INSTRUCTOR'S SOLUTIONS MANUAL

JUDITH A. PENNA

COLLEGE ALGEBRA GRAPHS & MODELS

SECOND EDITION

FUNDAMENTALS OF COLLEGE ALGEBRA GRAPHS & MODELS

Marvin L. Bittinger

Indiana University—Purdue University at Indianapolis

Judith A. Beecher

Indiana University—Purdue University at Indianapolis

David J. Ellenbogen

Community College of Vermont

Judith A. Penna

Indiana University—Purdue University at Indianapolis

Boston San Francisco New York
London Toronto Sydney Tokyo Singapore Madrid
Mexico City Munich Paris Cape Town Hong Kong Montreal

Reproduced by Addison Wesley Longman from camera-ready copy supplied by the author.

ISBN 0-201-70879-5

2 3 4 5 6 7 8 9 10 VG 04 03 02 01 00

Contents

Chapter G

Introduction to Graphs and the Graphing Calculator

1. To graph $(4, 0)$ we move from the origin 4 units to the right of the y-axis. Since the second coordinate is 0, we do not move up or down from the x-axis.

   To graph $(-3, -5)$ we move from the origin 3 units to the left of the y-axis. Then we move 5 units down from the x-axis.

   To graph $(-1, 4)$ we move from the origin 1 unit to the left of the y-axis. Then we move 4 units up from the x-axis.

   To graph $(0, 2)$ we do not move to the right or the left of the y-axis since the first coordinate is 0. From the origin we move 2 units up.

   To graph $(2, -2)$ we move from the origin 2 units to the right of the y-axis. Then we move 2 units down from the x-axis.

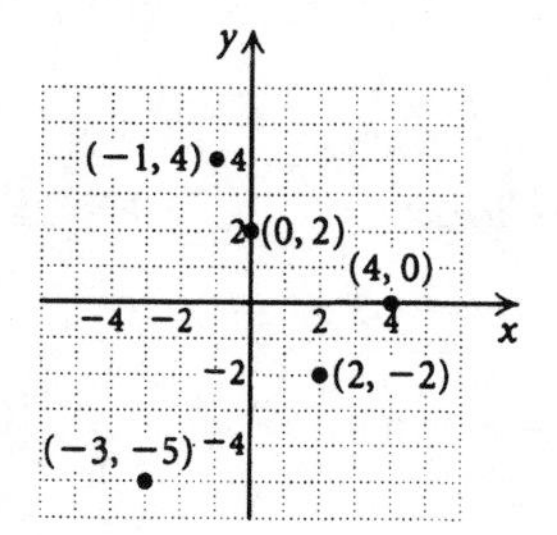

2.

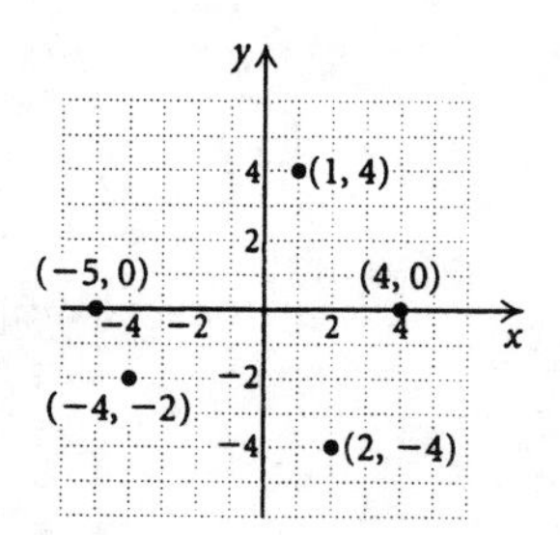

3. To graph $(-5, 1)$ we move from the origin 5 units to the left of the y-axis. Then we move 1 unit up from the x-axis.

   To graph $(5, 1)$ we move from the origin 5 units to the right of the y-axis. Then we move 1 unit up from the x-axis.

   To graph $(2, 3)$ we move from the origin 2 units to the right of the y-axis. Then we move 3 units up from the x-axis.

   To graph $(2, -1)$ we move from the origin 2 units to the right of the y-axis. Then we move 1 unit down from the x-axis.

   To graph $(0, 1)$ we do not move to the right or the left of the y-axis since the first coordinate is 0. From the origin we move 1 unit up.

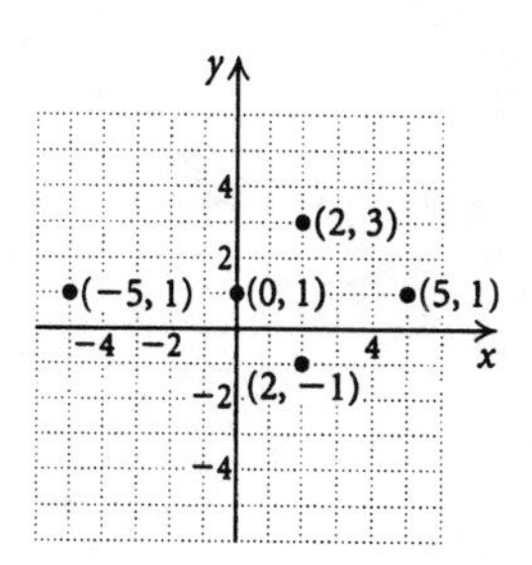

4.

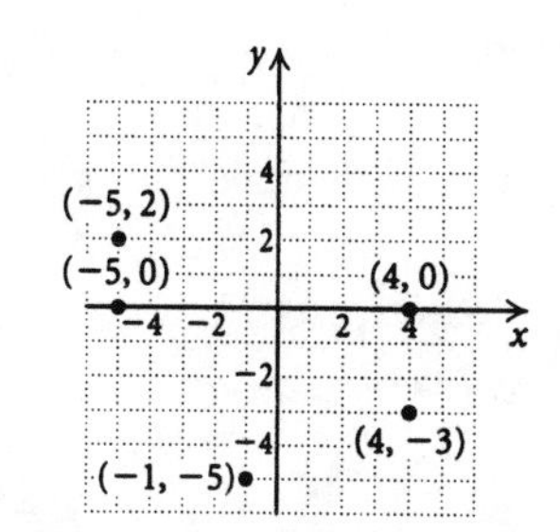

5. Enter the coordinates of the points in lists, set up a STAT PLOT, and then graph the points in the standard window. See the Graphing Calculator Manual that accompanies the text for the procedure.

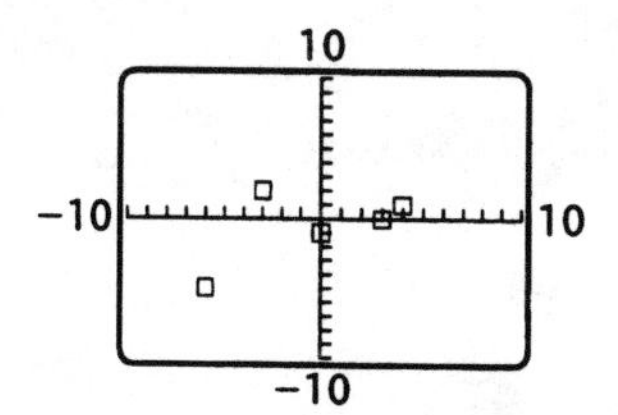

6.

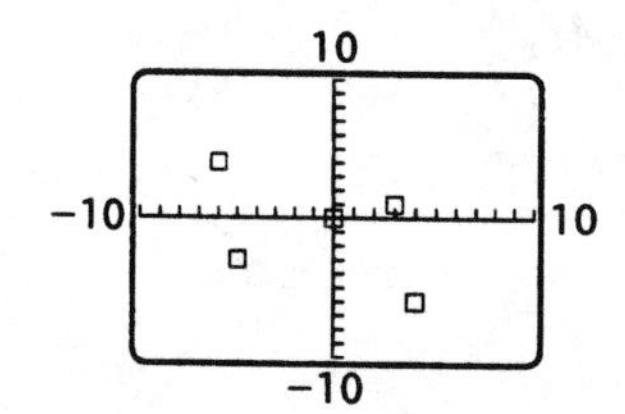

7. Enter the coordinates of the points in lists, set up a STAT PLOT, and then graph the points in the standard window.

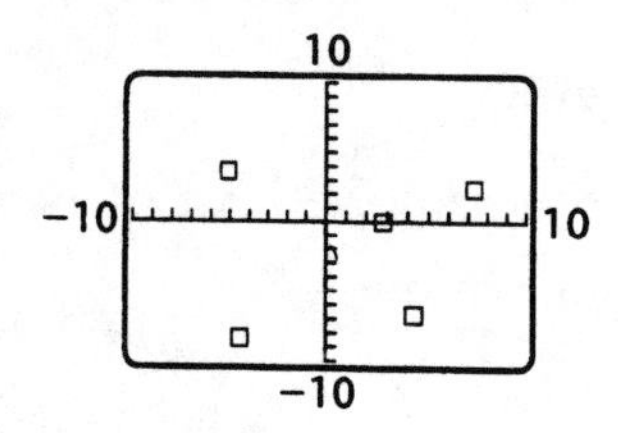

8.

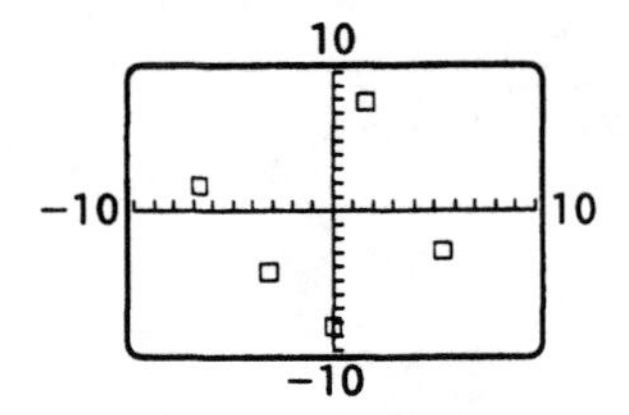

9. To determine whether $(1, -1)$ is a solution, substitute 1 for x and -1 for y.

$$\begin{array}{r|l} \multicolumn{2}{c}{y = 2x - 3} \\ \hline -1 & ?\ 2 \cdot 1 - 3 \\ & 2 - 3 \\ -1 & -1 \qquad \text{TRUE} \end{array}$$

The equation $-1 = -1$ is true, so $(1, -1)$ is a solution.

To determine whether $(0, 3)$ is a solution, substitute 0 for x and -3 for y.

$$\begin{array}{r|l} \multicolumn{2}{c}{y = 2x - 3} \\ \hline 3 & ?\ 2 \cdot 0 - 3 \\ & 0 - 3 \\ 3 & -3 \qquad \text{FALSE} \end{array}$$

The equation $3 = -3$ is false, so $(0, 3)$ is not a solution.

10. For $(2, 5)$:

$$\begin{array}{r|l} \multicolumn{2}{c}{y = 3x - 1} \\ \hline 5 & ?\ 3 \cdot 2 - 1 \\ & 6 - 1 \\ 5 & 5 \qquad \text{TRUE} \end{array}$$

$(2, 5)$ is a solution.

For $(-2, -5)$:

$$\begin{array}{r|l} \multicolumn{2}{c}{y = 3x - 1} \\ \hline -5 & ?\ 3(-2) - 1 \\ & -6 - 1 \\ -5 & -7 \qquad \text{FALSE} \end{array}$$

$(-2, -5)$ is not a solution.

11. To determine whether $\left(\frac{2}{3}, \frac{3}{4}\right)$ is a solution, substitute $\frac{2}{3}$ for x and $\frac{3}{4}$ for y.

$$\begin{array}{r|l} \multicolumn{2}{c}{6x - 4y = 1} \\ \hline 6 \cdot \frac{2}{3} - 4 \cdot \frac{3}{4} & ?\ 1 \\ 4 - 3 & \\ 1 & 1 \quad \text{TRUE} \end{array}$$

The equation $1 = 1$ is true, so $\left(\frac{2}{3}, \frac{3}{4}\right)$ is a solution.

To determine whether $\left(1, \frac{3}{2}\right)$ is a solution, substitute 1 for x and $\frac{3}{2}$ for y.

$$\begin{array}{r|l} \multicolumn{2}{c}{6x - 4y = 1} \\ \hline 6 \cdot 1 - 4 \cdot \frac{3}{2} & ?\ 1 \\ 6 - 6 & \\ 0 & 1 \quad \text{FALSE} \end{array}$$

The equation $0 = 1$ is false, so $\left(1, \frac{3}{2}\right)$ is not a solution.

12. For $(1.5, 2.6)$:

$$\begin{array}{r|l} \multicolumn{2}{c}{x^2 + y^2 = 9} \\ \hline (1.5)^2 + (2.6)^2 & ?\ 9 \\ 2.25 + 6.76 & \\ 9.01 & 9 \quad \text{FALSE} \end{array}$$

$(1.5, 2.6)$ is not a solution.

For $(-3, 0)$:

$$\begin{array}{r|l} \multicolumn{2}{c}{x^2 + y^2 = 9} \\ \hline (-3)^2 + 0^2 & ?\ 9 \\ 9 + 0 & \\ 9 & 9 \quad \text{TRUE} \end{array}$$

$(-3, 0)$ is a solution.

13. To determine whether $\left(-\frac{1}{2}, -\frac{4}{5}\right)$ is a solution, substitute $-\frac{1}{2}$ for a and $-\frac{4}{5}$ for b.

$$\begin{array}{r|l} \multicolumn{2}{c}{2a + 5b = 3} \\ \hline 2\left(-\frac{1}{2}\right) + 5\left(-\frac{4}{5}\right) & ?\ 3 \\ -1 - 4 & \\ -5 & 3 \quad \text{FALSE} \end{array}$$

The equation $-5 = 3$ is false, so $\left(-\frac{1}{2}, -\frac{4}{5}\right)$ is not a solution.

To determine whether $\left(0, \frac{3}{5}\right)$ is a solution, substitute 0 for a and $\frac{3}{5}$ for b.

$$\begin{array}{r|l} \multicolumn{2}{c}{2a + 5b = 3} \\ \hline 2 \cdot 0 + 5 \cdot \frac{3}{5} & ?\ 3 \\ 0 + 3 & \\ 3 & 3 \quad \text{TRUE} \end{array}$$

The equation $3 = 3$ is true, so $\left(0, \frac{3}{5}\right)$ is a solution.

14. For $\left(0, \frac{3}{2}\right)$:

$$\begin{array}{r|l} \multicolumn{2}{c}{3m + 4n = 6} \\ \hline 3 \cdot 0 + 4 \cdot \frac{3}{2} & ?\ 6 \\ 0 + 6 & \\ 6 & 6 \quad \text{TRUE} \end{array}$$

$\left(0, \frac{3}{2}\right)$ is a solution.

For $\left(\frac{2}{3}, 1\right)$:

$$\begin{array}{c|l} 3m + 4n & = 6 \\ \hline 3 \cdot \frac{2}{3} + 4 \cdot 1 & ?\ 6 \\ 2 + 4 & \\ 6 & 6 \ \text{TRUE} \end{array}$$

The equation $\left(\frac{2}{3}, 1\right)$ is true, so $\left(\frac{2}{3}, 1\right)$ is a solution.

15. To determine whether $(-0.75, 2.75)$ is a solution, substitute -0.75 for x and 2.75 for y.

$$\begin{array}{c|l} x^2 - y^2 & = 3 \\ \hline (-0.75)^2 - (2.75)^2 & ?\ 3 \\ 0.5625 - 7.5625 & \\ -7 & 3 \ \text{FALSE} \end{array}$$

The equation $-7 = 3$ is false, so $(-0.75, 2.75)$ is not a solution.

To determine whether $(2, -1)$ is a solution, substitute 2 for x and -1 for y.

$$\begin{array}{c|l} x^2 - y^2 & = 3 \\ \hline 2^2 - (-1)^2 & ?\ 3 \\ 4 - 1 & \\ 3 & 3 \ \text{TRUE} \end{array}$$

The equation $3 = 3$ is true, so $(2, -1)$ is a solution.

16. For $(2, -4)$:

$$\begin{array}{c|l} 5x + 2y^2 & = 70 \\ \hline 5 \cdot 2 + 2(-4)^2 & ?\ 70 \\ 10 + 2 \cdot 16 & \\ 10 + 32 & \\ 42 & 70 \ \text{FALSE} \end{array}$$

$(2, -4)$ is not a solution.

For $(4, -5)$:

$$\begin{array}{c|l} 5x + 2y^2 & = 70 \\ \hline 5 \cdot 4 + 2(-5)^2 & ?\ 70 \\ 20 + 2 \cdot 25 & \\ 20 + 50 & \\ 70 & 70 \ \text{TRUE} \end{array}$$

$(4, -5)$ is a solution.

17. First create a table of values as described in the Graphing Calculator Manual that accompanies the text.

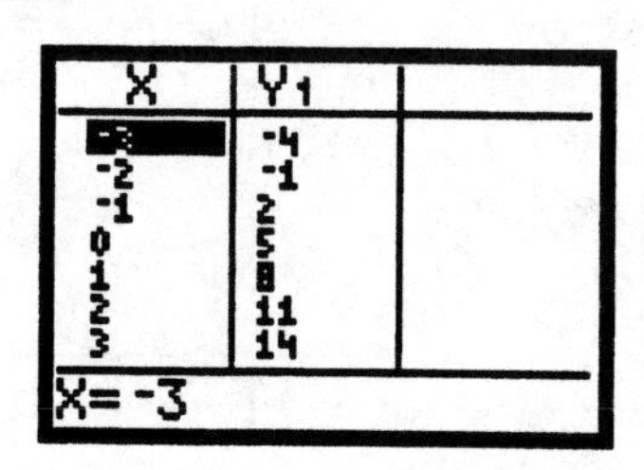

X	Y_1
-3	-4
-2	-1
-1	2
0	5
1	8
2	11
3	14

X= -3

Then plot the points in the table and draw the graph.

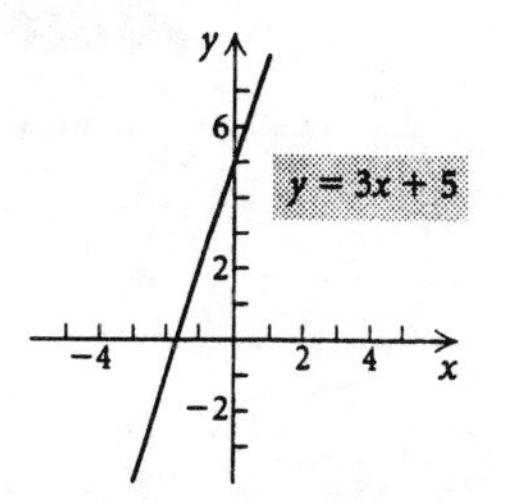

18.

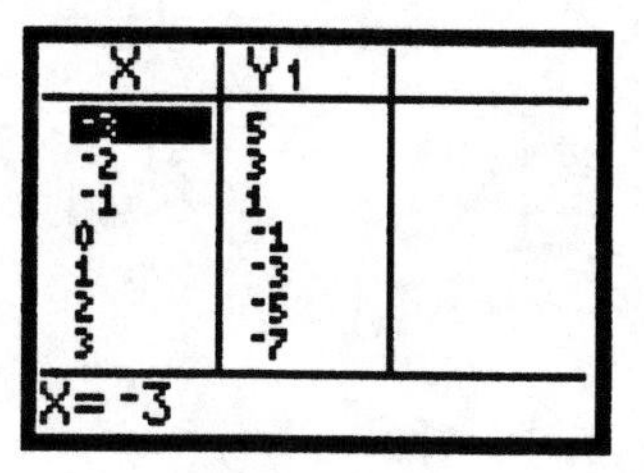

X	Y_1
-3	5
-2	3
-1	1
0	-1
1	-3
2	-5
3	-7

X= -3

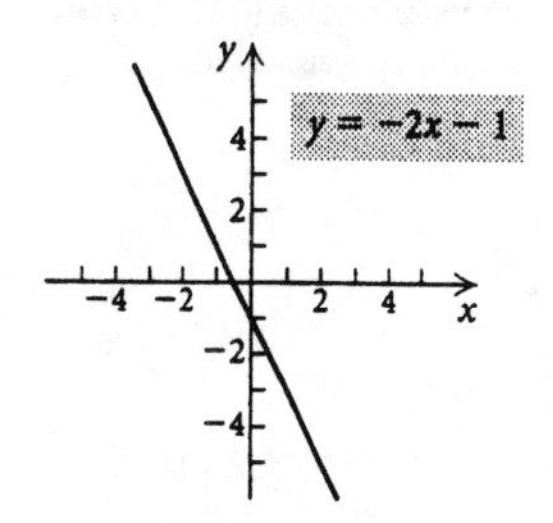

19. First create a table of values as described in the Graphing Calculator Manual that accompanies the text. Then plot the points in the table and draw the graph.

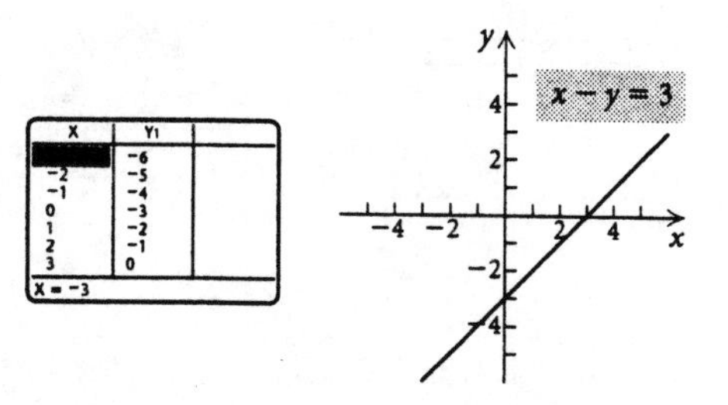

X	Y_1
-3	-6
-2	-5
-1	-4
0	-3
1	-2
2	-1
3	0

X = -3

20.

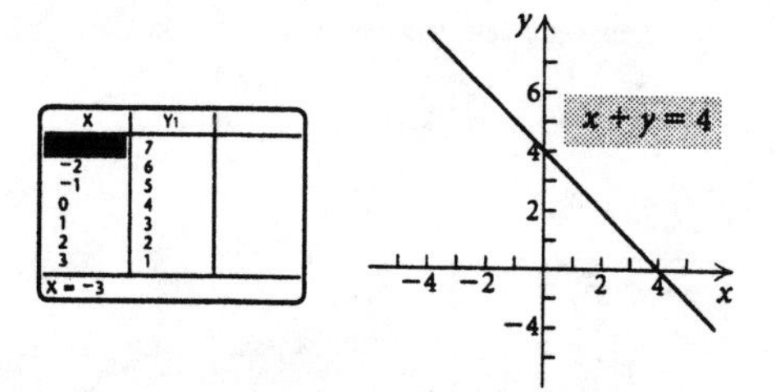

X	Y_1
-3	7
-2	6
-1	5
0	4
1	3
2	2
3	1

X = -3

21. First create a table of values as described in the Graphing Calculator Manual that accompanies the text. Then plot the points in the table and draw the graph.

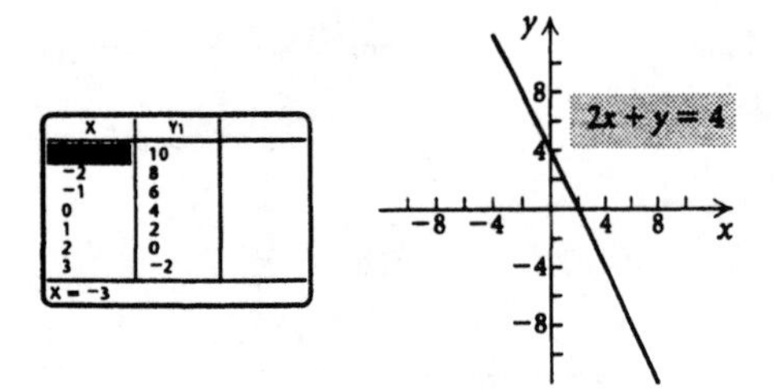

22.

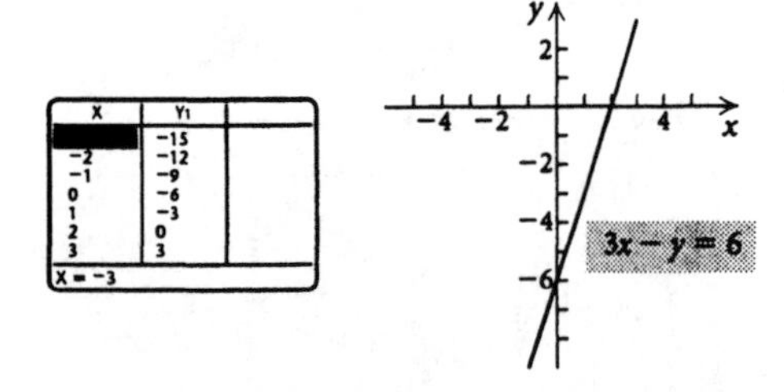

23. First create a table of values as described in the Graphing Calculator Manual that accompanies the text. Then plot the points in the table and draw the graph.

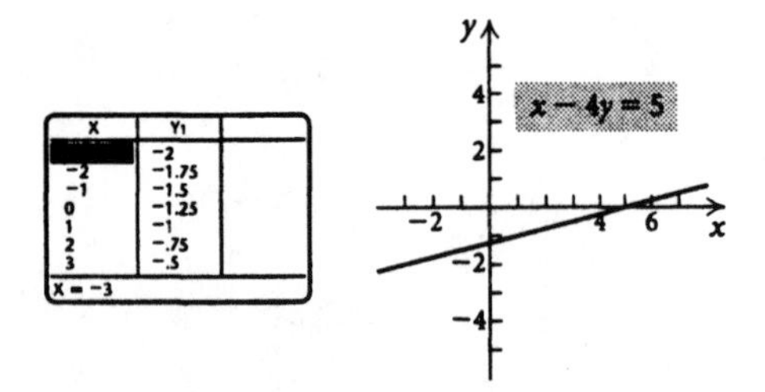

24.

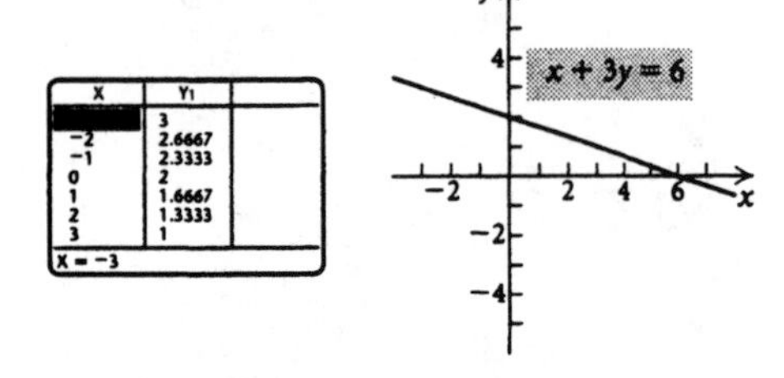

25. First create a table of values as described in the Graphing Calculator Manual that accompanies the text. Then plot the points in the table and draw the graph.

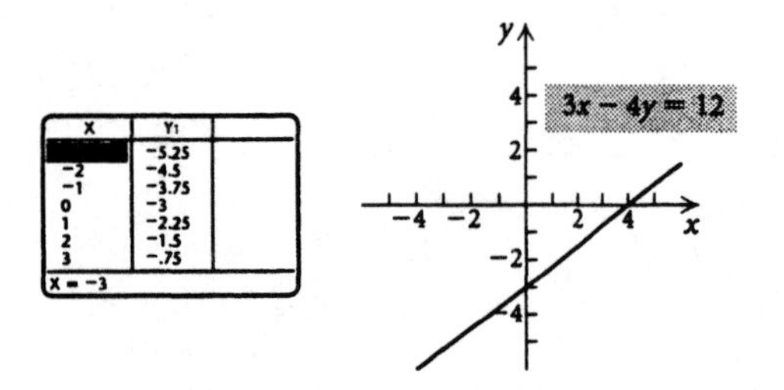

26.

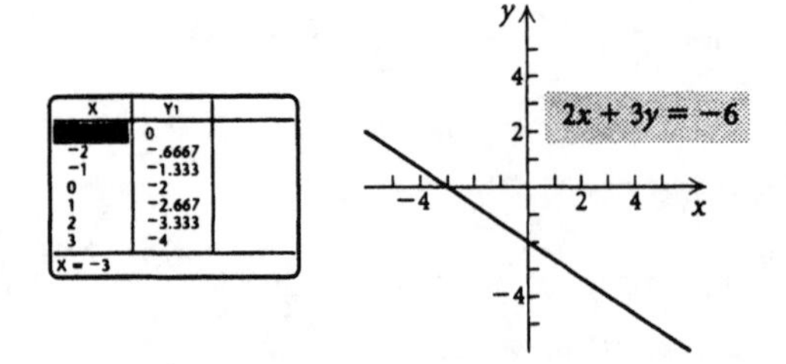

27. First create a table of values as described in the Graphing Calculator Manual that accompanies the text. Then plot the points in the table and draw the graph.

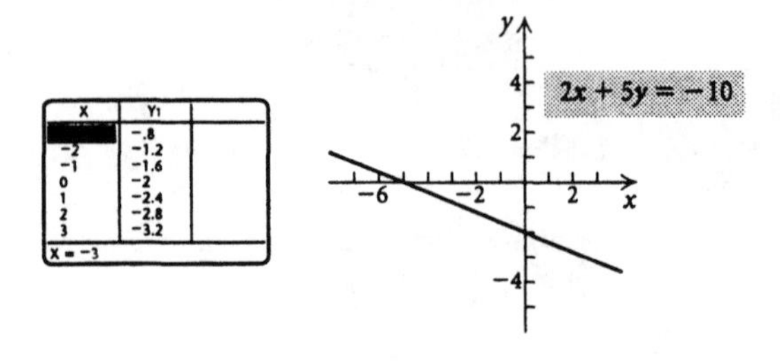

28.

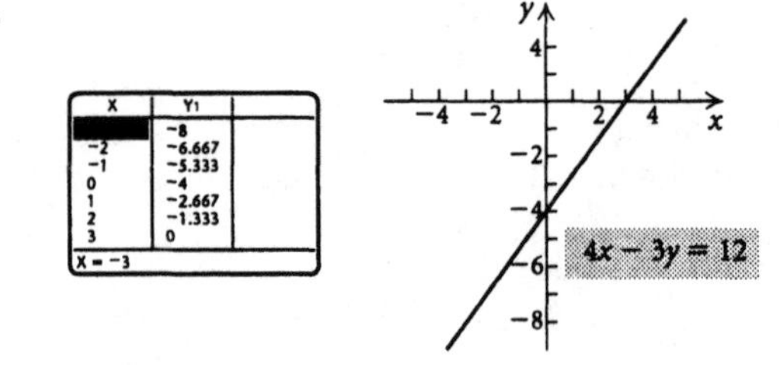

29. First create a table of values as described in the Graphing Calculator Manual that accompanies the text. Then plot the points in the table and draw the graph.

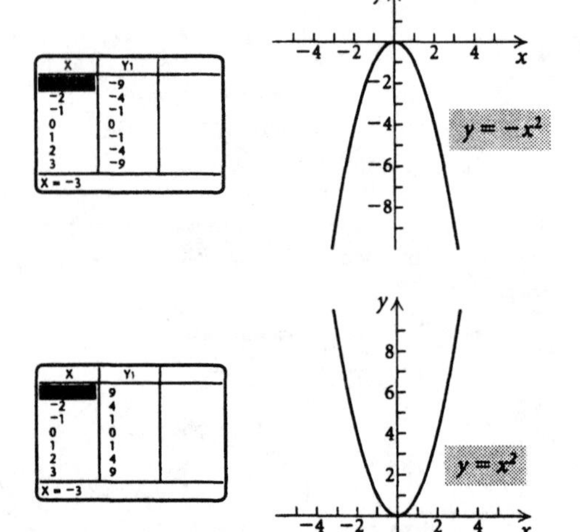

30.

31. First create a table of values as described in the Graphing Calculator Manual that accompanies the text. Then plot the points in the table and draw the graph.

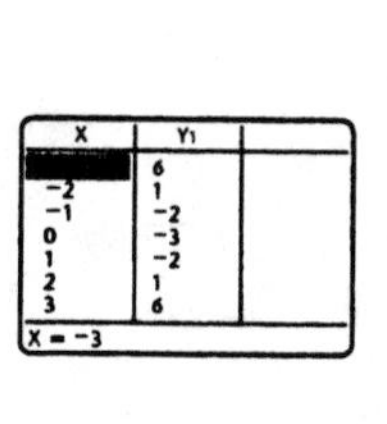

X	Y_1
-3	6
-2	1
-1	-2
0	-3
1	-2
2	1
3	6

X = -3

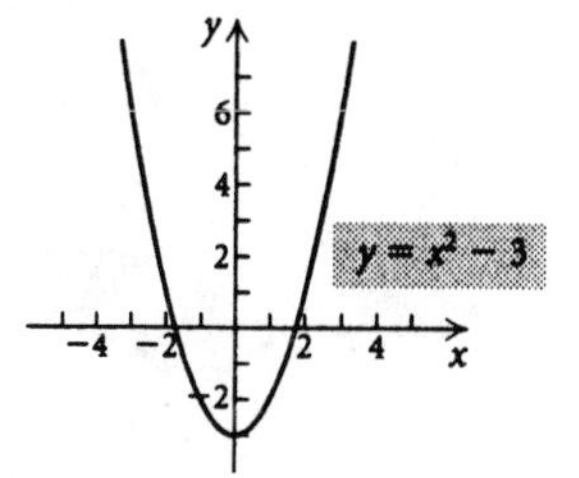

32.

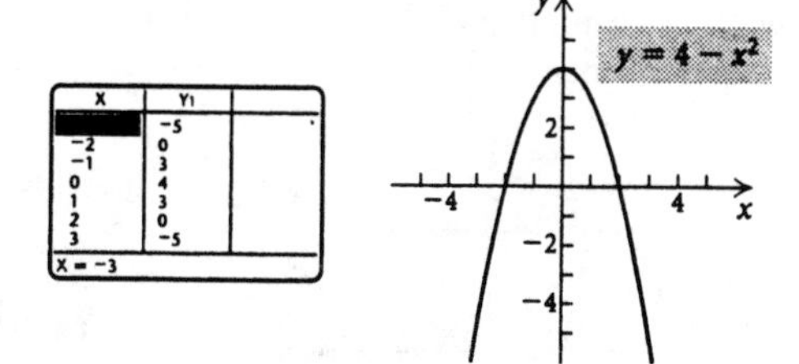

X	Y_1
-3	-5
-2	0
-1	3
0	4
1	3
2	0
3	-5

X = -3

33. First create a table of values as described in the Graphing Calculator Manual that accompanies the text. Then plot the points in the table and draw the graph.

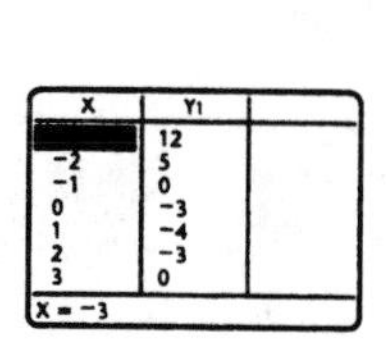

X	Y_1
-3	12
-2	5
-1	0
0	-3
1	-4
2	-3
3	0

X = -3

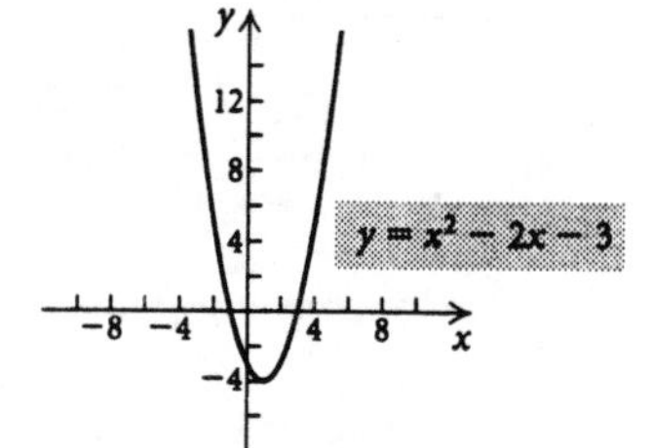

34.

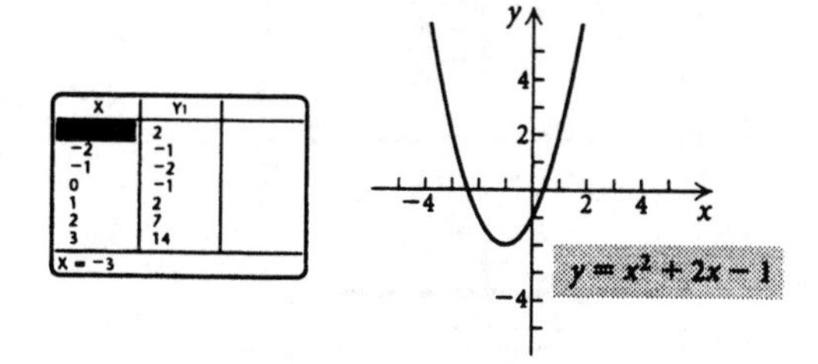

X	Y_1
-3	2
-2	-1
-1	-2
0	-1
1	2
2	7
3	14

X = -3

35. Graph (b) is the graph of $y = 3 - x$.

36. Graph (d) is the graph of $2x - y = 6$.

37. Graph (a) is the graph of $y = x^2 + 2x + 1$.

38. Graph (c) is the graph of $y = 4 - x^2$.

39. Enter the equation, select the standard window, and graph the equation as described in the Graphing Calculator Manual that accompanies the text.

$y = 2x + 1$

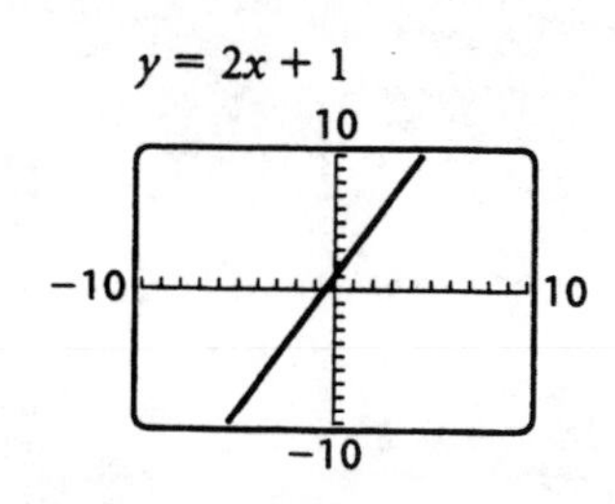

40. $y = 3x - 4$

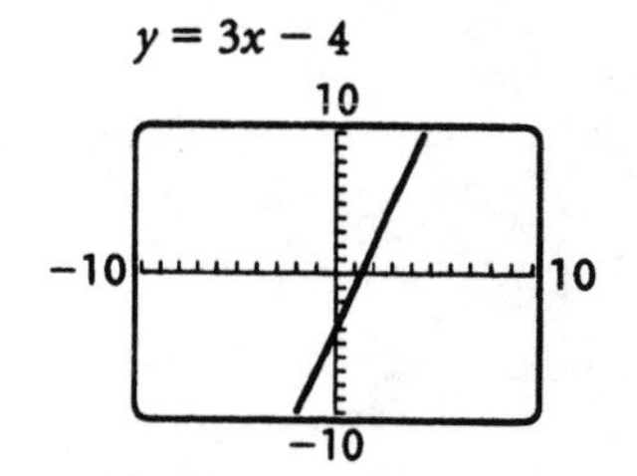

41. First solve the equation for y: $y = -4x+7$. Enter the equation in this form, select the standard window, and graph the equation as described in the Graphing Calculator Manual that accompanies the text.

$4x + y = 7$

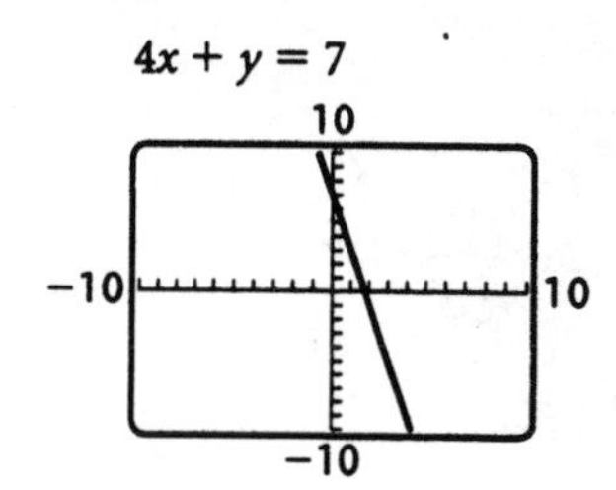

42. $5x + y = -8$

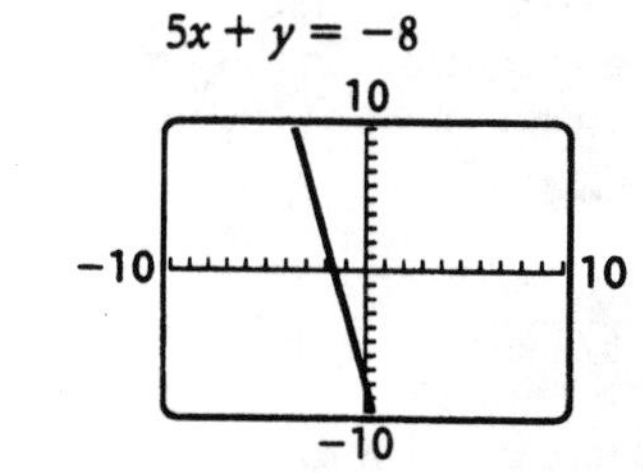

43. Enter the equation, select the standard window, and graph the equation as described in the Graphing Calculator Manual that accompanies the text.

$y = \frac{1}{3}x + 2$

44. $y = \frac{3}{2}x - 4$

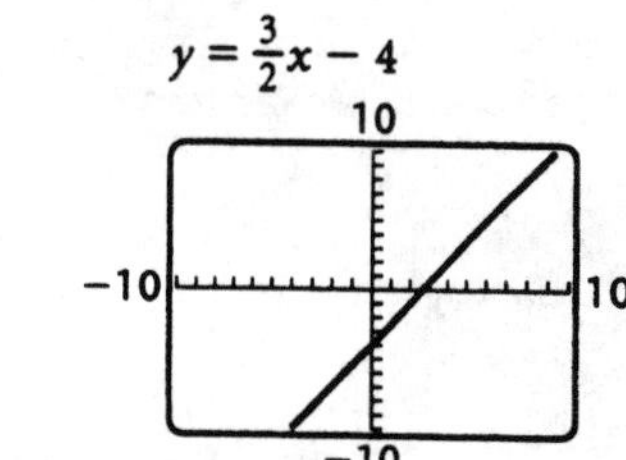

45. First solve the equation for y.

$$2x + 3y = -5$$
$$3y = -2x - 5$$
$$y = \frac{-2x - 5}{3}, \text{ or } \frac{1}{3}(-2x - 5)$$

Enter the equation in "$y =$" form, select the standard window, and graph the equation as described in the Graphing Calculator Manual that accompanies the text.

$2x + 3y = -5$

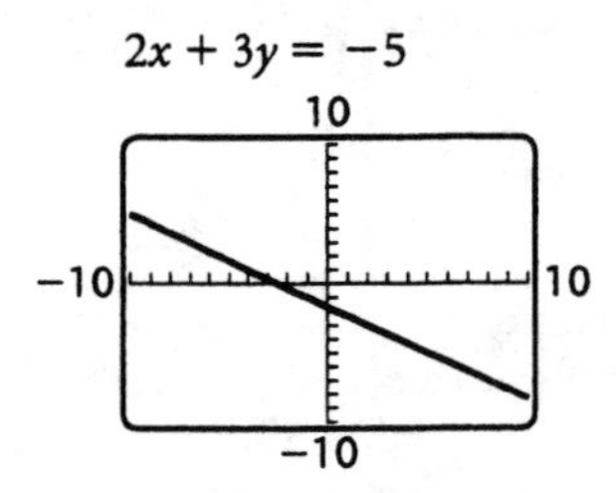

46. $3x + 4y = 1$

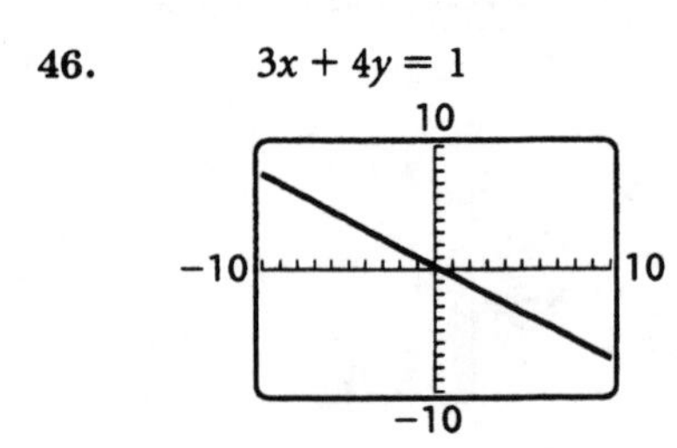

47. Enter the equation, select the standard window, and graph the equation as described in the Graphing Calculator Manual that accompanies the text.

$y = x^2 + 6$

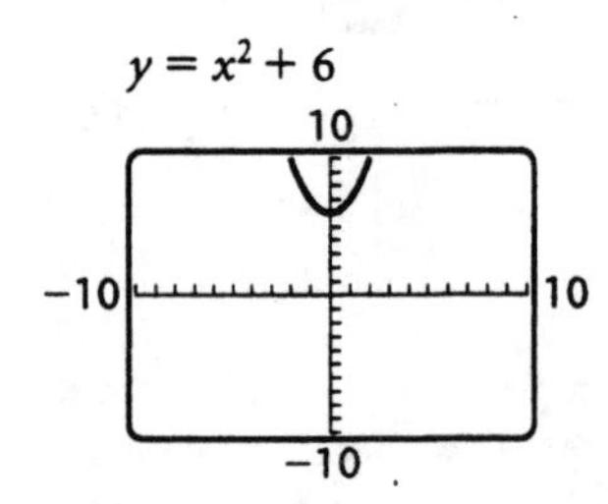

48. $y = x^2 - 8$

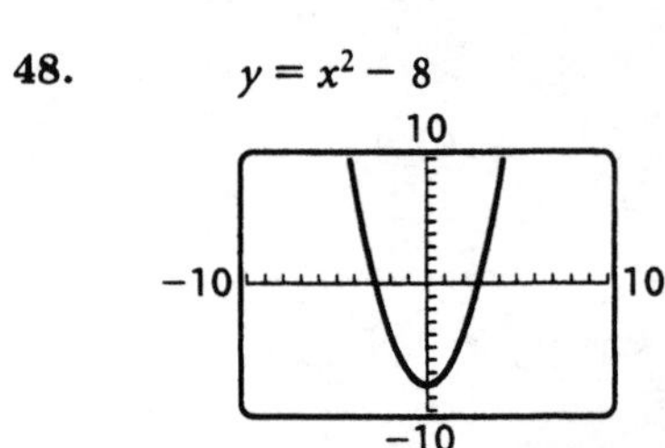

49. Enter the equation, select the standard window, and graph the equation as described in the Graphing Calculator Manual that accompanies the text.

$y = 2 - x^2$

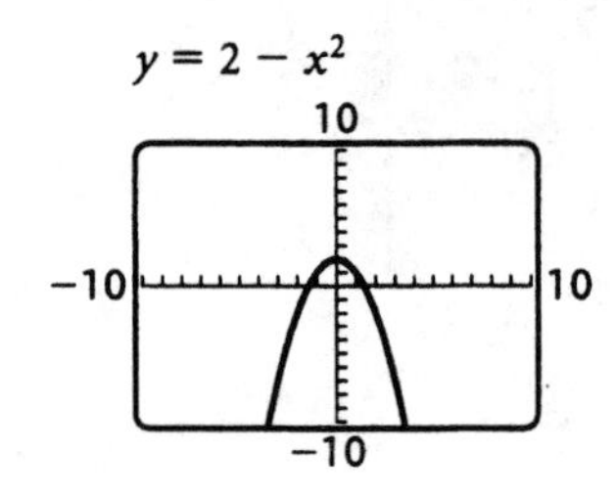

50. $y = 5 - x^2$

10

−10 10

−10

51. Enter the equation, select the standard window, and graph the equation as described in the Graphing Calculator Manual that accompanies the text.

$y = x^2 + 4x - 2$

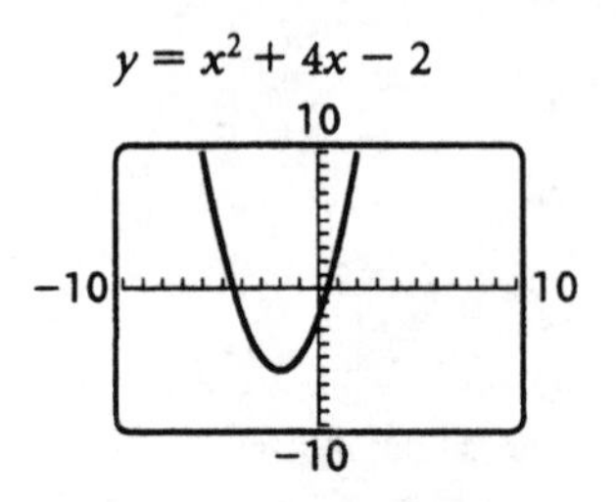

52. $y = x^2 - 5x + 3$

10

−10 10

−10

53. Standard window:

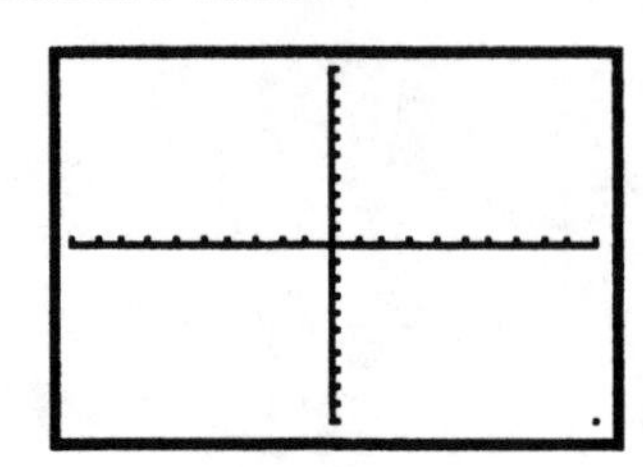

$[-25, 25, -25, 25]$, Xscl $= 5$, Yscl $= 5$

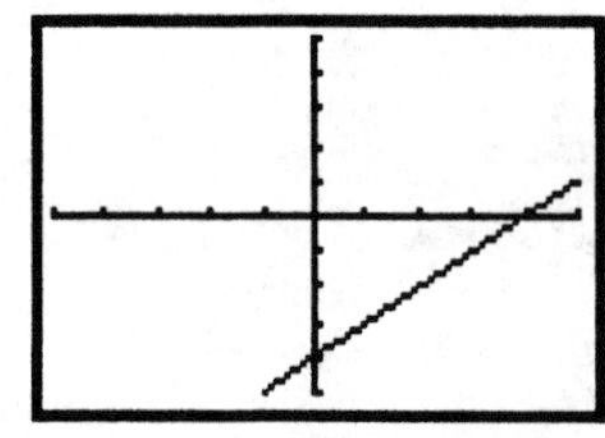

We see that $[-25, 25, -25, 25]$ is a better choice for this graph.

54. Standard window:

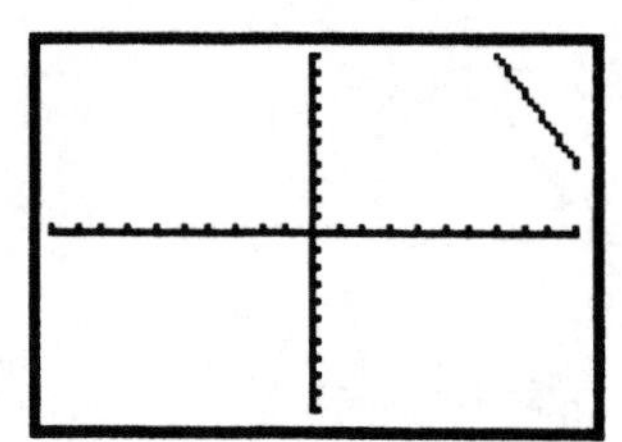

$[-15, 15, -10, 30]$, Xscl = 3, Yscl = 5

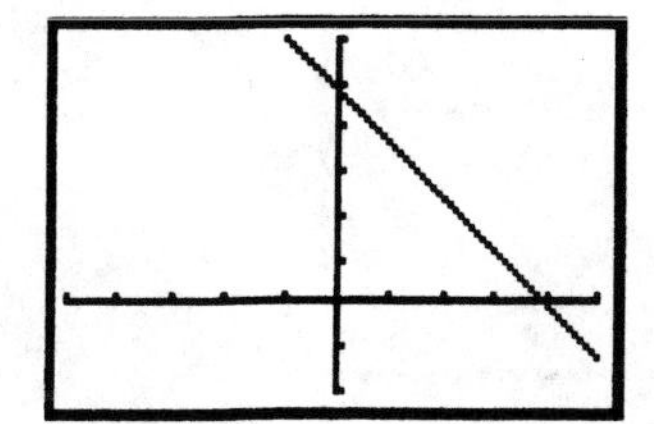

We see that $[-15, 15, -10, 30]$ is a better choice for this graph.

55. Standard window:

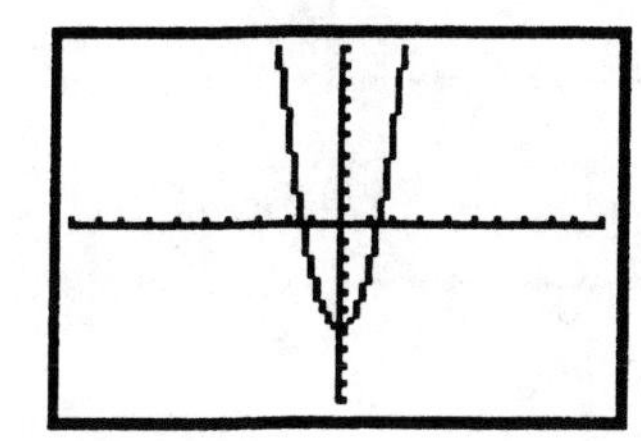

$[-4, 4, -4, 4]$

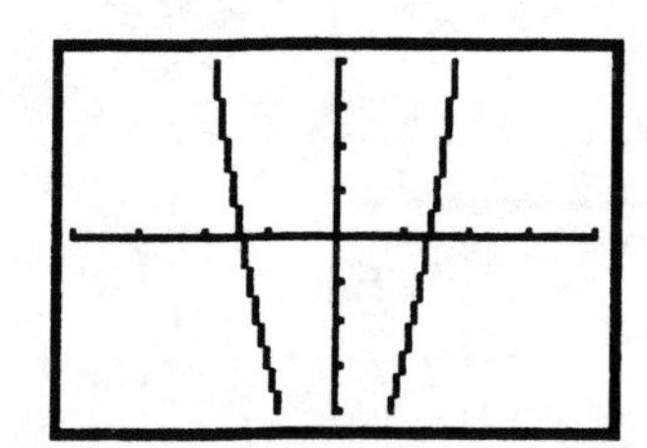

We see that the standard window is a better choice for this graph.

56. Standard window:

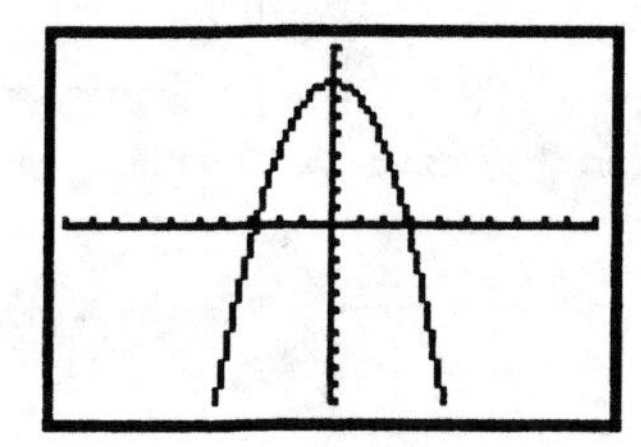

$[-3, 3, -3, 3]$

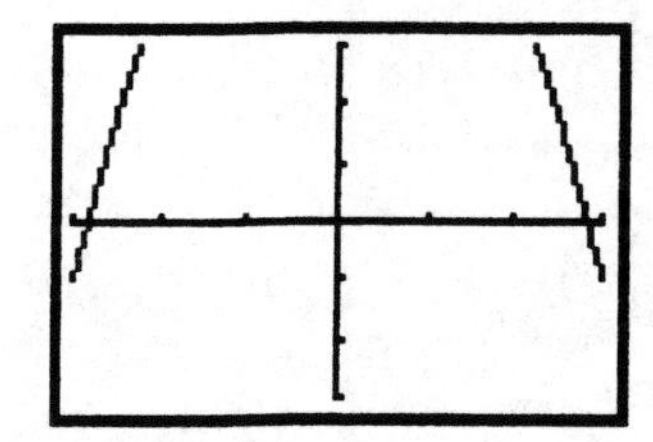

We see that the standard window is a better choice for this graph.

57. Standard window:

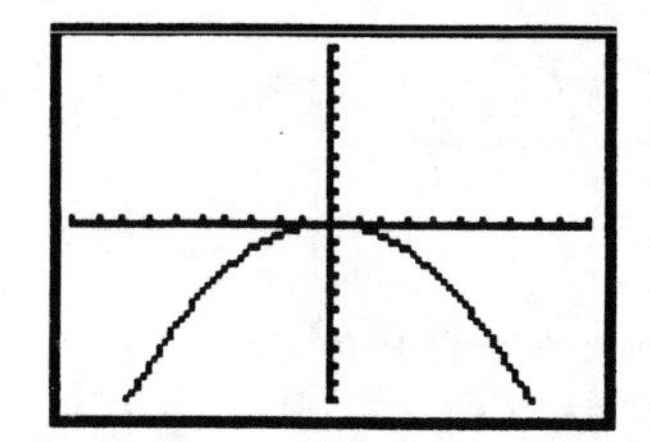

$[-1, 1, -0.3, 0.3]$, Xscl = 0.1, Yscl = 0.1

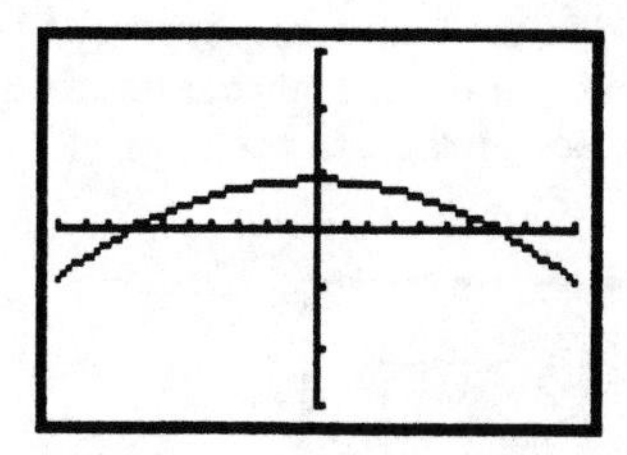

We see that $[-1, 1, -0.3, 0.3]$ is a better choice for this graph.

58. Standard window:

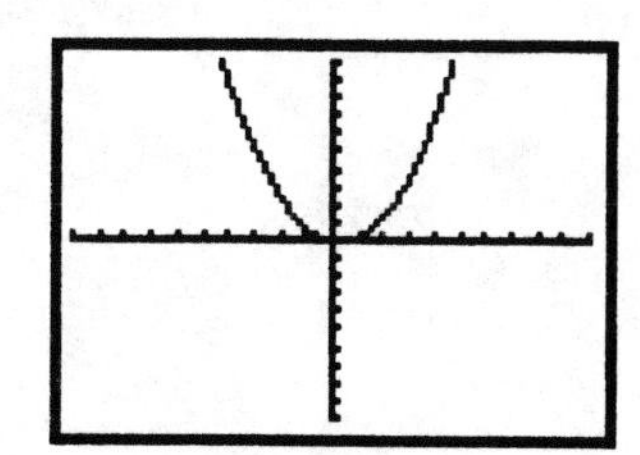

$[-0.8, 1.2, -0.5, 0.5]$, Xscl = 0.1, Yscl = 0.1

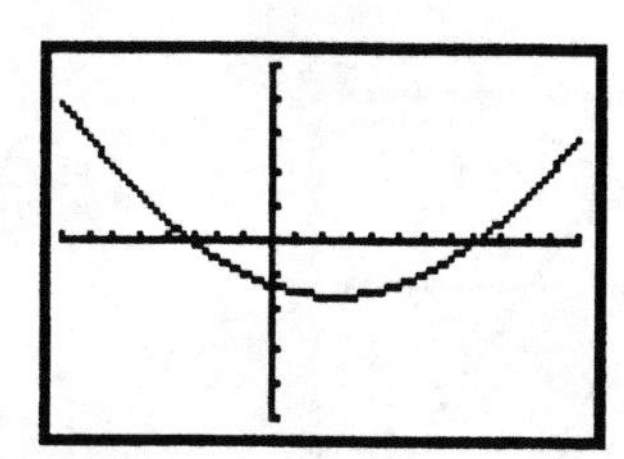

We see that $[-0.8, 1.2, -0.5, 0.5]$ is a better choice for this graph.

59. Graph the equations and use the INTERSECT feature as described in the Graphing Calculator Manual that accompanies the text.

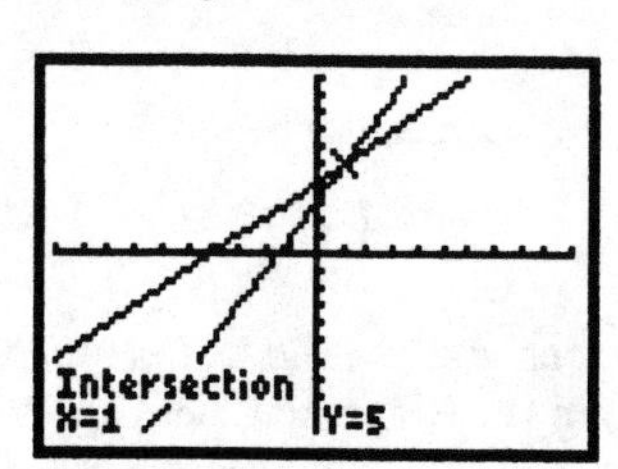

The point of intersection is $(1, 5)$.

60.

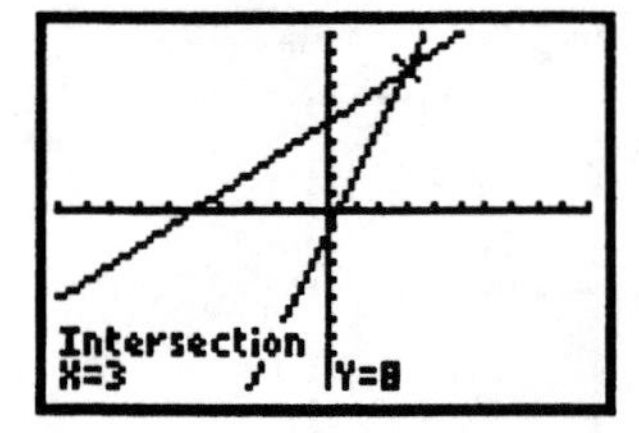

The point of intersection is $(3, 8)$.

61. Graph the equations and use the INTERSECT feature as described in the Graphing Calculator Manual that accompanies the text. The coordinates are rational numbers, so we can use the ▹Frac feature to find their exact values.

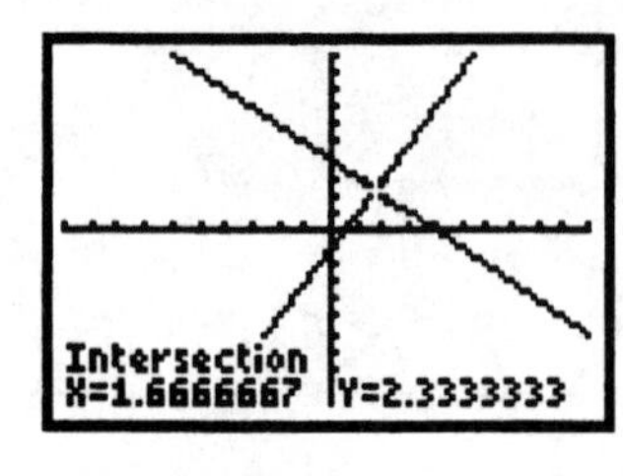

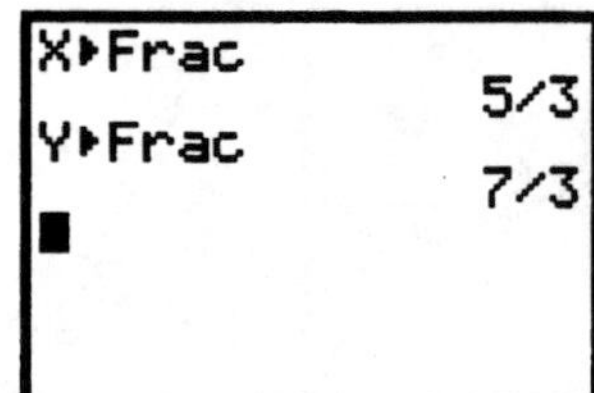

The point of intersection is $\left(\frac{5}{3}, \frac{7}{3}\right)$.

62.

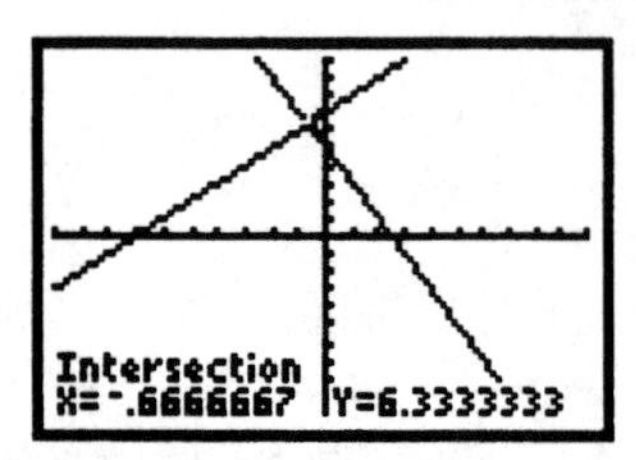

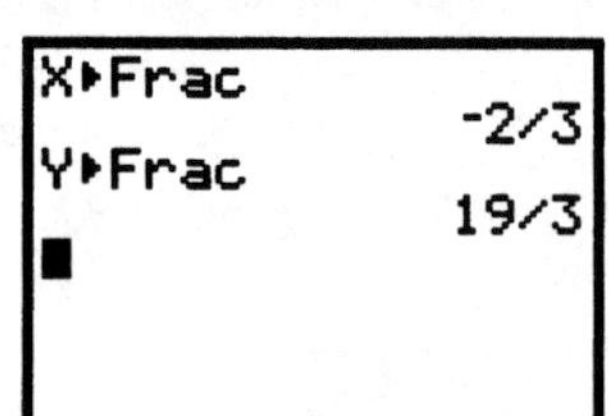

The point of intersection is $\left(-\frac{2}{3}, \frac{19}{3}\right)$.

63. Graph the equations and use the INTERSECT feature as described in the Graphing Calculator Manual that accompanies the text. The coordinates are rational numbers, so we can use the ▹Frac feature to find their exact values.

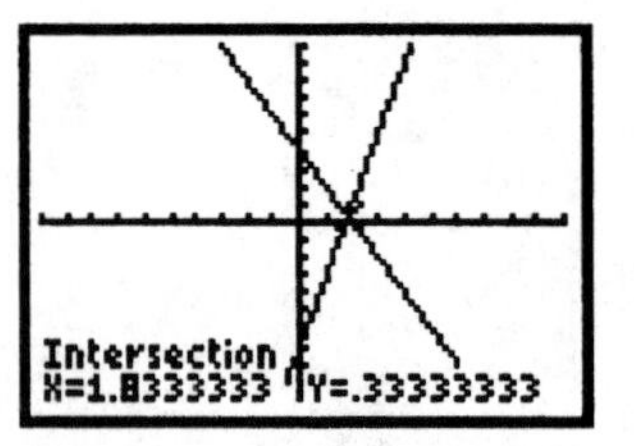

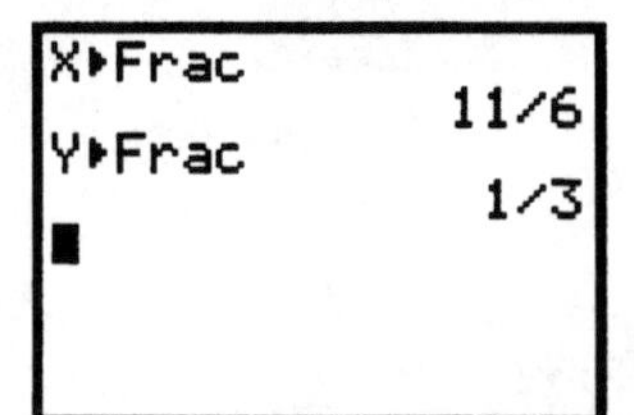

The point of intersection is $\left(\frac{11}{6}, \frac{1}{3}\right)$.

64.

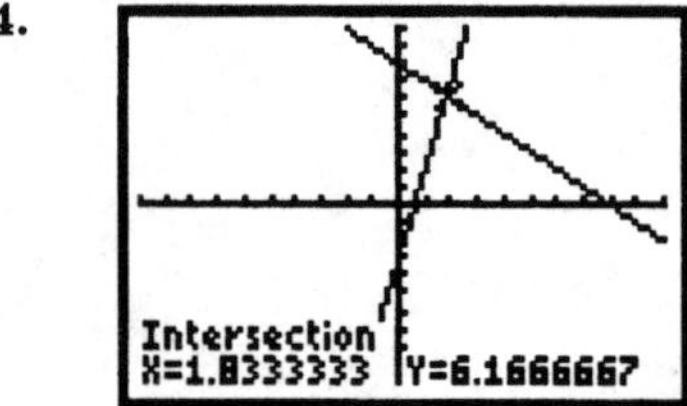

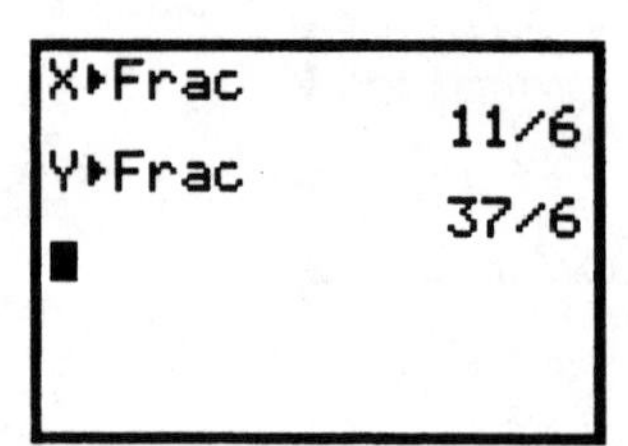

The point of intersection is $\left(\frac{11}{6}, \frac{37}{6}\right)$.

65. Graph the equations and use the INTERSECT feature as described in the Graphing Calculator Manual that accompanies the text.

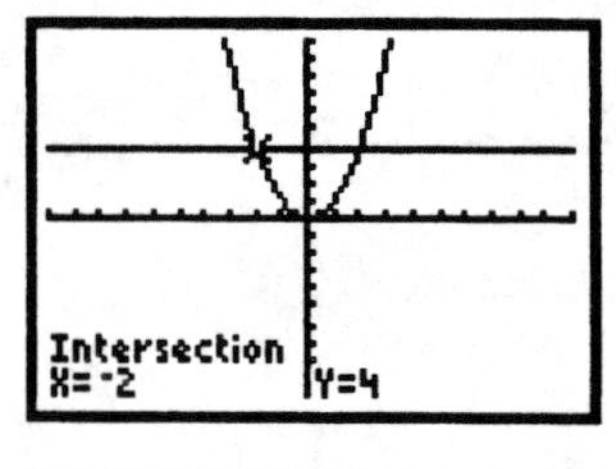

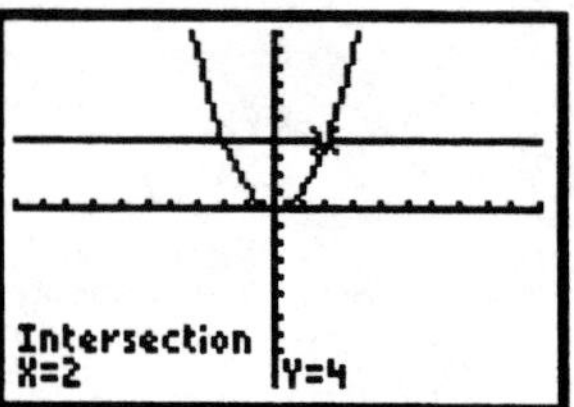

The points of intersection are $(-2, 4)$ and $(2, 4)$.

66.

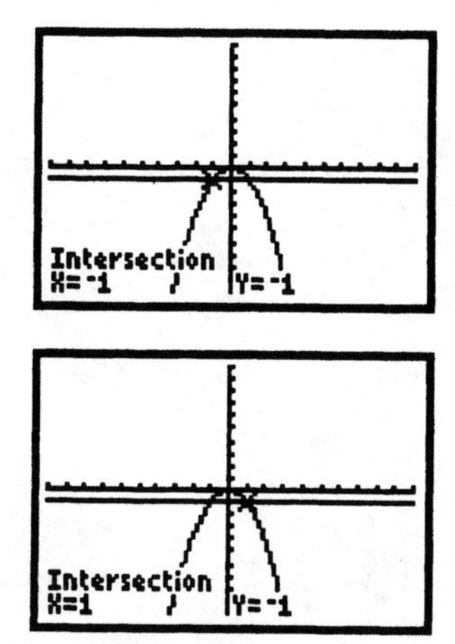

The points of intersection are $(-1, -1)$ and $(1, -1)$.

67. Graph the equations and use the INTERSECT feature as described in the Graphing Calculator Manual that accompanies the text.

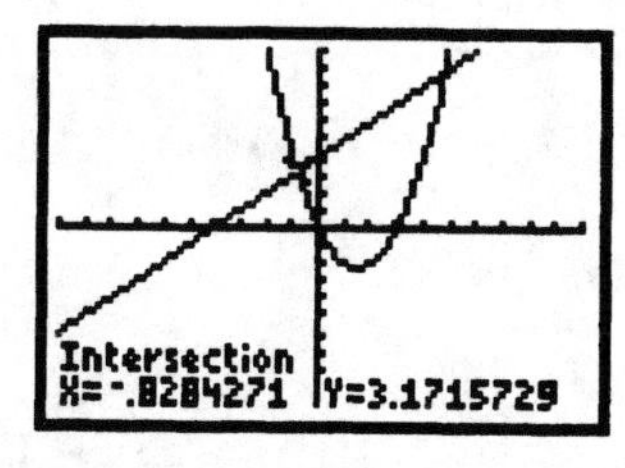

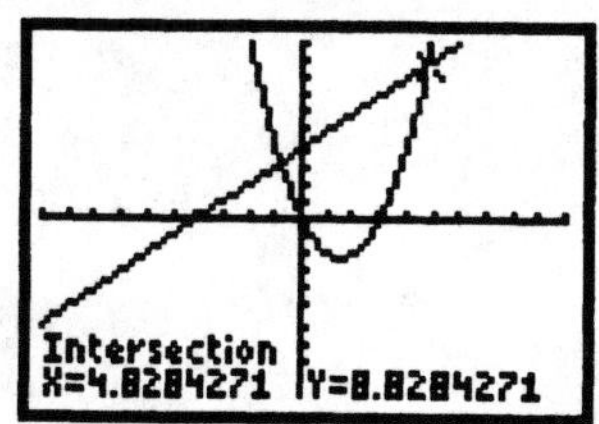

The coordinates are not rational numbers. We give their decimal approximations. The points of intersection are approximately $(-0.8284271, 3.1715729)$ and $(4.8284271, 8.8284271)$.

68.

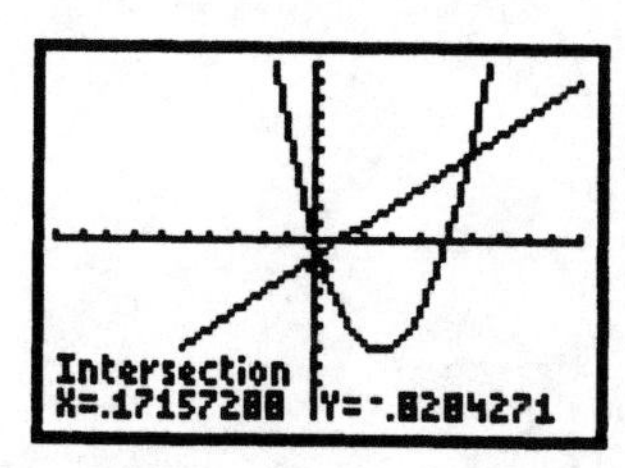

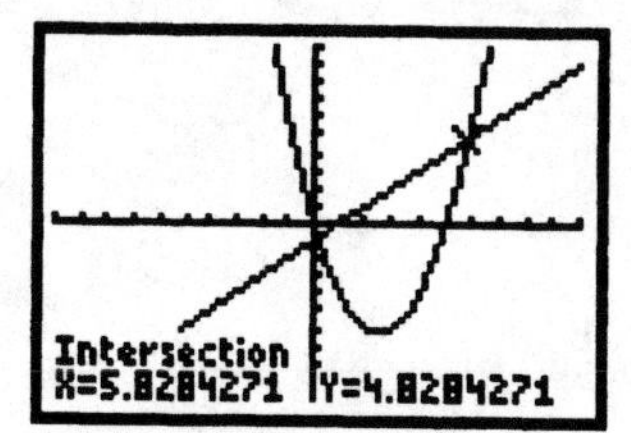

The coordinates of the points of intersection are approximately $(0.17157288, -0.8284271)$ and $(5.8284271, 4.8284271)$.

Chapter R

Basic Concepts of Algebra

Exercise Set R.1

1. Whole numbers: $\sqrt[3]{8}$, 0, 9, $\sqrt{25}$ $(\sqrt[3]{8} = 2, \sqrt{25} = 5)$

2. Integers: -12, $\sqrt[3]{8}$, 0, 9, $\sqrt{25}$ $(\sqrt[3]{8} = 2, \sqrt{25} = 5)$

3. Irrational numbers: $\sqrt{7}$, $5.242242224\ldots$, $-\sqrt{14}$, $\sqrt[5]{5}$, $\sqrt[3]{4}$

   (Although there is a pattern in $5.242242224\ldots$, there is no repeating block of digits.)

4. Natural numbers: $\sqrt[3]{8}$, 9, $\sqrt{25}$

5. Rational numbers: -12, $5.\overline{3}$, $-\dfrac{7}{3}$, $\sqrt[3]{8}$, 0, -1.96, 9, $4\dfrac{2}{3}$, $\sqrt{25}$, $\dfrac{5}{7}$

6. Real numbers: All of them

7. Answers may vary. Some examples are $-\dfrac{3}{4}$, $5.7\overline{6}$, $9\dfrac{1}{8}$, -1.067.

8. Answers may vary. Some examples are $\sqrt{5}$, $-\sqrt[3]{10}$, $6.191991999\ldots$.

9. Answers may vary. Some examples are -1, -5, -352.

10. Answers may vary. Some examples are $\dfrac{12}{5}$, -8, 0, $-4.\overline{5}$, $6\dfrac{5}{7}$, -14.059.

11. This is a closed interval, so we use brackets. Interval notation is $[-3, 3]$.

12. $(-4, 4)$

13. This is a half-open interval. We use a bracket on the left and a parenthesis on the right. Interval notation is $[-14, -11)$.

14. $(6, 20]$

15. This interval is of unlimited extent in the negative direction, and the endpoint -4 is included. Interval notation is $(-\infty, -4]$.

16. $(-5, \infty)$

17. This interval is of unlimited extent in the negative direction, and the endpoint 3.8 is not included. Interval notation is $(-\infty, 3.8)$.

18. $[\sqrt{3}, \infty)$

19. $\{x|x \neq 7\}$ is equivalent to $\{x|x < 7\} \cup \{x|x > 7\}$. Thus, interval notation is $(-\infty, 7) \cup (7, \infty)$.

20. $\{x|x \neq -3\}$ is equivalent to $\{x|x < -3\} \cup \{x|x > -3\}$. Thus, interval notation is $(-\infty, -3) \cup (-3, \infty)$.

21. The endpoints 0 and 5 are not included in the interval, so we use parentheses. Interval notation is $(0, 5)$.

22. $[-1, 2]$

23. The endpoint -9 is included in the interval, so we use a bracket before the -9. The endpoint -4 is not included, so we use a parenthesis after the -4. Interval notation is $[-9, -4)$.

24. $(-9, -5]$

25. Both endpoints are included in the interval, so we use brackets. Interval notation is $[x, x + h]$.

26. $(x, x + h]$

27. The endpoint p is not included in the interval, so we use a parenthesis before the p. The interval is of unlimited extent in the positive direction, so we use the infinity symbol ∞. Interval notation is (p, ∞).

28. $(-\infty, q]$

29. Since 6 is an element of the set of natural numbers, the statement is true.

30. True

31. Since 3.2 is not an element of the set of integers, the statement is false.

32. True

33. Since $-\dfrac{11}{5}$ is an element of the set of rational numbers, the statement is true.

34. False

35. Since $\sqrt{11}$ is an element of the set of real numbers, the statement is false.

36. False

37. Since 24 is an element of the set of whole numbers, the statement is false.

38. True

39. Since 1.089 is not an element of the set of irrational numbers, the statement is true.

40. True

41. Since every whole number is an integer, the statement is true.

42. False

43. Since every rational number is a real number, the statement is true.

44. True

45. Since there are real numbers that are not integers, the statement is false.

46. False

47. The sentence $6x = x6$ illustrates the commutative property of multiplication.

48. Associative property of addition

49. The sentence $-3 \cdot 1 = -3$ illustrates the multiplicative identity property.

50. Commutative property of addition

51. The sentence $5(ab) = (5a)b$ illustrates the associative property of multiplication.

52. Distributive property

53. The sentence $2(a+b) = (a+b)2$ illustrates the commutative property of multiplication.

54. Additive inverse property

55. The sentence $-6(m+n) = -6(n+m)$ illustrates the commutative property of addition.

56. Additive identity property

57. The sentence $8 \cdot \frac{1}{8} = 1$ illustrates the multiplicative inverse property.

58. Distributive property

59. The distance of -7.1 from 0 is 7.1, so $|-7.1| = 7.1$.

60. 0

61. The distance of $\frac{5}{4}$ from 0 is $\frac{5}{4}$, so $\left|\frac{5}{4}\right| = \frac{5}{4}$.

62. $\sqrt{3}$

63. $|-5-6| = |-11| = 11$, or
$|6-(-5)| = |6+5| = |11| = 11$

64. $|0-(-2.5)| = |2.5| = 2.5$, or
$|-2.5-0| = |-2.5| = 2.5$

65. $|-2-(-8)| = |-2+8| = |6| = 6$, or
$|-8-(-2)| = |-8+2| = |-6| = 6$

66. $\left|\frac{15}{8} - \frac{23}{12}\right| = \left|\frac{45}{24} - \frac{46}{24}\right| = \left|-\frac{1}{24}\right| = \frac{1}{24}$, or
$\left|\frac{23}{12} - \frac{15}{8}\right| = \left|\frac{46}{24} - \frac{45}{24}\right| = \left|\frac{1}{24}\right| = \frac{1}{24}$

67. $|12.1 - 6.7| = |5.4| = 5.4$, or
$|6.7 - 12.1| = |-5.4| = 5.4$

68. $|-3-(-14)| = |-3+14| = |11| = 11$, or
$|-14-(-3)| = |-14+3| = |-11| = 11$

69. Provide an example. For instance, $16 \div (8 \div 2) = 16 \div 4 = 4$, but $(16 \div 8) \div 2 = 2 \div 2 = 1$.

70. $\sqrt{a}$ is a rational number when a is the square of a rational number. That is, $\sqrt{a}$ is a rational number if there is a rational number c such that $a = c^2$.

71. Answers may vary. One such number is $0.124124412444\ldots$.

72. Answers may vary. Since $-\sqrt{2.01} \approx -1.418$ and $-\sqrt{2} \approx -1.414$, one such number is -1.415.

73. Answers may vary. Since $-\frac{1}{101} = 0.\overline{0099}$ and $-\frac{1}{100} = -0.01$, one such number is -0.00999.

74. Answers may vary. One such number is $\sqrt{5.995}$.

75. Since $1^2 + 3^2 = 10$, the hypotenuse of a right triangle with legs of lengths 1 unit and 3 units has a length of $\sqrt{10}$ units.

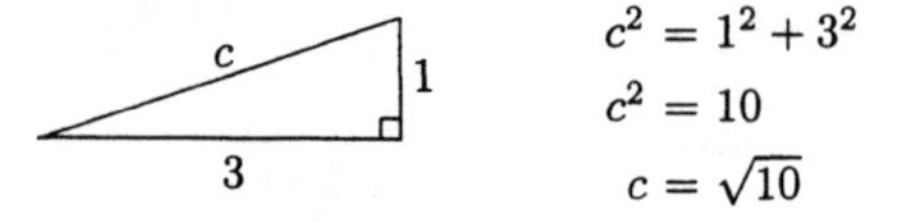

$c^2 = 1^2 + 3^2$
$c^2 = 10$
$c = \sqrt{10}$

Exercise Set R.2

1. $18^\circ = 1$ (For any nonzero real number, $a^0 = 1$.)

2. $\left(-\frac{4}{3}\right)^0 = 1$

3. $5^8 \cdot 5^{-6} = 5^{8+(-6)} = 5^2$, or 25

4. $6^2 \cdot 6^{-7} = 6^{2+(-7)} = 6^{-5}$, or $\frac{1}{6^5}$

5. $m^{-5} \cdot m^5 = m^{-5+5} = m^0 = 1$

6. $n^9 \cdot n^{-9} = n^{9+(-9)} = n^0 = 1$

7. $7^3 \cdot 7^{-5} \cdot 7 = 7^{3+(-5)+1} = 7^{-1}$, or $\frac{1}{7}$

8. $3^6 \cdot 3^{-5} \cdot 3^4 = 3^{6+(-5)+4} = 3^5$

9. $2x^3 \cdot 3x^2 = 2 \cdot 3 \cdot x^{3+2} = 6x^5$

10. $3y^4 \cdot 4y^3 = 3 \cdot 4 \cdot y^{4+3} = 12y^7$

11. $(5a^2b)(3a^{-3}b^4) = 5 \cdot 3 \cdot a^{2+(-3)} \cdot b^{1+4} = 15a^{-1}b^5$, or $\dfrac{15b^5}{a}$

12. $(4xy^2)(3x^{-4}y^5) = 4 \cdot 3 \cdot x^{1+(-4)} \cdot y^{2+5} = 12x^{-3}y^7$, or $\dfrac{12y^7}{x^3}$

13. $(2x)^3(3x)^2 = 2^3x^3 \cdot 3^2x^2 = 8 \cdot 9 \cdot x^{3+2} = 72x^5$

14. $(4y)^2(3y)^3 = 16y^2 \cdot 27y^3 = 432y^5$

15. $\dfrac{b^{40}}{b^{37}} = b^{40-37} = b^3$

16. $\dfrac{a^{39}}{a^{32}} = a^{39-32} = a^7$

17. $\dfrac{x^2y^{-2}}{x^{-1}y} = x^{2-(-1)}y^{-2-1} = x^3y^{-3}$, or $\dfrac{x^3}{y^3}$

18. $\dfrac{x^3y^{-3}}{x^{-1}y^2} = x^{3-(-1)}y^{-3-2} = x^4y^{-5}$, or $\dfrac{x^4}{y^5}$

19. $\dfrac{32x^{-4}y^3}{4x^{-5}y^8} = \dfrac{32}{4}x^{-4-(-5)}y^{3-8} = 8xy^{-5}$, or $\dfrac{8x}{y^5}$

20. $\dfrac{20a^5b^{-2}}{5a^7b^{-3}} = \dfrac{20}{5}a^{5-7}b^{-2-(-3)} = 4a^{-2}b$, or $\dfrac{4b}{a^2}$

21. $(2ab^2)^3 = 2^3a^3(b^2)^3 = 2^3a^3b^{2\cdot 3} = 8a^3b^6$

22. $(4xy^3)^2 = 4^2x^2(y^3)^2 = 16x^2y^6$

23. $(-2x^3)^4 = (-2)^4(x^3)^4 = (-2)^4x^{3\cdot 4} = 16x^{12}$

24. $(-3x^2)^4 = (-3)^4(x^2)^4 = 81x^8$

25. $(-5c^{-1}d^{-2})^{-2} = (-5)^{-2}c^{-1(-2)}d^{-2(-2)} = \dfrac{c^2d^4}{(-5)^2} = \dfrac{c^2d^4}{25}$

26. $(-4x^{-5}z^{-2})^{-3} = (-4)^{-3}(x^{-5})^{-3}(z^{-2})^{-3} = \dfrac{x^{15}z^6}{(-4)^3} = \dfrac{x^{15}z^6}{-64}$

27. $\left(\dfrac{24a^{10}b^{-8}c^7}{3a^6b^{-3}b^5}\right)^5 = (8a^4b^{-5}c^2)^5 = 8^5a^{20}b^{-25}c^{10}$, or $\dfrac{8^5a^{20}c^{10}}{b^{25}}$

28. $\left(\dfrac{125p^{12}q^{-14}r^{22}}{25p^8q^6r^{-15}}\right)^{-4} = (5p^4q^{-20}r^{37})^{-4} =$

$5^{-4}p^{-16}q^{80}r^{-148}$, or $\dfrac{q^{80}}{625p^{16}r^{148}}$

29. Convert 405,000 to scientific notation.

We want the decimal point to be positioned between the 4 and the first 0, so we move it 5 places to the left. Since 405,000 is greater than 10, the exponent must be positive.

$$405,000 = 4.05 \times 10^5$$

30. Position the decimal point 6 places to the left, between the 1 and the 6. Since 1,670,000 is greater than 10, the exponent must be positive.

$$1,670,000 = 1.67 \times 10^6$$

31. Convert 0.00000039 to scientific notation.

We want the decimal point to be positioned between the 3 and the 9, so we move it 7 places to the right. Since 0.00000039 is a number between 0 and 1 the exponent must be negative.

$$0.00000039 = 3.9 \times 10^{-7}$$

32. Position the decimal point 4 places to the right, between the 9 and the 2. Since 0.00092 is a number between 0 and 1, the exponent must be negative.

$$0.00092 = 9.2 \times 10^{-4}$$

33. Convert 0.000016 to scientific notation.

We want the decimal point to be positioned between the 1 and the 6, so we move it 5 places to the right. Since 0.000016 is a number between 0 and 1, the exponent must be negative.

$$0.000016\ \text{m}^3 = 1.6 \times 10^{-5}\ \text{m}^3$$

34. Position the decimal point 11 places to the left between the first 8 and the 2. Since 828,597,000,000 is greater than 10, the exponent must be positive.

$$\$828,597,000,000 = \$8.28597 \times 10^{11}$$

35. Convert 8.3×10^{-5} to decimal notation.

The exponent is negative, so the number is between 0 and 1. We move the decimal point 5 places to the left.

$$8.3 \times 10^{-5} = 0.000083$$

36. The exponent is positive, so the number is greater than 10. We move the decimal point 6 places to the right.

$$4.1 \times 10^6 = 4,100,000$$

37. Convert 2.07×10^7 to decimal notation.

The exponent is positive, so the number is greater than 10. We move the decimal point 7 places to the right.

$$2.07 \times 10^7 = 20,700,000$$

38. The exponent is negative, so the number is between 0 and 1. We move the decimal point 6 places to the left.

$$3.15 \times 10^{-6} = 0.00000315$$

39. Convert $\$1.1358 \times 10^{10}$ to decimal notation.

The exponent is positive, so the number is greater than 10. We move the decimal point 10 places to the right.

$$\$1.1358 \times 10^{10} = \$11,358,000,000$$

40. The exponent is negative, so the number is between 0 and 1. We move the decimal point 24 places to the left.

1.67×10^{-24} g $= 0.00000000000000000000000167$ g

41.
$$(3.1 \times 10^5)(4.5 \times 10^{-3})$$
$= (3.1 \times 4.5) \times (10^5 \times 10^{-3})$
$= 13.95 \times 10^2$ This is not scientific notation.
$= (1.395 \times 10) \times 10^2$
$= 1.395 \times 10^3$ Writing scientific notation

42. $(9.1 \times 10^{-17})(8.2 \times 10^3) = 74.62 \times 10^{-14}$
$= (7.462 \times 10) \times 10^{-14}$
$= 7.462 \times 10^{-13}$

43. $\dfrac{6.4 \times 10^{-7}}{8.0 \times 10^6} = \dfrac{6.4}{8.0} \times \dfrac{10^{-7}}{10^6}$
$= 0.8 \times 10^{-13}$ This is not scientific notation.
$= (8 \times 10^{-1}) \times 10^{-13}$
$= 8 \times 10^{-14}$ Writing scientific notation

44. $\dfrac{1.1 \times 10^{-40}}{2.0 \times 10^{-71}} = 0.55 \times 10^{31}$
$= (5.5 \times 10^{-1}) \times 10^{31}$
$= 5.5 \times 10^{30}$

45. The average cost per mile is the total cost divided by the number of miles.

$\dfrac{\$210 \times 10^6}{17.6}$
$= \dfrac{\$210 \times 10^6}{1.76 \times 10}$
$\approx \$119 \times 10^5$
$\approx (\$1.19 \times 10^2) \times 10^5$
$\approx \$1.19 \times 10^7$

The average cost per mile was about $\$1.19 \times 10^7$.

46. $\dfrac{\$6.152415 \times 10^9}{3.05 \times 10^7} \approx \$2.02 \times 10^2$

47. First find the number of seconds in 1 hour:

1 hour $= 1 \text{ hr} \times \dfrac{60 \text{ min}}{1 \text{ hr}} \times \dfrac{60 \text{ sec}}{1 \text{ min}} = 3600$ sec

The number of disintegrations produced in 1 hour is the number of disintegrations per second times the number of seconds in 1 hour.

37 billion $\times$ 3600
$= 37,000,000,000 \times 3600$
$= 3.7 \times 10^{10} \times 3.6 \times 10^3$ Writing scientific notation
$= (3.7 \times 3.6) \times (10^{10} \times 10^3)$
$= 13.32 \times 10^{13}$ Multiplying
$= (1.332 \times 10) \times 10^{13}$
$= 1.332 \times 10^{14}$

One gram of radium produces 1.332×10^{14} disintegrations in 1 hour.

48.
$$2\pi \times 93,000,000$$
$= 2\pi \times 9.3 \times 10^7$
≈ 58.43362336
$\approx (5.843362336 \times 10) \times 10^7$
$= 5.843362336 \times 10^8$ mi

49.
$$3 \cdot 2 - 4 \cdot 2^2 + 6(3-1)$$
$= 3 \cdot 2 - 4 \cdot 2^2 + 6 \cdot 2$ Working inside parentheses
$= 3 \cdot 2 - 4 \cdot 4 + 6 \cdot 2$ Evaluating 2^2
$= 6 - 16 + 12$ Multiplying
$= -10 + 12$ Adding in order
$= 2$ from left to right

50.
$$3[(2 + 4 \cdot 2^2) - 6(3-1)]$$
$= 3[(2 + 4 \cdot 4) - 6 \cdot 2]$
$= 3[(2 + 16) - 6 \cdot 2]$
$= 3[18 - 6 \cdot 2]$
$= 3[18 - 12]$
$= 3[6]$
$= 18$

51.
$$16 \div 4 \cdot 4 \div 2 \cdot 256$$
$= 4 \cdot 4 \div 2 \cdot 256$ Multiplying and dividing in order from left to right
$= 16 \div 2 \cdot 256$
$= 8 \cdot 256$
$= 2048$

52.
$$2^6 \cdot 2^{-3} \div 2^{10} \div 2^{-8}$$
$= 2^3 \div 2^{10} \div 2^{-8}$
$= 2^{-7} \div 2^{-8}$
$= 2$

53.
$$\frac{4(8-6)^2 - 4 \cdot 3 + 2 \cdot 8}{3^1 + 19^0}$$
$= \dfrac{4 \cdot 2^2 - 4 \cdot 3 + 2 \cdot 8}{3 + 1}$ Calculating in the numerator and in the denominator
$= \dfrac{4 \cdot 4 - 4 \cdot 3 + 2 \cdot 8}{4}$
$= \dfrac{16 - 12 + 16}{4}$
$= \dfrac{4 + 16}{4}$
$= \dfrac{20}{4}$
$= 5$

54. $\dfrac{[4(8-6)^2+4](3-2\cdot 8)}{2^2(2^3+5)}$

$= \dfrac{[4\cdot 2^2+4](3-16)}{2^2(8+5)}$

$= \dfrac{[4\cdot 4+4](-13)}{2^2\cdot 13}$

$= \dfrac{[16+4](-13)}{4\cdot 13}$

$= \dfrac{20(-13)}{52}$

$= \dfrac{-260}{52}$

$= -5$

55. Since interest is compounded semiannually, $n = 2$. Substitute \$2125 for P, 6.2% or 0.062 for i, 2 for n, and 5 for t in the compound interest formula.

$A = P\left(1+\dfrac{i}{n}\right)^{nt}$

$= \$2125\left(1+\dfrac{0.062}{2}\right)^{2\cdot 5}$ Substituting

$= \$2125(1+0.031)^{2\cdot 5}$ Dividing

$= \$2125(1.031)^{2\cdot 5}$ Adding

$= \$2125(1.031)^{10}$ Multiplying 2 and 5

$\approx \$2125(1.357021264)$ Evaluating the exponential expression

$\approx \$2883.670185$ Multiplying

$\approx \$2883.67$ Rounding to the nearest cent

56. $A = \$9550\left(1+\dfrac{0.054}{2}\right)^{2\cdot 7} \approx \$13,867.23$

57. Since interest is compounded quarterly, $n = 4$. Substitute \$6700 for P, 4.5% or 0.045 for i, 4 for n, and 6 for t in the compound interest formula.

$A = P\left(1+\dfrac{i}{n}\right)^{nt}$

$= \$6700\left(1+\dfrac{0.045}{4}\right)^{4\cdot 6}$ Substituting

$= \$6700(1+0.01125)^{4\cdot 6}$ Dividing

$= \$6700(1.01125)^{4\cdot 6}$ Adding

$= \$6700(1.01125)^{24}$ Multiplying 4 and 6

$\approx \$6700(1.307991226)$ Evaluating the exponential expression

$\approx \$8763.541217$ Multiplying

$\approx \$8763.54$ Rounding to the nearest cent

58. $A = \$4875\left(1+\dfrac{0.058}{4}\right)^{4\cdot 9} \approx \$8185.56$

59. Yes; find the results with parentheses and without them.

$4\cdot 25 \div (10-5) = 4\cdot 25\div 5 = 100\div 5 = 20,$

but $4\cdot 25\div 10-5 = 100\div 10-5 = 10-5 = 5.$

60. No; x^{-2}, or $\dfrac{1}{x^2}$ is positive for all $x < 0$ and x^{-1}, or $\dfrac{1}{x}$ is negative for all $x < 0$. Partial confirmation can be obtained by inspecting the graphs of $y_1 = x^{-2}$ and $y_2 = x^{-1}$ for $x < 0$.

61. $(x^t\cdot x^{3t})^2 = (x^{4t})^2 = x^{4t\cdot 2} = x^{8t}$

62. $(x^y\cdot x^{-y})^3 = (x^0)^3 = 1^3 = 1$

63. $(t^{a+x}\cdot t^{x-a})^4 = (t^{2x})^4 = t^{2x\cdot 4} = t^{8x}$

64. $(m^{x-b}\cdot n^{x+b})^x(m^bn^{-b})^x$

$= (m^{x^2-bx}n^{x^2+bx})(m^{bx}n^{-bx})$

$= m^{x^2}n^{x^2}$

65. $\left[\dfrac{(3x^ay^b)^3}{(-3x^ay^b)^2}\right]^2 = \left[\dfrac{27x^{3a}y^{3b}}{9x^{2a}y^{2b}}\right]^2$

$= \left[3x^ay^b\right]^2$

$= 9x^{2a}y^{2b}$

66. $\left[\left(\dfrac{x^r}{y^t}\right)^2\left(\dfrac{x^{2r}}{y^{4t}}\right)^{-2}\right]^{-3}$

$= \left[\left(\dfrac{x^{2r}}{y^{2t}}\right)\left(\dfrac{x^{-4r}}{y^{-8t}}\right)\right]^{-3}$

$= \left(\dfrac{x^{-2r}}{y^{-6t}}\right)^{-3}$

$= \dfrac{x^{6r}}{y^{18t}}$, or $x^{6r}y^{-18t}$

67. $P = \$98,000 - \$16,000 = \$82,000;$

$i = 8\frac{1}{2}\%$, or 0.085; $n = 12\cdot 25 = 300$

$M = P\left[\dfrac{\frac{i}{12}\left(1+\frac{i}{12}\right)^n}{\left(1+\frac{i}{12}\right)^n - 1}\right]$

$M = \$82,000\left[\dfrac{\frac{0.085}{12}\left(1+\frac{0.085}{12}\right)^{300}}{\left(1+\frac{0.085}{12}\right)^{300} - 1}\right]$

$\approx \$660.29$ Using a calculator

68. $P = \$124,000 - \$20,000 = \$104,000;$

$i = 7\frac{3}{4}\%$, or 0.0775; $n = 12\cdot 30 = 360$

$M = P\left[\dfrac{\frac{i}{12}\left(1+\frac{i}{12}\right)^n}{\left(1+\frac{i}{12}\right)^n - 1}\right]$

$M = \$104,000\left[\dfrac{\frac{0.0775}{12}\left(1+\frac{0.0775}{12}\right)^{360}}{\left(1+\frac{0.0775}{12}\right)^{360} - 1}\right]$

$\approx \$745.07$

69. $P = \$135,000 - \$18,000 = \$117,000$;

$i = 7\frac{1}{2}\%$, or 0.075; $n = 12 \cdot 20 = 240$

$$M = P\left[\frac{\frac{i}{12}\left(1+\frac{i}{12}\right)^n}{\left(1+\frac{i}{12}\right)^n - 1}\right]$$

$$M = \$117,000\left[\frac{\frac{0.075}{12}\left(1+\frac{0.075}{12}\right)^{240}}{\left(1+\frac{0.075}{12}\right)^{240} - 1}\right]$$

$\approx \$924.54$ Using a calculator

70. $P = \$151,000 - \$21,000 = \$130,000$;

$i = 8\frac{1}{4}\%$, or 0.0825; $n = 12 \cdot 25 = 300$

$$M = P\left[\frac{\frac{i}{12}\left(1+\frac{i}{12}\right)^n}{\left(1+\frac{i}{12}\right)^n - 1}\right]$$

$$M = \$130,000\left[\frac{\frac{0.0825}{12}\left(1+\frac{0.0825}{12}\right)^{300}}{\left(1+\frac{0.0825}{12}\right)^{300} - 1}\right]$$

$\approx \$1024.99$

71. Deselect the graph of y_3 and inspect the graphs of y_1 and y_2 in a suitable window such as $[-2, 2, -2, 2]$. The graph of y_1 lies on or above the graph of y_2 for $x \leq 1$.

72. Deselect the graph of y_2 and inspect the graphs of y_1 and y_3 in a suitable window such as $[-2, 2, -2, 2]$. The graph of y_3 lies above the graph of y_1 for $x < -1$ or $x > 1$.

Exercise Set R.3

1. $-5y^4 + 3y^3 + 7y^2 - y - 4 =$
$-5y^4 + 3y^3 + 7y^2 + (-y) + (-4)$

Terms: $-5y^4$, $3y^3$, $7y^2$, $-y$, -4

The degree of the term of highest degree, $-5y^4$, is 4. Thus, the degree of the polynomial is 4.

2. $2m^3 - m^2 - 4m + 11 = 2m^3 + (-m^2) + (-4m) + 11$

Terms: $2m^3$, $-m^2$, $-4m$, 11

The degree of the term of highest degree, $2m^3$, is 3. Thus, the degree of the polynomial is 3.

3. $3a^4b - 7a^3b^3 + 5ab - 2 = 3a^4b + (-7a^3b^3) + 5ab + (-2)$

Terms: $3a^4b$, $-7a^3b^3$, $5ab$, -2

The degrees of the terms are 5, 6, 2, and, 0, respectively, so the degree of the polynomial is 6.

4. $6p^3q^2 - p^2q^4 - 3pq^2 + 5 = 6p^3q^2 + (-p^2q^4) + (-3pq^2) + 5$

Terms: $6p^3q^2$, $-p^2q^4$, $-3pq^2$, 5

The degrees of the terms are 5, 6, 3, and 0, respectively, so the degree of the polynomial is 6.

5. $(5x^2y - 2xy^2 + 3xy - 5) + (-2x^2y - 3xy^2 + 4xy + 7)$
$= (5-2)x^2y + (-2-3)xy^2 + (3+4)xy + (-5+7)$
$= 3x^2y - 5xy^2 + 7xy + 2$

6. $2x^2y - 7xy^2 + 8xy + 5$

7. $(2x + 3y + z - 7) + (4x - 2y - z + 8) + (-3x + y - 2z - 4)$
$= (2+4-3)x + (3-2+1)y + (1-1-2)z + (-7+8-4)$
$= 3x + 2y - 2z - 3$

8. $7x^2 + 12xy - 2x - y - 9$

9. $(3x^2 - 2x - x^3 + 2) - (5x^2 - 8x - x^3 + 4)$
$= (3x^2 - 2x - x^3 + 2) + (-5x^2 + 8x + x^3 - 4)$
$= (3-5)x^2 + (-2+8)x + (-1+1)x^3 + (2-4)$
$= -2x^2 + 6x - 2$

10. $-4x^2 + 8xy - 5y^2 + 3$

11. $(x^4 - 3x^2 + 4x) - (3x^3 + x^2 - 5x + 3)$
$= (x^4 - 3x^2 + 4x) + (-3x^3 - x^2 + 5x - 3)$
$= x^4 - 3x^3 + (-3-1)x^2 + (4+5)x - 3$
$= x^4 - 3x^3 - 4x^2 + 9x - 3$

12. $2x^4 - 5x^3 - 5x^2 + 10x - 5$

13. $(a-b)(2a^3 - ab + 3b^2)$
$= (a-b)(2a^3) + (a-b)(-ab) + (a-b)(3b^2)$ Using the distributive property
$= 2a^4 - 2a^3b - a^2b + ab^2 + 3ab^2 - 3b^3$ Using the distributive property three more times
$= 2a^4 - 2a^3b - a^2b + 4ab^2 - 3b^3$ Collecting like terms

14. $(n+1)(n^2 - 6n - 4)$
$= (n+1)(n^2) + (n+1)(-6n) + (n+1)(-4)$
$= n^3 + n^2 - 6n^2 - 6n - 4n - 4$
$= n^3 - 5n^2 - 10n - 4$

15. $(x+5)(x-3)$
$= x^2 - 3x + 5x - 15$ Using FOIL
$= x^2 + 2x - 15$ Collecting like terms

16. $(y-4)(y+1) = y^2 + y - 4y - 4 = y^2 - 3y - 4$

17. $(2a+3)(a+5)$
$= 2a^2 + 10a + 3a + 15$ Using FOIL
$= 2a^2 + 13a + 15$ Collecting like terms

18. $(3b+1)(b-2) = 3b^2 - 6b + b - 2 = 3b^2 - 5b - 2$

19. Construct a table of values for $y_1 = (2x+3)(x+5)$ and $y_2 = 2x^2 + 13x + 15$. The values of y_1 and y_2 are the same for each given x-value, so the product appears to be correct.

20. Construct a table of values for $y_1 = (3x+1)(x-2)$ and $y_2 = 3x^2 - 5x - 2$. The values of y_1 and y_2 are the same for each given x-value, so the product appears to be correct.

21. $(2x+3y)(2x+y)$

$= 4x^2 + 2xy + 6xy + 3y^2$ Using FOIL

$= 4x^2 + 8xy + 3y^2$

22. $(2a-3b)(2a-b) = 4a^2 - 2ab - 6ab + 3b^2 = 4a^2 - 8ab + 3b^2$

23. $(y+5)^2$

$= y^2 + 2 \cdot y \cdot 5 + 5^2$

$[(A+B)^2 = A^2 + 2AB + B^2]$

$= y^2 + 10y + 25$

24. $(y+7)^2 = y^2 + 2 \cdot y \cdot 7 + 7^2 = y^2 + 14y + 49$

25. $(5x-3)^2$

$= (5x)^2 - 2 \cdot 5x \cdot 3 + 3^2$

$[(A-B)^2 = A^2 - 2AB + B^2]$

$= 25x^2 - 30x + 9$

26. $(3x-2)^2 = (3x)^2 - 2 \cdot 3x \cdot 2 + 2^2 = 9x^2 - 12x + 4$

27. Construct a table of values for $y_1 = (5x-3)^2$ and $y_2 = 25x^2 - 30x + 9$. The values of y_1 and y_2 are the same for each given x-value, so the product appears to be correct.

28. Construct a table of values for $y_1 = (3x-2)^2$ and $y_2 = 9x^2 - 12x + 4$. The values of y_1 and y_2 are the same for each given x-value, so the product appears to be correct.

29. $(2x+3y)^2$

$= (2x)^2 + 2(2x)(3y) + (3y)^2$

$[(A+B)^2 = A^2+2AB+B^2]$

$= 4x^2 + 12xy + 9y^2$

30. $(5x+2y)^2 = (5x)^2 + 2 \cdot 5x \cdot 2y + (2y)^2 = 25x^2 + 20xy + 4y^2$

31. $(2x^2-3y)^2$

$= (2x^2)^2 - 2(2x^2)(3y) + (3y)^2$

$[(A-B)^2 = A^2 - 2AB + B^2]$

$= 4x^4 - 12x^2y + 9y^2$

32. $(4x^2-5y)^2 = (4x^2)^2 - 2 \cdot 4x^2 \cdot 5y + (5y)^2 = 16x^4 - 40x^2y + 25y^2$

33. $(a+3)(a-3)$

$= a^2 - 3^2$ $[(A+B)(A-B) = A^2 - B^2]$

$= a^2 - 9$

34. $(b+4)(b-4) = b^2 - 4^2 = b^2 - 16$

35. Construct a table of values for $y_1 = (x+3)(x-3)$ and $y_2 = x^2 - 9$. The values of y_1 and y_2 are the same for each given x-value, so the product appears to be correct.

36. Construct a table of values for $y_1 = (x+4)(x-4)$ and $y_2 = x^2 - 16$. The values of y_1 and y_2 are the same for each given x-value, so the product appears to be correct.

37. $(3x-2y)(3x+2y)$

$= (3x)^2 - (2y)^2$ $[(A-B)(A+B) = A^2-B^2]$

$= 9x^2 - 4y^2$

38. $(3x+5y)(3x-5y) = (3x)^2 - (5y)^2 = 9x^2 - 25y^2$

39. $(2x+3y+4)(2x+3y-4)$

$= [(2x+3y)+4][(2x+3y)-4]$

$= (2x+3y)^2 - 4^2$

$= 4x^2 + 12xy + 9y^2 - 16$

40. $(5x+2y+3)(5x+2y-3) = (5x+2y)^2 - 3^2 = 25x^2 + 20xy + 4y^2 - 9$

41. $(x+1)(x-1)(x^2+1)$

$= (x^2-1)(x^2+1)$

$= x^4 - 1$

42. $(y-2)(y+2)(y^2+4)$

$= (y^2-4)(y^2+4)$

$= y^4 - 16$

43. No; if the leading coefficients of the polynomials are additive inverses, the degree of the sum is less than n. For example, the sum of the second degree polynomials x^2+x-1 and $-x^2+4$ is $x+3$, a first degree polynomial.

44. Algebraically: Choose specific values for A and B, $A \neq 0$, $B \neq 0$, and evaluate $(A+B)^2$ and A^2+B^2. For example, $(2+3)^2 = 5^2 = 25$, but $2^2+3^2 = 4+9 = 13$.

Graphically: For $A = x$ and $B \neq 0$, graph $y_1 = (x+B)^2$ and $y_2 = x^2 + B^2$ and observe that the graphs do not coincide. For example, let $B = 2$ and graph $y_1 = (x+2)^2$ and $y_2 = x^2 + 4$.

Numerically: For y_1 and y_2 as described in the graphical method above, use the TABLE feature and observe that $y_1 = y_2$ only for $x = 0$.

Geometrically: Show that the area of a square with side $A+B$ is not equal to A^2+B^2. See the figure below.

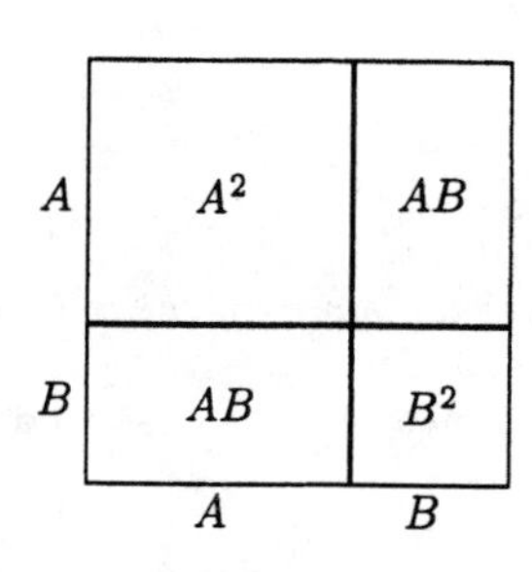

45. $(a^n + b^n)(a^n - b^n) = (a^n)^2 - (b^n)^2$
$= a^{2n} - b^{2n}$

46. $(t^a + 4)(t^a - 7) = (t^a)^2 - 7t^a + 4t^a - 28 =$
$t^{2a} - 3t^a - 28$

47. $(a^n + b^n)^2 = (a^n)^2 + 2 \cdot a^n \cdot b^n + (b^n)^2$
$= a^{2n} + 2a^n b^n + b^{2n}$

48. $(x^{3m} - t^{5n})^2 = (x^{3m})^2 - 2 \cdot x^{3m} \cdot t^{5n} + (t^{5n})^2 =$
$x^{6m} - 2x^{3m}t^{5n} + t^{10n}$

49. $(x-1)(x^2+x+1)(x^3+1)$
$= [(x-1)x^2 + (x-1)x + (x-1) \cdot 1](x^3+1)$
$= (x^3 - x^2 + x^2 - x + x - 1)(x^3+1)$
$= (x^3-1)(x^3+1)$
$= (x^3)^2 - 1^2$
$= x^6 - 1$

50. $[(2x-1)^2 - 1]^2$
$= [4x^2 - 4x + 1 - 1]^2$
$= [4x^2 - 4x]^2$
$= (4x^2)^2 - 2(4x^2)(4x) + (4x)^2$
$= 16x^4 - 32x^3 + 16x^2$

51. $(x^{a-b})^{a+b}$
$= x^{(a-b)(a+b)}$
$= x^{a^2-b^2}$

52. $(t^{m+n})^{m+n} \cdot (t^{m-n})^{m-n}$
$= t^{m^2+2mn+n^2} \cdot t^{m^2-2mn+n^2}$
$= t^{2m^2+2n^2}$

53. $(a+b+c)^2$
$= (a+b+c)(a+b+c)$
$= (a+b+c)(a) + (a+b+c)(b) + (a+b+c)(c)$
$= a^2 + ab + ac + ab + b^2 + bc + ac + bc + c^2$
$= a^2 + b^2 + c^2 + 2ab + 2ac + 2bc$

Exercise Set R.4

1. $2x - 10 = 2 \cdot x - 2 \cdot 5 = 2(x-5)$

2. $7y + 42 = 7 \cdot y + 7 \cdot 6 = 7(y+6)$

3. $3x^4 - 9x^2 = 3x^2 \cdot x^2 - 3x^2 \cdot 3 = 3x^2(x^2-3)$

4. $20y^2 - 5y^5 = 5y^2 \cdot 4 - 5y^2 \cdot y^3 = 5y^2(4 - y^3)$

5. $4a^2 - 12a + 16 = 4 \cdot a^2 - 4 \cdot 3a + 4 \cdot 4 = 4(a^2 - 3a + 4)$

6. $6n^2 + 24n - 18 = 6 \cdot n^2 + 6 \cdot 4n - 6 \cdot 3 = 6(n^2 + 4n - 3)$

7. $a(b-2) + c(b-2) = (b-2)(a+c)$

8. $a(x^2-3) - 2(x^2-3) = (x^2-3)(a-2)$

9. $x^3 + 3x^2 + 6x + 18$
$= x^2(x+3) + 6(x+3)$
$= (x+3)(x^2+6)$

10. $3x^3 + x^2 - 18x - 6$
$= x^2(3x+1) - 6(3x+1)$
$= (3x+1)(x^2-6)$

11. $y^3 - 3y^2 - 4y + 12$
$= y^2(y-3) - 4(y-3)$
$= (y-3)(y^2-4)$
$= (y-3)(y+2)(y-2)$

12. $p^3 - 2p^2 - 9p + 18$
$= p^2(p-2) - 9(p-2)$
$= (p-2)(p^2-9)$
$= (p-2)(p+3)(p-3)$

13. The graphs of $y_1 = x^3 + 3x^2 + 6x + 18$ and $y_2 = (x+3)(x^2+6)$ appear to coincide. A table of values also shows that the values of y_1 and y_2 are the same for the given x-values.

14. The graphs of $y_1 = 3x^3 + x^2 - 18x - 6$ and $y_2 = (3x+1)(x^2-6)$ appear to coincide. A table of values also shows that the values of y_1 and y_2 are the same for the given x-values.

15. $p^2 + 6p + 8$

We look for two numbers with a product of 8 and a sum of 6. By trial, we determine that they are 2 and 4.

$$p^2 + 6p + 8 = (p+2)(p+4)$$

16. Note that $(-5)(-2) = 10$ and $-5 + (-2) = -7$. Then
$$w^2 - 7w + 10 = (w-5)(w-2).$$

17. $2n^2 + 9n - 56$

We look for factors $(pn+q)(rn+s)$ for which $pn \cdot rn = 2n^2$ and $q \cdot s = -56$. When we multiply the inside terms, then the outside terms, and add, we must have $9n$. By trial, we determine the factorization:

$$2n^2 + 9n - 56 = (2n - 7)(n + 8)$$

18. $3y^2 + 7y - 20 = (3y - 5)(y + 4)$

19. $y^4 - 4y^2 - 21$

Think of this polynomial as $u^2 - 4u - 21$, where we have mentally substituted u for y^2. Then we look for factors of -21 whose sum is -4. By trial we determine the factors to be -7 and 3, so

$$u^2 - 4u - 21 = (u - 7)(u + 3).$$

Then, substituting y^2 for u, we obtain the factorization of the original trinomial.

$$y^4 - 4y^2 - 21 = (y^2 - 7)(y^2 + 3)$$

Neither of these factors can be factored further, so the factorization is complete.

20. $m^4 - m^2 - 90 = (m^2 - 10)(m^2 + 9)$

21. The graphs of $y_1 = x^2+6x+8$ and $y_2 = (x+2)(x+4)$ appear to coincide. A table of values also shows that the values of y_1 and y_2 are the same for the given x-values.

22. The graphs of $y_1 = x^2-7x+10$ and $y_2 = (x-5)(x-2)$ appear to coincide. A table of values also shows that the values of y_1 and y_2 are the same for the given x-values.

23. $9x^2 - 25 = (3x)^2 - 5^2$
$= (3x + 5)(3x - 5)$

24. $16x^2 - 9 = (4x - 3)(4x + 3)$

25. $4xy^4 - 4xz^2 = 4x(y^4 - z^2)$
$= 4x[(y^2)^2 - z^2]$
$= 4x(y^2 + z)(y^2 - z)$

26. $5x^4y - 5yz^4 = 5y(x^4 - z^4)$
$= 5y(x^2 + z^2)(x^2 - z^2)$
$= 5y(x^2 + z^2)(x + z)(x - z)$

27. $y^2 - 6y + 9 = y^2 - 2 \cdot y \cdot 3 + 3^2$
$= (y - 3)^2$

28. $x^2 + 8x + 16 = (x + 4)^2$

29. $1 - 8x + 16x^2 = 1^2 - 2 \cdot 1 \cdot 4x + (4x)^2$
$= (1 - 4x)^2$

30. $1 + 10x + 25x^2 = (1 + 5x)^2$

31. The graphs of $y_1 = 1 - 8x + 16x^2$ and $y_2 = (1 - 4x)^2$ appear to coincide. A table of values also shows that the values of y_1 and y_2 are the same for the given x-values.

32. The graphs of $y_1 = 1+10x+25x^2$ and $y_2 = (1+5x)^2$ appear to coincide. A table of values also shows that the values of y_1 and y_2 are the same for the given x-values.

33. $x^3 + 8 = x^3 + 2^3$
$= (x + 2)(x^2 - 2x + 4)$

34. $y^3 - 64 = (y - 4)(y^2 + 4y + 16)$

35. $m^3 - 1 = m^3 - 1^3$
$= (m - 1)(m^2 + m + 1)$

36. $n^3 + 216 = (n + 6)(n^2 - 6n + 36)$

37. The graphs of $y_1 = x^3 - 1$ and $y_2 = (x - 1)(x^2 + x + 1)$ appear to coincide. A table of values also shows that the values of y_1 and y_2 are the same for the given x-values.

38. The graphs of $y_1 = x^3 + 216$ and $y_2 = (x + 6)(x^2 - 6x + 36)$ appear to coincide. A table of values also shows that the values of y_1 and y_2 are the same for the given x-values.

39. $18a^2b - 15ab^2 = 3ab \cdot 6a - 3ab \cdot 5b$
$= 3ab(6a - 5b)$

40. $4x^2y + 12xy^2 = 4xy(x + 3y)$

41. $x^3 - 4x^2 + 5x - 20 = x^2(x - 4) + 5(x - 4)$
$= (x - 4)(x^2 + 5)$

42. $z^3 + 3z^2 - 3z - 9 = z^2(z + 3) - 3(z + 3)$
$= (z + 3)(z^2 - 3)$

43. $8x^2 - 32 = 8(x^2 - 4)$
$= 8(x + 2)(x - 2)$

44. $6y^2 - 6 = 6(y^2 - 1) = 6(y + 1)(y - 1)$

45. $4y^2 - 5$

There are no common factors. We might try to factor this polynomial as a difference of squares, but there is no integer which yields 5 when squared. Thus, the polynomial is prime.

46. There are no common factors and there is no integer which yields 7 when squared, so $16x^2 - 7$ is prime.

47. The graphs of $y_1 = x^3 - 4x^2 + 5x - 20$ and $y_2 = (x - 4)(x^2 + 5)$ appear to coincide. A table of values also shows that the values of y_1 and y_2 are the same for the given x-values.

48. The graphs of $y_1 = x^3 + 3x^2 - 3x - 9$ and $y_2 = (x + 3)(x^2 - 3)$ appear to coincide. A table of values also shows that the values of y_1 and y_2 are the same for the given x-values.

49. $x^2 + 9x + 20$

We look for two numbers with a product of 20 and a sum of 9. They are 4 and 5.

$$x^2 + 9x + 20 = (x+4)(x+5)$$

50. Note that $3(-2) = -6$ and $3 + (-2) = 1$. Then

$$y^2 + y - 6 = (y+3)(y-2).$$

51. $y^2 - 6y + 5$

We look for two numbers with a product of 5 and a sum of -6. They are -5 and -1.

$$y^2 - 6y + 5 = (y-5)(y-1)$$

52. Note that $-7(3) = -21$ and $-7 + 3 = -4$.

$$x^2 - 4x - 21 = (x-7)(x+3)$$

53. $2a^2 + 9a + 4$

We look for factors $(pa+q)(ra+s)$ for which $pa \cdot qa = 2a^2$ and $q \cdot s = 4$. When we multiply the inside terms, then the outside terms, and add, we must have $9a$. By trial, we determine the factorization:

$$2a^2 + 9a + 4 = (2a+1)(a+4)$$

54. $3b^2 - b - 2 = (3b+2)(b-1)$

55. $6x^2 + 7x - 3$

We look for factors $(px+q)(rx+s)$ for which $px \cdot rx = 6$ and $q \cdot s = -3$. When we multiply the inside terms, then the outside terms, and add, we must have $7x$. By trial, we determine the factorization:

$$6x^2 + 7x - 3 = (3x-1)(2x+3)$$

56. $8x^2 + 2x - 15 = (4x-5)(2x+3)$

57.
$$\begin{aligned} y^2 - 18y + 81 &= y^2 - 2 \cdot y \cdot 9 + 9^2 \\ &= (y-9)^2 \end{aligned}$$

58. $n^2 + 2n + 1 = (n+1)^2$

59.
$$\begin{aligned} x^2y^2 - 14xy + 49 &= (xy)^2 - 2 \cdot xy \cdot 7 + 7^2 \\ &= (xy-7)^2 \end{aligned}$$

60. $x^2y^2 - 16xy + 64 = (xy-8)^2$

61.
$$\begin{aligned} 4ax^2 + 20ax - 56a &= 4a(x^2 + 5x - 14) \\ &= 4a(x+7)(x-2) \end{aligned}$$

62.
$$\begin{aligned} 21x^2y + 2xy - 8y &= y(21x^2 + 2x - 8) \\ &= y(7x-4)(3x+2) \end{aligned}$$

63.
$$\begin{aligned} 3z^3 - 24 &= 3(z^3 - 8) \\ &= 3(z^3 - 2^3) \\ &= 3(z-2)(z^2 + 2z + 4) \end{aligned}$$

64.
$$\begin{aligned} 4t^3 + 108 &= 4(t^3 + 27) \\ &= 4(t+3)(t^2 - 3t + 9) \end{aligned}$$

65.
$$\begin{aligned} &16a^7b + 54ab^7 \\ &= 2ab(8a^6 + 27b^6) \\ &= 2ab[(2a^2)^3 + (3b^2)^3] \\ &= 2ab(2a^2 + 3b^2)(4a^4 - 6a^2b^2 + 9b^4) \end{aligned}$$

66.
$$\begin{aligned} &24a^2x^4 - 375a^8x \\ &= 3a^2x(8x^3 - 125a^6) \\ &= 3a^2x(2x - 5a^2)(4x^2 + 10a^2x + 25a^4) \end{aligned}$$

67. The graphs of $y_1 = 6x^2 + 7x - 3$ and $y_2 = (3x-1)(2x+3)$ appear to coincide. A table of values also shows that the values of y_1 and y_2 are the same for the given x-values.

68. The graphs of $y_1 = 8x^2 + 2x - 15$ and $y_2 = (4x-5)(2x+3)$ appear to coincide. A table of values also shows that the values of y_1 and y_2 are the same for the given x-values.

69. $A^2 + B^2$ can be factored when A and B have a common factor. For example, let $A = 2x$ and $B = 10$. Then $A^2 + B^2 = 4x^2 + 100 = 4(x^2 + 25)$.

70.
$$\begin{aligned} A^3 - B^3 &= A^3 + (-B)^3 \\ &= (A + (-B))(A^2 - A(-B) + (-B)^2) \\ &= (A-B)(A^2 + AB + B^2) \end{aligned}$$

71.
$$\begin{aligned} &y^4 - 84 + 5y^2 \\ &= y^4 + 5y^2 - 84 \\ &= u^2 + 5u - 84 && \text{Substituting } u \text{ for } y^2 \\ &= (u+12)(u-7) \\ &= (y^2+12)(y^2-7) && \text{Substituting } y^2 \text{ for } u \end{aligned}$$

72.
$$\begin{aligned} &11x^2 + x^4 - 80 \\ &= x^4 + 11x^2 - 80 \\ &= u^2 + 11u - 80 && \text{Substituting } u \text{ for } x^2 \\ &= (u+16)(u-5) \\ &= (x^2+16)(x^2-5) && \text{Substituting } x^2 \text{ for } u \end{aligned}$$

73.
$$\begin{aligned} y^2 - \frac{8}{49} + \frac{2}{7}y &= y^2 + \frac{2}{7}y - \frac{8}{49} \\ &= \left(y + \frac{4}{7}\right)\left(y - \frac{2}{7}\right) \end{aligned}$$

74.
$$\begin{aligned} t^2 - 0.27 + 0.6t &= t^2 + 0.6t - 0.27 \\ &= (t+0.9)(t-0.3) \end{aligned}$$

75.
$$\begin{aligned} &(x+h)^3 - x^3 \\ &= [(x+h) - x][(x+h)^2 + x(x+h) + x^2] \\ &= (x + h - x)(x^2 + 2xh + h^2 + x^2 + xh + x^2) \\ &= h(3x^2 + 3xh + h^2) \end{aligned}$$

76.
$$\begin{aligned} &(x+0.01)^2 - x^2 \\ &= (x + 0.01 + x)(x + 0.01 - x) \\ &= 0.01(2x + 0.01), \text{ or } 0.02(x + 0.005) \end{aligned}$$

77. $(y-4)^2+5(y-4)-24$
$= u^2+5u-24$ Substituting u for $y-4$
$= (u+8)(u-3)$
$= (y-4+8)(y-4-3)$ Substituting $y-4$ for u
$= (y+4)(y-7)$

78. $6(2p+q)^2-5(2p+q)-25$
$= 6u^2-5u-25$ Substituting u for $2p+q$
$= (3u+5)(2u-5)$
$= [3(2p+q)+5][2(2p+q)-5]$ Substituting $2p+q$ for u
$= (6p+3q+5)(4p+2q-5)$

79. $x^{2n}+5x^n-24=(x^n)^2+5x^n-24$
$= (x^n+8)(x^n-3)$

80. $4x^{2n}-4x^n-3=(2x^n-3)(2x^n+1)$

81. $x^2+ax+bx+ab=x(x+a)+b(x+a)$
$= (x+a)(x+b)$

82. $bdy^2+ady+bcy+ac$
$= dy(by+a)+c(by+a)$
$= (by+a)(dy+c)$

83. $25y^{2m}-(x^{2n}-2x^n+1)$
$= (5y^m)^2-(x^n-1)^2$
$= [5y^m+(x^n-1)][5y^m-(x^n-1)]$
$= (5y^m+x^n-1)(5y^m-x^n+1)$

84. $x^{6a}-t^{3b}=(x^{2a})^3-(t^b)^3=$
$(x^{2a}-t^b)(x^{4a}+x^{2a}t^b+t^{2b})$

85. $(y-1)^4-(y-1)^2$
$= (y-1)^2[(y-1)^2-1]$
$= (y-1)^2[y^2-2y+1-1]$
$= (y-1)^2(y^2-2y)$
$= y(y-1)^2(y-2)$

86. $x^6-2x^5+x^4-x^2+2x-1$
$= x^4(x^2-2x+1)-(x^2-2x+1)$
$= (x^2-2x+1)(x^4-1)$
$= (x-1)^2(x^2+1)(x^2-1)$
$= (x-1)^2(x^2+1)(x+1)(x-1)$
$= (x^2+1)(x+1)(x-1)^3$

Exercise Set R.5

1. Since $-\dfrac{3}{4}$ is defined for all real numbers, the domain is $\{x|x \text{ is a real number}\}$.

2. Since $8-x=0$ when $x=8$, the domain is $\{x|x \text{ is a real number } and\ x\neq 8\}$.

3. $\dfrac{3x-3}{x(x-1)}$

The denominator is 0 when the factor $x=0$ and also when $x-1=0$, or $x=1$. The domain is $\{x|x \text{ is a real number } and\ x\neq 0\ and\ x\neq 1\}$.

4. $\dfrac{(x^2-4)(x+1)}{(x+2)(x^2-1)}=\dfrac{(x^2-4)(x+1)}{(x+2)(x+1)(x-1)}$

$x+2=0$ when $x=-2$; $x+1=0$ when $x=-1$; $x-1=0$ when $x=1$. The domain is $\{x|x \text{ is a real number } and\ x\neq -2\ and\ x\neq -1\ and\ x\neq 1\}$.

5. We first factor the denominator completely.

$\dfrac{7x^2-28x+28}{(x^2-4)(x^2+3x-10)}=\dfrac{7x^2-28x+28}{(x+2)(x-2)(x+5)(x-2)}$

We see that $x+2=0$ when $x=-2$, $x-2=0$ when $x=2$, and $x+5=0$ when $x=-5$. Thus, the domain is $\{x|x \text{ is a real number } and\ x\neq -2\ and\ x\neq 2\ and\ x\neq -5\}$.

6. $\dfrac{7x^2+11x-6}{x(x^2-x-6)}=\dfrac{7x^2+11x-6}{x(x-3)(x+2)}$

The denominator is 0 when $x=0$ or when $x-3=0$ or when $x+2=0$. Now $x-3=0$ when $x=3$ and $x+2=0$ when $x=-2$. Thus, the domain is $\{x|x \text{ is a real number } and\ x\neq 0\ and\ x\neq 3\ and\ x\neq -2\}$.

7. Let $y_1=\dfrac{3x-3}{x(x-1)}$ and find the values of x for which the table entry is "ERROR."

X	Y1	Y2
−1	−3	
0	ERROR	
1	ERROR	
2	1.5	
3	1	
4	.75	
5	.6	

X = 0

8. Let $y_1=\dfrac{(x^2-4)(x+1)}{(x+2)(x^2-1)}$.

X	Y1
−4	1.2
−3	1.25
−2	ERROR
−1	ERROR
0	2
1	ERROR
2	0

X = 1

9. $\dfrac{x^2-y^2}{(x-y)^2}\cdot\dfrac{1}{x+y}$

$=\dfrac{(x^2-y^2)\cdot 1}{(x-y)^2(x+y)}$

$=\dfrac{(x+y)(x-y)\cdot 1}{(x-y)(x-y)(x+y)}$

$=\dfrac{1}{x-y}$

10. $\dfrac{r-s}{r+s}\cdot\dfrac{r^2-s^2}{(r-s)^2} = \dfrac{(r-s)(r^2-s^2)}{(r+s)(r-s)^2}$

$= \dfrac{(r-s)(r-s)(r+s)\cdot 1}{(r+s)(r-s)(r-s)}$

$= 1$

11. $\dfrac{x^2-2x-35}{2x^3-3x^2}\cdot\dfrac{4x^3-9x}{7x-49}$

$= \dfrac{(x-7)(x+5)(x)(2x+3)(2x-3)}{x\cdot x(2x-3)(7)(x-7)}$

$= \dfrac{(x+5)(2x+3)}{7x}$

12. $\dfrac{x^2+2x-35}{3x^3-2x^2}\cdot\dfrac{9x^3-4x}{7x+49}$

$= \dfrac{(x+7)(x-5)(x)(3x+2)(3x-2)}{x\cdot x(3x-2)(7)(x+7)}$

$= \dfrac{(x-5)(3x+2)}{7x}$

13. $\dfrac{a^2-a-6}{a^2-7a+12}\cdot\dfrac{a^2-2a-8}{a^2-3a-10}$

$= \dfrac{(a-3)(a+2)(a-4)(a+2)}{(a-4)(a-3)(a-5)(a+2)}$

$= \dfrac{a+2}{a-5}$

14. $\dfrac{a^2-a-12}{a^2-6a+8}\cdot\dfrac{a^2+a-6}{a^2-2a-24}$

$= \dfrac{(a-4)(a+3)(a+3)(a-2)}{(a-2)(a-4)(a-6)(a+4)}$

$= \dfrac{(a+3)^2}{(a-6)(a+4)}$

15. $\dfrac{m^2-n^2}{r+s}\div\dfrac{m-n}{r+s}$

$= \dfrac{m^2-n^2}{r+s}\cdot\dfrac{r+s}{m-n}$

$= \dfrac{(m+n)(m-n)(r+s)}{(r+s)(m-n)}$

$= m+n$

16. $\dfrac{a^2-b^2}{x-y}\div\dfrac{a+b}{x-y}$

$= \dfrac{a^2-b^2}{x-y}\cdot\dfrac{x-y}{a+b}$

$= \dfrac{(a+b)(a-b)(x-y)}{(x-y)(a+b)\cdot 1}$

$= a-b$

17. $\dfrac{3x+12}{2x-8}\div\dfrac{(x+4)^2}{(x-4)^2}$

$= \dfrac{3x+12}{2x-8}\cdot\dfrac{(x-4)^2}{(x+4)^2}$

$= \dfrac{3(x+4)(x-4)(x-4)}{2(x-4)(x+4)(x+4)}$

$= \dfrac{3(x-4)}{2(x+4)}$

18. $\dfrac{a^2-a-2}{a^2-a-6}\div\dfrac{a^2-2a}{2a+a^2}$

$= \dfrac{a^2-a-2}{a^2-a-6}\cdot\dfrac{2a+a^2}{a^2-2a}$

$= \dfrac{(a-2)(a+1)(a)(2+a)}{(a-3)(a+2)(a)(a-2)}$

$= \dfrac{a+1}{a-3}$

19. $\dfrac{x^2-y^2}{x^3-y^3}\cdot\dfrac{x^2+xy+y^2}{x^2+2xy+y^2}$

$= \dfrac{(x+y)(x-y)(x^2+xy+y^2)}{(x-y)(x^2+xy+y^2)(x+y)(x+y)}$

$= \dfrac{1}{x+y}\cdot\dfrac{(x+y)(x-y)(x^2+xy+y^2)}{(x+y)(x-y)(x^2+xy+y^2)}$

$= \dfrac{1}{x+y}\cdot 1$ Removing a factor of 1

$= \dfrac{1}{x+y}$

20. $\dfrac{c^3+8}{c^2-4}\div\dfrac{c^2-2c+4}{c^2-4c+4}$

$= \dfrac{c^3+8}{c^2-4}\cdot\dfrac{c^2-4c+4}{c^2-2c+4}$

$= \dfrac{(c+2)(c^2-2c+4)(c-2)(c-2)}{(c+2)(c-2)(c^2-2c+4)}$

$= \dfrac{(c+2)(c^2-2c+4)(c-2)}{(c+2)(c^2-2c+4)(c-2)}\cdot\dfrac{c-2}{1}$

$= c-2$

21. $\dfrac{(x-y)^2-z^2}{(x+y)^2-z^2}\div\dfrac{x-y+z}{x+y-z}$

$= \dfrac{(x-y)^2-z^2}{(x+y)^2-z^2}\cdot\dfrac{x+y-z}{x-y+z}$

$= \dfrac{(x-y+z)(x-y-z)(x+y-z)}{(x+y+z)(x+y-z)(x-y+z)}$

$= \dfrac{(x-y+z)(x+y-z)}{(x-y+z)(x+y-z)}\cdot\dfrac{x-y-z}{x+y+z}$

$= 1\cdot\dfrac{x-y-z}{x+y+z}$ Removing a factor of 1

$= \dfrac{x-y-z}{x+y+z}$

22. $\dfrac{(a+b)^2-9}{(a-b)^2-9}\cdot\dfrac{a-b-3}{a+b+3}$

$= \dfrac{(a+b+3)(a+b-3)(a-b-3)}{(a-b+3)(a-b-3)(a+b+3)}$

$= \dfrac{(a+b+3)(a-b-3)}{(a+b+3)(a-b-3)}\cdot\dfrac{a+b-3}{a-b+3}$

$= \dfrac{a+b-3}{a-b+3}$

23. The graphs of $y_1 = \dfrac{x^2-2x-35}{2x^3-3x^2}\cdot\dfrac{4x^3-9x}{7x-49}$ and $y_2 = \dfrac{(x+5)(2x+3)}{7x}$ appear to coincide. A table of values also shows that the values of y_1 and y_2 are the same for the given x-values.

24. The graphs of $y_1 = \frac{x^2+2x-35}{3x^3-2x^2} \cdot \frac{9x^3-4x}{7x+49}$ and $y_2 = \frac{(x-5)(3x+2)}{7x}$ appear to coincide. A table of values also shows that the values of y_1 and y_2 are the same for the given x-values.

25. $\frac{3}{2a+3} + \frac{2a}{2a+3}$

$= \frac{3+2a}{2a+3}$

$= 1$

26. $\frac{a-3b}{a+b} + \frac{a+5b}{a+b} = \frac{2a+2b}{a+b}$

$= \frac{2(a+b)}{1 \cdot (a+b)}$

$= 2$

27. $\frac{y}{y-1} + \frac{2}{1-y}$

$= \frac{y}{y-1} + \frac{-1}{-1} \cdot \frac{2}{1-y}$

$= \frac{y}{y-1} + \frac{-2}{y-1}$

$= \frac{y-2}{y-1}$

28. $\frac{a}{a-b} + \frac{b}{b-a} = \frac{a}{a-b} + \frac{-b}{a-b}$

$= \frac{a-b}{a-b}$

$= 1$

29. $\frac{x}{2x-3y} - \frac{y}{3y-2x}$

$= \frac{x}{2x-3y} - \frac{-1}{-1} \cdot \frac{y}{3y-2x}$

$= \frac{x}{2x-3y} - \frac{-y}{2x-3y}$

$= \frac{x+y}{2x-3y}$ $[x-(-y) = x+y]$

30. $\frac{3a}{3a-2b} - \frac{2a}{2b-3a} = \frac{3a}{3a-2b} - \frac{-2a}{3a-2b}$

$= \frac{5a}{3a-2b}$

31. $\frac{3}{x+2} + \frac{2}{x^2-4}$

$= \frac{3}{x+2} + \frac{2}{(x+2)(x-2)}$, LCD is $(x+2)(x-2)$

$= \frac{3}{x+2} \cdot \frac{x-2}{x-2} + \frac{2}{(x+2)(x-2)}$

$= \frac{3x-6}{(x+2)(x-2)} + \frac{2}{(x+2)(x-2)}$

$= \frac{3x-4}{(x+2)(x-2)}$

32. $\frac{5}{a-3} - \frac{2}{a^2-9}$

$= \frac{5}{a-3} - \frac{2}{(a+3)(a-3)}$, LCD is $(a+3)(a-3)$

$= \frac{5(a+3)-2}{(a+3)(a-3)}$

$= \frac{5a+15-2}{(a+3)(a-3)}$

$= \frac{5a+13}{(a+3)(a-3)}$

33. $\frac{y}{y^2-y-20} - \frac{2}{y+4}$

$= \frac{y}{(y+4)(y-5)} - \frac{2}{y+4}$, LCD is $(y+4)(y-5)$

$= \frac{y}{(y+4)(y-5)} - \frac{2}{y+4} \cdot \frac{y-5}{y-5}$

$= \frac{y}{(y+4)(y-5)} - \frac{2y-10}{(y+4)(y-5)}$

$= \frac{y-(2y-10)}{(y+4)(y-5)}$

$= \frac{y-2y+10}{(y+4)(y-5)}$

$= \frac{-y+10}{(y+4)(y-5)}$

34. $\frac{6}{y^2+6y+9} - \frac{5}{y+3}$

$= \frac{6}{(y+3)^2} - \frac{5}{y+3}$, LCD is $(y+3)^2$

$= \frac{6-5(y+3)}{(y+3)^2}$

$= \frac{6-5y-15}{(y+3)^2}$

$= \frac{-5y-9}{(y+3)^2}$

35. $\frac{3}{x+y}+\frac{x-5y}{x^2-y^2}$

$=\frac{3}{x+y}+\frac{x-5y}{(x+y)(x-y)}$, LCD is $(x+y)(x-y)$

$=\frac{3}{x+y}\cdot\frac{x-y}{x-y}+\frac{x-5y}{(x+y)(x-y)}$

$=\frac{3x-3y}{(x+y)(x-y)}+\frac{x-5y}{(x+y)(x-y)}$

$=\frac{4x-8y}{(x+y)(x-y)}$

36. $\frac{a^2+1}{a^2-1}-\frac{a-1}{a+1}$

$=\frac{a^2+1}{(a+1)(a-1)}-\frac{a-1}{a+1}$, LCD is $(a+1)(a-1)$

$=\frac{a^2+1-(a-1)(a-1)}{(a+1)(a-1)}$

$=\frac{a^2+1-a^2+2a-1}{(a+1)(a-1)}$

$=\frac{2a}{(a+1)(a-1)}$

37. $\frac{9x+2}{3x^2-2x-8}+\frac{7}{3x^2+x-4}$

$=\frac{9x+2}{(3x+4)(x-2)}+\frac{7}{(3x+4)(x-1)}$,
LCD is $(3x+4)(x-2)(x-1)$

$=\frac{9x+2}{(3x+4)(x-2)}\cdot\frac{x-1}{x-1}+\frac{7}{(3x+4)(x-1)}\cdot\frac{x-2}{x-2}$

$=\frac{9x^2-7x-2}{(3x+4)(x-2)(x-1)}+\frac{7x-14}{(3x+4)(x-1)(x-2)}$

$=\frac{9x^2-16}{(3x+4)(x-2)(x-1)}$

$=\frac{(3x+4)(3x-4)}{(3x+4)(x-2)(x-1)}$

$=\frac{3x-4}{(x-2)(x-1)}$

38. $\frac{3y}{y^2-7y+10}-\frac{2y}{y^2-8y+15}$

$=\frac{3y}{(y-2)(y-5)}-\frac{2y}{(y-5)(y-3)}$,
LCD is $(y-2)(y-5)(y-3)$

$=\frac{3y(y-3)-2y(y-2)}{(y-2)(y-5)(y-3)}$

$=\frac{3y^2-9y-2y^2+4y}{(y-2)(y-5)(y-3)}$

$=\frac{y^2-5y}{(y-2)(y-5)(y-3)}$

$=\frac{y(y-5)}{(y-2)(y-5)(y-3)}$

$=\frac{y}{(y-2)(y-3)}$

39. $\frac{5a}{a-b}+\frac{ab}{a^2-b^2}+\frac{4b}{a+b}$

$=\frac{5a}{a-b}+\frac{ab}{(a+b)(a-b)}+\frac{4b}{a+b}$,
LCD is $(a+b)(a-b)$

$=\frac{5a}{a-b}\cdot\frac{a+b}{a+b}+\frac{ab}{(a+b)(a-b)}+\frac{4b}{a+b}\cdot\frac{a-b}{a-b}$

$=\frac{5a^2+5ab}{(a+b)(a-b)}+\frac{ab}{(a+b)(a-b)}+\frac{4ab-4b^2}{(a+b)(a-b)}$

$=\frac{5a^2+10ab-4b^2}{(a+b)(a-b)}$

40. $\frac{6a}{a-b}+\frac{3b}{b-a}+\frac{5}{a^2-b^2}$

$=\frac{6a}{a-b}+\frac{-3b}{a-b}+\frac{5}{(a+b)(a-b)}$,
LCD is $(a+b)(a-b)$

$=\frac{6a(a+b)+3b(a+b)+5}{(a+b)(a-b)}$

$=\frac{6a^2+6ab+3ab+3b^2+5}{(a+b)(a-b)}$

$=\frac{6a^2+9ab+3b^2+5}{(a+b)(a-b)}$

41. $\frac{7}{x+2}-\frac{x+8}{4-x^2}+\frac{3x-2}{4-4x+x^2}$

$=\frac{7}{x+2}-\frac{x+8}{(2+x)(2-x)}+\frac{3x-2}{(2-x)^2}$,
LCD is $(2+x)(2-x)^2$

$=\frac{7}{2+x}\cdot\frac{(2-x)^2}{(2-x)^2}-\frac{x+8}{(2+x)(2-x)}\cdot\frac{2-x}{2-x}+\frac{3x-2}{(2-x)^2}\cdot\frac{2+x}{2+x}$

$=\frac{28-28x+7x^2-(16-6x-x^2)+3x^2+4x-4}{(2+x)(2-x)^2}$

$=\frac{28-28x+7x^2-16+6x+x^2+3x^2+4x-4}{(2+x)(2-x)^2}$

$=\frac{11x^2-18x+8}{(2+x)(2-x)^2}$, or $\frac{11x^2-18x+8}{(x+2)(x-2)^2}$

42. $\frac{6}{x+3}-\frac{x+4}{9-x^2}+\frac{2x-3}{9-6x+x^2}$

$=\frac{6}{x+3}-\frac{x+4}{(3+x)(3-x)}+\frac{2x-3}{(3-x)^2}$,
LCD is $(3+x)(3-x)^2$

$=\frac{6(3-x)^2-(x+4)(3-x)+(2x-3)(3+x)}{(3+x)(3-x)^2}$

$=\frac{54-36x+6x^2+x^2+x-12+2x^2+3x-9}{(3+x)(3-x)^2}$

$=\frac{33-32x+9x^2}{(3+x)(3-x)^2}$, or $\frac{9x^2-32x+33}{(x+3)(x-3)^2}$

43. $\frac{1}{x+1}+\frac{x}{2-x}+\frac{x^2+2}{x^2-x-2}$

$=\frac{1}{x+1}+\frac{x}{2-x}+\frac{x^2+2}{(x+1)(x-2)}$

$=\frac{1}{x+1}+\frac{-1}{-1}\cdot\frac{x}{2-x}+\frac{x^2+2}{(x+1)(x-2)}$

$=\frac{1}{x+1}+\frac{-x}{x-2}+\frac{x^2+2}{(x+1)(x-2)}$,

LCD is $(x+1)(x-2)$

$=\frac{1}{x+1}\cdot\frac{x-2}{x-2}+\frac{-x}{x-2}\cdot\frac{x+1}{x+1}+\frac{x^2+2}{(x+1)(x-2)}$

$=\frac{x-2}{(x+1)(x-2)}+\frac{-x^2-x}{(x+1)(x-2)}+\frac{x^2+2}{(x+1)(x-2)}$

$=\frac{x-2-x^2-x+x^2+2}{(x+1)(x-2)}$

$=\frac{0}{(x+1)(x-2)}$

$=0$

44. $\frac{x-1}{x-2}-\frac{x+1}{x+2}-\frac{x-6}{4-x^2}$

$=\frac{x-1}{x-2}-\frac{x+1}{x+2}-\frac{x-6}{(2+x)(2-x)}$

$=\frac{1-x}{2-x}-\frac{x+1}{x+2}-\frac{x-6}{(2+x)(2-x)}$,

LCD is $(2+x)(2-x)$

$=\frac{(1-x)(2+x)-(x+1)(2-x)-(x-6)}{(2+x)(2-x)}$

$=\frac{2-x-x^2+x^2-x-2-x+6}{(2+x)(2-x)}$

$=\frac{6-3x}{(2+x)(2-x)}$

$=\frac{3(2-x)}{(2+x)(2-x)}$

$=\frac{3}{2+x}$

45. The graphs of $y_1=\frac{3}{x+2}+\frac{2}{x^2-4}$ and $y_2=\frac{3x-4}{(x+2)(x-2)}$ appear to coincide. A table of values also shows that the values of y_1 and y_2 are the same for the given x-values.

46. The graphs of $y_1=\frac{5}{x-3}-\frac{2}{x^2-9}$ and $y_2=\frac{5x+13}{(x+3)(x-3)}$ appear to coincide. A table of values also shows that the values of y_1 and y_2 are the same for the given x-values.

47. $\frac{\frac{x^2-y^2}{xy}}{\frac{x-y}{y}}=\frac{x^2-y^2}{xy}\cdot\frac{y}{x-y}$

$=\frac{(x+y)(x-y)y}{xy(x-y)}$

$=\frac{x+y}{x}$

48. $\frac{\frac{a-b}{b}}{\frac{a^2-b^2}{ab}}=\frac{a-b}{b}\cdot\frac{ab}{(a+b)(a-b)}$

$=\frac{ab(a-b)}{b(a+b)(a-b)}$

$=\frac{a}{a+b}$

49. $\frac{a-a^{-1}}{a+a^{-1}}=\frac{a-\frac{1}{a}}{a+\frac{1}{a}}=\frac{a\cdot\frac{a}{a}-\frac{1}{a}}{a\cdot\frac{a}{a}+\frac{1}{a}}$

$=\frac{\frac{a^2-1}{a}}{\frac{a^2+1}{a}}$

$=\frac{a^2-1}{a}\cdot\frac{a}{a^2+1}$

$=\frac{a^2-1}{a^2+1}$

50. $\frac{a-\frac{a}{b}}{b-\frac{b}{a}}=\frac{\frac{ab-a}{b}}{\frac{ab-b}{a}}$

$=\frac{a(b-1)}{b}\cdot\frac{a}{b(a-1)}$

$=\frac{a^2(b-1)}{b^2(a-1)}$

51. $\frac{c+\frac{8}{c^2}}{1+\frac{2}{c}}=\frac{c\cdot\frac{c^2}{c^2}+\frac{8}{c^2}}{1\cdot\frac{c}{c}+\frac{2}{c}}$

$=\frac{\frac{c^3+8}{c^2}}{\frac{c+2}{c}}$

$=\frac{c^3+8}{c^2}\cdot\frac{c}{c+2}$

$=\frac{(c+2)(c^2-2c+4)c}{c\cdot c(c+2)}$

$=\frac{c^2-2c+4}{c}$

52. $\dfrac{x^{-1}+y^{-1}}{x^{-3}+y^{-3}} = \dfrac{\dfrac{1}{x}+\dfrac{1}{y}}{\dfrac{1}{x^3}+\dfrac{1}{y_3}}$

$= \dfrac{\left(\dfrac{1}{x}+\dfrac{1}{y}\right)(x^3y^3)}{\left(\dfrac{1}{x^3}+\dfrac{1}{y^3}\right)(x^3y^3)}$

$= \dfrac{x^2y^3+x^3y^2}{y^3+x^3}$

$= \dfrac{x^2y^2\cancel{(y+x)}}{\cancel{(y+x)}(y^2-yx+x^2)}$

$= \dfrac{x^2y^2}{y^2-yx+x^2}$

53. $\dfrac{x^2+xy+y^2}{\dfrac{x^2}{y}-\dfrac{y^2}{x}} = \dfrac{x^2+xy+y^2}{\dfrac{x^2}{y}\cdot\dfrac{x}{x}-\dfrac{y^2}{x}\cdot\dfrac{y}{y}}$

$= \dfrac{x^2+xy+y^2}{\dfrac{x^3-y^3}{xy}}$

$= (x^2+xy+y^2)\cdot\dfrac{xy}{x^3-y^3}$

$= \dfrac{(x^2+xy+y^2)(xy)}{(x-y)(x^2+xy+y^2)}$

$= \dfrac{x^2+xy+y^2}{x^2+xy+y^2}\cdot\dfrac{xy}{x-y}$

$= 1\cdot\dfrac{xy}{x-y}$

$= \dfrac{xy}{x-y}$

54. $\dfrac{\dfrac{a^2}{b}+\dfrac{b^2}{a}}{a^2-ab+b^2} = \dfrac{\dfrac{a^3+b^3}{ab}}{a^2-ab+b^2}$

$= \dfrac{(a+b)(a^2-ab+b^2)}{ab}\cdot\dfrac{1}{a^2-ab+b^2}$

$= \dfrac{a+b}{ab}\cdot\dfrac{a^2-ab+b^2}{a^2-ab+b^2}$

$= \dfrac{a+b}{ab}$

55. $\dfrac{\dfrac{x}{y}-\dfrac{y}{x}}{\dfrac{1}{y}+\dfrac{1}{x}} = \dfrac{\dfrac{x}{y}-\dfrac{y}{x}}{\dfrac{1}{y}+\dfrac{1}{x}}\cdot\dfrac{xy}{xy}$, LCM is xy

$= \dfrac{\left(\dfrac{x}{y}-\dfrac{y}{x}\right)(xy)}{\left(\dfrac{1}{y}+\dfrac{1}{x}\right)(xy)}$

$= \dfrac{x^2-y^2}{x+y}$

$= \dfrac{\cancel{(x+y)}(x-y)}{\cancel{(x+y)}\cdot 1}$

$= x-y$

56. $\dfrac{\dfrac{a}{b}-\dfrac{b}{a}}{\dfrac{1}{a}-\dfrac{1}{b}} = \dfrac{a^2-b^2}{b-a}$ Multiplying by $\dfrac{ab}{ab}$

$= \dfrac{(a+b)(a-b)}{b-a}$

$= \dfrac{(a+b)\cancel{(a-b)}}{-1\cdot\cancel{(a-b)}}$

$= -a-b$

57. $\dfrac{\dfrac{1}{x-3}+\dfrac{2}{x+3}}{\dfrac{3}{x-1}-\dfrac{4}{x+2}} = \dfrac{\dfrac{1}{x-3}\cdot\dfrac{x+3}{x+3}+\dfrac{2}{x+3}\cdot\dfrac{x-3}{x-3}}{\dfrac{3}{x-1}\cdot\dfrac{x+2}{x+2}-\dfrac{4}{x+2}\cdot\dfrac{x-1}{x-1}}$

$= \dfrac{\dfrac{x+3+2(x-3)}{(x-3)(x+3)}}{\dfrac{3(x+2)-4(x-1)}{(x-1)(x+2)}}$

$= \dfrac{\dfrac{x+3+2x-6}{(x-3)(x+3)}}{\dfrac{3x+6-4x+4}{(x-1)(x+2)}}$

$= \dfrac{\dfrac{3x-3}{(x-3)(x+3)}}{\dfrac{-x+10}{(x-1)(x+2)}}$

$= \dfrac{3x-3}{(x-3)(x+3)}\cdot\dfrac{(x-1)(x+2)}{-x+10}$

$= \dfrac{(3x-3)(x-1)(x+2)}{(x-3)(x+3)(-x+10)}$, or

$\dfrac{3(x-1)^2(x+2)}{(x-3)(x+3)(-x+10)}$

58. $\dfrac{\dfrac{5}{x+1}-\dfrac{3}{x-2}}{\dfrac{1}{x-5}+\dfrac{2}{x+2}} = \dfrac{\dfrac{5(x-2)-3(x+1)}{(x+1)(x-2)}}{\dfrac{x+2+2(x-5)}{(x-5)(x+2)}}$

$= \dfrac{\dfrac{5x-10-3x-3}{(x+1)(x-2)}}{\dfrac{x+2+2x-10}{(x-5)(x+2)}}$

$= \dfrac{\dfrac{2x-13}{(x+1)(x-2)}}{\dfrac{3x-8}{(x-5)(x+2)}}$

$= \dfrac{2x-13}{(x+1)(x-2)}\cdot\dfrac{(x-5)(x+2)}{3x-8}$

$= \dfrac{(2x-13)(x-5)(x+2)}{(x+1)(x-2)(3x-8)}$

59. $$\frac{\frac{a}{1-a}+\frac{1+a}{a}}{\frac{1-a}{a}+\frac{a}{1+a}} = \frac{\frac{a}{1-a}\cdot\frac{a}{a}+\frac{1+a}{a}\cdot\frac{1-a}{1-a}}{\frac{1-a}{a}\cdot\frac{1+a}{1+a}+\frac{a}{1+a}\cdot\frac{a}{a}}$$

$$= \frac{\frac{a^2+(1-a^2)}{a(1-a)}}{\frac{(1-a^2)+a^2}{a(1+a)}}$$

$$= \frac{1}{\not{a}(1-a)}\cdot\frac{\not{a}(1+a)}{1}$$

$$= \frac{1+a}{1-a}$$

60. $$\frac{\frac{1-x}{x}+\frac{x}{1+x}}{\frac{1+x}{x}+\frac{x}{1-x}} = \frac{\frac{1-x^2+x^2}{x(1+x)}}{\frac{1-x^2+x^2}{x(1-x)}}$$

$$= \frac{1}{x(1+x)}\cdot\frac{x(1-x)}{1}$$

$$= \frac{\not{x}(1-x)}{\not{x}(1+x)}$$

$$= \frac{1-x}{1+x}$$

61. $$\frac{\frac{1}{a^2}+\frac{2}{ab}+\frac{1}{b^2}}{\frac{1}{a^2}-\frac{1}{b^2}} = \frac{\frac{1}{a^2}+\frac{2}{ab}+\frac{1}{b^2}}{\frac{1}{a^2}-\frac{1}{b^2}}\cdot\frac{a^2b^2}{a^2b^2},$$

LCM is a^2b^2

$$= \frac{b^2+2ab+a^2}{b^2-a^2}$$

$$= \frac{(b+a)(b+a)}{(b+a)(b-a)}$$

$$= \frac{b+a}{b-a}$$

62. $$\frac{\frac{1}{x^2}-\frac{1}{y^2}}{\frac{1}{x^2}-\frac{2}{xy}+\frac{1}{y^2}} = \frac{y^2-x^2}{y^2-2xy+x^2}$$

Multiplying by $\frac{x^2y^2}{x^2y^2}$

$$= \frac{(y+x)(y-x)}{(y-x)(y-x)}$$

$$= \frac{y+x}{y-x}$$

63. When the least common denominator is used, the multiplication in the numerators is often simpler and there is usually less simplification required after the addition or subtraction is performed.

64. When there are three or more different binomial denominators, as in Exercise 57, Method 2 is usually preferable. Otherwise, some might prefer Method 1 while others will prefer Method 2.

65. $$\frac{(x+h)^2-x^2}{h} = \frac{x^2+2xh+h^2-x^2}{h}$$

$$= \frac{2xh+h^2}{h}$$

$$= \frac{\not{h}(2x+h)}{\not{h}\cdot 1}$$

$$= 2x+h$$

66. $$\frac{\frac{1}{x+h}-\frac{1}{x}}{h} = \frac{\frac{x-x-h}{x(x+h)}}{h}$$

$$= \frac{-h}{x(x+h)}\cdot\frac{1}{h}$$

$$= \frac{-1\cdot\not{h}}{x\not{h}(x+h)}$$

$$= \frac{-1}{x(x+h)}$$

67. $$\frac{(x+h)^3-x^3}{h} = \frac{x^3+3x^2h+3xh^2+h^3-x^3}{h}$$

$$= \frac{3x^2h+3xh^2+h^3}{h}$$

$$= \frac{\not{h}(3x^2+3xh+h^2)}{\not{h}\cdot 1}$$

$$= 3x^2+3xh+h^2$$

68. $$\frac{\frac{1}{(x+h)^2}-\frac{1}{x^2}}{h} = \frac{\frac{x^2-x^2-2xh-h^2}{x^2(x+h)^2}}{h}$$

$$= \frac{-2xh-h^2}{x^2(x+h)^2}\cdot\frac{1}{h}$$

$$= \frac{\not{h}(-2x-h)}{x^2\not{h}(x+h)^2}$$

$$= \frac{-2x-h}{x^2(x+h)^2}$$

69. $$\left[\frac{\frac{x+1}{x-1}+1}{\frac{x+1}{x-1}-1}\right]^5 = \left[\frac{\frac{(x+1)+(x-1)}{x-1}}{\frac{(x+1)-(x-1)}{x-1}}\right]^5$$

$$= \left[\frac{2x}{x-1}\cdot\frac{x-1}{2}\right]^5$$

$$= \left[\frac{\not{2}x(\not{x-1})}{1\cdot\not{2}(\not{x-1})}\right]^5$$

$$= x^5$$

70. $1+\cfrac{1}{1+\cfrac{1}{1+\cfrac{1}{1+\cfrac{1}{x}}}} = 1+\cfrac{1}{1+\cfrac{1}{1+\cfrac{1}{\frac{x+1}{x}}}}$

$= 1+\cfrac{1}{1+\cfrac{1}{1+\cfrac{x}{x+1}}}$

$= 1+\cfrac{1}{1+\cfrac{1}{\frac{2x+1}{x+1}}}$

$= 1+\cfrac{1}{1+\cfrac{x+1}{2x+1}}$

$= 1+\cfrac{1}{\frac{3x+2}{2x+1}}$

$= 1+\dfrac{2x+1}{3x+2}$

$= \dfrac{5x+3}{3x+2}$

71. $\dfrac{n(n+1)(n+2)}{2\cdot 3}+\dfrac{(n+1)(n+2)}{2}$

$= \dfrac{n(n+1)(n+2)}{2\cdot 3}+\dfrac{(n+1)(n+2)}{2}\cdot\dfrac{3}{3}$, LCD is $2\cdot 3$

$= \dfrac{n(n+1)(n+2)+3(n+1)(n+2)}{2\cdot 3}$

$= \dfrac{(n+1)(n+2)(n+3)}{2\cdot 3}$ Factoring the numerator by grouping

72. $\dfrac{n(n+1)(n+2)(n+3)}{2\cdot 3\cdot 4}+\dfrac{(n+1)(n+2)(n+3)}{2\cdot 3}$

$= \dfrac{n(n+1)(n+2)(n+3)+4(n+1)(n+2)(n+3)}{2\cdot 3\cdot 4}$, LCD is $2\cdot 3\cdot 4$

$= \dfrac{(n+1)(n+2)(n+3)(n+4)}{2\cdot 3\cdot 4}$

73. $\dfrac{x^2-9}{x^3+27}\cdot\dfrac{5x^2-15x+45}{x^2-2x-3}+\dfrac{x^2+x}{4+2x}$

$= \dfrac{(x+3)(x-3)(5)(x^2-3x+9)}{(x+3)(x^2-3x+9)(x-3)(x+1)}+\dfrac{x^2+x}{4+2x}$

$= \dfrac{(x+3)(x-3)(x^2-3x+9)}{(x+3)(x-3)(x^2-3x+9)}\cdot\dfrac{5}{x+1}+\dfrac{x^2+x}{4+2x}$

$= 1\cdot\dfrac{5}{x+1}+\dfrac{x^2+x}{4+2x}$

$= \dfrac{5}{x+1}+\dfrac{x^2+x}{2(2+x)}$

$= \dfrac{5\cdot 2(2+x)+(x^2+x)(x+1)}{2(x+1)(2+x)}$

$= \dfrac{20+10x+x^3+2x^2+x}{2(x+1)(2+x)}$

$= \dfrac{x^3+2x^2+11x+20}{2(x+1)(2+x)}$

74. $\dfrac{x^2+2x-3}{x^2-x-12}\div\dfrac{x^2-1}{x^2-16}-\dfrac{2x+1}{x^2+2x+1}$

$= \dfrac{x^2+2x-3}{x^2-x-12}\cdot\dfrac{x^2-16}{x^2-1}-\dfrac{2x+1}{x^2+2x+1}$

$= \dfrac{\cancel{(x+3)}\cancel{(x-1)}(x+4)\cancel{(x-4)}}{\cancel{(x-4)}\cancel{(x+3)}(x+1)\cancel{(x-1)}}-\dfrac{2x+1}{x^2+2x+1}$

$= \dfrac{x+4}{x+1}-\dfrac{2x+1}{(x+1)(x+1)}$

$= \dfrac{(x+4)(x+1)-(2x+1)}{(x+1)(x+1)}$

$= \dfrac{x^2+5x+4-2x-1}{(x+1)(x+1)}$

$= \dfrac{x^2+3x+3}{(x+1)^2}$

Exercise Set R.6

1. $\sqrt{(-11)^2}=|-11|=11$

2. $\sqrt{(-1)^2}=|-1|=1$

3. $\sqrt{16y^2}=\sqrt{(4y)^2}=|4y|=4|y|$

4. $\sqrt{36t^2}=|6t|=6|t|$

5. $\sqrt{(b+1)^2}=|b+1|$

6. $\sqrt{(2c-3)^2}=|2c-3|$

7. $\sqrt[3]{-27x^3}=\sqrt[3]{(-3x)^3}=-3x$

8. $\sqrt[3]{-8y^3}=-2y$

9. $\sqrt{x^2-4x+4} = \sqrt{(x-2)^2} = |x-2|$

10. $\sqrt{x^2+16x+64} = \sqrt{(x+8)^2} = |x+8|$

11. $\sqrt[5]{32} = \sqrt[5]{2^5} = 2$

12. $\sqrt[5]{-32} = -2$

13. $\sqrt{180} = \sqrt{36\cdot 5} = \sqrt{36}\cdot\sqrt{5} = 6\sqrt{5}$

14. $\sqrt{48} = \sqrt{16\cdot 3} = 4\sqrt{3}$

15. $\sqrt[3]{54} = \sqrt[3]{27\cdot 2} = \sqrt[3]{27}\cdot\sqrt[3]{2} = 3\sqrt[3]{2}$

16. $\sqrt[3]{135} = \sqrt[3]{27\cdot 5} = 3\sqrt[3]{5}$

17. $\sqrt{128c^2d^4} = \sqrt{64c^2d^4\cdot 2} = |8cd^2|\sqrt{2} = 8\sqrt{2}\,|c|d^2$

18. $\sqrt{162c^4d^6} = \sqrt{81c^4\cdot d^6\cdot 2} = 9c^2|d^3|\sqrt{2} =$
$9\sqrt{2}c^2d^2|d|$

19. The values of $y_1 = \sqrt{x^2-4x+4}$ and $y_2 = |x-2|$ appear to be the same for any value of x.

20. The values of $y_1 = \sqrt{x^2+16x+64}$ and $y_2 = |x+8|$ appear to be the same for any value of x.

21. $\sqrt{2x^3y}\sqrt{12xy} = \sqrt{24x^4y^2} = \sqrt{4x^4y^2\cdot 6} = 2x^2y\sqrt{6}$

22. $\sqrt{3y^4z}\sqrt{20z} = \sqrt{60y^4z^2} = \sqrt{4y^4z^2\cdot 15} = 2y^2z\sqrt{15}$

23. $\sqrt[3]{3x^2y}\sqrt[3]{36x} = \sqrt[3]{108x^3y} = \sqrt[3]{27x^3\cdot 4y} = 3x\sqrt[3]{4y}$

24. $\sqrt[5]{8x^3y^4}\sqrt[5]{4x^4y} = \sqrt[5]{32x^7y^5} = 2xy\sqrt[5]{x^2}$

25.
$$\begin{aligned}\sqrt[3]{2(x+4)}\sqrt[3]{4(x+4)^4} &= \sqrt[3]{8(x+4)^5}\\ &= \sqrt[3]{8(x+4)^3\cdot(x+4)^2}\\ &= 2(x+4)\sqrt[3]{(x+4)^2}\end{aligned}$$

26.
$$\begin{aligned}\sqrt[3]{4(x+1)^2}\sqrt[3]{18(x+1)^2} &= \sqrt[3]{72(x+1)^4}\\ &= \sqrt[3]{8(x+1)^3\cdot 9(x+1)}\\ &= 2(x+1)\sqrt[3]{9(x+1)}\end{aligned}$$

27. $\sqrt[6]{\dfrac{m^{12}n^{24}}{64}} = \sqrt[6]{\left(\dfrac{m^2n^4}{2}\right)^6} = \dfrac{m^2n^4}{2}$

28. $\sqrt[8]{\dfrac{m^{16}n^{24}}{2^8}} = \dfrac{m^2n^3}{2}$

29. $\dfrac{\sqrt[3]{40m}}{\sqrt[3]{5m}} = \sqrt[3]{\dfrac{40m}{5m}} = \sqrt[3]{8} = 2$

30. $\dfrac{\sqrt{40xy}}{\sqrt{8x}} = \sqrt{\dfrac{40xy}{8x}} = \sqrt{5y}$

31. $\dfrac{\sqrt[3]{3x^2}}{\sqrt[3]{24x^5}} = \sqrt[3]{\dfrac{3x^2}{24x^5}} = \sqrt[3]{\dfrac{1}{8x^3}} = \dfrac{1}{2x}$

32. $\dfrac{\sqrt{128a^2b^4}}{\sqrt{16ab}} = \sqrt{\dfrac{128a^2b^4}{16ab}} = \sqrt{8ab^3} =$
$\sqrt{4\cdot 2\cdot a\cdot b^2\cdot b} = 2b\sqrt{2ab}$

33.
$$\begin{aligned}\sqrt[3]{\frac{64a^4}{27b^3}} &= \sqrt[3]{\frac{64\cdot a^3\cdot a}{27\cdot b^3}}\\ &= \frac{\sqrt[3]{64a^3}\sqrt[3]{a}}{\sqrt[3]{27b^3}}\\ &= \frac{4a\sqrt[3]{a}}{3b}\end{aligned}$$

34. $\sqrt{\dfrac{9x^7}{16y^8}} = \dfrac{\sqrt{9\cdot x^6\cdot x}}{\sqrt{16\cdot y^8}} = \dfrac{3x^3\sqrt{x}}{4y^4}$

35.
$$\begin{aligned}\sqrt{\frac{7x^3}{36y^6}} &= \sqrt{\frac{7\cdot x^2\cdot x}{36\cdot y^6}}\\ &= \frac{\sqrt{x^2}\sqrt{7x}}{\sqrt{36y^6}}\\ &= \frac{x\sqrt{7x}}{6y^3}\end{aligned}$$

36. $\sqrt[3]{\dfrac{2yz}{250z^4}} = \sqrt[3]{\dfrac{y}{125z^3}} = \dfrac{\sqrt[3]{y}}{\sqrt[3]{125z^3}} = \dfrac{\sqrt[3]{y}}{5z}$

37.
$$\begin{aligned}9\sqrt{50}+6\sqrt{2} &= 9\sqrt{25\cdot 2}+6\sqrt{2}\\ &= 9\cdot 5\sqrt{2}+6\sqrt{2}\\ &= 45\sqrt{2}+6\sqrt{2}\\ &= (45+6)\sqrt{2}\\ &= 51\sqrt{2}\end{aligned}$$

38. $11\sqrt{27}-4\sqrt{3} = 33\sqrt{3}-4\sqrt{3} = 29\sqrt{3}$

39.
$$\begin{aligned}&8\sqrt{2x^2}-6\sqrt{20x}-5\sqrt{8x^2}\\ &= 8x\sqrt{2}-6\sqrt{4\cdot 5x}-5\sqrt{4x^2\cdot 2}\\ &= 8x\sqrt{2}-6\cdot 2\sqrt{5x}-5\cdot 2x\sqrt{2}\\ &= 8x\sqrt{2}-12\sqrt{5x}-10x\sqrt{2}\\ &= -2x\sqrt{2}-12\sqrt{5x}\end{aligned}$$

40.
$$\begin{aligned}&2\sqrt[3]{8x^2}+5\sqrt[3]{27x^2}-3\sqrt{x^3}\\ &= 4\sqrt[3]{x^2}+15\sqrt[3]{x^2}-3x\sqrt{x}\\ &= 19\sqrt[3]{x^2}-3x\sqrt{x}\end{aligned}$$

41.
$$\begin{aligned}&(\sqrt{3}-\sqrt{2})(\sqrt{3}+\sqrt{2})\\ &= (\sqrt{3})^2-(\sqrt{2})^2\\ &= 3-2\\ &= 1\end{aligned}$$

42. $(\sqrt{8}+2\sqrt{5})(\sqrt{8}-2\sqrt{5}) = 8-4\cdot 5 = -12$

43.
$$\begin{aligned}(1+\sqrt{3})^2 &= 1^2+2\cdot 1\cdot\sqrt{3}+(\sqrt{3})^2\\ &= 1+2\sqrt{3}+3\\ &= 4+2\sqrt{3}\end{aligned}$$

44. $(\sqrt{2}-5)^2 = 2-10\sqrt{2}+25 = 27-10\sqrt{2}$

45. The graphs of $y_1 = 8\sqrt{2x^2}-6\sqrt{20x}-5\sqrt{8x^2}$ and $y_2 = -2x\sqrt{2}-12\sqrt{5x}$ appear to coincide. A table of values also shows that the values of y_1 and y_2 are the same for the given x-values.

46. The graphs of $y_1 = 2\sqrt[3]{8x^2} + 5\sqrt[3]{27x^2} - 3\sqrt{x^3}$ and $y_2 = 19\sqrt[3]{x^2} - 3x\sqrt{x}$ appear to coincide. A table of values also shows that the values of y_1 and y_2 are the same for the given x-values.

47. We use the Pythagorean theorem to find b, the airplane's horizontal distance from the airport. We have $a = 3700$ and $c = 14,200$.

$$\begin{aligned} c^2 &= a^2 + b^2 \\ 14,200^2 &= 3700^2 + b^2 \\ 201,640,000 &= 13,690,000 + b^2 \\ 187,950,000 &= b^2 \\ 13,709.5 &\approx b \end{aligned}$$

The airplane is about 13,709.5 ft horizontally from the airport.

48.

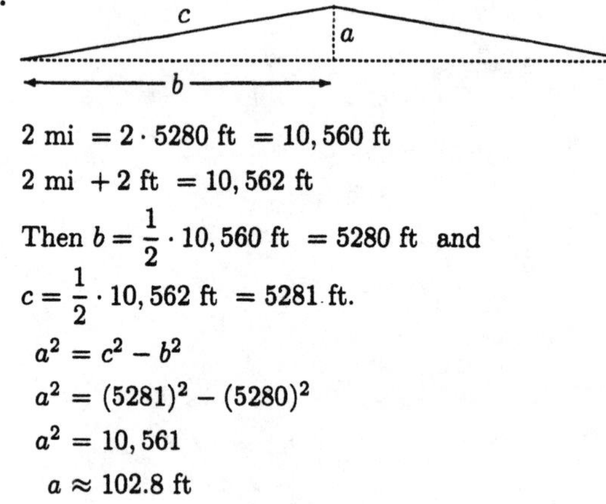

$2 \text{ mi} = 2 \cdot 5280 \text{ ft} = 10,560 \text{ ft}$

$2 \text{ mi} + 2 \text{ ft} = 10,562 \text{ ft}$

Then $b = \frac{1}{2} \cdot 10,560 \text{ ft} = 5280 \text{ ft}$ and

$c = \frac{1}{2} \cdot 10,562 \text{ ft} = 5281 \text{ ft}.$

$$\begin{aligned} a^2 &= c^2 - b^2 \\ a^2 &= (5281)^2 - (5280)^2 \\ a^2 &= 10,561 \\ a &\approx 102.8 \text{ ft} \end{aligned}$$

49. a) $h^2 + \left(\frac{a}{2}\right)^2 = a^2$ Pythagorean theorem

$$\begin{aligned} h^2 + \frac{a^2}{4} &= a^2 \\ h^2 &= \frac{3a^2}{4} \\ h &= \sqrt{\frac{3a^2}{4}} \\ h &= \frac{a}{2}\sqrt{3} \end{aligned}$$

b) Using the result of part (a) we have

$$\begin{aligned} A &= \frac{1}{2} \cdot \text{base} \cdot \text{height} \\ A &= \frac{1}{2} a \cdot \frac{a}{2}\sqrt{3} \quad \left(\frac{a}{2} + \frac{a}{2} = a\right) \\ A &= \frac{a^2}{4}\sqrt{3} \end{aligned}$$

50.
$$\begin{aligned} c^2 &= s^2 + s^2 \\ c^2 &= 2s^2 \\ c &= s\sqrt{2} \quad \text{Length of third side} \end{aligned}$$

51.

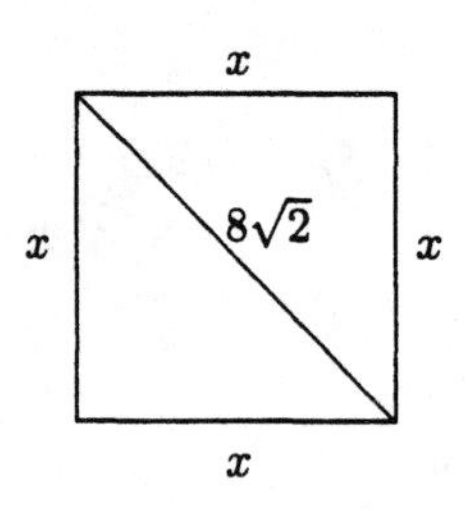

$$\begin{aligned} x^2 + x^2 &= (8\sqrt{2})^2 \quad \text{Pythagorean theorem} \\ 2x^2 &= 128 \\ x^2 &= 64 \\ x &= 8 \end{aligned}$$

52.

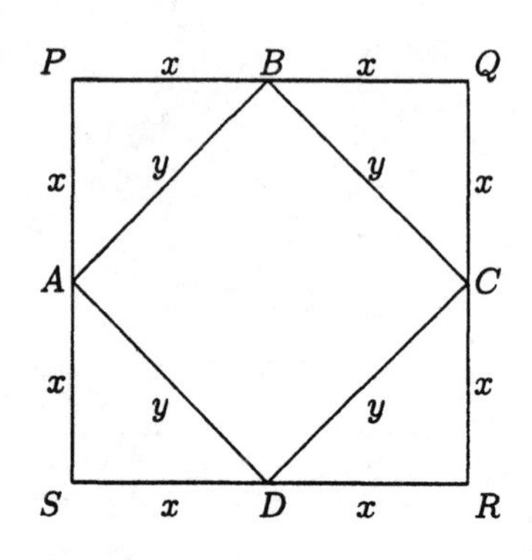

$$\begin{aligned} (2x)^2 &= 100 \\ 4x^2 &= 100 \\ x^2 &= 25 \\ x &= 5 \end{aligned}$$

$A = y^2 = x^2 + x^2 = 5^2 + 5^2 = 25 + 25 = 50 \text{ ft}^2$

53. $\sqrt{\frac{2}{3}} = \sqrt{\frac{2}{3} \cdot \frac{3}{3}} = \sqrt{\frac{6}{9}} = \frac{\sqrt{6}}{\sqrt{9}} = \frac{\sqrt{6}}{3}$

54. $\sqrt{\frac{3}{7}} = \sqrt{\frac{3}{7} \cdot \frac{7}{7}} = \sqrt{\frac{21}{49}} = \frac{\sqrt{21}}{7}$

55. $\frac{\sqrt[3]{5}}{\sqrt[3]{4}} = \frac{\sqrt[3]{5}}{\sqrt[3]{4}} \cdot \frac{\sqrt[3]{2}}{\sqrt[3]{2}} = \frac{\sqrt[3]{10}}{\sqrt[3]{8}} = \frac{\sqrt[3]{10}}{2}$

56. $\frac{\sqrt[3]{7}}{\sqrt[3]{25}} = \frac{\sqrt[3]{7}}{\sqrt[3]{25}} \cdot \frac{\sqrt[3]{5}}{\sqrt[3]{5}} = \frac{\sqrt[3]{35}}{\sqrt[3]{125}} = \frac{\sqrt[3]{35}}{5}$

57. $\sqrt[3]{\frac{16}{9}} = \sqrt[3]{\frac{16}{9} \cdot \frac{3}{3}} = \sqrt[3]{\frac{48}{27}} = \frac{\sqrt[3]{48}}{\sqrt[3]{27}} = \frac{\sqrt[3]{8 \cdot 6}}{3} = \frac{2\sqrt[3]{6}}{3}$

58. $\sqrt[3]{\frac{3}{5}} = \sqrt[3]{\frac{3}{5} \cdot \frac{25}{25}} = \sqrt[3]{\frac{75}{125}} = \frac{\sqrt[3]{75}}{5}$

59. $\dfrac{6}{3+\sqrt{5}} = \dfrac{6}{3+\sqrt{5}} \cdot \dfrac{3-\sqrt{5}}{3-\sqrt{5}}$
$= \dfrac{6(3-\sqrt{5})}{9-5}$
$= \dfrac{6(3-\sqrt{5})}{4}$
$= \dfrac{3(3-\sqrt{5})}{2} = \dfrac{9-3\sqrt{5}}{2}$

60. $\dfrac{2}{\sqrt{3}-1} = \dfrac{2}{\sqrt{3}-1} \cdot \dfrac{\sqrt{3}+1}{\sqrt{3}+1}$
$= \dfrac{\not{2}(\sqrt{3}+1)}{\not{2}\cdot 1}$
$= \sqrt{3}+1$

61. $\dfrac{6}{\sqrt{m}-\sqrt{n}} = \dfrac{6}{\sqrt{m}-\sqrt{n}} \cdot \dfrac{\sqrt{m}+\sqrt{n}}{\sqrt{m}+\sqrt{n}}$
$= \dfrac{6(\sqrt{m}+\sqrt{n})}{(\sqrt{m})^2-(\sqrt{n})^2}$
$= \dfrac{6\sqrt{m}+6\sqrt{n}}{m-n}$

62. $\dfrac{3}{\sqrt{v}+\sqrt{w}} = \dfrac{3}{\sqrt{v}+\sqrt{w}} \cdot \dfrac{\sqrt{v}-\sqrt{w}}{\sqrt{v}-\sqrt{w}} = \dfrac{3\sqrt{v}-3\sqrt{w}}{v-w}$

63. $\dfrac{\sqrt{12}}{5} = \dfrac{\sqrt{12}}{5} \cdot \dfrac{\sqrt{3}}{\sqrt{3}} = \dfrac{\sqrt{36}}{5\sqrt{3}} = \dfrac{6}{5\sqrt{3}}$

64. $\dfrac{\sqrt{50}}{3} = \dfrac{\sqrt{50}}{3} \cdot \dfrac{\sqrt{2}}{\sqrt{2}} = \dfrac{\sqrt{100}}{3\sqrt{2}} = \dfrac{10}{3\sqrt{2}}$

65. $\sqrt[3]{\dfrac{7}{2}} = \sqrt[3]{\dfrac{7}{2} \cdot \dfrac{49}{49}} = \sqrt[3]{\dfrac{343}{98}} = \dfrac{\sqrt[3]{343}}{\sqrt[3]{98}} = \dfrac{7}{\sqrt[3]{98}}$

66. $\sqrt[3]{\dfrac{2}{5}} = \sqrt[3]{\dfrac{2}{5} \cdot \dfrac{4}{4}} = \sqrt[3]{\dfrac{8}{20}} = \dfrac{2}{\sqrt[3]{20}}$

67. $\dfrac{\sqrt{11}}{\sqrt{3}} = \dfrac{\sqrt{11}}{\sqrt{3}} \cdot \dfrac{\sqrt{11}}{\sqrt{11}} = \dfrac{\sqrt{121}}{\sqrt{33}} = \dfrac{11}{\sqrt{33}}$

68. $\dfrac{\sqrt{5}}{\sqrt{2}} = \dfrac{\sqrt{5}}{\sqrt{2}} \cdot \dfrac{\sqrt{5}}{\sqrt{5}} = \dfrac{\sqrt{25}}{\sqrt{10}} = \dfrac{5}{\sqrt{10}}$

69. $\dfrac{9-\sqrt{5}}{3-\sqrt{3}} = \dfrac{9-\sqrt{5}}{3-\sqrt{3}} \cdot \dfrac{9+\sqrt{5}}{9+\sqrt{5}}$
$= \dfrac{9^2-(\sqrt{5})^2}{27+3\sqrt{5}-9\sqrt{3}-\sqrt{15}}$
$= \dfrac{81-5}{27+3\sqrt{5}-9\sqrt{3}-\sqrt{15}}$
$= \dfrac{76}{27+3\sqrt{5}-9\sqrt{3}-\sqrt{15}}$

70. $\dfrac{8-\sqrt{6}}{5-\sqrt{2}} = \dfrac{8-\sqrt{6}}{5-\sqrt{2}} \cdot \dfrac{8+\sqrt{6}}{8+\sqrt{6}}$
$= \dfrac{64-6}{40+5\sqrt{6}-8\sqrt{2}-\sqrt{12}}$
$= \dfrac{58}{40+5\sqrt{6}-8\sqrt{2}-2\sqrt{3}}$

71. $\dfrac{\sqrt{a}+\sqrt{b}}{3a} = \dfrac{\sqrt{a}+\sqrt{b}}{3a} \cdot \dfrac{\sqrt{a}-\sqrt{b}}{\sqrt{a}-\sqrt{b}}$
$= \dfrac{(\sqrt{a})^2-(\sqrt{b})^2}{3a(\sqrt{a}-\sqrt{b})}$
$= \dfrac{a-b}{3a\sqrt{a}-3a\sqrt{b}}$

72. $\dfrac{\sqrt{p}-\sqrt{q}}{1+\sqrt{q}} = \dfrac{\sqrt{p}-\sqrt{q}}{1+\sqrt{q}} \cdot \dfrac{\sqrt{p}+\sqrt{q}}{\sqrt{p}+\sqrt{q}}$
$= \dfrac{p-q}{\sqrt{p}+\sqrt{q}+\sqrt{pq}+q}$

73. $x^{3/4} = \sqrt[4]{x^3}$

74. $y^{2/5} = \sqrt[5]{y^2}$

75. $16^{3/4} = (16^{1/4})^3 = (\sqrt[4]{16})^3 = 2^3 = 8$

76. $4^{7/2} = (\sqrt{4})^7 = 2^7 = 128$

77. $125^{-1/3} = \dfrac{1}{125^{1/3}} = \dfrac{1}{\sqrt[3]{125}} = \dfrac{1}{5}$

78. $32^{-4/5} = (\sqrt[5]{32})^{-4} = 2^{-4} = \dfrac{1}{16}$

79. $a^{5/4}b^{-3/4} = \dfrac{a^{5/4}}{b^{3/4}} = \dfrac{\sqrt[4]{a^5}}{\sqrt[4]{b^3}} = \dfrac{a\sqrt[4]{a}}{\sqrt[4]{b^3}}$, or $a\sqrt[4]{\dfrac{a}{b^3}}$

80. $x^{2/5}y^{-1/5} = \sqrt[5]{\dfrac{x^2}{y}}$

81. $(\sqrt[4]{13})^5 = \sqrt[4]{13^5} = 13^{5/4}$, or $\sqrt[4]{13^5} = 13\sqrt[4]{13}$

82. $\sqrt[5]{17^3} = 17^{3/5}$

83. $\sqrt[3]{20^2} = 20^{2/3}$

84. $(\sqrt[5]{12})^4 = 12^{4/5}$

85. $\sqrt[3]{\sqrt{11}} = (\sqrt{11})^{1/3} = (11^{1/2})^{1/3} = 11^{1/6}$

86. $\sqrt[3]{\sqrt[4]{7}} = (7^{1/4})^{1/3} = 7^{1/12}$

87. $\sqrt{5}\sqrt[3]{5} = 5^{1/2} \cdot 5^{1/3} = 5^{1/2+1/3} = 5^{5/6}$

88. $\sqrt[3]{2}\sqrt{2} = 2^{1/3} \cdot 2^{1/2} = 2^{5/6}$

89. $\sqrt[5]{32^2} = 32^{2/5} = (32^{1/5})^2 = 2^2 = 4$

90. $\sqrt[3]{64^2} = 64^{2/3} = (64^{1/3})^2 = 4^2 = 16$

91. $(2a^{3/2})(4a^{1/2}) = 8a^{3/2+1/2} = 8a^2$

92. $(3a^{5/6})(8a^{2/3}) = 24a^{3/2} = 24a\sqrt{a}$

93. $\left(\dfrac{x^6}{9b^{-4}}\right)^{1/2} = \left(\dfrac{x^6}{3^2b^{-4}}\right)^{1/2} = \dfrac{x^3}{3b^{-2}}$, or $\dfrac{x^3b^2}{3}$

94. $\left(\dfrac{x^{2/3}}{4y^{-2}}\right)^{1/2} = \dfrac{x^{1/3}}{4^{1/2}y^{-1}} = \dfrac{\sqrt[3]{x}}{2y^{-1}}$, or $\dfrac{y\sqrt[3]{x}}{2}$

95. $\dfrac{x^{2/3}y^{5/6}}{x^{-1/3}y^{1/2}} = x^{2/3-(-1/3)}y^{5/6-1/2} = xy^{1/3} = x\sqrt[3]{y}$

96. $\dfrac{a^{1/2}b^{5/8}}{a^{1/4}b^{3/8}} = a^{1/4}b^{1/4} = \sqrt[4]{ab}$

97. The graphs of $y_1 = (2x^{3/2})(4x^{1/2})$ and $y_2 = 8x^2$ appear to coincide. A table of values also shows that the values of y_1 and y_2 are the same for the given x-values.

98. The graphs of $y_1 = (3x^{5/6})(8x^{2/3})$ and $y_2 = 24x\sqrt{x}$ appear to coincide. A table of values also shows that the values of y_1 and y_2 are the same for the given x-values.

99.
$$\begin{aligned}\sqrt[3]{6}\sqrt{2} = 6^{1/3}2^{1/2} &= 6^{2/6}2^{3/6}\\ &= (6^2 2^3)^{1/6}\\ &= \sqrt[6]{36\cdot 8}\\ &= \sqrt[6]{288}\end{aligned}$$

100. $\sqrt{2}\sqrt[4]{8} = 2^{1/2}(2^3)^{1/4} = 2^{1/2}2^{3/4} = 2^{5/4} = \sqrt[4]{2^5} = 2\sqrt[4]{2}$

101.
$$\begin{aligned}\sqrt[4]{xy}\sqrt[3]{x^2y} = (xy)^{1/4}(x^2y)^{1/3} &= (xy)^{3/12}(x^2y)^{4/12}\\ &= \left[(xy)^3(x^2y)^4\right]^{1/12}\\ &= \left[x^3y^3x^8y^4\right]^{1/12}\\ &= \sqrt[12]{x^{11}y^7}\end{aligned}$$

102. $\sqrt[3]{ab^2}\sqrt{ab} = (ab^2)^{1/3}(ab)^{1/2} = (ab^2)^{2/6}(ab)^{3/6} = \sqrt[6]{(ab^2)^2(ab)^3} = \sqrt[6]{a^5b^7} = b\sqrt[6]{a^5b}$

103.
$$\begin{aligned}\sqrt[3]{a^4\sqrt{a^3}} = \left(a^4\sqrt{a^3}\right)^{1/3} &= (a^4a^{3/2})^{1/3}\\ &= (a^{11/2})^{1/3}\\ &= a^{11/6}\\ &= \sqrt[6]{a^{11}}\\ &= a\sqrt[6]{a^5}\end{aligned}$$

104. $\sqrt{a^3\sqrt[3]{a^2}} = (a^3\cdot a^{2/3})^{1/2} = (a^{11/3})^{1/2} = a^{11/6} = \sqrt[6]{a^{11}} = a\sqrt[6]{a^5}$

105.
$$\begin{aligned}\frac{\sqrt{(a+x)^3}\sqrt[3]{(a+x)^2}}{\sqrt[4]{a+x}} &= \frac{(a+x)^{3/2}(a+x)^{2/3}}{(a+x)^{1/4}}\\ &= \frac{(a+x)^{26/12}}{(a+x)^{3/12}}\\ &= (a+x)^{23/12}\\ &= \sqrt[12]{(a+x)^{23}}\\ &= (a+x)\sqrt[12]{(a+x)^{11}}\end{aligned}$$

106. $\dfrac{\sqrt[4]{(x+y)^2}\sqrt[3]{x+y}}{\sqrt{(x+y)^3}} = \dfrac{(x+y)^{2/4}(x+y)^{1/3}}{(x+y)^{3/2}} = (x+y)^{-2/3} = \dfrac{1}{\sqrt[3]{(x+y)^2}}$

107. Algebraic explanation: Choose specific values for a and b ($a \neq 0$, $b \neq 0$) and show that $\sqrt{a+b} \neq \sqrt{a}+\sqrt{b}$. For example let $a = 9$ and $b = 16$. Then $\sqrt{9+16} = \sqrt{25} = 5$, but $\sqrt{9}+\sqrt{16} = 3+4 = 7$.

Graphical explanation: Let $a = x$ and $b =$ any real number except 0. Graph $y_1 = \sqrt{x+b}$ and $y_2 = \sqrt{x}+\sqrt{b}$ and observe that the graphs do not coincide. For example, when $b = 5$, observe that the graphs of $y_1 = \sqrt{x+5}$ and $y_2 = \sqrt{x}+\sqrt{5}$ do not coincide.

108. Observe that $26 > 25$, so $\sqrt{26} > \sqrt{25}$, or $\sqrt{26} > 5$. Then $10\sqrt{26} - 50 > 10\cdot 5 - 50$, or $10\sqrt{26} - 50 > 0$.

109.
$$\begin{aligned}&\sqrt{1+x^2} + \frac{1}{\sqrt{1+x^2}}\\ &= \sqrt{1+x^2}\cdot\frac{1+x^2}{1+x^2} + \frac{1}{\sqrt{1+x^2}}\cdot\frac{\sqrt{1+x^2}}{\sqrt{1+x^2}}\\ &= \frac{(1+x^2)\sqrt{1+x^2}}{1+x^2} + \frac{\sqrt{1+x^2}}{1+x^2}\\ &= \frac{(2+x^2)\sqrt{1+x^2}}{1+x^2}\end{aligned}$$

110.
$$\begin{aligned}&\sqrt{1-x^2} - \frac{x^2}{2\sqrt{1-x^2}}\\ &= \sqrt{1-x^2} - \frac{x^2\sqrt{1-x^2}}{2(1-x^2)} \quad \text{Rationalizing the denominator}\\ &= \frac{2(1-x^2)\sqrt{1-x^2} - x^2\sqrt{1-x^2}}{2(1-x^2)}\\ &= \frac{(2-3x^2)\sqrt{1-x^2}}{2(1-x^2)}\end{aligned}$$

111. $\left(\sqrt{a\sqrt{a}}\right)^{\sqrt{a}} = \left(a^{\sqrt{a}/2}\right)^{\sqrt{a}} = a^{a/2}$

112.
$$\begin{aligned}\frac{(2a^3b^{5/4}c^{1/7})^4}{(54a^{-2}b^{2/3}c^{6/5})^{-1/3}} &= \frac{16a^{12}b^5c^{4/7}}{54^{-1/3}a^{2/3}b^{-2/9}c^{-2/5}}\\ &= 16\sqrt[3]{54}a^{34/3}b^{47/9}c^{34/35}\\ &= 2^4\cdot 3\cdot 2^{1/3}a^{34/3}b^{47/9}c^{34/35}\\ &= 3\cdot 2^{13/3}a^{34/3}b^{47/9}c^{34/35}, \text{ or}\\ &\quad 48\cdot 2^{1/3}a^{34/3}b^{47/9}c^{34/35}\end{aligned}$$

113. Graph $y_1 = x^{1/2}$ and $y_2 = x^{1/3}$ in a window that shows the relative positions of the graphs. One suitable choice is $[-1, 4, -1, 3]$.

a) Find the first coordinates of the points of intersection of the graphs. They are 0 and 1, so $x^{1/2} = x^{1/3}$ for $x = 0$ or $x = 1$.

b) Find the values of x for which the graph of y_1 lies above the graph of y_2. They are $\{x|x > 1\}$.

c) Find the values of x for which the graph of y_1 lies below the graph of y_2. They are $\{x|0 < x < 1\}$.

Exercise Set R.7

1. $4x + 5 = 21$

$4x = 16$ Subtracting 5 on both sides

$x = 4$ Dividing by 4 on both sides

We can also graph $y_1 = 4x + 5$ and $y_2 = 21$ and find the first coordinate of the point of intersection.

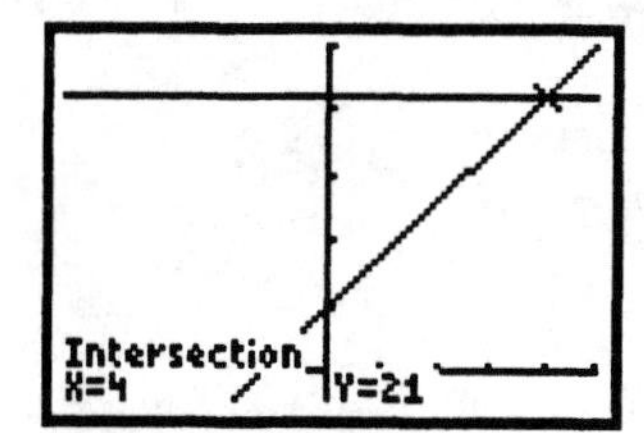

The solution is 4.

2. $2y - 1 = 3$

$2y = 4$

$y = 2$

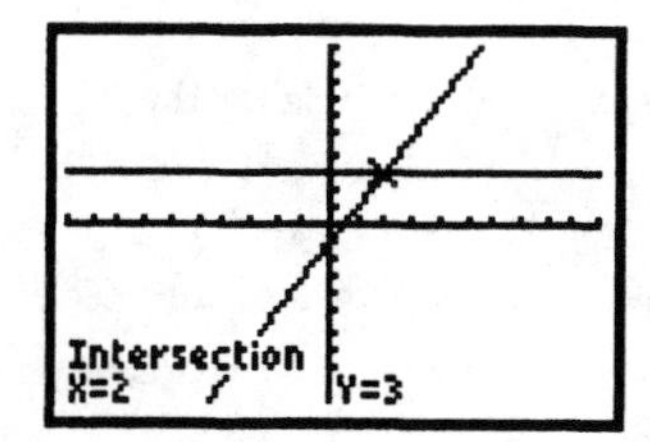

The solution is 2.

3. $y + 1 = 2y - 7$

$1 = y - 7$ Subtracting y on both sides

$8 = y$ Adding 7 on both sides

We can also graph $y_1 = x + 1$ and $y_2 = 2x - 7$ and find the first coordinate of the point of intersection.

The solution is 8.

4. $5 - 4x = x - 13$

$18 = 5x$

$\frac{18}{5} = x$, or

$3.6 = x$

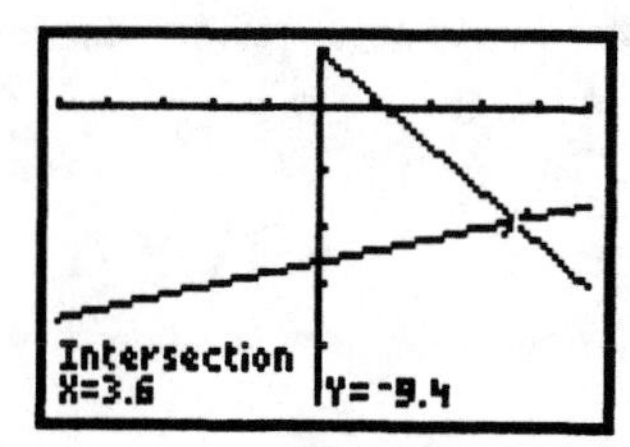

The solution is $\frac{18}{5}$, or 3.6.

5. $5x - 2 + 3x = 2x + 6 - 4x$

$8x - 2 = 6 - 2x$ Collecting like terms

$8x + 2x = 6 + 2$ Adding $2x$ and 2 on both sides

$10x = 8$ Collecting like terms

$x = \frac{8}{10}$ Dividing by 10 on both sides

$x = \frac{4}{5}$, or 0.8 Simplifying

We can also graph $y_1 = 5x - 2 + 3x$ and $y_2 = 2x + 6 - 4x$ and find the first coordinate of the point of intersection.

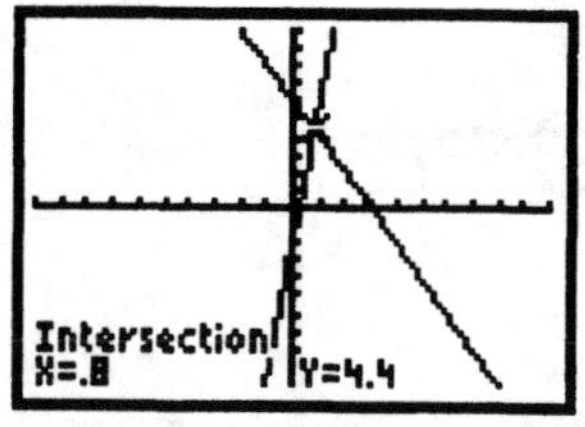

The solution is $\frac{4}{5}$, or 0.8.

6. $5x - 17 - 2x = 6x - 1 - x$

$3x - 17 = 5x - 1$

$-2x = 16$

$x = -8$

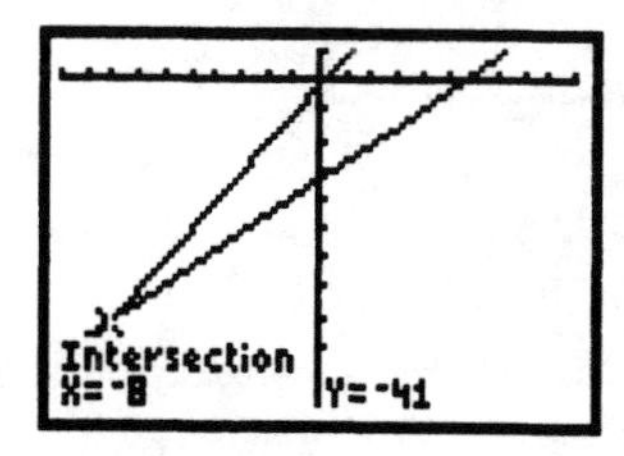

The solution is -8.

7. $7(3x + 6) = 11 - (x + 2)$

$21x + 42 = 11 - x - 2$ Using the distributive property

$21x + 42 = 9 - x$ Collecting like terms

$21x + x = 9 - 42$ Adding x and subtracting 42 on both sides

$22x = -33$ Collecting like terms

$x = -\frac{33}{22}$ Dividing by 22 on both sides

$x = -\frac{3}{2}$, or -1.5 Simplifying

We can also graph $y_1 = 7(3x + 6)$ and $y_2 = 11 - (x + 2)$ and find the first coordinate of the point of intersection.

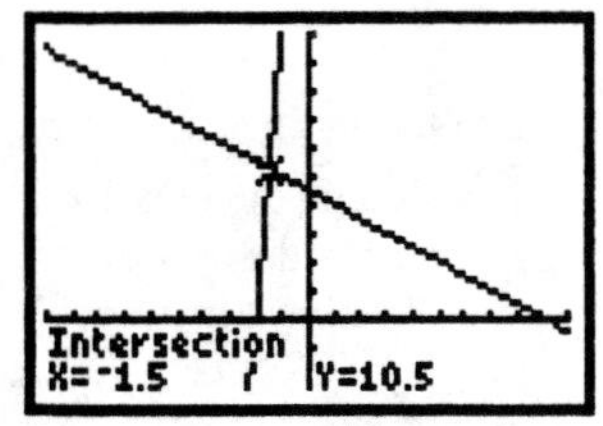

The solution is $-\frac{3}{2}$, or -1.5.

8. $4(5y+3) = 3(2y-5)$

$20y + 12 = 6y - 15$

$14y = -27$

$y = -\frac{27}{14}$

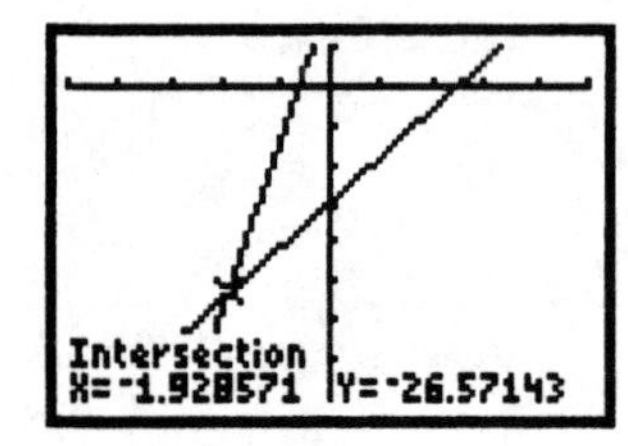

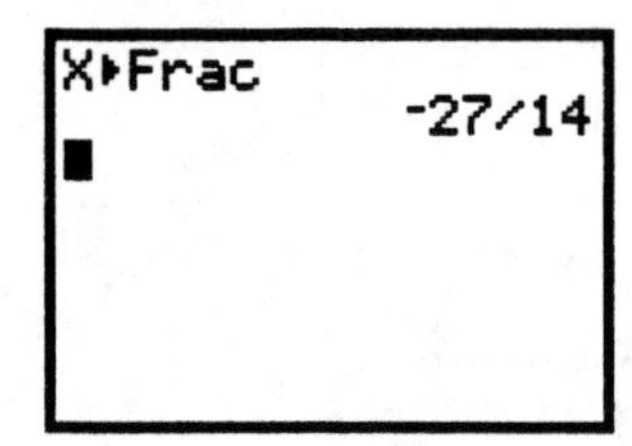

The solution is $-\frac{27}{14}$.

9. $(2x-3)(3x-2) = 0$

$2x - 3 = 0$ *or* $3x - 2 = 0$ Using the principle of zero products

$2x = 3$ *or* $3x = 2$

$x = \frac{3}{2}$ *or* $x = \frac{2}{3}$

We can also graph $y_1 = (2x-3)(3x-2)$ and $y_2 = 0$ and find the first coordinates of the points of intersection.

The solutions are $\frac{3}{2}$ and $\frac{2}{3}$.

10. $(5x-2)(2x+3) = 0$

$x = \frac{2}{5}$ *or* $x = -\frac{3}{2}$

We can also graph $y_1 = (5x-2)(2x+3)$ and $y_2 = 0$ and find the first coordinates of the points of intersection.

The solutions are $\frac{2}{5}$ and $-\frac{3}{2}$.

11. $3x^2 + x - 2 = 0$

$(3x-2)(x+1) = 0$ Factoring

$3x - 2 = 0$ *or* $x + 1 = 0$ Using the principle of zero products

$x = \frac{2}{3}$ *or* $x = -1$

We can also graph $y_1 = 3x^2 + x - 2$ and $y_2 = 0$ and find the first coordinates of the points of intersection.

The solutions are $\frac{2}{3}$ and -1.

12. $10x^2 - 16x + 6 = 0$

$2(5x-3)(x-1) = 0$

$x = \frac{3}{5}$ *or* $x = 1$

We can also graph $y_1 = 10x^2 - 16x + 6$ and $y_2 = 0$ and find the first coordinates of the points of intersection.

The solutions are $\frac{3}{5}$ and 1.

13. $4x^2 - 12 = 0$

$4x^2 = 12$

$x^2 = 3$

$x = \sqrt{3}$ *or* $x = -\sqrt{3}$ Using the principle of square roots

We can also graph $y_1 = 4x^2 - 12$ and $y_2 = 0$ and find the first coordinates of the points of intersection.

The solutions are $\sqrt{3}$ and $-\sqrt{3}$, or approximately 1.732 and -1.732.

14. $6x^2 = 36$

$x^2 = 6$

$x = \sqrt{6}$ *or* $x = -\sqrt{6}$

We can also graph $y_1 = 6x^2$ and $y_2 = 36$ and find the first coordinates of the points of intersection.

The solutions are $\sqrt{6}$ and $-\sqrt{6}$, or approximately 2.449 and -2.499.

15. $2x^2 = 6x$

$2x^2 - 6x = 0$ Subtracting $6x$ on both sides

$2x(x-3) = 0$

$2x = 0$ *or* $x - 3 = 0$

$x = 0$ *or* $x = 3$

We can also graph $y_1 = 2x^2$ and $y_2 = 6x$ and find the first coordinates of the points of intersection.

The solutions are 0 and 3.

16. $18x + 9x^2 = 0$

$9x(2+x) = 0$

$x = 0$ *or* $x = -2$

We can also graph $y_1 = 18x + 9x^2$ and $y_2 = 0$ and find the first coordinates of the points of intersection.

The solutions are -2 and 0.

17.
$$3y^3 - 5y^2 - 2y = 0$$
$$y(3y^2 - 5y - 2) = 0$$
$$y(3y+1)(y-2) = 0$$
$$y = 0 \text{ or } 3y+1 = 0 \text{ or } y-2 = 0$$
$$y = 0 \text{ or } y = -\frac{1}{3} \text{ or } y = 2$$

We can also graph $y_1 = 3x^3 - 5x^2 - 2x$ and $y_2 = 0$ and find the first coordinates of the points of intersection.

The solutions are $-\frac{1}{3}$, 0 and 2.

18.
$$3t^3 + 2t = 5t^2$$
$$3t^3 - 5t^2 + 2t = 0$$
$$t(t-1)(3t-2) = 0$$
$$t = 0 \text{ or } t = 1 \text{ or } t = \frac{2}{3}$$

We can also graph $y_1 = 3x^3 + 2x$ and $y_2 = 5x^2$ and find the first coordinates of the points of intersection.

The solutions are 0, $\frac{2}{3}$, and 1.

19.
$$7x^3 + x^2 - 7x - 1 = 0$$
$$x^2(7x+1) - (7x+1) = 0$$
$$(x^2-1)(7x+1) = 0$$
$$(x+1)(x-1)(7x+1) = 0$$
$$x+1 = 0 \text{ or } x-1 = 0 \text{ or } 7x+1 = 0$$
$$x = -1 \text{ or } x = 1 \text{ or } x = -\frac{1}{7}$$

We can also graph $y_1 = 7x^3 + x^2 - 7x - 1$ and $y_2 = 0$ and find the first coordinates of the points of intersection.

The solutions are -1, $-\frac{1}{7}$, and 1.

20.
$$3x^3 + x^2 - 12x - 4 = 0$$
$$x^2(3x+1) - 4(3x+1) = 0$$
$$(3x+1)(x^2-4) = 0$$
$$(3x+1)(x+2)(x-2) = 0$$
$$x = -\frac{1}{3} \text{ or } x = -2 \text{ or } x = 2$$

We can also graph $y_1 = 3x^3 + x^2 - 12x - 4$ and $y_2 = 0$ and find the first coordinates of the points of intersection.

The solutions are -2, $-\frac{1}{3}$, and 2.

21.
$$A = \frac{1}{2}bh$$
$$2A = bh \quad \text{Multiplying by 2 on both sides}$$
$$\frac{2A}{h} = b \quad \text{Dividing by } h \text{ on both sides}$$

22.
$$A = \pi r^2$$
$$\frac{A}{r^2} = \pi$$

23.
$$P = 2l + 2w$$
$$P - 2l = 2w \quad \text{Subtracting } 2l \text{ on both sides}$$
$$\frac{P-2l}{2} = w \quad \text{Dividing by 2 on both sides}$$

24.
$$A = P + Prt$$
$$A = P(1+rt)$$
$$\frac{A}{1+rt} = P$$

25.
$$A = \frac{1}{2}h(b_1 + b_2)$$
$$2A = h(b_1 + b_2) \quad \text{Multiplying by 2 on both sides}$$
$$\frac{2A}{b_1+b_2} = h \quad \text{Dividing by } b_1 + b_2 \text{ on both sides}$$

26.
$$A = \frac{1}{2}h(b_1 + b_2)$$
$$\frac{2A}{h} = b_1 + b_2$$
$$\frac{2A}{h} - b_1 = b_2, \text{ or}$$
$$\frac{2A - b_1h}{h} = b_2$$

27.
$$V = \frac{4}{3}\pi r^3$$
$$3V = 4\pi r^3 \quad \text{Multiplying by 3 on both sides}$$
$$\frac{3V}{4r^3} = \pi \quad \text{Dividing by } 4r^3 \text{ on both sides}$$

28.
$$V = \frac{4}{3}\pi r^3$$
$$\frac{3V}{4\pi} = r^3$$

29.
$$F = \frac{9}{5}C + 32$$
$$F - 32 = \frac{9}{5}C \quad \text{Subtracting 32 on both sides}$$
$$\frac{5}{9}(F - 32) = C \quad \text{Multiplying by } \frac{5}{9} \text{ on both sides}$$

30.
$$Ax + By = C$$
$$By = C - Ax$$
$$y = \frac{C - Ax}{B}$$

31. The graphs of $y_1 = x^2 - 6x + 9$ and $y_2 = 1 - x^4$ do not intersect.

32. When we want to know what principal should be invested at a given interest rate in order to have a given amount in an account at the end of a given period of time, it would be helpful to solve the formula for P. This would be particularly helpful when this computation is to be done multiple times with different values of A, r, and t.

33.
$$\begin{aligned}
2x - \{x - [3x - (6x + 5)]\} &= 4x - 1 \\
2x - \{x - [3x - 6x - 5]\} &= 4x - 1 \\
2x - \{x - [-3x - 5]\} &= 4x - 1 \\
2x - \{x + 3x + 5\} &= 4x - 1 \\
2x - \{4x + 5\} &= 4x - 1 \\
2x - 4x - 5 &= 4x - 1 \\
-2x - 5 &= 4x - 1 \\
-6x - 5 &= -1 \\
-6x &= 4 \\
x &= -\frac{2}{3}
\end{aligned}$$

We can also graph $y_1 = 2x - (x - (3x - (6x + 5)))$ and $y_2 = 4x - 1$ and find the first coordinate of the point of intersection.

The solution is $-\frac{2}{3}$.

34.
$$\begin{aligned}
14 - 2[3 + 5(x - 1)] &= 3\{x - 4[1 + 6(2 - x)]\} \\
14 - 2[3 + 5x - 5] &= 3\{x - 4[1 + 12 - 6x]\} \\
14 - 2[5x - 2] &= 3\{x - 4[13 - 6x]\} \\
14 - 10x + 4 &= 3\{x - 52 + 24x\} \\
18 - 10x &= 3\{25x - 52\} \\
18 - 10x &= 75x - 156 \\
174 &= 85x \\
\frac{174}{85} &= x
\end{aligned}$$

We can also graph $y_1 = 14 - 2(3 + 5(x - 1))$ and $y_2 = 3(x-4(1+6(2-x)))$ and find the first coordinate of the point of intersection.

The solution is $\frac{174}{85}$.

35.
$$\begin{aligned}
(x - 2)^3 &= x^3 - 2 \\
x^3 - 6x^2 + 12x - 8 &= x^3 - 2 \\
0 &= 6x^2 - 12x + 6 \\
0 &= 6(x^2 - 2x + 1) \\
0 &= 6(x - 1)(x - 1)
\end{aligned}$$

$x - 1 = 0$ *or* $x - 1 = 0$

$x = 1$ *or* $x = 1$

We can also graph $y_1 = (x - 2)^3$ and $y_2 = x^3 - 2$ and find the first coordinate of the point of intersection.

The solution is 1.

36.
$$\begin{aligned}
(x + 1)^3 &= (x - 1)^3 + 26 \\
x^3 + 3x^2 + 3x + 1 &= x^3 - 3x^2 + 3x - 1 + 26 \\
x^3 + 3x^2 + 3x + 1 &= x^3 - 3x^2 + 3x + 25 \\
6x^2 - 24 &= 0 \\
6(x^2 - 4) &= 0 \\
6(x + 2)(x - 2) &= 0
\end{aligned}$$

$x + 2 = 0$ *or* $x - 2 = 0$

$x = -2$ *or* $x = 2$

We can also graph $y_1 = (x + 1)^3$ and $y_2 = (x - 1)^3 + 26$ and find the first coordinates of the points of intersection.

The solutions are -2 and 2.

37.
$$\begin{aligned}
(6x^3 + 7x^2 - 3x)(x^2 - 7) &= 0 \\
x(6x^2 + 7x - 3)(x^2 - 7) &= 0 \\
x(3x - 1)(2x + 3)(x^2 - 7) &= 0
\end{aligned}$$

$x=0$ *or* $3x - 1=0$ *or* $2x + 3=0$ *or* $x^2 - 7 = 0$

$x=0$ *or* $x=\frac{1}{3}$ *or* $x=-\frac{3}{2}$ *or* $x = \sqrt{7}$ *or* $x = -\sqrt{7}$

We can also graph $y_1 = (6x^3 + 7x^2 - 3x)(x^2 - 7)$ and $y_2 = 0$ and find the first coordinates of the points of intersection.

The exact solutions are $-\sqrt{7}$, $-\frac{3}{2}$, 0, $\frac{1}{3}$, and $\sqrt{7}$.

38.
$$\begin{aligned}
\left(x - \frac{1}{5}\right)\left(x^2 - \frac{1}{4}\right) + \left(x - \frac{1}{5}\right)\left(x^2 + \frac{1}{8}\right) &= 0 \\
\left(x - \frac{1}{5}\right)\left(2x^2 - \frac{1}{8}\right) &= 0 \\
\left(x - \frac{1}{5}\right)(2)\left(x + \frac{1}{4}\right)\left(x - \frac{1}{4}\right) &= 0
\end{aligned}$$

$x = \frac{1}{5}$ *or* $x = -\frac{1}{4}$ *or* $x = \frac{1}{4}$

We can also graph $y_1 = \left(x - \frac{1}{5}\right)\left(x^2 - \frac{1}{4}\right) + \left(x - \frac{1}{5}\right)\left(x^2 + \frac{1}{8}\right)$ and $y_2 = 0$ and find the first coordinates of the points of intersection.

The solutions are $-\frac{1}{4}$, $\frac{1}{5}$, and $\frac{1}{4}$.

Chapter 1

Graphs, Functions, and Models

Exercise Set 1.1

1. This correspondence is a function, because each member of the domain corresponds to exactly one member of the range.

2. This correspondence is a function, because each member of the domain corresponds to exactly one member of the range.

3. This correspondence is a function, because each member of the domain corresponds to exactly one member of the range.

4. This correspondence is not a function, because there is a member of the domain (1) that corresponds to more than one member of the range (4 and 6).

5. This correspondence is not a function, because there is a member of the domain (m) that corresponds to more than one member of the range (A and B).

6. This correspondence is a function, because each member of the domain corresponds to exactly one member of the range.

7. This correspondence is a function, because each member of the domain corresponds to exactly one member of the range.

8. This correspondence is a function, because each member of the domain corresponds to exactly one member of the range.

9. This correspondence is a function, because each car has exactly one license number.

10. This correspondence is not a function, because we can safely assume that at least one person uses more than one doctor.

11. This correspondence is a function, because each member of the family has exactly one eye color.

12. This correspondence is not a function, because we can safely assume that at least one band member plays more than one instrument.

13. This correspondence is not a function, because at least one student will have more than one neighboring seat occupied by another student.

14. This correspondence is a function, because each bag has exactly one weight.

15. The relation is a function, because no two ordered pairs have the same first coordinate and different second coordinates.

    The domain is the set of all first coordinates: $\{2,3,4\}$.

    The range is the set of all second coordinates: $\{10,15,20\}$.

16. The relation is a function, because no two ordered pairs have the same first coordinate and different second coordinates.

    Domain: $\{3,5,7\}$

    Range: $\{1\}$

17. The relation is not a function, because the ordered pairs $(-2,1)$ and $(-2,4)$ have the same first coordinate and different second coordinates.

    The domain is the set of all first coordinates: $\{-7,-2,0\}$.

    The range is the set of all second coordinates: $\{3,1,4,7\}$.

18. The relation is not a function, because of each of the ordered pairs has the same first coordinate and different second coordinates.

    Domain: $\{1\}$

    Range: $\{3,5,7,9\}$

19. The relation is a function, because no two ordered pairs have the same first coordinate and different second coordinates.

    The domain is the set of all first coordinates: $\{-2,0,2,4,-3\}$.

    The range is the set of all second coordinates: $\{1\}$.

20. The relation is not a function, because the ordered pairs $(5,0)$ and $(5,-1)$ have the same first coordinates and different second coordinates. This is also true of the pairs $(3,-1)$ and $(3,-2)$.

    Domain: $\{5,3,0\}$

    Range: $\{0,-1,-2\}$

21. The point $(-1,2)$ is on the graph, so $f(-1)=2$; the point $(0,0)$ is on the graph, so $f(0)=0$; the point $(1,-2)$ is on the graph, so $f(1)=-2$.

22. The point $(-2,4)$ is on the graph, so $g(-2)=4$; the point $(0,-4)$ is on the graph, so $g(0)=-4$; the point $(2.4,-2.6176)$ is on the graph, so $g(2.4)=-2.6176$.

23. $g(x) = 3x^2 - 2x + 1$

a) $g(0) = 3 \cdot 0^2 - 2 \cdot 0 + 1 = 1$

b) $g(-1) = 3(-1)^2 - 2(-1) + 1 = 6$

c) $g(3) = 3 \cdot 3^2 - 2 \cdot 3 + 1 = 22$

d) $g(-x) = 3(-x)^2 - 2(-x) + 1 = 3x^2 + 2x + 1$

e) $g(1-t) = 3(1-t)^2 - 2(1-t) + 1 =$
$3(1-2t+t^2) - 2(1-t) + 1 = 3 - 6t + 3t^2 - 2 + 2t + 1 =$
$3t^2 - 4t + 2$

24. $f(x) = 5x^2 + 4x$

a) $f(0) = 5 \cdot 0^2 + 4 \cdot 0 = 0 + 0 = 0$

b) $f(-1) = 5(-1)^2 + 4(-1) = 5 - 4 = 1$

c) $f(3) = 5 \cdot 3^2 + 4 \cdot 3 = 45 + 12 = 57$

d) $f(t) = 5t^2 + 4t$

e) $f(t-1) = 5(t-1)^2 + 4(t-1) = 5t^2 - 6t + 1$

25. $g(x) = x^3$

a) $g(2) = 2^3 = 8$

b) $g(-2) = (-2)^3 = -8$

c) $g(-x) = (-x)^3 = -x^3$

d) $g(3y) = (3y)^3 = 27y^3$

e) $g(2+h) = (2+h)^3 = 8 + 12h + 6h^2 + h^3$

26. $f(x) = 2|x| + 3x$

a) $f(1) = 2|1| + 3 \cdot 1 = 2 + 3 = 5$

b) $f(-2) = 2|-2| + 3(-2) = 4 - 6 = -2$

c) $f(-x) = 2|-x| + 3(-x) = 2|x| - 3x$

d) $f(2y) = 2|2y| + 3 \cdot 2y = 4|y| + 6y$

e) $f(2-h) = 2|2-h| + 3(2-h) =$
$2|2-h| + 6 - 3h$

27. $g(x) = \dfrac{x-4}{x+3}$

a) $g(5) = \dfrac{5-4}{5+3} = \dfrac{1}{8}$

b) $g(4) = \dfrac{4-4}{4+7} = 0$

c) $g(-3) = \dfrac{-3-4}{-3+3} = \dfrac{-7}{0}$

Since division by 0 is not defined, $g(-3)$ does not exist.

d) $g(-16.25) = \dfrac{-16.25-4}{-16.25+3} = \dfrac{-20.25}{-13.25} = \dfrac{81}{53}$

e) $g(x+h) = \dfrac{x+h-4}{x+h+3}$

28. $f(x) = \dfrac{x}{2-x}$

a) $f(2) = \dfrac{2}{2-2} = \dfrac{2}{0}$

Since division by 0 is not defined, $f(2)$ does not exist.

b) $f(1) = \dfrac{1}{2-1} = 1$

c) $f(-16) = \dfrac{-16}{2-(-16)} = \dfrac{-16}{18} = -\dfrac{8}{9}$

d) $f(-x) = \dfrac{-x}{2-(-x)} = \dfrac{-x}{2+x}$

e) $f\left(-\dfrac{2}{3}\right) = \dfrac{-\frac{2}{3}}{2-\left(-\frac{2}{3}\right)} = \dfrac{-\frac{2}{3}}{\frac{8}{3}} = -\dfrac{1}{4}$

29. $g(x) = \dfrac{x}{\sqrt{1-x^2}}$

$g(0) = \dfrac{0}{\sqrt{1-0^2}} = \dfrac{0}{\sqrt{1}} = \dfrac{0}{1} = 0$

$g(-1) = \dfrac{-1}{\sqrt{1-(-1)^2}} = \dfrac{-1}{\sqrt{1-1}} = \dfrac{-1}{\sqrt{0}} = \dfrac{-1}{0}$

Since division by 0 is not defined, $g(-1)$ does not exist.

$g(5) = \dfrac{5}{\sqrt{1-5^2}} = \dfrac{5}{\sqrt{1-25}} = \dfrac{5}{\sqrt{-24}}$

Since $\sqrt{-24}$ is not defined as a real number, $g(5)$ does not exist as a real number.

$g\left(\dfrac{1}{2}\right) = \dfrac{\frac{1}{2}}{\sqrt{1-\left(\frac{1}{2}\right)^2}} = \dfrac{\frac{1}{2}}{\sqrt{1-\frac{1}{4}}} = \dfrac{\frac{1}{2}}{\sqrt{\frac{3}{4}}} =$

$\dfrac{\frac{1}{2}}{\frac{\sqrt{3}}{2}} = \dfrac{1}{2} \cdot \dfrac{2}{\sqrt{3}} = \dfrac{1 \cdot 2}{2\sqrt{3}} = \dfrac{1}{\sqrt{3}}$, or $\dfrac{\sqrt{3}}{3}$

30. $h(x) = x + \sqrt{x^2 - 1}$

$h(0) = 0 + \sqrt{0^2 - 1} = 0 + \sqrt{-1}$

Since $\sqrt{-1}$ is not defined as a real number, $h(0)$ does not exist as a real number.

$h(2) = 2 + \sqrt{2^2 - 1} = 2 + \sqrt{3}$

$h(-x) = -x + \sqrt{(-x)^2 - 1} = -x + \sqrt{x^2 - 1}$

31. We can substitute any real number for x. Thus, the domain is the set of all real numbers, or $(-\infty, \infty)$.

32. We can substitute any real number for x. Thus, the domain is the set of all real numbers, or $(-\infty, \infty)$.

33. The input 0 results in a denominator of 0. Thus, the domain is $\{x|x \neq 0\}$, or $(-\infty, 0) \cup (0, \infty)$.

34. The input 0 results in a denominator of 0. Thus, the domain is $\{x|x \neq 0\}$, or $(-\infty, 0) \cup (0, \infty)$.

35. We can substitute any real number in the numerator, but we must avoid inputs that make the denominator 0. We find these inputs.

$$2 - x = 0$$
$$2 = x$$

The domain is $\{x|x \neq 2\}$, or $(-\infty, 2) \cup (2, \infty)$.

36. We find the inputs that make the denominator 0:

$$x + 4 = 0$$
$$x = -4$$

The domain is $\{x|x \neq -4\}$, or $(-\infty, -4) \cup (-4, \infty)$.

37. We find the inputs that make the denominator 0:

$$x^2 - 4x - 5 = 0$$
$$(x - 5)(x + 1) = 0$$
$$x - 5 = 0 \quad or \quad x + 1 = 0$$
$$x = 5 \quad or \quad x = -1$$

The domain is $\{x|x \neq 5 \text{ } and \text{ } x \neq -1\}$, or $(-\infty, -1) \cup (-1, 5) \cup (5, \infty)$.

38. We can substitute any real number in the numerator. Find the inputs that make the denominator 0:

$$3x^2 - 10x - 8 = 0$$
$$(3x + 2)(x - 4) = 0$$
$$x = -\frac{2}{3} \quad or \quad x = 4$$

Domain: $\left\{x \middle| x \neq -\frac{2}{3} \text{ } and \text{ } x \neq 4\right\}$, or

$$\left(-\infty, -\frac{2}{3}\right) \cup \left(\frac{2}{3}, 4\right) \cup (4, \infty)$$

39.

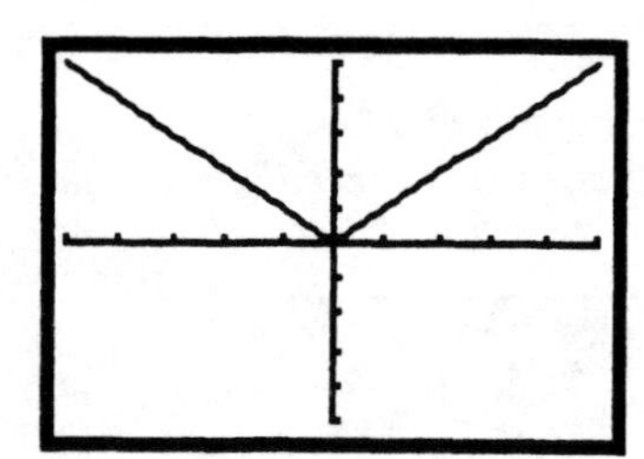

Domain: all real numbers, or $(-\infty, \infty)$

Range: $[0, \infty)$

40.

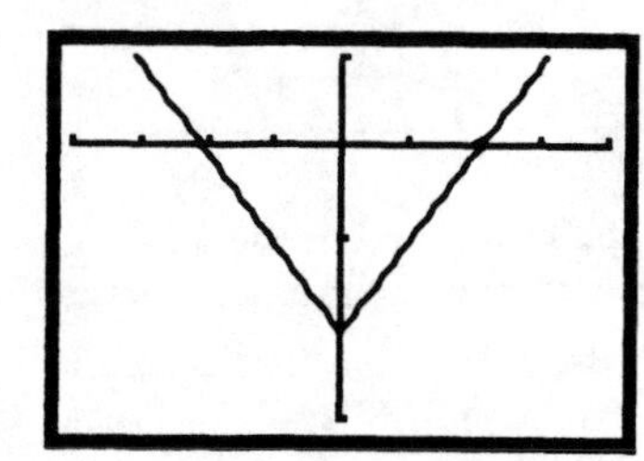

Domain: all real numbers, or $(-\infty, \infty)$

Range: $[-10.3, \infty)$

41.

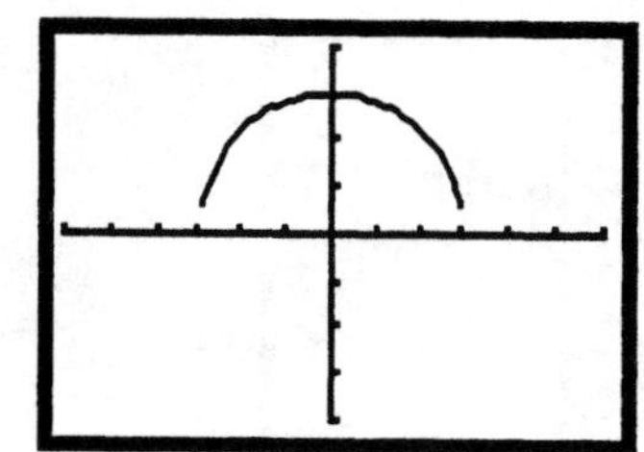

Domain: $[-3, 3]$

Range: $[0, 3]$

42.

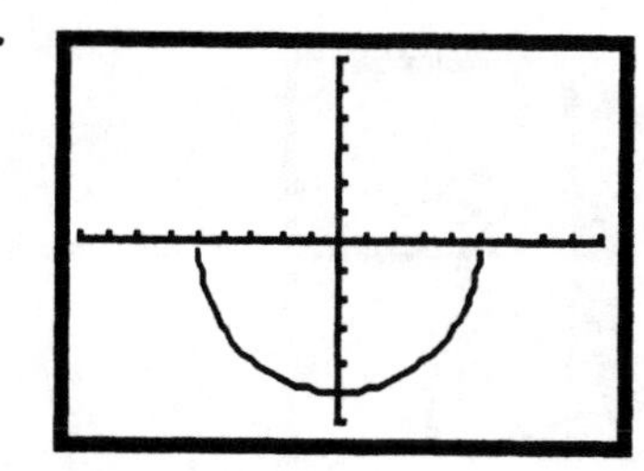

Domain: $[-5, 5]$

Range: $[-5, 0]$

43.

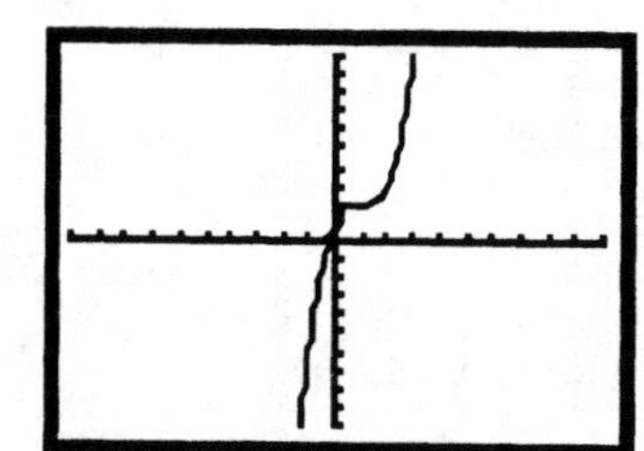

Domain: all real numbers, or $(-\infty, \infty)$

Range: all real numbers, or $(-\infty, \infty)$

44.

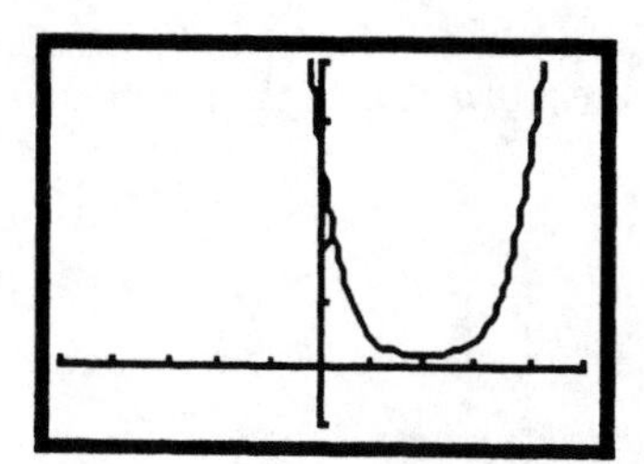

Domain: all real numbers, or $(-\infty, \infty)$

Range: $[1, \infty)$

45.

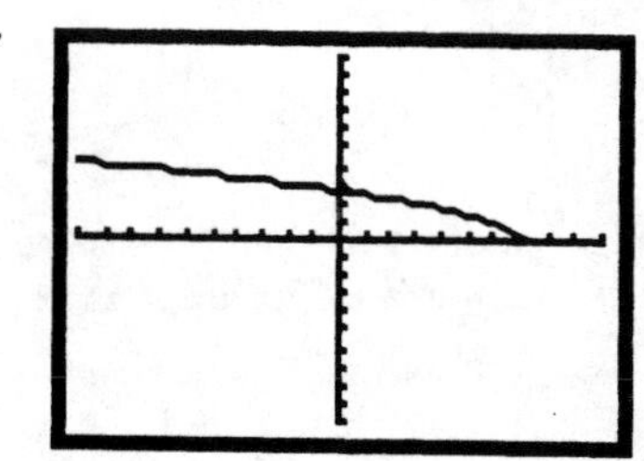

Domain: $(-\infty, 7]$

Range: $[0, \infty)$

46.

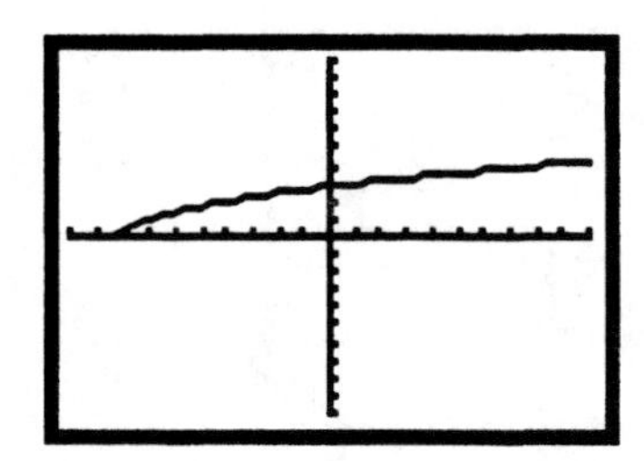

Domain: $[-8, \infty)$

Range: $[0, \infty)$

47.

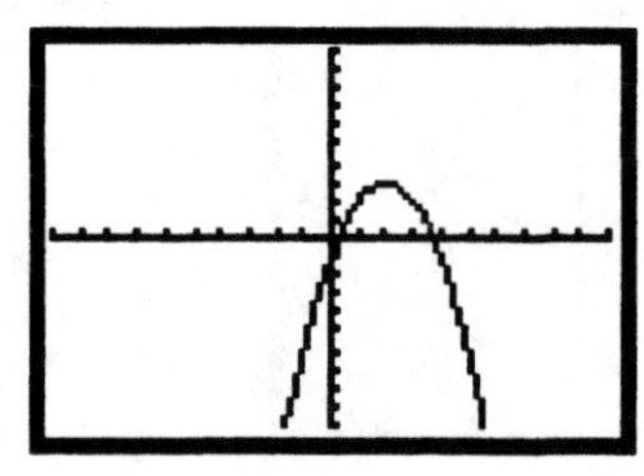

Domain: all real numbers, or $(-\infty, \infty)$

Range: $(-\infty, 3]$

48.

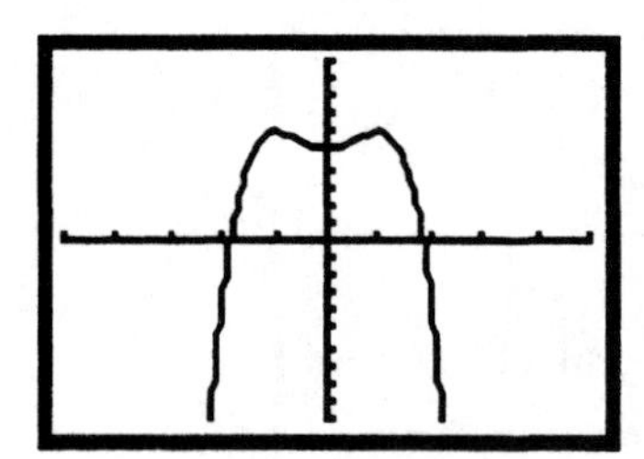

Domain: all real numbers, or $(-\infty, \infty)$

Range: $(-\infty, 6]$

49. This is not the graph of a function, because we can find a vertical line that crosses the graph more than once.

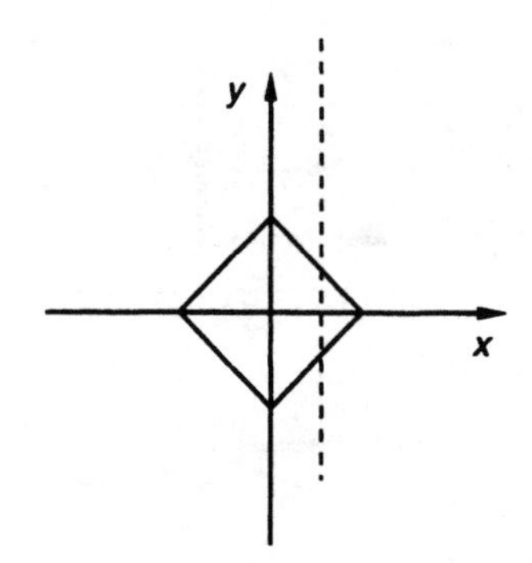

50. This is not the graph of a function, because we can find a vertical line that crosses the graph more than once.

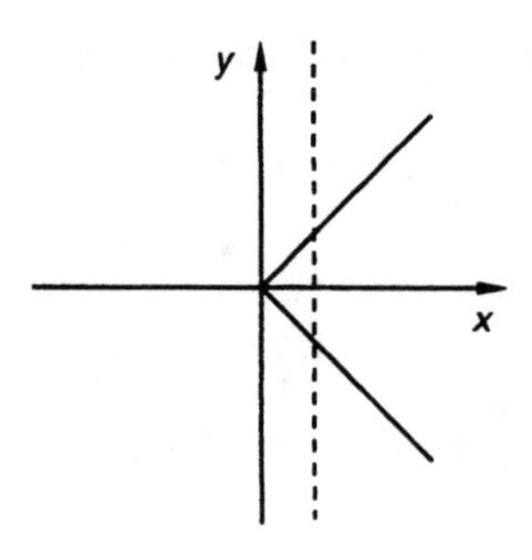

51. This is the graph of a function, because there is no vertical line that crosses the graph more than once.

52. This is the graph of a function, because there is no vertical line that crosses the graph more than once.

53. This is the graph of a function, because there is no vertical line that crosses the graph more than once.

54. This is the graph of a function, because there is no vertical line that crosses the graph more than once.

55. This is not the graph of a function, because we can find a vertical line that crosses the graph more than once.

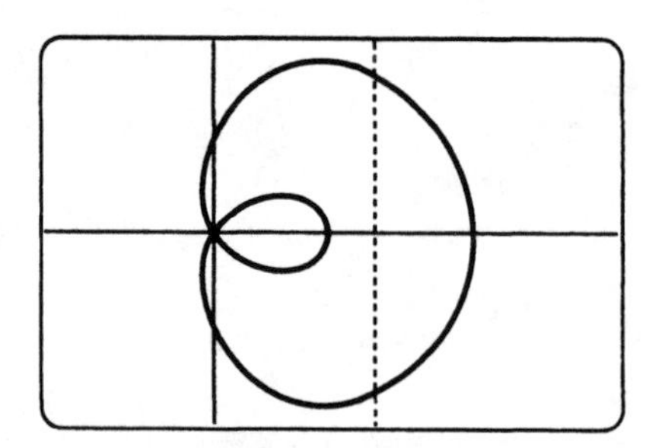

56. This is not the graph of a function, because we can find a vertical line that crosses the graph more than once.

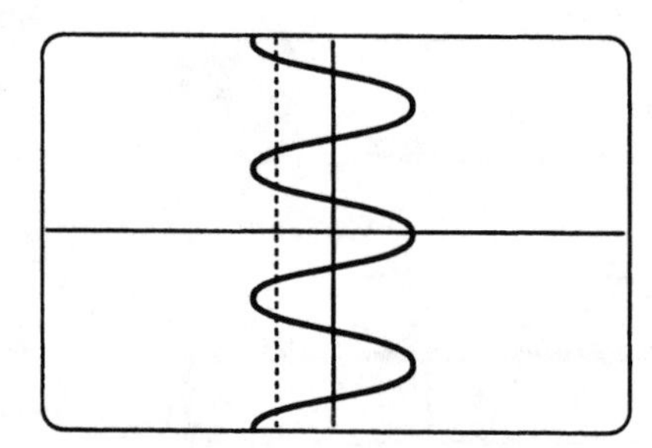

57. This is the graph of a function, because there is no vertical line that crosses the graph more than once.

The inputs on the x-axis that correspond to points on the graph extend from 0 to 5, inclusive. Thus, the domain is $\{x|0 \leq x \leq 5\}$, or $[0, 5]$.

The outputs on the y-axis extend from 0 to 3, inclusive. Thus, the range is $\{y|0 \leq y \leq 3\}$, or $[0, 3]$.

58. This is the graph of a function, because there is no vertical line that crosses the graph more than once.

The inputs on the x-axis that correspond to points on the graph extend from -3 up to but not including 5. Thus, the domain is $\{x| -3 \le x < 5\}$, or $[-3, 5)$.

The outputs on the y-axis extend from -4 up to but not including 1. Thus, the range is $\{y| -4 \le y < 1\}$, or $[-4, 1)$.

59. This is the graph of a function, because there is no vertical line that crosses the graph more than once.

The inputs on the x-axis that correspond to points on the graph extend from -2π to 2π inclusive. Thus, the domain is $\{x| -2\pi \le x \le 2\pi\}$, or $[-2\pi, 2\pi]$.

The outputs on the y-axis extend from -1 to 1, inclusive. Thus, the range is $\{y| -1 \le y \le 1\}$, or $[-1, 1]$.

60. This is the graph of a function, because there is no vertical line that crosses the graph more than once.

The inputs on the x-axis that correspond to points on the graph extend from -2 to 1, inclusive. Thus, the domain is $\{x| -2 \le x \le 1\}$, or $[-2, 1]$.

The outputs on the y-axis extend from -1 to 4, inclusive. Thus, the range is $\{y| -1 \le y \le 4\}$, or $[-1, 4]$.

61. $E(t) = 1000(100 - t) + 580(100 - t)^2$

a) $E(99.5) = 1000(100-99.5)+580(100-99.5)^2$
$= 1000(0.5) + 580(0.5)^2$
$= 500 + 580(0.25) = 500 + 145$
$= 645$ m above sea level

b) $E(100) = 1000(100 - 100) + 580(100 - 100)^2$
$= 1000 \cdot 0 + 580(0)^2 = 0 + 0$
$= 0$ m above sea level, or at sea level

62. a) $P(2003) = 0.1522(2003) - 298.592 \approx \$6.26$
$P(2010) = 0.1522(2010) - 298.592 = \$7.33$

b) Solve: $8 = 0.1522x - 298.592$
$x \approx 2014$

63. $T(0.5) = 0.5^{1.31} \approx 0.4$ acres
$T(10) = 10^{1.31} \approx 20.4$ acres
$T(20) = 20^{1.31} \approx 50.6$ acres
$T(100) = 100^{1.31} \approx 416.9$ acres
$T(200) = 200^{1.31} \approx 1033.6$ acres

64. A function is a correspondence between two sets in which each member of the first set corresponds to exactly one member of the second set.

65. The domain of a function is the set of all inputs of the function. The range is the set of all outputs. The range depends on the domain.

66.

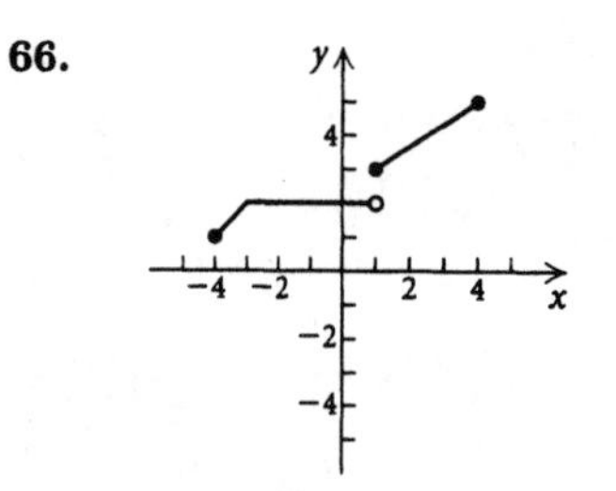

67. Answers may vary. Two possibilities are $f(x) = x$, $g(x) = x + 1$ and $f(x) = x^2$ and $g(x) = x^2 - 4$.

68.

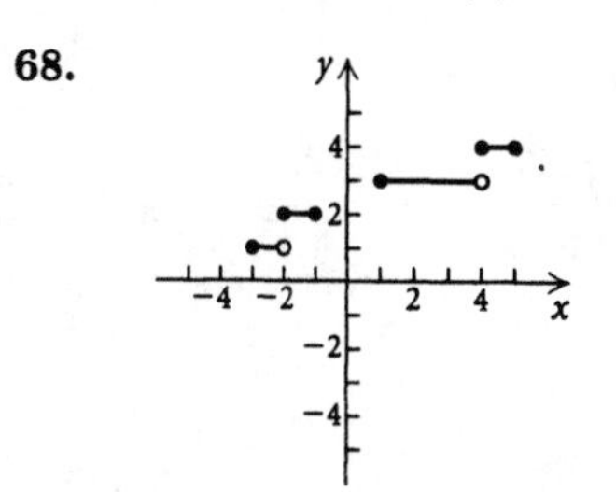

69. $f(x - 1) = 5x$
$f(6) = f(7 - 1) = 5 \cdot 7 = 35$

Exercise Set 1.2

1. a) Yes. Each input is 1 more than the one that precedes it.

b) Yes. Each output is 3 more than the one that precedes it.

c) Yes. Constant changes in inputs result in constant changes in outputs.

2. a) Yes. Each input is 10 more than the one that precedes it.

b) No. The change in the outputs varies.

c) No. Constant changes in inputs do not result in constant changes in outputs.

3. a) Yes. Each input is 15 more than the one that precedes it.

b) No. The change in the outputs varies.

c) No. Constant changes in inputs do not result in constant changes in outputs.

4. a) Yes. Each input is 2 more than the one that precedes it.

b) Yes. Each output is 4 less than the one that precedes it.

c) Yes. Constant changes in inputs result in constant changes in outputs.

5. Two points on the line are $(0, 3)$ and $(5, 0)$.

$$m = \frac{y_2 - y_1}{x_2 - x_1} = \frac{0 - 3}{5 - 0} = \frac{-3}{5}, \text{ or } -\frac{3}{5}$$

6. $m = \frac{0-(-3)}{-2-(-2)} = \frac{3}{0}$

The slope is not defined.

7. $m = \frac{y_2-y_1}{x_2-x_1} = \frac{3-3}{3-0} = \frac{0}{3} = 0$

8. $m = \frac{1-(-4)}{5-(-3)} = \frac{5}{8}$

9. $m = \frac{y_2-y_1}{x_2-x_1} = \frac{2-4}{-1-9} = \frac{-2}{-10} = \frac{1}{5}$

10. $m = \frac{-1-7}{5-(-3)} = \frac{-8}{8} = -1$

11. $m = \frac{y_2-y_1}{x_2-x_1} = \frac{6-(-9)}{-5-4} = \frac{15}{-9} = -\frac{5}{3}$

12. $m = \frac{-13-(-1)}{2-(-6)} = \frac{-12}{8} = -\frac{3}{2}$

13. $m = \frac{y_2-y_1}{x_2-x_1} = \frac{2-(-3)}{\pi-\pi} = \frac{5}{0}$

Since division by 0 is not defined, the slope is not defined.

14. $m = \frac{-4-(-4)}{0.56-\sqrt{2}} = \frac{0}{0.56-\sqrt{2}} = 0$

15. $m = \frac{y_2-y_1}{x_2-x_1} = \frac{(a+h)^2-a^2}{a+h-a} = \frac{a^2+2ah+h^2-a^2}{h} =$

$\frac{2ah+h^2}{h} = \frac{h(2a+h)}{h} = 2a+h$

16. $m = \frac{3(a+h)+1-(3a+1)}{a+h-a} =$

$\frac{3a+3h+1-3a-1}{h} = \frac{3h}{h} = 3$

17. $m = \frac{920.58}{13,740} = 0.067$

The road grade is 6.7%.

We find an equation of the line with slope 0.067 and containing the point $(13,740, 920.58)$:

$$y - 920.58 = 0.067(x - 13,740)$$
$$y - 920.58 = 0.067x - 920.58$$
$$y = 0.067x$$

18. Let h = the height of the end of the treadmill.

Solve: $0.08 = \frac{h}{5}$

$h = 0.4$ ft

19. We can use any two points on the graph to find the rate of change. We will use $(1991, 800)$ and $(1995, 1200)$.

$$\text{Rate of change} = \frac{\text{change in } y}{\text{change in } x} = \frac{1200-800}{1995-1991} = \frac{400}{4} = 100$$

The rate of change is \$100 per year.

20. $$\text{Rate of change} = \frac{\$17,634 - \$11,037}{1995-1986} = \frac{\$6597}{9} = \$733 \text{ per year}$$

21. First express $1\frac{1}{2}$ hr as 90 min. Then we have the points $(50, 10)$ and $(50 + 90, 25)$, or $(50, 10)$ and $(140, 25)$.

$$\text{Speed} = \text{average rate of change} = \frac{25-10}{140-50} = \frac{15}{90} = \frac{1}{6}$$

The speed is $\frac{1}{6}$ km per minute.

22. $$\text{Typing rate} = \frac{\frac{3}{4}-\frac{1}{6}}{6} = \frac{\frac{7}{12}}{6} = \frac{7}{72} \text{ of the paper per hour}$$

23. $y = \frac{3}{5}x - 7$

The equation is in the form $y = mx+b$ where $m = \frac{3}{5}$ and $b = -7$. Thus, the slope is $\frac{3}{5}$, and the y-intercept is $(0, -7)$.

24. $f(x) = -2x + 3$

Slope: -2; y-intercept: $(0, 3)$

25. $f(x) = 5 - \frac{1}{2}x$, or $f(x) = -\frac{1}{2}x + 5$

The second equation is in the form $y = mx+b$ where $m = -\frac{1}{2}$ and $b = 5$. Thus, the slope is $-\frac{1}{2}$ and the y-intercept is $(0, 5)$.

26. $y = 2 + \frac{3}{7}x$

Slope: $\frac{3}{7}$; y-intercept: $(0, 2)$

27. Solve the equation for y.

$$3x + 2y = 10$$
$$2y = -3x + 10$$
$$y = -\frac{3}{2}x + 5$$

Slope: $-\frac{3}{2}$; y-intercept: $(0, 5)$

28. $2x - 3y = 12$

$$-3y = -2x + 12$$
$$y = \frac{2}{3}x - 4$$

Slope: $\frac{2}{3}$; y-intercept: $(0, -4)$

29. Solve the equation for $f(x)$.

$$4x - 3f(x) - 15 = 6$$
$$-3f(x) = -4x + 21$$
$$f(x) = \frac{4}{3}x - 7$$

Slope: $\frac{4}{3}$; y-intercept: $(0, -7)$

30. $9 = 3 + 5x - 2f(x)$

$$-5x + 6 = -2f(x)$$
$$\frac{5}{2}x - 3 = f(x)$$

Slope: $\frac{5}{2}$; y-intercept: $(0, -3)$

31. Substitute $\frac{2}{9}$ for m, 0 for x_1, and 4 for y_1 in the point-slope equation.

$$y - y_1 = m(x - x_1)$$
$$y - 4 = \frac{2}{9}(x - 0)$$
$$y - 4 = \frac{2}{9}x$$
$$y = \frac{2}{9}x + 4 \quad \text{Slope-intercept equation}$$

32. $y - 5 = -\frac{3}{8}(x - 0)$

$$y = -\frac{3}{8}x + 5$$

33. Substitute -4 for m, 0 for x_1, and -7 for y_1 in the point-slope equation.

$$y - y_1 = m(x - x_1)$$
$$y - (-7) = -4(x - 0)$$
$$y + 7 = -4x$$
$$y = -4x - 7 \quad \text{Slope-intercept equation}$$

34. $y - (-6) = \frac{2}{7}(x - 0)$

$$y = \frac{2}{7}x - 6$$

35. $y - y_1 = m(x - x_1)$

$$y - \frac{3}{4} = -4.2(x - 0) \quad \text{Substituting}$$
$$y - \frac{3}{4} = -4.2x$$
$$y = -4.2x + \frac{3}{4} \quad \text{Slope-intercept equation}$$

36. $y - \left(-\frac{3}{2}\right) = -4(x - 0)$

$$y = -4x - \frac{3}{2}$$

37. $y - y_1 = m(x - x_1)$

$$y - 7 = \frac{2}{9}(x - 3) \quad \text{Substituting}$$
$$y - 7 = \frac{2}{9}x - \frac{2}{3}$$
$$y = \frac{2}{9}x + \frac{19}{3} \quad \text{Slope-intercept equation}$$

38. $y - 6 = -\frac{3}{8}(x - 5)$

$$y = -\frac{3}{8}x + \frac{63}{8}$$

39. $y - y_1 = m(x - x_1)$

$$y - (-2) = 3(x - 1)$$
$$y + 2 = 3x - 3$$
$$y = 3x - 5 \quad \text{Slope-intercept equation}$$

40. $y - 1 = -2(x - (-5))$

$$y = -2x - 9$$

41. $y - y_1 = m(x - x_1)$

$$y - (-1) = -\frac{3}{5}(x - (-4))$$
$$y + 1 = -\frac{3}{5}(x + 4)$$
$$y + 1 = -\frac{3}{5}x - \frac{12}{5}$$
$$y = -\frac{3}{5}x - \frac{17}{5} \quad \text{Slope-intercept equation}$$

42. $y - (-5) = \frac{2}{3}(x - (-4))$

$$y = \frac{2}{3}x - \frac{7}{3}$$

43. $m = \frac{-4 - 5}{2 - (-1)} = \frac{-9}{3} = -3$

Using the point $(-1, 5)$, we get

$$y - 5 = -3(x - (-1)), \text{ or } y - 5 = -3(x + 1).$$

Using the point $(2, -4)$, we get

$$y - (-4) = -3(x - 2), \text{ or } y + 4 = -3(x - 2).$$

In either case, the slope-intercept equation is $y = -3x + 2$.

44. $m = \frac{-11 - (-1)}{7 - 2} = \frac{-10}{5} = -2$

Using $(2, -1)$: $y - (-1) = -2(x - 2)$, or

$$y + 1 = -2(x - 2)$$

Using $(7, -11)$: $y - (-11) = -2(x - 7)$, or

$$y + 11 = -2(x - 7)$$

In either case, we have $y = -2x + 3$.

45. $m = \frac{4-0}{-1-7} = \frac{4}{-8} = -\frac{1}{2}$

Using the point $(7, 0)$, we get

$y - 0 = -\frac{1}{2}(x-7)$.

Using the point $(-1, 4)$, we get

$y - 4 = -\frac{1}{2}(x-(-1))$, or

$y - 4 = -\frac{1}{2}(x+1)$.

In either case, the slope-intercept equation is $y = -\frac{1}{2}x + \frac{7}{2}$.

46. $m = \frac{-5-7}{-1-(-3)} = \frac{-12}{2} = -6$

Using $(-3, 7)$: $y - 7 = -6(x-(-3))$, or

$y - 7 = -6(x+3)$

Using $(-1, -5)$: $y - (-5) = -6(x-(-1))$, or

$y + 5 = -6(x+1)$

In either case, we have $y = -6x - 11$.

47. We solve each equation for y.

$x + 2y = 5$ $\quad$ $2x + 4y = 8$

$y = -\frac{1}{2}x + \frac{5}{2}$ $\quad$ $y = -\frac{1}{2}x + 2$

We see that $m_1 = -\frac{1}{2}$ and $m_2 = -\frac{1}{2}$. Since the slopes are the same and the y-intercepts, $\frac{5}{2}$ and 2, are different, the lines are parallel.

48. $2x - 5y = -3$ $\quad$ $2x + 5y = 4$

$y = \frac{2}{5}x + \frac{3}{5}$ $\quad$ $y = -\frac{2}{5}x + \frac{4}{5}$

$m_1 = \frac{2}{5}$, $m_2 = -\frac{2}{5}$; $m_1 \neq m_2$; $m_1m_2 = -\frac{4}{25} \neq -1$

The lines are neither parallel nor perpendicular.

49. We solve each equation for y.

$y = 4x - 5$ $\quad$ $4y = 8 - x$

$y = -\frac{1}{4}x + 2$

We see that $m_1 = 4$ and $m_2 = -\frac{1}{4}$. Since $m_1m_2 = 4\left(-\frac{1}{4}\right) = -1$, the lines are perpendicular.

50. $y = 7 - x$,

$y = x + 3$

$m_1 = -1$, $m_2 = 1$; $m_1m_2 = -1 \cdot 1 = -1$

The lines are perpendicular.

51. $y = \frac{2}{7}x + 1$; $m = \frac{2}{7}$

The line parallel to the given line will have slope $\frac{2}{7}$. We use the point-slope equation for a line with slope $\frac{2}{7}$ and containing the point $(3, 5)$:

$$y - y_1 = m(x - x_1)$$
$$y - 5 = \frac{2}{7}(x-3)$$
$$y - 5 = \frac{2}{7}x - \frac{6}{7}$$
$$y = \frac{2}{7}x + \frac{29}{7} \quad \text{Slope-intercept form}$$

The slope of the line perpendicular to the given line is the opposite of the reciprocal of $\frac{2}{7}$, or $-\frac{7}{2}$. We use the point-slope equation for a line with slope $-\frac{7}{2}$ and containing the point $(3, 5)$:

$$y - y_1 = m(x - x_1)$$
$$y - 5 = -\frac{7}{2}(x-3)$$
$$y - 5 = -\frac{7}{2}x + \frac{21}{2}$$
$$y = -\frac{7}{2}x + \frac{31}{2} \quad \text{Slope-intercept form}$$

52. $f(x) = 2x + 9$

$m = 2$, $-\frac{1}{m} = -\frac{1}{2}$

Parallel line: $y - 6 = 2(x-(-1))$

$y = 2x + 8$

Perpendicular line: $y - 6 = -\frac{1}{2}(x-(-1))$

$y = -\frac{1}{2}x + \frac{11}{2}$

53. $y = -0.3x + 4.3$; $m = -0.3$

The line parallel to the given line will have slope -0.3. We use the point-slope equation for a line with slope -0.3 and containing the point $(-7, 0)$:

$$y - y_1 = m(x - x_1)$$
$$y - 0 = -0.3(x-(-7))$$
$$y = -0.3x - 2.1 \quad \text{Slope-intercept form}$$

The slope of the line perpendicular to the given line is the opposite of the reciprocal of -0.3, or $\frac{1}{0.3} = \frac{10}{3}$. We use the point-slope equation for a line with slope $\frac{10}{3}$ and containing the point $(-7, 0)$:

$$y - y_1 = m(x - x_1)$$
$$y - 0 = \frac{10}{3}(x-(-7))$$
$$y = \frac{10}{3}x + \frac{70}{3} \quad \text{Slope-intercept form}$$

54. $2x + y = -4$

$y = -2x - 4$

$m = -2$, $-\frac{1}{m} = \frac{1}{2}$

Parallel line: $y - (-5) = -2(x-(-4))$

$y = -2x - 13$

Perpendicular line: $y-(-5)=\frac{1}{2}(x-(-4))$

$$y=\frac{1}{2}x-3$$

55. $3x+4y=5$

$$4y=-3x+5$$

$$y=-\frac{3}{4}x+\frac{5}{4};\ m=-\frac{3}{4}$$

The line parallel to the given line will have slope $-\frac{3}{4}$. We use the point-slope equation for a line with slope $-\frac{3}{4}$ and containing the point $(3,-2)$:

$$y-y_1=m(x-x_1)$$

$$y-(-2)=-\frac{3}{4}(x-3)$$

$$y+2=-\frac{3}{4}x+\frac{9}{4}$$

$$y=-\frac{3}{4}x+\frac{1}{4}\quad \text{Slope-intercept form}$$

The slope of the line perpendicular to the given line is the opposite of the reciprocal of $-\frac{3}{4}$, or $\frac{4}{3}$. We use the point-slope equation for a line with slope $\frac{4}{3}$ and containing the point $(3,-2)$:

$$y-y_1=m(x-x_1)$$

$$y-(-2)=\frac{4}{3}(x-3)$$

$$y+2=\frac{4}{3}x-4$$

$$y=\frac{4}{3}x-6\quad \text{Slope-intercept form}$$

56. $y=4.2(x-3)+1$

$y=4.2x-11.6$

$m=4.2;\ -\frac{1}{m}=-\frac{1}{4.2}=-\frac{5}{21}$

Parallel line: $y-(-2)=4.2(x-8)$

$$y=4.2x-35.6$$

Perpendicular line: $y-(-2)=-\frac{5}{21}(x-8)$

$$y=-\frac{5}{21}x-\frac{2}{21}$$

57. $x=-1$ is the equation of a vertical line. The line parallel to the given line is a vertical line containing the point $(3,-3)$, or $x=3$.

The line perpendicular to the given line is a horizontal line containing the point $(3,-3)$, or $y=-3$.

58. $y=-1$ is a horizontal line.

Parallel line: $y=-5$

Perpendicular line: $x=4$

59. a) $W(h)=3.5h-110$

b)

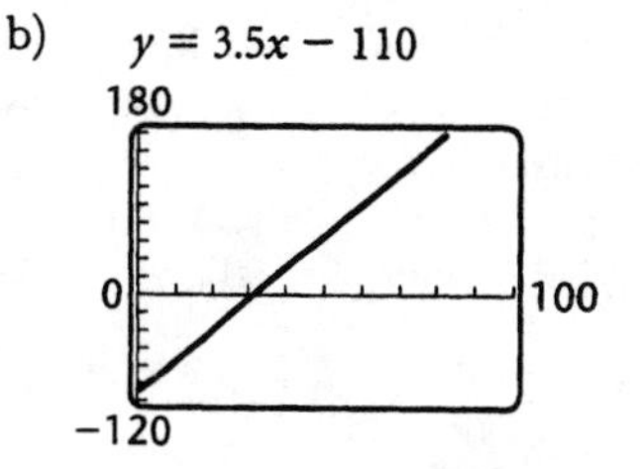

c) $W(62)=3.5(62)-110=217-110=107$ lb

d) Both the height and weight must be positive. Solving $h>0$ *and* $3.5h-110>0$, we find that the domain of the function is $\{h|h>31.43\}$, or $(31.43,\infty)$.

60. $P(d)=\frac{1}{33}d+1$

a)

$y=\frac{1}{33}x+1$

5

0

0

20

b) $P(0)=\frac{1}{33}\cdot 0+1=1$ atm

$P(5)=\frac{1}{33}\cdot 5+1=1\frac{5}{33}$ atm

$P(10)=\frac{1}{33}\cdot 10+1=1\frac{10}{33}$ atm

$P(33)=\frac{1}{33}\cdot 33+1=2$ atm

$P(200)=\frac{1}{33}\cdot 200+1=\frac{233}{33}$ atm, or $7\frac{2}{33}$ atm

c) The depth must be nonnegative, so the domain is $\{d|d\geq 0\}$, or $[0,\infty)$.

61. $D(F)=2F+115$

a)

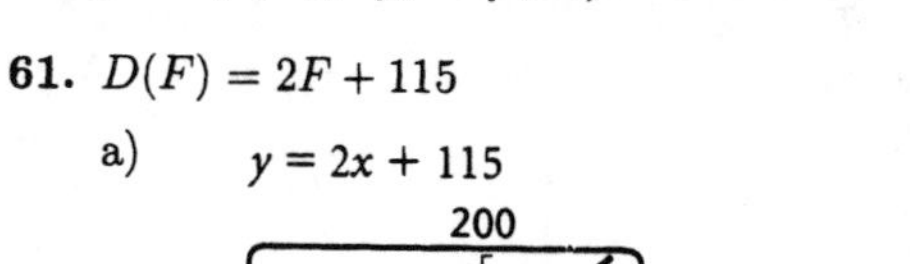

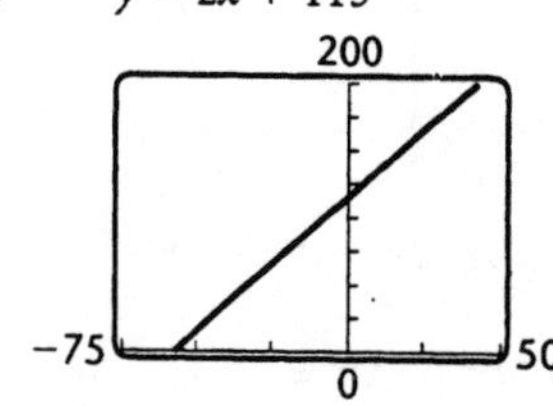

b) $D(0)=2\cdot 0+115=115$ ft

$D(-20)=2(-20)+115=-40+115=75$ ft

$D(10)=2\cdot 10+115=20+115=135$ ft

$D(32)=2\cdot 32+115=64+115=179$ ft

c) For $F<-57.5°$, $D(F)$ is negative; above 32°, ice doesn't form

62. a) $M(x)=2.89x+70.64$

$M(26)=2.89(26)+70.64=145.78$ cm

b) The length of the humerus must be positive, so the domain is $\{x|x > 0\}$, or $(0, \infty)$. Realistically, however, we might expect the length of the humerus to be between 20 cm and 60 cm, so the domain could be $\{x|20 \le x \le 60\}$, or $[20, 60]$.

63. a) $D(r) = \dfrac{11r+5}{10} = \dfrac{11}{10}r + \dfrac{5}{10}$

The slope is $\dfrac{11}{10}$.

For each mph faster the car travels, it takes $\dfrac{11}{10}$ ft longer to stop.

b)

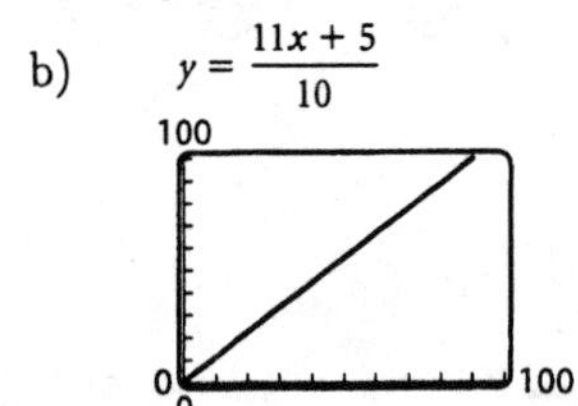

c) $D(5) = \dfrac{11 \cdot 5 + 5}{10} = \dfrac{60}{10} = 6$ ft

$D(10) = \dfrac{11 \cdot 10 + 5}{10} = \dfrac{115}{10} = 11.5$ ft

$D(20) = \dfrac{11 \cdot 20 + 5}{10} = \dfrac{225}{10} = 22.5$ ft

$D(50) = \dfrac{11 \cdot 50 + 5}{10} = \dfrac{555}{10} = 55.5$ ft

$D(65) = \dfrac{11 \cdot 65 + 5}{10} = \dfrac{720}{10} = 72$ ft

d) The speed cannot be negative. $D(0) = \dfrac{1}{2}$ which says that a stopped car travels $\dfrac{1}{2}$ ft before stopping. Thus, 0 is not in the domain. The speed can be positive, so the domain is $\{r|r > 0\}$, or $(0, \infty)$.

64. $V(t) = \$5200 - \$512.50t$

a)

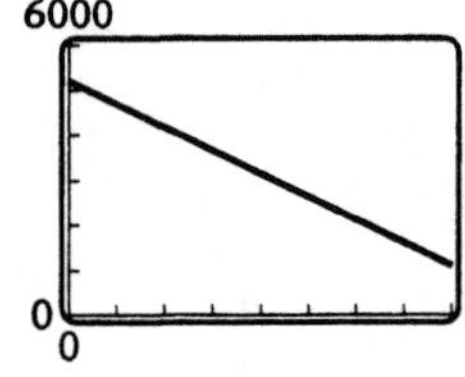

b) $V(0) = \$5200 - \$512.50(0) =$
$\$5200 - \$0 = \$5200$

$V(1) = \$5200 - \$512.50(1) =$
$\$5200 - \$512.50 = \$4687.50$

$V(2) = \$5200 - \$512.50(2) =$
$\$5200 - \$1025 = \$4175$

$V(3) = \$5200 - \$512.50(3) =$
$\$5200 - \$1537.50 = \$3662.50$

$V(8) = \$5200 - \$512.50(8) =$
$\$5200 - \$4100 = \$1100$

c) Since the time must be nonnegative and not more than 8 years, the domain is $[0, 8]$. The value starts at \$5200 and declines to \$1100, so the range is $[1100, 5200]$.

65. The rise represents "Height of corn in feet," and the run represents "Number of weeks after planting."

Slope $= \dfrac{\text{rise}}{\text{run}} = \dfrac{\text{Height of corn in feet}}{\text{Number of weeks after planting}}$

66. $\dfrac{\text{Distance walked in miles}}{\text{Number of hours spent walking}}$

67. $C(t) = 60 + 40t$

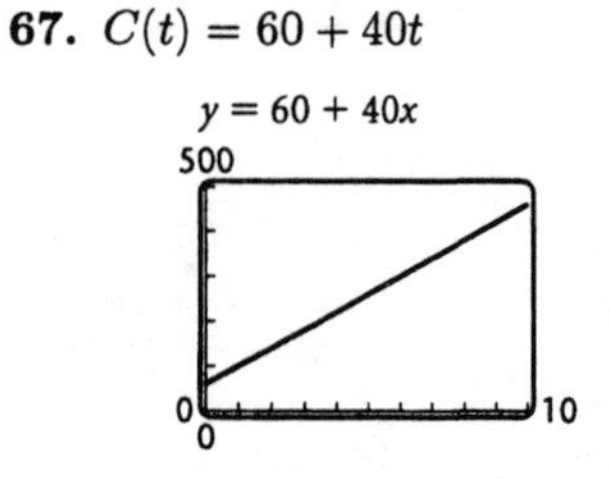

$C(6) = 60 + 40 \cdot 6 = \$300$

68. $C(t) = 35 + 30t$

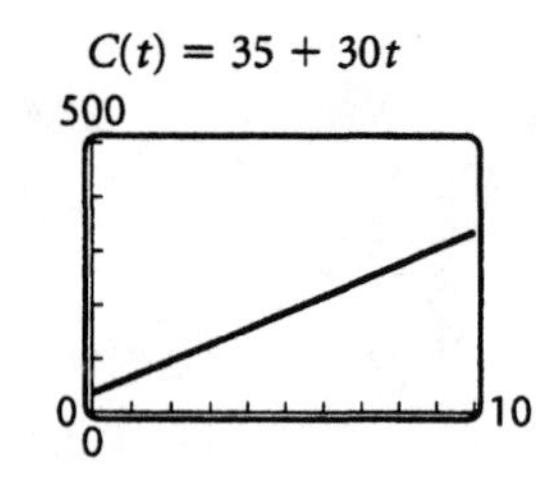

$C(8) = 35 + 30 \cdot 8 = \$275$

69. Let $x =$ the number of shirts produced.

$C(x) = 800 + 3x$

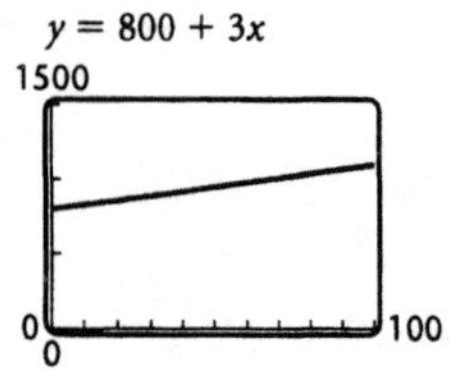

$C(75) = 800 + 3 \cdot 75 = \$1025$

70. Let $x =$ the number of rackets restrung.

$C(x) = 500 + 2x$

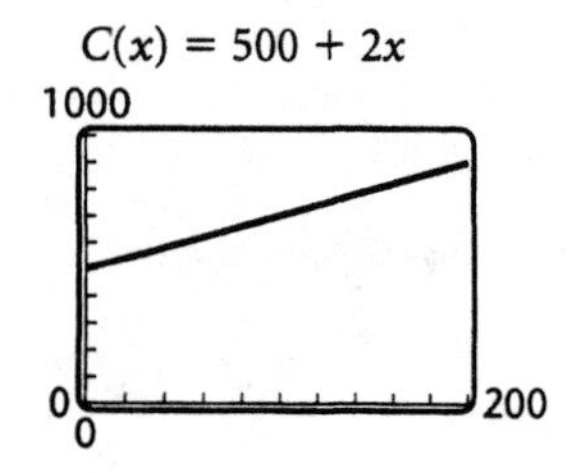

$C(150) = 500 + 2 \cdot 150 = \$800$

71. A vertical line ($x = a$) crosses the graph more than once.

72. The sign of the slope indicates the slant of a line. A line that slants up from left to right has positive slope because corresponding changes in x and y have the same sign. A line that slants down from left to right has negative slope, because corresponding changes in x and y have opposite signs. A horizontal line has zero slope, because there is no change in y for a given change in x. A vertical line has undefined slope, because there is no change in x for a given change in y and division by 0 is undefined. The larger the absolute value of slope, the steeper the line. This is because a larger absolute value corresponds to a greater change in y, compared to the change in x, than a smaller absolute value.

73. $f(x) = x^2 - 3x$

$f(-5) = (-5)^2 - 3(-5) = 25 + 15 = 40$

74. $f(5) = 5^2 - 3 \cdot 5 = 10$

75. $f(x) = x^2 - 3x$

$f(-a) = (-a)^2 - 3(-a) = a^2 + 3a$

76. $f(a+h) = (a+h)^2 - 3(a+h) = a^2 + 2ah + h^2 - 3a - 3h$

77. The slope of the line containing $(-3, k)$ and $(4, 8)$ is

$$\frac{8-k}{4-(-3)} = \frac{8-k}{7}.$$

The slope of the line containing $(5, 3)$ and $(1, -6)$ is

$$\frac{-6-3}{1-5} = \frac{-9}{-4} = \frac{9}{4}.$$

The slopes must be equal in order for the lines to be parallel:

$$\frac{8-k}{7} = \frac{9}{4}$$

$$32 - 4k = 63 \quad \text{Multiplying by 28}$$

$$-4k = 31$$

$$k = -\frac{31}{4}, \text{ or } -7.75$$

78. The slope of the line containing $(-1, 3)$ and $(2, 9)$ is

$$\frac{9-3}{2-(-1)} = \frac{6}{3} = 2.$$

Then the slope of the desired line is $-\frac{1}{2}$. We find the equation of that line:

$$y - 5 = -\frac{1}{2}(x - 4)$$

$$y - 5 = -\frac{1}{2}x + 2$$

$$y = -\frac{1}{2}x + 7$$

79. To express F as a function of C, we consider C to be the independent variable. First find the slope of the function containing the points $(0, 32)$ and $(100, 212)$.

$$\frac{32-212}{0-100} = \frac{-180}{-100} = \frac{9}{5}$$

Now find the function with slope $\frac{9}{5}$ and containing the point $(0, 32)$.

$$F - 32 = \frac{9}{5}(C - 0)$$

$$F - 32 = \frac{9}{5}C$$

$$F(C) = \frac{9}{5}C + 32$$

To express C as a function of F, we consider F to be the independent variable. First find the slope of the function containing the points $(32, 0)$ and $(212, 100)$.

$$\frac{100-0}{212-32} = \frac{100}{180} = \frac{5}{9}$$

Now find the function with slope $\frac{5}{9}$ and containing the point $(32, 0)$.

$$C - 0 = \frac{5}{9}(F - 32)$$

$$C(F) = \frac{5}{9}(F - 32)$$

80. False. For example, let $f(x) = x+1$. Then $f(c+d) = c+d+1$, but $f(c)+f(d) = c+1+d+1 = c+d+2$.

81. False. For example, let $f(x) = x + 1$. Then $f(cd) = cd+1$, but $f(c)f(d) = (c+1)(d+1) = cd+c+d+1 \neq cd+1$ for $c \neq -d$.

82. False. For example, let $f(x) = x + 1$. Then $f(kx) = kx + 1$, but $kf(x) = k(x+1) = kx + k \neq kx + 1$ for $k \neq 1$.

83. False. For example, let $f(x) = x+1$. Then $f(c-d) = c - d + 1$, but $f(c) - f(d) = c + 1 - (d + 1) = c - d$.

84. $3mx + b = 3(mx + b)$

$$3mx + b = 3mx + 3b$$

$$b = 3b$$

$$0 = 2b$$

$$0 = b$$

Thus, $f(x) = mx + 0$, or $f(x) = mx$.

85.

$$f(x) = mx + b$$

$$f(x+2) = f(x) + 2$$

$$m(x+2) + b = mx + b + 2$$

$$mx + 2m + b = mx + b + 2$$

$$2m = 2$$

$$m = 1$$

Thus, $f(x) = 1 \cdot x + b$, or $f(x) = x + b$.

Exercise Set 1.3

1. Yes. The rate of change of the occupancy rates generally seems to be constant, so the data might be modeled by a linear function.

2. Yes. Both the U.S. demand and the world demand have constant rates of change, so they might be modeled by linear functions.

3. No. The population figures seem to be rising more rapidly after 1993 than before (that is, the slope is not constant), so this data cannot be modeled by a linear function.

4. No. Each of the lines falls faster from 0 to 3 months than from 3 to 12 months (that is, the slopes are not constant), so they cannot be modeled by a linear function.

5. No. The data points fall faster from 0 to 2 than after 2 (that is, the rate of change is not constant), so they cannot be modeled by a linear function.

6. Yes. The rate of change seems to be constant, so the scatterplot might be modeled by a linear function.

7. Yes. The rate of change seems to be constant, so the scatterplot might be modeled by a linear function.

8. No. The data points rise, fall, and then rise again in a way that cannot be modeled by a linear function.

9. a) Using the linear regression feature on a grapher we get $y = 1304.844444x + 21,950.01111$, where $x =$ the number of years after 1980.

   In 2005, $x = 2005 - 1980 = 25$. Find y when $x = 25$.

   $y = 1304.844444(25) + 21,950.01111 \approx 54,571$

   There will be approximately 54,571 shopping centers in 2005.

   b) $r \approx 9890$; since r is close to 1, the regression line is a good fit.

10. a) $y = 230x + 5800$

    In 1998, $x = 10$: $y = 230(10) + 5800 = 8100$ thousand, or 8,100,000

    In 2000, $x = 12$: $y = 230(12) + 5800 = 8560$ thousand, or 8,560,000

    In 2010, $x = 22$: $y = 230(22) + 5800 = 10,860$ thousand, or 10,860,000

    b) $r = 0.9972$; since r is close to one, this is a good fit.

11. a) Using the linear regression feature on a grapher, we get $y = 645.7x + 9799.2$.

    In 2004-2005, $x = 7$:

    $y = 645.7(7) + 9799.2 = \$14,319.10$

    In 2006-2007, $x = 9$:

    $y = 645.7(9) + 9799.2 = \$15,610.50$

    In 2010-2011, $x = 13$:

    $y = 645.7(13) + 9799.2 = \$18,193.30$

    b) $r = 0.9994$; since this is close to 1, the regression line is a good fit.

    c) The president's assertion is about \$1000, or 6% higher than the estimate from the linear model.

12. a) $y = 1.2x + 21.4$

    In 2000, $x = 8$: $y = 1.2(8) + 21.4 = \$31.0$ billion

    In 2005, $x = 13$; $y = 1.2(13) + 21.4 =$ \$37.0 billion

    In 2010, $x = 18$: $y = 1.2(18) + 21.4 =$ \$43.0 billion

    b) $r = 0.9864$; since r is close to 1, this is a good fit.

    c) The owner's assertion is about \$14 billion, or 45%, higher than the estimate from the model.

13. a) Using the linear regression feature on a grapher, we get $M = 0.2H + 156$.

    b) For $H = 40$: $M = 0.2(40) + 156 = 164$ beats per minute

    For $H = 65$: $M = 0.2(65) + 156 = 169$ beats per minute

    For $H = 76$: $M = 0.2(76) + 156 \approx 171$ beats per minute

    For $H = 84$: $M = 0.2(84) + 156 \approx 173$ beats per minute

    c) $r = 1$; all the data points are on the regression line so it should be a good predictor.

14. a) $y = 0.072050673x + 81.99920823$

    b) For $x = 24$:
    $y = 0.072050673(24) + 81.99920823 \approx 84\%$

    For $x = 6$:
    $y = 0.072050673(6) + 81.99920823 \approx 82\%$

    For $x = 18$:
    $y = 0.072050673(18) + 81.99920823 \approx 83\%$

    c) $r = 0.0636$; since there is a very low correlation, the regression line is not a good predictor.

15. A change in one quantity does not necessarily mean that a corresponding change in the other quantity occurs.

16. Answers will vary.

17. $$m = \frac{y_2 - y_1}{x_2 - x_1} = \frac{-1-(-8)}{-5-2} = \frac{-1+8}{-7} = \frac{7}{-7} = -1$$

18. $m = \frac{-7-7}{5-5} = \frac{-14}{0}$

The slope is undefined.

19. Substitute $\frac{3}{4}$ for m and -5 for b in $y = mx + b$. We have $y = \frac{3}{4}x - 5$.

20. $m = \frac{-10-(-10)}{-2-2} = \frac{0}{-4} = 0$

$$y - (-10) = 0(x-2)$$
$$y + 10 = 0$$
$$y = -10$$

21. a) Using the home run data from the years 1939 through 1942, 1946 through 1951, and 1954 through 1960, we get $H = -0.4020073552x + 813.785167$. (Rounding the constants reduces the accuracy of any values calculated using this function.)

In 1943,
$y = -0.4020073552(1943) + 813.785167 \approx 33$.

In 1944,
$y = -0.4020073552(1944) + 813.785167 \approx 32$.

In 1945,
$y = -0.4020073552(1945) + 813.785167 \approx 32$.

In 1952,
$y = -0.4020073552(1952) + 813.785167 \approx 29$.

In 1953,
$y = -0.4020073552(1953) + 813.785167 \approx 29$.

The correlation coefficient is -0.3668. Since this is low, the function would not be a good predictor.

b) Adding the number of home runs predicted for the war years and the data in the table for the other years, we get 662 home runs. Thus, Ted Williams would not have broken Hank Aaron's home run record according to this function.

c) Using the RBI data for the years listed in part (a), we get $R = -3.353892124x + 6646.484217$. (Rounding the constants greatly reduces the accuracy of values found using this function.)

In 1943,
$R = -3.353892124(1943) + 6646.484217 \approx 130$.

In 1944,
$R = -3.353892124(1944) + 6646.484217 \approx 127$.

In 1945,
$R = -3.353892124(1945) + 6646.484217 \approx 123$.

In 1952,
$R = -3.353892124(1952) + 6646.484217 \approx 100$.

In 1953,
$R = -3.353892124(1953) + 6646.484217 \approx 96$.

The correlation coefficient is -0.7817. Since this is fairly high, the function would be an above-average predictor but not good enough to provide a high degree of confidence in its ability to predict.

d) Adding the number of RBIs predicted for the war years and the data in the table for the other years, we get 2378 RBIs. Thus, Ted Williams would have broken Hank Aaron's RBI record according to this function.

Exercise Set 1.4

1. a) The graph rises from left to right on $(-5, 1)$.

b) The graph drops from left to right on $(3, 5)$.

c) The graph neither rises nor drops on $(1, 3)$.

2. a) The graph rises from left to right on $(1, 3)$.

b) The graph drops from left to right on $(-5, 1)$.

c) The graph neither rises nor drops on $(3, 5)$.

3. a) The graph rises from left to right on $(-3, -1)$ and on (3,5).

b) The graph drops from left to right on $(1, 3)$.

c) The graph neither rises nor drops on $(-5, 3)$.

4. a) The graph rises from left to right on $(1, 2)$.

b) The graph drops from left to right on $(-5, -2)$, on $(-2, 1)$, and on $(3, 5)$.

c) The graph neither rises nor drops on $(2, 3)$.

5.

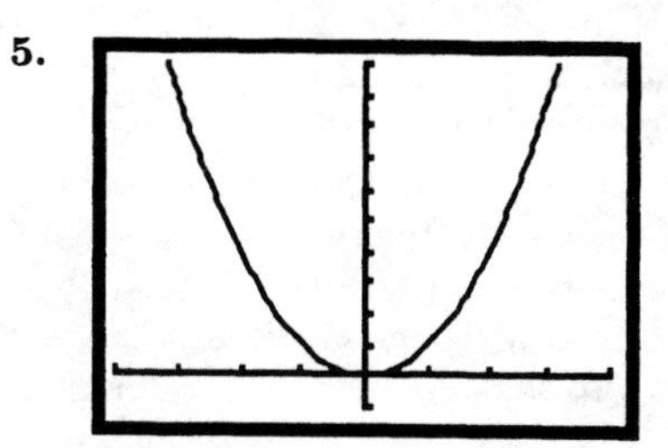

The graph is increasing on $(0, \infty)$ and decreasing on $(-\infty, 0)$. Using the MINIMUM feature we find that the relative minimum is 0 at $x = 0$. There are no relative maxima.

6.

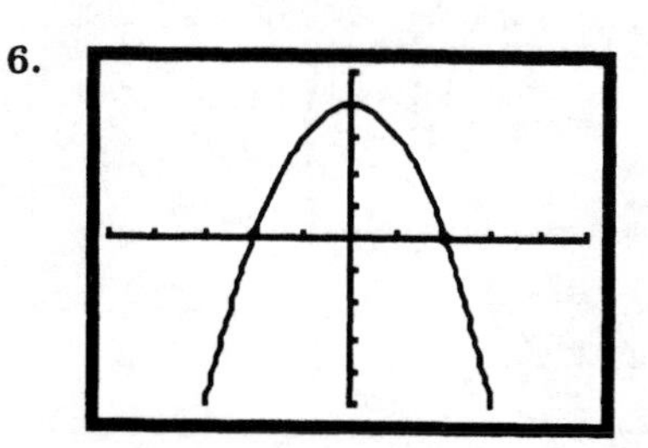

Increasing: $(-\infty, 0)$

Decreasing: $(0, \infty)$

Relative maximum: 4 at $x = 0$

Relative minima: none

7.

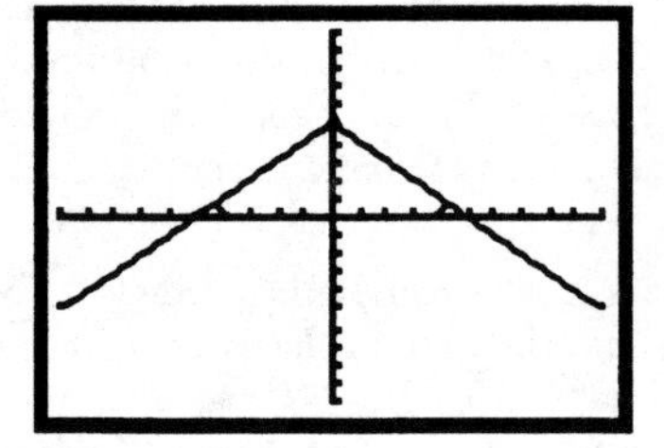

The graph is increasing on $(-\infty, 0)$ and decreasing on $(0, \infty)$. Using the MAXIMUM feature we find that the relative maximum is 5 at $x = 0$. There are no relative minima.

8.

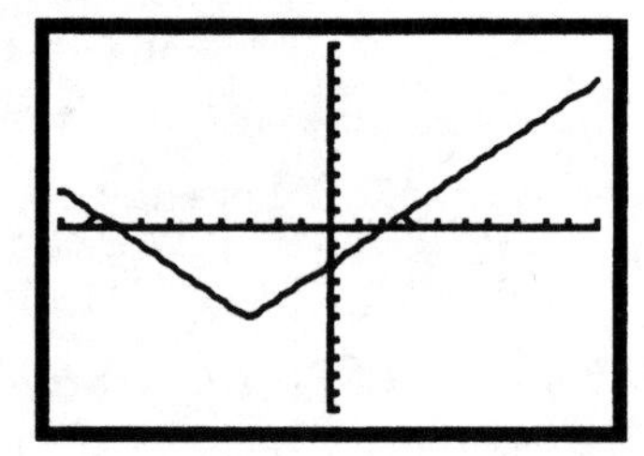

Increasing: $(-3, \infty)$

Decreasing: $(-\infty, -3)$

Relative maxima: none

Relative minimum: -5 at $x = -3$

9.

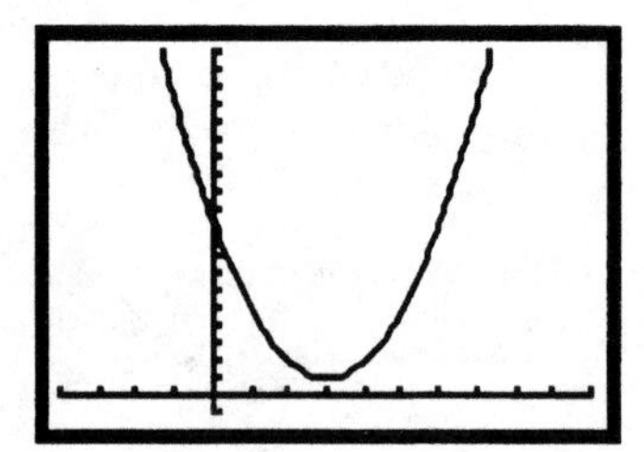

Beginning at the left side of the window, the graph first drops as we move to the right. We see that the function is decreasing on $(-\infty, 3)$. We then observe that the function is increasing on $(3, \infty)$. The MINIMUM feature also shows that the relative minimum is 1 at $x = 3$. There are no relative maxima.

10.

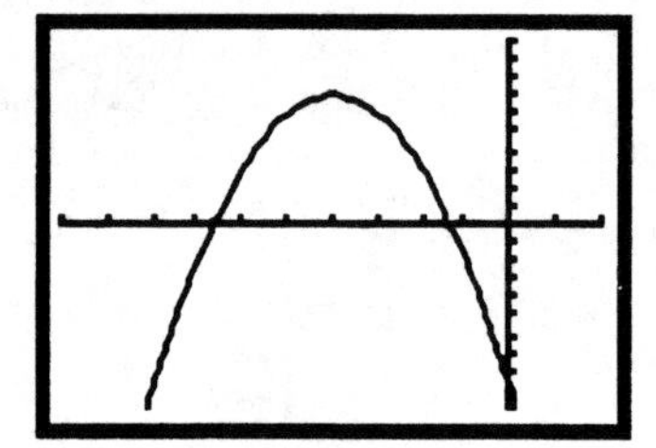

Increasing: $(-\infty, -4)$

Decreasing: $(-4, \infty)$

Relative maximum: 7 at $x = -4$

Relative minima: none

11.

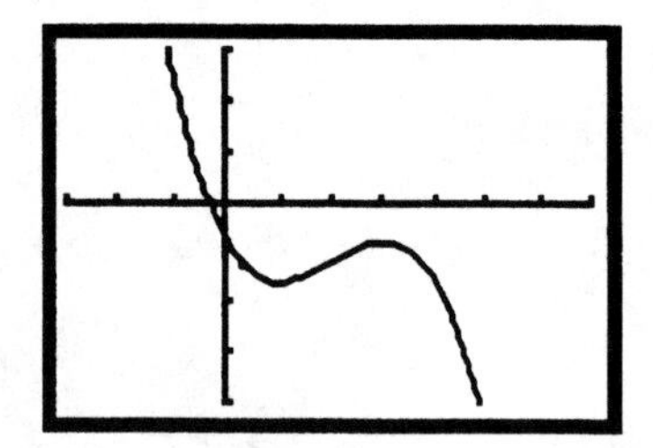

Beginning at the left side of the window, the graph first drops as we move to the right. We see that the function is decreasing on $(-\infty, 1)$. We then find that the function is increasing on $(1, 3)$ and decreasing again on $(3, \infty)$. The MAXIMUM and MINIMUM features also show that the relative maximum is -4 at $x = 3$ and the relative minimum is -8 at $x = 1$.

12.

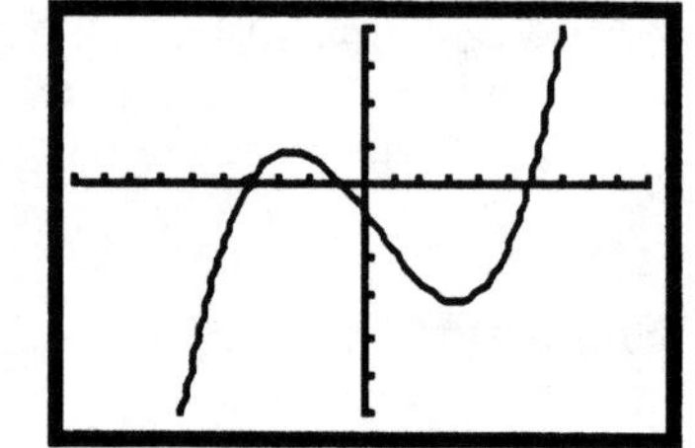

Increasing: $(-\infty, -2.573)$, $(3.239, \infty)$

Decreasing: $(-2.573, 3.239)$

Relative maximum: 4.134 at $x = -2.573$

Relative minimum: -15.497 at $x = 3.239$

13.

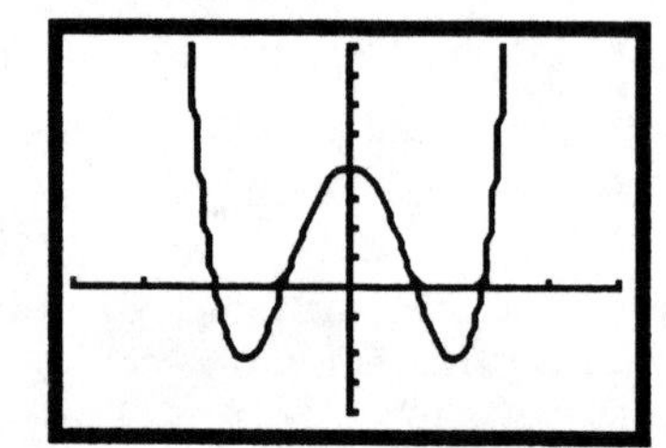

We find that the function is increasing on $(-1.552.0)$ and on $(1.552, \infty)$ and decreasing on $(-\infty, -1.552)$ and on $(0, 1.552)$. The relative maximum is 4.07 at $x = 0$ and the relative minima are -2.314 at $x = -1.552$ and -2.314 at $x = 1.552$.

14.

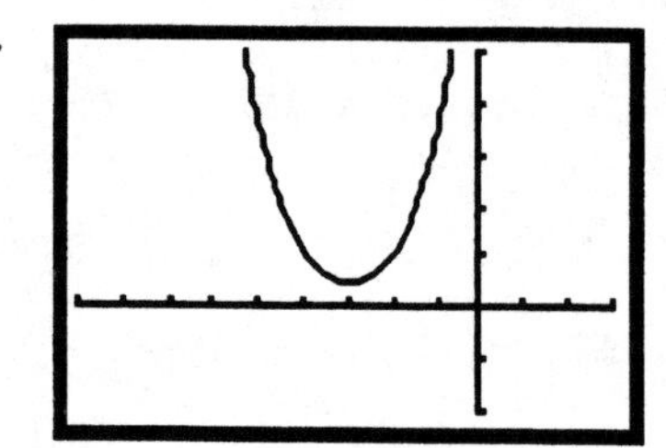

Increasing: $(-3, \infty)$

Decreasing: $(-\infty, -3)$

Relative maxima: none

Relative minimum: 9.78 at $x = -3$

15. a) $y = -0.1x^2 + 1.2x + 98.6$

110

0 12

90

b) Using the MAXIMUM feature we find that the relative maximum is 102.2 at $t = 6$.

c) Using the result in part (b), we know that the patient's temperature was the highest at $t = 6$, or 6 days after the onset of the illness and that the highest temperature was 102.2°F.

16. a)

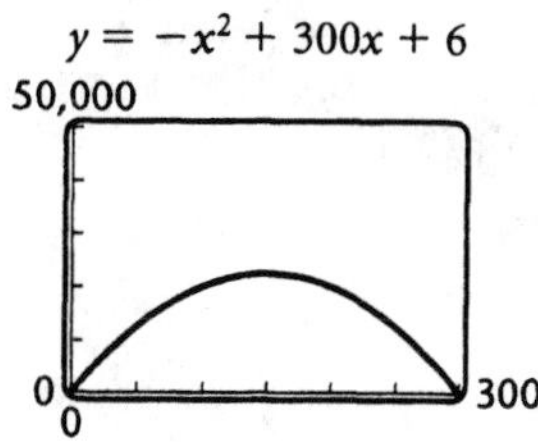

b) $22,506$ at $a = 150$

c) The greatest number of games will be sold when \$150 thousand is spent on advertising. For that amount, 22,506 games will be sold.

17. Graph $y = \dfrac{8x}{x^2+1}$.

Increasing: $(-1, 1)$

Decreasing: $(-\infty, -1)$, $(1, \infty)$

18. Graph $y = \dfrac{-4}{x^2+1}$.

Increasing: $(0, \infty)$

Decreasing: $(-\infty, 0)$

19. Graph $y = x\sqrt{4-x^2}$, for $-2 \le x \le 2$.

Increasing: $(-1.414, 1.414)$

Decreasing: $(-2, -1.414)$, $(1.414, 2)$

20. Graph $y = -0.8x\sqrt{9-x^2}$, for $-3 \le x \le 3$.

Increasing: $(-3, -2.121)$, $(2.121, 3)$

Decreasing: $(-2.121, 2.121)$

21. After t minutes, the balloon has risen $120t$ ft. We use the Pythagorean theorem.

$$[d(t)]^2 = (120t)^2 + (400)^2$$
$$d(t) = \sqrt{(120t)^2 + (400)^2}$$

We only considered the positive square root since distance must be nonnegative.

22. Use the Pythagorean theorem.

$$[h(d)]^2 + (3700)^2 = d^2$$
$$[h(d)]^2 = d^2 - (3700)^2$$
$$h(d) = \sqrt{d^2 - (3700)^2} \quad \text{Taking the positive square root}$$

23. If x = the length of the rectangle, in meters, then the width is $\dfrac{48-2x}{2}$, or $24 - x$. We use the formula Area = length $\times$ width:

$$A(x) = x(24 - x)$$
$$A(x) = 24x - x^2$$

24. Let h = the height of the flag, in inches. Then the length of the base $= 2h - 7$.

$$A(h) = \frac{1}{2}(2h-7)(h)$$
$$A(h) = h^2 - \frac{7}{2}h$$

25. Let w = the width of the rectangle. Then the length $= \dfrac{40-2w}{2}$, or $20 - w$. Divide the rectangle into quadrants as shown below.

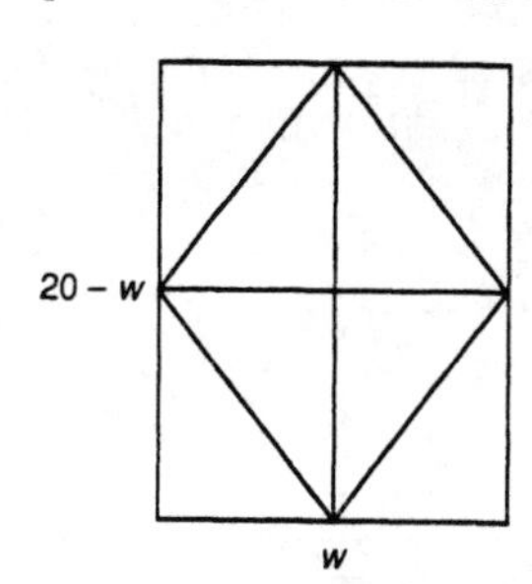

In each quadrant there are two congruent triangles. One triangle is part of the rhombus and both are part of the rectangle. Thus, in each quadrant the area of the rhombus is one-half the area of the rectangle. Then, in total, the area of the rhombus is one-half the area of the rectangle.

$$A(w) = \frac{1}{2}(20-w)(w)$$
$$A(w) = 10w - \frac{w^2}{2}$$

26. Let w = the width, in feet. Then the length $= \dfrac{16-2w}{2}$, or $8 - w$.

$$A(w) = (8-w)w$$
$$A(w) = 8w - w^2$$

27. We will use similar triangles, expressing all distances in feet. $\left(6 \text{ in.} = \dfrac{1}{2} \text{ ft}, \ s \text{ in.} = \dfrac{s}{12} \text{ ft, and } d \text{ yd} = 3d \text{ ft}\right)$ We have

$$\frac{3d}{7} = \frac{\frac{1}{2}}{\frac{s}{12}}$$

$$\frac{s}{12} \cdot 3d = 7 \cdot \frac{1}{2}$$

$$\frac{sd}{4} = \frac{7}{2}$$

$$d = \frac{4}{s} \cdot \frac{7}{2}, \text{ so}$$

$$d(s) = \frac{14}{s}.$$

28. The volume of the tank is the sum of the volume of a sphere with radius r and a right circular cylinder with radius r and height 6 ft.

$$V(r) = \frac{4}{3}\pi r^3 + 6\pi r^2$$

29. a) If the length $= x$ feet, then the width $= 30 - x$ feet.

$$A(x) = x(30 - x)$$
$$A(x) = 30x - x^2$$

b) The length of the rectangle must be positive and less than 30 ft, so the domain of the function is $\{x|0 < x < 30\}$, or $(0, 30)$.

c) $y = 30x - x^2$

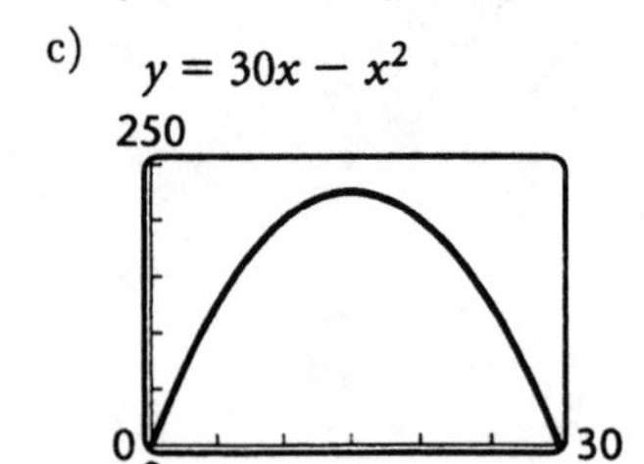

d) Using the MAXIMUM feature, we find that the maximum area occurs when $x = 15$. Then the dimensions that yield the maximum area are length $= 15$ ft and width $= 30 - 15$, or 15 ft.

30. a) $A(x) = x(360 - 3x)$, or $360x - 3x^2$

b) The domain is $\left\{x \middle| 0 < x < \frac{360}{3}\right\}$, or $\{x|0 < x < 120\}$, or $(0, 120)$.

c) $y = 360x - 3x^2$

15,000

0 0 120

d) The maximum value occurs when $x = 60$ so the width of each corral should be 60 yd and the total length of the two corrals should be $360 - 3 \cdot 60$, or 180 yd.

31. a) When a square with sides of length x are cut from each corner, the length of each of the remaining sides of the piece of cardboard is $12 - 2x$. Then the dimensions of the box are x by $12 - 2x$ by $12 - 2x$. We use the formula Volume = length × width × height to find the volume of the box:

$$V(x) = (12 - 2x)(12 - 2x)(x)$$
$$V(x) = (144 - 48x + 4x^2)(x)$$
$$V(x) = 144x - 48x^2 + 4x^3$$

This can also be expressed as $V(x) = 4x(x - 6)^2$.

b) The length of the sides of the square corners that are cut out must be positive and less than half the length of a side of the piece of cardboard. Thus, the domain of the function is $\{x|0 < x < 6\}$, or $(0, 6)$.

c) $y = 4x^3 - 48x^2 + 144x$

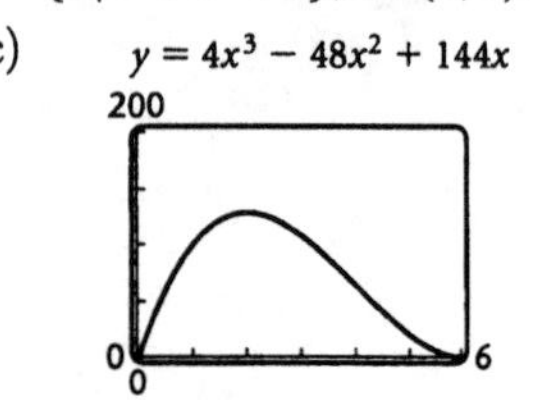

d) Using the MAXIMUM feature, we find that the maximum value of the volume occurs when $x = 2$. When $x = 2$, $12 - 2x = 12 - 2 \cdot 2 = 8$, so the dimensions that yield the maximum volume are 8 cm by 8 cm by 2 cm.

32. a) $V(x) = 8x(14 - 2x)$, or $112x - 16x^2$

b) The domain is $\left\{x \middle| 0 < x < \frac{14}{2}\right\}$, or $\{x|0 < x < 7\}$, or $(0, 7)$.

c) $y = 112x - 16x^2$

250

0 0 7

d) The maximum occurs when $x = 3.5$, so the file should be 3.5 in. tall.

33. a) The length of a diameter of the circle (and a diagonal of the rectangle) is $2 \cdot 8$, or 16 ft. Let $l =$ the length of the rectangle. Use the Pythagorean theorem to write l as a function of x.

$$x^2 + l^2 = 16^2$$
$$x^2 + l^2 = 256$$
$$l^2 = 256 - x^2$$
$$l = \sqrt{256 - x^2}$$

Since the length must be positive, we considered only the positive square root.

Use the formula Area = length × width to find the area of the rectangle:

$$A(x) = x\sqrt{256 - x^2}$$

b) The width of the rectangle must be positive and less than the diameter of the circle. Thus, the domain of the function is $\{x|0 < x < 16\}$, or $(0, 16)$.

c) 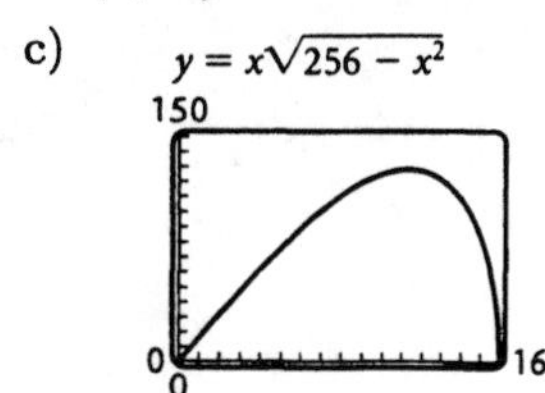

d) Using the MAXIMUM feature, we find that the maximum area occurs when x is about 11.314. When $x \approx 11.314$, $\sqrt{256 - x^2} \approx \sqrt{256 - (11.314)^2} \approx 11.313$. Thus, the dimensions that maximize the area are about 11.314 ft by 11.313 ft. (Answers may vary slightly due to rounding differences.)

34. a) Let $h(x)$ = the height of the box.

$$320 = x \cdot x \cdot h(x)$$

$$\frac{320}{x^2} = h(x)$$

Area of the bottom: x^2

Area of each side: $x\left(\frac{320}{x^2}\right)$, or $\frac{320}{x}$

Area of the top: x^2

$$C(x) = 1.5x^2 + +4(2.5)\left(\frac{320}{x}\right) + 1 \cdot x^2$$

$$C(x) = 2.5x^2 + \frac{3200}{x}$$

b) The length of the base must be positive, so the domain of the function is $\{x|x > 0\}$, or $(0, \infty)$.

c) 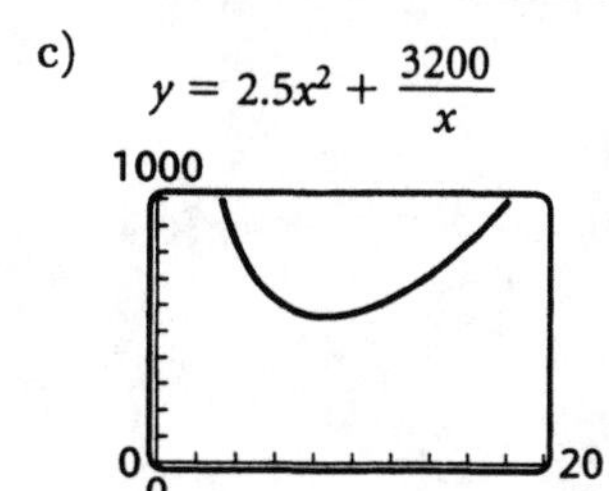

d) Using the MIMIMUM feature, we find that the minimum cost occurs when $x \approx 8.618$. Thus, the dimensions that minimize the cost are about 8.618 ft by 8.618 ft by $\frac{320}{(8.618)^2}$, or about 4.309 ft.

35. $f(x) = \begin{cases} \frac{1}{2}x, & \text{for } x < 0, \\ x + 3, & \text{for } x \geq 0 \end{cases}$

We create the graph in two parts. Graph $f(x) = \frac{1}{2}x$ for inputs x less than 0. The graph $f(x) = x + 3$ for inputs x greater than or equal to 0.

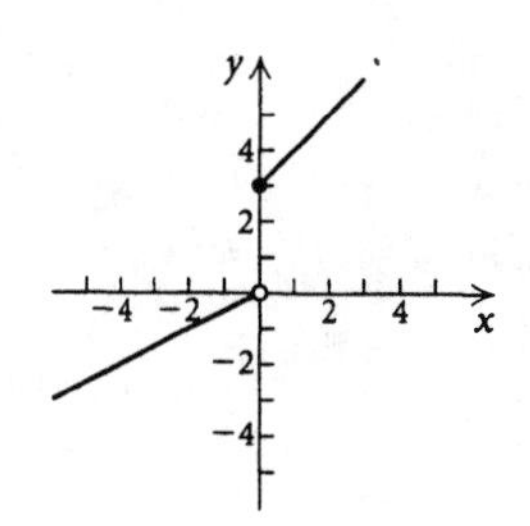

36. $f(x) = \begin{cases} -\frac{1}{3}x + 2, & \text{for } x \leq 0, \\ x - 5, & \text{for } x > 0 \end{cases}$

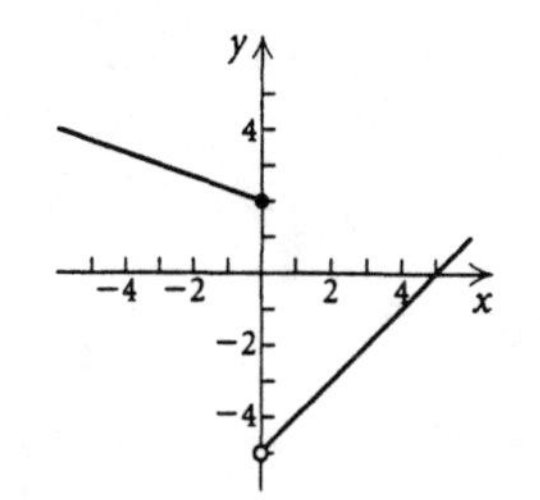

37. $f(x) = \begin{cases} -\frac{3}{4}x + 2, & \text{for } x < 4, \\ -1, & \text{for } x \geq 4 \end{cases}$

We create the graph in two parts. Graph $f(x) = -\frac{3}{4}x + 2$ for inputs x less than 4. The graph $f(x) = -1$ for inputs x greater than or equal to 4.

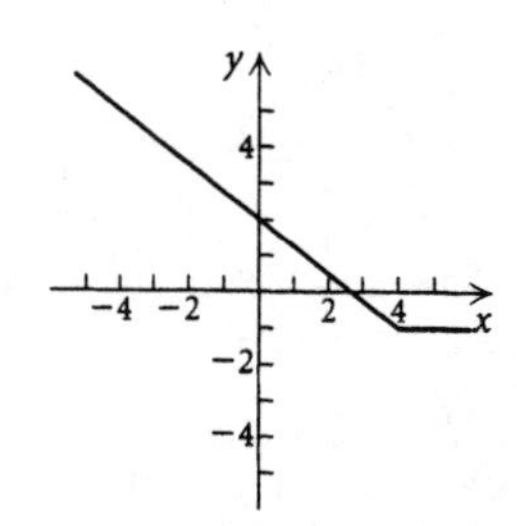

38. $f(x) = \begin{cases} 4, & \text{for } x \leq -2, \\ x + 1, & \text{for } -2 < x < 3 \\ -x, & \text{for } x \geq 3 \end{cases}$

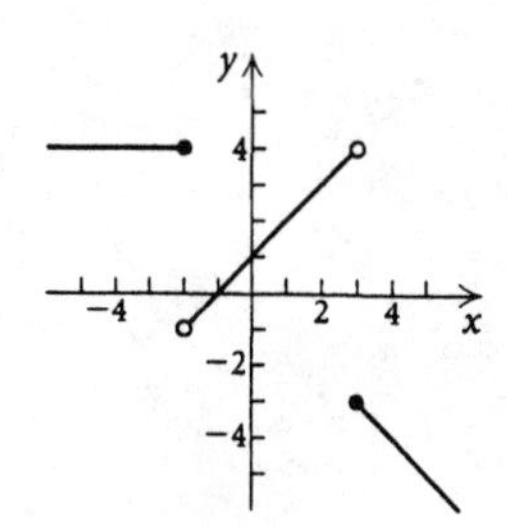

39. $f(x) = \begin{cases} x + 1, & \text{for } x \leq -3, \\ -1, & \text{for } -3 < x < 4 \\ \frac{1}{2}x, & \text{for } x \geq 4 \end{cases}$

We create the graph in three parts. Graph $f(x) = x + 1$ for inputs x less than or equal to -3. Graph $f(x) = -1$ for inputs greater than -3 and less than 4. Then graph $f(x) = \frac{1}{2}x$ for inputs greater than or equal to 4.

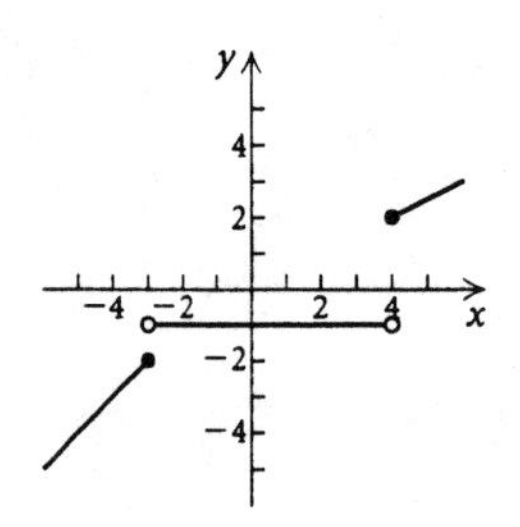

40. $f(x) = \begin{cases} \dfrac{x^2 - 9}{x + 3}, & \text{for } x \neq -3, \\ 5, & \text{for } x = -3 \end{cases}$

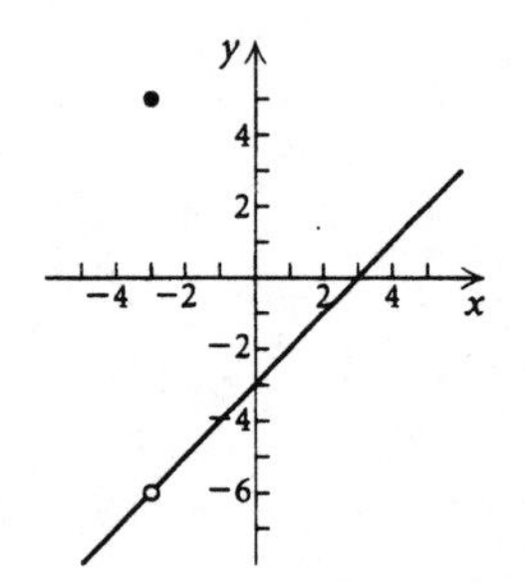

41. $f(x) = \begin{cases} 2, & \text{for } x = 5, \\ \dfrac{x^2 - 25}{x - 5}, & \text{for } x \neq 5 \end{cases}$

When $x \neq 5$, the denominator of $(x^2 - 25)/(x - 5)$ is nonzero so we can simplify:

$$\frac{x^2 - 25}{x - 5} = \frac{(x + 5)(x - 5)}{x - 5} = x + 5.$$

Thus, $f(x) = x + 5$, for $x \neq 5$.

The graph of this part of the function consists of a line with a "hole" at the point $(5, 10)$, indicated by an open dot. At $x = 5$, we have $f(5) = 2$, so the point $(5, 2)$ is plotted below the open dot.

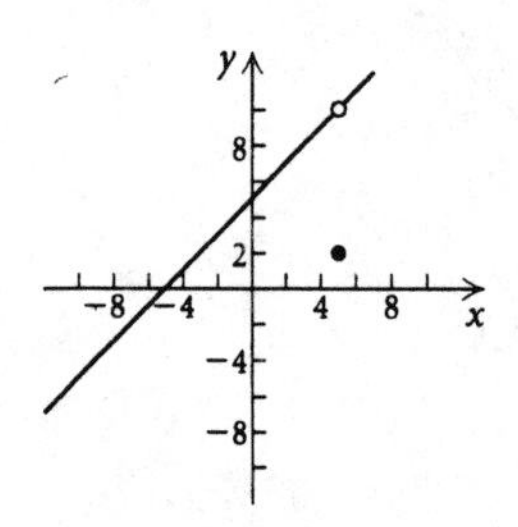

42. $f(x) = \begin{cases} \dfrac{x^2 + 3x + 2}{x + 1}, & \text{for } x \neq -1, \\ 7, & \text{for } x = -1 \end{cases}$

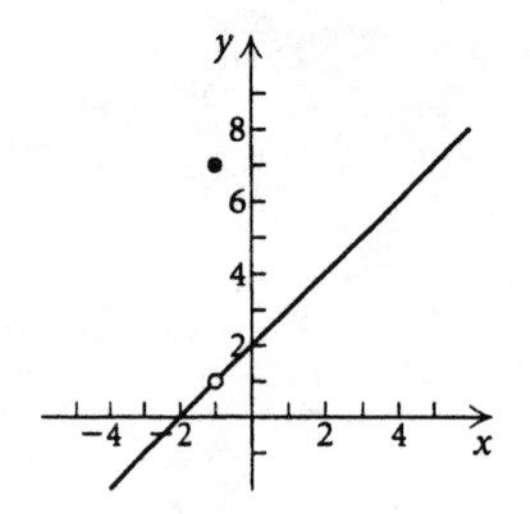

43. $f(x) = \text{int}(x)$

See Example 7.

44. $f(x) = 2\,\text{int}(x)$

This function can be defined by a piecewise function with an infinite number of statements:

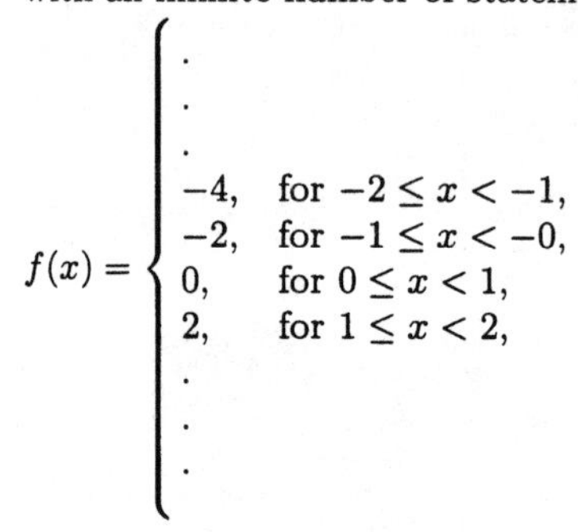

$$f(x) = \begin{cases} \cdot \\ \cdot \\ \cdot \\ -4, & \text{for } -2 \leq x < -1, \\ -2, & \text{for } -1 \leq x < -0, \\ 0, & \text{for } 0 \leq x < 1, \\ 2, & \text{for } 1 \leq x < 2, \\ \cdot \\ \cdot \\ \cdot \end{cases}$$

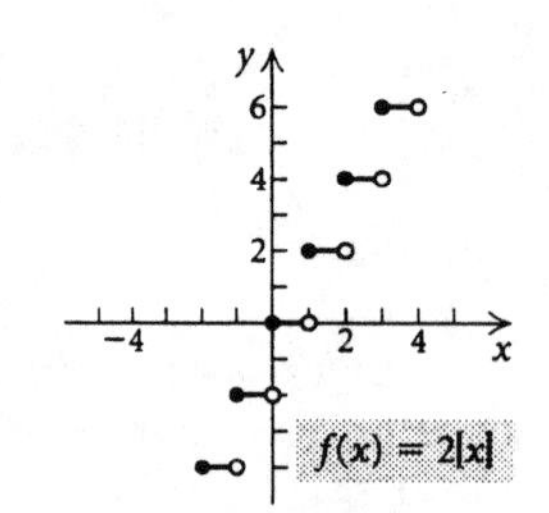

45. $f(x) = 1 + \text{int}(x)$

This function can be defined by a piecewise function with an infinite number of statements:

$$f(x) = \begin{cases} \cdot \\ \cdot \\ \cdot \\ -1, & \text{for } -2 \leq x < -1, \\ 0, & \text{for } -1 \leq x < -0, \\ 1, & \text{for } 0 \leq x < 1, \\ 2, & \text{for } 1 \leq x < 2, \\ \cdot \\ \cdot \\ \cdot \end{cases}$$

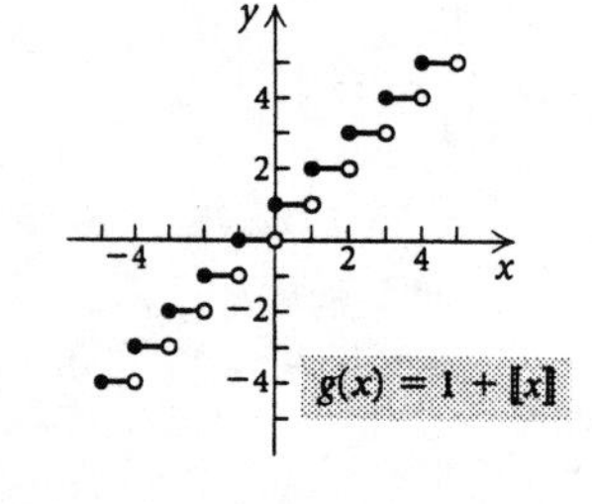

46. $f(x) = \frac{1}{2}\,\text{int}(x) - 2$

This function can be defined by a piecewise function with an infinite number of statements:

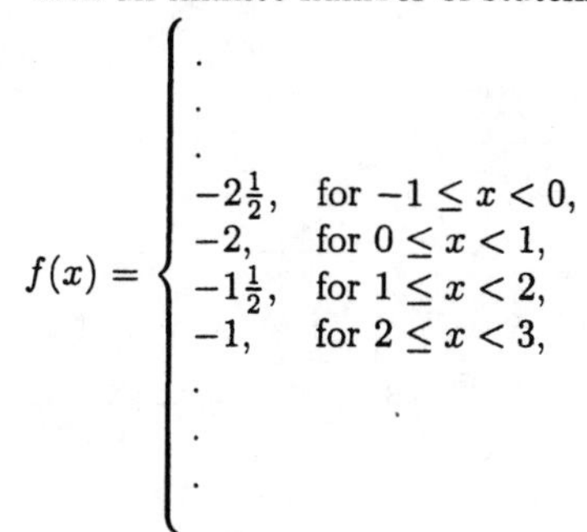

$$f(x) = \begin{cases} \vdots \\ -2\frac{1}{2}, & \text{for } -1 \le x < 0, \\ -2, & \text{for } 0 \le x < 1, \\ -1\frac{1}{2}, & \text{for } 1 \le x < 2, \\ -1, & \text{for } 2 \le x < 3, \\ \vdots \end{cases}$$

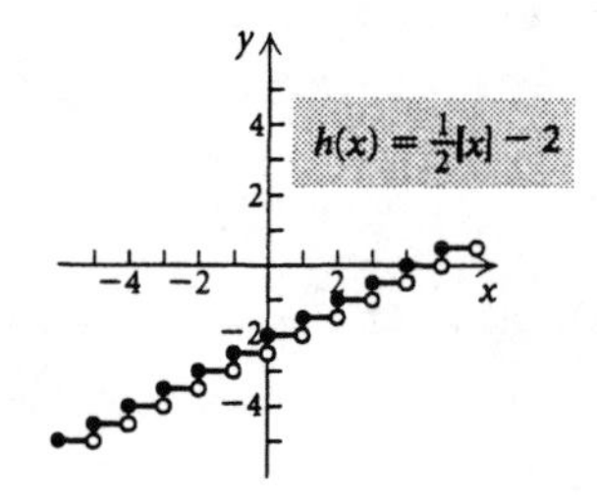

47. Use DOT mode.

$$y = \begin{cases} \sqrt[3]{x}, & \text{for } x \leqslant -1, \\ x^2 - 3x, & \text{for } -1 < x < 4, \\ \sqrt{x-4}, & \text{for } x \geqslant 4 \end{cases}$$

10

−10 10

−10

48. Use DOT mode.

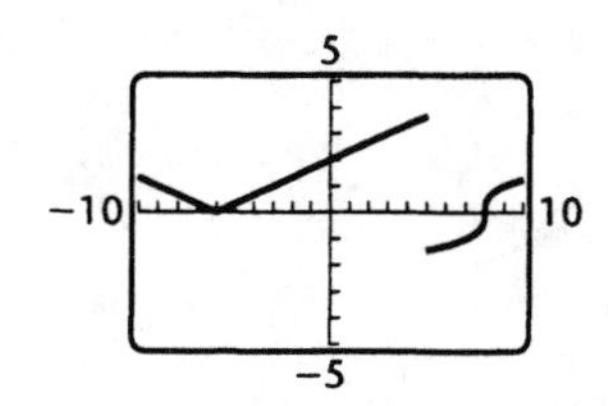

49. $(f+g)(5) = f(5) + g(5)$
$= (5^2 - 3) + (2 \cdot 5 + 1)$
$= 25 - 3 + 10 + 1$
$= 33$

50. $(f-g)(3) = f(3) - g(3)$
$= (3^2 - 3) - (2 \cdot 3 + 1)$
$= 6 - 7$
$= -1$

51. $(f-g)(-1) = f(-1) - g(-1)$
$= ((-1)^2 - 3) - (2(-1) + 1)$
$= -2 - (-1) = -2 + 1$
$= -1$

52. $(fg)(0) = f(0) \cdot g(0)$
$= (0^2 - 3)(2 \cdot 0 + 1)$
$= -3(1) = -3$

53. $(fg)(2) = f(2) \cdot g(2)$
$= (2^2 - 3)(2 \cdot 2 + 1)$
$= 1 \cdot 5 = 5$

54. $(fg)(-2) = f(-2) \cdot g(-2)$
$= [(-2)^2 - 3)][2(-2) + 1]$
$= 1 \cdot (-3) = -3$

55. $(f/g)(-1) = \dfrac{f(-1)}{g(-1)}$
$= \dfrac{(-1)^2 - 3}{2(-1) + 1}$
$= \dfrac{-2}{-1} = 2$

56. $(f/g)\left(-\frac{1}{2}\right) = \dfrac{f\left(-\frac{1}{2}\right)}{g\left(-\frac{1}{2}\right)}$
$= \dfrac{\left(-\frac{1}{2}\right)^2 - 3}{2\left(-\frac{1}{2}\right) + 1}$
$= \dfrac{-\frac{11}{4}}{0}$

Since division by 0 is not defined, $(f/g)\left(-\frac{1}{2}\right)$ does not exist.

57. $(fg)\left(-\frac{1}{2}\right) = f\left(-\frac{1}{2}\right) \cdot g\left(-\frac{1}{2}\right)$
$= \left[\left(-\frac{1}{2}\right)^2 - 3\right]\left[2\left(-\frac{1}{2}\right) + 1\right]$
$= -\frac{11}{4} \cdot 0 = 0$

58. $(f/g)(-\sqrt{3}) = \dfrac{f(-\sqrt{3})}{g(-\sqrt{3})}$
$= \dfrac{(-\sqrt{3})^2 - 3}{2(-\sqrt{3}) + 1}$
$= \dfrac{0}{-2\sqrt{3} + 1} = 0$

59. $(f/g)(\sqrt{3}) = \dfrac{f(\sqrt{3})}{g(\sqrt{3})}$
$= \dfrac{(\sqrt{3})^2 - 3}{2\sqrt{3}+1}$
$= \dfrac{3-3}{2\sqrt{3}+1}$
$= \dfrac{0}{2\sqrt{3}+1} = 0$

60. $(f-g)(0) = f(0) - g(0)$
$= (0^2 - 3) - (2\cdot 0 + 1)$
$= -3 - 1 = -4$

61. $f(x) = x - 3$, $g(x) = \sqrt{x+4}$

a) Any number can be an input in f, so the domain of f is the set of all real numbers.

The inputs of g must be nonnegative, so we have $x + 4 \geq 0$, or $x \geq -4$. Thus, the domain of g is $[-4, \infty)$.

The domain of $f+g$, $f-g$, and fg is the set of all numbers in the domains of both f and g. This is $[-4, \infty)$.

The domain of ff is the domain of f, or the set of all real numbers.
The domain of f/g is the set of all numbers in the domains of f and g, excluding those for which $g(x) = 0$. Since $g(-4) = 0$, the domain of f/g is $(-4, \infty)$.

The domain of g/f is the set of all numbers in the domains of g and f, excluding those for which $f(x) = 0$. Since $f(3) = 0$, the domain of g/f is $[-4, 3) \cup (3, \infty)$.

b) $(f+g)(x) = f(x) + g(x) = x - 3 + \sqrt{x+4}$
$(f-g)(x) = f(x) - g(x) = x - 3 - \sqrt{x+4}$
$(fg)(x) = f(x)\cdot g(x) = (x-3)\sqrt{x+4}$
$(ff)(x) = [f(x)]^2 = (x-3)^2 = x^2 - 6x + 9$
$(f/g)(x) = \dfrac{f(x)}{g(x)} = \dfrac{x-3}{\sqrt{x+4}}$
$(g/f)(x) = \dfrac{g(x)}{f(x)} = \dfrac{\sqrt{x+4}}{x-3}$

62. $f(x) = x^2 - 1$, $g(x) = 2x + 5$

a) The domain of f and of g is the set of all real numbers. Then the domain of $f+g$, $f-g$, fg and ff is the set of all real numbers. Since $g\left(-\dfrac{5}{2}\right) = 0$, the domain of f/g is the set of all real numbers except $-\dfrac{5}{2}$. Since $f(1) = 0$ and $f(-1) = 0$, the domain of g/f is the set of all real numbers except 1 and -1.

b) $(f+g)(x) = x^2 - 1 + 2x + 5 = x^2 + 2x + 4$
$(f-g)(x) = x^2 - 1 - (2x+5) = x^2 - 2x - 6$
$(fg)(x) = (x^2-1)(2x+5) = 2x^3+5x^2-2x-5$
$(ff)(x) = (x^2-1)^2 = x^4 - 2x^2 + 1$
$(f/g)(x) = \dfrac{x^2-1}{2x+5}$
$(g/f)(x) = \dfrac{2x+5}{x^2-1}$

63. $f(x) = x^3$, $g(x) = 2x^2 + 5x - 3$

a) Since any number can be an input for either f or g, the domain of f, g, $f+g$, $f-g$, fg, and ff is the set of all real numbers.

Since $g(-3) = 0$ and $g\left(\dfrac{1}{2}\right) = 0$, the domain of f/g is the set of all real numbers except -3 and $\dfrac{1}{2}$.

Since $f(0) = 0$, the domain of g/f is the set of all real numbers except 0.

b) $(f+g)(x) = f(x) + g(x) = x^3 + 2x^2 + 5x - 3$
$(f-g)(x) = f(x) - g(x) = x^3 - (2x^2+5x-3) = x^3 - 2x^2 - 5x + 3$
$(fg)(x) = f(x)\cdot g(x) = x^3(2x^2+5x-3) = 2x^5 + 5x^4 - 3x^3$
$(ff)(x) = [f(x)]^2 = (x^3)^2 = x^6$
$(f/g)(x) = \dfrac{f(x)}{g(x)} = \dfrac{x^3}{2x^2+5x-3}$
$(g/f)(x) = \dfrac{g(x)}{f(x)} = \dfrac{2x^2+5x-3}{x^3}$

64. $f(x) = x^2$, $g(x) = \sqrt{x}$

a) The domain of f is the set of all real numbers, and the domain of g is $[0, \infty)$. Then the domain of $f+g$, $f-g$, and fg is $[0, \infty)$. The domain of ff is the set of all real numbers. The domain of f/g and g/f is $(0, \infty)$.

b) $(f+g)(x) = x^2 + \sqrt{x}$
$(f-g)(x) = x^2 - \sqrt{x}$
$(fg)(x) = x^2\sqrt{x}$
$(ff)(x) = (x^2)^2 = x^4$
$(f/g)(x) = \dfrac{x^2}{\sqrt{x}}$
$(g/f)(x) = \dfrac{\sqrt{x}}{x^2}$

65. $f(x) = x^2 - 4$, $g(x) = x^2 + 2$
$(f+g)(x) = f(x) + g(x) = x^2 - 4 + x^2 + 2 = 2x^2 - 2$
$(f-g)(x) = f(x) - g(x) = x^2 - 4 - (x^2+2) = x^2 - 4 - x^2 - 2 = -6$
$(fg)(x) = f(x)\cdot g(x) = (x^2-4)(x^2+2) = x^4 - 2x^2 - 8$
$(f/g)(x) = \dfrac{f(x)}{g(x)} = \dfrac{x^2-4}{x^2+2}$

66. $f(x) = x^2 + 2,\ g(x) = 4x - 7$

$(f + g)(x) = x^2 + 2 + 4x - 7 = x^2 + 4x - 5$

$(f - g)(x) = x^2 + 2 - (4x - 7) = x^2 - 4x + 9$

$(fg)(x) = (x^2 + 2)(4x - 7) = 4x^3 - 7x^2 + 8x - 14$

$(f/g)(x) = \dfrac{x^2 + 2}{4x - 7}$

67. $f(x) = \sqrt{x - 7},\ g(x) = x^2 - 25$

$(f + g)(x) = f(x) + g(x) = \sqrt{x - 7} + x^2 - 25$

$(f - g)(x) = f(x) - g(x) = \sqrt{x - 7} - (x^2 - 25) =$
$\sqrt{x - 7} - x^2 + 25$

$(fg)(x) = f(x) \cdot g(x) = \sqrt{x - 7}(x^2 - 25)$

$(f/g)(x) = \dfrac{f(x)}{g(x)} = \dfrac{\sqrt{x - 7}}{x^2 - 25}$

68. $f(x) = \sqrt{x - 1},\ g(x) = \sqrt{3x - 8}$

$(f + g)(x) = \sqrt{x - 1} + \sqrt{3x - 8}$

$(f - g)(x) = \sqrt{x - 1} - \sqrt{3x - 8}$

$(fg)(x) = \sqrt{x - 1}\sqrt{3x - 8} = \sqrt{3x^2 - 11x + 8}$

$(f/g)(x) = \dfrac{\sqrt{x - 1}}{\sqrt{3x - 8}}$

69. From the graph, we see that the domain of F is $[0, 9]$ and the domain of G is $[3, 10]$. The domain of $F + G$ is the set of numbers in the domains of both F and G. This is $[3, 9]$.

70. The domain of $F - G$ and FG is the set of numbers in the domains of both F and G. (See Exercise 69.) This is $[3, 9]$.

The domain of F/G is the set of numbers in the domains of both F and G, excluding those for which $G = 0$. Since $G > 0$ for all values of x in its domain, the domain of F/G is $[3, 9]$.

71. The domain of G/F is the set of numbers in the domains of both F and G (See Exercise 69.), excluding those for which $F = 0$. Since $F(6) = 0$ and $F(8) = 0$, the domain of G/F is $[3, 6) \cup (6, 8) \cup (8, 9]$.

72. $(F + G)(x) = F(x) + G(x)$

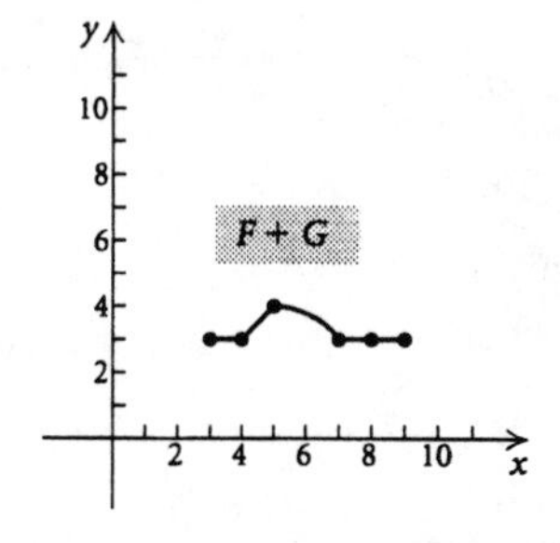

73.

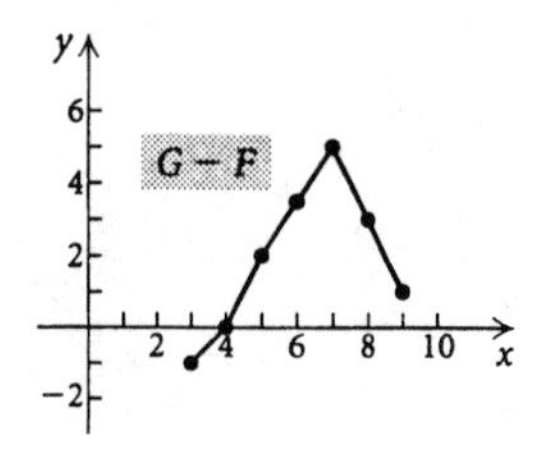

74.

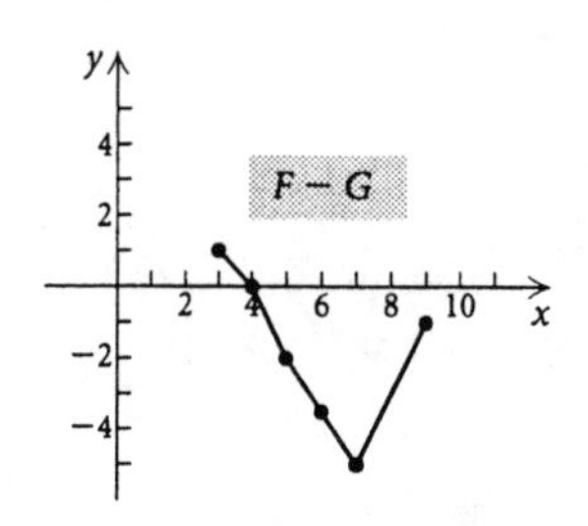

75. a) $P(x) = R(x) - C(x) = 60x - 0.4x^2 - (3x + 13) =$
$60x - 0.4x^2 - 3x - 13 = -0.4x^2 + 57x - 13$

b) $R(100) = 60\cdot100 - 0.4(100)^2 = 6000 - 0.4(10,000) =$
$6000 - 4000 = 2000$

$C(100) = 3 \cdot 100 + 13 = 300 + 13 = 313$

$P(100) = R(100) - C(100) = 2000 - 313 = 1687$

c)

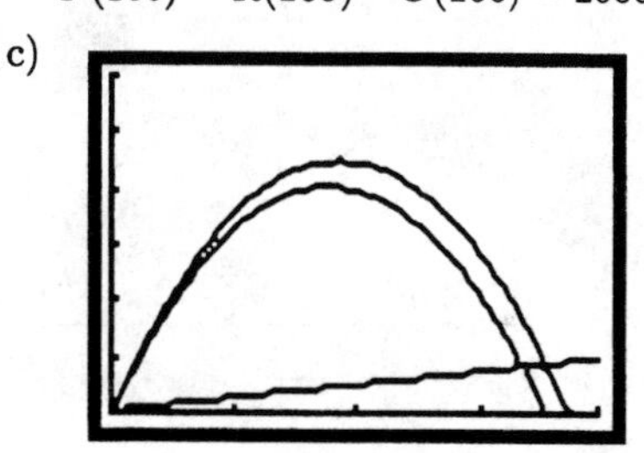

76. a) $P(x) = 200x - x^2 - (5000 + 8x) =$
$200x - x^2 - 5000 - 8x = -x^2 + 192x - 5000$

b) $R(175) = 200(175) - 175^2 = 4375$

$C(175) = 5000 + 8 \cdot 175 = 6400$

$P(175) = R(175) - C(175) = 4375 - 6400 =$
-2025

(We could also use the function found in part (a) to find $P(175)$.)

c)

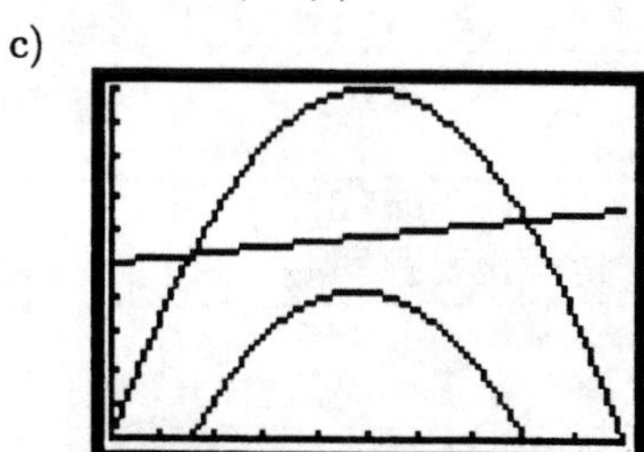

77. $f(x) = 3x^2 - 2x + 1$

$f(x + h) = 3(x + h)^2 - 2(x + h) + 1 =$

$3(x^2 + 2xh + h^2) - 2(x + h) + 1 =$

$3x^2 + 6xh + 3h^2 - 2x - 2h + 1$

$f(x) = 3x^2 - 2x + 1$

$\frac{f(x+h) - f(x)}{h} =$

$\frac{(3x^2 + 6xh + 3h^2 - 2x - 2h + 1) - (3x^2 - 2x + 1)}{h} =$

$\frac{3x^2 + 6xh + 3h^2 - 2x - 2h + 1 - 3x^2 + 2x - 1}{h} =$

$\frac{6xh + 3h^2 - 2h}{h} = \frac{h(6x + 3h - 2)}{h \cdot 1} =$

$\frac{h}{h} \cdot \frac{6x + 3h - 2}{1} = 6x + 3h - 2$

78. $f(x) = 5x^2 + 4x$

$\frac{f(x+h)-f(x)}{h} = \frac{(5x^2+10xh+5h^2+4x+4h)-(5x^2+4x)}{h} =$

$\frac{10xh + 5h^2 + 4h}{h} = 10x + 5h + 4$

79. $f(x) = x^3$

$f(x+h) = (x+h)^3 = x^3 + 3x^2h + 3xh^2 + h^3$

$f(x) = x^3$

$\frac{f(x+h) - f(x)}{h} = \frac{x^3 + 3x^2h + 3xh^2 + h^3 - x^3}{h} =$

$\frac{3x^2h + 3xh^2 + h^3}{h} = \frac{h(3x^2 + 3xh + h^2)}{h \cdot 1} =$

$\frac{h}{h} \cdot \frac{3x^2 + 3xh + h^2}{1} = 3x^2 + 3xh + h^2$

80. $f(x) = 2|x| + 3x$

$\frac{f(x+h)-f(x)}{h} = \frac{(2|x+h|+3x+3h)-(2|x|+3x)}{h} =$

$\frac{2|x+h| - 2|x| + 3h}{h}$

81. $f(x) = \frac{x-4}{x+3}$

$\frac{f(x+h) - f(x)}{h} = \frac{\frac{x+h-4}{x+h+3} - \frac{x-4}{x+3}}{h} =$

$\frac{\frac{x+h-4}{x+h+3} - \frac{x-4}{x+3}}{h} \cdot \frac{(x+h+3)(x+3)}{(x+h+3)(x+3)} =$

$\frac{(x+h-4)(x+3) - (x-4)(x+h+3)}{h(x+h+3)(x+3)} =$

$\frac{x^2+hx-4x+3x+3h-12-(x^2+hx+3x-4x-4h-12)}{h(x+h+3)(x+3)} =$

$\frac{x^2 + hx - x + 3h - 12 - x^2 - hx + x + 4h + 12}{h(x+h+3)(x+3)} =$

$\frac{7h}{h(x+h+3)(x+3)} = \frac{h}{h} \cdot \frac{7}{(x+h+3)(x+3)} =$

$\frac{7}{(x+h+3)(x+3)}$

82. $f(x) = \frac{x}{2-x}$

$\frac{f(x+h) - f(x)}{h} = \frac{\frac{x+h}{2-(x+h)} - \frac{x}{2-x}}{h} =$

$\frac{\frac{(x+h)(2-x) - x(2-x-h)}{(2-x-h)(2-x)}}{h} =$

$\frac{\frac{2x - x^2 + 2h - hx - 2x + x^2 + hx}{(2-x-h)(2-x)}}{h} =$

$\frac{\frac{2h}{(2-x-h)(2-x)}}{h} =$

$\frac{2h}{(2-x-h)(2-x)} \cdot \frac{1}{h} = \frac{2}{(2-x-h)(2-x)}$

83. Some possibilities are outdoor temperature during a 24 hour period, sales of a new product, and temperature during an illness.

84. For continuous functions, relative extrema occur at points for which the function changes from increasing to decreasing or vice versa.

85. $f(x) = -2x - 3$

We can substitute any real number for x, so the domain is the set of all real numbers, or $(-\infty, \infty)$.

86. $g(x) = \frac{1}{(x-3)^2}$

The input 3 results in a denominator of 0, so the domain is $\{x|x \neq 3\}$, or $(-\infty, 3) \cup (3, \infty)$.

87. $h(x) = \frac{x}{|x+5|}$

We can substitute any real number in the numerator, but the input -5 results in a denominator of 0. Thus, the domain is $\{x|x \neq -5\}$, or $(-\infty, -5) \cup (-5, \infty)$.

88. $f(x) = -5$

We can substitute any real number for x, so the domain is the set of all real numbers, or $(-\infty, \infty)$.

89. Graph $y = x^4 + 4x^3 - 36x^2 - 160x + 400$

Increasing: $(-5, -2)$, $(4, \infty)$

Decreasing: $(-\infty, -5)$, $(-2, 4)$

Relative maximum: 560 at $x = -2$

Relative minima: 425 at $x = -5$, -304 at $x = 4$

90. Graph $y = 3.22x^5 - 5.208x^3 - 11$

Increasing: $(-\infty, -0.985)$, $(0.985, \infty)$

Decreasing: $(-0.985, 0.985)$

Relative maximum: -9.008 at $x = -0.985$

Relative minimum: -12.992 at $x = 0.985$

91. a) The function $C(t)$ can be defined piecewise.

$$C(t) = \begin{cases} 2, & \text{for } 0 < t < 1, \\ 4, & \text{for } 1 \le t < 2, \\ 6, & \text{for } 2 \le t < 3, \\ \cdot \\ \cdot \\ \cdot \end{cases}$$

We graph this function.

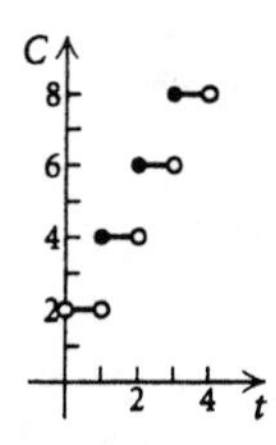

b) From the definition of the function in part (a), we see that it can be written as

$$C(t) = 2[\text{int}(t) + 1],\ t > 0.$$

92. If $\text{int}(x + 2) = -3$, then $-3 \le x + 2 < -2$, or $-5 \le x < -4$. The possible inputs for x are $\{x| -5 \le x < -4\}$.

93. If $[\text{int}(x)]^2 = 25$, then $\text{int}(x) = -5$ or $\text{int}(x) = 5$. For $-5 \le x < -4$, $\text{int}(x) = -5$. For $5 \le x < 6$, $\text{int}(x) = 5$. Thus, the possible inputs for x are $\{x| -5 \le x < -4 \text{ or } 5 \le x < 6\}$.

94. a) The distance from A to S is $4 - x$.

Using the Pythagorean theorem, we find that the distance from S to C is $\sqrt{1 + x^2}$.

Then $C(x) = 3000(4 - x) + 5000\sqrt{1 + x^2}$, or $12,000 - 3000x + 5000\sqrt{1 + x^2}$.

b) Graph $y = 12,000 - 3000x + 5000\sqrt{1 + x^2}$ in a window such as $[0, 5, 10,000, 20,000]$, Xscl = 1, Yscl = 1000. Using the MINIMUM feature, we find that cost is minimized when $x = 0.75$, so the line should come to shore 0.75 mi from B.

95. a) We add labels to the drawing in the text.

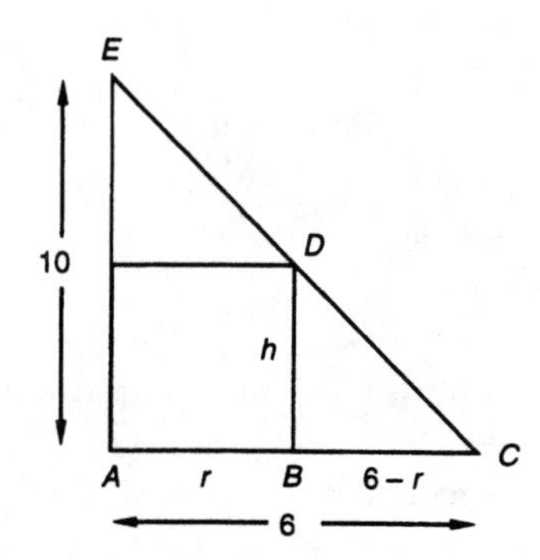

We write a proportion involving the lengths of the sides of the similar triangles BCD and ACE. Then we solve it for h.

$$\frac{h}{6 - r} = \frac{10}{6}$$

$$h = \frac{10}{6}(6 - r) = \frac{5}{3}(6 - r)$$

$$h = \frac{30 - 5r}{3}$$

Thus, $h(r) = \dfrac{30 - 5r}{3}$.

b) $V = \pi r^2 h$

$V(r) = \pi r^2 \left(\dfrac{30 - 5r}{3}\right)$ Substituting for h

c) We first express r in terms of h.

$$h = \frac{30 - 5r}{3}$$

$$3h = 30 - 5r$$

$$5r = 30 - 3h$$

$$r = \frac{30 - 3h}{5}$$

$$V = \pi r^2 h$$

$$V(h) = \pi\left(\frac{30 - 3h}{5}\right)^2 h$$

Substituting for r

We can also write $V(h) = \pi h\left(\dfrac{30 - 3h}{5}\right)^2$.

Exercise Set 1.5

1. If the graph were folded on the x-axis, the parts above and below the x-axis would not coincide, so the graph is not symmetric with respect to the x-axis.

If the graph were folded on the y-axis, the parts to the left and right of the y-axis would coincide, so the graph is symmetric with respect to the y-axis.

If the graph were rotated 180°, the resulting graph would not coincide with the original graph, so it is not symmetric with respect to the origin.

2. If the graph were folded on the x-axis, the parts above and below the x-axis would not coincide, so the graph is not symmetric with respect to the x-axis.

If the graph were folded on the y-axis, the parts to the left and right of the y-axis would coincide, so the graph is symmetric with respect to the y-axis.

If the graph were rotated 180°, the resulting graph would not coincide with the original graph, so it is not symmetric with respect to the origin.

3. If the graph were folded on the x-axis, the parts above and below the x-axis would coincide, so the graph is symmetric with respect to the x-axis.

If the graph were folded on the y-axis, the parts to the left and right of the y-axis would not coincide, so the graph is not symmetric with respect to the y-axis.

If the graph were rotated 180°, the resulting graph would not coincide with the original graph, so it is not symmetric with respect to the origin.

4. If the graph were folded on the x-axis, the parts above and below the x-axis would not coincide, so the graph is not symmetric with respect to the x-axis.

If the graph were folded on the y-axis, the parts to the left and right of the y-axis would not coincide, so the graph is not symmetric with respect to the y-axis.

If the graph were rotated 180°, the resulting graph would coincide with the original graph, so it is symmetric with respect to the origin.

5. If the graph were folded on the x-axis, the parts above and below the x-axis would not coincide, so the graph is not symmetric with respect to the x-axis.

If the graph were folded on the y-axis, the parts to the left and right of the y-axis would not coincide, so the graph is not symmetric with respect to the y-axis.

If the graph were rotated 180°, the resulting graph would coincide with the original graph, so it is symmetric with respect to the origin.

6. If the graph were folded on the x-axis, the parts above and below the x-axis would coincide, so the graph is symmetric with respect to the x-axis.

If the graph were folded on the y-axis, the parts to the left and right of the y-axis would coincide, so the graph is symmetric with respect to the y-axis.

If the graph were rotated 180°, the resulting graph would coincide with the original graph, so it is symmetric with respect to the origin.

7.

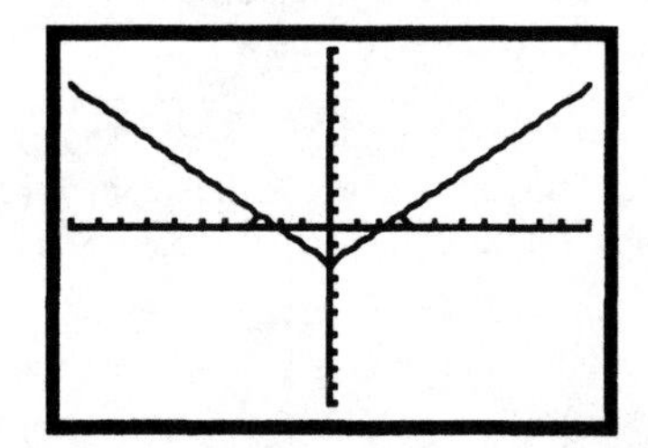

The graph is symmetric with respect to the y-axis. It is not symmetric with respect to the x-axis or the origin.

Test algebraically for symmetry with respect to the x-axis:

$y = |x| - 2$ Original equation

$-y = |x| - 2$ Replacing y by $-y$

$y = -|x| + 2$ Simplifying

The last equation is not equivalent to the original equation, so the graph is not symmetric with respect to the x-axis.

Test algebraically for symmetry with respect to the y-axis:

$y = |x| - 2$ Original equation

$y = |-x| - 2$ Replacing x by $-x$

$y = |x| - 2$ Simplifying

The last equation is equivalent to the original equation, so the graph is symmetric with respect to the y-axis.

Test algebraically for symmetry with respect to the origin:

$y = |x| - 2$ Original equation

$-y = |-x| - 2$ Replacing x by $-x$ and y by $-y$

$-y = |x| - 2$ Simplifying

$y = -|x| + 2$

The last equation is not equivalent to the original equation, so the graph is not symmetric with respect to the origin.

8.

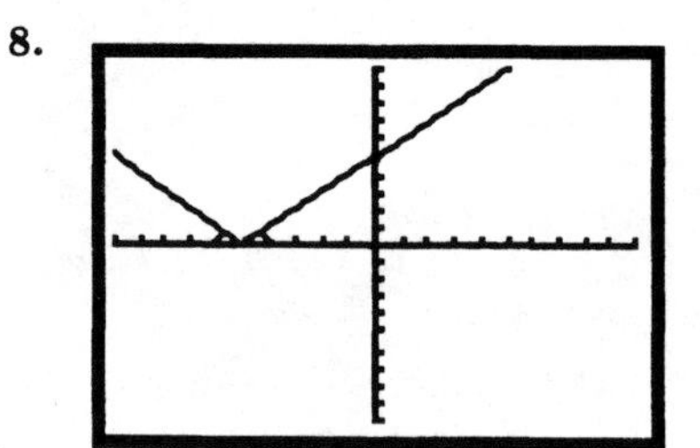

The graph is not symmetric with respect to the x-axis, the y-axis, or the origin.

Test algebraically for symmetry with respect to the x-axis:

$y = |x + 5|$ Original equation

$-y = |x + 5|$ Replacing y by $-y$

$y = -|x + 5|$ Simplifying

The last equation is not equivalent to the original equation, so the graph is not symmetric with respect to the x-axis.

Test algebraically for symmetry with respect to the y-axis:

$y = |x + 5|$ Original equation

$y = |-x + 5|$ Replacing x by $-x$

The last equation is not equivalent to the original equation, so the graph is not symmetric with respect to the y-axis.

Test algebraically for symmetry with respect to the origin:

$y = |x + 5|$ Original equation

$-y = |-x + 5|$ Replacing x by $-x$ and y by $-y$

$y = -|-x + 5|$ Simplifying

The last equation is not equivalent to the original equation, so the graph is not symmetric with respect to the origin.

9.

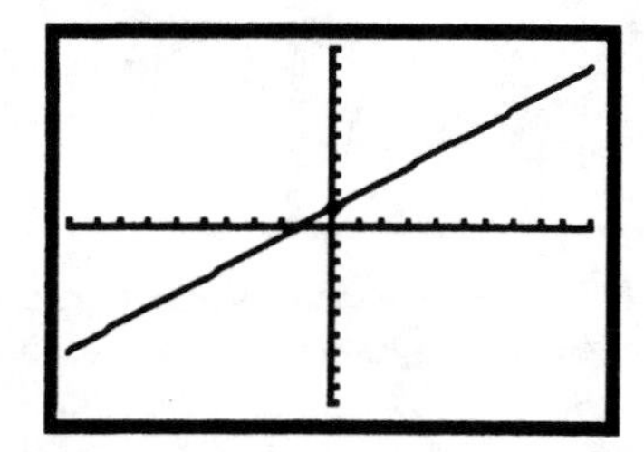

The graph is not symmetric with respect to the x-axis, the y-axis, or the origin.

Test algebraically for symmetry with respect to the x-axis:

$$\begin{aligned} 5y &= 4x+5 && \text{Original equation} \\ 5(-y) &= 4x+5 && \text{Replacing } y \text{ by } -y \\ -5y &= 4x+5 && \text{Simplifying} \\ 5y &= -4x-5 && \end{aligned}$$

The last equation is not equivalent to the original equation, so the graph is not symmetric with respect to the x-axis.

Test algebraically for symmetry with respect to the y-axis:

$$\begin{aligned} 5y &= 4x+5 && \text{Original equation} \\ 5y &= 4(-x)+5 && \text{Replacing } x \text{ by } -x \\ 5y &= -4x+5 && \text{Simplifying} \end{aligned}$$

The last equation is not equivalent to the original equation, so the graph is not symmetric with respect to the y-axis.

Test algebraically for symmetry with respect to the origin:

$$\begin{aligned} 5y &= 4x+5 && \text{Original equation} \\ 5(-y) &= 4(-x)+5 && \text{Replacing } x \text{ by } -x \text{ and } y \text{ by } -y \\ -5y &= -4x+5 && \text{Simplifying} \\ 5y &= 4x-5 && \end{aligned}$$

The last equation is not equivalent to the original equation, so the graph is not symmetric with respect to the origin.

10.

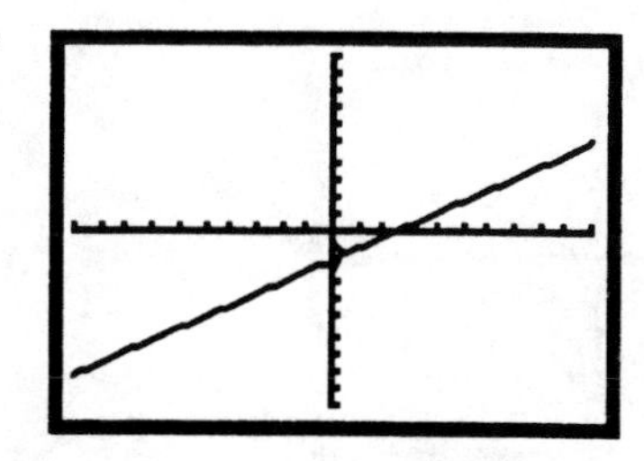

The graph is not symmetric with respect to the x-axis, the y-axis, or the origin.

Test algebraically for symmetry with respect to the x-axis:

$$\begin{aligned} 2x-5 &= 3y && \text{Original equation} \\ 2x-5 &= 3(-y) && \text{Replacing } y \text{ by } -y \\ -2x+5 &= 3y && \text{Simplifying} \end{aligned}$$

The last equation is not equivalent to the original equation, so the graph is not symmetric with respect to the x-axis.

Test algebraically for symmetry with respect to the y-axis:

$$\begin{aligned} 2x-5 &= 3y && \text{Original equation} \\ 2(-x)-5 &= 3y && \text{Replacing } x \text{ by } -x \\ -2x-5 &= 3y && \text{Simplifying} \end{aligned}$$

The last equation is not equivalent to the original equation, so the graph is not symmetric with respect to the y-axis.

Test algebraically for symmetry with respect to the origin:

$$\begin{aligned} 2x-5 &= 3y && \text{Original equation} \\ 2(-x)-5 &= 3(-y) && \text{Replacing } x \text{ by } -x \text{ and } y \text{ by } -y \\ -2x-5 &= -3y && \text{Simplifying} \\ 2x+5 &= 3y && \end{aligned}$$

The last equation is not equivalent to the original equation, so the graph is not symmetric with respect to the origin.

11.

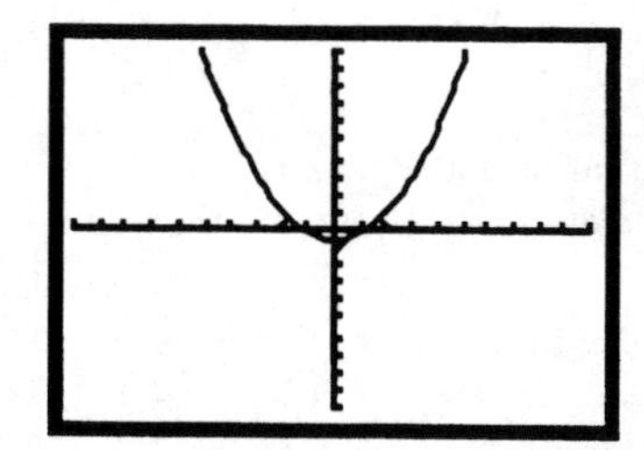

The graph is symmetric with respect to the y-axis. It is not symmetric with respect to the x-axis or the origin.

Test algebraically for symmetry with respect to the x-axis:

$$\begin{aligned} 5y &= 2x^2-3 && \text{Original equation} \\ 5(-y) &= 2x^2-3 && \text{Replacing } y \text{ by } -y \\ -5y &= 2x^2-3 && \text{Simplifying} \\ 5y &= -2x^2+3 && \end{aligned}$$

The last equation is not equivalent to the original equation, so the graph is not symmetric with respect to the x-axis.

Test algebraically for symmetry with respect to the y-axis:

$$\begin{aligned} 5y &= 2x^2-3 && \text{Original equation} \\ 5y &= 2(-x)^2-3 && \text{Replacing } x \text{ by } -x \\ 5y &= 2x^2-3 && \end{aligned}$$

The last equation is equivalent to the original equation, so the graph is symmetric with respect to the y-axis.

Test algebraically for symmetry with respect to the origin:

$5y = 2x^2 - 3$	Original equation
$5(-y) = 2(-x)^2 - 3$	Replacing x by $-x$ and y by $-y$
$-5y = 2x^2 - 3$	Simplifying
$5y = -2x^2 + 3$	

The last equation is not equivalent to the original equation, so the graph is not symmetric with respect to the origin.

12.

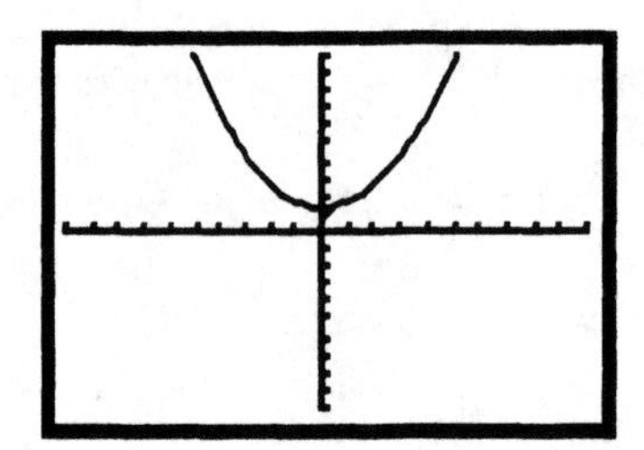

The graph is symmetric with respect to the y-axis. It is not symmetric with respect to the x-axis or the origin.

Test algebraically for symmetry with respect to the x-axis:

$x^2 + 4 = 3y$	Original equation
$x^2 + 4 = 3(-y)$	Replacing y by $-y$
$-x^2 - 4 = 3y$	Simplifying

The last equation is not equivalent to the original equation, so the graph is not symmetric with respect to the x-axis.

Test algebraically for symmetry with respect to the y-axis:

$x^2 + 4 = 3y$	Original equation
$(-x)^2 + 4 = 3y$	Replacing x by $-x$
$x^2 + 4 = 3y$	

The last equation is equivalent to the original equation, so the graph is symmetric with respect to the y-axis.

Test algebraically for symmetry with respect to the origin:

$x^2 + 4 = 3y$	Original equation
$(-x)^2 + 4 = 3(-y)$	Replacing x by $-x$ and y by $-y$
$x^2 + 4 = -3y$	Simplifying
$-x^2 - 4 = 3y$	

The last equation is not equivalent to the original equation, so the graph is not symmetric with respect to the origin.

13.

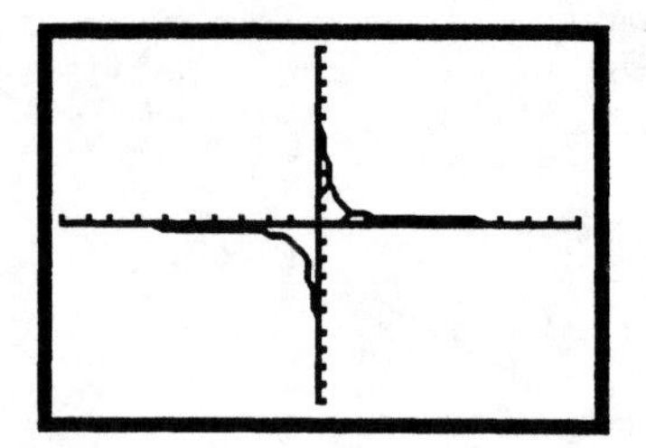

The graph is not symmetric with respect to the x-axis or the y-axis. It is symmetric with respect to the origin.

Test algebraically for symmetry with respect to the x-axis:

$y = \frac{1}{x}$	Original equation
$-y = \frac{1}{x}$	Replacing y by $-y$
$y = -\frac{1}{x}$	Simplifying

The last equation is not equivalent to the original equation, so the graph is not symmetric with respect to the x-axis.

Test algebraically for symmetry with respect to the y-axis:

$y = \frac{1}{x}$	Original equation
$y = \frac{1}{-x}$	Replacing x by $-x$
$y = -\frac{1}{x}$	Simplifying

The last equation is not equivalent to the original equation, so the graph is symmetric with respect to the y-axis.

Test algebraically for symmetry with respect to the origin:

$y = \frac{1}{x}$	Original equation
$-y = \frac{1}{-x}$	Replacing x by $-x$ and y by $-y$
$y = \frac{1}{x}$	Simplifying

The last equation is equivalent to the original equation, so the graph is symmetric with respect to the origin.

14.

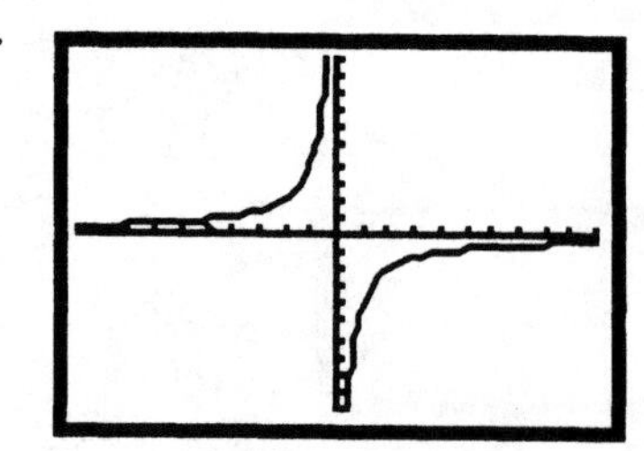

The graph is not symmetric with respect to the x-axis or the y-axis. It is symmetric with respect to the origin.

Test algebraically for symmetry with respect to the x-axis:

$$y = -\frac{4}{x} \quad \text{Original equation}$$
$$-y = -\frac{4}{x} \quad \text{Replacing } y \text{ by } -y$$
$$y = \frac{4}{x} \quad \text{Simplifying}$$

The last equation is not equivalent to the original equation, so the graph is not symmetric with respect to the x-axis.

Test algebraically for symmetry with respect to the y-axis:

$$y = -\frac{4}{x} \quad \text{Original equation}$$
$$y = -\frac{4}{-x} \quad \text{Replacing } x \text{ by } -x$$
$$y = \frac{4}{x} \quad \text{Simplifying}$$

The last equation is not equivalent to the original equation, so the graph is not symmetric with respect to the y-axis.

Test algebraically for symmetry with respect to the origin:

$$y = -\frac{4}{x} \quad \text{Original equation}$$
$$-y = -\frac{4}{-x} \quad \text{Replacing } x \text{ by } -x \text{ and } y \text{ by } -y$$
$$y = -\frac{4}{x} \quad \text{Simplifying}$$

The last equation is equivalent to the original equation, so the graph is symmetric with respect to the origin.

15. Test for symmetry with respect to the x-axis:

$$5x - 5y = 0 \quad \text{Original equation}$$
$$5x - 5(-y) = 0 \quad \text{Replacing } y \text{ by } -y$$
$$5x + 5y = 0 \quad \text{Simplifying}$$

The last equation is not equivalent to the original equation, so the graph is not symmetric with respect to the x-axis.

Test for symmetry with respect to the y-axis:

$$5x - 5y = 0 \quad \text{Original equation}$$
$$5(-x) - 5y = 0 \quad \text{Replacing } x \text{ by } -x$$
$$-5x - 5y = 0 \quad \text{Simplifying}$$
$$5x + 5y = 0$$

The last equation is not equivalent to the original equation, so the graph is not symmetric with respect to the y-axis.

Test for symmetry with respect to the origin:

$$5x - 5y = 0 \quad \text{Original equation}$$
$$5(-x) - 5(-y) = 0 \quad \text{Replacing } x \text{ by } -x \text{ and } y \text{ by } -y$$
$$-5x + 5y = 0 \quad \text{Simplifying}$$
$$5x - 5y = 0$$

The last equation is equivalent to the original equation, so the graph is symmetric with respect to the origin.

16. Test for symmetry with respect to the x-axis:

$$6x + 7y = 0 \quad \text{Original equation}$$
$$6x + 7(-y) = 0 \quad \text{Replacing } y \text{ by } -y$$
$$6x - 7y = 0 \quad \text{Simplifying}$$

The last equation is not equivalent to the original equation, so the graph is not symmetric with respect to the x-axis.

Test for symmetry with respect to the y-axis:

$$6x + 7y = 0 \quad \text{Original equation}$$
$$6(-x) + 7y = 0 \quad \text{Replacing } x \text{ by } -x$$
$$6x - 7y = 0 \quad \text{Simplifying}$$

The last equation is not equivalent to the original equation, so the graph is not symmetric with respect to the y-axis.

Test for symmetry with respect to the origin:

$$6x + 7y = 0 \quad \text{Original equation}$$
$$6(-x) + 7(-y) = 0 \quad \text{Replacing } x \text{ by } -x \text{ and } y \text{ by } -y$$
$$6x + 7y = 0 \quad \text{Simplifying}$$

The last equation is equivalent to the original equation, so the graph is symmetric with respect to the origin.

17. Test for symmetry with respect to the x-axis:

$$3x^2 - 2y^2 = 3 \quad \text{Original equation}$$
$$3x^2 - 2(-y)^2 = 3 \quad \text{Replacing } y \text{ by } -y$$
$$3x^2 - 2y^2 = 3 \quad \text{Simplifying}$$

The last equation is equivalent to the original equation, so the graph is symmetric with respect to the x-axis.

Test for symmetry with respect to the y-axis:

$$3x^2 - 2y^2 = 3 \quad \text{Original equation}$$
$$3(-x)^2 - 2y^2 = 3 \quad \text{Replacing } x \text{ by } -x$$
$$3x^2 - 2y^2 = 3 \quad \text{Simplifying}$$

The last equation is equivalent to the original equation, so the graph is symmetric with respect to the y-axis.

Test for symmetry with respect to the origin:

$$3x^2 - 2y^2 = 3 \quad \text{Original equation}$$
$$3(-x)^2 - 2(-y)^2 = 3 \quad \text{Replacing } x \text{ by } -x \text{ and } y \text{ by } -y$$
$$3x^2 - 2y^2 = 3 \quad \text{Simplifying}$$

The last equation is equivalent to the original equation, so the graph is symmetric with respect to the origin.

18. Test for symmetry with respect to the x-axis:

$$5y = 7x^2 - 2x \quad \text{Original equation}$$
$$5(-y) = 7x^2 - 2x \quad \text{Replacing } y \text{ by } -y$$
$$5y = -7x^2 + 2x \quad \text{Simplifying}$$

The last equation is not equivalent to the original equation, so the graph is not symmetric with respect to the x-axis.

Test for symmetry with respect to the y-axis:

$5y = 7x^2 - 2x$ Original equation

$5y = 7(-x)^2 - 2(-x)$ Replacing x by $-x$

$5y = 7x^2 + 2x$ Simplifying

The last equation is not equivalent to the original equation, so the graph is not symmetric with respect to the y-axis.

Test for symmetry with respect to the origin:

$5y = 7x^2 - 2x$ Original equation

$5(-y) = 7(-x)^2 - 2(-x)$ Replacing x by $-x$ and y by $-y$

$-5y = 7x^2 + 2x$ Simplifying

$5y = -7x^2 - 2x$

The last equation is not equivalent to the original equation, so the graph is not symmetric with respect to the origin.

19. Test for symmetry with respect to the x-axis:

$y = |2x|$ Original equation

$-y = |2x|$ Replacing y by $-y$

$y = -|2x|$ Simplifying

The last equation is not equivalent to the original equation, so the graph is not symmetric with respect to the x-axis.

Test for symmetry with respect to the y-axis:

$y = |2x|$ Original equation

$y = |2(-x)|$ Replacing x by $-x$

$y = |-2x|$ Simplifying

$y = |2x|$

The last equation is equivalent to the original equation, so the graph is symmetric with respect to the y-axis.

Test for symmetry with respect to the origin:

$y = |2x|$ Original equation

$-y = |2(-x)|$ Replacing x by $-x$ and y by $-y$

$-y = |-2x|$ Simplifying

$-y = |2x|$

$y = -|2x|$

The last equation is not equivalent to the original equation, so the graph is not symmetric with respect to the origin.

20. Test for symmetry with respect to the x-axis:

$y^3 = 2x^2$ Original equation

$(-y)^3 = 2x^2$ Replacing y by $-y$

$-y^3 = 2x^2$ Simplifying

$y^3 = -2x^2$

The last equation is not equivalent to the original equation, so the graph is not symmetric with respect to the x-axis.

Test for symmetry with respect to the y-axis:

$y^3 = 2x^2$ Original equation

$y^3 = 2(-x)^2$ Replacing x by $-x$

$y^3 = 2x^2$ Simplifying

The last equation is equivalent to the original equation, so the graph is symmetric with respect to the y-axis.

Test for symmetry with respect to the origin:

$y^3 = 2x^2$ Original equation

$(-y)^3 = 2(-x)^2$ Replacing x by $-x$ and y by $-y$

$-y^3 = 2x^2$ Simplifying

$y^3 = -2x^2$

The last equation is not equivalent to the original equation, so the graph is not symmetric with respect to the origin.

21. Test for symmetry with respect to the x-axis:

$2x^4 + 3 = y^2$ Original equation

$2x^4 + 3 = (-y)^2$ Replacing y by $-y$

$2x^4 + 3 = y^2$ Simplifying

The last equation is equivalent to the original equation, so the graph is symmetric with respect to the x-axis.

Test for symmetry with respect to the y-axis:

$2x^4 + 3 = y^2$ Original equation

$2(-x)^4 + 3 = y^2$ Replacing x by $-x$

$2x^4 + 3 = y^2$ Simplifying

The last equation is equivalent to the original equation, so the graph is symmetric with respect to the y-axis.

Test for symmetry with respect to the origin:

$2x^4 + 3 = y^2$ Original equation

$2(-x)^4 + 3 = (-y)^2$ Replacing x by $-x$ and y by $-y$

$2x^4 + 3 = y^2$ Simplifying

The last equation is equivalent to the original equation, so the graph is symmetric with respect to the origin.

22. Test for symmetry with respect to the x-axis:

$2y^2 = 5x^2 + 12$ Original equation

$2(-y)^2 = 5x^2 + 12$ Replacing y by $-y$

$2y^2 = 5x^2 + 12$ Simplifying

The last equation is equivalent to the original equation, so the graph is symmetric with respect to the x-axis.

Test for symmetry with respect to the y-axis:

$2y^2 = 5x^2 + 12$ Original equation

$2y^2 = 5(-x)^2 + 12$ Replacing x by $-x$

$2y^2 = 5x^2 + 12$ Simplifying

The last equation is equivalent to the original equation, so the graph is symmetric with respect to the y-axis.

Test for symmetry with respect to the origin:

$2y^2 = 5x^2 + 12$ Original equation

$2(-y)^2 = 5(-x)^2 + 12$ Replacing x by $-x$ and y by $-y$

$2y^2 = 5x^2 + 12$ Simplifying

The last equation is equivalent to the original equation, so the graph is symmetric with respect to the origin.

23. Test for symmetry with respect to the x-axis:

$3y^3 = 4x^3 + 2$ Original equation

$3(-y)^3 = 4x^3 + 2$ Replacing y by $-y$

$-3y^3 = 4x^3 + 2$ Simplifying

$3y^3 = -4x^3 - 2$

The last equation is not equivalent to the original equation, so the graph is not symmetric with respect to the x-axis.

Test for symmetry with respect to the y-axis:

$3y^3 = 4x^3 + 2$ Original equation

$3y^3 = 4(-x)^3 + 2$ Replacing x by $-x$

$3y^3 = -4x^3 + 2$ Simplifying

The last equation is not equivalent to the original equation, so the graph is not symmetric with respect to the y-axis.

Test for symmetry with respect to the origin:

$3y^3 = 4x^3 + 2$ Original equation

$3(-y)^3 = 4(-x)^3 + 2$ Replacing x by $-x$ and y by $-y$

$-3y^3 = -4x^3 + 2$ Simplifying

$3y^3 = 4x^3 - 2$

The last equation is not equivalent to the original equation, so the graph is not symmetric with respect to the origin.

24. Test for symmetry with respect to the x-axis:

$3x = |y|$ Original equation

$3x = |-y|$ Replacing y by $-y$

$3x = |y|$ Simplifying

The last equation is equivalent to the original equation, so the graph is symmetric with respect to the x-axis.

Test for symmetry with respect to the y-axis:

$3x = |y|$ Original equation

$3(-x) = |y|$ Replacing x by $-x$

$-3x = |y|$ Simplifying

The last equation is not equivalent to the original equation, so the graph is not symmetric with respect to the y-axis.

Test for symmetry with respect to the origin:

$3x = |y|$ Original equation

$3(-x) = |-y|$ Replacing x by $-x$ and y by $-y$

$-3x = |y|$ Simplifying

The last equation is not equivalent to the original equation, so the graph is not symmetric with respect to the origin.

25. Test for symmetry with respect to the x-axis:

$xy = 12$ Original equation

$x(-y) = 12$ Replacing y by $-y$

$-xy = 12$ Simplifying

$xy = -12$

The last equation is not equivalent to the original equation, so the graph is not symmetric with respect to the x-axis.

Test for symmetry with respect to the y-axis:

$xy = 12$ Original equation

$-xy = 12$ Replacing x by $-x$

$xy = -12$ Simplifying

The last equation is not equivalent to the original equation, so the graph is not symmetric with respect to the y-axis.

Test for symmetry with respect to the origin:

$xy = 12$ Original equation

$-x(-y) = 12$ Replacing x by $-x$ and y by $-y$

$xy = 12$ Simplifying

The last equation is equivalent to the original equation, so the graph is symmetric with respect to the origin.

26. Test for symmetry with respect to the x-axis:

$xy - x^2 = 3$ Original equation

$x(-y) - x^2 = 3$ Replacing y by $-y$

$xy + x^2 = -3$ Simplifying

The last equation is not equivalent to the original equation, so the graph is not symmetric with respect to the x-axis.

Test for symmetry with respect to the y-axis:

$xy - x^2 = 3$ Original equation

$-xy - (-x)^2 = 3$ Replacing x by $-x$

$xy + x^2 = -3$ Simplifying

The last equation is not equivalent to the original equation, so the graph is not symmetric with respect to the y-axis.

Test for symmetry with respect to the origin:

$xy - x^2 = 3$ Original equation

$-x(-y) - (-x)^2 = 3$ Replacing x by $-x$ and y by $-y$

$xy - x^2 = 3$ Simplifying

The last equation is equivalent to the original equation, so the graph is symmetric with respect to the origin.

27. The graph is symmetric with respect to the y-axis, so the function is even.

28. The graph is symmetric with respect to the y-axis, so the function is even.

29. The graph is symmetric with respect to the origin, so the function is odd.

30. The graph is not symmetric with respect to either the y-axis or the origin, so the function is neither even nor odd.

31. The graph is not symmetric with respect to either the y-axis or the origin, so the function is neither even nor odd.

32. The graph is not symmetric with respect to either the y-axis or the origin, so the function is neither even nor odd.

33. $f(x) = -3x^3 + 2x$

$f(-x) = -3(-x)^3 + 2(-x) = 3x^3 - 2x$

$-f(x) = -(-3x^3 + 2x) = 3x^3 - 2x$

$f(-x) = -f(x)$, so f is odd.

34. $f(x) = 7x^3 + 4x - 2$

$f(-x) = 7(-x)^3 + 4(-x) - 2 = -7x^3 - 4x - 2$

$-f(x) = -(7x^3 + 4x - 2) = -7x^3 - 4x + 2$

$f(x) \neq f(-x)$, so f is not even.

$f(-x) \neq -f(x)$, so f is not odd.

Thus, $f(x) = 7x^3 + 4x - 2$ is neither even nor odd.

35. $f(x) = 5x^2 + 2x^4 - 1$

$f(-x) = 5(-x)^2 + 2(-x)^4 - 1 = 5x^2 + 2x^4 - 1$

$f(x) = f(-x)$, so f is even.

36. $f(x) = x + \dfrac{1}{x}$

$f(-x) = -x + \dfrac{1}{-x} = -x - \dfrac{1}{x}$

$-f(x) = -\left(x + \dfrac{1}{x}\right) = -x - \dfrac{1}{x}$

$f(-x) = -f(x)$, so f is odd.

37. $f(x) = x^{17}$

$f(-x) = (-x)^{17} = -x^{17}$

$-f(x) = -x^{17}$

$f(-x) = -f(x)$, so f is odd.

38. $f(x) = \sqrt[3]{x}$

$f(-x) = \sqrt[3]{-x} = -\sqrt[3]{x}$

$-f(x) = -\sqrt[3]{x}$

$f(-x) = -f(x)$, so f is odd.

39. $f(x) = \dfrac{1}{x^2}$

$f(-x) = \dfrac{1}{(-x)^2} = \dfrac{1}{x^2}$

$f(x) = f(-x)$, so f is even.

40. $f(x) = x - |x|$

$f(-x) = (-x) - |(-x)| = -x - |x|$

$-f(x) = -(x - |x|) = -x + |x|$

$f(x) \neq f(-x)$, so f is not even.

$f(-x) \neq -f(x)$, so f is not odd.

Thus, $f(x) = x - |x|$ is neither even nor odd.

41. $f(x) = 8$

$f(-x) = 8$

$f(x) = f(-x)$, so f is even.

42. $f(x) = \sqrt{x^2 + 1}$

$f(-x) = \sqrt{(-x)^2 + 1} = \sqrt{x^2 + 1}$

$f(x) = f(-x)$, so f is even.

43. Think of the graph of $g(x) = x^2$. Since $f(x) = g(x) + 1$, the graph of $f(x) = x^2 + 1$ is the graph of $g(x) = x^2$ shifted up 1 unit.

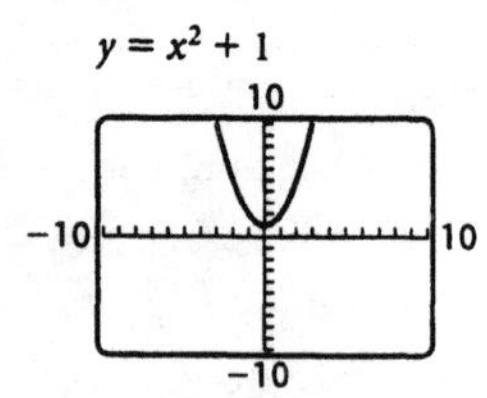

44. Think of the graph of $f(x) = |x|$. Since $g(x) = f(3x)$, the graph of $g(x) = |3x|$ is the graph of $f(x) = |x|$ shrunk horizontally by dividing each x-coordinate by 3 $\left(\text{or multiplying each } x\text{-coordinate by } \dfrac{1}{3}\right)$.

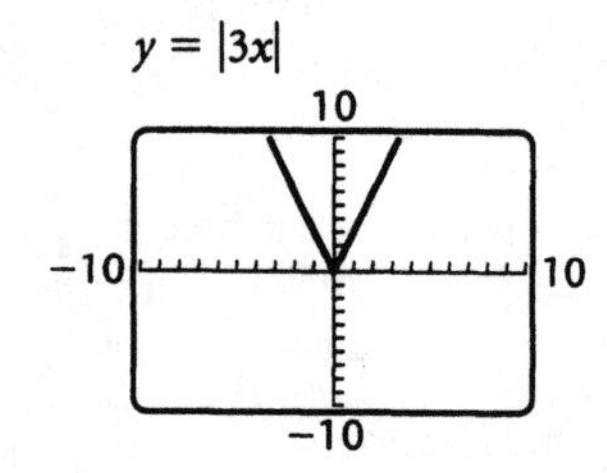

45. Think of the graph of $f(x) = x^3$. Since $g(x) = f(x + 5)$, the graph of $g(x) = (x + 5)^3$ is the graph of $f(x) = x^3$ shifted left 5 units.

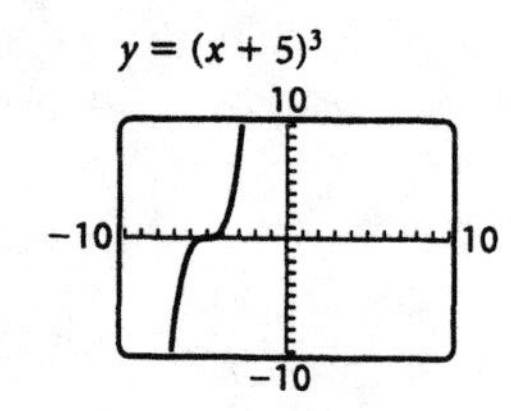

46. Think of the graph of $g(x) = \sqrt[3]{x}$. Since $f(x) = \frac{1}{2}g(x)$, the graph of $f(x) = \frac{1}{2}\sqrt[3]{x}$ is the graph of $g(x) = \sqrt[3]{x}$ shrunk vertically by multiplying each y-coordinate by $\frac{1}{2}$.

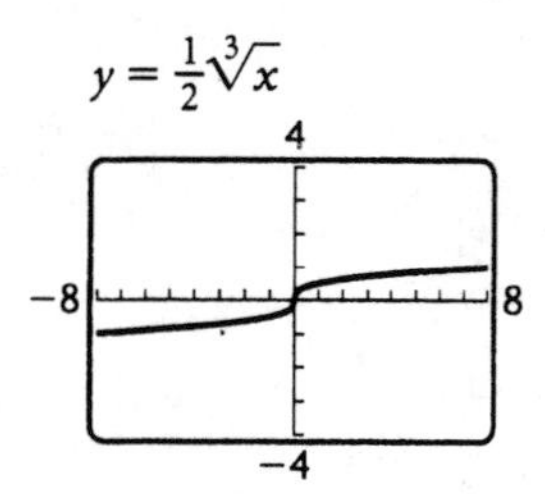

47. Think of the graph of $g(x) = x^2$. Since $f(x) = -g(x)$, the graph of $f(x) = -x^2$ is the graph of $g(x) = x^2$ reflected across the x-axis.

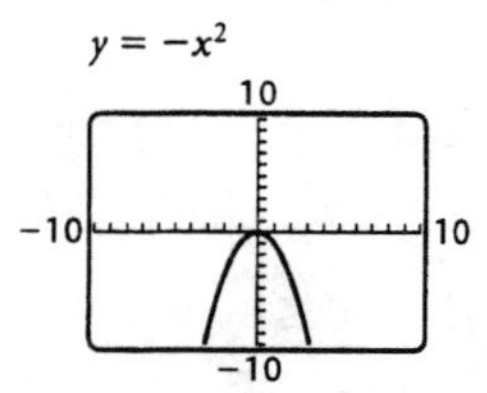

48. Think of the graph of $g(x) = |x|$. Since $f(x) = g(x-3) - 4$, the graph of $f(x) = |x-3| - 4$ is the graph of $g(x) = |x|$ shifted right 3 units and down 4 units.

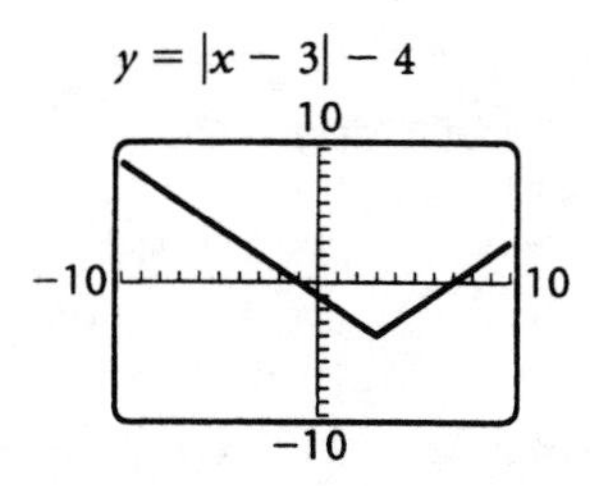

49. Think of the graph of $g(x) = \sqrt{x}$. Since $f(x) = 3g(x) - 5$, the graph of $f(x) = 3\sqrt{x} - 5$ is the graph of $g(x) = \sqrt{x}$ stretched vertically by multiplying each y-coordinate by 3 and then shifted down 5 units.

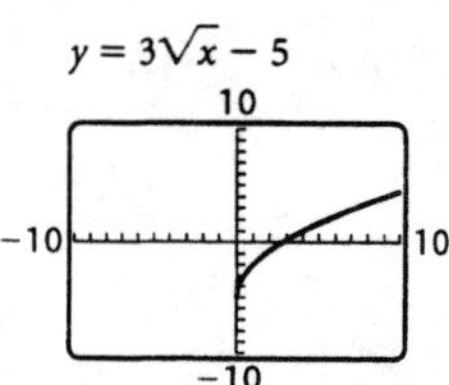

50. Think of the graph of $g(x) = \frac{1}{x}$. Since $f(x) = 5 - g(x)$, or $f(x) = -g(x) + 5$, the graph of $f(x) = 5 - \frac{1}{x}$ is the graph of $g(x) = \frac{1}{x}$ reflected across the x-axis and then shifted up 5 units.

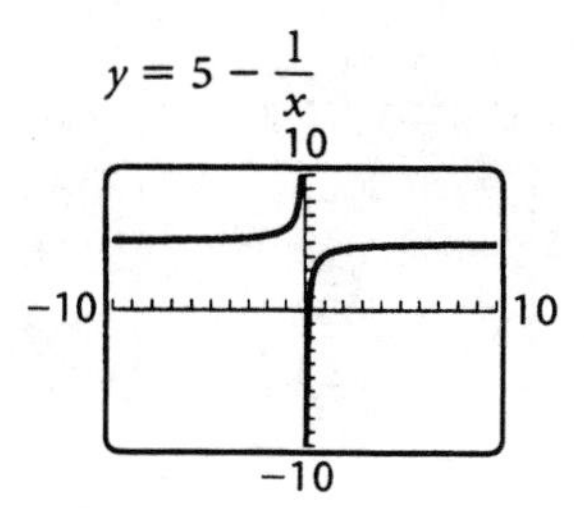

51. Think of the graph of $f(x) = |x|$. Since $g(x) = f\left(\frac{1}{3}x\right) - 4$, the graph of $g(x) = \left|\frac{1}{3}x\right| - 4$ is the graph of $f(x) = |x|$ stretched horizontally by multiplying each x-coordinate by 3 and then shifted down 4 units.

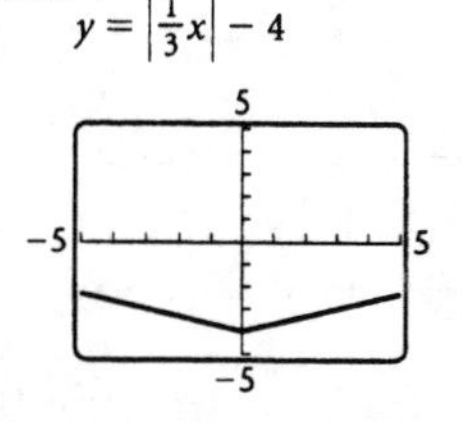

52. Think of the graph of $g(x) = x^3$. Since $f(x) = \frac{2}{3}g(x) - 4$, the graph of $f(x) = \frac{2}{3}x^3 - 4$ is the graph of $g(x) = x^3$ shrunk vertically by multiplying each y-coordinate by $\frac{2}{3}$ and then shifted down 4 units.

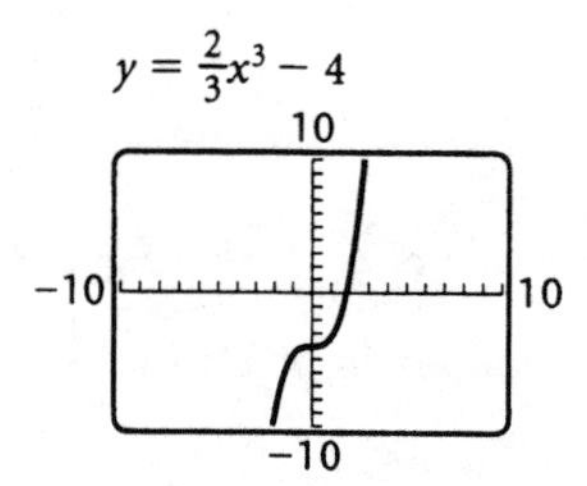

53. Think of the graph of $g(x) = x^2$. Since $f(x) = g(x+5) - 4$, the graph of $f(x) = (x+5)^2 - 4$ is the graph of $g(x) = x^2$ shifted left 5 units and then down 4 units.

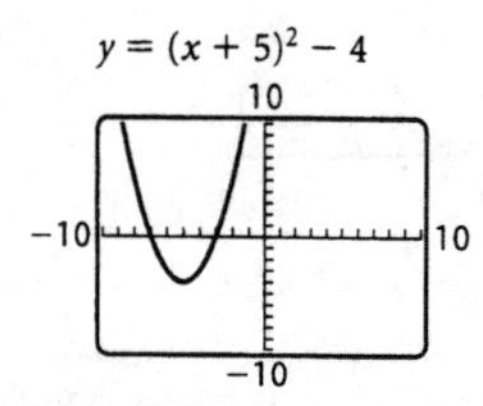

54. Think of the graph of $g(x) = x^3$. Since $f(x) = g(-x) - 5$, the graph of $f(x) = (-x)^3 - 5$ is the graph of $g(x) = x^3$ reflected across the y-axis and shifted down 5 units.

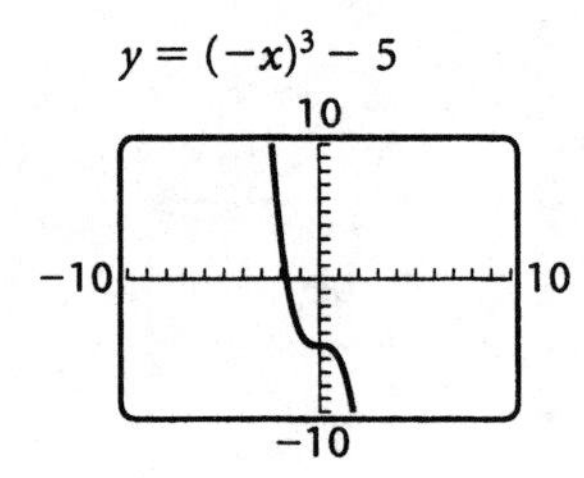

55. Think of the graph of $g(x) = x^2$. Since $f(x) = -\frac{1}{4}g(x-5)$, the graph of $f(x) = -\frac{1}{4}(x-5)^2$ is the graph of $g(x) = x^2$ shifted right 5 units, shrunk vertically by multiplying each y-coordinate by $\frac{1}{4}$, and reflected across the x-axis.

$y = -\frac{1}{4}(x-5)^2$

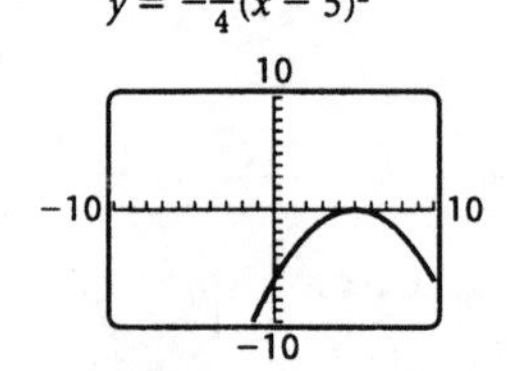

56. Think of the graph of $f(x) = \sqrt{x}$. Since $g(x) = f(-x)+5$, the graph of $g(x) = \sqrt{-x}+5$ is the graph of $f(x) = \sqrt{x}$ reflected across the y-axis and shifted up 5 units.

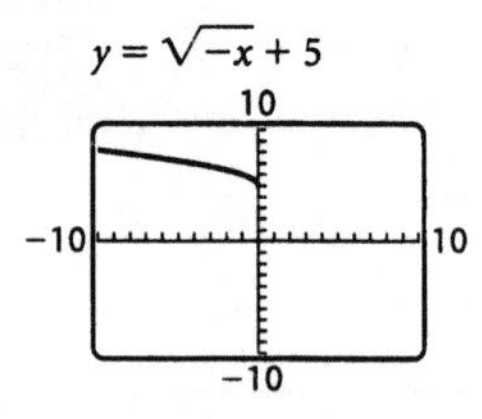

57. Think of the graph of $g(x) = \frac{1}{x}$. Since $f(x) = g(x+3)+2$, the graph of $f(x) = \frac{1}{x+3}+2$ is the graph of $g(x) = \frac{1}{x}$ shifted left 3 units and up 2 units.

$y = \frac{1}{x+3} + 2$

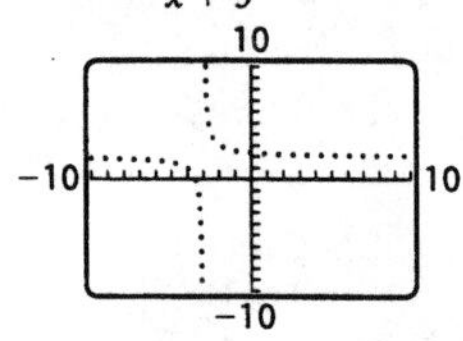

58. Think of the graph of $g(x) = x^2$. Since $f(x) = 3g(x+4) - 3$, the graph of $f(x) = 3(x+4)^2 - 3$ is the graph of $g(x) = x^2$ shifted left 4 units, stretched vertically by multiplying each y-coordinate by 3, and then shifted down 3 units.

$y = 3(x+4)^2 - 3$

10

−10 10

−10

59. Shape: $h(x) = x^2$

Turn $h(x)$ upside-down (that is, reflect it across the x-axis): $g(x) = -h(x) = -x^2$

Shift $g(x)$ right 8 units: $f(x) = g(x-8) = -(x-8)^2$

60. Shape: $h(x) = \sqrt{x}$

Shift $h(x)$ left 6 units: $g(x) = h(x+6) = \sqrt{x+6}$

Shift $g(x)$ down 5 units: $f(x) = g(x)-5 = \sqrt{x+6}-5$

61. Shape: $h(x) = |x|$

Shift $h(x)$ left 7 units: $g(x) = h(x+7) = |x+7|$

Shift $g(x)$ up 2 units: $f(x) = g(x)+2 = |x+7|+2$

62. Shape: $h(x) = x^3$

Turn $h(x)$ upside-down (that is, reflect it across the x-axis): $g(x) = -h(x) = -x^3$

Shift $g(x)$ right 5 units: $f(x) = g(x-5) = -(x-5)^3$

63. Shape: $h(x) = \frac{1}{x}$

Shrink $h(x)$ vertically by a factor of $\frac{1}{2}$ $\left(\text{that is, multiply each function value by } \frac{1}{2}\right)$:

$g(x) = \frac{1}{2}h(x) = \frac{1}{2}\cdot\frac{1}{x}$, or $\frac{1}{2x}$

Shift $g(x)$ down 3 units: $f(x) = g(x) - 3 = \frac{1}{2x} - 3$

64. Shape: $h(x) = x^2$

Shift $h(x)$ right 6 units: $g(x) = h(x-6) = (x-6)^2$

Shift $g(x)$ up 2 units: $f(x) = g(x)+2 = (x-6)^2+2$

65. Shape: $m(x) = x^2$

Turn $m(x)$ upside-down (that is, reflect it across the x-axis): $h(x) = -m(x) = -x^2$

Shift $h(x)$ right 3 units: $g(x) = h(x-3) = -(x-3)^2$

Shift $g(x)$ up 4 units: $f(x) = g(x)+4 = -(x-3)^2+4$

66. Shape: $h(x) = |x|$

Stretch $h(x)$ horizontally by a factor of 2 $\left(\text{that is, multiply each } x\text{-value by } \frac{1}{2}\right)$: $g(x) = h\left(\frac{1}{2}x\right) = \left|\frac{1}{2}x\right|$

Shift $g(x)$ down 5 units: $f(x) = g(x) - 5 = \left|\frac{1}{2}x\right| - 5$

67. Shape: $m(x) = \sqrt{x}$

Reflect $m(x)$ across the y-axis: $h(x) = m(-x) = \sqrt{-x}$

Shift $h(x)$ left 2 units: $g(x) = h(x+2) = \sqrt{-(x+2)}$

Shift $g(x)$ down 1 unit: $f(x) = g(x) - 1 = \sqrt{-(x+2)} - 1$

68. Shape: $h(x) = \frac{1}{x}$

Reflect $h(x)$ across the x-axis: $g(x) = -h(x) = -\frac{1}{x}$

Shift $g(x)$ up 1 unit: $f(x) = g(x) + 1 = -\frac{1}{x} + 1$

69. Shape: $h(x) = x^3$

Shift $h(x)$ left 4 units: $g(x) = h(x+4) = (x+4)^3$

Shrink $g(x)$ vertically by a factor of 0.83 (that is, multiply each function value by 0.83): $f(x) = 0.83g(x) = 0.83(x+4)^3$

70. Each y-coordinate is multiplied by $\frac{1}{2}$. We plot and connect

$(-4, 0)$, $(-3, -1)$, $(-1, -1)$, $(2, 1.5)$, and $(5, 0)$.

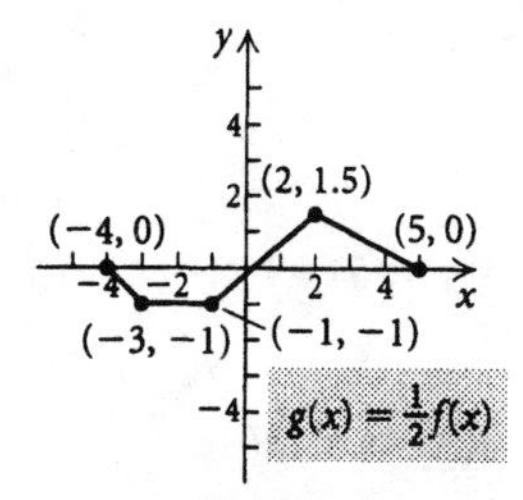

71. Each y-coordinate is multiplied by -2. We plot and connect $(-4, 0)$, $(-3, 4)$, $(-1, 4)$, $(2, -6)$, and $(5, 0)$.

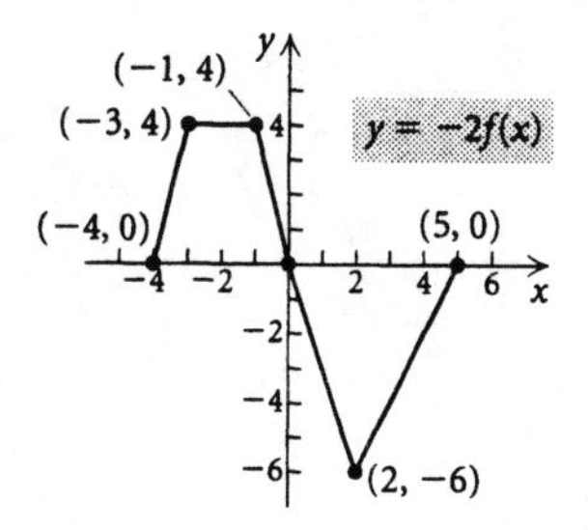

72. The graph is shrunk horizontally by a factor of 2. That is, each x-coordinate is divided by 2 $\left(\text{or multiplied by } \frac{1}{2}\right)$. We plot and connect $(-2, 0)$, $(-1.5, -2)$, $(-0.5, -2)$, $(1, 3)$, and $(2.5, 0)$.

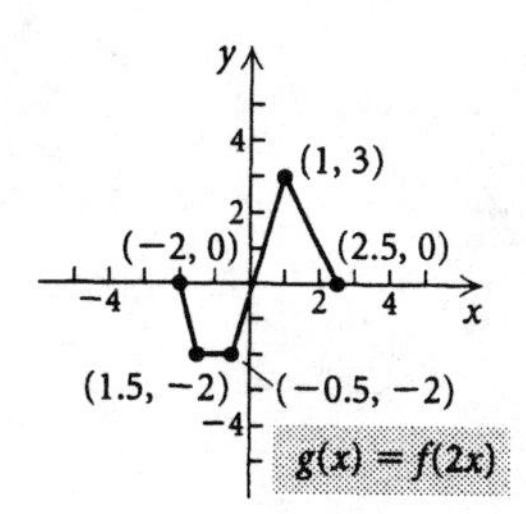

73. The graph is reflected across the y-axis and stretched horizontally by a factor of 2. That is, each x-coordinate is multiplied by -2 $\left(\text{or divided by } -\frac{1}{2}\right)$. We plot and connect $(8, 0)$, $(6, -2)$, $(2, -2)$, $(-4, 3)$, and $(-10, 0)$.

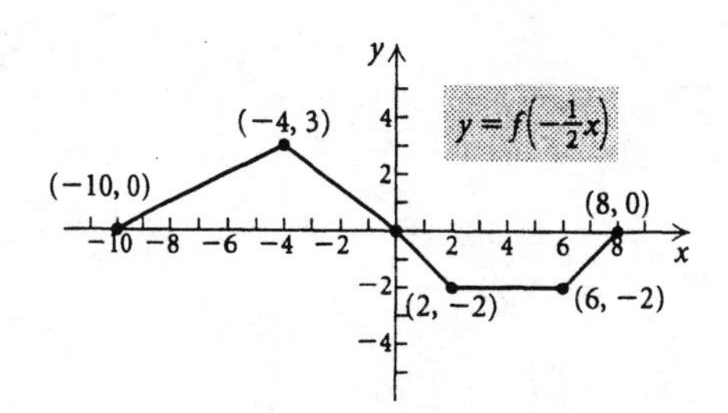

74. The graph is shifted left 1 unit so each x-coordinate is decreased by 1. The graph is also reflected across the x-axis, stretched vertically by a factor of 3, and shifted down 4 units. Thus, each y-coordinate is multiplied by -3 and then decreased by 4. We plot and connect $(-5, -4)$, $(-4, 2)$, $(-2, 2)$, $(1, -13)$, and $(4, -4)$.

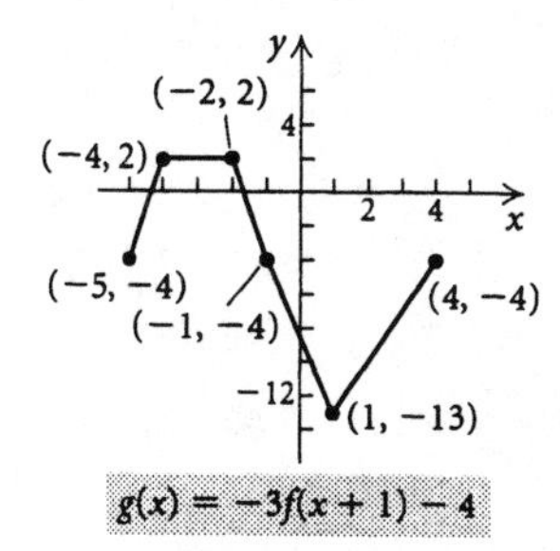

75. The graph is shifted right 1 unit so each x-coordinate is increased by 1. The graph is also reflected across the x-axis, shrunk vertically by a factor of 2, and shifted up 3 units. Thus, each y-coordinate is multiplied by $-\frac{1}{2}$ and then increased by 3. We plot and connect $(-3, 3)$, $(-2, 4)$, $(0, 4)$, $(3, 1.5)$, and $(6, 3)$.

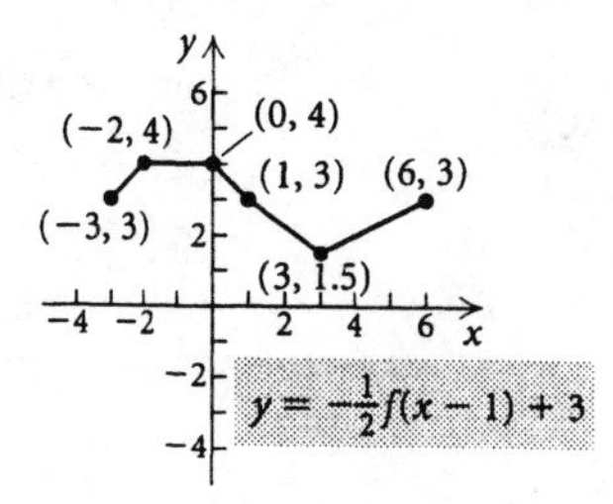

76. $g(x) = f(x) + 3$

The graph of $g(x)$ is the graph of $f(x)$ shifted up 3 units. This is graph h.

77. $g(x) = f(-x) + 3$

The graph of $g(x)$ is the graph of $f(x)$ reflected across the y-axis and shifted up 3 units. This is graph f.

78. $g(x) = -f(-x)$

The graph of $g(x)$ is the graph of $f(x)$ reflected across the x-axis and the y-axis. This is graph a.

79. $g(x) = -f(x) + 3$

The graph of $g(x)$ is the graph of $f(x)$ reflected across the x-axis and shifted up 3 units. This is graph f.

80. $g(x) = \frac{1}{3}f(x) - 3$

The graph of $g(x)$ is the graph of $f(x)$ shrunk vertically by a factor of 3 $\left(\text{that is, each } y\text{-coordinate is multiplied by } \frac{1}{3}\right)$ and then shifted down 3 units. This is graph e.

81. $g(x) = \frac{1}{3}f(x - 2)$

The graph of $g(x)$ is the graph of $f(x)$ shrunk vertically by a factor of 3 $\left(\text{that is, each } y\text{-coordinate is multiplied by } \frac{1}{3}\right)$ and then shifted right 2 units. This is graph d.

82. $g(x) = -f(x + 2)$

The graph of $g(x)$ is the graph $f(x)$ reflected across the x-axis and shifted left 2 units. This is graph b.

83. $g(x) = \frac{1}{3}f(x + 2)$

The graph of $g(x)$ is the graph of $f(x)$ shrunk vertically by a factor of 3 $\left(\text{that is, each } y\text{-coordinate is multiplied by } \frac{1}{3}\right)$ and then shifted left 2 units. This is graph c.

84. $f(-x) = \frac{1}{4}(-x)^4 + \frac{1}{5}(-x)^3 - 81(-x)^2 - 17 = \frac{1}{4}x^4 - \frac{1}{5}x^3 - 81x^2 - 17 \neq g(x)$

85. $f(-x) = 2(-x)^4 - 35(-x)^3 + 3(-x) - 5 = 2x^4 + 35x^3 - 3x - 5 = g(x)$

86. Each y-coordinate of the graph of $f(x) = x^3 - 3x^2$ is multiplied by $\frac{1}{2}$. A formula for the transformed function is $h(x) = \frac{1}{2}f(x)$, or $h(x) = \frac{1}{2}(x^3 - 3x^2)$.

87. The graph of $f(x) = x^3 - 3x^2$ is shifted up 2 units. A formula for the transformed function is $g(x) = f(x) + 2$, or $g(x) = x^3 - 3x^2 + 2$.

88. The graph of $f(x) = x^3 - 3x^2$ is shifted right 2 units and up 1 unit. A formula for the transformed function is $m(x) = f(x - 2) + 1$, or $m(x) = (x - 2)^3 - 3(x - 2)^2 + 1$.

89. The graph of $f(x) = x^3 - 3x^2$ is shifted left 1 unit. A formula for the transformed function is $k(x) = f(x + 1)$, or $k(x) = (x + 1)^3 - 3(x + 1)^2$.

90. If all of the exponents are even numbers, then $f(x)$ is an even function. If $a_0 = 0$ and all of the exponents are odd numbers, then $f(x)$ is an odd function.

91. The graph of $f(x) = 0$ is symmetric with respect to the x-axis, the y-axis, and the origin. This function is both even and odd. In general, a graph that is symmetric with respect to the x-axis fails the vertical line test, so a function cannot be symmetric with respect to the x-axis. (The exception is $f(x) = 0$.)

92. The graph of $f(x) = |x^2 - 9|$ looks like the graph of $g(x) = x^2 - 9$ with the points with negative y-coordinates reflected across the x-axis.

93. For every point (x, y) on the graph of $y = f(x)$, its reflection across the y-axis $(-x, y)$ is on the graph of $y = f(-x)$.

94. $f(x) = 4x^3 - 5x$

a) $f(2) = 4 \cdot 2^3 - 5 \cdot 2 = 4 \cdot 8 - 5 \cdot 2 = 32 - 10 = 22$

b) $f(-2) = 4(-2)^3 - 5(-2) = 4(-8) - 5(-2) = -32 + 10 = -22$

c) $f(a) = 4a^3 - 5a$

d) $f(-a) = 4(-a)^3 - 5(-a) = 4(-a^3) - 5(-a) = -4a^3 + 5a$

95. $f(x) = 5x^2 - 7$

a) $f(-3) = 5(-3)^2 - 7 = 5 \cdot 9 - 7 = 45 - 7 = 38$

b) $f(3) = 5 \cdot 3^2 - 7 = 5 \cdot 9 - 7 = 45 - 7 = 38$

c) $f(a) = 5a^2 - 7$

d) $f(-a) = 5(-a)^2 - 7 = 5a^2 - 7$

96.

$$2x - 9y + 1 = 0$$
$$2x + 1 = 9y$$
$$\frac{2}{9}x + \frac{1}{9} = y$$

Slope: $\frac{2}{9}$; y-intercept: $\left(0, \frac{1}{9}\right)$

97. First find the slope of the given line.

$$8x - y = 10$$
$$8x = y + 10$$
$$8x - 10 = y$$

The slope of the given line is 8. The slope of a line perpendicular to this line is the opposite of the reciprocal of 8, or $-\frac{1}{8}$.

$$y - y_1 = m(x - x_1)$$
$$y - 1 = -\frac{1}{8}[x - (-1)]$$
$$y - 1 = -\frac{1}{8}(x + 1)$$
$$y - 1 = -\frac{1}{8}x - \frac{1}{8}$$
$$y = -\frac{1}{8}x + \frac{7}{8}$$

98.

$$f(x) = \frac{x^2 + 1}{x^3 + 1}$$
$$f(-x) = \frac{(-x)^2 + 1}{(-x)^3 + 1} = \frac{x^2 + 1}{-x^3 + 1}$$
$$-f(x) = -\frac{x^2 + 1}{x^3 + 1}$$

Since $f(x) \neq f(-x)$, f is not even.

Since $f(-x) \neq -f(x)$, f is not odd.

Thus, $f(x) = \dfrac{x^2+1}{x^3+1}$ is neither even nor odd.

99. $f(x) = x\sqrt{10-x^2}$

$f(-x) = -x\sqrt{10-(-x)^2} = -x\sqrt{10-x^2}$

$-f(x) = -x\sqrt{10-x^2}$

Since $f(-x) = -f(x)$, f is odd.

100. If the graph were folded on the x-axis, the parts above and below the x-axis would coincide, so the graph is symmetric with respect to the x-axis.

If the graph were folded on the y-axis, the parts to the left and right of the y-axis would not coincide, so the graph is not symmetric with respect to the y-axis.

If the graph were rotated 180°, the resulting graph would not coincide with the original graph, so it is not symmetric with respect to the origin.

101. If the graph were folded on the x-axis, the parts above and below the x-axis would coincide, so the graph is symmetric with respect to the x-axis.

If the graph were folded on the y-axis, the parts to the left and right of the y-axis would coincide, so the graph is symmetric with respect to the y-axis.

If the graph were rotated 180°, the resulting graph would coincide with the original graph, so it is symmetric with respect to the origin.

102. If the graph were folded on the x-axis, the parts above and below the x-axis would not coincide, so the graph is not symmetric with respect to the x-axis.

If the graph were folded on the y-axis, the parts to the left and right of the y-axis would not coincide, so the graph is not symmetric with respect to the y-axis.

If the graph were rotated 180°, the resulting graph would coincide with the original graph, so it is symmetric with respect to the origin.

103. If the graph were folded on the x-axis, the parts above and below the x-axis would coincide, so the graph is symmetric with respect to the x-axis.

If the graph were folded on the y-axis, the parts to the left and right of the y-axis would not coincide, so the graph is not symmetric with respect to the y-axis.

If the graph were rotated 180°, the resulting graph would not coincide with the original graph, so it is not symmetric with respect to the origin.

104. Call the transformed function $g(x)$.

Then $g(5) = 4 - f(-3) = 4 - f(5-8)$,

$g(8) = 4 - f(0) = 4 - f(8-8)$,

and $g(11) = 4 - f(3) = 4 - f(11-8)$.

Thus $g(x) = 4 - f(x-8)$, or $g(x) = 4 - |x-8|$.

105. Think of the graph of $g(x) = \text{int}(x)$. Since $f(x) = g\left(x - \dfrac{1}{2}\right)$, the graph of $f(x) = \text{int}\left(x - \dfrac{1}{2}\right)$ is the graph of $g(x) = \text{int}(x)$ shifted right $\dfrac{1}{2}$ unit. The domain is the set of all real numbers; the range is the set of all integers.

106. This function can be defined piecewise as follows:

$$f(x) = \begin{cases} -(\sqrt{x}-1), & \text{for } 0 \le x < 1, \\ \sqrt{x}-1, & \text{for } x \ge 1, \end{cases}$$

Think of the graph of $g(x) = \sqrt{x}$. First shift it down 1 unit. Then reflect across the x-axis the portion of the graph for which $0 < x < 1$. The domain and range are both the set of nonnegative real numbers, or $[0, \infty)$.

107. On the graph of $y = 2f(x)$ each y-coordinate of $y = f(x)$ is multiplied by 2, so $(3, 4\cdot 2)$, or $(3, 8)$ is on the transformed graph.

On the graph of $y = 2 + f(x)$, each y-coordinate of $y = f(x)$ is increased by 2 (shifted up 2 units), so $(3, 4+2)$, or $(3, 6)$ is on the transformed graph.

On the graph of $y = f(2x)$, each x-coordinate of $y = f(x)$ is multiplied by $\dfrac{1}{2}$ (or divided by 2), so $\left(\dfrac{1}{2}\cdot 3, 4\right)$, or $\left(\dfrac{3}{2}, 4\right)$ is on the transformed graph.

108. Using a grapher we find that the zeros are -2.582, 0, and 2.582.

The graph of $y = f(x-3)$ is the graph of $y = f(x)$ shifted right 3 units. Thus we shift each of the zeros of $f(x)$ 3 units right to find the zeros of $f(x-3)$. They are $-2.582+3$, or 0.418; $0+3$, or 3; and $2.582+3$, or 5.582.

The graph of $y = f(x+8)$ is the graph of $y = f(x)$ shifted 8 units left. Thus we shift each of the zeros of $f(x)$ 8 units left to find the zeros of $f(x+8)$. They are $-2.582-8$, or -10.582; $0-8$, or -8; and $2.582-8$, or -5.418.

109. $f(2-3) = f(-1) = 5$, so $b = 5$.

(The graph of $y = f(x-3)$ is the graph of $y = f(x)$ shifted right 3 units, so the point $(-1, 5)$ on $y = f(x)$ is transformed to the point $(-1+3, 5)$, or $(2, 5)$ on $y = f(x-3)$.)

110. Let $f(x) = g(x) = x$. Now f and g are odd functions, but $(fg)(x) = x^2 = (fg)(-x)$. Thus, the product is even, so the statement is false.

111. Let $f(x)$ and $g(x)$ be even functions. Then by definition, $f(x) = f(-x)$ and $g(x) = g(-x)$. Thus, $(f+g)(x) = f(x) + g(x) = f(-x) + g(-x) = (f+g)(-x)$ and $f+g$ is even. The statement is true.

112. Let $f(x)$ be an even function, and let $g(x)$ be an odd function. By definition $f(x) = f(-x)$ and $g(-x) = -g(x)$, or $g(x) = -g(-x)$. Then $fg(x) = f(x) \cdot g(x) = f(-x) \cdot [-g(-x)] = -f(-x) \cdot g(-x) = -fg(-x)$, and fg is odd. The statement is true.

113. See the answer section in the text.

114. $O(-x) = \dfrac{f(-x) - f(-(-x))}{2} = \dfrac{f(-x) - f(x)}{2}$,

$-O(x) = -\dfrac{f(x) - f(-x)}{2} = \dfrac{f(-x) - f(x)}{2}$. Thus, $O(-x) = -O(x)$ and O is odd.

115. a), b) See the answer section in the text.

Exercise Set 1.6

1.
$$y = kx$$
$$54 = k \cdot 12$$
$$\frac{54}{12} = k, \text{ or } k = \frac{9}{2}$$

The variation constant is $\frac{9}{2}$, or 4.5. The equation of variation is $y = \frac{9}{2}x$, or $y = 4.5x$.

2.
$$y = kx$$
$$0.1 = k(0.2)$$
$$\frac{1}{2} = k \quad \text{Variation constant}$$

Equation of variation: $y = \frac{1}{2}x$, or $y = 0.5x$.

3.
$$y = \frac{k}{x}$$
$$3 = \frac{k}{12}$$
$$36 = k$$

The variation constant is 36. The equation of variation is $y = \frac{36}{x}$.

4.
$$y = \frac{k}{x}$$
$$12 = \frac{k}{5}$$
$$60 = k \quad \text{Variation constant}$$

Equation of variation: $y = \frac{60}{x}$

5.
$$y = kx$$
$$1 = k \cdot \frac{1}{4}$$
$$4 = k$$

The variation constant is 4. The equation of variation is $y = 4x$.

6.
$$y = \frac{k}{x}$$
$$0.1 = \frac{k}{0.5}$$
$$0.05 = k \quad \text{Variation constant}$$

Equation of variation: $y = \frac{0.05}{x}$

7.
$$y = \frac{k}{x}$$
$$32 = \frac{k}{\frac{1}{8}}$$
$$\frac{1}{8} \cdot 32 = k$$
$$4 = k$$

The variation constant is 4. The equation of variation is $y = \frac{4}{x}$.

8.
$$y = kx$$
$$3 = k \cdot 33$$
$$\frac{1}{11} = k \quad \text{Variation constant}$$

Equation of variation: $y = \frac{1}{11}x$

9.
$$y = kx$$
$$\frac{3}{4} = k \cdot 2$$
$$\frac{1}{2} \cdot \frac{3}{4} = k$$
$$\frac{3}{8} = k$$

The variation constant is $\frac{3}{8}$. The equation of variation is $y = \frac{3}{8}x$.

10.
$$y = \frac{k}{x}$$
$$\frac{1}{5} = \frac{k}{35}$$
$$7 = k \quad \text{Variation constant}$$

Equation of variation: $y = \frac{7}{x}$

11.
$$y = \frac{k}{x}$$
$$1.8 = \frac{k}{0.3}$$
$$0.54 = k$$

The variation constant is 0.54. The equation of variation is $y = \frac{0.54}{x}$.

12.
$$y = kx$$
$$0.9 = k(0.4)$$
$$\frac{9}{4} = k \quad \text{Variation constant}$$

Equation of variation: $y = \frac{9}{4}x$, or $y = 2.25x$

13. $T = \dfrac{k}{P}$ T varies inversely as P.

$5 = \dfrac{k}{7}$ Substituting

$35 = k$ Variation constant

$T = \dfrac{35}{P}$ Equation of variation

$T = \dfrac{35}{10}$ Substituting

$T = 3.5$

It will take 10 bricklayers 3.5 hr to complete the job.

14. $A = kG$

$9.66 = k \cdot 9$

$\dfrac{161}{150} = k$

$A = \dfrac{161}{150}G$

$A = \dfrac{161}{150} \cdot 4$

$A \approx \$4.29$

15. Let F = the number of grams of fat and w = the weight.

$F = kw$ F varies directly as w.

$60 = k \cdot 120$ Substituting

$\dfrac{60}{120} = k$, or Solving for k

$\dfrac{1}{2} = k$ Variation constant

$F = \dfrac{1}{2}w$ Equation of variation

$F = \dfrac{1}{2} \cdot 180$ Substituting

$F = 90$

The maximum daily fat intake for a person weighing 180 lb is 90 g.

16. $t = \dfrac{k}{r}$

$5 = \dfrac{k}{80}$

$400 = r$

$t = \dfrac{400}{r}$

$t = \dfrac{400}{70}$

$t = \dfrac{40}{7}$, or $5\dfrac{5}{7}$ hr

17. $W = \dfrac{k}{L}$ W varies inversely as L.

$1200 = \dfrac{k}{8}$ Substituting

$9600 = k$ Variation constant

$W = \dfrac{9600}{L}$ Equation of variation

$W = \dfrac{9600}{14}$ Substituting

$W = 685\dfrac{5}{7}$

A 14-m beam can support $685\dfrac{5}{7}$ kg, or about 686 kg.

18. $N = kP$

$31 = k \cdot 18,137,226$ Substituting

$\dfrac{31}{18,137,226} = k$ Variation constant

$N = \dfrac{31}{18,137,226}P$

$N = \dfrac{31}{18,137,226} \cdot 3,892,644$ Substituting

$N \approx 6$ (Rounding down)

Colorado has 6 representatives.

19. $M = kE$ M varies directly as E.

$38 = k \cdot 95$ Substituting

$\dfrac{2}{5} = k$ Variation constant

$M = \dfrac{2}{5}E$ Equation of variation

$M = \dfrac{2}{5} \cdot 100$ Substituting

$M = 40$

A 100-lb person would weigh 40 lb on Mars.

20. $t = \dfrac{k}{r}$

$45 = \dfrac{k}{600}$

$27,000 = k$

$t = \dfrac{27,000}{r}$

$t = \dfrac{27,000}{1000}$

$t = 27$ min

21. $d = km$ d varies directly as m.

$40 = k \cdot 3$ Substituting

$\dfrac{40}{3} = k$ Variation constant

$d = \dfrac{40}{3}m$ Equation of variation

$d = \dfrac{40}{3} \cdot 5 = \dfrac{200}{3}$ Substituting

$d = 66\dfrac{2}{3}$

A 5-kg mass will stretch the spring $66\dfrac{2}{3}$ cm.

22.
$$\begin{aligned} f &= kF \\ 6.3 &= k \cdot 150 \\ 0.042 &= k \\ f &= 0.042F \\ f &= 0.042(80) \\ f &= 3.36 \end{aligned}$$

23.
$$\begin{aligned} P &= \frac{k}{W} && P \text{ varies inversely as } W. \\ 330 &= \frac{k}{3.2} && \text{Substituting} \\ 1056 &= k && \text{Variation constant} \\ P &= \frac{1056}{W} && \text{Equation of variation} \\ 550 &= \frac{1056}{W} && \text{Substituting} \\ 550W &= 1056 && \text{Multiplying by } W \\ W &= \frac{1056}{550} && \text{Dividing by 550} \\ W &= 1.92 && \text{Simplifying} \end{aligned}$$

A tone with a pitch of 550 vibrations per second has a wavelength of 1.92 ft.

24.
$$\begin{aligned} L &= kP \\ 385 &= k \cdot 12,500 \\ 0.0308 &= k \\ L &= 0.0308P \\ L &= 0.0308(250,000,000) \\ L &= 7,700,000 \text{ tons} \end{aligned}$$

25.
$$\begin{aligned} y &= \frac{k}{x^2} \\ 0.15 &= \frac{k}{(0.1)^2} && \text{Substituting} \\ 0.15 &= \frac{k}{0.01} \\ 0.15(0.01) &= k \\ 0.0015 &= k \end{aligned}$$

The equation of variation is $y = \frac{0.0015}{x^2}$.

26.
$$\begin{aligned} y &= \frac{k}{x^2} \\ 6 &= \frac{k}{3^2} \\ 54 &= k \\ y &= \frac{54}{x^2} \end{aligned}$$

27.
$$\begin{aligned} y &= kx^2 \\ 0.15 &= k(0.1)^2 && \text{Substituting} \\ 0.15 &= 0.01k \\ \frac{0.15}{0.01} &= k \\ 15 &= k \end{aligned}$$

The equation of variation is $y = 15x^2$.

28.
$$\begin{aligned} y &= kx^2 \\ 6 &= k \cdot 3^2 \\ \frac{2}{3} &= k \\ y &= \frac{2}{3}x^2 \end{aligned}$$

29.
$$\begin{aligned} y &= kxz \\ 56 &= k \cdot 7 \cdot 8 && \text{Substituting} \\ 56 &= 56k \\ 1 &= k \end{aligned}$$

The equation of variation is $y = xz$.

30.
$$\begin{aligned} y &= \frac{kx}{z} \\ 4 &= \frac{k \cdot 12}{15} \\ 5 &= k \\ y &= \frac{5x}{z} \end{aligned}$$

31.
$$\begin{aligned} y &= kxz^2 \\ 105 &= k \cdot 14 \cdot 5^2 && \text{Substituting} \\ 105 &= 350k \\ \frac{105}{350} &= k \\ \frac{3}{10} &= k \end{aligned}$$

The equation of variation is $y = \frac{3}{10}xz^2$.

32.
$$\begin{aligned} y &= k \cdot \frac{xz}{w} \\ \frac{3}{2} &= k \cdot \frac{2 \cdot 3}{4} \\ 1 &= k \\ y &= \frac{xz}{w} \end{aligned}$$

33.
$$\begin{aligned} y &= k\frac{xz}{wp} \\ \frac{3}{28} &= k\frac{3 \cdot 10}{7 \cdot 8} && \text{Substituting} \\ \frac{3}{28} &= k \cdot \frac{30}{56} \\ \frac{3}{28} \cdot \frac{56}{30} &= k \\ \frac{1}{5} &= k \end{aligned}$$

The equation of variation is $y = \dfrac{xz}{5wp}$.

34.
$$y = k \cdot \frac{xz}{w^2}$$
$$\frac{12}{5} = k \cdot \frac{16 \cdot 3}{5^2}$$
$$\frac{5}{4} = k$$
$$y = \frac{5xz}{4w^2}$$

35.
$$I = \frac{k}{d^2}$$
$$90 = \frac{k}{5^2} \quad \text{Substituting}$$
$$90 = \frac{k}{25}$$
$$2250 = k$$

The equation of variation is $I = \dfrac{2250}{d^2}$.

Substitute 40 for I and find d.
$$40 = \frac{2250}{d^2}$$
$$40d^2 = 2250$$
$$d^2 = 56.25$$
$$d = 7.5$$

The distance from 5 m to 7.5 m is $7.5 - 5$, or 2.5 m, so it is 2.5 m further to a point where the intensity is 40 W/m^2.

36.
$$D = kAv$$
$$222 = k \cdot 37.8 \cdot 40$$
$$\frac{37}{252} = k$$
$$D = \frac{37}{252}Av$$
$$430 = \frac{37}{252} \cdot 51v$$
$$v \approx 57.4 \text{ mph}$$

37.
$$d = kr^2$$
$$200 = k \cdot 60^2 \quad \text{Substituting}$$
$$200 = 3600k$$
$$\frac{200}{3600} = k$$
$$\frac{1}{18} = k$$

The equation of variation is $d = \dfrac{1}{18}r^2$.

Substitute 72 for d and find r.
$$72 = \frac{1}{18}r^2$$
$$1296 = r^2$$
$$36 = r$$

A car can travel 36 mph and still stop in 72 ft.

38.
$$W = \frac{k}{d^2}$$
$$220 = \frac{k}{(3978)^2}$$
$$3,481,386,480 = k$$
$$W = \frac{3,481,386,480}{d^2}$$
$$W = \frac{3,481,386,480}{(3978 + 200)^2}$$
$$W \approx 199 \text{ lb}$$

39. $E = \dfrac{kR}{I}$

We first find k.
$$3.18 = \frac{k \cdot 71}{201} \quad \text{Substituting}$$
$$3.18\left(\frac{201}{71}\right) = k \quad \text{Multiplying by } \frac{201}{71}$$
$$9 \approx k$$

The equation of variation is $E = \dfrac{9R}{I}$.

Substitute 3.18 for E and 300 for I and solve R.
$$3.18 = \frac{9R}{300}$$
$$3.18\left(\frac{300}{9}\right) = R \quad \text{Multiplying by } \frac{300}{9}$$
$$106 = R$$

Shawn Estes would have given up 106 earned runs if he had pitched 300 innings.

40.
$$V = \frac{kT}{P}$$
$$231 = \frac{k \cdot 42}{20}$$
$$110 = k$$
$$V = \frac{110T}{P}$$
$$V = \frac{110 \cdot 30}{15}$$
$$V = 220 \text{ cm}^3$$

41. Let $y(x) = kx^2$. Then $y(2x) = k(2x)^2 = k \cdot 4x^2 = 4 \cdot kx^2 = 4 \cdot y(x)$. Thus, doubling x causes y to be quadrupled.

42. Let $y = k_1x$ and $x = \dfrac{k_2}{z}$. Then $y = k_1 \cdot \dfrac{k_2}{z}$, or $y = \dfrac{k_1k_2}{z}$, so y varies inversely as z.

43.

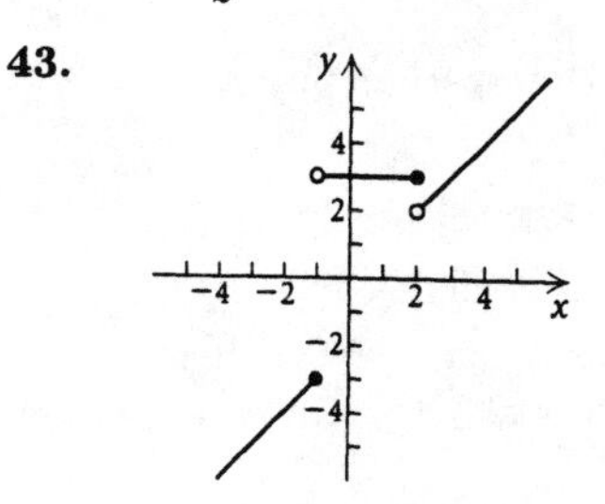

44. Test for symmetry with respect to the x-axis.

$y = 3x^4 - 3$ Original equation

$-y = 3x^4 - 3$ Replacing y by $-y$

$y = -3x^4 + 3$ Simplifying

The last equation is not equivalent to the original equation, so the graph is not symmetric with respect to the x-axis.

Test for symmetry with respect to the y-axis.

$y = 3x^4 - 3$ Original equation

$y = 3(-x)^4 - 3$ Replacing x by $-x$

$y = 3x^4 - 3$ Simplifying

The last equation is equivalent to the original equation, so the graph is symmetric with respect to the y-axis.

Test for symmetry with respect to the origin:

$y = 3x^4 - 3$

$-y = 3(-x)^4 - 3$ Replacing x by $-x$ and y by $-y$

$-y = 3x^4 - 3$

$y = -3x^4 + 3$ Simplifying

The last equation is not equivalent to the original equation, so the graph is not symmetric with respect to the origin.

45. Test for symmetry with respect to the x-axis.

$y^2 = x$ Original equation

$(-y)^2 = x$ Replacing y by $-y$

$y^2 = x$ Simplifying

The last equation is equivalent to t he original equation, so the graph is symmetric with respect to the x-axis.

Test for symmetry with respect to the y-axis:

$y^2 = x$ Original equation

$y^2 = -x$ Replacing x by $-x$

The last equation is not equivalent to the original equation, so the graph is not symmetric with respect to the y-axis.

Test for symmetry with respect to the origin:

$y^2 = x$ Original equation

$(-y)^2 = -x$ Replacing x by $-x$ and y by $-y$

$y^2 = -x$ Simplifying

The last equation is not equivalent to the original equation, so the graph is not symmetric with respect to the origin.

46. Test for symmetry with respect to the x-axis:

$2x - 5y = 0$ Original equation

$2x - 5(-y) = 0$ Replacing y by $-y$

$2x + 5y = 0$ Simplifying

The last equation is not equivalent to the original equation, so the graph is not symmetric with respect to the x-axis.

Test for symmetry with respect to the y-axis:

$2x - 5y = 0$ Original equation

$2(-x) - 5y = 0$ Replacing x by $-x$

$-2x - 5y = 0$ Simplifying

The last equation is not equivalent to the original equation, so the graph is not symmetric with respect to the y-axis.

Test for symmetry with respect to the origin:

$2x - 5y = 0$ Original equation

$2(-x) - 5(-y) = 0$ Replacing x by $-x$ and y by $-y$

$-2x + 5y = 0$

$2x - 5y = 0$ Simplifying

The last equation is equivalent to the original equation, so the graph is symmetric with respect to the origin.

47. Let V represent the volume and p represent the price of a jar of peanut butter.

$V = kp$ V varies directly as p.

$\pi\left(\frac{3}{2}\right)^2(4) = k(1.2)$ Substituting

$7.5\pi = k$ Variation constant

$V = 7.5\pi p$ Equation of variation

$\pi(3)^2(6) = 7.5\pi p$ Substituting

$7.2 = p$

The bigger jar should cost $7.20.

48. a) $7xy = 14$

$y = \frac{2}{x}$

Inversely

b) $x - 2y = 12$

$y = \frac{x}{2} - 6$

Neither

c) $-2x + y = 0$

$y = 2x$

Directly

d) $x = \frac{3}{4}y$

$y = \frac{4}{3}x$

Directly

e) $\frac{x}{y} = 2$

$y = \frac{1}{2}x$

Directly

49. We are told $A = kd^2$, and we know $A = \pi r^2$ so we have:

$kd^2 = \pi r^2$

$kd^2 = \pi\left(\frac{d}{2}\right)^2 \qquad r = \frac{d}{2}$

$kd^2 = \frac{\pi d^2}{4}$

$k = \frac{\pi}{4} \qquad$ Variation constant

50. $Q = \frac{kp^2}{q^3}$

Q varies directly as the square of p and inversely as the cube of q.

Exercise Set 1.7

1. Either point can be considered as (x_1, y_1).

$d = \sqrt{(4-5)^2 + (6-9)^2}$
$= \sqrt{(-1)^2 + (-3)^2} = \sqrt{10} \approx 3.162$

2. $d = \sqrt{(-3-2)^2 + (7-11)^2} = \sqrt{41} \approx 6.403$

3. Either point can be considered as (x_1, y_1).

$d = \sqrt{(6-9)^2 + (-1-5)^2}$
$= \sqrt{(-3)^2 + (-6)^2} = \sqrt{45} \approx 6.708$

4. $d = \sqrt{(-4-(-1))^2 + (-7-3)^2} = \sqrt{109} \approx 10.440$

5. Either point can be considered as (x_1, y_1).

$d = \sqrt{(-\sqrt{6}-\sqrt{3})^2 + (0-(-\sqrt{5}))^2}$
$= \sqrt{6 + 2\sqrt{18} + 3 + 5} = \sqrt{14 + 2\sqrt{9 \cdot 2}}$
$= \sqrt{14 + 2 \cdot 3\sqrt{2}} = \sqrt{14 + 6\sqrt{2}} \approx 4.742$

6. $d = \sqrt{(-\sqrt{2}-0)^2 + (1-\sqrt{7})^2} = \sqrt{2+1-2\sqrt{7}+7} = \sqrt{10-2\sqrt{7}} \approx 2.170$

7. First we find the length of the diameter:

$d = \sqrt{(-3-9)^2 + (-1-4)^2}$
$= \sqrt{(-12)^2 + (-5)^2} = \sqrt{169} = 13$

The length of the radius is one-half the length of the diameter, or $\frac{1}{2}(13)$, or 6.5.

8. Radius $= \sqrt{(-3-0)^2 + (5-1)^2} = \sqrt{25} = 5$

Diameter $= 2 \cdot 5 = 10$

9. First we find the distance between each pair of points.

For $(-4, 5)$ and $(6, 1)$:

$d = \sqrt{(-4-6)^2 + (5-1)^2}$
$= \sqrt{(-10)^2 + 4^2} = \sqrt{116}$

For $(-4, 5)$ and $(-8, -5)$:

$d = \sqrt{(-4-(-8))^2 + (5-(-5))^2}$
$= \sqrt{4^2 + 10^2} = \sqrt{116}$

For $(6, 1)$ and $(-8, -5)$:

$d = \sqrt{(6-(-8))^2 + (1-(-5))^2}$
$= \sqrt{14^2 + 6^2} = \sqrt{232}$

Since $(\sqrt{116})^2 + (\sqrt{116})^2 = (\sqrt{232})^2$, the points could be the vertices of a right triangle.

10. For $(-3, 1)$ and $(2, -1)$:

$d = \sqrt{(-3-2)^2 + (1-(-1))^2} = \sqrt{29}$

For $(-3, 1)$ and $(6, 9)$:

$d = \sqrt{(-3-6)^2 + (1-9)^2} = \sqrt{145}$

For $(2, -1)$ and $(6, 9)$:

$d = \sqrt{(2-6)^2 + (-1-9)^2} = \sqrt{116}$

Since $(\sqrt{29})^2 + (\sqrt{116})^2 = (\sqrt{145})^2$, the points could be the vertices of a right triangle.

11. First we find the distance between each pair of points.

For $(-3, 4)$ and $(0, 5)$:

$d = \sqrt{(-3-0)^2 + (4-5)^2}$
$= \sqrt{(-3)^2 + (-1)^2} = \sqrt{10}$

For $(-3, 4)$ and $(3, -4)$:

$d = \sqrt{(-3-3)^2 + (4-(-4))^2}$
$= \sqrt{(-6)^2 + 8^2} = \sqrt{100} = 10$

For $(0, 5)$ and $(3, -4)$:

$d = \sqrt{(0-3)^2 + (5-(-4))^2}$
$= \sqrt{(-3)^2 + 9^2} = \sqrt{90}$

Since $(\sqrt{10})^2 + (\sqrt{90})^2 = 10^2$, the points are vertices of a right triangle.

12. See the graph of this rectangle in Exercise 19.

The segments with endpoints $(-3, 4)$, $(2, -1)$ and $(5, 2)$, $(0, 7)$ are one pair of opposite sides. We find the length of each of these sides.

For $(-3, 4)$, $(2, -1)$:

$d = \sqrt{(-3-2)^2 + (4-(-1))^2} = \sqrt{50}$

For $(5, 2)$, $(0, 7)$:

$d = \sqrt{(5-0)^2 + (2-7)^2} = \sqrt{50}$

The segments with endpoints $(2, -1)$, $(5, 2)$ and $(0, 7)$, $(-3, 4)$ are the second pair of opposite sides. We find their lengths.

For $(2, -1)$, $(5, 2)$:

$d = \sqrt{(2-5)^2 + (-1-2)^2} = \sqrt{18}$

For $(0, 7)$, $(-3, 4)$:

$d = \sqrt{(0-(-3))^2 + (7-4)^2} = \sqrt{18}$

The endpoints of the diagonals are $(-3, 4)$, $(5, 2)$ and $(2, -1)$, $(0, 7)$. We find the length of each.

For $(-3,4)$, $(5,2)$:

$$d=\sqrt{(-3-5)^2+(4-2)^2}=\sqrt{68}$$

For $(2,-1)$, $(0,7)$:

$$d=\sqrt{(2-0)^2+(-1-7)^2}=\sqrt{68}$$

The opposite sides of the quadrilateral are the same length and the diagonals are the same length, so the quadrilateral is a rectangle.

13. We use the midpoint formula.

$$\left(\frac{4+5}{2},\frac{6+9}{2}\right)=\left(\frac{9}{2},\frac{15}{2}\right)$$

14. $\left(\frac{-3+2}{2},\frac{7+11}{2}\right)=\left(-\frac{1}{2},9\right)$

15. We use the midpoint formula.

$$\left(\frac{6+9}{2},\frac{-1+5}{2}\right)=\left(\frac{15}{2},2\right)$$

16. $\left(\frac{-4+(-1)}{2},\frac{-7+3}{2}\right)=\left(-\frac{5}{2},-2\right)$

17. We use the midpoint formula.

$$\left(\frac{\sqrt{3}+(-\sqrt{6})}{2},\frac{-\sqrt{5}+0}{2}\right)=\left(\frac{\sqrt{3}-\sqrt{6}}{2},-\frac{\sqrt{5}}{2}\right)$$

18. $\left(-\frac{\sqrt{2}+0}{2},\frac{1+\sqrt{7}}{2}\right)=\left(-\frac{\sqrt{2}}{2},\frac{1+\sqrt{7}}{2}\right)$

19.

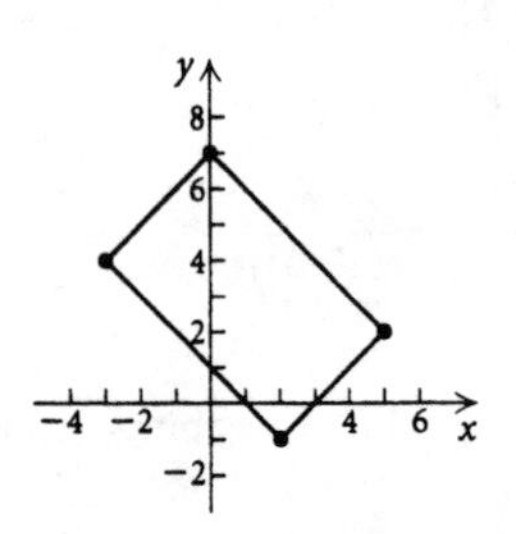

For the side with vertices $(-3,4)$ and $(2,-1)$:

$$\left(\frac{-3+2}{2},\frac{4+(-1)}{2}\right)=\left(-\frac{1}{2},\frac{3}{2}\right)$$

For the sides with vertices $(2,-1)$ and $(5,2)$:

$$\left(\frac{2+5}{2},\frac{-1+2}{2}\right)=\left(\frac{7}{2},\frac{1}{2}\right)$$

For the sides with vertices $(5,2)$ and $(0,7)$:

$$\left(\frac{5+0}{2},\frac{2+7}{2}\right)=\left(\frac{5}{2},\frac{9}{2}\right)$$

For the sides with vertices $(0,7)$ and $(-3,4)$:

$$\left(\frac{0+(-3)}{2},\frac{7+4}{2}\right)=\left(-\frac{3}{2},\frac{11}{2}\right)$$

For the quadrilateral whose vertices are the points found above, the diagonals have endpoints

$\left(-\frac{1}{2},\frac{3}{2}\right),\left(\frac{5}{2},\frac{9}{2}\right)$ and $\left(\frac{7}{2},\frac{1}{2}\right),\left(-\frac{3}{2},\frac{11}{2}\right)$.

We find the length of each of these diagonals.

For $\left(-\frac{1}{2},\frac{3}{2}\right),\left(\frac{5}{2},\frac{9}{2}\right)$:

$$d=\sqrt{\left(-\frac{1}{2}-\frac{5}{2}\right)^2+\left(\frac{3}{2}-\frac{9}{2}\right)^2}$$

$$=\sqrt{(-3)^2+(-3)^2}=\sqrt{18}$$

For $\left(\frac{7}{2},\frac{1}{2}\right),\left(-\frac{3}{2},\frac{11}{2}\right)$:

$$d=\sqrt{\left(\frac{7}{2}-\left(-\frac{3}{2}\right)\right)^2+\left(\frac{1}{2}-\frac{11}{2}\right)^2}$$

$$=\sqrt{5^2+(-5)^2}=\sqrt{50}$$

Since the diagonals do not have the same lengths, the midpoints are not vertices of a rectangle.

20.

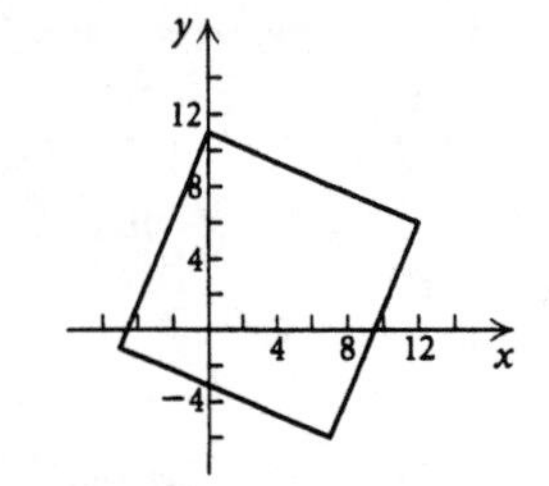

For the side with vertices $(-5,-1)$ and $(7,-6)$:

$$\left(\frac{-5+7}{2},\frac{-1+(-6)}{2}\right)=\left(1,-\frac{7}{2}\right)$$

For the side with vertices $(7,-6)$ and $(12,6)$:

$$\left(\frac{7+12}{2},\frac{-6+6}{2}\right)=\left(\frac{19}{2},0\right)$$

For the side with vertices $(12,6)$ and $(0,11)$:

$$\left(\frac{12+0}{2},\frac{6+11}{2}\right)=\left(6,\frac{17}{2}\right)$$

For the side with vertices $(0,11)$ and $(-5,-1)$:

$$\left(\frac{0+(-5)}{2},\frac{11+(-1)}{2}\right)=\left(-\frac{5}{2},5\right)$$

For the quadrilateral whose vertices are the points found above, one pair of opposite sides has endpoints $\left(1,-\frac{7}{2}\right),\left(\frac{19}{2},0\right)$ and $\left(6,\frac{17}{2}\right),\left(-\frac{5}{2},5\right)$. The length of each of these sides is $\frac{\sqrt{338}}{2}$. The other pair of opposite sides has endpoints $\left(\frac{19}{2},0\right),\left(6,\frac{17}{2}\right)$ and $\left(-\frac{5}{2},5\right),\left(1,-\frac{7}{2}\right)$.

The length of each of these sides is also $\frac{\sqrt{338}}{2}$. The endpoints of the diagonals of the quadrilateral are $\left(1,-\frac{7}{2}\right),\left(6,\frac{17}{2}\right)$ and $\left(\frac{19}{2},0\right),\left(-\frac{5}{2},5\right)$. The length of each diagonal is 13. Since the four sides of the quadrilateral are the same length and the diagonals are the same length, the midpoints are vertices of a square.

21. We use the midpoint formula.

$$\left(\frac{\sqrt{7}+\sqrt{2}}{2}, \frac{-4+3}{2}\right) = \left(\frac{\sqrt{7}+\sqrt{2}}{2}, -\frac{1}{2}\right)$$

22. $\left(\frac{-3+1}{2}, \frac{\sqrt{5}+\sqrt{2}}{2}\right) = \left(-1, \frac{\sqrt{5}+\sqrt{2}}{2}\right)$

23. Square the viewing window. For the graph shown, one possibility is $[-12, 9, -4, 10]$.

24. Square the viewing window. For the window shown, one possibility is $[-10, 20, -15, 5]$.

25.
$$\begin{aligned}(x-h)^2 + (y-k)^2 &= r^2\\ (x-2)^2 + (y-3)^2 &= \left(\frac{5}{3}\right)^2 \quad \text{Substituting}\\ (x-2)^2 + (y-3)^2 &= \frac{25}{9}\end{aligned}$$

26.
$$\begin{aligned}(x-4)^2 + (y-5)^2 &= (4.1)^2\\ (x-4)^2 + (y-5)^2 &= 16.81\end{aligned}$$

27. The length of a radius is the distance between $(-1, 4)$ and $(3, 7)$:

$$\begin{aligned}r &= \sqrt{(-1-3)^2 + (4-7)^2}\\ &= \sqrt{(-4)^2 + (-3)^2} = \sqrt{25} = 5\end{aligned}$$

$$\begin{aligned}(x-h)^2 + (y-k)^2 &= r^2\\ [x-(-1)]^2 + (y-4)^2 &= 5^2\\ (x+1)^2 + (y-4)^2 &= 25\end{aligned}$$

28. Find the length of a radius:

$$r = \sqrt{(6-1)^2 + (-5-7)^2} = \sqrt{169} = 13$$

$$\begin{aligned}(x-6)^2 + [y-(-5)]^2 &= 13^2\\ (x-6)^2 + (y+5)^2 &= 169\end{aligned}$$

29. The center is the midpoint of the diameter:

$$\left(\frac{7+(-3)}{2}, \frac{13+(-11)}{2}\right) = (2, 1)$$

Use the center and either endpoint of the diameter to find the length of a radius. We use the point $(7, 13)$:

$$\begin{aligned}r &= \sqrt{(7-2)^2 + (13-1)^2}\\ &= \sqrt{5^2 + 12^2} = \sqrt{169} = 13\end{aligned}$$

$$\begin{aligned}(x-h)^2 + (y-k)^2 &= r^2\\ (x-2)^2 + (y-1)^2 &= 13^2\\ (x-2)^2 + (y-1)^2 &= 169\end{aligned}$$

30. The points $(-9, 4)$ and $(-1, -2)$ are opposite vertices of the square and hence endpoints of a diameter of the circle. We use these points to find the center and radius.

Center: $\left(\frac{-9+(-1)}{2}, \frac{4+(-2)}{2}\right) = (-5, 1)$

Radius: $\frac{1}{2}\sqrt{(-9-(-1))^2 + (4-(-2))^2} = \frac{1}{2}\cdot 10 = 5$

$$\begin{aligned}[x-(-5)]^2 + (y-1)^2 &= 5^2\\ (x+5)^2 + (y-1)^2 &= 25\end{aligned}$$

31. Since the center is 2 units to the left of the y-axis and the circle is tangent to the y-axis, the length of a radius is 2.

$$\begin{aligned}(x-h)^2 + (y-k)^2 &= r^2\\ [x-(-2)]^2 + (y-3)^2 &= 2^2\\ (x+2)^2 + (y-3)^2 &= 4\end{aligned}$$

32. Since the center is 5 units below the x-axis and the circle is tangent to the x-axis, the length of a radius is 5.

$$\begin{aligned}(x-4)^2 + [y-(-5)]^2 &= 5^2\\ (x-4)^2 + (y+5)^2 &= 25\end{aligned}$$

33.
$$\begin{aligned}x^2 + y^2 &= 4\\ (x-0)^2 + (y-0)^2 &= 2^2\end{aligned}$$

Center: $(0, 0)$; radius: 2

$x^2 + y^2 = 4$

$y_1 = \sqrt{4-x^2}$, $y_2 = -\sqrt{4-x^2}$

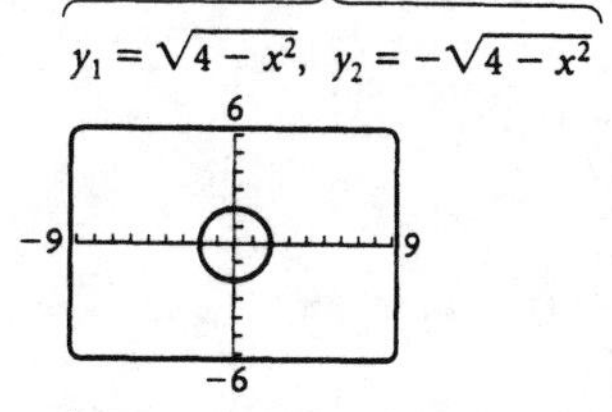

34.
$$\begin{aligned}x^2 + y^2 &= 81\\ (x-0)^2 + (y-0)^2 &= 9^2\end{aligned}$$

Center: $(0, 0)$; radius: 9

$x^2 + y^2 = 81$

35.
$$\begin{aligned}x^2 + (y-3)^2 &= 16\\ (x-0)^2 + (y-3)^2 &= 4^2\end{aligned}$$

Center: $(0, 3)$; radius: 4

$x^2 + (y-3)^2 = 16$

$y_1 = 3 + \sqrt{16-x^2}$, $y_2 = 3 - \sqrt{16-x^2}$

36.
$$\begin{aligned}(x+2)^2 + y^2 &= 100\\ [x-(-2)]^2 + (y-0)^2 &= 10^2\end{aligned}$$

Center: $(-2, 0)$; radius: 10

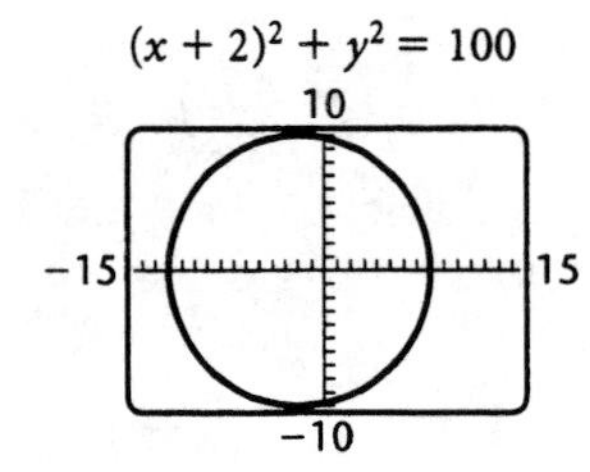

37. $(x-1)^2+(y-5)^2=36$

$(x-1)^2+(y-5)^2=6^2$

Center: $(1,5)$; radius: 6

$(x-1)^2+(y-5)^2=36$

$y_1=5+\sqrt{36-(x-1)^2},\ y_2=5-\sqrt{36-(x-1)^2}$

38. $(x-7)^2+(y+2)^2=25$

$(x-7)^2+[y-(-2)]^2=5^2$

Center: $(7,-2)$; radius: 5

$(x-7)^2+(y+2)^2=25$

39. $(x+4)^2+(y+5)^2=9$

$[x-(-4)]^2+[y-(-5)]^2=3^2$

Center: $(-4,-5)$; radius: 3

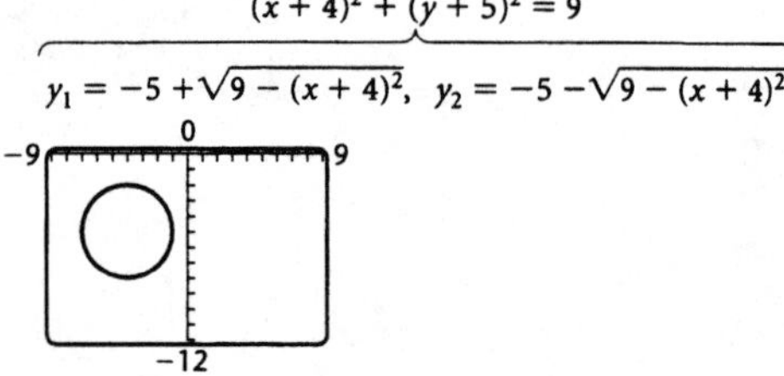

40. $(x+1)^2+(y-2)^2=64$

$[x-(-1)]^2+(y-2)^2=8^2$

Center: $(-1,2)$; radius: 8

$(x+1)^2+(y-2)^2=64$

41. The Pythagorean theorem is used to derive the distance formula, and the distance formula is used to derive the equation of a circle in standard form.

42. Let $A=(a,b)$ and $B=(c,d)$. The coordinates of a point C one-half of the way from A to B are $\left(\frac{a+c}{2},\frac{b+d}{2}\right)$. A point D that is one-half of the way from C to B is $\frac{1}{2}+\frac{1}{2}\cdot\frac{1}{2}$, or $\frac{3}{4}$ of the way from A to B. Its coordinates are $\left(\frac{\frac{a+c}{2}+c}{2},\frac{\frac{b+d}{2}+d}{2}\right)$, or $\left(\frac{a+3c}{4},\frac{b+3d}{4}\right)$. Then a point E that is one-half of the way from D to B is $\frac{3}{4}+\frac{1}{2}\cdot\frac{1}{4}$, or $\frac{7}{8}$ of the way from A to B. Its coordinates are $\left(\frac{\frac{a+3c}{4}+c}{2},\frac{\frac{b+3d}{4}+d}{2}\right)$, or $\left(\frac{a+7c}{8},\frac{b+7d}{8}\right)$.

43. The equation $y=-\frac{3}{5}x+4$ is in the form $y=mx+b$. The slope is $-\frac{3}{5}$ and the y-intercept is $(0,4)$.

44.
$$2x-3y=15$$
$$-3y=-2x+15$$
$$y=\frac{2}{3}x-5$$

Slope: $\frac{2}{3}$; y-intercept: $(0,-5)$

45. Substitute 3 for m and -1 for b in the equation $y=mx+b$. We have $y=3x-1$.

46. $m=\frac{1-3}{4-(-2)}=\frac{-2}{6}=-\frac{1}{3}$
$$y-1=-\frac{1}{3}(x-4)$$
$$y-1=-\frac{1}{3}x+\frac{4}{3}$$
$$y=-\frac{1}{3}x+\frac{7}{3}$$

47. Use the distance formula. Either point can be considered as (x_1,y_1).
$$d=\sqrt{(a+h-a)^2+(\sqrt{a+h}-\sqrt{a})^2}$$
$$=\sqrt{h^2+a+h-2\sqrt{a^2+ah}+a}$$
$$=\sqrt{h^2+2a+h-2\sqrt{a^2+ah}}$$

Next we use the midpoint formula.
$$\left(\frac{a+a+h}{2},\frac{\sqrt{a}+\sqrt{a+h}}{2}\right)=\left(\frac{2a+h}{2},\frac{\sqrt{a}+\sqrt{a+h}}{2}\right)$$

48. Use the distance formula:

$$d = \sqrt{(a+h-a)^2 + \left(\frac{1}{a+h} - \frac{1}{a}\right)^2} =$$

$$\sqrt{h^2 + \left(\frac{-h}{a(a+h)}\right)^2} = \sqrt{h^2 + \frac{h^2}{a^2(a+h)^2}} =$$

$$\sqrt{\frac{h^2a^2(a+h)^2 + h^2}{a^2(a+h)^2}} = \sqrt{\frac{h^2(a^2(a+h)^2+1)}{a^2(a+h)^2}} =$$

$$\left|\frac{h}{a(a+h)}\right|\sqrt{a^2(a+h)^2+1}$$

Find the midpoint:

$$\left(\frac{a+a+h}{2}, \frac{\frac{1}{a}+\frac{1}{a+h}}{2}\right) = \left(\frac{2a+h}{2}, \frac{2a+h}{2a(a+h)}\right)$$

49. First use the formula for the area of a circle to find r^2:

$$A = \pi r^2$$
$$36\pi = \pi r^2$$
$$36 = r^2$$

Then we have:

$$(x-h)^2 + (y-k)^2 = r^2$$
$$(x-2)^2 + [y-(-7)]^2 = 36$$
$$(x-2)^2 + (y+7)^2 = 36$$

50. $C = 2\pi r$

$10\pi = 2\pi r$

$5 = r$

Then $[x-(-5)]^2+(y-8)^2 = 5^2$, or $(x+5)^2+(y-8)^2 = 25$.

51. Label the drawing with additional information and lettering.

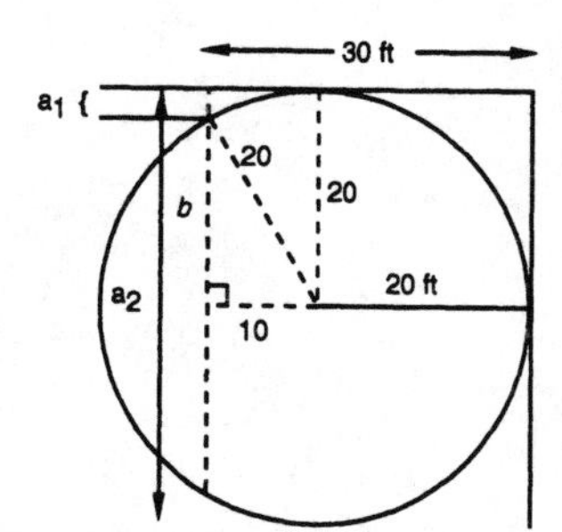

Find b using the Pythagorean theorem.

$$b^2 + 10^2 = 20^2$$
$$b^2 + 100 = 400$$
$$b^2 = 300$$
$$b = 10\sqrt{3}$$
$$b \approx 17.3$$

Find a_1:

$a_1 = 20 - b \approx 20 - 17.3 \approx 2.7$ ft

Find a_2:

$a_2 = 2b + a_1 \approx 2(17.3) + 2.7 \approx 37.3$ ft

52. a) When the circle is positioned on a coordinate system as shown in the text, the center lies on the y-axis and is equidistant from $(-4,0)$ and $(0,2)$.

Let $(0,y)$ be the coordinates of the center.

$$\sqrt{(-4-0)^2+(0-y)^2} = \sqrt{(0-0)^2+(2-y)^2}$$
$$4^2 + y^2 = (2-y)^2$$
$$16 + y^2 = 4 - 4y + y^2$$
$$12 = -4y$$
$$-3 = y$$

The center of the circle is $(0,-3)$.

b) Use the point $(-4,0)$ and the center $(0,-3)$ to find the radius.

$$(-4-0)^2 + [0-(-3)]^2 = r^2$$
$$25 = r^2$$
$$5 = r$$

The radius is 5 ft.

53.

$$x^2 + y^2 = 1$$

$\left(\frac{\sqrt{3}}{2}\right)^2 + \left(-\frac{1}{2}\right)^2$	? 1
$\frac{3}{4} + \frac{1}{4}$	
1	1 TRUE

$\left(\frac{\sqrt{3}}{2}, -\frac{1}{2}\right)$ lies on the unit circle.

54.

$$x^2 + y^2 = 1$$

$0^2 + (-1)^2$	? 1
1	1 TRUE

$(0,-1)$ lies on the unit circle.

55.

$$x^2 + y^2 = 1$$

$\left(\frac{\sqrt{2}}{2}\right)^2 + \left(\frac{\sqrt{2}}{2}\right)^2$	? 1
$\frac{2}{4} + \frac{2}{4}$	
1	1 TRUE

$\left(\frac{\sqrt{2}}{2}, \frac{\sqrt{2}}{2}\right)$ lies on the unit circle.

56.

$$x^2 + y^2 = 1$$

$\left(\frac{1}{2}\right)^2 + \left(-\frac{\sqrt{3}}{2}\right)^2$	? 1
$\frac{1}{4} + \frac{3}{4}$	
1	1 TRUE

$\left(\frac{1}{2}, -\frac{\sqrt{3}}{2}\right)$ lies on the unit circle.

57. Let $(0, y)$ be the required point. We set the distance between $(-2, 0)$ and $(0, y)$ equal to the distance between $(4, 6)$ and $(0, y)$ and solve for y.

$$\sqrt{[0-(-2)]^2 + (y-0)^2} = \sqrt{(0-4)^2 + (y-6)^2}$$

$$\sqrt{4+y^2} = \sqrt{16+y^2-12y+36}$$

$$4+y^2 = 16+y^2-12y+36$$

Squaring both sides

$$-48 = -12y$$

$$4 = y$$

The point is $(0, 4)$.

58. The coordinates of P are $\left(\frac{b}{2}, \frac{h}{2}\right)$ by the midpoint formula. By the distance formula, each of the distances from P to $(0, h)$, from P to $(0, 0)$, and from P to $(b, 0)$ is $\frac{\sqrt{b^2+h^2}}{2}$.

59. a), b) See the answer section in the text.

Chapter 2

Functions and Equations: Zeros and Solutions

Exercise Set 2.1

1. a) The graph crosses the x-axis at $(4, 0)$. This is the x-intercept.

   b) The zero of the function is the first coordinate of the x-intercept. It is 4.

2. a) $(5, 0)$

   b) 5

3. $x + 5 = 0$ Setting $f(x) = 0$

   $x + 5 - 5 = 0 - 5$ Subtracting 5 on both sides

   $x = -5$

   The zero of the function is -5.

4. $5x + 20 = 0$

   $5x = -20$

   $x = -4$

5. $-x + 18 = 0$ Setting $f(x) = 0$

   $-x + 18 + x = 0 + x$ Adding x on both sides

   $18 = x$

   The zero of the function is 18.

6. $8 + x = 0$

   $x = -8$

7. $16 - x = 0$ Setting $f(x) = 0$

   $16 - x + x = 0 + x$ Adding x on both sides

   $16 = x$

   The zero of the function is 16.

8. $-2x + 7 = 0$

   $-2x = -7$

   $x = \frac{7}{2}$, or 3.5

9. $f(x) = x + 12$

   Graph the function in a window that displays the x-intercept. We use $[-15, 5, -10, 10]$. Then use the Zero feature.

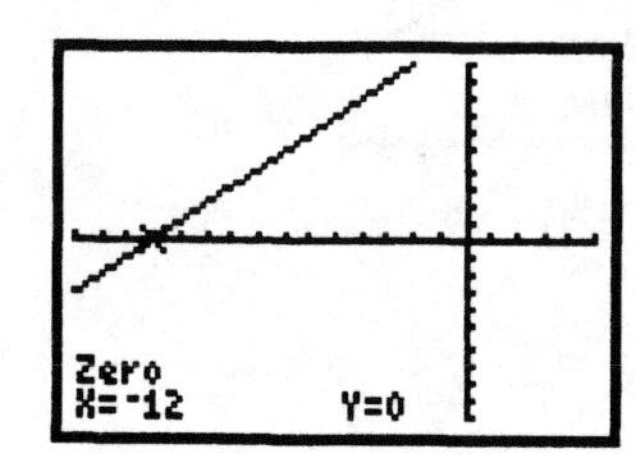

The zero of the function is -12.

10. $f(x) = 8x + 2$

    Graph the function in a window that displays the x-intercept. We use $[-2, 2, -2, 2]$. Then use the Zero feature.

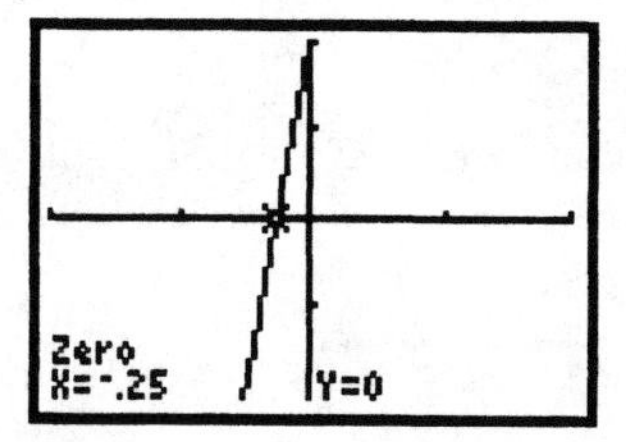

The zero of the function is -0.25.

11. $f(x) = -x + 6$

    Graph the function in a window that displays the x-intercept. We use the standard window. Then use the Zero feature.

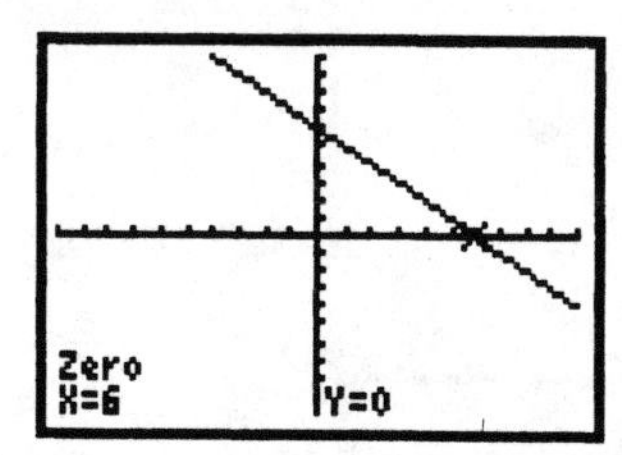

The zero of the function is 6.

12. $f(x) = 4 + x$

    Graph the function in a window that displays the x-intercept. We use the standard window. Then use the Zero feature.

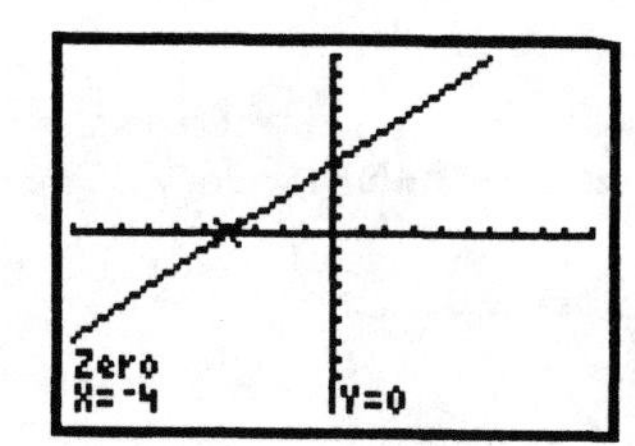

The zero of the function is -4.

13. $f(x) = 20 - x$

    Graph the function in a window that displays the x-intercept. We use $[-5, 25, -10, 10]$, Xscl = 5. Then use the Zero feature.

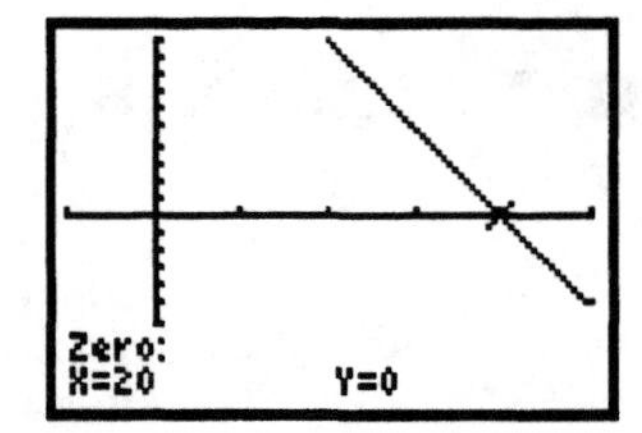

The zero of the function is 20.

14. $f(x) = -3x + 13$

Graph the function in a window that displays the x-intercept. We use the standard window. Then use the Zero feature.

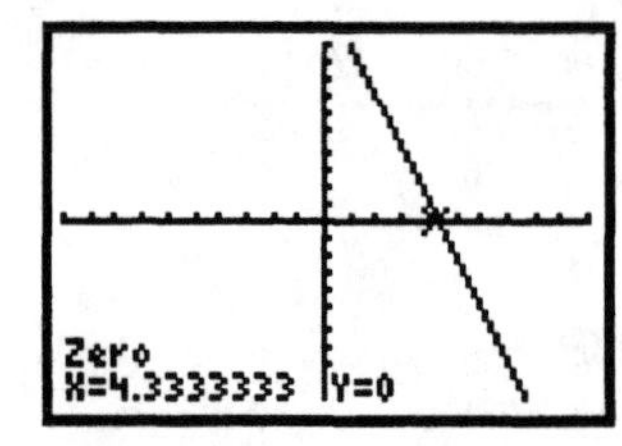

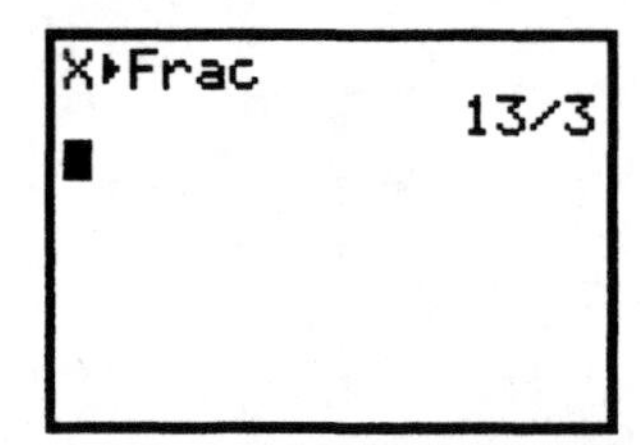

The zero of the function is $4.\overline{3}$, or $\frac{13}{3}$.

15. $f(x) = x - 6$

Algebraic solution:

$x - 6 = 0$ Setting $f(x) = 0$

$x = 6$ Adding 6 on both sides

Graphical solution:

Graph the function in a window that displays the x-intercept. We use the standard window. Then use the Zero feature.

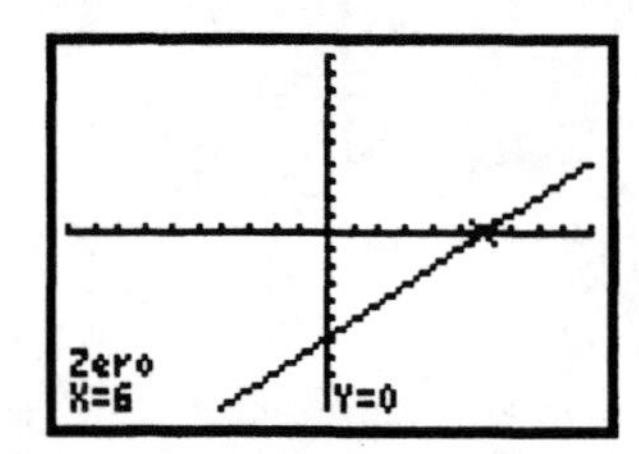

The zero of the function is 6.

16. $f(x) = 3x - 9$

Algebraic solution:

$3x - 9 = 0$

$3x = 9$

$x = 3$

Graphical solution:

Graph the function in a window that displays the x-intercept. We use the standard window. Then use the Zero feature.

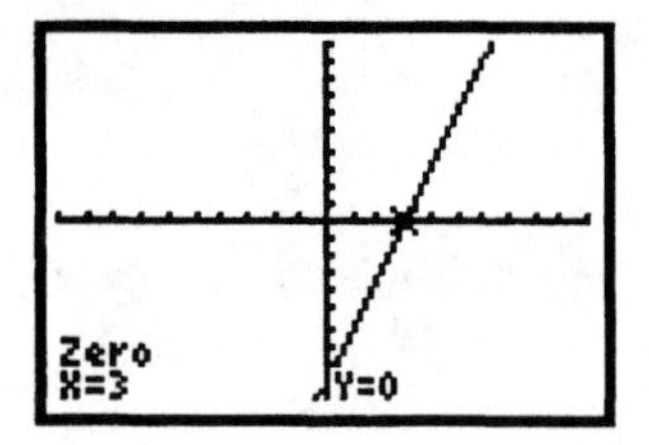

The zero of the function is 3.

17. $f(x) = x + 15$

Algebraic solution:

$-x + 15 = 0$ Setting $f(x) = 0$

$15 = x$ Adding x on both sides

Graphical solution:

Graph the function in a window that displays the x-intercept. We use $[-5, 20, -10, 10]$. Then use the Zero feature.

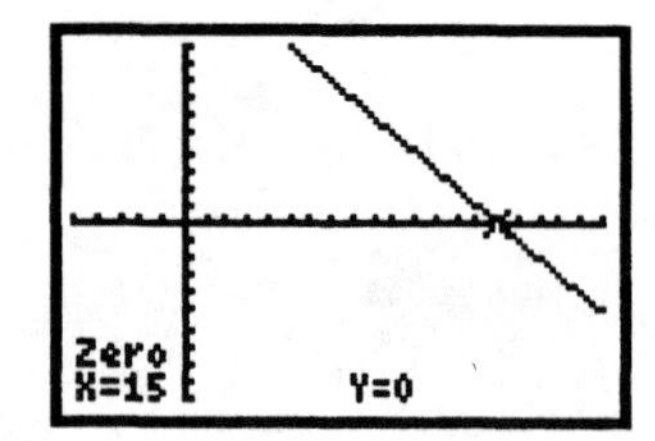

The zero of the function is 15.

18. $f(x) = 4 - x$

Algebraic solution:

$4 - x = 0$

$4 = x$

Graphical solution:

Graph the function in a window that displays the x-intercept. We use the standard window. Then use the Zero feature.

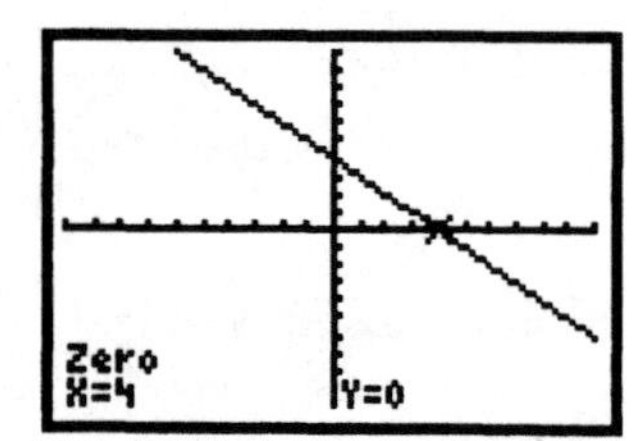

The zero of the function is 4.

19. $4x + 3 = 0$

$4x = -3$ Subtracting 3

$x = -\frac{3}{4}$ Dividing by 4

The solution is $-\frac{3}{4}$.

20.
$$\begin{aligned} 3x - 16 &= 0 \\ 3x &= 16 \\ x &= \frac{16}{3} \end{aligned}$$

21.
$$\begin{aligned} 2x + 7 &= x + 3 \\ x + 7 &= 3 && \text{Subtracting } x \\ x &= -4 && \text{Subtracting } 7 \end{aligned}$$

The solution is -4.

22.
$$\begin{aligned} 5x - 4 &= 2x + 5 \\ 3x - 4 &= 5 \\ 3x &= 9 \\ x &= 3 \end{aligned}$$

23.
$$\begin{aligned} 3(x + 1) &= 5 - 2(3x + 4) \\ 3x + 3 &= 5 - 6x - 8 && \text{Removing parentheses} \\ 3x + 3 &= -6x - 3 && \text{Collecting like terms} \\ 9x + 3 &= -3 && \text{Adding } 6x \\ 9x &= -6 && \text{Subtracting } 3 \\ x &= -\frac{2}{3} && \text{Dividing by } 9 \end{aligned}$$

24.
$$\begin{aligned} 4(3x + 2) - 7 &= 3(x - 2) \\ 12x + 8 - 7 &= 3x - 6 \\ 12x + 1 &= 3x - 6 \\ 9x + 1 &= -6 \\ 9x &= -7 \\ x &= -\frac{7}{9} \end{aligned}$$

25. $5x - 8 = 0$

The solution of this equation is the zero of the function $f(x) = 5x - 8$. Graph the function in a window that displays the x-intercept. We use the standard window. Then use the Zero feature.

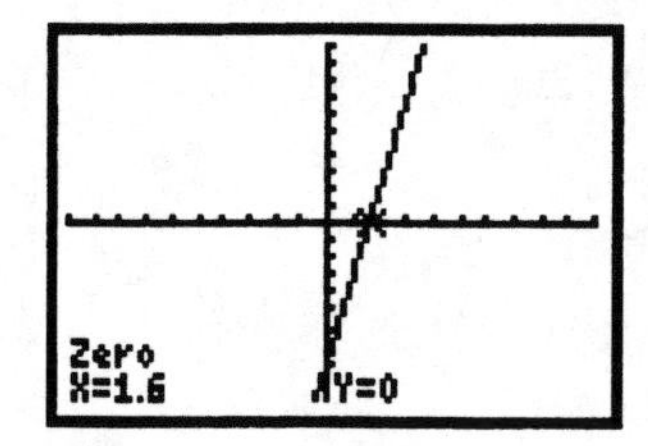

The solution is 1.6.

26. $6x + 5 = 0$

The solution of this equation is the zero of the function $f(x) = 6x + 5$. Graph the function in a window that displays the x-intercept. We use $[-5, 5, -5, 5]$. Then use the Zero feature.

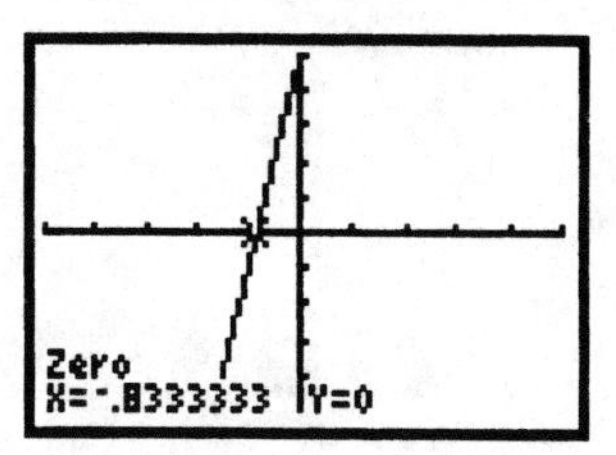

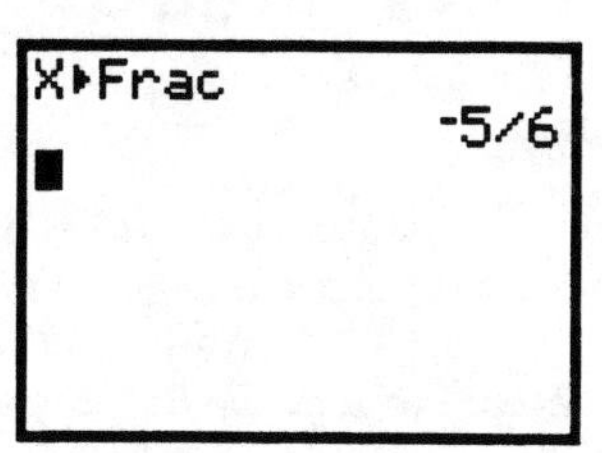

The solution is $-0.8\overline{3}$, or $-\frac{5}{6}$.

27. $3x - 5 = 2x + 1$

Graph $y_1 = 3x - 5$ and $y_2 = 2x + 1$ in a window that displays the point of intersection of the graphs. We use $[-10, 10, -5, 20]$, Yscl $= 5$. Then use the Intersect feature to find the first coordinate of the point of intersection.

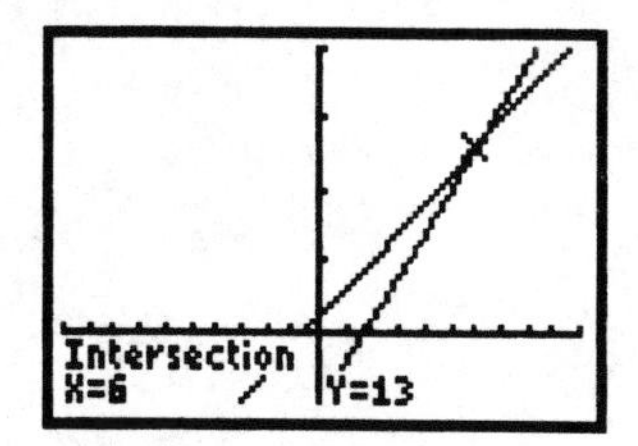

The solution is 6.

28. $4x + 3 = 2x - 7$

Graph $y_1 = 4x + 3$ and $y_2 = 2x - 7$ in a window that displays the point of intersection of the graphs. We use the $[-10, 10, -25, 5]$, Yscl $= 5$. Then use the Intersect feature to find the first coordinate of the point of intersection.

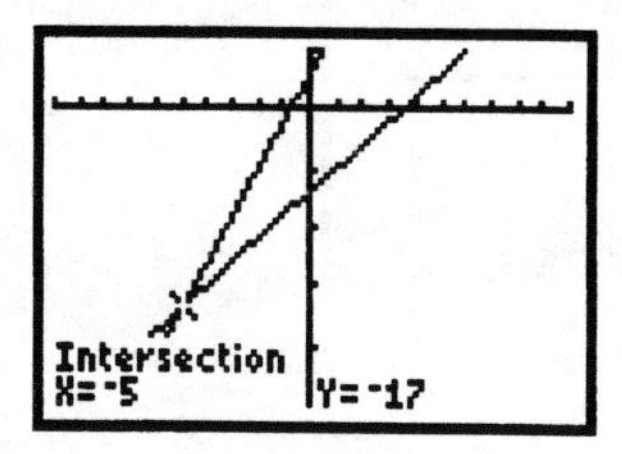

The solution is -5.

29. $2(x - 4) = 3 - 5(2x + 1)$

Graph $y_1 = 2(x - 4)$ and $y_2 = 3 - 5(2x + 1)$ in a window that displays the point of intersection of the graphs. We use the standard window. Then use the Intersect feature to find the first coordinate of the point of intersection.

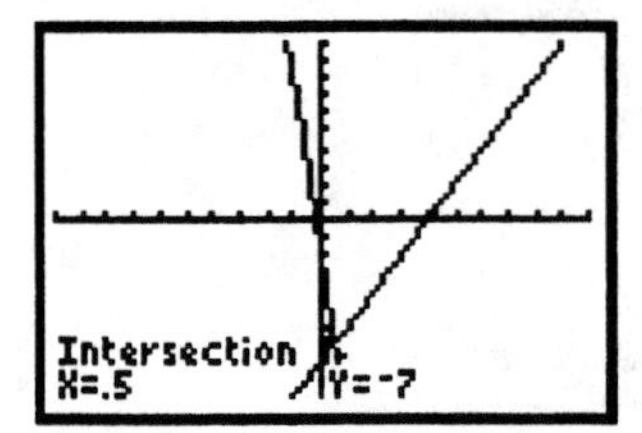

The solution is 0.5.

30. $3(2x-5)+4=2(4x+3)$

Graph $y_1 = 3(2x-5)+4$ and $y_2 = 2(4x+3)$ in a window that displays the point of intersection of the graphs. We use $[-15, 5, -80, 5]$, Yscl $= 10$. Then use the Intersect feature to find the first coordinate of the point of intersection.

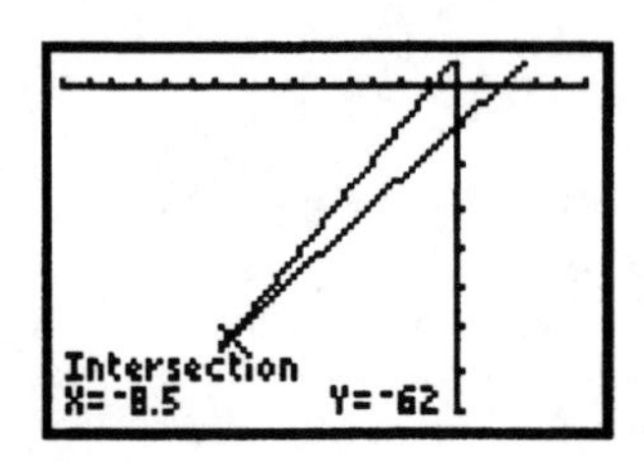

The solution is -8.5.

31. $8x+25=0$

Algebraic solution:

$$\begin{aligned} 8x+25 &= 0 \\ 8x &= -25 && \text{Subtracting 25} \\ x &= -\frac{25}{8}, \text{ or } -3.125 && \text{Dividing by 8} \end{aligned}$$

Graphical solution:

The solution of this equation is the zero of the function $f(x) = 8x+25$. Graph the function in a window that displays the x-intercept. We use the standard window. Then use the Zero feature.

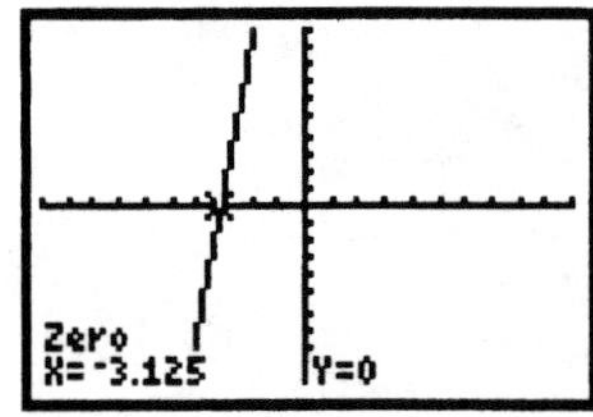

The solution is $-\frac{25}{8}$, or -3.125.

32. $7-9x=0$

Algebraic solution:

$$\begin{aligned} 7-9x &= 0 \\ 7 &= 9x \\ \frac{7}{9} &= x \end{aligned}$$

Graphical solution:

The solution of this equation is the zero of the function $f(x) = 7-9x$. Graph the function in a window that displays the x-intercept. We use $[-5, 5, -5, 5]$. Then use the Zero feature.

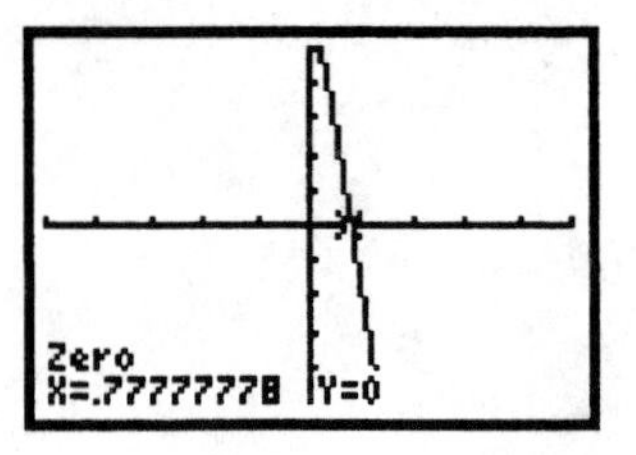

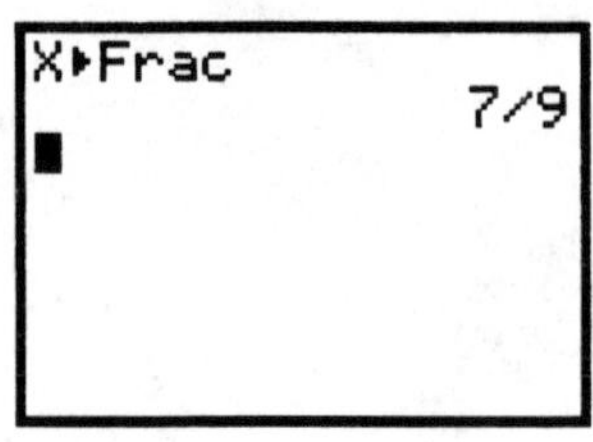

The solution is $0.\overline{7}$, or $\frac{7}{9}$.

33. $6x-5=3x+10$

Algebraic solution:

$$\begin{aligned} 6x-5 &= 3x+10 \\ 3x-5 &= 10 && \text{Subtracting } 3x \\ 3x &= 15 && \text{Adding 5} \\ x &= 5 && \text{Dividing by 3} \end{aligned}$$

Graphical solution:

Graph $y_1 = 6x-5$ and $y_2 = 3x+10$ in a window that displays the point of intersection of the graphs. We use $[-10, 10, -5, 35]$, Yscl $= 5$. Then use the Intersect feature to find the first coordinate of the point of intersection.

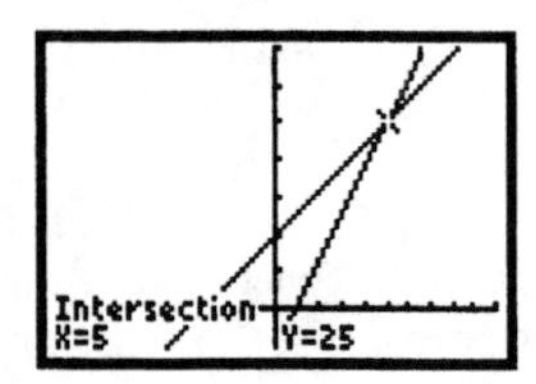

The solution is 5.

34. $2(3x-1)=3-4(x+2)$

Algebraic solution:

$$\begin{aligned} 2(3x-1) &= 3-4(x+2) \\ 6x-2 &= 3-4x-8 \\ 6x-2 &= -4x-5 \\ 10x-2 &= -5 \\ 10x &= -3 \\ x &= -\frac{3}{10}, \text{ or } -0.3 \end{aligned}$$

Graphical solution:

Graph $y_1 = 2(3x-1)$ and $y_2 = 3-4(x+2)$ in a window that displays the point of intersection of the graphs. We use $[-2, 2, -5, 2]$. Then use the Intersect feature to find the first coordinate of the point of intersection.

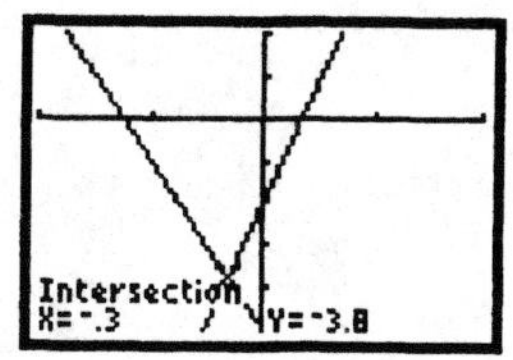

The solution is $-\frac{3}{10}$, or -0.3.

35. ***Familiarize.*** Let $x =$ the number of foreign tourists who visited the United States in 1998, in millions.

Translate. We reword the problem and translate to an equation.

Number of foreign tourists in the U.S. plus 22.9 is number of foreign tourists in France.

$$x + 22.9 = 70$$

Carry out. We will solve algebraically.

$$x + 22.9 = 70$$
$$x = 47.1 \quad \text{Subtracting 22.9}$$

Check. $47.1 + 22.9 = 70$, so the answer checks.

State. In 1998, 47.1 million foreign tourists visited the United States.

36. Let $m =$ the number of miles a courier in Amsterdam walks on a route.

Solve: $6m = 11$

$m = \frac{11}{6}$ mi

37. ***Familiarize.*** Let $c =$ the number of catalogs each household received.

Translate.

Number of households times number of catalogs per household is 14 billion.

$$112,000,000 \cdot c = 14,000,000,000$$

Carry out. We will solve algebraically.

$$112,000,000c = 14,000,000,000$$
$$c = 125$$

Check. If each of 112 million households receives 125 catalogs, then $112,000,000 \cdot 125$, or 14,000,000,000 catalogs are received. The answer checks.

State. Each household received 125 catalogs.

38. Let $d =$ the annual delay in 1982, in hours.

Solve: $8d = 32$

$d = 4$ hours

39. ***Familiarize.*** Let $p =$ the number of Braille pages published in 1977, in millions.

Translate.

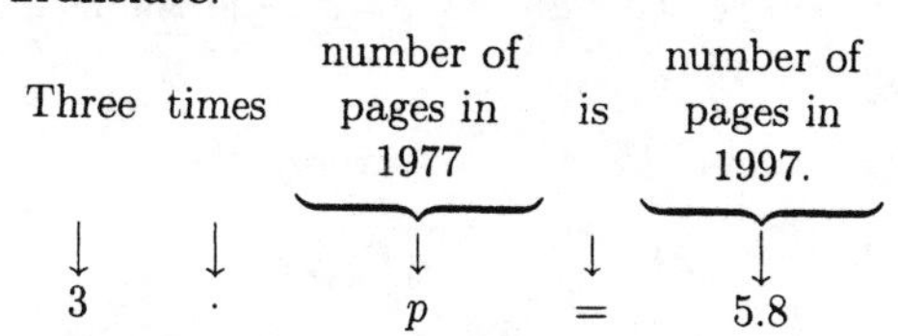

Carry out. We will solve graphically. Graph $y_1 = 3x$ and $y_2 = 5.8$ and find the first coordinate of the point of intersection of the graphs.

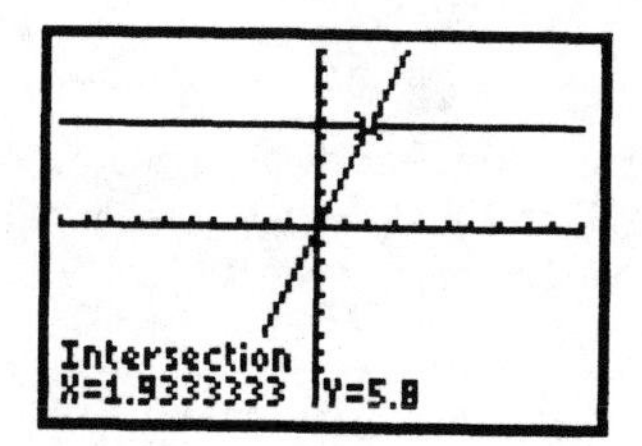

The possible solution is approximately 1.93 million.

Check. $3(1.93 \text{ million}) = 5.79 \text{ million} \approx 5.8$ million, so the answer checks. (Recall that we rounded the possible solution to 2 decimal places.)

State. Approximately 1.93 million Braille pages were published in 1977.

40. Let $x =$ the amount spent for domestic long-distance fax service, in millions of dollars.

Solve: $\frac{1}{3}x + x = 13$

$x = 9.75$, so \$9.75 million is spent on domestic long-distance fax service and $\frac{1}{3}(\$9.75)$, or \$3.25 million is spent on local fax service.

41. ***Familiarize.*** Using the labels on the drawing in the text, we let $w =$ the width of the test plot and $w + 25 =$ the length. Recall that for a rectangle, Perimeter $= 2 \cdot \text{length} + 2 \cdot \text{width}$.

Translate.

Perimeter = 2 · length + 2 · width

$$322 = 2(w + 25) + 2 \cdot w$$

Carry out. We solve the equation.

$$322 = 2(w + 25) + 2 \cdot w$$
$$322 = 2w + 50 + 2w$$
$$322 = 4w + 50$$
$$272 = 4w$$
$$68 = w$$

We could also find the first coordinate of the point of intersection of $y_1 = 322$ and $y_2 = 2(x + 25) + 2x$ using the Intersect feature.

When $w = 68$, then $w + 25 = 68 + 25 = 93$.

Check. The length is 25 m more than the width: $93 = 68 + 25$. The perimeter is $2 \cdot 93 + 2 \cdot 68$, or $186 + 136$, or 322 m. The answer checks.

State. The length is 93 m; the width is 68 m.

42. Let w = the width of the garden.

Solve: $2 \cdot 2w + 2 \cdot w = 39$

$w = 6.5$, so the width is 6.5 m, and the length is 2(6.5), or 13 m.

43. ***Familiarize.*** We make a drawing.

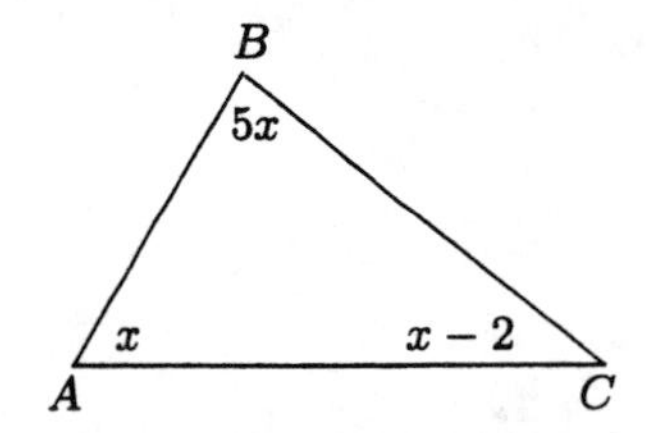

We let x = the measure of angle A. Then $5x$ = the measure of angle B, and $x - 2$ = the measure of angle C. The sum of the angle measures is 180°.

Translate.

Measure of angle A + Measure of angle B + Measure of angle C = 180

$$x + 5x + x - 2 = 180$$

Carry out. We solve the equation.

$$x + 5x + x - 2 = 180$$
$$7x - 2 = 180$$
$$7x = 182$$
$$x = 26$$

We could also find the first coordinate of the point of intersection of $y_1 = x + 5x + x - 2$ and $y_2 = 180$ using the Intersect feature.

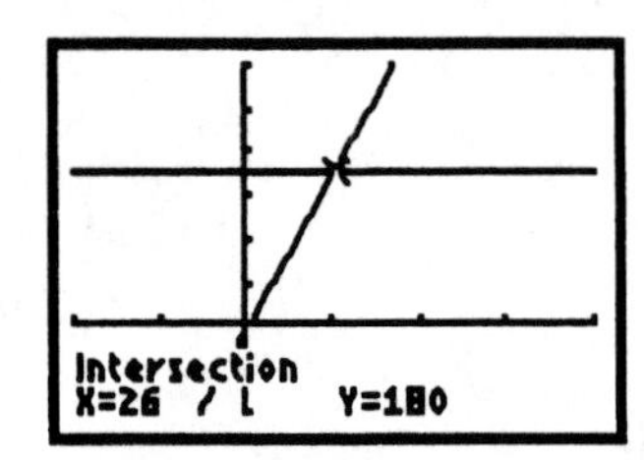

If $x = 26$, then $5x = 5 \cdot 26$, or 130, and $x - 2 = 26 - 2$, or 24.

Check. The measure of angle B, 130°, is five times the measure of angle A, 26°. The measure of angle C, 24°, is 2° less than the measure of angle A, 26°. The sum of the angle measures is 26° + 130° + 24°, or 180°. The answer checks.

State. The measure of angles A, B, and C are 26°, 130°, and 24°, respectively.

44. Let x = the measure of angle A.

Solve. $x + 2x + x + 20 = 180$

$x = 40°$, so the measure of angle A is 40°; the measure of angle B is $2 \cdot 40°$, or 80°; and the measure of angle C is 40° + 20°, or 60°.

45. ***Familiarize.*** Let n = the former number of enrollees.

Translate. We will express 48.4% as 0.484.

Former number of enrollees + 48.4% of Former number of enrollees is 416,300.

$$n + 0.484 \cdot n = 416,300$$

Carry out. We solve the equation.

$$n + 0.484n = 416,300$$
$$1.484n = 416,300 \quad \text{Collecting like terms}$$
$$n \approx 280,526$$

We could also find the first coordinate of the point of intersection of $y_1 = x + 0.484x$ and $y_2 = 416,300$ using the Intersect feature.

Check. 48.4% of 280,526 ≈ 135,775 and 280,526 + 135,775 = 416,301 ≈ 416,300. The answer checks.

State. The former number of enrollees was about 280,526.

46. Let c = the cost of the professionally handled move.

Solve: $0.52c = 1035$

$c \approx \$1990$

47. ***Familiarize.*** We make a drawing. Let r = the speed of the Central Railway freight train. Then $r + 14$ = the speed of the Amtrak passenger train. Also let t = the time the train travels.

Passenger train
r mph t hr 330 mi

Passenger train
$r + 14$ mph t hr 400 mi

We can also organize the information in a table.

$d = r \cdot t$

	Distance	Rate	Time
Freight train	330	r	t
Passenger train	400	$r + 14$	t

Translate. Using the formula $d = rt$ in each row of the table, we get two equations.

$$330 = rt \quad \text{and} \quad 400 = (r + 14)t$$

Solve each equation for t.

$$\frac{330}{r} = t \quad \text{and} \quad \frac{400}{r + 14} = t$$

Thus, we have the equation

$$\frac{330}{r} = \frac{400}{r + 14}.$$

Carry out. We solve the equation.

$$\frac{330}{r} = \frac{400}{r+14}, \text{ LCD is } r(r+14)$$

$$r(r+14)\cdot\frac{330}{r} = r(r+14)\cdot\frac{400}{r+14}$$

$$330(r+14) = 400r$$

$$330r + 4620 = 400r$$

$$4620 = 70r$$

$$66 = r$$

We could also find the first coordinate of the point of intersection of $y_1 = \frac{330}{x}$ and $y_2 = \frac{400}{x+14}$ using the Intersect feature.

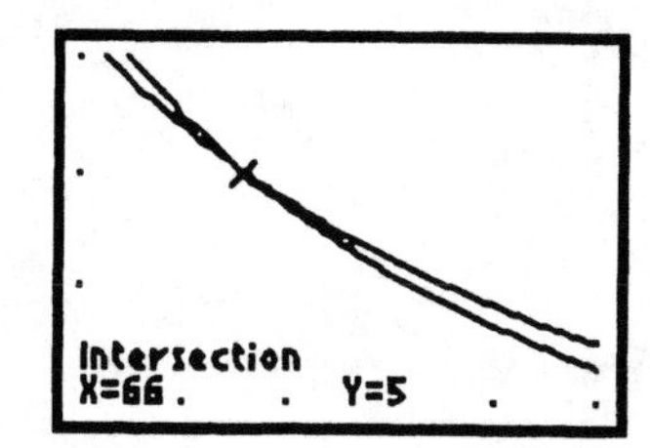

When $r = 66$, $r + 14 = 66 + 14 = 80$.

Check. The freight train travels 330 mi in 330/66, or 5 hr and the passenger train travels 400 mi in 400/80, or 5 hr. Since the times are the same the answer checks.

State. The speed of the Central Railway freight train is 66 mph, and the speed of the Amtrak passenger train is 80 mph.

48. Let t = the time the private airplane travels.

	Distance	Rate	Time
Private airplane	d	180	t
Jet	d	900	$t-2$

From the table we have the following equations:

$$d = 180t \quad\text{and}\quad d = 900(t-2)$$

Solve: $180t = 900(t-2)$

$t = 2.5$

In 2.5 hr the private airplane travels 180(2.5), or 450 km. This is the distance from the airport at which it is overtaken by the jet.

49. ***Familiarize.*** Let w = the number of pounds of Jocelyn's body weight that is water.

Translate.

50% of body weight is water.

$$0.5 \times 135 = w$$

Carry out.

$$0.5 \times 135 = w$$

$$67.5 = w$$

Check. Since 50% of 138 is 67.5, the answer checks.

State. 67.5 lb of Jocelyn's boyd weight is water.

50. Let w = the number of pounds of Reggie's body weight that is water.

Solve: $0.6 \times 186 = w$

$w = 111.6$ lb

51. ***Familiarize.*** Let n = the number of Americans who had cardiovascular procedures in 1996, in millions. Then 9% more than this number is $n + 9\%n$, or $n + 0.09n$, or $1.09n$. This is the number of procedures performed in 1999.

Translate.

The number of procedures in 1999 is 6 million.

$$1.09n = 6$$

Carry out. We will solve graphically. Graph $y_1 = 1.09x$ and $y_2 = 6$ and find the first coordinate of the point of intersection of the graphs.

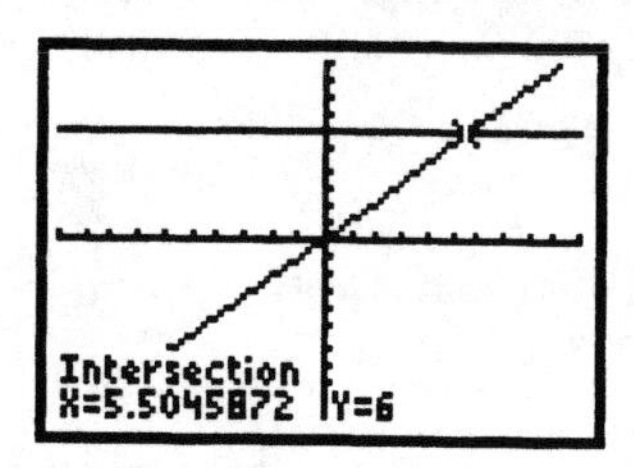

The possible solution is approximately 5.5 million.

Check. 9% of 5.5 is 0.495 and $5.5 + 0.495 = 5.995 \approx 6$. The answer checks.

State. Approximately 5.5 million Americans had cardiovascular procedures in 1996.

52. Let c = the cost of a 30-sec commercial on NBC in the fall of 1997.

Solve: $c - 0.102c = 168,000$

$c \approx \$187,082$

53. ***Familiarize.*** Let c = the number of 30-sec commercials NBC must sell on *ER* in order to break even on a episode of the show.

Translate.

Cost per commercial times number of commercials is $13 million.

$$565,000 \times c = 13,000,000$$

Carry out. We will solve algebraically.

$$565,000c = 13,000,000$$

$$c \approx 23$$

Check. If 23 commercials are sold at \$565,000 each, then $23(\$565,000) = \$12,995,000 \approx \$13,000,000$ is raised. The answer checks.

State. NBC must sell 23 30-sec commercials on *ER* in order to break even on the episode.

54. Let p = the percent of book buyers who are 18-24 years old.

Solve: $15 = 2.5p$

$p = 6\%$

55. ***Familiarize.*** Let x = the number of years it will take Horseshoe Falls to migrate one-fourth mile upstream. We will express one-fourth mile in feet:

$$\frac{1}{4}\text{ mi} \times \frac{5280\text{ ft}}{1\text{ mi}} = 1320\text{ ft}$$

Translate.

$$\underbrace{\text{Rate per year}}_{\downarrow\atop 2}\ \ \underset{\downarrow\atop \cdot}{\text{times}}\ \ \underbrace{\text{number of years}}_{\downarrow\atop x}\ \ \underset{\downarrow\atop =}{\text{is}}\ \ \underbrace{1320\text{ ft.}}_{\downarrow\atop 1320}$$

Carry out. We will solve algebraically.

$2x = 1320$

$x = 660$

Check. At a rate of 2 ft per year, in 660 yr the falls will migrate $2 \cdot 660$, or 1320 ft. The answer checks.

State. It will take Horseshoe Falls 660 yr to migrate one-fourth mile upstream.

56. Express one-half mile in inches:

$$\frac{1}{2}\text{ mi} \times \frac{5280\text{ ft}}{1\text{ mi}} \times \frac{12\text{ in.}}{1\text{ ft}} = 31,680\text{ in.}$$

Let n = the number of inches the volcano rises in a year.

Solve: $50,000n = 31,680$

$n = 0.6336$ in.

57. The graph of $f(x) = mx + b$, $m \neq 0$, is a straight line that is not horizontal. The graph of such a line intersects the x-axis exactly once. Thus, the function has exactly one zero.

58. If $m = 0$, then $f(x) = mx + b$ has the form $f(x) = b$. The graph of this function is a horizontal line. If $b \neq 0$, the graph does not intersect the x-axis and hence the function has no zero. If $b = 0$, then the function is $f(x) = 0$ and it has infinitely many zeros. Thus, we must include the restriction $m \neq 0$ in the given statement.

59. First find the slope of the given line.

$$3x + 4y = 7$$
$$4y = -3x + 7$$
$$y = -\frac{3}{4}x + \frac{7}{4}$$

The slope is $-\frac{3}{4}$. Now write a slope-intersect equation of the line containing $(-1, 4)$ with slope $-\frac{3}{4}$.

$$y - 4 = -\frac{3}{4}[x - (-1)]$$
$$y - 4 = -\frac{3}{4}(x + 1)$$
$$y - 4 = -\frac{3}{4}x - \frac{3}{4}$$
$$y = -\frac{3}{4}x + \frac{13}{4}$$

60. Enter the data and find the equation of the regression line. We have $y = 4.07x + 39.36$. When $x = 9$, $y \approx 76.0$ million.

61. $f(x) = 2x - 1$, $g(x) = x^2 - 5$

$(f + g)(x) = f(x) + g(x) = (2x - 1) + (x^2 - 5) = x^2 + 2x - 6$

62. $(fg)(-1) = f(-1)g(-1) = [2(-1) - 1][(-1)^2 - 5] = (-3)(-4) = 12$

63. $f(x) = 7 - \frac{3}{2}x = -\frac{3}{2}x + 7$

The function can be written in the form $y = mx + b$, so it is a linear function.

64. $f(x) = \frac{3}{2x} + 5$ cannot be written in the form $f(x) = mx + b$, so it is not a linear function.

65. $f(x) = x^2 + 1$ cannot be written in the form $f(x) = mx + b$, so it is not a linear function.

66. $f(x) = \frac{3}{4}x - (2.4)^2$ is in the form $f(x) = mx + b$, so it is a linear function.

Exercise Set 2.2

1.
$$\begin{aligned} &(-5 + 3i) + (7 + 8i) \\ &= (-5 + 7) + (3i + 8i) \quad \text{Collecting the real parts and the imaginary parts} \\ &= 2 + (3 + 8)i \\ &= 2 + 11i \end{aligned}$$

2. $(-6 - 5i) + (9 + 2i) = (-6 + 9) + (-5i + 2i) = 3 - 3i$

3.
$$\begin{aligned} &(4 - 9i) + (1 - 3i) \\ &= (4 + 1) + (-9i - 3i) \quad \text{Collecting the real parts and the imaginary parts} \\ &= 5 + (-9 - 3)i \\ &= 5 - 12i \end{aligned}$$

4. $(7 - 2i) + (4 - 5i) = (7 + 4) + (-2i - 5i) = 11 - 7i$

5.
$$\begin{aligned} (3 + \sqrt{-16}) + (2 + \sqrt{-25}) &= (3 + 4i) + (2 + 5i) \\ &= (3 + 2) + (4i + 5i) \\ &= 5 + 9i \end{aligned}$$

6. $(7-\sqrt{-36})+(2+\sqrt{-9})=(7-6i)+(2+3i)=$ $(7+2)+(-6i+3i)=9-3i$

7. $(10+7i)-(5+3i)$

$=(10-5)+(7i-3i)$ The 5 and the $3i$ are both being subtracted.

$=5+4i$

8. $(-3-4i)-(8-i)=(-3-8)+[-4i-(-i)]=$ $-11-3i$

9. $(13+9i)-(8+2i)$

$=(13-8)+(9i-2i)$ The 8 and the $2i$ are both being subtracted.

$=5+7i$

10. $(-7+12i)-(3-6i)=(-7-3)+[12i-(-6i)]=$ $-10+18i$

11. $(1+3i)(1-4i)$

$=1-4i+3i-12i^2$ Using FOIL

$=1-4i+3i-12(-1)$ $i^2=-1$

$=1-i+12$

$=13-i$

12. $(1-2i)(1+3i)=1+3i-2i-6i^2=1+i+6=$ $7+i$

13. $(2+3i)(2+5i)$

$=4+10i+6i+15i^2$ Using FOIL

$=4+10i+6i-15$ $i^2=-1$

$=-11+16i$

14. $(3-5i)(8-2i)=24-6i-40i+10i^2=$ $24-6i-40i-10=14-46i$

15. $7i(2-5i)$

$=14i-35i^2$ Using the distributive law

$=14i+35$ $i^2=-1$

$=35+14i$ Writing in the form $a+bi$

16. $3i(6+4i)=18i+12i^2=18i-12=-12+18i$

17. $(3+\sqrt{-16})(2+\sqrt{-25})$

$=(3+4i)(2+5i)$

$=6+15i+8i+20i^2$

$=6+15i+8i-20$ $i^2=-1$

$=-14+23i$

18. $(7-\sqrt{-16})(2+\sqrt{-9})=(7-4i)(2+3i)=$ $14+21i-8i-12i^2=14+21i-8i+12=$ $26+13i$

19. $(5-4i)(5+4i)=5^2-(4i)^2$

$=25-16i^2$

$=25+16$ $i^2=-1$

$=41$

20. $(5+9i)(5-9i)=25-81i^2=25+81=106$

21. $(4+2i)^2$

$=16+2\cdot4\cdot2i+(2i)^2$ Recall $(A+B)^2=A^2+2AB+B^2$

$=16+16i+4i^2$

$=16+16i-4$ $i^2=-1$

$=12+16i$

22. $(5-4i)^2=25-40i+16i^2=25-40i-16=$ $9-40i$

23. $(-2+7i)^2$

$=(-2)^2+2(-2)(7i)+(7i)^2$ Recall $(A+B)^2=A^2+2AB+B^2$

$=4-28i+49i^2$

$=4-28i-49$ $i^2=-1$

$=-45-28i$

24. $(-3+2i)^2=9-12i+4i^2=9-12i-4=5-12i$

25. $\dfrac{2+\sqrt{3}i}{5-4i}$

$=\dfrac{2+\sqrt{3}i}{5-4i}\cdot\dfrac{5+4i}{5+4i}$ $5+4i$ is the conjugate of the divisor.

$=\dfrac{(2+\sqrt{3}i)(5+4i)}{(5-4i)(5+4i)}$

$=\dfrac{10+8i+5\sqrt{3}i+4\sqrt{3}i^2}{25-16i^2}$

$=\dfrac{10+8i+5\sqrt{3}i-4\sqrt{3}}{25+16}$ $i^2=-1$

$=\dfrac{10-4\sqrt{3}+(8+5\sqrt{3})i}{41}$

$=\dfrac{10-4\sqrt{3}}{41}+\dfrac{8+5\sqrt{3}}{41}i$ Writing in the form $a+bi$

26. $\dfrac{\sqrt{5}+3i}{1-i}=\dfrac{\sqrt{5}+3i}{1-i}\cdot\dfrac{1+i}{1+i}$

$=\dfrac{\sqrt{5}+\sqrt{5}i+3i+3i^2}{1-i^2}$

$=\dfrac{\sqrt{5}+\sqrt{5}i+3i-3}{1+1}$

$=\dfrac{\sqrt{5}-3}{2}+\dfrac{\sqrt{5}+3}{2}i$

27. $\dfrac{4+i}{-3-2i}$

$= \dfrac{4+i}{-3-2i} \cdot \dfrac{-3+2i}{-3+2i}$ $\quad -3+2i$ is the conjugate of the divisor.

$= \dfrac{(4+i)(-3+2i)}{(-3-2i)(-3+2i)}$

$= \dfrac{-12+5i+2i^2}{9-4i^2}$

$= \dfrac{-12+5i-2}{9+4}$ $\quad i^2 = -1$

$= \dfrac{-14+5i}{13}$

$= -\dfrac{14}{13} + \dfrac{5}{13}i$ $\quad$ Writing in the form $a+bi$

28. $\dfrac{5-i}{-7+2i} = \dfrac{5-i}{-7+2i} \cdot \dfrac{-7-2i}{-7-2i}$

$= \dfrac{-35-3i+2i^2}{49-4i^2}$

$= \dfrac{-35-3i-2}{49+4}$

$= -\dfrac{37}{53} - \dfrac{3}{53}i$

29. $\dfrac{3}{5-11i}$

$= \dfrac{3}{5-11i} \cdot \dfrac{5+11i}{5+11i}$ $\quad 5-11i$ is the conjugate of $5+11i$.

$= \dfrac{3(5+11i)}{(5-11i)(5+11i)}$

$= \dfrac{15+33i}{25-121i^2}$

$= \dfrac{15+33i}{25+121}$ $\quad i^2 = -1$

$= \dfrac{15+33i}{146}$

$= \dfrac{15}{146} + \dfrac{33}{146}i$ $\quad$ Writing in the form $a+bi$

30. $\dfrac{i}{2+i} = \dfrac{i}{2+i} \cdot \dfrac{2-i}{2-i}$

$= \dfrac{2i-i^2}{4-i^2}$

$= \dfrac{2i+1}{4+1}$

$= \dfrac{1}{5} + \dfrac{2}{5}i$

31. $\dfrac{1+i}{(1-i)^2}$

$= \dfrac{1+i}{1-2i+i^2}$

$= \dfrac{1+i}{1-2i-1}$ $\quad i^2 = -1$

$= \dfrac{1+i}{-2i}$

$= \dfrac{1+i}{-2i} \cdot \dfrac{2i}{2i}$ $\quad 2i$ is the conjugate of $-2i$.

$= \dfrac{(1+i)(2i)}{(-2i)(2i)}$

$= \dfrac{2i+2i^2}{-4i^2}$

$= \dfrac{2i-2}{4}$ $\quad i^2 = -1$

$= -\dfrac{2}{4} + \dfrac{2}{4}i$

$= -\dfrac{1}{2} + \dfrac{1}{2}i$

32. $\dfrac{1-i}{(1+i)^2} = \dfrac{1-i}{1+2i+i^2}$

$= \dfrac{1-i}{1+2i-1}$

$= \dfrac{1-i}{2i}$

$= \dfrac{1-i}{2i} \cdot \dfrac{-2i}{-2i}$

$= \dfrac{-2i+2i^2}{-4i^2}$

$= \dfrac{-2i-2}{4}$

$= -\dfrac{1}{2} - \dfrac{1}{2}i$

33. $\dfrac{4-2i}{1+i} + \dfrac{2-5i}{1+i}$

$= \dfrac{6-7i}{1+i}$ $\quad$ Adding

$= \dfrac{6-7i}{1+i} \cdot \dfrac{1-i}{1-i}$ $\quad 1-i$ is the conjugate of $1+i$.

$= \dfrac{(6-7i)(1-i)}{(1+i)(1-i)}$

$= \dfrac{6-13i+7i^2}{1-i^2}$

$= \dfrac{6-13i-7}{1+1}$ $\quad i^2 = -1$

$= \dfrac{-1-13i}{2}$

$= -\dfrac{1}{2} - \dfrac{13}{2}i$

34. $\dfrac{3+2i}{1-i}+\dfrac{6+2i}{1-i}=\dfrac{9+4i}{1-i}$

$=\dfrac{9+4i}{1-i}\cdot\dfrac{1+i}{1+i}$

$=\dfrac{9+13i+4i^2}{1-i^2}$

$=\dfrac{9+13i-4}{1+1}$

$=\dfrac{5}{2}+\dfrac{13}{2}i$

35. $i^{11}=i^{10}\cdot i=(i^2)^5\cdot i=(-1)^5\cdot i=-1\cdot i=-i$

36. $i^7=i^6\cdot i=(i^2)^3\cdot i=(-1)^3\cdot i=-1\cdot i=-i$

37. $i^{35}=i^{34}\cdot i=(i^2)^{17}\cdot i=(-1)^{17}\cdot i=-1\cdot i=-i$

38. $i^{24}=(i^2)^{12}=(-1)^{12}=1$

39. $i^{64}=(i^2)^{32}=(-1)^{32}=1$

40. $i^{42}=(i^2)^{21}=(-1)^{21}=-1$

41. $(-i)^{71}=(-1\cdot i)^{71}=(-1)^{71}\cdot i^{71}=-i^{70}\cdot i=-(i^2)^{35}\cdot i=-(-1)^{35}\cdot i=-(-1)i=i$

42. $(-i)^6=i^6=(i^2)^3=(-1)^3=-1$

43. $(5i)^4=5^4\cdot i^4=625(i^2)^2=625(-1)^2=625\cdot 1=625$

44. $(2i)^5=32i^5=32\cdot i^4\cdot i=32(i^2)^2\cdot i=32(-1)^2\cdot i=32\cdot 1\cdot i=32i$

45. The sum of two imaginary numbers is not always an imaginary number. For example, $(2+i)+(3-i)=5$, a real number.

46. The product of two imaginary numbers is not always an imaginary number. For example, $i\cdot i=i^2=-1$, a real number.

47. First find the slope of the given line.

$$3x-6y=7$$
$$-6y=-3x+7$$
$$y=\frac{1}{2}x-\frac{7}{6}$$

The slope is $\frac{1}{2}$. The slope of the desired line is the opposite of the reciprocal of $\frac{1}{2}$, or -2. Write a slope-intercept equation of the line containing $(3,-5)$ with slope -2.

$$y-(-5)=-2(x-3)$$
$$y+5=-2x+6$$
$$y=-2x+1$$

48. $(f-g)(x)=f(x)-g(x)=x^2+4-(3x+5)=x^2-3x-1$

49. $(f/g)(2)=\dfrac{f(2)}{g(2)}=\dfrac{2^2+4}{3\cdot 2+5}=\dfrac{4+4}{6+5}=\dfrac{8}{11}$

50. $\dfrac{f(x+h)-f(x)}{h}$

$=\dfrac{(x+h)^2-3(x+h)+4-(x^2-3x+4)}{h}$

$=\dfrac{x^2+2xh+h^2-3x-3h+4-x^2+3x-4}{h}$

$=\dfrac{2xh+h^2-3h}{h}$

$=\dfrac{h(2x+h-3)}{h}$

$=2x+h-3$

51. $(a+bi)+(a-bi)=2a$, a real number. Thus, the statement is true.

52. $(a+bi)+(c+di)=(a+c)+(b+d)i$. The conjugate of this sum is $(a+c)-(b+d)i=a+c-bi-di=(a-bi)+(c-di)$, the sum of the conjugates of the individual complex numbers. Thus, the statement is true.

53. $(a+bi)(c+di)=(ac-bd)+(ad+bc)i$. The conjugate of the product is $(ac-bd)-(ad+bc)i=(a-bi)(c-di)$, the product of the conjugates of the individual complex numbers. Thus, the statement is true.

54. $\dfrac{1}{z}=\dfrac{1}{a+bi}\cdot\dfrac{a-bi}{a-bi}=\dfrac{a}{a^2+b^2}+\dfrac{-b}{a^2+b^2}i$

55. $z\bar{z}=(a+bi)(a-bi)=a^2-b^2i^2=a^2+b^2$

56.
$$z+6\bar{z}=7$$
$$a+bi+6(a-bi)=7$$
$$a+bi+6a-6bi=7$$
$$7a-5bi=7$$

Then $7a=7$, so $a=1$, and $-5b=0$, so $b=0$. Thus, $z=1$.

Exercise Set 2.3

1. a) The graph crosses the x-axis at $(-4,0)$ and at $(2,0)$. These are the x-intercepts.

b) The zeros of the function are the first coordinates of the x-intercepts of the graph. They are -4 and 2.

2. a) $(-5,0)$, $(3,0)$

b) -5, 3

3.

$x^2+6x=7$

$x^2+6x+9=7+9$ Completing the square: $\frac{1}{2}\cdot 6=3$ and $3^2=9$

$(x+3)^2=16$ Factoring

$x+3=\pm 4$ Using the principle of square roots

$x=-3\pm 4$

$x=-3-4$ *or* $x=-3+4$

$x=-7$ *or* $x=1$

The solutions are -7 and 1.

4. $x^2 + 8x = -15$

$x^2+8x+16 = -15+16$ ($\frac{1}{2}\cdot 8 = 4$ and $4^2 = 16$)

$(x+4)^2 = 1$

$x+4 = \pm 1$

$x = -4 \pm 1$

$x = -4-1$ *or* $x = -4+1$

$x = -5$ *or* $x = -3$

The solutions are -5 and -3.

5. $x^2 = 8x - 9$

$x^2 - 8x = -9$ Subtracting $8x$

$x^2 - 8x + 16 = -9 + 16$ Completing the square: $\frac{1}{2}(-8) = -4$ and $(-4)^2 = 16$

$(x-4)^2 = 7$ Factoring

$x - 4 = \pm\sqrt{7}$ Using the principle of square roots

$x = 4 \pm \sqrt{7}$

The solutions are $4 - \sqrt{7}$ and $4 + \sqrt{7}$, or $4 \pm \sqrt{7}$.

6. $x^2 = 22 + 10x$

$x^2 - 10x = 22$

$x^2 - 10x + 25 = 22 + 25$ ($\frac{1}{2}(-10) = -5$ and $(-5)^2 = 25$)

$(x-5)^2 = 47$

$x - 5 = \pm\sqrt{47}$

$x = 5 \pm \sqrt{47}$

The solutions are $5 - \sqrt{47}$ and $5 + \sqrt{47}$, or $5 \pm \sqrt{47}$.

7. $x^2 + 8x + 25 = 0$

$x^2 + 8x = -25$ Subtracting 25

$x^2 + 8x + 16 = -25 + 16$ Completing the square: $\frac{1}{2}\cdot 8 = 4$ and $4^2 = 16$

$(x+4)^2 = -9$ Factoring

$x + 4 = \pm 3i$ Using the principle of square roots

$x = -4 \pm 3i$

The solutions are $-4 - 3i$ and $-4 + 3i$, or $-4 \pm 3i$.

8. $x^2 + 6x + 13 = 0$

$x^2 + 6x = -13$

$x^2 + 6x + 9 = -13 + 9$ ($\frac{1}{2}\cdot 6 = 3$ and $3^2 = 9$)

$(x+3)^2 = -4$

$x + 3 = \pm 2i$

$x = -3 \pm 2i$

The solution are $-3 - 2i$ and $-3 + 2i$, or $-3 \pm 2i$.

9. $3x^2 + 5x - 2 = 0$

$3x^2 + 5x = 2$ Adding 2

$x^2 + \frac{5}{3}x = \frac{2}{3}$ Dividing by 3

$x^2 + \frac{5}{3}x + \frac{25}{36} = \frac{2}{3} + \frac{25}{36}$ Completing the square: $\frac{1}{2}\cdot\frac{5}{3} = \frac{5}{6}$ and $(\frac{5}{6})^2 = \frac{25}{36}$

$\left(x + \frac{5}{6}\right)^2 = \frac{49}{36}$ Factoring and simplifying

$x + \frac{5}{6} = \pm\frac{7}{6}$ Using the principle of square roots

$x = -\frac{5}{6} \pm \frac{7}{6}$

$x = -\frac{5}{6} - \frac{7}{6}$ *or* $x = -\frac{5}{6} + \frac{7}{6}$

$x = -\frac{12}{6}$ *or* $x = \frac{2}{6}$

$x = -2$ *or* $x = \frac{1}{3}$

The solutions are -2 and $\frac{1}{3}$.

10. $2x^2 - 5x - 3 = 0$

$2x^2 - 5x = 3$

$x^2 - \frac{5}{2}x = \frac{3}{2}$

$x^2 - \frac{5}{2}x + \frac{25}{16} = \frac{3}{2} + \frac{25}{16}$ ($\frac{1}{2}(-\frac{5}{2}) = -\frac{5}{4}$ and $(-\frac{5}{4})^2 = \frac{25}{16}$)

$\left(x - \frac{5}{4}\right)^2 = \frac{49}{16}$

$x - \frac{5}{4} = \pm\frac{7}{4}$

$x = \frac{5}{4} \pm \frac{7}{4}$

$x = \frac{5}{4} - \frac{7}{4}$ *or* $x = \frac{5}{4} + \frac{7}{4}$

$x = -\frac{1}{2}$ *or* $x = 3$

The solutions are $-\frac{1}{2}$ and 3.

11. $x^2 - 2x = 15$

$x^2 - 2x - 15 = 0$

$(x-5)(x+3) = 0$ Factoring

$x - 5 = 0$ *or* $x + 3 = 0$

$x = 5$ *or* $x = -3$

The solutions are 5 and -3.

12. $x^2 + 4x = 5$

$x^2 + 4x - 5 = 0$

$(x+5)(x-1) = 0$

$$x+5=0 \quad or \quad x-1=0$$
$$x=-5 \quad or \quad x=1$$

The solutions are -5 and 1.

13. $5m^2+3m=2$

$5m^2+3m-2=0$

$(5m-2)(m+1)=0$ Factoring

$5m-2=0 \quad or \quad m+1=0$

$m=\frac{2}{5} \quad or \quad m=-1$

The solutions are $\frac{2}{5}$ and -1.

14. $2y^2-3y-2=0$

$(2y+1)(y-2)=0$

$2y+1=0 \quad or \quad y-2=0$

$y=-\frac{1}{2} \quad or \quad y=2$

The solutions are $-\frac{1}{2}$ and 2.

15. $3x^2+6=10x$

$3x^2-10x+6=0$

We use the quadratic formula. Here $a=3$, $b=-10$, and $c=6$.

$$x=\frac{-b\pm\sqrt{b^2-4ac}}{2a}$$
$$=\frac{-(-10)\pm\sqrt{(-10)^2-4\cdot 3\cdot 6}}{2\cdot 3} \quad \text{Substituting}$$
$$=\frac{10\pm\sqrt{28}}{6}=\frac{10\pm 2\sqrt{7}}{6}$$
$$=\frac{2(5\pm\sqrt{7})}{2\cdot 3}=\frac{5\pm\sqrt{7}}{3}$$

The solutions are $\frac{5-\sqrt{7}}{3}$ and $\frac{5+\sqrt{7}}{3}$, or $\frac{5\pm\sqrt{7}}{3}$.

16. $3t^2+8t+3=0$

$$t=\frac{-8\pm\sqrt{8^2-4\cdot 3\cdot 3}}{2\cdot 3}$$
$$=\frac{-8\pm\sqrt{28}}{6}=\frac{-8\pm 2\sqrt{7}}{6}$$
$$=\frac{2(-4\pm\sqrt{7})}{2\cdot 3}=\frac{-4\pm\sqrt{7}}{3}$$

The solutions are $\frac{-4-\sqrt{7}}{3}$ and $\frac{-4+\sqrt{7}}{3}$, or $\frac{-4\pm\sqrt{7}}{3}$.

17. $x^2+x+2=0$

We use the quadratic formula. Here $a=1$, $b=1$, and $c=2$.

$$x=\frac{-b\pm\sqrt{b^2-4ac}}{2a}$$
$$=\frac{-1\pm\sqrt{1^2-4\cdot 1\cdot 2}}{2\cdot 1} \quad \text{Substituting}$$
$$=\frac{-1\pm\sqrt{-7}}{2}$$
$$=\frac{-1\pm\sqrt{7}i}{2}=-\frac{1}{2}\pm\frac{\sqrt{7}}{2}i$$

The solutions are $-\frac{1}{2}-\frac{\sqrt{7}}{2}i$ and $-\frac{1}{2}+\frac{\sqrt{7}}{2}i$, or $-\frac{1}{2}\pm\frac{\sqrt{7}}{2}i$.

18. $x^2+1=x$

$x^2-x+1=0$

$$x=\frac{-(-1)\pm\sqrt{(-1)^2-4\cdot 1\cdot 1}}{2\cdot 1}$$
$$=\frac{1\pm\sqrt{-3}}{2}=\frac{1\pm\sqrt{3}i}{2}$$
$$=\frac{1}{2}\pm\frac{\sqrt{3}}{2}i$$

The solutions are $\frac{1}{2}-\frac{\sqrt{3}}{2}i$ and $\frac{1}{2}+\frac{\sqrt{3}}{2}i$, or $\frac{1}{2}\pm\frac{\sqrt{3}}{2}i$.

19. $5t^2-8t=3$

$5t^2-8t-3=0$

We use the quadratic formula. Here $a=5$, $b=-8$, and $c=-3$.

$$t=\frac{-b\pm\sqrt{b^2-4ac}}{2a}$$
$$=\frac{-(-8)\pm\sqrt{(-8)^2-4\cdot 5(-3)}}{2\cdot 5}$$
$$=\frac{8\pm\sqrt{124}}{10}=\frac{8\pm 2\sqrt{31}}{10}$$
$$=\frac{2(4\pm\sqrt{31})}{2\cdot 5}=\frac{4\pm\sqrt{31}}{5}$$

The solutions are $\frac{4-\sqrt{31}}{5}$ and $\frac{4+\sqrt{31}}{5}$, or $\frac{4\pm\sqrt{31}}{5}$.

20. $5x^2+2=x$

$5x^2-x+2=0$

$$x=\frac{-(-1)\pm\sqrt{(-1)^2-4\cdot 5\cdot 2}}{2\cdot 5}$$
$$=\frac{1\pm\sqrt{-39}}{10}=\frac{1\pm\sqrt{39}i}{10}$$
$$=\frac{1}{10}\pm\frac{\sqrt{39}}{10}i$$

The solutions are $\frac{1}{10}-\frac{\sqrt{39}}{10}i$ and $\frac{1}{10}+\frac{\sqrt{39}}{10}i$, or

$\frac{1}{10} \pm \frac{\sqrt{39}}{10}i.$

21. $3x^2 + 4 = 5x$

$3x^2 - 5x + 4 = 0$

We use the quadratic formula. Here $a = 3$, $b = -5$, and $c = 4$.

$$x = \frac{-b \pm \sqrt{b^2 - 4ac}}{2a}$$

$$= \frac{-(-5) \pm \sqrt{(-5)^2 - 4 \cdot 3 \cdot 4}}{2 \cdot 3}$$

$$= \frac{5 \pm \sqrt{-23}}{6} = \frac{5 \pm \sqrt{23}i}{6}$$

$$= \frac{5}{6} \pm \frac{\sqrt{23}}{6}i$$

The solutions are $\frac{5}{6} - \frac{\sqrt{23}}{6}i$ and $\frac{5}{6} + \frac{\sqrt{23}}{6}i$, or $\frac{5}{6} \pm \frac{\sqrt{23}}{6}i.$

22. $2t^2 - 5t = 1$

$2t^2 - 5t - 1 = 0$

$$t = \frac{-(-5) \pm \sqrt{(-5)^2 - 4 \cdot 2(-1)}}{2 \cdot 2}$$

$$= \frac{5 \pm \sqrt{33}}{4}$$

The solutions are $\frac{5 - \sqrt{33}}{4}$ and $\frac{5 + \sqrt{33}}{4}$, or $\frac{5 \pm \sqrt{33}}{4}$.

23. $4x^2 = 8x + 5$

$4x^2 - 8x - 5 = 0$

$a = 4,\ b = -8,\ c = -5$

$b^2 - 4ac = (-8)^2 - 4 \cdot 4(-5) = 144$

Since $b^2 - 4ac > 0$, there are no imaginary solutions.

24. $4x^2 - 12x + 9 = 0$

$b^2 - 4ac = (-12)^2 - 4 \cdot 4 \cdot 9 = 0$

There are no imaginary solutions.

25. $x^2 + 3x + 4 = 0$

$a = 1,\ b = 3,\ c = 4$

$b^2 - 4ac = 3^2 - 4 \cdot 1 \cdot 4 = -7$

Since $b^2 - 4ac < 0$, there are imaginary solutions.

26. $x^2 - 2x + 4 = 0$

$b^2 - 4ac = (-2)^2 - 4 \cdot 1 \cdot 4 = -12 < 0$

There are imaginary solutions.

27. $5t^2 - 7t = 0$

$a = 5,\ b = -7,\ c = 0$

$b^2 - 4ac = (-7)^2 - 4 \cdot 5 \cdot 0 = 49$

Since $b^2 - 4ac > 0$, there are no imaginary solutions.

28. $5t^2 - 4t = 11$

$5t^2 - 4t - 11 = 0$

$b^2 - 4ac = (-4)^2 - 4 \cdot 5(-11) = 236 > 0$

There are no imaginary solutions.

29. Graph $y = x^2 - 8x + 12$ and use the Zero feature twice.

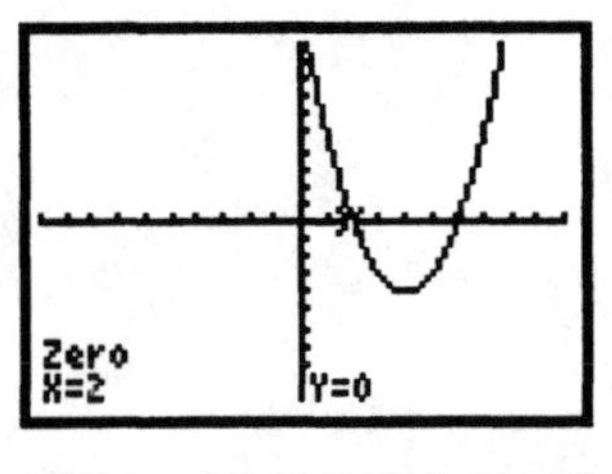

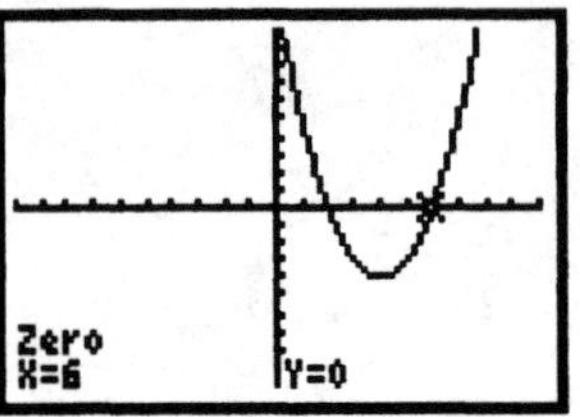

The solutions are 2 and 6.

30. Graph $y = 5x^2 + 42x + 16$ and use the Zero feature twice.

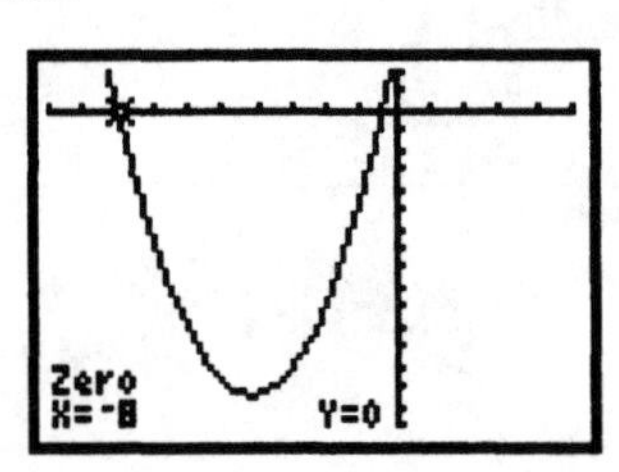

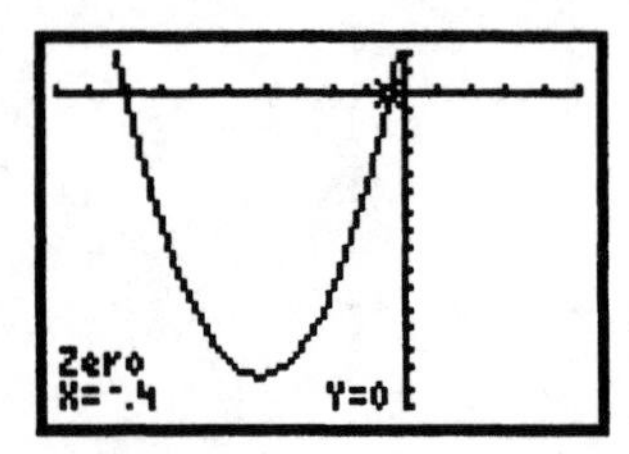

The solutions are -8 and -0.4.

31. Graph $y = 7x^2 - 43x + 6$ and use the Zero feature twice.

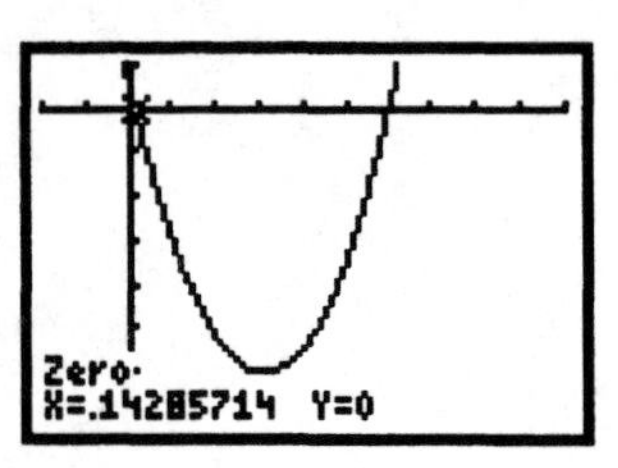

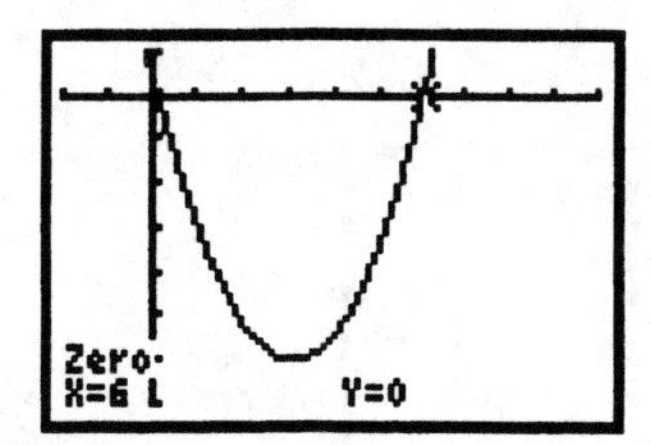

One solution is approximately 0.143 and the other is 6.

32. Graph $y = 10x^2 - 23x + 12$ and use the Zero feature twice.

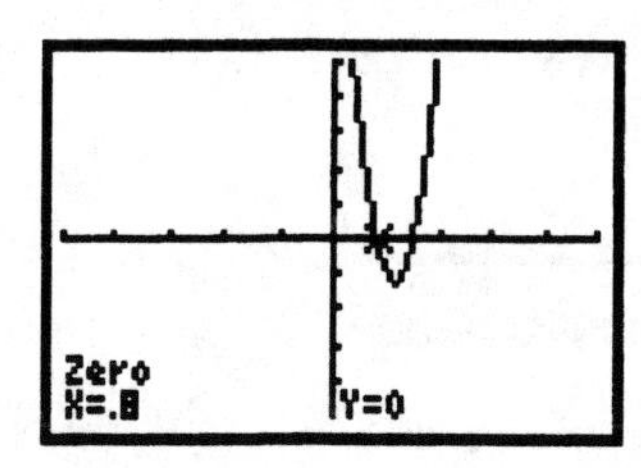

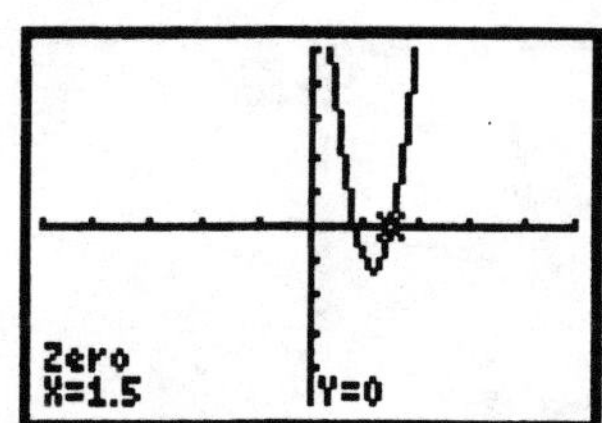

The solutions are 0.8 and 1.5.

33. Graph $y_1 = 6x+1$ and $y_2 = 4x^2$ and use the Intersect feature twice.

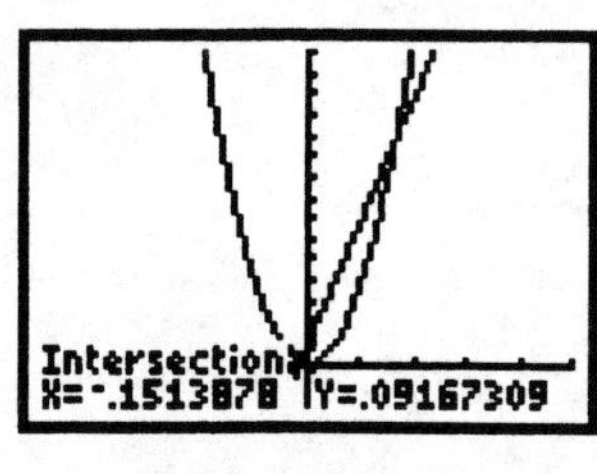

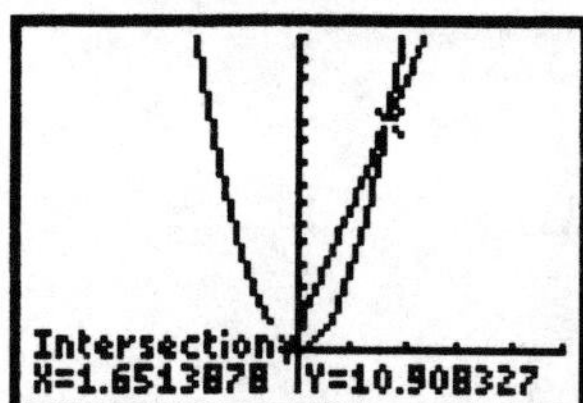

The solutions are approximately -0.151 and 1.651.

34. Graph $y_1 = 3x^2+5x$ and $y_2 = 3$ and use the Intersect feature twice.

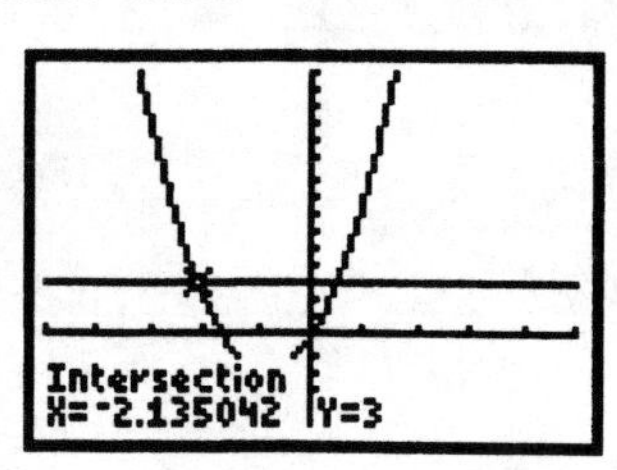

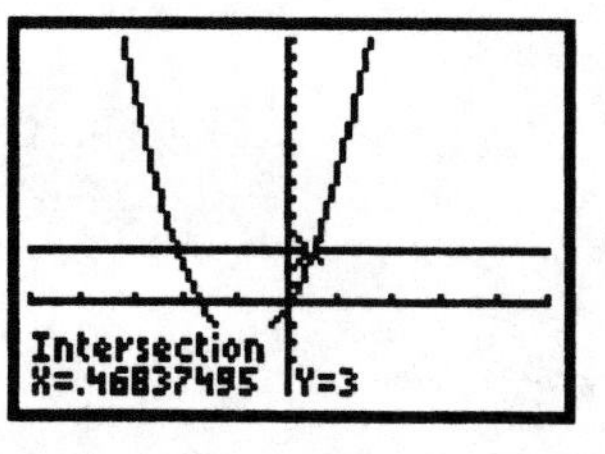

The solutions are approximately -2.135 and 0.468.

35. Graph $y_1 = 2x^2-4$ and $y_2 = 5x$ and use the Intersect feature twice.

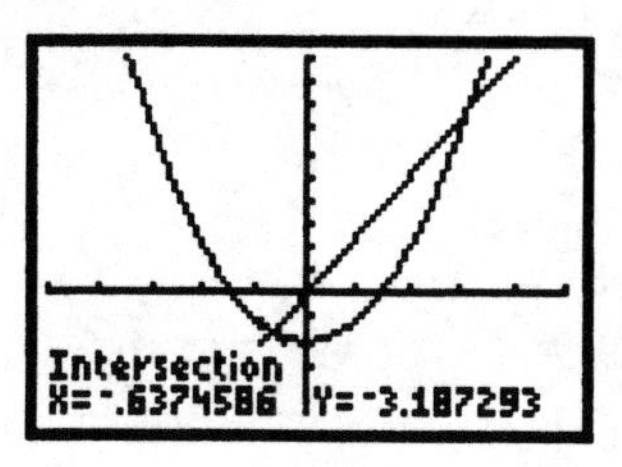

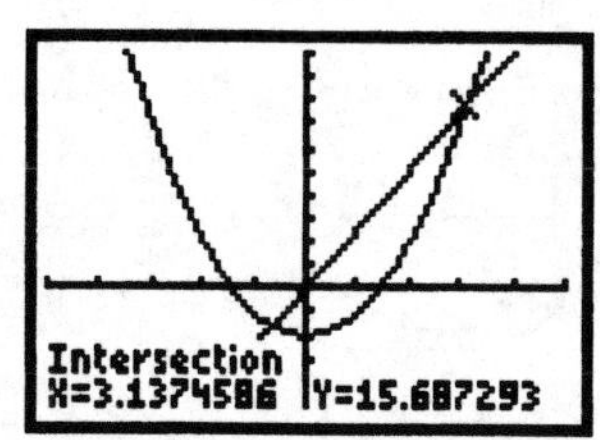

The solutions are approximately -0.637 and 3.137.

36. Graph $y_1 = 4x^2-2$ and $y_2 = 3x$ and use the Intersect feature twice.

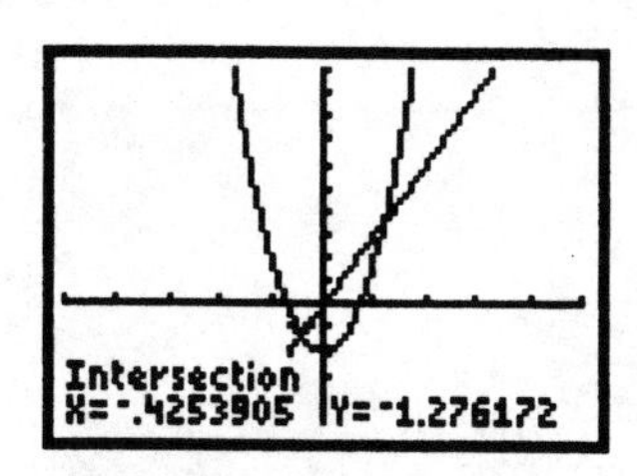

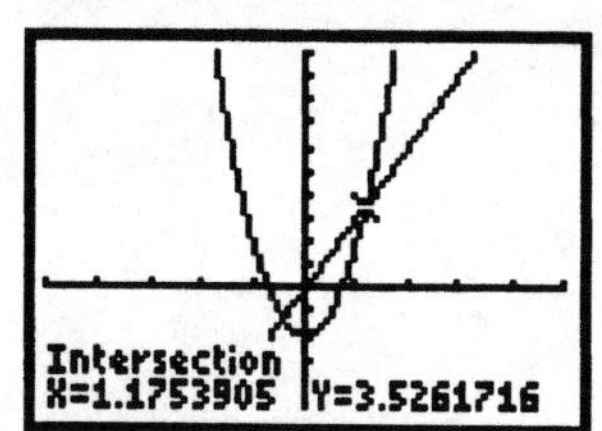

The solutions are approximately -0.425 and 1.175.

37.
$$x^2 + 6x + 5 = 0 \quad \text{Setting } f(x) = 0$$
$$(x+5)(x+1) = 0 \quad \text{Factoring}$$
$$x+5 = 0 \quad or \quad x+1 = 0$$
$$x = -5 \quad or \quad x = -1$$

The solutions are -5 and -1.

38.
$$x^2 - x - 2 = 0$$
$$(x+1)(x-2) = 0$$

$x+1=0 \quad or \quad x-2=0$

$x=-1 \quad or \quad x=2$

The solutions are -1 and 2.

39. $x^2-3x-3=0$

$a=1,\ b=-3,\ c=-3$

$$x=\frac{-b\pm\sqrt{b^2-4ac}}{2a}$$

$$=\frac{-(-3)\pm\sqrt{(-3)^2-4\cdot1\cdot(-3)}}{2\cdot1}$$

$$=\frac{3\pm\sqrt{9+12}}{2}$$

$$=\frac{3\pm\sqrt{21}}{2}$$

The solutions are $\frac{3-\sqrt{21}}{2}$ and $\frac{3+\sqrt{21}}{2}$, or $\frac{3\pm\sqrt{21}}{2}$.

40. $3x^2+8x+2=0$

$$x=\frac{-8\pm\sqrt{8^2-4\cdot3\cdot2}}{2\cdot3}$$

$$=\frac{-8\pm\sqrt{40}}{6}=\frac{-8\pm2\sqrt{10}}{6}$$

$$=\frac{-4\pm\sqrt{10}}{3}$$

The solutions are $\frac{-4-\sqrt{10}}{3}$ and $\frac{-4+\sqrt{10}}{3}$, or $\frac{-4\pm\sqrt{10}}{3}$.

41. $x^2-5x+1=0$

$a=1,\ b=-5,\ c=1$

$$x=\frac{-b\pm\sqrt{b^2-4ac}}{2a}$$

$$=\frac{-(-5)\pm\sqrt{(-5)^2-4\cdot1\cdot1}}{2\cdot1}$$

$$=\frac{5\pm\sqrt{25-4}}{2}$$

$$=\frac{5\pm\sqrt{21}}{2}$$

The solutions are $\frac{5-\sqrt{21}}{2}$ and $\frac{5+\sqrt{21}}{2}$, or $\frac{5\pm\sqrt{21}}{2}$.

42. $x^2-3x-7=0$

$$x=\frac{-(-3)\pm\sqrt{(-3)^2-4\cdot1\cdot(-7)}}{2\cdot1}$$

$$=\frac{3\pm\sqrt{37}}{2}$$

The solutions are $\frac{3-\sqrt{37}}{2}$ and $\frac{3+\sqrt{37}}{2}$, or $\frac{3\pm\sqrt{37}}{2}$.

43. Graph $y=x^2+2x-5$ and use the Zero feature twice.

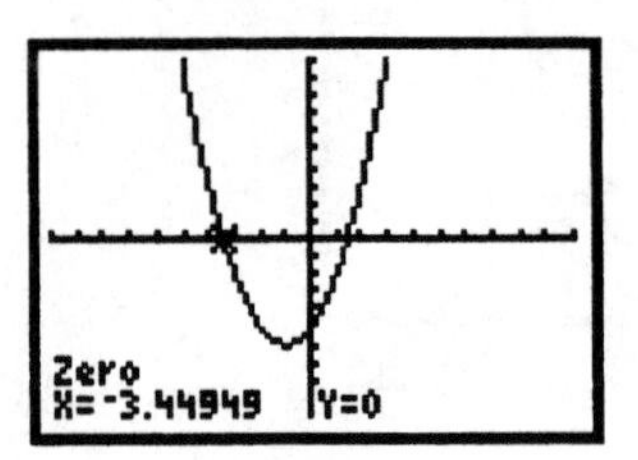

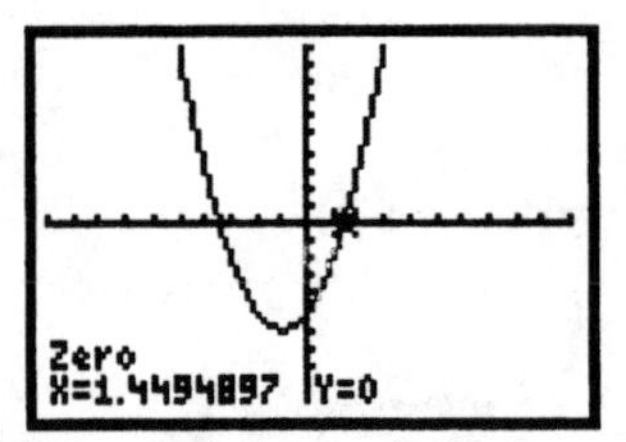

The zeros are approximately -3.449 and 1.449.

44. Graph $y=x^2-x-4$ and use the Zero feature twice.

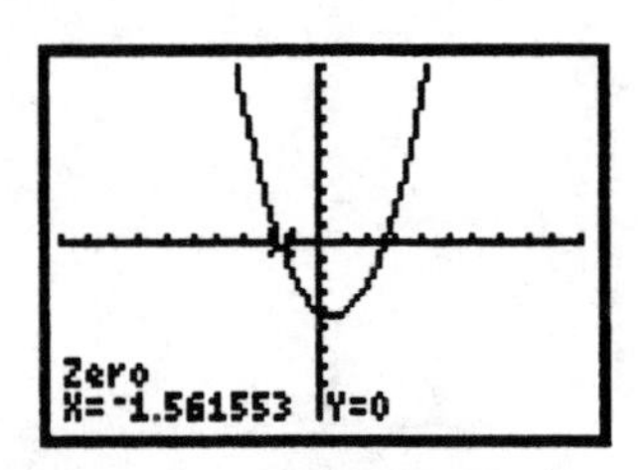

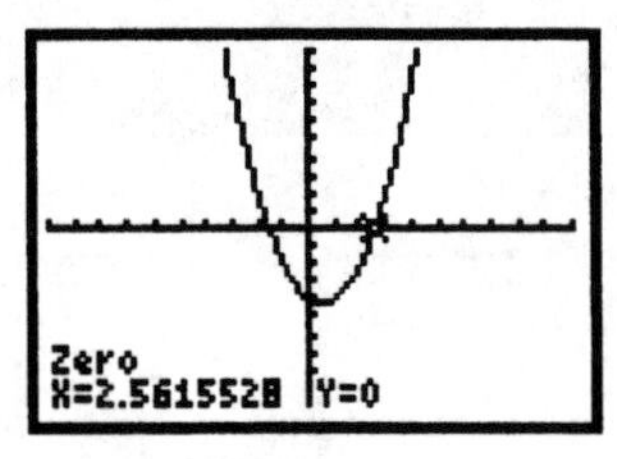

The zeros are approximately -1.562 and 2.562.

45. Graph $y=3x^2+2x-4$ and use the Zero feature twice.

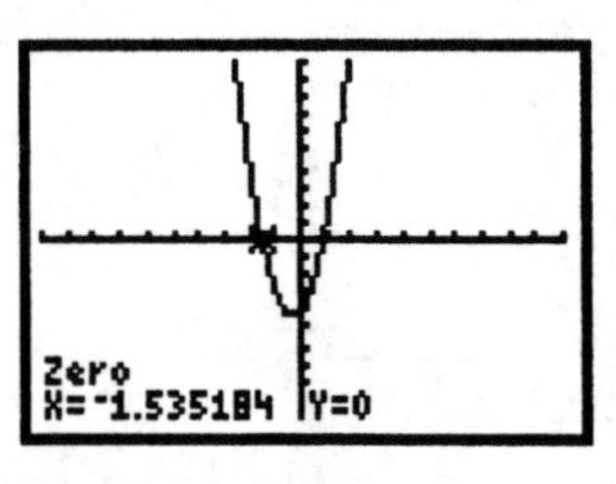

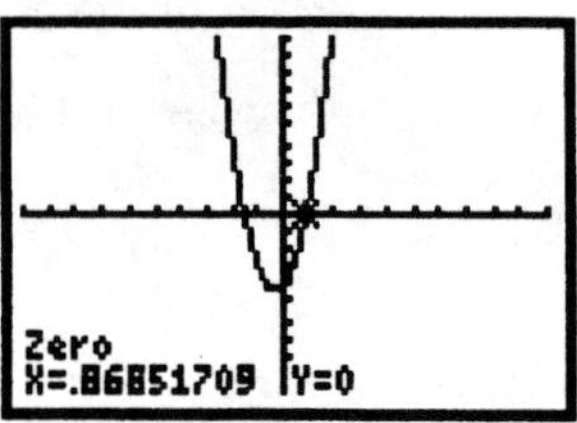

The zeros are approximately -1.535 and 0.869.

46. Graph $y = 9x^2 - 8x - 7$ and use the Zero feature twice.

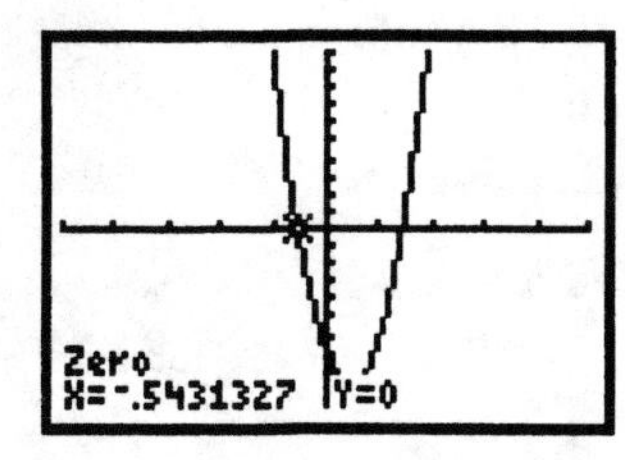

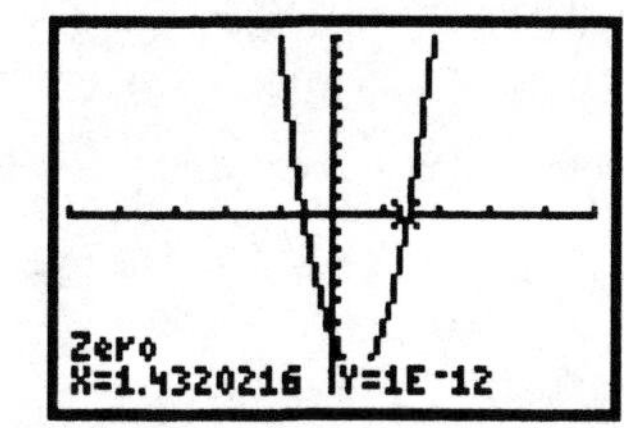

The zeros are approximately -0.543 and 1.432.

47. Graph $y = 5.02x^2 - 4.19x - 2.057$ and use the Zero feature twice.

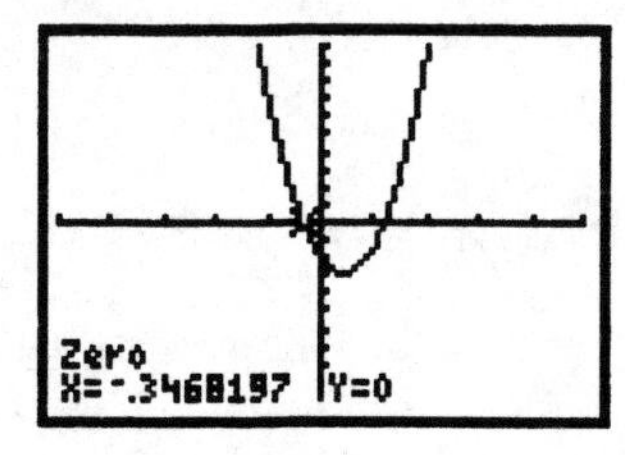

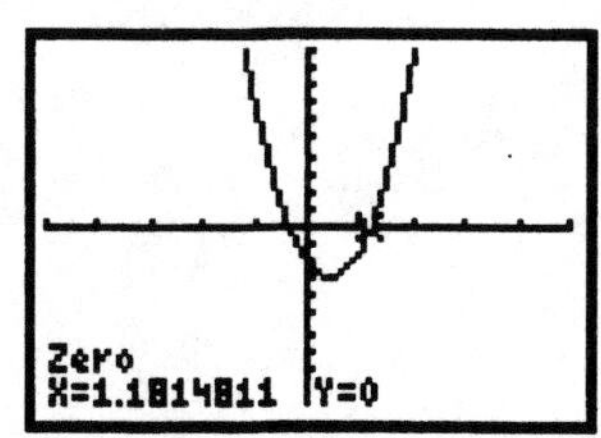

The zeros are approximately -0.347 and 1.181.

48. Graph $y = 1.21x^2 - 2.34x - 5.63$ and use the Zero feature twice.

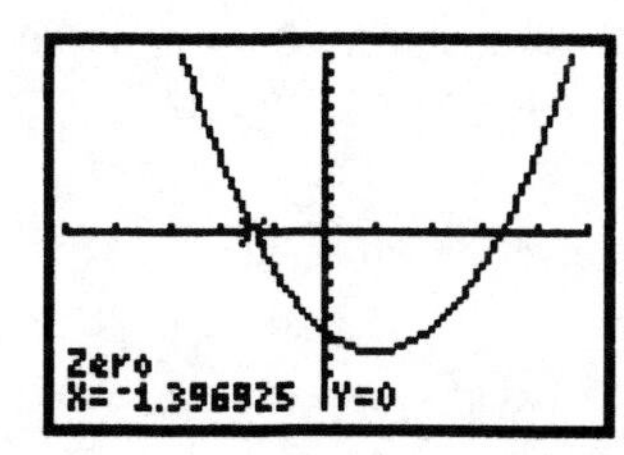

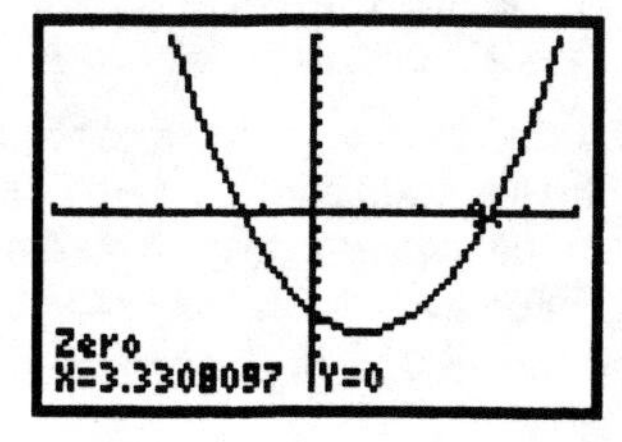

The zeros are approximately -1.397 and 3.331.

49. $x^4 - 3x^2 + 2 = 0$

Let $u = x^2$.

$u^2 - 3u + 2 = 0$ Substituting u for x^2

$(u-1)(u-2) = 0$

$u - 1 = 0$ *or* $u - 2 = 0$

$u = 1$ *or* $u = 2$

Now substitute x^2 for u and solve for x.

$x^2 = 1$ *or* $x^2 = 2$

$x = \pm 1$ *or* $x = \pm\sqrt{2}$

The solutions are -1, 1, $-\sqrt{2}$, and $\sqrt{2}$.

50. $x^4 + 3 = 4x^2$

$x^4 - 4x^2 + 3 = 0$

Let $u = x^2$.

$u^2 - 4u + 3 = 0$ Substituting u for x^2

$(u-1)(u-3) = 0$

$u - 1 = 0$ *or* $u - 3 = 0$

$u = 1$ *or* $u = 3$

Substitute x^2 for u and solve for x.

$x^2 = 1$ *or* $x^2 = 3$

$x = \pm 1$ *or* $x = \pm\sqrt{3}$

The solutions are -1, 1, $-\sqrt{3}$, and $\sqrt{3}$.

51. $x - 3\sqrt{x} - 4 = 0$

Let $u = \sqrt{x}$.

$u^2 - 3u - 4 = 0$ Substituting u for $\sqrt{x}$

$(u+1)(u-4) = 0$

$u + 1 = 0$ *or* $u - 4 = 0$

$u = -1$ *or* $u = 4$

Now substitute $\sqrt{x}$ for u and solve for x.

$\sqrt{x} = -1$ *or* $\sqrt{x} = 4$

No solution $\quad x = 16$

Note that $\sqrt{x}$ must be nonnegative, so $\sqrt{x} = -1$ has no solution. The number 16 checks and is the solution. The solution is 16.

52. $2x - 9\sqrt{x} + 4 = 0$

Let $u = \sqrt{x}$.

$2u^2 - 9u + 4 = 0$ Substituting u for $\sqrt{x}$

$(2u-1)(u-4) = 0$

$2u - 1 = 0$ *or* $u - 4 = 0$

$u = \frac{1}{2}$ *or* $u = 4$

Substitute $\sqrt{x}$ for u and solve for u.

$\sqrt{x} = \frac{1}{2}$ *or* $\sqrt{x} = 4$

$x = \frac{1}{4}$ *or* $x = 16$

Both numbers check. The solutions are $\frac{1}{4}$ and 16.

53. $m^{2/3} - 2m^{1/3} - 8 = 0$

Let $u = m^{1/3}$.

$u^2 - 2u - 8 = 0$ Substituting u for $m^{1/3}$

$(u+2)(u-4) = 0$

$u+2 = 0$ *or* $u-4 = 0$

$u = -2$ *or* $u = 4$

Now substitute $m^{1/3}$ for u and solve for m.

$m^{1/3} = -2$ *or* $m^{1/3} = 4$

$(m^{1/3})^3 = (-2)^3$ *or* $(m^{1/3})^3 = 4^3$ Using the principle of powers

$m = -8$ *or* $m = 64$

The solutions are -8 and 64.

54. $t^{2/3} + t^{1/3} - 6 = 0$

Let $u = t^{1/3}$.

$u^2 + u - 6 = 0$

$(u+3)(u-2) = 0$

$u+3 = 0$ *or* $u-2 = 0$

$u = -3$ *or* $u = 2$

Substitute $t^{1/3}$ for u and solve for t.

$t^{1/3} = -3$ *or* $t^{1/3} = 2$

$t = -27$ *or* $t = 8$

The solutions are -27 and 8.

55. $(2x-3)^2 - 5(2x-3) + 6 = 0$

Let $u = 2x - 3$.

$u^2 - 5u + 6 = 0$ Substituting u for $2x - 3$

$(u-2)(u-3) = 0$

$u-2 = 0$ *or* $u-3 = 0$

$u = 2$ *or* $u = 3$

Now substitute $2x - 3$ for u and solve for x.

$2x - 3 = 2$ *or* $2x - 3 = 3$

$2x = 5$ *or* $2x = 6$

$x = \frac{5}{2}$ *or* $x = 3$

The solutions are $\frac{5}{2}$ and 3.

56. $(3x+2)^2 + 7(3x+2) - 8 = 0$

Let $u = 3x + 2$.

$u^2 + 7u - 8 = 0$ Substituting u for $3x + 2$

$(u+8)(u-1) = 0$

$u+8 = 0$ *or* $u-1 = 0$

$u = -8$ *or* $u = 1$

Substitute $3x + 2$ for u and solve for x.

$3x + 2 = -8$ *or* $3x + 2 = 1$

$3x = -10$ *or* $3x = -1$

$x = -\frac{10}{3}$ *or* $x = -\frac{1}{3}$

The solutions are $-\frac{10}{3}$ and $-\frac{1}{3}$.

57. $(2t^2+t)^2 - 4(2t^2+t) + 3 = 0$

Let $u = 2t^2 + t$.

$u^2 - 4u + 3 = 0$ Substituting u for $2t^2 + t$

$(u-1)(u-3) = 0$

$u-1 = 0$ *or* $u-3 = 0$

$u = 1$ *or* $u = 3$

Now substitute $2t^2 + t$ for u and solve for t.

$2t^2 + t = 1$ *or* $2t^2 + t = 3$

$2t^2 + t - 1 = 0$ *or* $2t^2 + t - 3 = 0$

$(2t-1)(t+1) = 0$ *or* $(2t+3)(t-1) = 0$

$2t-1=0$ *or* $t+1=0$ *or* $2t+3=0$ *or* $t-1 = 0$

$t=\frac{1}{2}$ *or* $t=-1$ *or* $t=-\frac{3}{2}$ *or* $t = 1$

The solutions are $\frac{1}{2}$, -1, $-\frac{3}{2}$ and 1.

58. $12 = (m^2-5m)^2 + (m^2-5m)$

$0 = (m^2-5m)^2 + (m^2-5m) - 12$

Let $u = m^2 - 5m$.

$0 = u^2 + u - 12$ Substituting u for $m^2 - 5m$

$0 = (u+4)(u-3)$

$u+4 = 0$ *or* $u-3 = 0$

$u = -4$ *or* $u = 3$

Substitute $m^2 - 5m$ for u and solve for m.

$m^2-5m = -4$ *or* $m^2-5m = 3$

$m^2-5m+4 = 0$ *or* $m^2-5m-3 = 0$

$(m-1)(m-4) = 0$ *or*

$$m = \frac{-(-5) \pm \sqrt{(-5)^2-4\cdot 1(-3)}}{2\cdot 1}$$

$m = 1$ *or* $m = 4$ *or* $m = \frac{5 \pm \sqrt{37}}{2}$

The solutions are 1, 4, $\frac{5-\sqrt{37}}{2}$, and $\frac{5+\sqrt{37}}{2}$, or 1, 4, and $\frac{5 \pm \sqrt{37}}{2}$.

59. ***Familiarize and Translate***. We will use the formula $s = 16t^2$, substituting 2120 for s.

$2120 = 16t^2$

Carry out. We solve the equation.

$2120 = 16t^2$

$132.5 = t^2$ Dividing by 16 on both sides

$11.5 \approx t$ Taking the square root on both sides

We could also find the first coordinates of the points of intersection of the graphs of $y_1 = 2120$ and $y_2 = 16x^2$. Note that since a negative result has no meaning in this application, we only need to consider the point of intersection in the first quadrant.

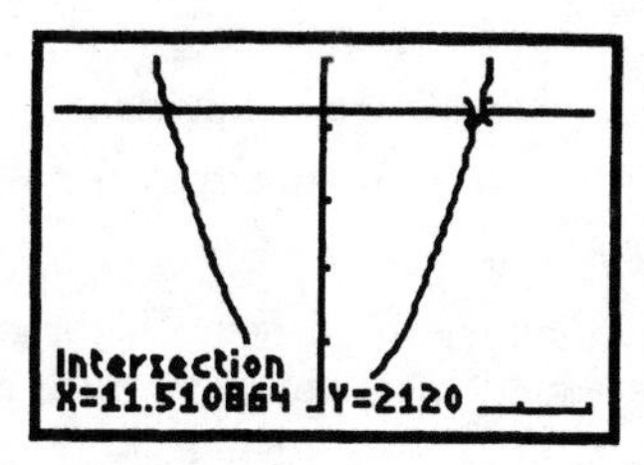

The possible solution is approximately 11.5.

Check. When $t = 11.5$, $s = 16(11.5)^2 = 2116 \approx 2120$. The answer checks.

State. It would take an object about 11.5 sec to reach the ground.

60. Solve: $2063 = 16t^2$

$t \approx 11.4$ sec

61. ***Familiarize***. Let w = the width of the rug. Then $w + 1$ = the length.

Translate. We use the Pythagorean equation.

$$w^2 + (w+1)^2 = 5^2$$

Carry out. We solve the equation.

$$w^2 + (w+1)^2 = 5^2$$
$$w^2 + w^2 + 2w + 1 = 25$$
$$2w^2 + 2w + 1 = 25$$
$$2w^2 + 2w - 24 = 0$$
$$2(w+4)(w-3) = 0$$
$$w + 4 = 0 \quad or \quad w - 3 = 0$$
$$w = -4 \quad or \quad w = 3$$

Since the width cannot be negative, we consider only 3. When $w = 3$, $w + 1 = 3 + 1 = 4$.

Check. The length, 4 ft, is 1 ft more than the width, 3 ft. The length of a diagonal of a rectangle with width 3 ft and length 4 ft is $\sqrt{3^2+4^2} = \sqrt{9+16} = \sqrt{25} = 5$. The answer checks.

State. The length is 4 ft, and the width is 3 ft.

62. Let x = the length of the longer leg.

Solve: $x^2 + (x-7)^2 = 13^2$

$x = -5$ *or* $x = 12$

Only 12 has meaning in the original problem. The length of one leg is 12 cm, and the length of the other leg is $12 - 7$, or 5 cm.

63. ***Familiarize***. Let n = the smaller number. Then $n + 5$ = the larger number.

Translate.

$$\underbrace{\text{The product of the numbers}}_{n(n+5)} \text{ is } 36.$$
$$n(n+5) = 36$$

Carry out.

$$n(n+5) = 36$$
$$n^2 + 5n = 36$$
$$n^2 + 5n - 36 = 0$$
$$(n+9)(n-4) = 0$$
$$n + 9 = 0 \quad or \quad n - 4 = 0$$
$$n = -9 \quad or \quad n = 4$$

If $n = -9$, then $n + 5 = -9 + 5 = -4$. If $n = 4$, then $n + 5 = 4 + 5 = 9$.

Check. The number -4 is 5 more than -9 and $(-4)(-9) = 36$, so the pair -9 and -4 check. The number 9 is 5 more than 4 and $9 \cdot 4 = 36$, so the pair 4 and 9 also check.

State. The numbers are -9 and -4 or 4 and 9.

64. Let n = the larger number.

Solve: $n(n-6) = 72$

$n = -6$ *or* $n = 12$

When $n = -6$, then $n - 6 = -6 - 6 = -12$, so one pair of numbers is -6 and -12. When $n = 12$, then $n - 6 = 12 - 6 = 6$, so the other pair of numbers is 6 and 12.

65 ***Familiarize***. We add labels to the drawing in the text.

We let x represent the length of a side of the square in each corner. Then the length and width of the resulting base are represented by $20 - 2x$ and $10 - 2x$, respectively. Recall that for a rectangle, Area = length × width.

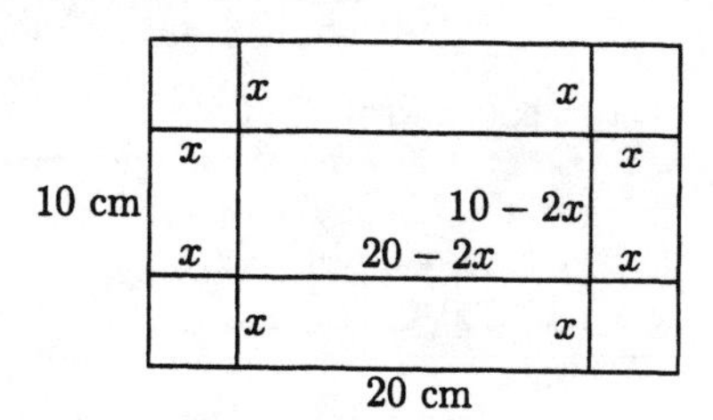

Translate.

$$\underbrace{\text{The area of the base}}_{(20-2x)(10-2x)} \text{ is } \underbrace{96 \text{ cm}^2}_{96}.$$
$$(20-2x)(10-2x) = 96$$

Carry out. We solve the equation.

$$200 - 60x + 4x^2 = 96$$
$$4x^2 - 60x + 104 = 0$$
$$x^2 - 15x + 26 = 0$$
$$(x-13)(x-2) = 0$$
$$x - 13 = 0 \quad or \quad x - 2 = 0$$
$$x = 13 \quad or \quad x = 2$$

We could also find the first coordinates of the points of intersection of $y_1 = (20-2x)(10-2x)$ and $y_2 = 96$ using the Intersect feature.

Check. When $x = 13$, both $20 - 2x$ and $10 - 2x$ are negative numbers, so we only consider $x = 2$. When $x = 2$, then $20 - 2x = 20 - 2 \cdot 2 = 16$ and $10 - 2x = 10 - 2 \cdot 2 = 6$, and the area of the base is $16 \cdot 6$, or 96 cm^2. The answer checks.

State. The length of the sides of the squares is 2 cm.

66. Let $w =$ the width of the frame.

Solve: $(32 - 2w)(28 - 2w) = 192$

$w = 8$ or $w = 22$

Only 8 has meaning in the original problem. The width of the frame is 8 cm.

67. $f(x) = 4 - 5x = -5x + 4$

The function can be written in the form $y = mx + b$, so it is a linear function.

68. $f(x) = 4 - 5x^2 = -5x^2 + 4$

The function can be written in the form $f(x) = ax^2 + bx + c$, $a \neq 0$, so it is a quadratic function.

69. $f(x) = 7x^2$

The function is in the form $f(x) = ax^2 + bx + c$, $a \neq 0$, so it is a quadratic function.

70. $f(x) = 23x + 6$

The function is in the form $f(x) = mx + b$, so it is a linear function.

71. $f(x) = 1.2x - (3.6)^2$

The function is in the form $f(x) = mx + b$, so it is a linear function.

72. $f(x) = 2 - x - x^2 = -x^2 - x + 2$

The function can be written in the form $f(x) = ax^2 + bx + c$, $a \neq 0$, so it is a quadratic function.

73. No; consider the quadratic formula $x = \dfrac{-b \pm \sqrt{b^2 - 4ac}}{2a}$. If $b^2 - 4ac = 0$, then $x = \dfrac{-b}{2a}$, so there is one real zero. If $b^2 - 4ac > 0$, then $\sqrt{b^2 - 4ac}$ is a real number and there are two real zeros. If $b^2 - 4ac < 0$, then $\sqrt{b^2 - 4ac}$ is an imaginary number and there are two imaginary zeros. Thus, a quadratic function cannot have one real zero and one imaginary zero.

74. Use the discriminant. If $b^2 - 4ac < 0$, there are no x-intercepts. If $b^2 - 4ac = 0$, there is one x-intercept. If $b^2 - 4ac > 0$, there are two x-intercepts.

75. Test for symmetry with respect to the x-axis:

$$\begin{aligned} 3x^2 + 4y^2 &= 5 && \text{Original equation} \\ 3x^2 + 4(-y)^2 &= 5 && \text{Replacing } y \text{ by } -y \\ 3x^2 + 4y^2 &= 5 && \text{Simplifying} \end{aligned}$$

The last equation is equivalent to the original equation, so the graph is symmetric with respect to the x-axis.

Test for symmetry with respect to the y-axis:

$$\begin{aligned} 3x^2 + 4y^2 &= 5 && \text{Original equation} \\ 3(-x)^2 + 4y^2 &= 5 && \text{Replacing } x \text{ by } -x \\ 3x^2 + 4y^2 &= 5 && \text{Simplifying} \end{aligned}$$

The last equation is equivalent to the original equation, so the equation is symmetric with respect to the y-axis.

Test for symmetry with respect to the origin:

$$\begin{aligned} 3x^2 + 4y^2 &= 5 && \text{Original equation} \\ 3(-x)^2 + 4(-y)^2 &= 5 && \text{Replacing } x \text{ by } -x \\ & && \text{and } y \text{ by } -y \\ 3x^2 + 4y^2 &= 5 && \text{Simplifying} \end{aligned}$$

The last equation is equivalent to the original equation, so the equation is symmetric with respect to the origin.

76. Test for symmetry with respect to the x-axis:

$$\begin{aligned} y^3 &= 6x^2 && \text{Original equation} \\ (-y)^3 &= 6x^2 && \text{Replacing } y \text{ by } -y \\ -y^3 &= 6x && \text{Simplifying} \end{aligned}$$

The last equation is not equivalent to the original equation, so the graph is not symmetric with respect to the x-axis.

Test for symmetry with respect to the y-axis:

$$\begin{aligned} y^3 &= 6x^2 && \text{Original equation} \\ y^3 &= 6(-x)^2 && \text{Replacing } x \text{ by } -x \\ y^3 &= 6x^2 && \text{Simplifying} \end{aligned}$$

The last equation is equivalent to the original equation, so the equation is symmetric with respect to the y-axis.

Test for symmetry with respect to the origin:

$$\begin{aligned} y^3 &= 6x^2 && \text{Original equation} \\ (-y)^3 &= 6(-x)^2 && \text{Replacing } x \text{ by } -x \\ & && y \text{ by } -y \\ -y^3 &= 6x^2 && \text{Simplifying} \end{aligned}$$

The last equation is not equivalent to the original equation, so the graph is not symmetric with respect to the origin.

77.

$$\begin{aligned} f(x) &= 2x^3 - x \\ f(-x) &= 2(-x)^3 - (-x) = -2x^3 + x \\ -f(x) &= -2x^3 + x \\ f(x) &\neq f(-x) \text{ so } f \text{ is not even} \\ f(-x) &= -f(x), \text{ so } f \text{ is odd.} \end{aligned}$$

78.

$$\begin{aligned} f(x) &= 4x^2 + 2x - 3 \\ f(-x) &= 4(-x)^2 + 2(-x) - 3 = 4x^2 - 2x - 3 \\ -f(x) &= -4x^2 - 2x + 3 \\ f(x) &\neq f(-x) \text{ so } f \text{ is not even} \\ f(-x) &\neq -f(x), \text{ so } f \text{ is not odd.} \end{aligned}$$

Thus $f(x) = 4x^2 + 2x - 3$ is neither even nor odd.

79. $x^2 + x - \sqrt{2} = 0$

$$x = \frac{-b \pm \sqrt{b^2 - 4ac}}{2a}$$

$$= \frac{-1 \pm \sqrt{1^2 - 4 \cdot 1(-\sqrt{2})}}{2 \cdot 1} = \frac{-1 \pm \sqrt{1 + 4\sqrt{2}}}{2}$$

The solutions are $\dfrac{-1 \pm \sqrt{1 + 4\sqrt{2}}}{2}$.

80. $x^2 + \sqrt{5}x - \sqrt{3} = 0$

Use the quadratic formula. Here $a = 1$, $b = \sqrt{5}$, and $c = -\sqrt{3}$.

$$x = \frac{-b \pm \sqrt{b^2 - 4ac}}{2a}$$

$$= \frac{-\sqrt{5} \pm \sqrt{(\sqrt{5})^2 - 4 \cdot 1(-\sqrt{3})}}{2 \cdot 1}$$

$$= \frac{-\sqrt{5} \pm \sqrt{5 + 4\sqrt{3}}}{2}$$

The solutions are $\dfrac{-\sqrt{5} \pm \sqrt{5 + 4\sqrt{3}}}{2}$.

81.

$$2t^2 + (t-4)^2 = 5t(t-4) + 24$$
$$2t^2 + t^2 - 8t + 16 = 5t^2 - 20t + 24$$
$$0 = 2t^2 - 12t + 8$$
$$0 = t^2 - 6t + 4 \quad \text{Dividing by 2}$$

Use the quadratic formula.

$$t = \frac{-b \pm \sqrt{b^2 - 4ac}}{2a}$$

$$= \frac{-(-6) \pm \sqrt{(-6)^2 - 4 \cdot 1 \cdot 4}}{2 \cdot 1}$$

$$= \frac{6 \pm \sqrt{20}}{2} = \frac{6 \pm 2\sqrt{5}}{2}$$

$$= \frac{2(3 \pm \sqrt{5})}{2} = 3 \pm \sqrt{5}$$

The solutions are $3 \pm \sqrt{5}$.

82.

$$9t(t+2) - 3t(t-2) = 2(t+4)(t+6)$$
$$9t^2 + 18t - 3t^2 + 6t = 2t^2 + 20t + 48$$
$$4t^2 + 4t - 48 = 0$$
$$4(t+4)(t-3) = 0$$

$t + 4 = 0$ *or* $t - 3 = 0$

$t = -4$ *or* $t = 3$

The solutions are -4 and 3.

83. $\sqrt{x-3} - \sqrt[4]{x-3} = 2$

Substitute u for $\sqrt[4]{x-3}$.

$$u^2 - u - 2 = 0$$
$$(u-2)(u+1) = 0$$

$u - 2 = 0$ *or* $u + 1 = 0$

$u = 2$ *or* $u = -1$

Substitute $\sqrt[4]{x-3}$ for u and solve for x.

$\sqrt[4]{x-3} = 2$ *or* $\sqrt[4]{x-3} = 1$

$x - 3 = 16$ No solution

$x = 19$

The value checks. The solution is 19.

84. $x^6 - 28x^3 + 27 = 0$

Substitute u for x^3.

$$u^2 - 28u + 27 = 0$$
$$(u-27)(u-1) = 0$$

$u = 27$ or $u = 1$

Substitute x^3 for u and solve for x.

$x^3 = 27$ *or* $x^3 = 1$

$x^3 - 27 = 0$ *or* $x^3 - 1 = 0$

$(x-3)(x^2+3x+9) = 0$ *or* $(x-1)(x^2+x+1) = 0$

Using the principle of zero products and, where necessary, the quadratic formula, we find that the solutions are $3, -\frac{3}{2} \pm \frac{3\sqrt{3}}{2}i, 1$, and $-\frac{1}{2} \pm \frac{\sqrt{3}}{2}i$.

85.

$$\left(y + \frac{2}{y}\right)^2 + 3y + \frac{6}{y} = 4$$

$$\left(y + \frac{2}{y}\right)^2 + 3\left(y + \frac{2}{y}\right) - 4 = 0$$

Substitute u for $y + \dfrac{2}{y}$.

$$u^2 + 3u - 4 = 0$$
$$(u+4)(u-1) = 0$$

$u = -4$ *or* $u = 1$

Substitute $y + \dfrac{2}{y}$ for u and solve for y.

$y + \dfrac{2}{y} = -4$ *or* $y + \dfrac{2}{y} = 1$

$y^2 + 2 = -4y$ *or* $y^2 + 2 = y$

$y^2 + 4y + 2 = 0$ *or* $y^2 - y + 2 = 0$

$y = \dfrac{-4 \pm \sqrt{4^2 - 4 \cdot 1 \cdot 2}}{2 \cdot 1}$ *or*

$y = \dfrac{-(-1) \pm \sqrt{(-1)^2 - 4 \cdot 1 \cdot 2}}{2 \cdot 1}$

$y = \dfrac{-4 \pm \sqrt{8}}{2}$ *or* $y = \dfrac{1 \pm \sqrt{-7}}{2}$

$y = \dfrac{-4 \pm 2\sqrt{2}}{2}$ *or* $y = \dfrac{1 \pm \sqrt{7}i}{2}$

$y = -2 \pm \sqrt{2}$ *or* $y = \dfrac{1}{2} \pm \dfrac{\sqrt{7}}{2}i$

The solutions are $-2 \pm \sqrt{2}$ and $\dfrac{1}{2} \pm \dfrac{\sqrt{7}}{2}i$.

86.
$$x^2+3x+1-\sqrt{x^2+3x+1}=8$$
$$x^2+3x+1-\sqrt{x^2+3x+1}-8=0$$
$$u^2-u-8=0$$
$$u=\frac{1+\sqrt{33}}{2}\ or\ u=\frac{1-\sqrt{33}}{2}$$
$$\sqrt{x^2+3x+1}=\frac{1+\sqrt{33}}{2}\ or$$
$$\sqrt{x^2+3x+1}=\frac{1-\sqrt{33}}{2}$$
$$x^2+3x+1=\frac{34+2\sqrt{33}}{4}\ or$$
$$x^2+3x+1=\frac{34-2\sqrt{33}}{4}$$
$$x^2+3x+\frac{-15-\sqrt{33}}{2}=0\ or$$
$$x^2+3x+\frac{-15+\sqrt{33}}{2}=0$$
$$x=\frac{-3\pm\sqrt{39+2\sqrt{33}}}{2}\ or$$
$$x=\frac{-3\pm\sqrt{39-2\sqrt{33}}}{2}$$

Only $\frac{-3\pm\sqrt{39+2\sqrt{33}}}{2}$ checks. The solutions are $\frac{-3\pm\sqrt{39+2\sqrt{33}}}{2}$.

87. $\frac{1}{2}at+v_0t+x_0=0$

Use the quadratic formula. Here $a=\frac{1}{2}a$, $b=v_0$, and $c=x_0$.

$$t=\frac{-v_0\pm\sqrt{(v_0)^2-4\cdot\frac{1}{2}a\cdot x_0}}{2\cdot\frac{1}{2}a}$$
$$t=\frac{-v_0\pm\sqrt{v_0^2-2ax_0}}{a}$$

Exercise Set 2.4

1. a) The minimum function value occurs at the vertex, so the vertex is $(-0.4999992, -2.25)$, or about $(-0.5, -2.25)$.

b) The line of symmetry is a vertical line through the vertex. It is $x=-0.4999992$, or about $x=-0.5$.

c) The minimum value of the function is -2.25.

2. a) $(-0.4999994, 6.25)$, or about $(-0.5, 6.25)$

b) $x=-0.4999994$, or about $x=-0.5$

c) Maximum: 6.25

3. $f(x)=x^2-8x+12$ 16 completes the square for x^2-8x

$=x^2-8x+16-16+12$ Adding $16-16$ on the right side

$=(x^2-8x+16)-16+12$

$=(x-4)^2-4$ Factoring and simplifying

$=(x-4)^2+(-4)$ Writing in the form $f(x)=a(x-h)^2+k$

a) Vertex: $(4,-4)$

b) Line of symmetry: $x=4$

c) Minimum value: -4

4. $g(x)=x^2+7x-8$

$=x^2+7x+\frac{49}{4}-\frac{49}{4}-8$ $\left(\frac{1}{2}\cdot 7=\frac{7}{2}\right.$ and $\left.\left(\frac{7}{2}\right)^2=\frac{49}{4}\right)$

$=\left(x+\frac{7}{2}\right)^2-\frac{81}{4}$

$=\left[x-\left(-\frac{7}{2}\right)\right]^2+\left(-\frac{81}{4}\right)$

a) Vertex: $\left(-\frac{7}{2},-\frac{81}{4}\right)$

b) Line of symmetry: $x=-\frac{7}{2}$

c) Minimum value: $-\frac{81}{4}$

5. $f(x)=x^2-7x+12$ $\frac{49}{4}$ completes the square for x^2-7x

$=x^2-7x+\frac{49}{4}-\frac{49}{4}+12$ Adding $\frac{49}{4}-\frac{49}{4}$ on the right side

$=\left(x^2-7x+\frac{49}{4}\right)-\frac{49}{4}+12$

$=\left(x-\frac{7}{2}\right)^2-\frac{1}{4}$ Factoring and simplifying

$=\left(x-\frac{7}{2}\right)^2+\left(-\frac{1}{4}\right)$ Writing in the form $f(x)=a(x-h)^2+k$

a) Vertex: $\left(\frac{7}{2},-\frac{1}{4}\right)$

b) Line of symmetry: $x=\frac{7}{2}$

c) Minimum value: $-\frac{1}{4}$

6. $g(x) = x^2 - 5x + 6$

$= x^2 - 5x + \frac{25}{4} - \frac{25}{4} + 6 \quad \left(\frac{1}{2}(-5) = -\frac{5}{2}\right.$ and $\left.\left(-\frac{5}{2}\right)^2 = \frac{25}{4}\right)$

$= \left(x - \frac{5}{2}\right)^2 - \frac{1}{4}$

$= \left(x - \frac{5}{2}\right)^2 + \left(-\frac{1}{4}\right)$

a) Vertex: $\left(\frac{5}{2}, -\frac{1}{4}\right)$

b) Line of symmetry: $x = \frac{5}{2}$

c) Minimum value: $-\frac{1}{4}$

7. $g(x) = 2x^2 + 6x + 8$

$= 2(x^2 + 3x) + 8$ Factoring 2 out of the first two terms

$= 2\left(x^2 + 3x + \frac{9}{4} - \frac{9}{4}\right) + 8$ Adding $\frac{9}{4} - \frac{9}{4}$ inside the parentheses

$= 2\left(x^2 + 3x + \frac{9}{4}\right) - 2 \cdot \frac{9}{4} + 8$ Removing $-\frac{9}{4}$ from within the parentheses

$= 2\left(x + \frac{3}{2}\right)^2 + \frac{7}{2}$ Factoring and simplifying

$= 2\left[x - \left(-\frac{3}{2}\right)\right]^2 + \frac{7}{2}$

a) Vertex: $\left(-\frac{3}{2}, \frac{7}{2}\right)$

b) Line of symmetry: $x = -\frac{3}{2}$

c) Minimum value: $\frac{7}{2}$

8. $f(x) = 2x^2 - 10x + 14$

$= 2(x^2 - 5x) + 14$

$= 2\left(x^2 - 5x + \frac{25}{4} - \frac{25}{4}\right) + 14$

$= 2\left(x^2 - 5x + \frac{25}{4}\right) - 2 \cdot \frac{25}{4} + 14$

$= 2\left(x - \frac{5}{2}\right)^2 + \frac{3}{2}$

a) Vertex: $\left(\frac{5}{2}, \frac{3}{2}\right)$

b) Line of symmetry: $x = \frac{5}{2}$

c) Minimum value: $\frac{3}{2}$

9. $g(x) = -2x^2 + 2x + 1$

$= -2(x^2 - x) + 1$ Factoring -2 out of the first two terms

$= -2\left(x^2 - x + \frac{1}{4} - \frac{1}{4}\right) + 1$ Adding $\frac{1}{4} - \frac{1}{4}$ inside the parentheses

$= -2\left(x^2 - x + \frac{1}{4}\right) - 2\left(-\frac{1}{4}\right) + 1$

Removing $-\frac{1}{4}$ from within the parentheses

$= -2\left(x - \frac{1}{2}\right)^2 + \frac{3}{2}$

a) Vertex: $\left(\frac{1}{2}, \frac{3}{2}\right)$

b) Line of symmetry: $x = \frac{1}{2}$

c) Maximum value: $\frac{3}{2}$

10. $f(x) = -3x^2 - 3x + 1$

$= -3(x^2 + x) + 1$

$= -3\left(x^2 + x + \frac{1}{4} - \frac{1}{4}\right) + 1$

$= -3\left(x^2 + x + \frac{1}{4}\right) - 3\left(-\frac{1}{4}\right) + 1$

$= -3\left(x + \frac{1}{2}\right)^2 + \frac{7}{4}$

$= -3\left[x - \left(-\frac{1}{2}\right)\right]^2 + \frac{7}{4}$

a) Vertex: $\left(-\frac{1}{2}, \frac{7}{4}\right)$

b) Line of symmetry: $x = -\frac{1}{2}$

c) Maximum value: $\frac{7}{4}$

11. The graph of $y = (x + 3)^2$ has vertex $(-3, 0)$ and opens up. It is graph (f).

12. The graph of $y = -(x - 4)^2 + 3$ has vertex $(4, 3)$ and opens down. It is graph (e).

13. The graph of $y = 2(x - 4)^2 - 1$ has vertex $(4, -1)$ and opens up. It is graph (b).

14. The graph of $y = x^2 - 3$ has vertex $(0, -3)$ and opens up. It is graph (g).

15. The graph of $y = -\frac{1}{2}(x + 3)^2 + 4$ has vertex $(-3, 4)$ and opens down. It is graph (h).

16. The graph of $y = (x - 3)^2$ has vertex $(3, 0)$ and opens up. It is graph (a).

17. The graph of $y = -(x + 3)^2 + 4$ has vertex $(-3, 4)$ and opens down. It is graph (c).

18. The graph of $y = 2(x-1)^2 - 4$ has vertex $(1, -4)$ and opens up. It is graph (d).

19. $f(x) = x^2 - 6x + 5$

a) The x-coordinate of the vertex is
$$-\frac{b}{2a} = -\frac{-6}{2 \cdot 1} = 3.$$
Since $f(3) = 3^2 - 6 \cdot 3 + 5 = -4$, the vertex is $(3, -4)$.

b) Since $a = 1 > 0$, the graph opens up so the second coordinate of the vertex, -4, is the minimum value of the function.

c) The range is $[-4, \infty)$.

d) Since the graph opens up, function values decrease to the left of the vertex and increase to the right of the vertex. Thus, $f(x)$ is increasing on $(3, \infty)$ and decreasing on $(-\infty, 3)$.

20. $f(x) = x^2 + 4x - 5$

a) $-\frac{b}{2a} = -\frac{4}{2 \cdot 1} = -2$

$f(-2) = (-2)^2 + 4(-2) - 5 = -9$

The vertex is $(-2, -9)$.

b) Since $a = 1 > 0$, the graph opens up. The minimum value of $f(x)$ is -9.

c) Range: $[-9, \infty)$

d) Increasing: $(-2, \infty)$; decreasing: $(-\infty, -2)$

21. $f(x) = 2x^2 + 4x - 16$

a) The x-coordinate of the vertex is
$$-\frac{b}{2a} = -\frac{4}{2 \cdot 2} = -1.$$
Since $f(-1) = 2(-1)^2 + 4(-1) - 16 = -18$, the vertex is $(-1, -18)$.

b) Since $a = 2 > 0$, the graph opens up so the second coordinate of the vertex, -18, is the minimum value of the function.

c) The range is $[-18, \infty)$.

d) Since the graph opens up, function values decrease to the left of the vertex and increase to the right of the vertex. Thus, $f(x)$ is increasing on $(-1, \infty)$ and decreasing on $(-\infty, -1)$.

22. $f(x) = \frac{1}{2}x^2 - 3x + \frac{5}{2}$

a) $-\frac{b}{2a} = -\frac{-3}{2 \cdot \frac{1}{2}} = 3$

$f(3) = \frac{1}{2} \cdot 3^2 - 3 \cdot 3 + \frac{5}{2} = -2$

The vertex is $(3, -2)$.

b) Since $a = \frac{1}{2} > 0$, the graph opens up. The minimum value of $f(x)$ is -2.

c) Range: $[-2, \infty)$

d) Increasing: $(3, \infty)$; decreasing: $(-\infty, 3)$

23. $f(x) = -\frac{1}{2}x^2 + 5x - 8$

a) The x-coordinate of the vertex is
$$-\frac{b}{2a} = -\frac{5}{2\left(-\frac{1}{2}\right)} = 5.$$
Since $f(5) = -\frac{1}{2} \cdot 5^2 + 5 \cdot 5 - 8 = \frac{9}{2}$, the vertex is $\left(5, \frac{9}{2}\right)$.

b) Since $a = -\frac{1}{2} < 0$, the graph opens down so the second coordinate of the vertex, $\frac{9}{2}$, is the maximum value of the function.

c) The range is $\left(-\infty, \frac{9}{2}\right]$.

d) Since the graph opens down, function values increase to the left of the vertex and decreases to the right of the vertex. Thus, $f(x)$ is increasing on $(-\infty, 5)$ and decreasing on $(5, \infty,)$.

24. $f(x) = -2x^2 - 24x - 64$

a) $-\frac{b}{2a} = -\frac{-24}{2(-2)} = -6.$

$f(-6) = -2(-6)^2 - 24(-6) - 64 = 8$

The vertex is $(-6, 8)$.

b) Since $a = -2 < 0$, the graph opens down. The maximum value of $f(x)$ is 8.

c) Range: $(-\infty, 8]$

d) Increasing: $(-\infty, -6)$; decreasing: $(-6, \infty)$

25. $f(x) = 3x^2 + 6x + 5$

a) The x-coordinate of the vertex is
$$-\frac{b}{2a} = -\frac{6}{2 \cdot 3} = -1.$$
Since $f(-1) = 3(-1)^2 + 6(-1) + 5 = 2$, the vertex is $(-1, 2)$.

b) Since $a = 3 > 0$, the graph opens up so the second coordinate of the vertex, 2, is the minimum value of the function.

c) The range is $[2, \infty)$.

d) Since the graph opens up, function values decrease to the left of the vertex and increase to the right of the vertex. Thus, $f(x)$ is increasing on $(-1, \infty)$ and decreasing on $(-\infty, -1)$.

26. $f(x) = -3x^2 + 24x - 49$

a) $-\frac{b}{2a} = -\frac{24}{2(-3)} = 4.$

$f(4) = -3 \cdot 4^2 + 24 \cdot 4 - 49 = -1$

The vertex is $(4, -1)$.

b) Since $a = -3 < 0$, the graph opens down. The maximum value of $f(x)$ is -1.

c) Range: $(-\infty, -1]$

d) Increasing: $(-\infty, 4)$; decreasing: $(4, \infty)$

27. ***Familiarize.*** Using the label in the text, we let $x =$ the height of the file. Then the length $= 10$ and the width $= 18 - 2x$.

Translate. Since the volume of a rectangular solid is length $\times$ width $\times$ height we have

$$V(x) = 10(18 - 2x)x, \text{ or } -20x^2 + 180x.$$

Carry out. Since $V(x)$ is a quadratic function with $a = -20 < 0$, the maximum function value occurs at the vertex of the graph of the function. The first coordinate of the vertex is

$$-\frac{b}{2a} = -\frac{180}{2(-20)} = 4.5.$$

Check. When $x = 4.5$, then $18-2x = 9$ and $V(x) = 10 \cdot 9(4.5)$, or 405. As a partial check, we can find $V(x)$ for a value of x less than 4.5 and for a value of x greater than 4.5. For instance, $V(4.4) = 404.8$ and $V(4.6) = 404.8$. Since both of these values are less than 405, our result appears to be correct. We could also examine a table of values for $V(x)$ and/or examine its graph.

State. The file should be 4.5 in. tall in order to maximize the volume.

28. Let $w =$ the width of the garden. Then the length $= 32-2w$ and the area is given by $A(w) = (32-2w)w$, or $-2w^2 + 32w$. The maximum function value occurs at the vertex of the graph of $A(w)$. The first coordinate of the vertex is

$$-\frac{b}{2a} = -\frac{32}{2(-2)} = 8.$$

When $w = 8$, then $32-2w = 16$ and the area is $16 \cdot 8$, or 128 ft^2. A garden with dimensions 8 ft by 16 ft yields this area.

29. ***Familiarize.*** Let $b =$ the length of the base of the triangle. Then the height $= 20 - b$.

Translate. Since the area of a triangle is $\frac{1}{2} \times$ base $\times$ height, we have

$$A(b) = \frac{1}{2}b(20 - b), \text{ or } -\frac{1}{2}b^2 + 10b.$$

Carry out. Since $A(b)$ is a quadratic function with $a = -\frac{1}{2} < 0$, the maximum function value occurs at the vertex of the graph of the function. The first coordinate of the vertex is

$$-\frac{b}{2a} = -\frac{10}{2\left(-\frac{1}{2}\right)} = 10.$$

When $b = 10$, then $20 - b = 20 - 10 = 10$, and the area is $\frac{1}{2} \cdot 10 \cdot 10 = 50$ cm^2.

Check. As a partial check, we can find $A(b)$ for a value of b less than 10 and for a value of b greater than 10. For instance, $V(9.9) = 49.995$ and $V(10.1) = 49.995$. Since both of these values are less than 50, our result appears to be correct. We could also examine a table of values for $A(b)$ and/or examine its graph.

State. The area is a maximum when the base and the height are both 10 cm.

30. Let $b =$ the length of the base. Then $69 - b =$ the height and $A(b) = b(69 - b)$, or $-b^2 + 69b$. The maximum function value occurs at the vertex of the graph of $A(b)$. The first coordinate of the vertex is

$$-\frac{b}{2a} = -\frac{69}{2(-1)} = 34.5.$$

When $b = 34.5$, then $69 - b = 34.5$. The area is a maximum when the base and height are both 34.5 cm.

31. ***Familiarize.*** We let $s =$ the height of the elevator shaft, $t_1 =$ the time it takes the screwdriver to reach the bottom of the shaft, and $t_2 =$ the time it takes the sound to reach the top of the shaft.

Translate. We know that $t_1 + t_2 = 5$. Using the information in Example 4 we also know that

$$s = 16t_1^2, \quad \textit{or} \quad t_1 = \frac{\sqrt{x}}{4} \text{ and}$$

$$s = 1100t_2, \quad \textit{or} \quad t_2 = \frac{s}{1100}.$$

Then $\frac{\sqrt{s}}{4} + \frac{s}{1100} = 5.$

Carry out. We solve the last equation above.

$$\frac{\sqrt{s}}{4} + \frac{s}{1100} = 5$$

$$275\sqrt{s} + s = 5500 \quad \text{Multiplying by 1100}$$

$$2 + 275\sqrt{s} - 5500 = 0$$

Let $u = \sqrt{s}$ and substitute.

$$u^2 + 275u - 5500 = 0$$

$$u = \frac{-b + \sqrt{b^2 - 4ac}}{2a} \quad \text{We only want the positive solution.}$$

$$= \frac{-275 + \sqrt{275^2 - 4 \cdot 1(-5500)}}{2 \cdot 1}$$

$$= \frac{-275 + \sqrt{97{,}625}}{2} \approx 18.725$$

Since $u \approx 18.725$, we have $\sqrt{s} = 18.725$, so $s \approx 350.6$.

Check. If $s \approx 350.6$, then $t_1 = \frac{\sqrt{s}}{4} = \frac{\sqrt{350.6}}{4} \approx 4.68$ and $t_2 = \frac{s}{1100} = \frac{350.6}{1100} \approx 0.32$, so $t_1 + t_2 = 4.68 + 0.32 = 5$.

The result checks.

State. The elevator shaft is about 350.6 ft tall.

32. Let $s =$ the height of the cliff, $t_1 =$ the time it takes the balloon to hit the ground, and $t_2 =$ the time it takes for the sound to reach the top of the cliff. Then we have

$t_1 + t_2 = 3,$

$s = 16t_1^2, \quad or \quad t_1 = \frac{\sqrt{s}}{4}$, and

$s = 1100t_2, \quad or \quad t_2 = \frac{s}{1100}$, so

$\frac{\sqrt{s}}{4} + \frac{s}{1100} = 3.$

Solving the last equation, we find that $s \approx 132.7$ ft.

33. $C(x) = 0.1x^2 - 0.7x + 2.425$

Since $C(x)$ is a quadratic function with $a = 0.1 > 0$, a minimum function value occurs at the vertex of the graph of $C(x)$. The first coordinate of the vertex is

$$-\frac{b}{2a} = -\frac{-0.7}{2(0.1)} = 3.5.$$

Thus, 3.5 hundred, or 350 bicycles should be built to minimize the average cost per bicycle.

34. $P(x) = R(x) - C(x)$

$P(x) = 5x - (0.001x^2 + 1.2x + 60)$

$P(x) = -0.001x^2 + 3.8x - 60$

Since $P(x)$ is a quadratic function with $a = -0.001 < 0$, a maximum function value occurs at the vertex of the graph of the function. The first coordinate of the vertex is

$$-\frac{b}{2a} = -\frac{3.8}{2(-0.001)} = 1900.$$

$P(1900) = -0.001(1900)^2 + 3.8(1900) - 60 = 3550$

Thus, the maximum profit is \$3550. It occurs when 1900 units are sold.

35. $P(x) = R(x) - C(x)$

$P(x) = (50x - 0.5x^2) - (10x + 3)$

$P(x) = -0.5x^2 + 40x - 3$

Since $P(x)$ is a quadratic function with $a = -0.5 < 0$, a maximum function value occurs at the vertex of the graph of the function. The first coordinate of the vertex is

$$-\frac{b}{2a} = -\frac{40}{2(-0.5)} = 40.$$

$P(40) = -0.5(40)^2 + 40 \cdot 40 - 3 = 797$

Thus, the maximum profit is \$797. It occurs when 40 units are sold.

36. $P(x) = R(x) - C(x)$

$P(x) = 20x - 0.1x^2 - (4x + 2)$

$P(x) = -0.1x^2 + 16x - 2$

Since $P(x)$ is a quadratic function with $a = -0.1 < 0$, a maximum function value occurs at the vertex of the graph of the function. The first coordinate of the vertex is

$$-\frac{b}{2a} = -\frac{16}{2(-0.1)} = 80.$$

$P(80) = -0.1(80)^2 + 16(80) - 2 = 638$

Thus, the maximum profit is \$638. It occurs when 80 units are sold.

37. ***Familiarize***. Using the labels on the drawing in the text, we let x = the width of each corral and $240 - 3x$ = the total length of the corrals.

Translate. Since the area of a rectangle is length × width, we have

$$A(x) = (240 - 3x)x = -3x^2 + 240x.$$

Carry out. Since $A(x)$ is a quadratic function with $a = -3 < 0$, the maximum function value occurs at the vertex of the graph of $A(x)$. The first coordinate of the vertex is

$$-\frac{b}{2a} = -\frac{240}{2(-3)} = 40.$$

$A(40) = -3(40)^2 + 240(40) = 4800$

Check. As a partial check we can find $A(x)$ for a value of x less than 40 and for a value of x greater than 40. For instance, $A(39.9) = 4799.97$ and $A(40.1) = 4799.97$. Since both of these values are less than 4800, our result appears to be correct. We could also examine a table of values for $A(x)$ and/or examine its graph.

State. The largest total area that can be enclosed is 4800 yd^2.

38. $\frac{1}{2} \cdot 2\pi x + 2x + 2y = 24$, so $y = 12 - \frac{\pi x}{2} - x$.

$$A(x) = \frac{1}{2} \cdot \pi x^2 + 2x\left(12 - \frac{\pi x}{2} - x\right)$$

$$A(x) = \frac{\pi x^2}{2} + 24x - \pi x^2 - 2x^2$$

$$A(x) = 24x - \frac{\pi x^2}{2} - 2x^2, \text{ or } 24x - \left(\frac{\pi}{2} + 2\right)x^2$$

Since $A(x)$ is a quadratic function with $a = -\left(\frac{\pi}{2} + 2\right) < 0$, the maximum function value occurs at the vertex of the graph of $A(x)$. The first coordinate of the vertex is

$$\frac{-b}{2a} = -\frac{24}{2\left[-\left(\frac{\pi}{2} + 2\right)\right]} = \frac{24}{\pi + 4}.$$

When $x = \frac{24}{\pi + 4}$, then $y = \frac{24}{\pi + 4}$. Thus, the maximum amount of light will enter when the dimensions of the rectangular part of the window are $2x$ by y, or $\frac{48}{\pi + 4}$ ft by $\frac{24}{\pi + 4}$ ft, or approximately 6.72 ft by 3.36 ft.

39. Answers will vary. The problem could be similar to Example 5 or Exercises 27 through 38.

40. Completing the square was used in Section 2.3 to solve quadratic equations. It was used again in this section to write quadratic functions in the form $f(x) = a(x - h)^2 + k$.

41. The x-intercepts of $g(x)$ are also $(x_1, 0)$ and $(x_2, 0)$. This is true because $f(x)$ and $g(x)$ have the same zeros. Consider $g(x) = 0$, or $-ax^2 - bx - c = 0$.

Multiplying by -1 on both sides, we get an equivalent equation $ax^2 + bx + c = 0$, or $f(x) = 0$.

42. $$\frac{f(x+h) - f(x)}{h} = \frac{3(x+h) - 7 - (3x - 7)}{h}$$
$$= \frac{3x + 3h - 7 - 3x + 7}{h}$$
$$= \frac{3h}{h} = 3$$

43. $f(x) = 2x^2 - x + 4$
$$f(x+h) = 2(x+h)^2 - (x+h) + 4$$
$$= 2(x^2 + 2xh + h^2) - (x+h) + 4$$
$$= 2x^2 + 4xh + 2h^2 - x - h - 4$$
$$\frac{f(x+h) - f(x)}{h}$$
$$= \frac{2x^2 + 4xh + 2h^2 - x - h - 4 - (2x^2 - x + 4)}{h}$$
$$= \frac{2x^2 + 4xh + 2h^2 - x - h - 4 - 2x^2 + x - 4}{h}$$
$$= \frac{4xh + 2h^2 - h}{h} = \frac{h(4x + 2h - 1)}{h}$$
$$= 4x + 2h - 1$$

44.

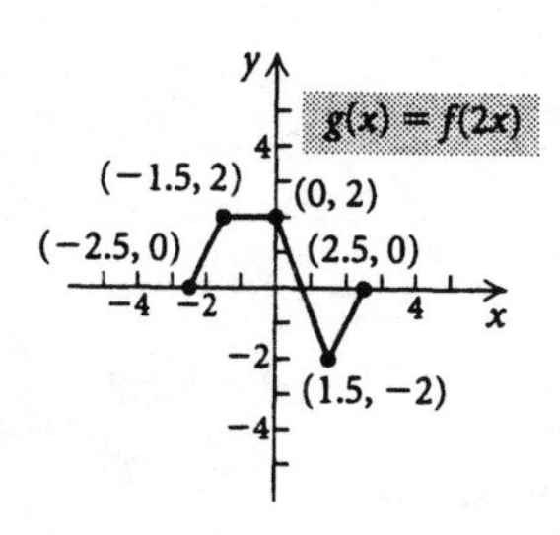

45. The graph of $f(x)$ is stretched vertically and reflected across the x-axis.

(3, 4)
(−5, 0)
(5, 0)
(0, −4)
(−3, −4)
g(x) = −2f(x)

46. a)
$$kx^2 - 2x + k = 0$$
$$k(-3)^2 - 2(-3) + k = 0 \quad \text{Substituting } -3 \text{ for } x$$
$$9k + 6 + k = 0$$
$$10k = -6$$
$$k = -\frac{3}{5}$$

b) $$-\frac{3}{5}x^2 - 2x - \frac{3}{5} = 0 \quad \text{Substituting } -\frac{3}{5} \text{ for } k$$
$$3x^2 + 10x + 3 = 0 \quad \text{Multiplying by } -5$$
$$(3x+1)(x+3) = 0$$
$$3x + 1 = 0 \quad \text{or} \quad x + 3 = 0$$
$$3x = -1 \quad \text{or} \quad x = -3$$
$$x = -\frac{1}{3} \quad \text{or} \quad x = -3$$

The other solution is $-\frac{1}{3}$.

47. a)
$$kx^2 - 17x + 33 = 0$$
$$k(3)^2 - 17(3) + 33 = 0 \quad \text{Substituting 3 for } x$$
$$9k - 51 + 33 = 0$$
$$9k = 18$$
$$k = 2$$

b) $$2x^2 - 17x + 33 = 0 \quad \text{Substituting 2 for } k$$
$$(2x - 11)(x - 3) = 0$$
$$2x - 11 = 0 \quad \text{or} \quad x - 3 = 0$$
$$x = \frac{11}{2} \quad \text{or} \quad x = 3$$

The other solution is $\frac{11}{2}$.

48. a)
$$x^2 - (6+3i)x + k = 0$$
$$3^2 - (6+3i) \cdot 3 + k = 0 \quad \text{Substituting 3 for } x$$
$$9 - 18 - 9i + k = 0$$
$$k = 9 + 9i$$

b) $$x^2 - (6+3i)x + 9 + 9i = 0$$
$$x = \frac{-[-(6+3i)] \pm \sqrt{[-(6+3i)]^2 - 4(1)(9+9i)}}{2 \cdot 1}$$
$$x = \frac{6 + 3i \pm \sqrt{36 + 36i - 9 - 36 - 36i}}{2}$$
$$x = \frac{6 + 3i \pm \sqrt{-9}}{2} = \frac{6 + 3i \pm 3i}{2}$$
$$x = \frac{6 + 3i + 3i}{2} \quad \text{or} \quad x = \frac{6 + 3i - 3i}{2}$$
$$x = \frac{6 + 6i}{2} \quad \text{or} \quad x = \frac{6}{2}$$
$$x = 3 + 3i \quad \text{or} \quad x = 3$$

The other solution is $3 + 3i$.

49. a)
$$(1+i)^2 - k(1+i) + 2 = 0 \quad \text{Substituting } 1 + i \text{ for } x$$
$$1 + 2i - 1 - k - ki + 2 = 0$$
$$2 + 2i = k + ki$$
$$2(1+i) = k(1+i)$$
$$2 = k$$

b) $x^2 - 2x + 2 = 0$ Substituting 2 for k

$$x = \frac{-(-2) \pm \sqrt{(-2)^2 - 4 \cdot 1 \cdot 2}}{2 \cdot 1}$$
$$= \frac{2 \pm \sqrt{-4}}{2}$$
$$= \frac{2 \pm 2i}{2} = 1 \pm i$$

The other solution is $1 - i$.

50. $f(x) = -4x^2 + bx + 3$

The x-coordinate of the vertex of $f(x)$ is $-\frac{b}{2(-4)}$, or $\frac{b}{8}$. Now we find b such that $f\left(\frac{b}{8}\right) = 50$.

$$-4\left(\frac{b}{8}\right)^2 + b \cdot \frac{b}{8} + 3 = 50$$
$$-\frac{b^2}{16} + \frac{b^2}{8} + 3 = 50$$
$$\frac{b^2}{16} = 47$$
$$b^2 = 16 \cdot 47$$
$$b = \pm\sqrt{16 \cdot 47}$$
$$b = \pm 4\sqrt{47}$$

51. $f(x) = -0.2x^2 - 3x + c$

The x-coordinate of the vertex of $f(x)$ is $-\frac{b}{2a} = -\frac{-3}{2(-0.2)} = -7.5$. Now we find c such that $f(-7.5) = -225$.

$$-0.2(-7.5)^2 - 3(-7.5) + c = -225$$
$$-11.25 + 22.5 + c = -225$$
$$c = -236.25$$

52. $f(x) = a(x - h)^2 + k$

$1 = a(-3 - 4)^2 - 5$, so $a = \frac{6}{49}$. Then $f(x) = \frac{6}{49}(x - 4)^2 - 5$.

53. $y = (|x| - 5)^2 - 3$

30, −10, 10, −10

54. First we find the radius r of a circle with circumference x:

$$2\pi r = x$$
$$r = \frac{x}{2\pi}$$

Then we find the length s of a side of a square with perimeter $24 - x$:

$$4s = 24 - x$$
$$s = \frac{24 - x}{4}$$

Then S = area of circle + area of square

$$S = \pi r^2 + s^2$$
$$S(x) = \pi\left(\frac{x}{2\pi}\right)^2 + \left(\frac{24 - x}{4}\right)^2$$
$$S(x) = \left(\frac{1}{4\pi} + \frac{1}{16}\right)x^2 - 3x + 36$$

Since $S(x)$ is a quadratic function with $a = \frac{1}{4\pi} + \frac{1}{16} > 0$, the minimum function value occurs at the vertex of the graph of $S(x)$. The first coordinate of the vertex is

$$-\frac{b}{2a} = -\frac{-3}{2\left(\frac{1}{4\pi} + \frac{1}{16}\right)} = \frac{24\pi}{4 + \pi}.$$

Then the string should be cut so that one piece is $\frac{24\pi}{4 + \pi}$ in., or about 10.56 in. The other piece will be $24 - \frac{24\pi}{4 + \pi}$, or $\frac{96}{4 + \pi}$ in., or about 13.44 in.

Exercise Set 2.5

1. The data points rise and then fall, suggesting that a quadratic function $f(x) = ax^2 + bx + c$, $a < 0$, might fit the data. The answer is (c).

2. The data points rise at a steady rate, so a linear function $f(x) = mx + b$ might fit the data. The answer is (a).

3. The data points fall, then rise, and then fall again. Thus, neither a linear nor a quadratic function fits the data. The answer is (d).

4. The data points fall and then rise, suggesting that a quadratic function $f(x) = ax^2 + bx + c$, $a > 0$, might fit the data. The answer is (b).

5. The data points fall at a steady rate, so a linear function $f(x) = mx + b$ might fit the data. The answer is (a).

6. The data points rise, then fall, and then rise again. Thus, neither a linear nor a quadratic function fits the data. The answer is (d).

7. a)

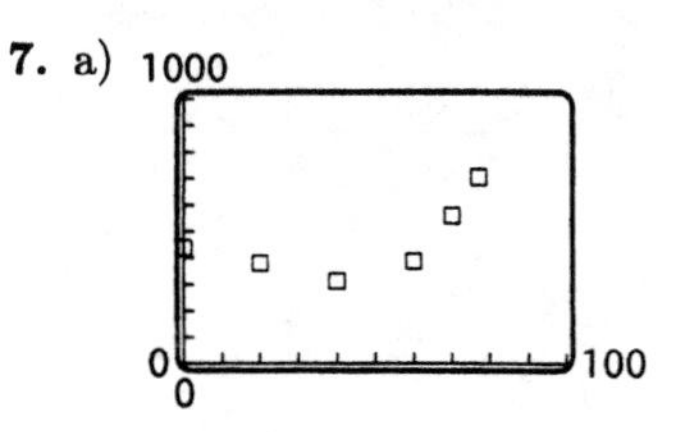

b) The data points fall and then rise, suggesting that a quadratic function might fit the data.

c) Using the quadratic regression feature on a grapher, we have $f(x) = 0.1729980538x^2 - 10.70013387x + 465.7917864$.

d)

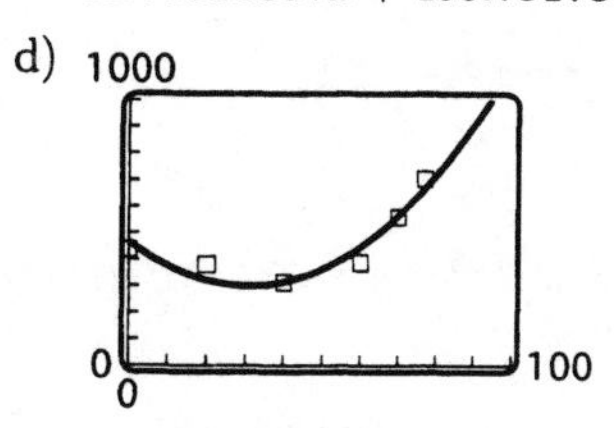

e) In 2004, $x = 2004 - 1920$, or 84; $f(84) \approx 788$, so we predict that there will be about 788 morning newspapers in 2004.

In 2007, $x = 2007 - 1920$, or 87; $f(87) \approx 844$, so we predict that there will be about 844 morning newspapers in 2007.

8. a)

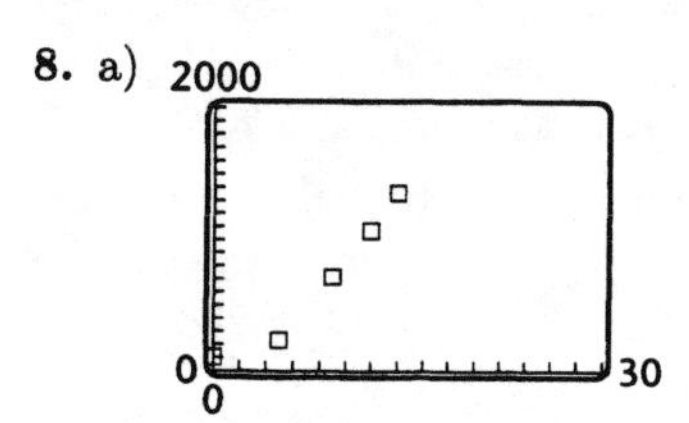

b) Quadratic

c) $f(x) = 5.674698286x^2 + 12.96730487x + 75.57847395$

d)

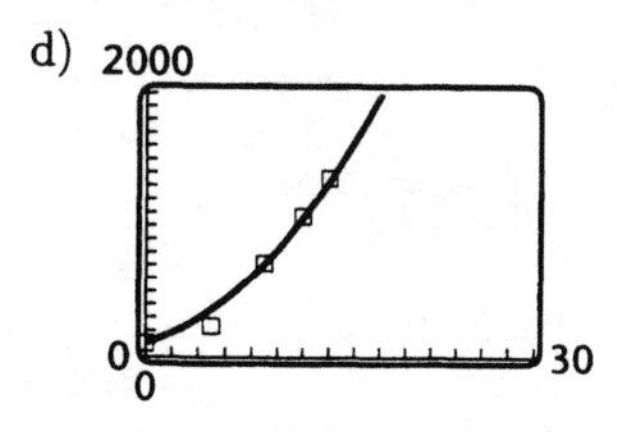

e) $f(22) \approx 3107.4$, so we predict that about 3107.4 thousand children will be home-educated in 2005.

$f(27) \approx 4562.6$, so we predict that about 4562.6 thousand children will be home-educated in 2010.

9. a)

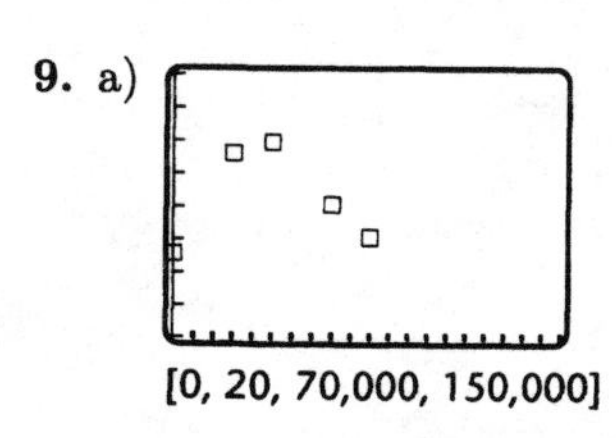

[0, 20, 70,000, 150,000]

b) The data points rise and then fall, suggesting that a quadratic function might fit the data.

c) Using the quadratic regression feature of a grapher, we have $f(x) = -1226.47072x^2 + 12,279.05135x + 97,260.77349$.

d)

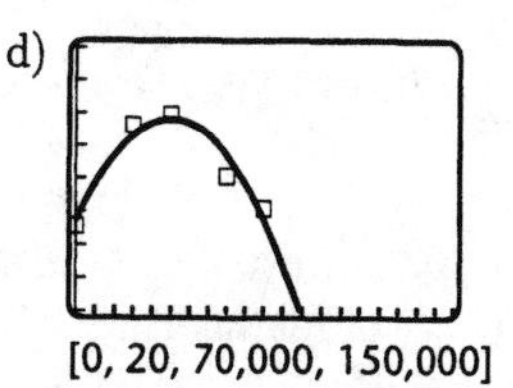

[0, 20, 70,000, 150,000]

e) In 2003, $x = 2003 - 1990$, or 13; $f(13) \approx 49,615$, so we predict that the assets of the Hospital Insurance program will be about $49,615 million in 2003.

In 2007, $x = 2007 - 1990$, or 17; $f(17) \approx 5491$, so we predict that the assets of the Hospital Insurance program will be about $5491 million in 2007.

10. a)

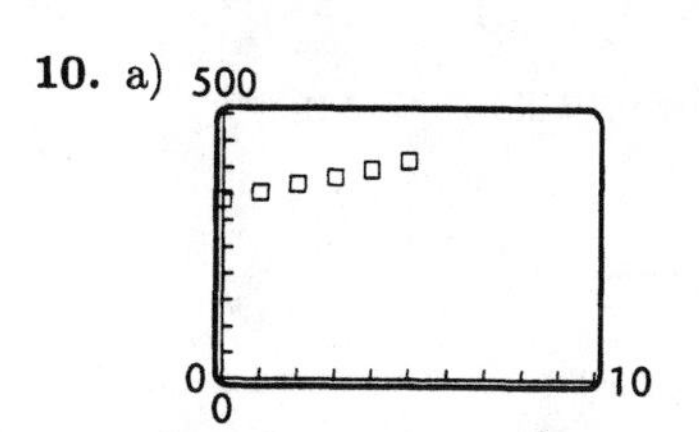

b) Linear

c) $f(x) = 14.56857143x + 339.4285714$

d)

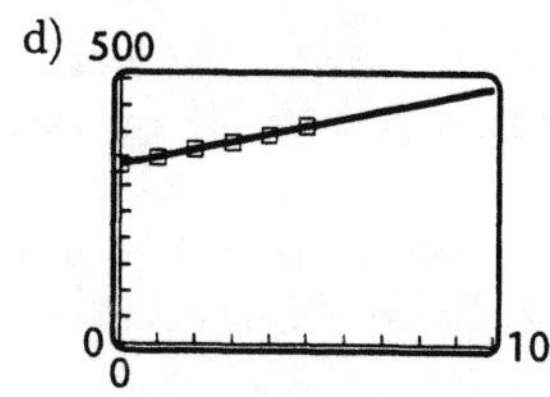

e) $f(9) \approx 470.5$, so we predict that the assets of the Social Security Trust Fund will be about $470.5 billion in 2004.

$f(15) \approx 558.0$, so we predict that the assets of the Social Security Trust Fund will be about $558.0 billion in 2010.

11. a)

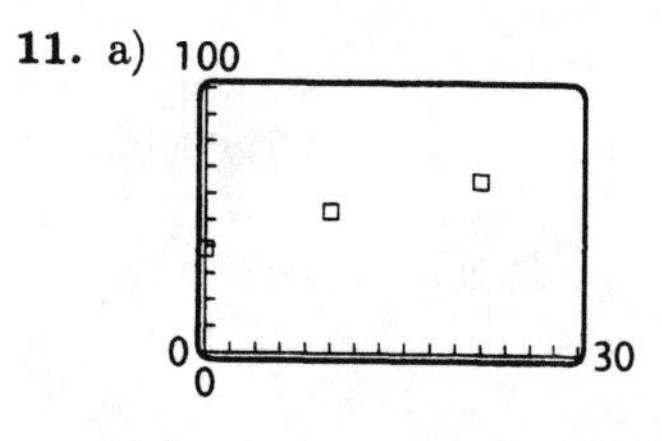

b) It appears that either a linear or a quadratic function could be fit to these data. We will use a quadratic function.

c) Using the quadratic regression feature on a grapher, we have $f(x) = -0.0223484848x^2 + 1.673484848x + 39$.

d)

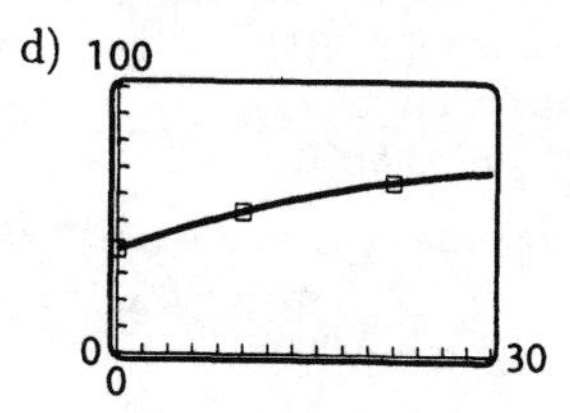

e) In 2003, $x = 2003 - 1975$, or 28; $f(28) \approx 68.3$, so we predict that 68.3% of mothers in the labor force in 2003 will have children under 6 years old.

In 2010, $x = 2010 - 1975$, or 35; $f(35) \approx 70.2$, so we predict that 70.2% of mothers in the labor force in 2010 will have children under 6 years old.

12. a)

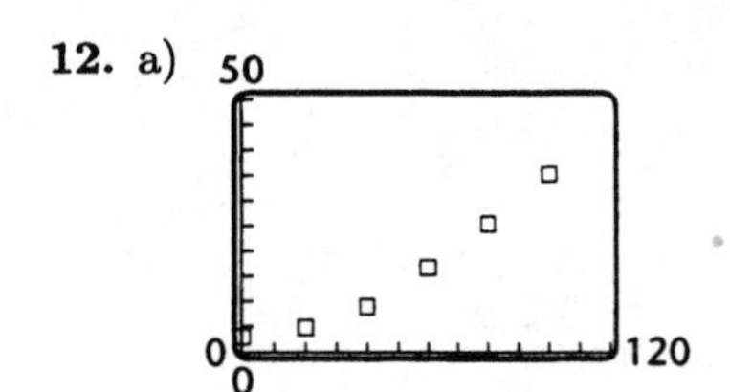

b) Quadratic

c) $f(x) = 0.002625x^2 + 0.0667857143x + 2.785714286$

d)

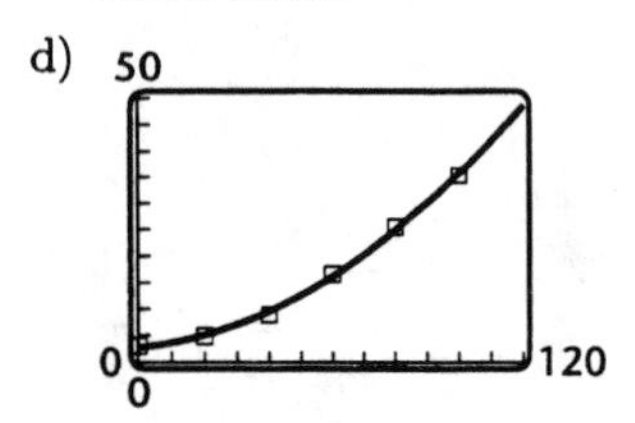

e) $f(106) \approx 39.4$, so we predict that there will be about 39.4 million Americans age 65 and over in 2006.

$f(111) \approx 42.5$, so we predict that there will be about 42.5 million Americans age 65 and over in 2011.

13. a)

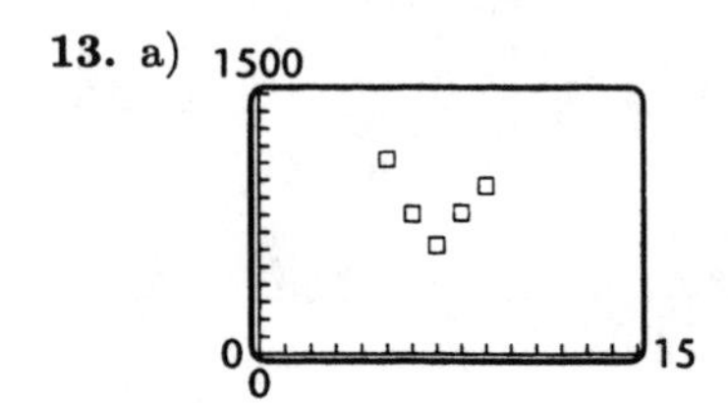

b) The data points fall and then rise, suggesting that a quadratic function might fit the data.

c) Using the quadratic regression feature on a grapher, we have $f(x) = 93.28571429x^2 - 1336x + 5460.828571$.

d)

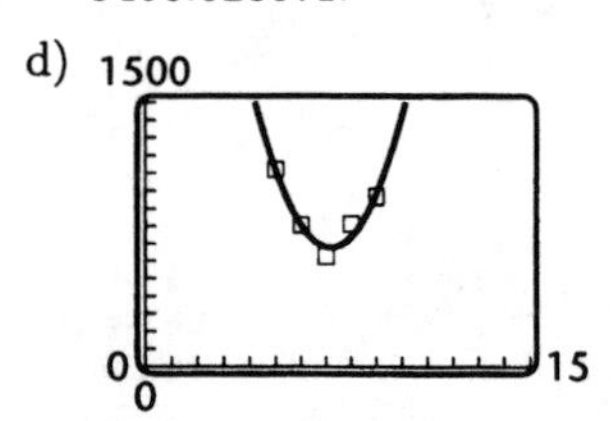

e) $f(7.5) \approx 688$, so we predict that the death rate of males who slept for an average of 7.5 hr per day was about 688 per 100,000.

$f(10) \approx 1429$, so we predict that the death rate of males who slept for an average of 10 hr per day was about 1429 per 100,000.

14. a)

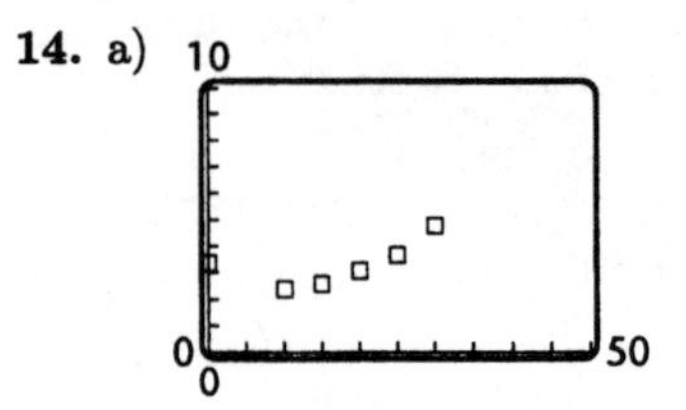

b) Quadratic

c) $f(x) = 0.0067142857x^2 - 0.1531428571x + 3.367857143$

d)

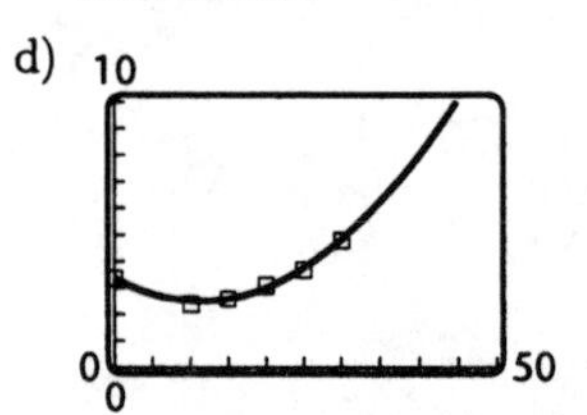

e) $f(34) \approx 5.9$, so we predict that the federal outlay for basic science research, in terms of 1992 dollars, will be about \$5.9 billion in 2004.

$f(38) \approx 7.2$, so we predict that the federal outlay for basic science research, in terms of 1992 dollars, will be about \$7.2 billion in 2008.

15. a) Using the quadratic regression feature of a grapher, we have $f(x) = -0.8630952381x^2 + 12.125x + 36.76785714$.

b) $f(5.5) \approx 77$ yr; $f(8.5) \approx 77$ yr

16. a) $f(x) = -0.4772727273x^2 + 9.856060606x + 26.37878788$

b) $f(10.5) \approx 77$ yr; $f(15.5) \approx 64$ yr

17. It would be unreasonable to think that a person could sleep an average of 0 hours or 24 hours per day. In addition, it would not be possible to sleep for more than 24 hours per day. Thus, function values like $f(0)$, $f(24)$, and $f(25)$ have no meaning.

18. Compare some function values with actual data.

19. $g(x) = \dfrac{x+5}{x-1}$

$g(-2) = \dfrac{-2+5}{-2-1} = \dfrac{3}{-3} = -1$

20. $g(1) = \dfrac{1+5}{1-1} = \dfrac{6}{0}$

$g(1)$ does not exist.

21. $g(x) = \dfrac{x+5}{x-1}$

$g(-5) = \dfrac{-5+5}{-5-1} = \dfrac{0}{-6} = 0$

22. When $x = 1$, the denominator is 0, so the domain is $\{x|x \neq 1\}$, or $(-\infty, 1) \cup (1, \infty)$.

23. a)

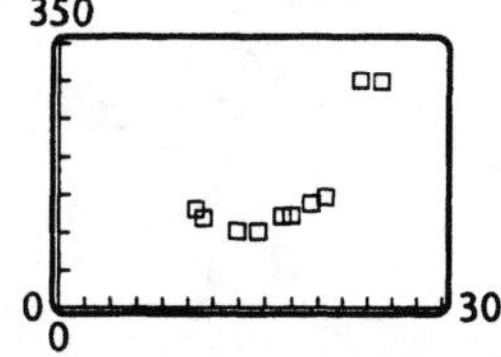

b) The data points fall and then rise, suggesting that a quadratic function might fit the data.

c) Using the quadratic regression feature on a grapher, we have $f(x) = 2.031904548x^2 - 59.04179598x + 527.2818092$.

d)

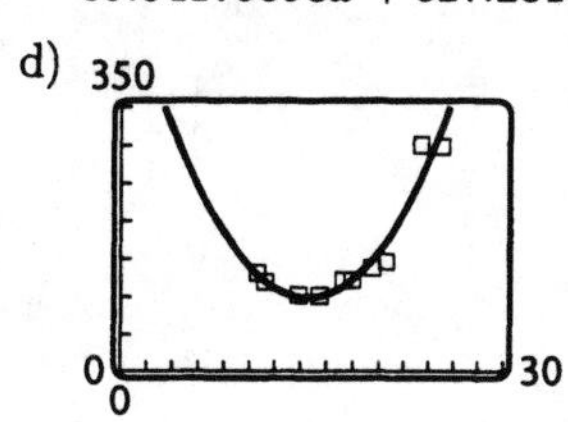

e) Graph $y_1 = f(x)$ and $y_2 = 180$ and find the first coordinate of the point of intersection of the graphs. We find that about \$20.9 million should be allotted for the advertising budget.

Exercise Set 2.6

1. $\frac{1}{4}+\frac{1}{5}=\frac{1}{t}$, LCD is $20t$

$$20t\left(\frac{1}{4}+\frac{1}{5}\right) = 20t\cdot\frac{1}{t}$$
$$20t\cdot\frac{1}{4}+20t\cdot\frac{1}{5} = 20t\cdot\frac{1}{t}$$
$$5t+4t = 20$$
$$9t = 20$$
$$t = \frac{20}{9}$$

Check:

$\frac{1}{4}+\frac{1}{5}=\frac{1}{t}$		
$\frac{1}{4}+\frac{1}{5}$?	$\frac{1}{\frac{20}{9}}$	
$\frac{5}{20}+\frac{4}{20}$	$1\cdot\frac{9}{20}$	
$\frac{9}{20}$	$\frac{9}{20}$	TRUE

The solution is $\frac{20}{9}$.

2. $\frac{1}{3}-\frac{5}{6}=\frac{1}{x}$, LCD is $6x$

$2x-5x = 6$ Multiplying by $6x$

$$-3x = 6$$
$$x = -2$$

-2 checks. The solution is -2.

3. $\frac{x+2}{4}-\frac{x-1}{5} = 15$, LCD is 20

$$20\left(\frac{x+2}{4}-\frac{x-1}{5}\right) = 20\cdot 15$$
$$5(x+2)-4(x-1) = 300$$
$$5x+10-4x+4 = 300$$
$$x+14 = 300$$
$$x = 286$$

The solution is 286.

4. $\frac{t+1}{3}-\frac{t-1}{2} = 1$, LCD is 6

$2t+2-3t+3 = 6$ Multiplying by 6

$$-t = 1$$
$$t = -1$$

The solution is -1.

5. $\frac{1}{2}+\frac{2}{x}=\frac{1}{3}+\frac{3}{x}$, LCD is $6x$

$$6x\left(\frac{1}{2}+\frac{2}{x}\right) = 6x\left(\frac{1}{3}+\frac{3}{x}\right)$$
$$3x+12 = 2x+18$$
$$3x-2x = 18-12$$
$$x = 6$$

Check:

$\frac{1}{2}+\frac{2}{x}=\frac{1}{3}+\frac{3}{x}$		
$\frac{1}{2}+\frac{2}{6}$?	$\frac{1}{3}+\frac{3}{6}$	
$\frac{1}{2}+\frac{1}{3}$	$\frac{1}{3}+\frac{1}{2}$	TRUE

The solution is 6.

6. $\frac{1}{t}+\frac{1}{2t}+\frac{1}{3t} = 5$, LCD is $6t$

$6+3+2 = 30t$ Multiplying by $6t$

$$11 = 30t$$
$$\frac{11}{30} = t$$

$\frac{11}{30}$ checks. The solution is $\frac{11}{30}$.

7. $$\frac{3x}{x+2}+\frac{6}{x} = \frac{12}{x^2+2x}$$

$\frac{3x}{x+2}+\frac{6}{x} = \frac{12}{x(x+2)}$, LCD is $x(x+2)$

$$x(x+2)\left(\frac{3x}{x+2}+\frac{6}{x}\right) = x(x+2)\cdot\frac{12}{x(x+2)}$$
$$3x\cdot x+6(x+2) = 12$$
$$3x^2+6x+12 = 12$$
$$3x^2+6x = 0$$
$$3x(x+2) = 0$$
$$3x = 0 \quad or \quad x+2 = 0$$
$$x = 0 \quad or \quad x = -2$$

Neither 0 nor -2 checks, so the equation has no solution.

8. $\dfrac{5x}{x-4} - \dfrac{20}{x} = \dfrac{80}{x^2-4x}$

$\dfrac{5x}{x-4} - \dfrac{20}{x} = \dfrac{80}{x(x-4)}$, LCD is $x(x-4)$

$5x^2 - 20x + 80 = 80$ Multiplying by $x(x-4)$

$5x^2 - 20x = 0$

$5x(x-4) = 0$

$x = 0$ or $x = 4$

Neither 0 nor 4 checks. There is no solution.

9. $\dfrac{4}{x^2-1} - \dfrac{2}{x-1} = \dfrac{3}{x+1}$,

LCD is $(x+1)(x-1)$

$(x+1)(x-1)\left(\dfrac{4}{(x+1)(x-1)} - \dfrac{2}{x-1}\right) =$

$(x+1)(x-1) \cdot \dfrac{3}{x+1}$

$4 - 2(x+1) = 3(x-1)$

$4 - 2x - 2 = 3x - 3$

$2 - 2x = 3x - 3$

$2 + 3 = 3x + 2x$

$5 = 5x$

$1 = x$

Check:

$$\frac{4}{x^2-1} - \frac{2}{x-1} = \frac{3}{x+1}$$

$$\frac{4}{1^2-1} - \frac{2}{1-1} \;?\; \frac{3}{1+1}$$

$$\frac{4}{0} - \frac{2}{0} \;\Big|\; \frac{3}{2}$$

Division by zero is undefined.

There is no solution.

10. $\dfrac{3y+5}{y^2+5y} + \dfrac{y+4}{y+5} = \dfrac{y+1}{y}$

$\dfrac{3y+5}{y(y+5)} + \dfrac{y+4}{y+5} = \dfrac{y+1}{y}$, LCD is $y(y+5)$

$3y + 5 + y^2 + 4y = y^2 + 6y + 5$

Multiplying by $y(y+5)$

$y = 0$

0 does not check. There is no solution.

11. $\dfrac{490}{x^2-49} = \dfrac{5x}{x-7} - \dfrac{35}{x+7}$

$\dfrac{490}{(x+7)(x-7)} = \dfrac{5x}{x-7} - \dfrac{35}{x+7}$,

LCD is $(x+7)(x-7)$

$(x+7)(x-7)\left(\dfrac{490}{(x+7)(x-7)}\right) =$

$(x+7)(x-7)\left(\dfrac{5x}{x-7} - \dfrac{35}{x+7}\right)$

$490 = 5x(x+7) - 35(x-7)$

$490 = 5x^2 + 35x - 35x + 245$

$0 = 5x^2 - 245$

$0 = 5(x+7)(x-7)$

$x + 7 = 0$ or $x - 7 = 0$

$x = -7$ or $x = 7$

Neither -7 nor 7 checks, so the equation has no solution.

12. $\dfrac{3}{m+2} + \dfrac{2}{m} = \dfrac{4m-4}{m^2-4}$

$\dfrac{3}{m+2} + \dfrac{2}{m} = \dfrac{4m-4}{(m+2)(m-2)}$,

LCD is $m(m+2)(m-2)$

$3m^2 - 6m + 2m^2 - 8 = 4m^2 - 4m$

Multiplying by $m(m+2)(m-2)$

$m^2 - 2m - 8 = 0$

$(m-4)(m+2) = 0$

$m = 4$ or $m = -2$

Only 4 checks. The solution is 4.

13. $\dfrac{1}{x-6} - \dfrac{1}{x} = \dfrac{6}{x^2-6x}$

$\dfrac{1}{x-6} - \dfrac{1}{x} = \dfrac{6}{x(x-6)}$, LCD is $x(x-6)$

$x(x-6)\left(\dfrac{1}{x-6} - \dfrac{1}{x}\right) = x(x-6) \cdot \dfrac{6}{x(x-6)}$

$x - (x-6) = 6$

$x - x + 6 = 6$

$6 = 6$

We get an equation that is true for all real numbers. Note, however, that when $x = 6$ or $x = 0$, division by 0 occurs in the original equation. Thus, the solution set is $\{x|x$ is a real number *and* $x \neq 6$ *and* $x \neq 0\}$, or $(-\infty, 0) \cup (0, 6) \cup (6, \infty)$.

14. $\dfrac{8}{x^2-4} = \dfrac{x}{x-2} - \dfrac{2}{x+2}$

$\dfrac{8}{(x+2)(x-2)} = \dfrac{x}{x-2} - \dfrac{2}{x+2}$,

LCD is $(x+2)(x-2)$

$8 = x^2 + 2x - 2x + 4$

Multiplying by $(x+2)(x-2)$

$0 = x^2 - 4$

$0 = (x+2)(x-2)$

$x = -2$ or $x = 2$

Neither -2 nor 2 checks. There is no solution.

15. $$\frac{8}{x^2-2x+4} = \frac{x}{x+2} + \frac{24}{x^3+8},$$

LCD is $(x+2)(x^2-2x+4)$

$$(x+2)(x^2-2x+4) \cdot \frac{8}{x^2-2x+4} =$$
$$(x+2)(x^2-2x+4)\left(\frac{x}{x+2}+\frac{24}{(x+2)(x^2-2x+4)}\right)$$
$$8(x+2) = x(x^2-2x+4)+24$$
$$8x+16 = x^3-2x^2+4x+24$$
$$0 = x^3-2x^2-4x+8$$
$$0 = x^2(x-2)-4(x-2)$$
$$0 = (x-2)(x^2-4)$$
$$0 = (x-2)(x+2)(x-2)$$

$x-2=0$ or $x+2=0$ or $x-2=0$

$x=2$ or $x=-2$ or $x=2$

Only 2 checks. The solution is 2.

16. $$\frac{18}{x^2-3x+9} - \frac{x}{x+3} = \frac{81}{x^3+27},$$

LCD is $(x+3)(x^2-3x+9)$

$18x+54-x^3+3x^2-9x = 81$ Multiplying by $(x+3)(x^2-3x+9)$

$$-x^3+3x^2+9x-27 = 0$$
$$-x^2(x-3)+9(x-3) = 0$$
$$(x-3)(9-x^2) = 0$$
$$(x-3)(3+x)(3-x) = 0$$

$x=3$ or $x=-3$

Only 3 checks. The solution is 3.

17. $$\sqrt{3x-4} = 1$$
$$(\sqrt{3x-4})^2 = 1^2$$
$$3x-4 = 1$$
$$3x = 5$$
$$x = \frac{5}{3}$$

Check:

$\sqrt{3x-4} = 1$	
$\sqrt{3\cdot\frac{5}{3}-4}$?	1
$\sqrt{5-4}$	
$\sqrt{1}$	
1	1 TRUE

The solution is $\frac{5}{3}$.

18. $$\sqrt[3]{2x+1} = -5$$
$$2x+1 = -125$$
$$2x = -126$$
$$x = -63$$

The answer checks. The solution is -63.

19. $$\sqrt[4]{x^2-1} = 1$$
$$(\sqrt[4]{x^2-1})^4 = 1^4$$
$$x^2-1 = 1$$
$$x^2 = 2$$
$$x = \pm\sqrt{2}$$

Check:

$\sqrt[4]{x^2-1} = 1$	
$\sqrt[4]{(\pm\sqrt{2})^2-1}$?	1
$\sqrt[4]{2-1}$	
$\sqrt[4]{1}$	
1	1 TRUE

The solutions are $\pm\sqrt{2}$.

20. $$\sqrt{m+1}-5 = 8$$
$$\sqrt{m+1} = 13$$
$$m+1 = 169$$
$$m = 168$$

The answer checks. The solution is 168.

21. $$\sqrt{y-1}+4 = 0$$
$$\sqrt{y-1} = -4$$

The principal square root is never negative. Thus, there is no solution.

If we do not observe the above fact, we can continue and reach the same answer.

$$(\sqrt{y-1})^2 = (-4)^2$$
$$y-1 = 16$$
$$y = 17$$

Check:

$\sqrt{y-1}+4 = 0$	
$\sqrt{17-1}+4$?	0
$\sqrt{16}+4$	
$4+4$	
8	0 FALSE

Since 17 does not check, there is no solution.

22. $$\sqrt[5]{3x+4} = 2$$
$$3x+4 = 32$$
$$3x = 28$$
$$x = \frac{28}{3}$$

The answer checks. The solution is $\frac{28}{3}$.

23. $$\sqrt[3]{6x+9}+8 = 5$$
$$\sqrt[3]{6x+9} = -3$$
$$(\sqrt[3]{6x+9})^3 = (-3)^3$$
$$6x+9 = -27$$
$$6x = -36$$
$$x = -6$$

Check:

$\sqrt[3]{6x+9}+8=5$		
$\sqrt[3]{6(-6)+9}+8$	? 5	
$\sqrt[3]{-27}+8$		
$-3+8$		
5	5	TRUE

The solution is -6.

24. $\sqrt{6x+7}=x+2$

$$6x+7=x^2+4x+4$$
$$0=x^2-2x-3$$
$$0=(x-3)(x+1)$$

$x=3$ or $x=-1$

Both values check. The solutions are 3 and -1.

25. $\sqrt{x-3}+\sqrt{x+2}=5$

$$\sqrt{x+2}=5-\sqrt{x-3}$$
$$(\sqrt{x+2})^2=(5-\sqrt{x-3})^2$$
$$x+2=25-10\sqrt{x-3}+(x-3)$$
$$x+2=22-10\sqrt{x-3}+x$$
$$10\sqrt{x-3}=20$$
$$\sqrt{x-3}=2$$
$$(\sqrt{x-3})^2=2^2$$
$$x-3=4$$
$$x=7$$

Check:

$\sqrt{x-3}+\sqrt{x+2}=5$		
$\sqrt{7-3}+\sqrt{7+2}$	? 5	
$\sqrt{4}+\sqrt{9}$		
$2+3$		
5	5	TRUE

The solution is 7.

26. $\sqrt{x}-\sqrt{x-5}=1$

$$\sqrt{x}=\sqrt{x-5}+1$$
$$x=x-5+2\sqrt{x-5}+1$$
$$4=2\sqrt{x-5}$$
$$2=\sqrt{x-5}$$
$$4=x-5$$
$$9=x$$

The answer checks. The solution is 9.

27. $\sqrt{3x-5}+\sqrt{2x+3}+1=0$

$$\sqrt{3x-5}+\sqrt{2x+3}=-1$$

The principal square root is never negative. Thus the sum of two principal square roots cannot equal -1. There is no solution.

28. $\sqrt{2m-3}=\sqrt{m+7}-2$

$$2m-3=m+7-4\sqrt{m+7}+4$$
$$m-14=-4\sqrt{m+7}$$
$$m^2-28m+196=16m+112$$
$$m^2-44m+84=0$$
$$(m-2)(m-42)=0$$

$m=2$ or $m=42$

Only 2 checks. The solution is 2.

29. $\sqrt{x}-\sqrt{3x-3}=1$

$$\sqrt{x}=\sqrt{3x-3}+1$$
$$(\sqrt{x})^2=(\sqrt{3x-3}+1)^2$$
$$x=(3x-3)+2\sqrt{3x-3}+1$$
$$2-2x=2\sqrt{3x-3}$$
$$1-x=\sqrt{3x-3}$$
$$(1-x)^2=(\sqrt{3x-3})^2$$
$$1-2x+x^2=3x-3$$
$$x^2-5x+4=0$$
$$(x-4)(x-1)=0$$

$x=4$ or $x=1$

The number 4 does not check, but 1 does. The solution is 1.

30. $\sqrt{2x+1}-\sqrt{x}=1$

$$\sqrt{2x+1}=\sqrt{x}+1$$
$$2x+1=x+2\sqrt{x}+1$$
$$x=2\sqrt{x}$$
$$x^2=4x$$
$$x^2-4x=0$$
$$x(x-4)=0$$

$x=0$ or $x=4$

Both values check. The solutions are 0 and 4.

31. $\sqrt{2y-5}-\sqrt{y-3}=1$

$$\sqrt{2y-5}=\sqrt{y-3}+1$$
$$(\sqrt{2y-5})^2=(\sqrt{y-3}+1)^2$$
$$2y-5=(y-3)+2\sqrt{y-3}+1$$
$$y-3=2\sqrt{y-3}$$
$$(y-3)^2=(2\sqrt{y-3})^2$$
$$y^2-6y+9=4(y-3)$$
$$y^2-6y+9=4y-12$$
$$y^2-10y+21=0$$
$$(y-7)(y-3)=0$$

$y=7$ or $y=3$

Both numbers check. The solutions are 7 and 3.

32. $\sqrt{4p+5}+\sqrt{p+5}=3$

$$\begin{aligned}\sqrt{4p+5}&=3-\sqrt{p+5}\\4p+5&=9-6\sqrt{p+5}+p+5\\3p-9&=-6\sqrt{p+5}\\p-3&=-2\sqrt{p+5}\\p^2-6p+9&=4p+20\\p^2-10p-11&=0\\(p-11)(p+1)&=0\\p=11 \text{ or } p&=-1\end{aligned}$$

Only -1 checks. The solution is -1.

33. $x^{1/3}=-2$

$(x^{1/3})^3=(-2)^3 \qquad (x^{1/3}=\sqrt[3]{x})$

$x=-8$

The value checks. The solution is -8.

34. $t^{1/5}=2$

$t=32$

The value checks. The solution is 32.

35. $t^{1/4}=3$

$(t^{1/4})^4=3^4 \qquad (t^{1/4}=\sqrt[4]{t})$

$t=81$

The value checks. The solution is 81.

36. $m^{1/2}=-7$

The principal square root is never negative. There is no solution.

37. $|x|=7$

The solutions are those numbers whose distance from 0 on a number line is 7. They are -7 and 7. That is,

$x=-7$ *or* $x=7$.

The solutions are -7 and 7.

38. $|x|=4.5$

$x=-4.5$ *or* $x=4.5$

The solutions are -4.5 and 4.5.

39. $|x|=-10.7$

The absolute value of a number is nonnegative. Thus, the equation has no solution.

40. $|x|=-\frac{3}{5}$

The absolute value of a number is nonnegative. Thus, there is no solution.

41. $|x-1|=4$

$x-1=-4$ or $x-1=4$

$x=-3$ or $x=5$

The solutions are -3 and 5.

42. $|x-7|=5$

$x-7=-5$ or $x-7=5$

$x=2$ or $x=12$

The solutions are 2 and 12.

43. $|3x|=1$

$3x=-1$ or $3x=1$

$x=-\frac{1}{3}$ or $x=\frac{1}{3}$

The solutions are $-\frac{1}{3}$ and $\frac{1}{3}$.

44. $|5x|=4$

$5x=-4$ or $5x=4$

$x=-\frac{4}{5}$ or $x=\frac{4}{5}$

The solutions are $-\frac{4}{5}$ and $\frac{4}{5}$.

45. $|x|=0$

The distance of 0 from 0 on a number line is 0. That is,

$x=0$.

The solution is 0.

46. $|6x|=0$

$6x=0$

$x=0$

The solution is 0.

47. $|3x+2|=1$

$3x+2=-1$ or $3x+2=1$

$3x=-3$ or $3x=-1$

$x=-1$ or $x=-\frac{1}{3}$

The solutions are -1 and $-\frac{1}{3}$.

48. $|7x-4|=8$

$7x-4=-8$ or $7x-4=8$

$7x=-4$ or $7x=12$

$x=-\frac{4}{7}$ or $x=\frac{12}{7}$

The solutions are $-\frac{4}{7}$ and $\frac{12}{7}$.

49. $\left|\frac{1}{2}x-5\right|=17$

$\frac{1}{2}x-5=-17$ or $\frac{1}{2}x-5=17$

$\frac{1}{2}x=-12$ or $\frac{1}{2}x=22$

$x=-24$ or $x=44$

The solutions are -24 and 44.

50. $\left|\frac{1}{3}x - 4\right| = 13$

$\frac{1}{3}x - 4 = -13$ or $\frac{1}{3}x - 4 = 13$

$\frac{1}{3}x = -9$ or $\frac{1}{3}x = 17$

$x = -27$ or $x = 51$

The solutions are -27 and 51.

51. $|x-1| + 3 = 6$

$|x-1| = 3$

$x - 1 = -3$ or $x - 1 = 3$

$x = -2$ or $x = 4$

The solutions are -2 and 4.

52. $|x+2| - 5 = 9$

$|x+2| = 14$

$x + 2 = -14$ or $x + 2 = 14$

$x = -16$ or $x = 12$

The solutions are -16 and 12.

53. $\frac{P_1V_1}{T_1} = \frac{P_2V_2}{T_2}$

$P_1V_1T_2 = P_2V_2T_1$ Multiplying by T_1T_2 on both sides

$\frac{P_1V_1T_2}{P_2V_2} = T_1$ Dividing by P_2V_2 on both sides

54. $\frac{1}{F} = \frac{1}{m} + \frac{1}{p}$

$mp = Fp + Fm$

$mp = F(p+m)$

$\frac{mp}{p+m} = F$

55. $\frac{1}{R} = \frac{1}{R_1} + \frac{1}{R_2}$

$RR_1R_2 \cdot \frac{1}{R} = RR_1R_2\left(\frac{1}{R_1} + \frac{1}{R_2}\right)$

Multiplying by RR_1R_2 on both sides

$R_1R_2 = RR_2 + RR_1$

$R_1R_2 - RR_2 = RR_1$ Subtracting RR_2 on both sides

$R_2(R_1 - R) = RR_1$ Factoring

$R_2 = \frac{RR_1}{R_1 - R}$ Dividing by $R_1 - R$ on both sides

56. $A = P(1+i)^2$

$\frac{A}{P} = (1+i)^2$

$\sqrt{\frac{A}{P}} = 1 + i$

$\sqrt{\frac{A}{P}} - 1 = i$

57. $\frac{1}{F} = \frac{1}{m} + \frac{1}{p}$

$Fmp \cdot \frac{1}{F} = Fmp\left(\frac{1}{m} + \frac{1}{p}\right)$ Multiplying by Fmp on both sides

$mp = Fp + Fm$

$mp - Fp = Fm$ Subtracting Fp on both sides

$p(m - F) = Fm$ Factoring

$p = \frac{Fm}{m - F}$ Dividing by $m - F$ on both sides

58. When both sides of an equation are multiplied by the LCD, the resulting equation might not be equivalent to the original equation. One or more of the possible solutions of the resulting equation might make a denominator of the original equation 0.

59. When both sides of an equation are raised to an even power, the resulting equation might not be the equivalent to the original equation. For example, the solution set of $x = -2$ is $\{-2\}$, but the solution set of $x^2 = (-2)^2$, or $x^2 = 4$, is $\{-2, 2\}$.

60. Graph $y = -3x + 9$ and use the Zero feature.

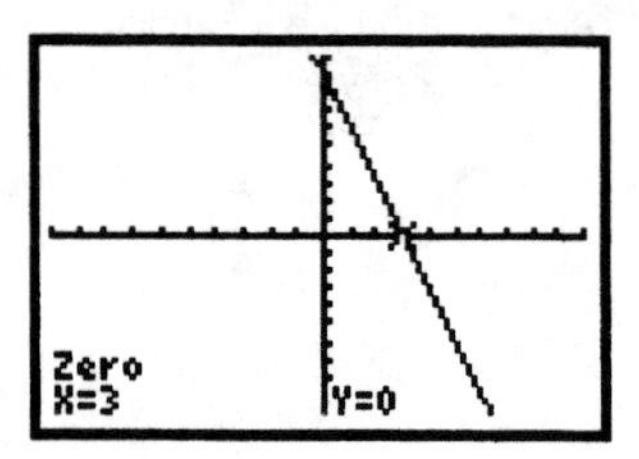

The zero is 3.

61. Graph $y = 15 - 2x$ and use the Zero feature.

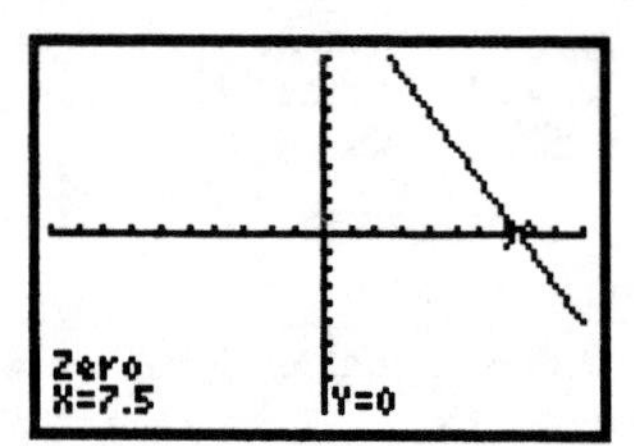

The zero is 7.5.

62. The amount of the increase is $27.7 - 25.1$, or 2.6 lb. Let $p =$ the percent of increase.

Solve: $p = \frac{2.6}{25.1}$

$p \approx 0.104$, or 10.4%

63. ***Familiarize.*** Let a = the number of adults who passed the GED test in 1997.

Translate.

Number of adults who passed the test in 1998	was	25,000	more than	number who passed the test in 1997.
$\downarrow$	$\downarrow$	$\downarrow$	$\downarrow$	$\downarrow$
506,000	$=$	25,000	$+$	a

Carry out. We will solve algebraically.

$$506,000 = 25,000 + a$$

$$481,000 = a \quad \text{Subtracting 25,000}$$

Check. 25,000 more than 481,000 is 25,000 + 481,000, or 506,000. The answer checks.

State. In 1997, 481,000 adults passed the GED test.

64. $\dfrac{x+3}{x+2} - \dfrac{x+4}{x+3} = \dfrac{x+5}{x+4} - \dfrac{x+6}{x+5}$,

LCD is $(x+2)(x+3)(x+4)(x+5)$

$x^4 + 15x^3 + 83x^2 + 201x + 180 - x^4 - 15x^3 -$
$82x^2 - 192x - 160 = x^4 + 15x^3 + 81x^2 + 185x +$
$150 - x^4 - 15x^3 - 80x^2 - 180x - 144$

$$x^2 + 9x + 20 = x^2 + 5x + 6$$
$$4x = -14$$
$$x = -\frac{7}{2}$$

The number $-\dfrac{7}{2}$ checks. The solution is $-\dfrac{7}{2}$.

65.
$$(x-3)^{2/3} = 2$$
$$[(x-3)^{2/3}]^3 = 2^3$$
$$(x-3)^2 = 8$$
$$x^2 - 6x + 9 = 8$$
$$x^2 - 6x + 1 = 0$$

$a = 1$, $b = -6$, $c = 1$

$$x = \frac{-b \pm \sqrt{b^2 - 4ac}}{2a}$$
$$= \frac{-(-6) \pm \sqrt{(-6)^2 - 4 \cdot 1 \cdot 1}}{2 \cdot 1}$$
$$= \frac{6 \pm \sqrt{32}}{2} = \frac{6 \pm 4\sqrt{2}}{2}$$
$$= \frac{2(3 \pm 2\sqrt{2})}{2} = 3 \pm 2\sqrt{2}$$

Both values check. The solutions are $3 \pm 2\sqrt{2}$.

66.
$$\sqrt{15 + \sqrt{2x+80}} = 5$$
$$\left(\sqrt{15 + \sqrt{2x+80}}\right)^2 = 5^2$$
$$15 + \sqrt{2x+80} = 25$$
$$\sqrt{2x+80} = 10$$
$$(\sqrt{2x+80})^2 = 10^2$$
$$2x + 80 = 100$$
$$2x = 20$$
$$x = 10$$

This number checks. The solution is 10.

67.
$$\sqrt{x+5} + 1 = \frac{6}{\sqrt{x+5}}, \quad \text{LCD is } \sqrt{x+5}$$
$$x + 5 + \sqrt{x+5} = 6 \quad \text{Multiplying by } \sqrt{x+5}$$
$$\sqrt{x+5} = 1 - x$$
$$x + 5 = 1 - 2x + x^2$$
$$0 = x^2 - 3x - 4$$
$$0 = (x-4)(x+1)$$

$x = 4$ or $x = -1$

Only -1 checks. The solution set is -1.

68. $x^{2/3} = x + 1$

Find the first coordinate of the point of intersection of $y_1 = (x^{1/3})^2$ and $y_2 = x + 1$. The solution is approximately -0.430.

Exercise Set 2.7

1.
$$x + 6 < 5x - 6$$
$$6 + 6 < 5x - x \quad \text{Subtracting } x \text{ and adding 6 on both sides}$$
$$12 < 4x$$
$$\frac{12}{4} < x \quad \text{Dividing by 4 on both sides}$$
$$3 < x$$

This inequality could also be solved as follows:

$$x + 6 < 5x - 6$$
$$x - 5x < -6 - 6 \quad \text{Subtracting } 5x \text{ and 6 on both sides}$$
$$-4x < -12$$
$$x > \frac{-12}{-4} \quad \text{Dividing by } -4 \text{ on both sides and reversing the inequality symbol}$$
$$x > 3$$

The solution set is $\{x|x > 3\}$, or $(3, \infty)$.

2.
$$3 - x < 4x + 7$$
$$-5x < 4$$
$$x > -\frac{4}{5}$$

The solution set is $\left\{x\middle|x > -\frac{4}{5}\right\}$, or $\left(-\frac{4}{5}, \infty\right)$.

3. $3x - 3 + 2x \geq 1 - 7x - 9$

$5x - 3 \geq -7x - 8$ Collecting like terms

$5x + 7x \geq -8 + 3$ Adding $7x$ and 3 on both sides

$12x \geq -5$

$x \geq -\frac{5}{12}$ Dividing by 12 on both sides

The solution set is $\left\{x \middle| x \geq -\frac{5}{12}\right\}$, or $\left[-\frac{5}{12}, \infty\right)$.

4. $5y - 5 + y \leq 2 - 6y - 8$

$6y - 5 \leq -6y - 6$

$12y \leq -1$

$y \leq -\frac{1}{12}$

The solution set is $\left\{y \middle| y \leq -\frac{1}{12}\right\}$, or $\left(-\infty, -\frac{1}{12}\right]$.

5. $14 - 5y \leq 8y - 8$

$14 + 8 \leq 8y + 5y$

$22 \leq 13y$

$\frac{22}{13} \leq y$

This inequality could also be solved as follows:

$14 - 5y \leq 8y - 8$

$-5y - 8y \leq -8 - 14$

$-13y \leq -22$

$y \geq \frac{22}{13}$ Dividing by -13 on both sides and reversing the inequality symbol

The solution set is $\left\{y \middle| y \geq \frac{22}{13}\right\}$, or $\left[\frac{22}{13}, \infty\right)$.

6. $8x - 7 < 6x + 3$

$2x < 10$

$x < 5$

The solution set is $\{x|x < 5\}$, or $(-\infty, 5)$.

7. $-\frac{3}{4}x \geq -\frac{5}{8} + \frac{2}{3}x$

$\frac{5}{8} \geq \frac{3}{4}x + \frac{2}{3}x$

$\frac{5}{8} \geq \frac{9}{12}x + \frac{8}{12}x$

$\frac{5}{8} \geq \frac{17}{12}x$

$\frac{12}{17} \cdot \frac{5}{8} \geq \frac{12}{17} \cdot \frac{17}{12}x$

$\frac{15}{34} \geq x$

The solution set is $\left\{x \middle| x \leq \frac{15}{34}\right\}$, or $\left(-\infty, \frac{15}{34}\right]$.

8. $-\frac{5}{6}x \leq \frac{3}{4} + \frac{8}{3}x$

$-\frac{21}{6}x \leq \frac{3}{4}$

$x \geq -\frac{3}{14}$

The solution set is $\left\{x \middle| x \geq -\frac{3}{14}\right\}$, or $\left[-\frac{3}{14}, \infty\right)$.

9. $4x(x - 2) < 2(2x - 1)(x - 3)$

$4x(x - 2) < 2(2x^2 - 7x + 3)$

$4x^2 - 8x < 4x^2 - 14x + 6$

$-8x < -14x + 6$

$-8x + 14x < 6$

$6x < 6$

$x < \frac{6}{6}$

$x < 1$

The solution set is $\{x|x < 1\}$, or $(-\infty, 1)$.

10. $(x + 1)(x + 2) > x(x + 1)$

$x^2 + 3x + 2 > x^2 + x$

$2x > -2$

$x > -1$

The solution set is $\{x|x > -1\}$, or $(-1, \infty)$.

11. The graph of $y_1 = 4x(x - 2)$ lies below the graph of $y_2 = 2(2x - 1)(x - 3)$ for $x < 1$.

12. The graph of $y_1 = (x+1)(x+2)$ lies above the graph of $y_2 = x(x + 1)$ for $x > -1$.

13. $-2 \leq x + 1 < 4$

$-3 \leq x < 3$ Subtracting 1

The solution set is $[-3, 3)$.

14. $-3 < x + 2 \leq 5$

$-5 < x \leq 3$

$(-5, 3]$

15. $5 \leq x - 3 \leq 7$

$8 \leq x \leq 10$ Adding 3

The solution set is $[8, 10]$.

16. $-1 < x - 4 < 7$

$3 < x < 11$

$(3, 11)$

17. $-3 \leq x + 4 \leq 3$

$-7 \leq x \leq -1$ Subtracting 4

The solution set is $[-7, -1]$.

18. $-5 < x + 2 < 15$

$-7 < x < 13$

$(-7, 13)$

19. $-2 < 2x + 1 < 5$

$-3 < 2x < 4$ Adding -1

$-\frac{3}{2} < x < 2$ Multiplying by $\frac{1}{2}$

The solution set is $\left(-\frac{3}{2}, 2\right)$.

20. $-3 \le 5x + 1 \le 3$

$-4 \le 5x \le 2$

$-\frac{4}{5} \le x \le \frac{2}{5}$

$\left[-\frac{4}{5}, \frac{2}{5}\right]$

21. $-4 \le 6 - 2x < 4$

$-10 \le -2x < -2$ Adding -6

$5 \ge x > 1$ Multiplying by $-\frac{1}{2}$

or $1 < x \le 5$

The solution set is $(1, 5]$.

22. $-3 < 1 - 2x \le 3$

$-4 < -2x \le 2$

$2 > x \ge -1$

$[-1, 2)$

23. $-5 < \frac{1}{2}(3x + 1) \le 7$

$-10 < 3x + 1 \le 14$ Multiplying by 2

$-11 < 3x \le 13$ Adding -1

$-\frac{11}{3} < x \le \frac{13}{3}$ Multiplying by $\frac{1}{3}$

The solution set is $\left(-\frac{11}{3}, \frac{13}{3}\right]$

24. $\frac{2}{3} \le -\frac{4}{5}(x - 3) < 1$

$-\frac{5}{6} \ge x - 3 > -\frac{5}{4}$

$\frac{13}{6} \ge x > \frac{7}{4}$

$\left(\frac{7}{4}, \frac{13}{6}\right]$

25. $3x \le -6$ *or* $x - 1 > 0$

$x \le -2$ *or* $x > 1$

The solution set is $(-\infty, -2] \cup (1, \infty)$.

26. $2x < 8$ *or* $x + 3 \ge 10$

$x < 4$ *or* $x \ge 7$

$(-\infty, 4) \cup [7, \infty)$

27. $2x + 3 \le -4$ *or* $2x + 3 \ge 4$

$2x \le -7$ *or* $2x \ge 1$

$x \le -\frac{7}{2}$ *or* $x \ge \frac{1}{2}$

The solution set is $\left(-\infty, -\frac{7}{2}\right] \cup \left[\frac{1}{2}, \infty\right)$.

28. $3x - 1 < -5$ *or* $3x - 1 > 5$

$3x < -4$ *or* $3x > 6$

$x < -\frac{4}{3}$ *or* $x > 2$

$\left(-\infty, -\frac{4}{3}\right) \cup (2, \infty)$

29. $2x - 20 < -0.8$ *or* $2x - 20 > 0.8$

$2x < 19.2$ *or* $2x > 20.8$

$x < 9.6$ *or* $x > 10.4$

The solution set is $(-\infty, 9.6) \cup (10.4, \infty)$.

30. $5x + 11 \le -4$ *or* $5x + 11 \ge 4$

$5x \le -15$ *or* $5x \ge -7$

$x \le -3$ *or* $x \ge -\frac{7}{5}$

$(-\infty, -3] \cup \left[-\frac{7}{5}, \infty\right)$

31. $x + 14 \le -\frac{1}{4}$ *or* $x + 14 \ge \frac{1}{4}$

$x \le -\frac{57}{4}$ *or* $x \ge -\frac{55}{4}$

The solution set is $\left(-\infty, -\frac{57}{4}\right] \cup \left[-\frac{55}{4}, \infty\right)$.

32. $x - 9 < -\frac{1}{2}$ *or* $x - 9 > \frac{1}{2}$

$x < \frac{17}{2}$ *or* $x > \frac{19}{2}$

$\left(-\infty, \frac{17}{2}\right) \cup \left(\frac{19}{2}, \infty\right)$

33. The graph of $y_1 = 6 - 2x$ lies both on or above the graph of $y_2 = -4$ and below the graph of $y_3 = 4$ for $1 < x \le 5$.

34. The graph of $y_1 = 1 - 2x$ lies both above the graph of $y_2 = -3$ and on or below the graph of $y_3 = 3$ for $-1 \le x < 2$.

35. $|x| < 7$

To solve we look for all numbers x whose distance from 0 is less than 7. These are the numbers between -7 and 7. That is, $-7 < x < 7$. The solution set and its graph are as follows:

$(-7, 7)$

−7 0 7

36. $|x| \le 4.5$

$-4.5 \le x \le 4.5$

The solution set is $[-4.5, 4.5]$.

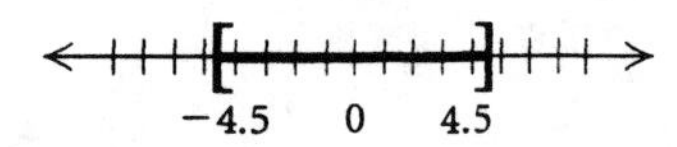

37. $|x| \ge 4.5$

To solve we look for all numbers x whose distance from 0 is greater than or equal to 4.5. That is, $x \le -4.5$ or $x \ge 4.5$. The solution set and its graph are as follows.

$\{x|x \le -4.5 \text{ or } x \ge 4.5\}$, or $(-\infty, -4.5] \cup [4.5, \infty)$

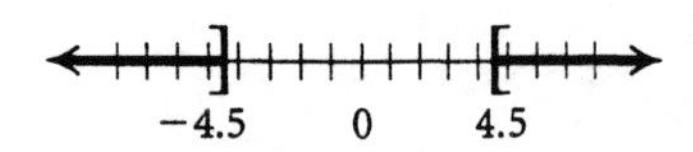

38. $|x| > 7$

$x < -7$ or $x > 7$

The solution set is $(-\infty, -7) \cup (7, \infty)$.

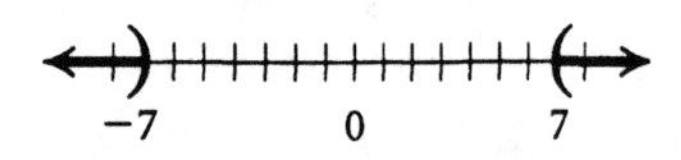

39. $|x+8| < 9$

$-9 < x+8 < 9$

$-17 < x < 1$ Subtracting 8

The solution set is $(-17, 1)$.

40. $|x+6| < 10$

$-10 \le x+6 \le 10$

$-16 \le x \le 4$

The solution set is $[-16, 4]$.

41. $|x+8| \ge 9$

$x+8 \le -9$ *or* $x+8 \ge 9$

$x \le -17$ *or* $x \ge 1$ Subtracting 8

The solution set is $(-\infty, -17] \cup [1, \infty)$.

42. $|x+6| > 10$

$x+6 < -10$ *or* $x+6 > 10$

$x < -16$ *or* $x > 4$

The solution set is $(-\infty, -16) \cup (4, \infty)$.

43. $\left|x - \frac{1}{4}\right| < \frac{1}{2}$

$-\frac{1}{2} < x - \frac{1}{4} < \frac{1}{2}$

$-\frac{1}{4} < x < \frac{3}{4}$ Adding $\frac{1}{4}$

The solution set is $\left(-\frac{1}{4}, \frac{3}{4}\right)$.

44. $|x - 0.5| \le 0.2$

$-0.2 \le x - 0.5 \le 0.2$

$0.3 \le x \le 0.7$

The solution set is $[0.3, 0.7]$.

45. $|3x| < 1$

$-1 < 3x < 1$

$-\frac{1}{3} < x < \frac{1}{3}$ Dividing by 3

The solution set is $\left(-\frac{1}{3}, \frac{1}{3}\right)$.

46. $|5x| \le 4$

$-4 \le 5x \le 4$

$-\frac{4}{5} \le x \le \frac{4}{5}$

The solution set is $\left[-\frac{4}{5}, \frac{4}{5}\right]$.

47. $|2x+3| \le 9$

$-9 \le 2x+3 \le 9$

$-12 \le 2x \le 6$ Subtracting 3

$-6 \le x \le 3$ Dividing by 2

The solution set is $[-6, 3]$.

48. $|3x+4| < 13$

$-13 < 3x+4 < 13$

$-17 < 3x < 9$

$-\frac{17}{3} < x < 3$

The solution set is $\left(-\frac{17}{3}, 3\right)$.

49. $|x-5| > 0.1$

$x - 5 < -0.1$ *or* $x - 5 > 0.1$

$x < 4.9$ *or* $x > 5.1$ Adding 5

The solution set is $(-\infty, 4.9) \cup (5.1, \infty)$.

50. $|x-7| \ge 0.4$

$x - 7 \le -0.4$ *or* $x - 7 \ge 0.4$

$x \le 6.6$ *or* $x \ge 7.4$

The solution set is $(-\infty, 6.6] \cup [7.4, \infty)$.

51. $|6 - 4x| \le 8$

$-8 \le 6 - 4x \le 8$

$-14 \le -4x \le 2$ Subtracting 6

$\frac{14}{4} \ge x \ge -\frac{2}{4}$ Dividing by -4 and reversing the inequality symbols

$\frac{7}{2} \ge x \ge -\frac{1}{2}$ Simplifying

The solution set is $\left[-\frac{1}{2}, \frac{7}{2}\right]$.

52. $|5-2x|>10$

$$5-2x<-10 \quad or \quad 5-2x>10$$
$$-2x<-15 \quad or \quad -2x>5$$
$$x>\frac{15}{2} \quad or \quad x<-\frac{5}{2}$$

The solution set is $\left(-\infty,-\frac{5}{2}\right)\cup\left(\frac{15}{2},\infty\right)$.

53. $\left|x+\frac{2}{3}\right|\le\frac{5}{3}$

$$-\frac{5}{3}\le x+\frac{2}{3}\le\frac{5}{3}$$
$$-\frac{7}{3}\le x\le 1 \quad \text{Subtracting } \frac{2}{3}$$

The solution set is $\left[-\frac{7}{3},1\right]$.

54. $\left|x+\frac{3}{4}\right|<\frac{1}{4}$

$$-\frac{1}{4}<x+\frac{3}{4}<\frac{1}{4}$$
$$-1<x<-\frac{1}{2}$$

The solution set is $\left(-1,-\frac{1}{2}\right)$.

55. $\left|\frac{2x+1}{3}\right|>5$

$$\frac{2x+1}{3}<-5 \quad or \quad \frac{2x+1}{3}>5$$
$$2x+1<-15 \quad or \quad 2x+1>15 \quad \text{Multiplying by 3}$$
$$2x<-16 \quad or \quad 2x>14 \quad \text{Subtracting 1}$$
$$x<-8 \quad or \quad x>7 \quad \text{Dividing by 2}$$

The solution set is $\{x|x<-8 \text{ or } x>7\}$, or $(-\infty,-8)\cup(7,\infty)$.

56. $\left|\frac{2x-1}{3}\right|\ge\frac{5}{6}$

$$\frac{2x-1}{3}\le-\frac{5}{6} \quad or \quad \frac{2x-1}{3}\ge\frac{5}{6}$$
$$2x-1\le-\frac{5}{2} \quad or \quad 2x-1\ge\frac{5}{2}$$
$$2x\le-\frac{3}{2} \quad or \quad 2x\ge\frac{7}{2}$$
$$x\le-\frac{3}{4} \quad or \quad x\ge\frac{7}{4}$$

The solution set is $\left(-\infty,-\frac{3}{4}\right]\cup\left[\frac{7}{4},\infty\right)$.

57. $|2x-4|<-5$

Since $|2x-4|\ge 0$ for all x, there is no x such that $|2x-4|$ would be less than -5. There is no solution.

58. $|3x+5|<0$

$|3x+5|\ge 0$ for all x, so there is no solution.

59. The graph of $y_1=|x+8|$ lies on or above the graph of $y_2=9$ for $x\le-17$ or $x\ge 1$.

60. The graph of $y_1=|x+6|$ lies above the graph of $y_2=10$ for $x<-16$ or $x>4$.

61. Absolute value is nonnegative.

62. $|x|\ge 0>p$ for any real number x.

63. The graphs of $y_1=-\frac{1}{2}x+1$ and $y_2=|x-5|$ do not intersect.

64. $(-1+2i)+(3-i)=(-1+3)+(2-1)i=2+i$

65.
$$\begin{aligned}(6-4i)-(-4+3i)&=6-4i+4-3i\\&=(6+4)+(-4-3)i\\&=10-7i\end{aligned}$$

66.
$$\begin{aligned}(2-3i)(5+i)&=10+2i-15i-3i^2\\&=10+2i-15i+3\\&=13-13i\end{aligned}$$

67.
$$\begin{aligned}\frac{5+2i}{3-4i}&=\frac{5+2i}{3-4i}\cdot\frac{3+4i}{3+4i}\\&=\frac{(5+2i)(3+4i)}{(3-4i)(3+4i)}\\&=\frac{15+20i+6i+8i^2}{9-16i^2}\\&=\frac{15+20i+6i-8}{9+16}\\&=\frac{7+26i}{25}\\&=\frac{7}{25}+\frac{26}{25}i\end{aligned}$$

68. $x\le 3x-2\le 2-x$

$$x\le 3x-2 \quad and \quad 3x-2\le 2-x$$
$$-2x\le-2 \quad and \quad 4x\le 4$$
$$x\ge 1 \quad and \quad x\le 1$$

The solution is 1.

69. $2x\le 5-7x<7+x$

$$2x\le 5-7x \quad and \quad 5-7x<7+x$$
$$9x\le 5 \quad and \quad -8x<2$$
$$x\le\frac{5}{9} \quad and \quad x>-\frac{1}{4}$$

The solution set is $\left(-\frac{1}{4},\frac{5}{9}\right]$.

70. $|x+2|\le|x-5|$

Divide the set of real numbers into three intervals: $(-\infty,-2)$, $[-2,5)$, and $[5,\infty,)$.

Find the solution set of $|x+2| \leq |x-5|$ in each interval. Then find the union of the three solution sets.

If $x < -2$, then $|x+2| = -(x+2)$ and $|x-5| = -(x-5)$.

Solve: $x < -2$ *and* $-(x+2) \leq -(x-5)$

$x < -2$ *and* $-x-2 \leq -x+5$

$x < -2$ *and* $-2 \leq 5$

The solution set for this interval is $(-\infty, -2)$.

If $-2 \leq x < 5$, then $|x+2| = x+2$ and $|x-5| = -(x-5)$.

Solve: $-2 \leq x < 5$ *and* $x+2 \leq -(x-5)$

$-2 \leq x < 5$ *and* $x+2 \leq -x+5$

$-2 \leq x < 5$ *and* $2x \leq 3$

$-2 \leq x < 5$ *and* $x \leq \frac{3}{2}$

The solution set for this interval is $\left[-2, \frac{3}{2}\right]$.

If $x \geq 5$, then $|x+2| = x+2$ and $|x-5| = x-5$.

Solve: $x \geq 5$ *and* $x+2 \leq x-5$

$x \geq 5$ *and* $2 \leq -5$

The solution set for this interval is $\emptyset$.

The union of the above three solution set is $\left(-\infty, \frac{3}{2}\right]$. This is the solution set of $|x+2| \leq |x-5|$.

71. $|3x-1| > 5x-2$

$3x-1 < -(5x-2)$ *or* $3x-1 > 5x-2$

$3x-1 < -5x+2$ *or* $1 > 2x$

$8x < 3$ *or* $\frac{1}{2} > x$

$x < \frac{3}{8}$ *or* $\frac{1}{2} > x$

The solution set is $\left(-\infty, \frac{3}{8}\right) \cup \left(-\infty, \frac{1}{2}\right)$. This is equivalent to $\left(-\infty, \frac{1}{2}\right)$.

72. $|x| + |x+1| < 10$

If $x < -1$, then $|x| = -x$ and $|x+1| = -(x+1)$ and we have:

$x < -1$ *and* $-x + [-(x+1)] < 10$

$x < -1$ *and* $-x-x-1 < 10$

$x < -1$ *and* $-2x-1 < 10$

$x < -1$ *and* $-2x < 10$

$x < -1$ *and* $x > -\frac{11}{2}$

The solution set for this interval is $\left(-\frac{11}{2}, -1\right)$.

If $-1 \leq x < 0$, then $|x| = -x$ and $|x+1| = x+1$ and we have:

$-1 \leq x$ *and* $-x+x+1 < 10$

$-1 \leq x$ *and* $1 < 10$

The solution set for this interval is $[-1, 0]$.

If $x \geq 0$, then $|x| = x$ and $|x+1| = x+1$ and we have:

$x \geq 0$ *and* $x+x+1 < 10$

$x \geq 0$ *and* $2x+1 < 10$

$x \geq 0$ *and* $2x < 9$

$x \geq 0$ *and* $x < \frac{9}{2}$

The solution set for this interval is $\left[0, \frac{9}{2}\right)$.

The union of the three solution sets above is $\left(-\frac{11}{2}, \frac{9}{2}\right)$. This is the solution set of $|x| + |x+1| < 10$.

73. $|p-4| + |p+4| < 8$

If $p < -4$, then $|p-4| = -(p-4)$ and $|p+4| = -(p+4)$.

Solve: $-(p-4) + [-(p+4)] < 8$

$-p+4-p-4 < 8$

$-2p < 8$

$p > -4$

Since this is false for all values of p in the interval $(-\infty, -4)$ there is no solution in this interval.

If $p \geq -4$, then $|p+4| = p+4$.

Solve: $|p-4| + p + 4 < 8$

$|p-4| < 4-p$

$p-4 > -(4-p)$ *and* $p-4 < 4-p$

$p-4 > p-4$ *and* $2p < 8$

$-4 > -4$ *and* $p < 4$

Since $-4 > -4$ is false for all values of p, there is no solution in the interval $[-4, \infty)$.

Thus, $|p-4| + |p+4| < 8$ has no solution.

74. $|x-3| + |2x+5| > 6$

Divide the set of real numbers into three intervals: $\left(-\infty, -\frac{5}{2}\right)$, $\left[-\frac{5}{2}, 3\right)$, and $[3, \infty)$.

Find the solution set of $|x-3| + |2x+5| > 6$ in each interval. Then find the union of the three solution sets.

If $x < -\frac{5}{2}$, then $|x-3| = -(x-3)$ and $|2x+5| = -(2x+5)$.

Solve: $x < -\frac{5}{2}$ *and* $-(x-3) + [-(2x+5)] > 6$

$x < -\frac{5}{2}$ *and* $-x+3-2x-5 > 6$

$x < -\frac{5}{2}$ *and* $-3x > 8$

$x < -\frac{5}{2}$ *and* $x < -\frac{8}{3}$

The solution set in this interval is $\left(-\infty, -\frac{8}{3}\right)$.

If $-\frac{5}{2} \le x < 3$, then $|x-3| = -(x-3)$ and $|2x+5| = 2x+5$.

Solve: $-\frac{5}{2} \le x < 3$ *and* $-(x-3)+2x+5 > 6$

$-\frac{5}{2} \le x < 3$ *and* $-x+3+2x+5 > 6$

$-\frac{5}{2} \le x < 3$ *and* $x > -2$

The solution set in this interval is $(-2, 3)$.

If $x \ge 3$, then $|x-3| = x-3$ and $|2x+5| = 2x+5$.

Solve: $x \ge 3$ *and* $x-3+2x+5 > 6$

$x \ge 3$ *and* $3x > 4$

$x \ge 3$ *and* $x > \frac{4}{3}$

The solution set in this interval is $[3, \infty)$.

The union of the above solution sets is $\left(-\infty, -\frac{8}{3}\right) \cup (-2, \infty)$. This is the solution set of $|x-3| + |2x+5| > 6$.

Chapter 3

Polynomial and Rational Functions

Exercise Set 3.1

1. $g(x) = \frac{1}{2}x^3 - 10x + 8$

   The degree of the polynomial is 3, so the polynomial is cubic. The leading term is $\frac{1}{2}x^3$ and the leading coefficient is $\frac{1}{2}$.

2. $f(x) = 15x^2 - 10 + 0.11x^4 - 7x^3 =$
   $0.11x^4 - 7x^3 + 15x^2 - 10$

   The degree of the polynomial is 4, so the polynomial is quartic. The leading term is $0.11x^4$ and the leading coefficient is 0.11.

3. $h(x) = 0.9x - 0.13$

   The degree of the polynomial is 1, so the polynomial is linear. The leading term is $0.9x$ and the leading coefficient is 0.9.

4. $f(x) = -6 = -6x^0$

   The degree of the polynomial is 0, so the polynomial is constant. The leading term and leading coefficient are both -6.

5. $g(x) = 305x^4 + 4021$

   The degree of the polynomial is 4, so the polynomial is quartic. The leading term is $305x^4$ and the leading coefficient is 305.

6. $h(x) = 2.4x^3 + 5x^2 - x + \frac{7}{8}$

   The degree of the polynomial is 3, so the polynomial is cubic. The leading term is $2.4x^3$ and the leading coefficient is 2.4.

7. $f(x) = \frac{1}{4}x^2 - 5$

   The leading term is $\frac{1}{4}x^2$. The sign of the leading coefficient, $\frac{1}{4}$, is positive and the dgree, 2, is even, so we would choose either graph (b) or graph (d). Note also that $f(0) = -5$, so the y-intercept is $(0, -5)$. Thus, graph (d) is the graph of this function.

8. $f(x) = -0.5x^6 - x^5 + 4x^4 - 5x^3 - 7x^2 + x - 3$

   The leading term is $-0.5x^6$. The sign of the leading coefficient, -0.5, is negative and the degree, 6, is even. Thus, graph (a) is the graph of this function.

9. $f(x) = x^5 - x^4 + x^2 + 4$

   The leading term is x^5. The sign of the leading coefficient, 1, is positive and the degree, 5, is odd. Thus, graph (f) is the graph of this function.

10. $f(x) = -\frac{1}{3}x^3 - 4x^2 + 6x + 42$

    The leading term is $-\frac{1}{3}x^3$. The sign of the leading coefficient, $-\frac{1}{3}$, is negative and the degree, 3, is odd, so we would choose either graph (c) or graph (e). Note also that $f(0) = 42$, so the y-intercept is $(0, 42)$. Thus, graph (c) is the graph of this function.

11. $f(x) = x^4 - 2x^3 + 12x^2 + x - 20$

    The leading term is x^4. The sign of the leading coefficient, 1, is positive and the degree, 4, is even, so we would choose either graph (b) or graph (d). Note also that $f(0) = -20$, so the y-intercept is $(0, -20)$. Thus, graph (b) is the graph of this function.

12. $f(x) = -0.3x^7 + 0.11x^6 - 0.25x^5 + x^4 + x^3 - 6x - 5$

    The leading term is $-0.3x^7$. The sign of the leading coefficient, -0.3, is negative and the degree, 7, is odd, so we would choose either graph (c) or graph (e). Note also that $f(0) = -5$, so the y-intercept is $(0, -5)$. Thus, graph (e) is the graph of this function.

13. Graph $y = x^3 - 3x - 1$ and use the Zero feature three times. The real-number zeros are about -1.532, -0.347, and 1.879.

14. Graph $y = x^3 + 3x^2 - 9x - 13$ and use the Zero feature three times. The real-number zeros are about -4.378, -1.167, and 2.545.

15. Graph $y = x^4 - 2x^2$ and use the Zero feature three times. The real-number zeros are about -1.414, 0, and 1.414.

16. Graph $y = x^4 - 2x^3 - 5.6$ and use the Zero feature twice. The real-number zeros are about -1.205 and 2.403.

17. Graph $y = x^3 - x$ and use the Zero feature three times. The real-number zeros are -1, 0, and 1.

18. Graph $y = 2x^3 - x^2 - 14x - 10$ and use the Zero feature three times. The real-number zeros are about -1.831, -0.856, and 3.188.

19. Graph $y = x^8 + 8x^7 - 28x^6 - 56x^5 + 70x^4 + 56x^3 - 28x^2 - 8x + 1$ and use the Zero feature eight times. The real-number zeros are about -10.153, -1.871, -0.821, -0.303, 0.098, 0.535, 1.219, and 3.297.

20. Graph $y = x^6 - 10x^5 + 13x^3 - 4x^2 - 5$ and use the Zero feature twice. The real-number zeros are -1.281 and 9.871.

21. $g(x) = x^3 - 1.2x + 1$

Graph the function and use the Zero, Maximum, and Minimum features.

Zero: -1.368

Relative maximum: 1.506 when $x \approx -0.632$

Relative minimum: 0.494 when $x \approx 0.632$

Range: $(-\infty, \infty)$

22. $g(x) = 5x - 14$

Graph the function and use the Zero feature.

Zero: 2.8

There are no relative maxima or minima.

Range: $(-\infty, \infty)$

23. $h(x) = -\frac{1}{2}x^4 + 3x^3 - 5x^2 + 3x + 6$

Graph the function and use the Zero, Maximum, and Minimum features.

Zeros: -0.720, 4.089

Relative maxima: 6.59375 when $x = 0.5$, 10.5 when $x = 3$

Relative minimum: 6.5 when $x = 1$

Range: $(-\infty, 10.5]$

24. $f(x) = -3.5x^3 - x^2 + 24.5x - 4$

Graph the function and use the Zero, Maximum, and Minimum features.

Zeros: -2.867, 0.165, 2.416

Relative maximum: 18.756 at $x \approx 1.435$

Relative minimum: -31.435 at $x \approx -1.626$

Range: $(-\infty, \infty)$

25. $f(x) = x^6 - 3.8$

Graph the function and use the Zero and Minimum features.

Zeros: -1.249, 1.249

There is no relative maximum.

Relative minimum: -3.8 when $x = 0$

Range: $[-3.8, \infty)$

26. $f(x) = x^5 - 3x^4 + 6x + 6$

Graph the function and use the Zero, Maximum, and Minimum features.

Zero: -0.774

Relative maximum: 10.032 when $x \approx 0.936$

Relative minimum: 0.211 when $x \approx 2.302$

Range: $(-\infty, \infty)$

27. $g(x) = 12 - 3.14x$

Graph the function and use the Zero feature.

Zero: 3.822

There are no relative maxima or minima.

Range: $(-\infty, \infty)$

28. $h(x) = 2x^3 - x^4 + 20$

Graph the function and use the Zero and Maximum features.

Zeros: -1.748, 2.857

Relative maximum: 21.688 when $x = 1.5$

There is no relative minimum.

Range: $(-\infty, 21.688]$

29. $f(x) = x^2 + 10x - x^5$

Graph the function and use the Zero, Maximum, and Minimum features.

Zeros: -1.697, 0, 1.856

Relative maximum: 11.012 when $x \approx 1.258$

Relative minimum: -8.183 when $x \approx -1.116$

Range: $(-\infty, \infty)$

30. $g(x) = 2x^2 - 5.8x + 1$

Graph the function and use the Zero and Minimum features.

Zeros: 0.184, 2.716

There is no relative maximum.

Relative minimum: -3.205 when $x \approx 1.450$

Range: $[-3.205, \infty)$

31. $h(x) = -x^5 + 4x^3 - x$

Graph the function and use the Zero, Maximum, and Minimum features.

Zeros: -1.932, -0.518, 0, 0.518, 1.932

Relative maxima: 0.195 when $x \approx -0.294$, 4.414 when $x \approx 1.521$

Relative minima: -4.414 when $x \approx -1.521$, -0.195 when $x \approx 0.294$

Range: $(-\infty, \infty)$

32. $f(x) = 2x^4 - 5.6x^2 + 10$

Graph the function and use the Zero, Maximum, and Minimum features.

There are no real-number zeros.

Relative maximum: 10 when $x = 0$

Relative minima: 6.08 when $x \approx -1.183$ and when $x \approx 1.183$

Range: $[6.08, \infty)$

33. $f(-5) = (-5)^3 + 3(-5)^2 - 9(-5) - 13 = -18$

$f(-4) = (-4)^3 + 3(-4)^2 - 9(-4) - 13 = 7$

By the intermediate value theorem, since $f(-5)$ and $f(-4)$ have opposite signs then $f(x)$ has a zero between -5 and -4.

34. $f(2) = 2^3 + 3 \cdot 2^2 - 9 \cdot 2 - 13 = -11$

$f(3) = 3^3 + 3 \cdot 3^2 - 9 \cdot 3 - 13 = 14$

By the intermediate value theorem, since $f(2)$ and $f(3)$ have opposite signs then $f(x)$ has a zero between 2 and 3.

35.
$$V = 48T^2$$
$$36 = 48T^2 \quad \text{Substituting}$$
$$0.75 = T^2$$
$$0.866 \approx T \quad \text{We only want the positive solution.}$$

Anfernee Hardaway's hang time is 0.866 sec.

36. Solve $294 = 4.9t^2 + 34.3t$.

$t = 5$ sec

37. First find the number of games played.

$N(x) = x^2 - x$

$N(9) = 9^2 - 9 = 72$

Now multiply the number of games by the cost per game to find the total cost.

$72 \cdot 45 = 3240$

It will cost \$3240 to play the entire schedule.

38. a) $P(15) = 0.015(15)^3 = 50.625$ watts per hour

b) Solve $120 = 0.015v^3$.

$v = 20$ mph

39. $A = P(1+i)^t$

a)
$$3610 = 2560(1+i)^2 \quad \text{Substituting}$$
$$\frac{3610}{2560} = (1+i)^2$$
$$\pm 1.1875 = 1 + i \quad \text{Taking the square root on both sides}$$
$$-1 \pm 1.1875 = i$$
$$-1 - 1.1875 = i \quad or \quad -1 + 1.1875 = i$$
$$-2.1875 = i \quad or \quad 0.1875 = i$$

Only the positive result has meaning in this application. The interest rate is 0.1875, or 18.75%.

b)
$$13{,}310 = 10{,}000(1+i)^3 \quad \text{Substituting}$$
$$\frac{13{,}310}{10{,}000} = (1+i)^3$$
$$1.10 \approx 1 + i \quad \text{Taking the cube root on both sides}$$
$$0.10 \approx i$$

The interest rate is 0.1, or 10%.

40. 5 ft, 7 in. = 67 in.

$$W(67) = \left(\frac{67}{12.3}\right)^3 \approx 162 \text{ lb}$$

41. The sales drop and then rise. This suggests that a quadratic function that opens up might fit the data. Thus, we choose (b).

42. The sales rise steadily. Thus, we choose (a), a linear function.

43. The sales rise and then drop. This suggests that a quadratic function that opens down might fit the data. Thus, we choose (c).

44. The sales rise, then fall, and then rise again. This suggests that a polynomial function that is neither linear nor quadratic fits the data. Thus, we choose (d).

45. The sales fall steadily. Thus, we choose (a), a linear function.

46. The sales rise at an increasing rate. This suggests that a quadratic function that opens up might fit the data. Thus, we choose (b). (It is also possible that there is no polynomial function that fits the data.)

47. The rate falls and then rises. This suggests that a quadratic function that opens up might fit the data. Thus, we choose (b).

48. The rate rises, then falls, and then rises again. This suggests that a polynomial function that is neither linear nor quadratic fits the data. Thus, we choose (d).

49. a)

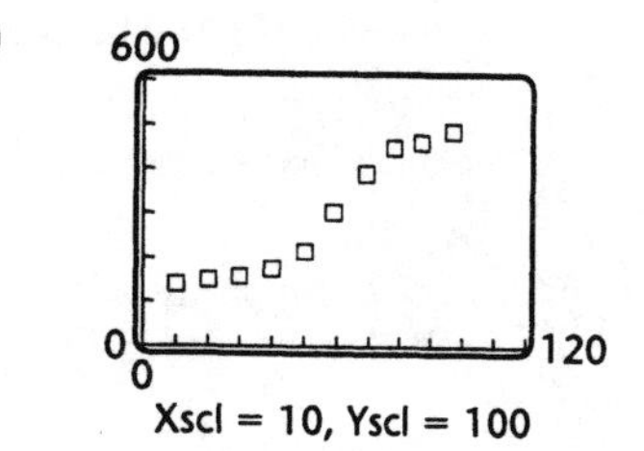

b) Linear: $y = 4.717948718x + 37.93076923$; $r^2 \approx 0.9306$

Quadratic: $y = 0.0281175273x^2 + 1.711977591x + 96.76946268$; $R^2 \approx 0.9504$

Cubic: $y = -0.0016285158x^3 + 0.2897410836x^2 - 10.07476028x + 227.1101167$; $R^2 \approx 0.9872$

Power: $y = 22.9451257x^{0.6396947491}$; $r^2 \approx 0.8142$

The value of R^2 is highest for the cubic function so we determine that this function fits the data best.

c)

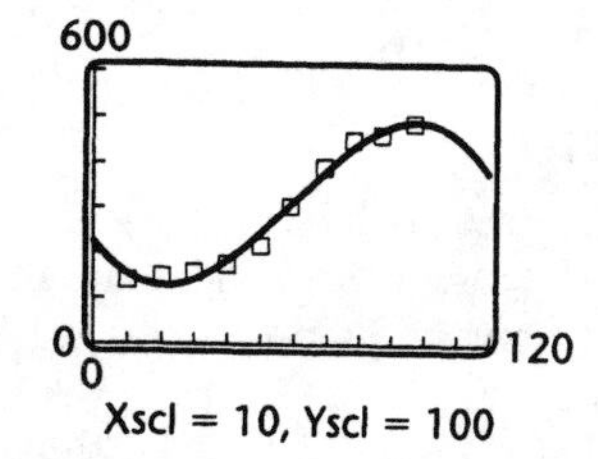

d) In 2000, $x = 2000 - 1900 = 100$; in 2010, $x = 2010 - 1900 - 110$.

For the linear function: $f(100) \approx 510$ and $f(110) \approx 557$.

For the quadratic function: $f(100) \approx 549$ and $f(110) \approx 625$.

For the cubic function: $f(100) \approx 489$ and $f(110) \approx 457$.

For the power function: $f(100) \approx 437$ and $f(110) \approx 464$.

It is reasonable to assume that the average acreage will continue to increase. For this reason we rule out the predictions made using the cubic and power functions. The rate of increase appears to be more like that predicted by the linear function than the quadratic function. Thus, we say that the prediction of 510 acres and 557 acres given by the linear function are the most realistic. Answers may vary.

50. a)

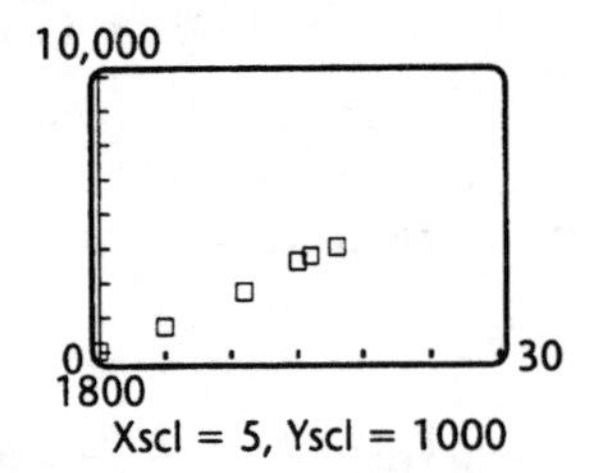

b) Quadratic: $y = -0.1043930032x^2 + 184.2372482x + 1806.309436$; $R^2 \approx 0.99757$

Cubic: $y = -0.1383359141x^3 + 3.639259836x^2 + 159.5283577 + 1820.903401$; $R^2 \approx 0.99787$

Quartic: $y = -0.0949406858x^4 + 3.37618171x^3 - 36.52854211x^2 + 295.5893298x + 1809.08903$; $R^2 \approx 0.99996$

The value of R^2 is highest for the quartic function so we determine that this function fits the data best.

c)

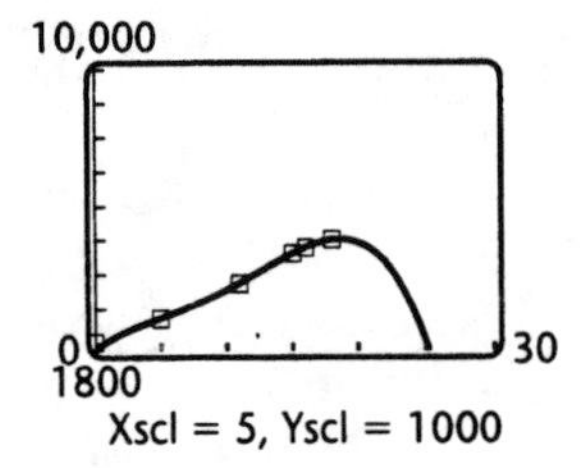

d) For the quadratic function: $f(20) \approx \$5449$ and $f(30) \approx \$7240$.

For the cubic function: $f(20) \approx \$5361$ and $f(30) \approx \$6147$

For the quartic function: $f(20) \approx \$4928$ and $f(30) \approx -\$7944$.

The quartic function does not give a realistic estimate for the cost in 2010. The rate of increase in the cost appears to be more like that estimated by the cubic function than the quadratic function. Thus, we say that the estimates given by the cubic function are the most realistic. Answers may vary.

51. a)

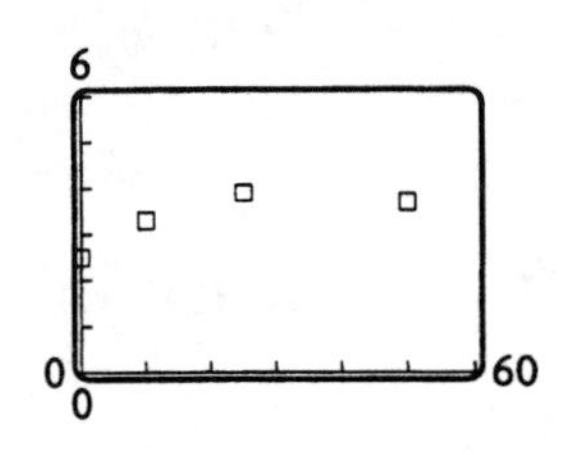

b) Linear: $y = 0.0216740088x + 2.889427313$; $r^2 \approx 0.5795$

Quadratic: $y = -0.001301354x^2 + 0.0887195358x + 2.513926499$; $R^2 \approx 0.9988$

Cubic: $y = 0.000008x^3 - 0.00188x^2 + 0.98x + 2.5$; $R^2 = 1$

The value of R^2 is highest for the cubic function. In fact, $R^2 = 1$ for this function so it fits the data perfectly.

c)

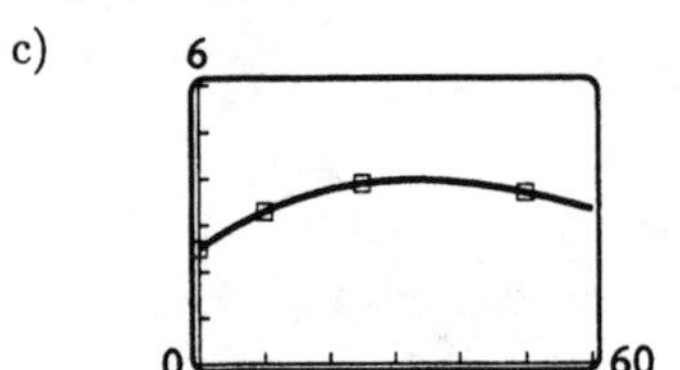

d) In 2060, $x = 2060 - 1985 = 75$.

For the linear function: $f(75) \approx 4.5$.

For the quadratic function: $f(75) \approx 1.8$.

For the cubic function: $f(75) \approx 2.65$.

Since the estimated concentration in 2035 is lower than the concentration in 2010, it would seem reasonable to assume that the level would continue to drop so we would not use the linear prediction. The rate of decrease appears to be more like that estimated by the cubic function than the quadratic function. Thus, we say that the estimate of 2.65 ppb given by the cubic function is most realistic. Answers may vary.

52. a)

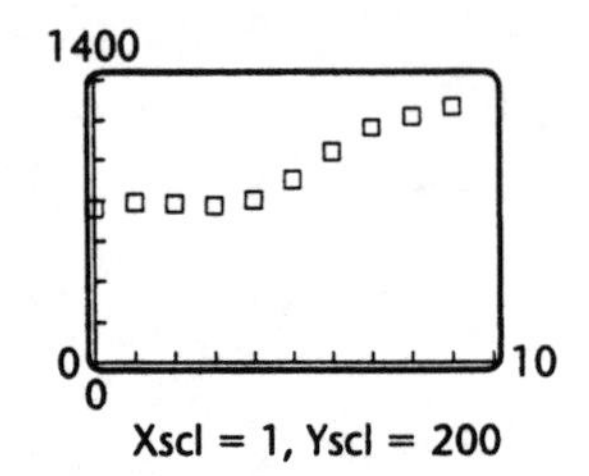

b) Linear: $y = 62.43636364x + 668.6363636$; $r^2 \approx 0.87806$

Quadratic: $y = 7.465909091x^2 - 4.756818182x + 758.2272727$; $R^2 \approx 0.95841$

Cubic: $y = -1.431429681x^3 + 26.79020979x^2 - 70.7457265x + 794.2993007$; $R^2 \approx 0.97569$

Quartic: $y = -0.6566142191x^4 + 10.38762626x^3 - 39.52782634x^2 + 47.44483294x + 765.9335664$; $R^2 \approx 0.99508$

The value of R^2 is highest for the quartic function, so we determine that it fits the data best.

c) 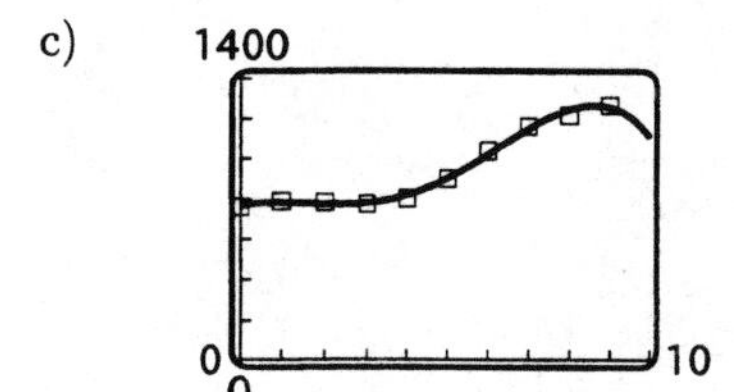

d) For the linear function: $f(14) \approx \$1543$ billion.

For the quadratic function: $f(14) \approx \$2155$ billion.

For the cubic function: $f(14) \approx \$1127$ billion.

For the quartic function: $f(14) \approx -\$3038$.

We rule out the quartic function since $f(14)$ is a negative number. It seems reasonable to assume that he debt will continue to rise so we also rule out the cubic function. Arguments could be made for both the linear and the quadratic functions. Since R^2 is higher for the quadratic function, we say that it gives the most realistic estimate. Answers may vary.

53. We get $y = -.4767604618x^3 + 120.2087302x^2 - 591.5532468x + 3710$, where x is the number of years after 1979. In 2000 (when $x = 21$), $y \approx 147$. A graph of the model shows that the number of deaths continues to decline each year after 1993 until it reaches 0 about 21.1 years after 1979, or in 2001, and then becomes negative. Clearly, this is not a realistic model for these years.

54. The value of R^2 does not necessarily reflect the accuracy of the fit of a regression equation beyond the domain of the data. See Exercises 49, 50, 52, and 53.

55. The range of a polynomial with an odd degree is $(-\infty, \infty)$. The range of a polynomial with an even degree is $[s, \infty)$ for some real numbers s if $a_n > 0$ and is $(-\infty, s]$ for some real number s if $a_n < 0$.

56. $d = \sqrt{(-2-4)^2 + (-4-2)^2} = \sqrt{36+36} = \sqrt{72} = 6\sqrt{2}$

57.
$$\begin{aligned} d &= \sqrt{(x_2 - x_1)^2 + (y_2 - y_1)^2} \\ &= \sqrt{[-1-(-5)]^2 + (0-3)^2} \\ &= \sqrt{4^2 + (-3)^2} = \sqrt{16+9} \\ &= \sqrt{25} = 5 \end{aligned}$$

58. The center of the circle is the midpoint of a segment that is a diameter:

$$\left(\frac{-6+(-2)}{2}, \frac{5+1}{2}\right) = (-4, 3).$$

The length of a radius is the distance from the center to one of the endpoints of the diameter:

$$\begin{aligned} r &= \sqrt{[-6-(-4)]^2 + (5-3)^2} \\ &= \sqrt{4+4} = \sqrt{8} \\ &= 2\sqrt{2} \end{aligned}$$

59.
$$\begin{aligned} (x-3)^2 + (y+5)^2 &= 49 \\ (x-3)^2 + [y-(-5)]^2 &= 7^2 \end{aligned}$$

Center: $(3, -5)$; radius: 7

60. ***Familiarize***. We will use the compound interest formula. The \$2000 deposit is invested for two years and grows to an amount A_1 given by $A_1 = 2000(1+i)^2$. The \$1200 deposit is invested for one year and grows to an amount A_2 given by $A_2 = 1200(1+i)$.

Translate. There is a total of \$3573.80 in both accounts at the end of the second year, so we have

$A_1 + A_2 = 3573.80$, or

$2000(1+i)^2 + 1200(1+i) = 3573.80$.

Carry out. We solve the equation.

$2000(1+i)^2 + 1200(1+i) - 3573.8 = 0$

Substitute u for $1+i$.

$2000u^2 + 1200u - 3573.8 = 0$

Using the quadratic formula we find that $u = 1.07$ or $u = -3.34$. Only the positive value has meaning in this application. Then since $u = 1.07$, we have $1 + i = 1.07$, or $i = 0.07$.

Check. At an interest rate of 0.07, or 7%, in two years \$2000 would grow to $2000(1+0.07)^2$, or \$2289.80. In one year \$1200 would grow to $1200(1+0.07)$, or \$1284. Now \$2289.80 + \$1284 = \$3573.80, so the result checks.

State. The interest rate is 7%.

Exercise Set 3.2

1. $f(x) = x^3 - 9x^2 + 14x + 24$

$f(4) = 4^3 - 9 \cdot 4^2 + 14 \cdot 4 + 24 = 0$

Since $f(4) = 0$, 4 is a zero of $f(x)$.

$f(5) = 5^3 - 9 \cdot 5^2 + 14 \cdot 5 + 24 = -6$

Since $f(5) \neq 0$, 5 is not a zero of $f(x)$.

$f(-2) = (-2)^3 - 9(-2)^2 + 14(-2) + 24 = -48$

Since $f(-2) \neq 0$, -2 is not a zero of $f(x)$.

2. $f(x) = 2x^3 - 3x^2 + x - 1$

$f(2) = 2 \cdot 2^3 - 3 \cdot 2^2 + 2 - 1 = 5$

$f(2) \neq 0$, so 2 is not a zero of $f(x)$.

$f(3) = 2 \cdot 3^3 - 3 \cdot 3^2 + 3 - 1 = 29$

$f(3) \neq 0$, so 3 is not a zero of $f(x)$.

$f(-1) = 2(-1)^3 - 3(-1)^2 + (-1) - 1 = -7$

$f(-1) \neq 0$, so -1 is not a zero of $f(x)$.

3. We divide to determine whether each binomial is a factor of $f(x)$.

a)
$$\begin{array}{r} x^2-5x-6 \\ x-4\,\overline{)\,x^3-9x^2+14x+24} \\ \underline{x^3-4x^2} \\ -5x^2+14x \\ \underline{-5x^2+20x} \\ -6x+24 \\ \underline{-6x+24} \\ 0 \end{array}$$

Since the remainder is 0, $x-4$ is a factor of $f(x)$.

b)
$$\begin{array}{r} x^2-4x-6 \\ x-5\,\overline{)\,x^3-9x^2+14x+24} \\ \underline{x^3-5x^2} \\ -4x^2+14x \\ \underline{-4x^2+20x} \\ -6x+24 \\ \underline{-6x+30} \\ -6 \end{array}$$

Since the remainder is not 0, $x-5$ is not a factor of $f(x)$.

c)
$$\begin{array}{r} x^2-11x+36 \\ x+2\,\overline{)\,x^3-9x^2+14x+24} \\ \underline{x^3+2x^2} \\ -11x^2+14x \\ \underline{-11x^2-22x} \\ 36x+24 \\ \underline{36x+72} \\ -48 \end{array}$$

Since the remainder is not 0, $x+2$ is not a factor of $f(x)$.

4. We divide to determine whether each binomial is a factor of $f(x)$.

a)
$$\begin{array}{r} 2x^2+x+3 \\ x-2\,\overline{)\,2x^3-3x^2+x-1} \\ \underline{2x^3-4x^2} \\ x^2+x \\ \underline{x^2-2x} \\ 3x-1 \\ \underline{3x-6} \\ 5 \end{array}$$

Since the remainder is not 0, $x-2$ is not a factor of $f(x)$.

b)
$$\begin{array}{r} 2x^2+3x+10 \\ x-3\,\overline{)\,2x^3-3x^2+x-1} \\ \underline{2x^3-6x^2} \\ 3x^2+x \\ \underline{3x^2-9x} \\ 10x-1 \\ \underline{10x-30} \\ 29 \end{array}$$

Since the remainder is not 0, $x-3$ is not a factor of $f(x)$.

c)
$$\begin{array}{r} 2x^2-5x+6 \\ x+1\,\overline{)\,2x^3-3x^2+x-1} \\ \underline{2x^3+2x^2} \\ -5x^2+x \\ \underline{-5x^2-5x} \\ 6x-1 \\ \underline{6x+6} \\ -7 \end{array}$$

Since the remainder is not 0, $x+1$ is not a factor of $f(x)$.

5.
$$\begin{array}{r} x^2-2x+4 \\ x+2\,\overline{)\,x^3+0x^2+0x-8} \\ \underline{x^3+2x^2} \\ -2x^2+0x \\ \underline{-2x^2-4x} \\ 4x-8 \\ \underline{4x+8} \\ -16 \end{array}$$

$x^3-8=(x+2)(x^2-2x+4)-16$

6. See the work in Exercise 4(b).

$2x^3-3x^2+x-1=(x-3)(2x^2+3x+10)+29$

7.
$$\begin{array}{r} x^2+5 \\ x^2+4\,\overline{)\,x^4+9x^2+20} \\ \underline{x^4+4x^2} \\ 5x^2+20 \\ \underline{5x^2+20} \\ 0 \end{array}$$

$x^4+9x^2+20=(x^2+4)(x^2+5)+0$

8.
$$\begin{array}{r} x^2-x+1 \\ x^2+x+1\,\overline{)\,x^4+0x^3+x^2+0x+2} \\ \underline{x^4+x^3+x^2} \\ -x^3+0x \\ \underline{-x^3-x^2-x} \\ x^2+x+2 \\ \underline{x^2+x+1} \\ 1 \end{array}$$

$x^4+x^2+2=(x^2+x+1)(x^2-x+1)+1$

9. $(2x^4+7x^3+x-12)\div(x+3)$

$=(2x^4+7x^3+0x^2+x-12)\div[x-(-3)]$

$$\begin{array}{r|rrrrr} -3 & 2 & 7 & 0 & 1 & -12 \\ & & -6 & -3 & 9 & -30 \\ \hline & 2 & 1 & -3 & 10 & -42 \end{array}$$

The quotient is $2x^3+x^2-3x+10$. The remainder is -42.

10.
$$\begin{array}{r|rrrr} 2 & 1 & -7 & 13 & 3 \\ & & 2 & -10 & 6 \\ \hline & 1 & -5 & 3 & 9 \end{array}$$

$Q(x)=x^2-5x+3$, $R(x)=9$

11. $(x^3-2x^2-8)\div(x+2)$

$=(x^3-2x^2+0x-8)\div[x-(-2)]$

$$\begin{array}{r|rrrr} -2 & 1 & -2 & 0 & -8 \\ & & -2 & 8 & -16 \\ \hline & 1 & -4 & 8 & -24 \end{array}$$

The quotient is $x^2 - 4x + 8$. The remainder is -24.

12.
$$\begin{array}{r|rrrr} 2 & 1 & 0 & -3 & 10 \\ & & 2 & 4 & 2 \\ \hline & 1 & 2 & 1 & 12 \end{array}$$

$Q(x) = x^2 + 2x + 1$, $R(x) = 12$

13. $(x^4 - 1) \div (x - 1)$

$= (x^4 + 0x^3 + 0x^2 + 0x - 1) \div (x - 1)$

$$\begin{array}{r|rrrrr} 1 & 1 & 0 & 0 & 0 & -1 \\ & & 1 & 1 & 1 & 1 \\ \hline & 1 & 1 & 1 & 1 & 0 \end{array}$$

The quotient is $x^3 + x^2 + x + 1$. The remainder is 0.

14.
$$\begin{array}{r|rrrrrr} -2 & 1 & 0 & 0 & 0 & 0 & 32 \\ & & -2 & 4 & -8 & 16 & -32 \\ \hline & 1 & -2 & 4 & -8 & 16 & 0 \end{array}$$

$Q(x) = x^4 - 2x^3 + 4x^2 - 8x + 16$, $R(x) = 0$

15. $(2x^4 + 3x^2 - 1) \div \left(x - \frac{1}{2}\right)$

$(2x^4 + 0x^3 + 3x^2 + 0x - 1) \div \left(x - \frac{1}{2}\right)$

$$\begin{array}{r|rrrrr} \frac{1}{2} & 2 & 0 & 3 & 0 & -1 \\ & & 1 & \frac{1}{2} & \frac{7}{4} & \frac{7}{8} \\ \hline & 2 & 1 & \frac{7}{2} & \frac{7}{4} & -\frac{1}{8} \end{array}$$

The quotient is $2x^3 + x^2 + \frac{7}{2}x + \frac{7}{4}$. The remainder is $-\frac{1}{8}$.

16.
$$\begin{array}{r|rrrrr} \frac{1}{4} & 3 & 0 & -2 & 0 & 2 \\ & & \frac{3}{4} & \frac{3}{16} & -\frac{29}{64} & -\frac{29}{256} \\ \hline & 3 & \frac{3}{4} & -\frac{29}{16} & -\frac{29}{64} & \frac{483}{256} \end{array}$$

$Q(x) = 3x^3 + \frac{3}{4}x^2 - \frac{29}{16}x - \frac{29}{64}$, $R(x) = \frac{483}{256}$

17. $(x^4 - y^4) \div (x - y)$

$= (x^4 + 0x^3 + 0x^2 + 0x - y^4) \div (x - y)$

$$\begin{array}{r|rrrrr} y & 1 & 0 & 0 & 0 & -y^4 \\ & & y & y^2 & y^3 & y^4 \\ \hline & 1 & y & y^2 & y^3 & 0 \end{array}$$

The quotient is $x^3 + x^2y + xy^2 + y^3$. The remainder is 0.

18.
$$\begin{array}{r|rrrr} -i & 1 & 3i & -4i & -2 \\ & & -i & 2 & -4-2i \\ \hline & 1 & 2i & 2-4i & -6-2i \end{array}$$

$Q(x) = x^2 + 2ix + (2 - 4i)$, $R(x) = -6 - 2i$

19. $f(x) = x^3 - 6x^2 + 11x - 6$

Find $f(1)$.

$$\begin{array}{r|rrrr} 1 & 1 & -6 & 11 & -6 \\ & & 1 & -5 & 6 \\ \hline & 1 & -5 & 6 & 0 \end{array}$$

$f(1) = 0$

Find $f(-2)$.

$$\begin{array}{r|rrrr} -2 & 1 & -6 & 11 & -6 \\ & & -2 & 16 & -54 \\ \hline & 1 & -8 & 27 & -60 \end{array}$$

$f(-2) = -60$

Find $f(3)$.

$$\begin{array}{r|rrrr} 3 & 1 & -6 & 11 & -6 \\ & & 3 & -9 & 6 \\ \hline & 1 & -3 & 2 & 0 \end{array}$$

$f(3) = 0$

20.
$$\begin{array}{r|rrrr} -3 & 1 & 7 & -12 & -3 \\ & & -3 & -12 & 72 \\ \hline & 1 & 4 & -24 & 69 \end{array}$$

$f(-3) = 69$

$$\begin{array}{r|rrrr} -2 & 1 & 7 & -12 & -3 \\ & & -2 & -10 & 44 \\ \hline & 1 & 5 & -22 & 41 \end{array}$$

$f(-2) = 41$

$$\begin{array}{r|rrrr} 1 & 1 & 7 & -12 & -3 \\ & & 1 & 8 & -4 \\ \hline & 1 & 8 & -4 & -7 \end{array}$$

$f(1) = -7$

21. $f(x) = 2x^5 - 3x^4 + 2x^3 - x + 8$

Find $f(20)$.

$$\begin{array}{r|rrrrrr} 20 & 2 & -3 & 2 & 0 & -1 & 8 \\ & & 40 & 740 & 14,840 & 296,800 & 5,935,980 \\ \hline & 2 & 37 & 742 & 14,840 & 296,799 & 5,935,988 \end{array}$$

$f(20) = 5,935,988$

Find $f(-3)$.

$$\begin{array}{r|rrrrrr} -3 & 2 & -3 & 2 & 0 & -1 & 8 \\ & & -6 & 27 & -87 & 261 & -780 \\ \hline & 2 & -9 & 29 & -87 & 260 & -772 \end{array}$$

$f(-3) = -772$

22.
$$\begin{array}{r|rrrrrr} -10 & 1 & -10 & 20 & 0 & -5 & -100 \\ & & -10 & 200 & -2200 & 22,000 & -219,950 \\ \hline & 1 & -20 & 220 & -2200 & 21,995 & -220,050 \end{array}$$

$f(-10) = -220,050$

$$\begin{array}{r|rrrrrr} 5 & 1 & -10 & 20 & 0 & -5 & -100 \\ & & 5 & -25 & -25 & -125 & -650 \\ \hline & 1 & -5 & -5 & -25 & -130 & -750 \end{array}$$

$f(5) = -750$

23. $f(x) = x^4 - 16$

Find $f(2)$.

$$\begin{array}{r|rrrrr} 2 & 1 & 0 & 0 & 0 & -16 \\ & & 2 & 4 & 8 & 16 \\ \hline & 1 & 2 & 4 & 8 & 0 \end{array}$$

$f(2) = 0$

Find $f(-2)$.

$$\begin{array}{r|rrrrr} -2 & 1 & 0 & 0 & 0 & -16 \\ & & -2 & 4 & -8 & 16 \\ \hline & 1 & -2 & 4 & -8 & 0 \end{array}$$

$f(-2) = 0$

Find $f(3)$.

$$\begin{array}{r|rrrrr} 3 & 1 & 0 & 0 & 0 & -16 \\ & & 3 & 9 & 27 & 81 \\ \hline & 1 & 3 & 9 & 27 & 65 \end{array}$$

$f(3) = 65$

Find $f(1 - \sqrt{2})$.

$$\begin{array}{r|rrrrr} 1-\sqrt{2} & 1 & 0 & 0 & 0 & -16 \\ & & 1-\sqrt{2} & 3-2\sqrt{2} & 7-5\sqrt{2} & 17-12\sqrt{2} \\ \hline & 1 & 1-\sqrt{2} & 3-2\sqrt{2} & 7-5\sqrt{2} & 1-12\sqrt{2} \end{array}$$

$f(1 - \sqrt{2}) = 1 - 12\sqrt{2}$

24.

$$\begin{array}{r|rrrrrr} 2 & 1 & 0 & 0 & 0 & 0 & 32 \\ & & 2 & 4 & 8 & 16 & 32 \\ \hline & 1 & 2 & 4 & 8 & 16 & 64 \end{array}$$

$f(2) = 64$

$f(-2) = 0$ (See Exercise 14.)

$$\begin{array}{r|rrrrrr} 3 & 1 & 0 & 0 & 0 & 0 & 32 \\ & & 3 & 9 & 27 & 81 & 243 \\ \hline & 1 & 3 & 9 & 27 & 81 & 275 \end{array}$$

$f(3) = 275$

$$\begin{array}{rrrrrr} \multicolumn{1}{r|}{2+3i} \\ 1 & 0 & 0 & 0 & 0 & 32 \\ & 2+3i & -5+12i & -46+9i & -119-120i & 122-597i \\ \hline 1 & 2+3i & -5+12i & -46+9i & -119-120i & 154-597i \end{array}$$

$f(2 + 3i) = 154 - 597i$

25. $f(x) = 3x^3 + 5x^2 - 6x + 18$

If -3 is a zero of $f(x)$, then $f(-3) = 0$. Find $f(-3)$ using synthetic division.

$$\begin{array}{r|rrrr} -3 & 3 & 5 & -6 & 18 \\ & & -9 & 12 & -18 \\ \hline & 3 & -4 & 6 & 0 \end{array}$$

Since $f(-3) = 0$, -3 is a zero of $f(x)$.

If 2 is a zero of $f(x)$, then $f(2) = 0$. Find $f(2)$ using synthetic division.

$$\begin{array}{r|rrrr} 2 & 3 & 5 & -6 & 18 \\ & & 6 & 22 & 32 \\ \hline & 3 & 11 & 16 & 50 \end{array}$$

Since $f(2) \neq 0$, 2 is not a zero of $f(x)$.

26.

$$\begin{array}{r|rrrr} -4 & 3 & 11 & -2 & 8 \\ & & -12 & 4 & -8 \\ \hline & 3 & -1 & 2 & 0 \end{array}$$

$f(-4) = 0$, so -4 is a zero of $f(x)$.

$$\begin{array}{r|rrrr} 2 & 3 & 11 & -2 & 8 \\ & & 6 & 34 & 64 \\ \hline & 3 & 17 & 32 & 72 \end{array}$$

$f(2) \neq 0$, so 2 is not a zero of $f(x)$.

27. $f(x) = x^3 - \dfrac{7}{2}x^2 + x - \dfrac{3}{2}$

If -3 is a zero of $f(x)$, then $f(-3) = 0$. Find $f(-3)$ using synthetic division.

$$\begin{array}{r|rrrr} -3 & 1 & -\frac{7}{2} & 1 & -\frac{3}{2} \\ & & -3 & \frac{39}{2} & -\frac{123}{2} \\ \hline & 1 & -\frac{13}{2} & \frac{41}{2} & -63 \end{array}$$

Since $f(-3) \neq 0$, -3 is not a zero of $f(x)$.

If $\dfrac{1}{2}$ is a zero of $f(x)$, then $f\left(\dfrac{1}{2}\right) = 0$.

Find $f\left(\dfrac{1}{2}\right)$ using synthetic division.

$$\begin{array}{r|rrrr} \frac{1}{2} & 1 & -\frac{7}{2} & 1 & -\frac{3}{2} \\ & & \frac{1}{2} & -\frac{3}{2} & -\frac{1}{4} \\ \hline & 1 & -3 & -\frac{1}{2} & -\frac{7}{4} \end{array}$$

Since $f\left(\dfrac{1}{2}\right) \neq 0$, $\dfrac{1}{2}$ is not a zero of $f(x)$.

28.

$$\begin{array}{r|rrrr} i & 1 & 2 & 1 & 2 \\ & & i & -1+2i & -2 \\ \hline & 1 & 2+i & 2i & 0 \end{array}$$

$f(i) = 0$, so i is a zero of $f(x)$.

$$\begin{array}{r|rrrr} -i & 1 & 2 & 1 & 2 \\ & & -i & -1-2i & -2 \\ \hline & 1 & 2-i & -2i & 0 \end{array}$$

$f(-i) = 0$, so $-i$ is a zero of $f(x)$.

$$\begin{array}{r|rrrr} -2 & 1 & 2 & 1 & 2 \\ & & -2 & 0 & -2 \\ \hline & 1 & 0 & 1 & 0 \end{array}$$

$f(-2) = 0$, so -2 is a zero of $f(x)$.

29. $f(x) = x^3 + 4x^2 + x - 6$

Try $x - 1$. Use synthetic division to see whether $f(1) = 0$.

$$\begin{array}{r|rrrr} 1 & 1 & 4 & 1 & -6 \\ & & 1 & 5 & 6 \\ \hline & 1 & 5 & 6 & 0 \end{array}$$

Since $f(1) = 0$, $x - 1$ is a factor of $f(x)$. Thus $f(x) = (x - 1)(x^2 + 5x + 6)$.

Factoring the trinomial we get

$f(x) = (x - 1)(x + 2)(x + 3)$.

To solve the equation $f(x) = 0$, use the principle of zero products.

$(x-1)(x+2)(x+3) = 0$

$x-1=0$ *or* $x+2=0$ *or* $x+3=0$

$x=1$ *or* $x=-2$ *or* $x=-3$

The solutions are 1, -2, and -3.

30.
$$\begin{array}{r|rrrr} 2 & 1 & 5 & -2 & -24 \\ & & 2 & 14 & 24 \\ \hline & 1 & 7 & 12 & 0 \end{array}$$

$f(x) = (x-2)(x^2+7x+12)$

$= (x-2)(x+3)(x+4)$

The solutions are 2, -3, and -4.

31. $f(x) = x^3 - 6x^2 + 3x + 10$

Try $x-1$. Use synthetic division to see whether $f(1) = 0$.

$$\begin{array}{r|rrrr} 1 & 1 & -6 & 3 & 10 \\ & & 1 & -5 & -2 \\ \hline & 1 & -5 & -2 & 8 \end{array}$$

Since $f(1) \neq 0$, $x-1$ is not a factor of $P(x)$.

Try $x+1$. Use synthetic division to see whether $f(-1) = 0$.

$$\begin{array}{r|rrrr} -1 & 1 & -6 & 3 & 10 \\ & & -1 & 7 & -10 \\ \hline & 1 & -7 & 10 & 0 \end{array}$$

Since $f(-1) = 0$, $x+1$ is a factor of $f(x)$.

Thus $f(x) = (x+1)(x^2-7x+10)$.

Factoring the trinomial we get

$f(x) = (x+1)(x-2)(x-5)$.

To solve the equation $f(x) = 0$, use the principle of zero products.

$(x+1)(x-2)(x-5) = 0$

$x+1=0$ *or* $x-2=0$ *or* $x-5=0$

$x=-1$ *or* $x=2$ *or* $x=5$

The solutions are -1, 2, and 5.

32.
$$\begin{array}{r|rrrr} 1 & 1 & 2 & -13 & 10 \\ & & 1 & 3 & -10 \\ \hline & 1 & 3 & -10 & 0 \end{array}$$

$f(x) = (x-1)(x^2+3x-10)$

$= (x-1)(x-2)(x+5)$

The solutions are 1, 2, and -5.

33. $f(x) = x^3 - x^2 - 14x + 24$

Try $x+1$, $x-1$, and $x+2$. Using synthetic division we find that $f(-1) \neq 0$, $f(1) \neq 0$ and $f(-2) \neq 0$. Thus $x+1$, $x-1$, and $x+2$, are not factors of $f(x)$.

Try $x-2$. Use synthetic division to see whether $P(2) = 0$.

$$\begin{array}{r|rrrr} 2 & 1 & -1 & -14 & 24 \\ & & 2 & 2 & -24 \\ \hline & 1 & 1 & -12 & 0 \end{array}$$

Since $f(2) = 0$, $x-2$ is a factor of $f(x)$. Thus $f(x) = (x-2)(x^2+x-12)$.

Factoring the trinomial we get

$f(x) = (x-2)(x+4)(x-3)$

To solve the equation $f(x) = 0$, use the principle of zero products.

$(x-2)(x+4)(x-3) = 0$

$x-2=0$ *or* $x+4=0$ *or* $x-3=0$

$x=2$ *or* $x=-4$ *or* $x=3$

The solutions are 2, -4, and 3.

34.
$$\begin{array}{r|rrrr} 2 & 1 & -3 & -10 & 24 \\ & & 2 & -2 & -24 \\ \hline & 1 & -1 & -12 & 0 \end{array}$$

$f(x) = (x-2)(x^2-x-12)$

$= (x-2)(x-4)(x+3)$

The solutions are 2, 4, and -3.

35. $f(x) = x^4 - x^3 - 19x^2 + 49x - 30$

Try $x-1$. Use synthetic division to see whether $f(1) = 0$.

$$\begin{array}{r|rrrrr} 1 & 1 & -1 & -19 & 49 & -30 \\ & & 1 & 0 & -19 & 30 \\ \hline & 1 & 0 & -19 & 30 & 0 \end{array}$$

Since $f(1) = 0$, $x-1$ is a factor of $f(x)$. Thus $f(x) = (x-1)(x^3-19x+30)$.

We continue to use synthetic division to factor $g(x) = x^3-19x+30$. Trying $x-1$, $x+1$, and $x+2$ we find that $g(1) \neq 0$, $g(-1) \neq 0$, and $g(-2) \neq 0$. Thus $x-1$, $x+1$, and $x+2$ are not factors of $x^3-19x+30$. Try $x-2$.

$$\begin{array}{r|rrrr} 2 & 1 & 0 & -19 & 30 \\ & & 2 & 4 & -30 \\ \hline & 1 & 2 & -15 & 0 \end{array}$$

Since $g(2) = 0$, $x-2$ is a factor of $x^3-19x+30$.

Thus $f(x) = (x-1)(x-2)(x^2+2x-15)$.

Factoring the trinomial we get

$f(x) = (x-1)(x-2)(x-3)(x+5)$.

To solve the equation $f(x) = 0$, use the principle of zero products.

$(x-1)(x-2)(x-3)(x+5) = 0$

$x-1=0$ *or* $x-2=0$ *or* $x-3=0$ *or* $x+5=0$

$x=1$ *or* $x=2$ *or* $x=3$ *or* $x=-5$

The solutions are 1, 2, 3, and -5.

36.
$$\begin{array}{r|rrrrr} -1 & 1 & 11 & 41 & 61 & 30 \\ & & -1 & -10 & -31 & -30 \\ \hline & 1 & 10 & 31 & 30 & 0 \end{array}$$

$$\begin{array}{r|rrrr} -2 & 1 & 10 & 31 & 30 \\ & & -2 & -16 & -30 \\ \hline & 1 & 8 & 15 & 0 \end{array}$$

$$f(x) = (x+1)(x+2)(x^2+8x+15)$$
$$= (x+1)(x+2)(x+3)(x+5)$$

The solutions are -1, -2, -3, and -5.

37. It is usually faster to use a grapher.

38. No; the polynomial cannot have more than n linear factors.

39. $2x - 7 = 5x + 8$

$-7 = 3x + 8$ Subtracting $2x$ on both sides

$-15 = 3x$ Subtracting 8 on both sides

$-5 = x$ Dividing by 3 on both sides

We could also graph $y_1 = 2x-7$ and $y_2 = 5x+8$ and use the Intersect feature to find the first coordinate of the point of intersection of the graphs.

The solution is -5.

40. $2x^2 + 12 = 5x$

$$2x^2 - 5x + 12 = 0$$
$$x = \frac{-(-5) \pm \sqrt{(-5)^2 - 4 \cdot 2 \cdot 12}}{2 \cdot 2}$$
$$= \frac{5 \pm \sqrt{-71}}{4}$$
$$= \frac{5 \pm i\sqrt{71}}{4}$$

41. $7x^2 + 4x = 3$

$$7x^2 + 4x - 3 = 0$$
$$(7x-3)(x+1) = 0$$
$$7x - 3 = 0 \quad or \quad x + 1 = 0$$
$$7x = 3 \quad or \quad x = -1$$
$$x = \frac{3}{7} \quad or \quad x = -1$$

We could also graph $y_1 = 7x^2 + 4x$ and $y_2 = 3$ and use the Intersect feature to find the first coordinate of the point of intersection of the graphs.

The solutions are $\frac{3}{7}$ and -1.

42. Let b and h represent the length of the base and the height of the triangle, respectively.

$b + h = 30$, so $b = 30 - h$.

$$A = \frac{1}{2}bh = \frac{1}{2}(30-h)h = -\frac{1}{2}h^2 + 15h$$

Find the value of h for which A is a maximum:

$$h = \frac{-15}{2(-1/2)} = 15$$

When $h = 15$, $b = 30 - 15 = 15$.

The area is a maximum when the base and the height are each 15 in.

43. a) -4, -3, 2, and 5 are zeros of the function, so $x+4$, $x+3$, $x-2$, and $x-5$ are factors.

b) We first write the product of the factors:

$P(x) = (x+4)(x+3)(x-2)(x-5)$

Note that $P(0) = 4 \cdot 3(-2)(-5) > 0$ and the graph shows a positive y-intercept, so this function is a correct one.

c) Yes; two examples are $f(x) = c \cdot P(x)$ for any non-zero constant c and $g(x) = (x-a)P(x)$.

d) No; only the function in part (b) has the given graph.

44. a) -5, -3, 4, 6, and 7 are zeros of the function, so $x+5$, $x+3$, $x-4$, $x-6$, and $x-7$ are factors.

b) We first write the product of the factors:

$F(x) = (x+5)(x+3)(x-4)(x-6)(x-7)$

Note that $F(0) = 5 \cdot 3(-4)(-6)(-7) < 0$, but that the y-intercept of the graph is positive. Thus we must reflect $F(x)$ across the x-axis to obtain a function $P(x)$ with the given graph. We have:

$$P(x) = -F(x)$$
$$P(x) = -(x+5)(x+3)(x-4)(x-6)(x-7)$$

c) Yes; two examples are $f(x) = c \cdot P(x)$, for any non-zero constant c, and $g(x) = (x-a)P(x)$.

d) No; only the function in part (b) has the given graph.

45. Divide $x^3 - kx^2 + 3x + 7k$ by $x + 2$.

$$\begin{array}{r|rrrr} -2 & 1 & -k & 3 & 7k \\ & & -2 & 2k+4 & -4k-14 \\ \hline & 1 & -k-2 & 2k+7 & 3k-14 \end{array}$$

Thus $P(-2) = 3k - 14$.

We know that if $x + 2$ is a factor of $f(x)$, then $f(-2) = 0$.

We solve $0 = 3k - 14$ for k.

$$0 = 3k - 14$$
$$\frac{14}{3} = k$$

46. Divide $x^2 + kx + 4$ by $x - 1$.

$$\begin{array}{r|rrr} 1 & 1 & k & 4 \\ & & 1 & k+1 \\ \hline & 1 & k+1 & k+5 \end{array}$$

The remainder is $k + 5$.

Divide $x^2 + kx + 4$ by $x + 1$.

$$\begin{array}{r|rrr} -1 & 1 & k & 4 \\ & & -1 & -k+1 \\ \hline & 1 & k-1 & -k+5 \end{array}$$

The remainder is $-k + 5$.

Let $k + 5 = -k + 5$ and solve for k.

$$k + 5 = -k + 5$$
$$2k = 0$$
$$k = 0$$

47. $y = \frac{1}{13}x^3 - \frac{1}{14}x$

$y = x\left(\frac{1}{13}x^2 - \frac{1}{14}\right)$

We use the principle of zero products to find the zeros of the polynomial.

$x = 0 \quad or \quad \frac{1}{13}x^2 - \frac{1}{14} = 0$

$x = 0 \quad or \quad \frac{1}{13}x^2 = \frac{1}{14}$

$x = 0 \quad or \quad x^2 = \frac{13}{14}$

$x = 0 \quad or \quad x = \pm\sqrt{\frac{13}{14}}$

$x = 0 \quad or \quad x \approx \pm 0.9636$

Only 0 and 0.9636 are in the interval $[0, 2]$.

48. $\frac{6x^2}{x^2+11} + \frac{60}{x^3 - 7x^2 + 11x - 77} = \frac{1}{x-7}$,

LCM is $(x^2+11)(x-7)$

$$(x^2+11)(x-7)\left[\frac{6x^2}{x^2+11} + \frac{60}{(x^2+11)(x-7)}\right] = (x^2+11)(x-7)\cdot\frac{1}{x-7}$$

$6x^2(x-7) + 60 = x^2 + 11$

$6x^3 - 42x^2 + 60 = x^2 + 11$

$6x^3 - 43x^2 + 49 = 0$

Use synthetic division to find factors of $f(x) = 6x^3 - 43x^2 + 49$.

$$\begin{array}{r|rrrr} -1 & 6 & -43 & 0 & 49 \\ & & -6 & 49 & -49 \\ \hline & 6 & -49 & 49 & 0 \end{array}$$

Then we have:

$(x+1)(6x^2 - 49x + 49) = 0$

$(x+1)(6x-7)(x-7) = 0$

$x = -1$ or $x = \frac{7}{6}$ or $x = 7$

Only -1 and $\frac{7}{6}$ check.

49. $\frac{2x^2}{x^2-1} + \frac{4}{x+3} = \frac{32}{3x^2 - x - 3}$,

LCM is $(x+1)(x-1)(x+3)$

$$(x+1)(x-1)(x+3)\left[\frac{2x^2}{(x+1)(x-1)} + \frac{4}{x+3}\right] = (x+1)(x-1)(x+3)\cdot\frac{32}{(x+1)(x-1)(x+3)}$$

$2x^2(x+3) + 4(x+1)(x-1) = 32$

$2x^3 + 6x^2 + 4x^2 - 4 = 32$

$2x^3 + 10x^2 - 36 = 0$

$x^3 + 5x^2 - 18 = 0$

Using synthetic division and several trials, we find that -3 is a factor of $f(x) = x^3 + 5x^2 - 18$:

$$\begin{array}{r|rrrr} -3 & 1 & 5 & 0 & -18 \\ & & -3 & -6 & 18 \\ \hline & 1 & 2 & -6 & 0 \end{array}$$

Then we have:

$(x+3)(x^2 + 2x - 6) = 0$

$x + 3 = 0 \quad or \quad x^2 + 2x - 6 = 0$

$x = -3 \quad or \quad x = -1 \pm \sqrt{7}$

Only $-1 \pm \sqrt{7}$ check.

50. The graphs have different x-intercepts, so the zeros change when a coefficient is changed.

51. Answers may vary. One possibility is $P(x) = x^{15} - x^{14}$.

52.
$$\begin{array}{r|rrr} 3+2i & 1 & -4 & -2 \\ & & 3+2i & -7+4i \\ \hline & 1 & -1+2i & -9+4i \end{array}$$

The answer is $x - 1 + 2i$, R $-9 + 4i$.

53.
$$\begin{array}{r|rrr} i & 1 & -3 & 7 \\ & & i & -3i-1 \\ \hline & 1 & -3+i & 6-3i \end{array} \qquad (i^2 = -1)$$

The answer is $x - 3 + i$, R $6 - 3i$.

Exercise Set 3.3

1. $f(x) = (x+3)^2(x-1) = (x+3)(x+3)(x-1)$

The factor $x+3$ occurs twice. Thus the zero -3 has a multiplicity of two.

The factor $x-1$ occurs only one time. Thus the zero 1 has a multiplicity of one.

2. 3, Multiplicity 2; -4, Multiplicity 3:

0, Multiplicity 4

3. $f(x) = x^3(x-1)^2(x+4)$

$x \cdot x \cdot x(x-1)(x-1)(x+4) = 0$

The factor x occurs three times. Thus the zero 0 has a multiplicity of three.

The factor $x-1$ occurs twice. Thus the zero 1 has a multiplicity of two.

The factor $x+4$ occurs only one time. Thus the zero -4 has a multiplicity of one.

4. $f(x) = (x^2 - 5x + 6)^2$

$= [(x-3)(x-2)]^2$

$= (x-3)^2(x-2)^2$

3, Multiplicity 2; 2, Multiplicity 2

5. $f(x) = x^4 - 4x^2 + 3$

We factor as follows:

$P(x) = (x^2 - 3)(x^2 - 1)$

$= (x - \sqrt{3})(x + \sqrt{3})(x - 1)(x + 1)$

The zeros of the polynomial are $\sqrt{3}$, $-\sqrt{3}$, 1, and -1. Each has multiplicity of one.

6. $f(x) = x^4 - 10x^2 + 9$
$= (x^2 - 9)(x^2 - 1)$
$= (x + 3)(x - 3)(x + 1)(x - 1)$
± 3, ± 1; each multiplicity of one.

7. $f(x) = x^3 + 3x^2 - x - 3$
We factor by grouping:
$P(x) = x^2(x + 3) - (x + 3)$
$= (x^2 - 1)(x + 3)$
$= (x - 1)(x + 1)(x + 3)$
The zeros of the polynomial are 1, -1, and -3. Each has multiplicity of one.

8. $f(x) = x^3 - x^2 - 2x + 2$
$= x^2(x - 1) - 2(x - 1)$
$= (x^2 - 2)(x - 1)$
$= (x - \sqrt{2})(x + \sqrt{2})(x - 1)$
$\sqrt{2}$, $-\sqrt{2}$, 1; each has multiplicity of 1.

9. Find a polynomial function of degree 3 with -2, 3, and 5 as zeros.
Such a polynomial has factors $x + 2$, $x - 3$, and $x - 5$, so we have
$f(x) = a_n(x + 2)(x - 3)(x - 5)$.
The number a_n can be any nonzero number. The simplest polynomial will be obtained if we let it be 1. Multiplying the factors, we obtain
$f(x) = (x + 2)(x - 3)(x - 5)$
$= (x^2 - x - 6)(x - 5)$
$= x^3 - 6x^2 - x + 30$

10. $f(x) = (x - 2)(x - i)(x + i)$
$= (x - 2)(x^2 + 1)$
$= x^3 - 2x^2 + x - 2$

11. Find a polynomial function of degree 3 with -3, $2i$, and $-2i$ as zeros.
Such a polynomial has factors $x + 3$, $x - 2i$, and $x + 2i$, so we have
$f(x) = a_n(x + 3)(x - 2i)(x + 2i)$.
The number a_n can be any nonzero number. The simplest polynomial will be obtained if we let it be 1. Multiplying the factors, we obtain
$f(x) = (x + 3)(x - 2i)(x + 2i)$
$= (x + 3)(x^2 + 4)$
$= x^3 + 3x^2 + 4x + 12$

12. $f(x) = [x - (1 + 4i)][x - (1 - 4i)](x + 1)$
$= (x^2 - 2x + 17)(x + 1)$
$= x^3 - x^2 + 15x + 17$

13. Find a polynomial function of degree 3 with $\sqrt{2}$, $-\sqrt{2}$, and $\sqrt{3}$ as zeros.
Such a polynomial has factors $x - \sqrt{2}$, $x + \sqrt{2}$, and $x - \sqrt{3}$, so we have
$f(x) = a_n(x - \sqrt{2})(x + \sqrt{2})(x - \sqrt{3})$.
The number a_n can be any nonzero number. The simplest polynomial will be obtained if we let it be 1. Multiplying the factors, we obtain
$f(x) = (x - \sqrt{2})(x + \sqrt{2})(x - \sqrt{3})$
$= (x^2 - 2)(x - \sqrt{3})$
$= x^3 - \sqrt{3}x^2 - 2x + 2\sqrt{3}$

14. $f(x) = (x + 3)(x)\left(x - \frac{1}{2}\right)$
$= (x^2 + 3x)(\left(x - \frac{1}{2}\right)$
$= x^3 + \frac{5}{2}x^2 - \frac{3}{2}x$

15. A polynomial function of degree 5 has at most 5 zeros. Since 5 zeros are given, these are all of the zeros of the desired function.
$f(x) = (x + 1)^3(x - 0)(x - 1)$
$= (x^3 + 3x^2 + 3x + 1)(x^2 - x)$
$= x^5 + 2x^4 - 2x^2 - x$

16. $f(x) = (x + 2)(x - 3)^2(x + 1)$
$= x^4 - 3x^3 - 7x^2 + 15x + 18$

17. A polynomial function $f(x)$ of degree 5 has at most 5 zeros. Three of the zeros are 6, $-3 + 4i$, and $4 - \sqrt{5}$. Since $f(x)$ has rational coefficients we know that the conjugates of $-3 + 4i$ and $4 - \sqrt{5}$, or $-3 - 4i$ and $4 + \sqrt{5}$, are also zeros.

18. The conjugate of $1 - i$, or $1 + i$, is the other zero.

19. Find a polynomial function of lowest degree with rational coefficients that has $1 + i$ and 2 as some of its zeros.
$1 - i$ is also a zero.
Thus the polynomial function is
$f(x) = a_n(x - 2)[x - (1 + i)][x - (1 - i)]$.
If we let $a_n = 1$, we obtain
$f(x) = (x - 2)[(x - 1) - i][(x - 1) + i]$
$= (x - 2)[(x - 1)^2 - i^2]$
$= (x - 2)(x^2 - 2x + 1 + 1)$
$= (x - 2)(x^2 - 2x + 2)$
$= x^3 - 4x^2 + 6x - 4$

20. $f(x) = [x - (2 - i)][x - (2 + i)](x + 1)$
$= x^3 - 3x^2 + x + 5$

21. Find a polynomial function of lowest degree with rational coefficients that has $-4i$ and 5 as some of its zeros.

$4i$ is also a zero.

Thus the polynomial function is

$f(x) = a_n(x-5)(x+4i)(x-4i)$.

If we let $a_n = 1$, we obtain

$$\begin{aligned} f(x) &= (x-5)[x^2-(4i)^2] \\ &= (x-5)(x^2+16) \\ &= x^3-5x^2+16x-80 \end{aligned}$$

22. $f(x)$

$$\begin{aligned} &= [x-(2-\sqrt{3})][x-(2+\sqrt{3})][x-(1+i)][x-(1-i)] \\ &= x^4-6x^3+11x^2-10x+2 \end{aligned}$$

23. Find a polynomial function of lowest degree with rational coefficients that has $\sqrt{5}$ and $-3i$ as some of its zeros.

$-\sqrt{5}$ and $3i$ are also zeros.

Thus the polynomial function is

$f(x) = a_n(x-\sqrt{5})(x+\sqrt{5})(x+3i)(x-3i)$.

If we let $a_n = 1$, we obtain

$$\begin{aligned} f(x) &= (x^2-5)(x^2+9) \\ &= x^4+4x^2-45 \end{aligned}$$

24. $f(x) = (x+\sqrt{2})(x-\sqrt{2})(x-4i)(x+4i)$
$= x^4+14x^2-32$

25. If $-i$ is a zero of $f(x) = x^4-5x^3+7x^2-5x+6$, i is also a zero. Thus $x+i$ and $x-i$ are factors of the polynomial. Since $(x+i)(x-i) = x^2+1$, we know that $f(x) =$ $(x^2+1)\cdot Q(x)$. Divide $x^4-5x^3+7x^2-5x+6$ by x^2+1.

$$\begin{array}{r|l} & x^2 - 5x + 6 \\ x^2+1 & x^4 - 5x^3 + 7x^2 - 5x + 6 \\ & x^4 \qquad\quad + x^2 \\ \hline & -5x^3 + 6x^2 - 5x \\ & -5x^3 \qquad\quad - 5x \\ \hline & 6x^2 \qquad + 6 \\ & 6x^2 \qquad + 6 \\ \hline & 0 \end{array}$$

Thus

$$\begin{aligned} x^4-5x^3+7x^2-5x+6 &= (x+i)(x-i)(x^2-5x+6) \\ &= (x+i)(x-i)(x-2)(x-3) \end{aligned}$$

Using the principle of zero products we find the other zeros to be i, 2, and 3.

26. $(x-2i)$ and $(x+2i)$ are both factors of $P(x) = x^4-16$.

$(x-2i)(x+2i) = x^2+4$

$$\begin{array}{r|l} & x^2 - 4 \\ x^2+4 & x^4 + 0x^2 - 16 \\ & x^4 + 4x^2 \\ \hline & -4x^2 - 16 \\ & -4x^2 - 16 \\ \hline & 0 \end{array}$$

$$\begin{aligned} (x-2i)(x+2i)(x^2-4) &= 0 \\ (x-2i)(x+2i)(x+2)(x-2) &= 0 \end{aligned}$$

The other zeros are $-2i$, -2, and 2.

27. $x^3-6x^2+13x-20 = 0$

If 4 is a zero, then $x-4$ is a factor. Use synthetic division to find another factor.

$$\begin{array}{r|rrrr} 4 & 1 & -6 & 13 & -20 \\ & & 4 & -8 & 20 \\ \hline & 1 & -2 & 5 & 0 \end{array}$$

$(x-4)(x^2-2x+5) = 0$

$x-4 = 0$ *or* $x^2-2x+5 = 0$ Principle of zero products

$x = 4$ *or* $x = \dfrac{2\pm\sqrt{4-20}}{2}$ Quadratic formula

$x = 4$ *or* $x = \dfrac{2\pm 4i}{2} = 1\pm 2i$

The other zeros are $1+2i$ and $1-2i$.

28.

$$\begin{array}{r|rrrr} 2 & 1 & 0 & 0 & -8 \\ & & 2 & 4 & 8 \\ \hline & 1 & 2 & 4 & 0 \end{array}$$

$(x-2)(x^2+2x+4) = 0$

$x = 2$ *or* $x = -1\pm\sqrt{3}i$

The other zeros are $-1+\sqrt{3}i$ and $-1-\sqrt{3}i$.

29. $f(x) = x^5-3x^2+1$

According to the rational zeros theorem, any rational zero of f must be of the form p/q, where p is a factor of the constant term, 1, and q is a factor of the coefficient of x^5, 1.

$$\frac{\text{Possibilities for } p}{\text{Possibilities for } q} : \frac{\pm 1}{\pm 1}$$

Possibilities for p/q: $1, -1$

30. $f(x) = x^7+37x^5-6x^2+12$

$$\frac{\text{Possibilities for } p}{\text{Possibilities for } q} : \frac{\pm 1, \pm 2, \pm 3, \pm 4, \pm 6, \pm 12}{\pm 1}$$

Possibilities for p/q: $1, -1, 2, -2, 3, -3, 4, -4, 6, -6, 12, -12$

31. $f(x) = 15x^6+47x^2+2$

According to the rational zeros theorem, any rational zero of f must be of the form p/q, where p is a factor of 2 and q is a factor of 15.

$$\frac{\text{Possibilities for } p}{\text{Possibilities for } q} : \frac{\pm 1, \pm 2}{\pm 1, \pm 3, \pm 5, \pm 15}$$

Possibilities for p/q: $1, -1, 2, -2, \frac{1}{3}, -\frac{1}{3}, \frac{2}{3}, -\frac{2}{3}, \frac{1}{5}, -\frac{1}{5}, \frac{2}{5}, -\frac{2}{5}, \frac{1}{15}, -\frac{1}{15}, \frac{2}{15}, -\frac{2}{15}$

32. $f(x) = 10x^{25}+3x^{17}-35x+6$

$$\frac{\text{Possibilities for } p}{\text{Possibilities for } q} : \frac{\pm 1, \pm 2, \pm 3, \pm 6}{\pm 1, \pm 2, \pm 5, \pm 10}$$

Possibilities for p/q: $1, -1, 2, -2, 3, -3, 6, -6,$
$\frac{1}{2}, -\frac{1}{2}, \frac{3}{2}, -\frac{3}{2}, \frac{1}{5}, -\frac{1}{5}, \frac{2}{5}, -\frac{2}{5}, \frac{3}{5}, -\frac{3}{5}, \frac{6}{5}, -\frac{6}{5}, \frac{1}{10}, -\frac{1}{10}, \frac{3}{10}, -\frac{3}{10}$

33. $f(x) = x^3 + 3x^2 - 2x - 6$

a) $\dfrac{\text{Possibilities for } p}{\text{Possibilities for } q} : \dfrac{\pm 1, \pm 2, \pm 3, \pm 6}{\pm 1}$

Possibilities for p/q: $1, -1, 2, -2, 3, -3, 6, -6$

From the graph of $y = x^3 + 3x^2 - 2x - 6$, we see that, of the possibilities above, only -3 might be a zero. We use synthetic division to determine whether -3 is indeed a zero.

$$\begin{array}{r|rrrr} -3 & 1 & 3 & -2 & -6 \\ & & -3 & 0 & 6 \\ \hline & 1 & 0 & -2 & 0 \end{array}$$

Then we have $f(x) = (x + 3)(x^2 - 2)$.

We find the other zeros:

$$x^2 - 2 = 0$$
$$x^2 = 2$$
$$x = \pm\sqrt{2}.$$

There is only one rational zero, -3. The other zeros are $\pm\sqrt{2}$. (Note that we could have used factoring by grouping to find this result.)

b) $f(x) = (x + 3)(x - \sqrt{2})(x + \sqrt{2})$

34. $f(x) = x^3 - x^2 - 3x + 3$

a) $\dfrac{\text{Possibilities for } p}{\text{Possibilities for } q} : \dfrac{\pm 1, \pm 3}{\pm 1}$

Possibilities for p/q: $1, -1, 3, -3$

From the graph of $y = x^3 - x^2 - 3x + 3$, we see that, of the possibilities above, only 1 might be a zero.

$$\begin{array}{r|rrrr} 1 & 1 & -1 & -3 & 3 \\ & & 1 & 0 & -3 \\ \hline & 1 & 0 & -3 & 0 \end{array}$$

$f(x) = (x - 1)(x^2 - 3)$

Now $x^2 - 3 = 0$ for $x = \pm\sqrt{3}$. Thus, there is only one rational zero, 1. The other zeros are $\pm\sqrt{3}$. (Note that we would have used factoring by grouping to find this result.)

b) $f(x) = (x - 1)(x - \sqrt{3})(x + \sqrt{3})$

35. $f(x) = x^3 - 3x + 2$

a) $\dfrac{\text{Possibilities for } p}{\text{Possibilities for } q} : \dfrac{\pm 1, \pm 2}{\pm 1}$

Possibilities for p/q: $1, -1, 2, -2$

From the graph of $y = x^3 - 3x + 2$, we see that, of the possibilities above, -2 and 1 might be a zeros. We use synthetic division to determine whether -2 is a zero.

$$\begin{array}{r|rrrr} -2 & 1 & 0 & -3 & 2 \\ & & -2 & 4 & -2 \\ \hline & 1 & -2 & 1 & 0 \end{array}$$

Then we have $f(x) = (x + 2)(x^2 - 2x + 1) = (x + 2)(x - 1)^2$.

Now $(x - 1)^2 = 0$ for $x = 1$. Thus, the rational zeros are -2 and 1. (The zero 1 has a multiplicity of 2.) These are the only zeros.

b) $f(x) = (x + 2)(x - 1)^2$

36. $f(x) = x^3 - 2x + 4$

a) $\dfrac{\text{Possibilities for } p}{\text{Possibilities for } q} : \dfrac{\pm 1, \pm 2, \pm 4}{\pm 1}$

Possibilities for p/q: $1, -1, 2, -2, 4, -4$

From the graph of $y = x^3 - 2x + 4$, we see that, of the possibilities above, only -2 might be a zero.

$$\begin{array}{r|rrrr} -2 & 1 & 0 & -2 & 4 \\ & & -2 & 4 & -4 \\ \hline & 1 & -2 & 2 & 0 \end{array}$$

$f(x) = (x + 2)(x^2 - 2x + 2)$

Using the quadratic formula, we find that the other zeros are $1 \pm i$. The only rational zero is -2. The other zeros are $1 \pm i$.

b) $f(x) = (x + 2)[x - (1 + i)][x - (1 - i)]$
$= (x + 2)(x - 1 - i)(x - 1 + i)$

37. $f(x) = x^3 - 5x^2 + 11x + 17$

a) $\dfrac{\text{Possibilities for } p}{\text{Possibilities for } q} : \dfrac{\pm 1, \pm 17}{\pm 1}$

Possibilities for p/q: $1, -1, 17, -17$

From the graph of $y = x^3 - 5x^2 + 11x + 17$, we see that, of the possibilities above, we see that only -1 might be a zero. We use synthetic division to determine whether -1 is indeed a zero.

$$\begin{array}{r|rrrr} -1 & 1 & -5 & 11 & 17 \\ & & -1 & 6 & -17 \\ \hline & 1 & -6 & 17 & 0 \end{array}$$

Then we have $f(x) = (x + 1)(x^2 - 6x + 17)$. We use the quadratic formula to find the other zeros.

$$x^2 - 6x + 17 = 0$$

$$x = \frac{-(-6) \pm \sqrt{(-6)^2 - 4 \cdot 1 \cdot 17}}{2 \cdot 1}$$
$$= \frac{6 \pm \sqrt{-32}}{2} = \frac{6 \pm 4\sqrt{2}i}{2}$$
$$= 3 \pm 2\sqrt{2}i$$

The only rational zero is -1. The other zeros are $3 \pm 2\sqrt{2}i$.

b) $f(x) = (x + 1)[x - (3 + 2\sqrt{2}i)][x - (3 - 2\sqrt{2}i)]$
$= (x + 1)(x - 3 - 2\sqrt{2}i)(x - 3 + 2\sqrt{2}i)$

38. $f(x) = 2x^3 + 7x^2 + 2x - 8$

a) $\dfrac{\text{Possibilities for } p}{\text{Possibilities for } q} : \dfrac{\pm1, \pm2, \pm4, \pm8}{\pm1, \pm2}$

Possibilities for p/q: $1, -1, 2, -2, 4, -4, 8, -8, \frac{1}{2}, -\frac{1}{2}$

From the graph of $y = 2x^3 + 7x^2 + 2x - 8$, we see that, of the possibilities above, only -2 and 1 might be a zeros.

$$\begin{array}{r|rrrr} -2 & 2 & 7 & 2 & -8 \\ & & -4 & -6 & 8 \\ \hline & 2 & 3 & -4 & 0 \end{array}$$

$f(x) = (x+2)(2x^2 + 3x - 4)$

Using the quadratic formula, we find that the other zeros are $\dfrac{-3 \pm \sqrt{41}}{4}$. The only rational zero is -2. The other zeros are $\dfrac{-3 \pm \sqrt{41}}{4}$.

b) $f(x) = (x+2)\left(x - \dfrac{-3+\sqrt{41}}{4}\right)\left(x - \dfrac{-3-\sqrt{41}}{4}\right)$

39. $f(x) = 5x^4 - 4x^3 + 19x^2 - 16x - 4$

a) $\dfrac{\text{Possibilities for } p}{\text{Possibilities for } q} : \dfrac{\pm1, \pm2, \pm4}{\pm1, \pm5}$

Possibilities for p/q: $1, -1, 2, -2, 4, -4, \frac{1}{5}, -\frac{1}{5}, \frac{2}{5}, -\frac{2}{5}, \frac{4}{5}, -\frac{4}{5}$

From the graph of $y = 5x^4 - 4x^3 + 19x^2 - 16x - 4$, we see that, of the possibilities above, only $-\frac{2}{5}$, $-\frac{1}{5}$ and 1 might be zeros. We use synthetic division to determine whether 1 is a zero.

$$\begin{array}{r|rrrrr} 1 & 5 & -4 & 19 & -16 & -4 \\ & & 5 & 1 & 20 & 4 \\ \hline & 5 & 1 & 20 & 4 & 0 \end{array}$$

Then we have

$$\begin{aligned} f(x) &= (x-1)(5x^3 + x^2 + 20x + 4) \\ &= (x-1)[x^2(5x+1) + 4(5x+1)] \\ &= (x-1)(5x+1)(x^2+4). \end{aligned}$$

We find the other zeros:

$$\begin{aligned} 5x + 1 = 0 \quad &or \quad x^2 + 4 = 0 \\ 5x = -1 \quad &or \quad x^2 = -4 \\ x = -\frac{1}{5} \quad &or \quad x = \pm 2i \end{aligned}$$

The rational zeros are $-\frac{1}{5}$ and 1. The other zeros are $\pm 2i$.

b) From part (a) we see that

$f(x) = (5x+1)(x-1)(x+2i)(x-2i)$.

40. $f(x) = 3x^4 - 4x^3 + x^2 + 6x - 2$

a) $\dfrac{\text{Possibilities for } p}{\text{Possibilities for } q} : \dfrac{\pm1, \pm2}{\pm1, \pm3}$

Possibilities for p/q: $1, -1, 2, -2, \frac{1}{3}, -\frac{1}{3}, \frac{2}{3}, -\frac{2}{3}$

From the graph of $y = 3x^4 - 4x^3 + x^2 + 6x - 2$, we see that, of the possibilities above, only -1 and $\frac{1}{3}$ might be zeros.

$$\begin{array}{r|rrrrr} -1 & 3 & -4 & 1 & 6 & -2 \\ & & -3 & 7 & -8 & 2 \\ \hline & 3 & -7 & 8 & -2 & 0 \end{array}$$

$$\begin{array}{r|rrrr} \frac{1}{3} & 3 & -7 & 8 & -2 \\ & & 1 & -2 & 2 \\ \hline & 3 & -6 & 6 & 0 \end{array}$$

$$\begin{aligned} f(x) &= (x+1)\left(x - \frac{1}{3}\right)(3x^2 - 6x + 6) \\ &= (x+1)\left(x - \frac{1}{3}\right)(3)(x^2 - 2x + 2) \end{aligned}$$

Using the quadratic formula, we find that the other zeros are $1 \pm i$.

The rational zeros are -1 and $\frac{1}{3}$. The other zeros are $1 \pm i$.

b)
$$\begin{aligned} f(x) &= 3(x+1)\left(x - \frac{1}{3}\right)[x - (1+i)][x - (1-i)] \\ &= (x+1)(3x-1)(x-1-i)(x-1+i) \end{aligned}$$

41. $f(x) = x^4 - 3x^3 - 20x^2 - 24x - 8$

a) $\dfrac{\text{Possibilities for } p}{\text{Possibilities for } q} : \dfrac{\pm1, \pm2, \pm4, \pm8}{\pm1}$

Possibilities for p/q: $1, -1, 2, -2, 4, -4, 8, -8$

From the graph of $y = x^4 - 3x^3 - 20x^2 - 24x - 8$, we see that, of the possibilities above, only -2 and -1 might be zeros. We use synthetic division to determine if -2 is a zero.

$$\begin{array}{r|rrrrr} -2 & 1 & -3 & -20 & -24 & -8 \\ & & -2 & 10 & 20 & 8 \\ \hline & 1 & -5 & -10 & -4 & 0 \end{array}$$

We see that -2 is a zero. Now we determine whether -1 is a zero.

$$\begin{array}{r|rrrr} -1 & 1 & -5 & -10 & -4 \\ & & -1 & 6 & 4 \\ \hline & 1 & -6 & -4 & 0 \end{array}$$

Then we have $f(x) = (x+2)(x+1)(x^2 - 6x - 4)$. Use the quadratic formula to find the other zeros.

$x^2 - 6x - 4 = 0$

$$\begin{aligned} x &= \frac{-(-6) \pm \sqrt{(-6)^2 - 4 \cdot 1 \cdot (-4)}}{2 \cdot 1} \\ &= \frac{6 \pm \sqrt{52}}{2} = \frac{6 \pm 2\sqrt{13}}{2} \\ &= 3 \pm \sqrt{13} \end{aligned}$$

The rational zeros are -2 and -1. The other zeros are $3 \pm \sqrt{13}$.

b) $f(x) = (x+2)(x+1)[x-(3+\sqrt{13})][x-(3-\sqrt{13})]$
$= (x+2)(x+1)(x-3-\sqrt{13})(x-3+\sqrt{13})$

42. $f(x) = x^4 + 5x^3 - 27x^2 + 31x - 10$

a) $\dfrac{\text{Possibilities for } p}{\text{Possibilities for } q} : \dfrac{\pm 1, \pm 2, \pm 5, \pm 10}{\pm 1}$

Possibilities for p/q: $1, -1, 2, -2, 5, -5, 10, -10$

From the graph of $y = x^4 + 5x^3 - 27x^2 + 31x - 10$, we see that, of the possibilities above, only 1 and 2 might be zeros.

$$\begin{array}{r|rrrrr} 1 & 1 & 5 & -27 & 31 & -10 \\ & & 1 & 6 & -21 & 10 \\ \hline & 1 & 6 & -21 & 10 & 0 \end{array}$$

$$\begin{array}{r|rrrr} 2 & 1 & 6 & -21 & 10 \\ & & 2 & 16 & -10 \\ \hline & 1 & 8 & -5 & 0 \end{array}$$

$f(x) = (x-1)(x-2)(x^2 + 8x - 5)$

Using the quadratic formula, we find that the other zeros are $-4 \pm \sqrt{21}$.

The rational zeros are 1 and 2. The other zeros are $-4 \pm \sqrt{21}$.

b) $f(x) = (x-1)(x-2)[x-(4+\sqrt{21})][x-(4-\sqrt{21})]$
$= (x-1)(x-2)(x-4-\sqrt{21})(x-4+\sqrt{21})$

43. $f(x) = x^3 - 4x^2 + 2x + 4$

a) $\dfrac{\text{Possibilities for } p}{\text{Possibilities for } q} : \dfrac{\pm 1, \pm 2, \pm 4}{\pm 1}$

Possibilities for p/q: $1, -1, 2, -2, 4, -4$

From the graph of $y = x^3 - 4x^2 + 2x + 4$, we see that, of the possibilities above, only -1, 1, and 2 might be zeros. Synthetic division shows that neither -1 nor 1 is a zero. Try 2.

$$\begin{array}{r|rrrr} 2 & 1 & -4 & 2 & 4 \\ & & 2 & -4 & -4 \\ \hline & 1 & -2 & -2 & 0 \end{array}$$

Then we have $f(x) = (x-2)(x^2 - 2x - 2)$. Use the quadratic formula to find the other zeros.

$x^2 - 2x - 2 = 0$

$$x = \frac{-(-2) \pm \sqrt{(-2)^2 - 4 \cdot 1 \cdot (-2)}}{2 \cdot 1}$$
$$= \frac{2 \pm \sqrt{12}}{2} = \frac{2 \pm 2\sqrt{3}}{2}$$
$$= 1 \pm \sqrt{3}$$

The only rational zero is 2. The other zeros are $1 \pm \sqrt{3}$.

b) $f(x) = (x-2)[x-(1+\sqrt{3})][x-(1-\sqrt{3})]$
$= (x-2)(x-1-\sqrt{3})(x-1+\sqrt{3})$

44. $f(x) = x^3 - 8x^2 + 17x - 4$

a) $\dfrac{\text{Possibilities for } p}{\text{Possibilities for } q} : \dfrac{\pm 1, \pm 2, \pm 4}{\pm 1}$

Possibilities for p/q: $1, -1, 2, -2, 4, -4$

From the graph of $y = x^3 - 8x^2 + 17x - 4$, we see that, of the possibilities above, only 4 might be a zero.

$$\begin{array}{r|rrrr} 4 & 1 & -8 & 17 & -4 \\ & & 4 & -16 & 4 \\ \hline & 1 & -4 & 1 & 0 \end{array}$$

$f(x) = (x-4)(x^2 - 4x + 1)$

Using the quadratic formula, we find that the other zeros are $2 \pm \sqrt{3}$.

The only rational zero is 4. The other zeros are $2 \pm \sqrt{3}$.

b) $f(x) = (x-4)[x-(2+\sqrt{3})][x-(2-\sqrt{3})]$
$= (x-4)(x-2-\sqrt{3})(x-2+\sqrt{3})$

45. $f(x) = x^3 + 8$

a) $\dfrac{\text{Possibilities for } p}{\text{Possibilities for } q} : \dfrac{\pm 1, \pm 2, \pm 4, \pm 8}{\pm 1}$

Possibilities for p/q: $1, -1, 2, -2, 4, -4, 8, -8$

From the graph of $y = x^3 + 8$, we see that, of the possibilities above, only -2 might be a zero. We use synthetic division to see if it is.

$$\begin{array}{r|rrrr} -2 & 1 & 0 & 0 & 8 \\ & & -2 & 4 & -8 \\ \hline & 1 & -2 & 4 & 0 \end{array}$$

We have $f(x) = (x+2)(x^2 - 2x + 4)$. Use the quadratic formula to find the other zeros.

$x^2 - 2x + 4 = 0$

$$x = \frac{-(-2) \pm \sqrt{(-2)^2 - 4 \cdot 1 \cdot 4}}{2 \cdot 1}$$
$$= \frac{2 \pm \sqrt{-12}}{2} = \frac{2 \pm 2\sqrt{3}i}{2}$$
$$= 1 \pm \sqrt{3}i$$

The only rational zero is -2. The other zeros are $1 \pm \sqrt{3}i$.

b) $f(x) = (x+2)[x-(1+\sqrt{3}i)][x-(1-\sqrt{3}i)]$
$= (x+2)(x-1-\sqrt{3}i)(x-1+\sqrt{3}i)$

46. $f(x) = x^3 - 8$

a) As in Exercise 43, the possibilities for p/q are 1, -1, 2, -2, 4, -4, 8, and -8.

From the graph of $y = x^3 - 8$, we see that, of the possibilities above, only 2 might be a zero.

$$\begin{array}{r|rrrr} 2 & 1 & 0 & 0 & -8 \\ & & 2 & 4 & 8 \\ \hline & 1 & 2 & 4 & 0 \end{array}$$

$f(x) = (x-2)(x^2 + 2x + 4)$

Using the quadratic formula, we find that the other zeros are $-1 \pm \sqrt{3}i$.

The only rational zero is 2. The other zeros are $-1 \pm \sqrt{3}i$.

b) $f(x) = (x-2)[x-(-1+\sqrt{3}i)][x-(-1-\sqrt{3}i)]$
$= (x-2)(x+1-\sqrt{3}i)(x+1+\sqrt{3}i)$

47. $f(x) = \frac{1}{3}x^3 - \frac{1}{2}x^2 - \frac{1}{6}x + \frac{1}{6}$
$= \frac{1}{6}(2x^3 - 3x^2 - x + 1)$

a) The second form of the equation is equivalent to the first and has the advantage of having integer coefficients. Thus, we can use the rational zeros theorem for $g(x) = 2x^3 - 3x^2 - x + 1$. The zeros of $g(x)$ are the same as the zeros of $f(x)$. We find the zeros of $g(x)$.

$\frac{\text{Possibilities for } p}{\text{Possibilities for } q} : \frac{\pm 1}{\pm 1, \pm 2}$

Possibilities for p/q: $1, -1, \frac{1}{2}, -\frac{1}{2}$

From the graph of $y = 2x^3 - 3x^2 - x + 1$, we see that, of the possibilities above, only $-\frac{1}{2}$ and $\frac{1}{2}$ might be zeros. Synthetic division shows that $-\frac{1}{2}$ is not a zero. Try $\frac{1}{2}$.

$$\begin{array}{r|rrrr} \frac{1}{2} & 2 & -3 & -1 & 1 \\ & & 1 & -1 & -1 \\ \hline & 2 & -2 & -2 & 0 \end{array}$$

We have $g(x) = \left(x - \frac{1}{2}\right)(2x^2 - 2x - 2) = \left(x - \frac{1}{2}\right)(2)(x^2 - x - 1)$. Use the quadratic formula to find the other zeros.

$x^2 - x - 1 = 0$

$$x = \frac{-(-1) \pm \sqrt{(-1)^2 - 4 \cdot 1 \cdot (-1)}}{2 \cdot 1}$$
$$= \frac{1 \pm \sqrt{5}}{2}$$

The only rational zero is $\frac{1}{2}$. The other zeros are $\frac{1 \pm \sqrt{5}}{2}$.

b) $f(x) = \frac{1}{6}g(x)$
$= \frac{1}{6}\left(x - \frac{1}{2}\right)(2)\left[x - \frac{1+\sqrt{5}}{2}\right]\left[x - \frac{1-\sqrt{5}}{2}\right]$
$= \frac{1}{3}\left(x - \frac{1}{2}\right)\left(x - \frac{1+\sqrt{5}}{2}\right)\left(x - \frac{1-\sqrt{5}}{2}\right)$

48. $f(x) = \frac{2}{3}x^3 - \frac{1}{2}x^2 + \frac{2}{3}x - \frac{1}{2}$
$= \frac{1}{6}(4x^3 - 3x^2 + 4x - 3)$

a) Find the zeros of $g(x) = 4x^3 - 3x^2 + 4x - 3$.

$\frac{\text{Possibilities for } p}{\text{Possibilities for } q} : \frac{\pm 1, \pm 3}{\pm 1, \pm 2, \pm 4}$

Possibilities for p/q: $1, -1, 3, -3, \frac{1}{2}, -\frac{1}{2}, \frac{3}{2}, -\frac{3}{2}, \frac{1}{4}, -\frac{1}{4}, \frac{3}{4}, -\frac{3}{4}$

From the graph of $y = 4x^3 - 3x^2 + 4x - 3$, we see that, of the possibilities above, only $\frac{1}{2}$, $\frac{3}{4}$, and 1 might be zeros. Synthetic division shows that $\frac{1}{2}$ is not a zero. Try $\frac{3}{4}$.

$$\begin{array}{r|rrrr} \frac{3}{4} & 4 & -3 & 4 & -3 \\ & & 3 & 0 & 3 \\ \hline & 4 & 0 & 4 & 0 \end{array}$$

$g(x) = \left(x - \frac{3}{4}\right)(4x^2 + 4) = \left(x - \frac{3}{4}\right)(4)(x^2 + 1)$

Now $x^2 + 1 = 0$ when $x = \pm i$. Thus, the only rational zero is $\frac{3}{4}$. The other zeros are $\pm i$. (Note that we could have used factoring by grouping to find this result.)

b) $f(x) = \frac{1}{6}g(x)$
$= \frac{1}{6}\left(x - \frac{3}{4}\right)(4)(x + i)(x - i)$
$= \frac{2}{3}\left(x - \frac{3}{4}\right)(x + i)(x - i)$

49. $f(x) = x^4 + 32$

According to the rational zeros theorem, the possible rational zeros are $\pm 1, \pm 2, \pm 4, \pm 8, \pm 16$, and ± 32. The graph of $y = x^4 + 32$ has no x-intercepts, so $f(x)$ has no real-number zeros and hence no rational zeros.

50. $f(x) = x^6 + 8$

Possible rational zeros: ± 1, ± 2, ± 4, ± 8

The graph of $y = x^6 + 8$ has no x-intercepts, so $f(x)$ has no real-number zeros and hence no rational zeros.

51. $f(x) = x^3 - x^2 - 4x + 3$

According to the rational zeros theorem, the possible rational zeros are ± 1 and ± 3. The graph of $y = x^3 - x^2 - 4x + 3$ shows that none of these is a zero. Thus, there are no rational zeros.

52. $f(x) = 2x^3 + 3x^2 + 2x + 3$

Possible rational zeros: ± 1, ± 3, $\pm\frac{1}{2}$, $\pm\frac{3}{2}$

The graph of $y = 2x^3 + 3x^2 + 2x + 3$ shows that, of these possibilities, only $-\frac{3}{2}$ might be a zero.

$$\begin{array}{r|rrrr} -\frac{3}{2} & 2 & 3 & 2 & 3 \\ & & -3 & 0 & -3 \\ \hline & 2 & 0 & 2 & 0 \end{array}$$

$$f(x) = \left(x + \frac{3}{2}\right)(2x^2 + 2) = \left(x + \frac{3}{2}\right)(2)(x^2 + 1)$$

Since $g(x) = x^2 + 1$ has no real-number zeros, the only rational zero is $-\frac{3}{2}$. (We could have used factoring by grouping to find this result.)

53. $f(x) = x^4 + 2x^3 + 2x^2 - 4x - 8$

According to the rational zeros theorem, the possible rational zeros are ± 1, ± 2, ± 4, and ± 8. The graph of $y = x^4 + 2x^3 + 2x^2 - 4x - 8$ shows that none of the possibilities is a zero. Thus, there are no rational zeros.

54. $f(x) = x^4 + 6x^3 + 17x^2 + 36x + 66$

Possible rational zeros: $\pm 1, \pm 2, \pm 3, \pm 6, \pm 11, \pm 22, \pm 33, \pm 66$

The graph of $y = x^4 + 6x^3 + 17x^2 + 36x + 66$ has no x-intercepts, so $f(x)$ has no real-number zeros and hence no rational zeros.

55. $f(x) = x^5 - 5x^4 + 5x^3 + 15x^2 - 36x + 20$

According to the rational zeros theorem, the possible rational zeros are ± 1, ± 2, ± 4, ± 5, ± 10, and ± 20. The graph of $y = x^5 - 5x^4 + 5x^3 + 15x^2 - 36x + 20$ shows that, of these possibilities, only -2, 1 and 2 might be zeros. We try -2.

$$\begin{array}{r|rrrrrr} -2 & 1 & -5 & 5 & 15 & -36 & 20 \\ & & -2 & 14 & -38 & 46 & -20 \\ \hline & 1 & -7 & 19 & -23 & 10 & 0 \end{array}$$

Thus, -2 is a zero. Now try 1.

$$\begin{array}{r|rrrrr} 1 & 1 & -7 & 19 & -23 & 10 \\ & & 1 & -6 & 13 & 10 \\ \hline & 1 & -6 & 13 & -10 & 0 \end{array}$$

1 is also a zero. Try 2.

$$\begin{array}{r|rrrr} 2 & 1 & -6 & 13 & -10 \\ & & 2 & -8 & 10 \\ \hline & 1 & -4 & 5 & 0 \end{array}$$

2 is also a zero.

We have $f(x) = (x + 2)(x - 1)(x - 2)(x^2 - 4x + 5)$. The discriminant of $x^2 - 4x + 5$ is $(-4)^2 - 4 \cdot 1 \cdot 5$, or $4 < 0$, so $x^2 - 4x + 5$ has two nonreal zeros. Thus, the rational zeros are -2, 1, and 2.

56. $f(x) = x^5 - 3x^4 - 3x^3 + 9x^2 - 4x + 12$

Possible rational zeros: ± 1, ± 2, ± 3, ± 4, ± 6, ± 12. The graph of $y = x^5 - 3x^4 - 3x^3 + 9x^2 - 4x + 12$ shows that, of these possibilities, only -2, 2 and 3 might be zeros.

$$\begin{array}{r|rrrrrr} -2 & 1 & -3 & -3 & 9 & -4 & 12 \\ & & -2 & 10 & -14 & 10 & -12 \\ \hline & 1 & -5 & 7 & -5 & 6 & 0 \end{array}$$

-2 is a zero.

$$\begin{array}{r|rrrrr} 2 & 1 & -5 & 7 & -5 & 6 \\ & & 2 & -6 & 2 & -6 \\ \hline & 1 & -3 & 1 & -3 & 0 \end{array}$$

2 is a zero.

$$\begin{array}{r|rrrr} 3 & 1 & -3 & 1 & -3 \\ & & 3 & 0 & 3 \\ \hline & 1 & 0 & 1 & 0 \end{array}$$

3 is also a zero.

$f(x) = (x + 2)(x - 2)(x - 3)(x^2 + 1)$

Since $g(x) = x^2 + 1$ has no real-number zeros, the rational zeros are -2, 2, and 3.

57. No; since imaginary zeros of polynomials with rational coefficients occur in conjugate pairs, a third-degree polynomial with rational coefficients can have at most two imaginary zeros. Thus, there must be at least one real zero.

58. Yes; let c be a zero of $P(x)$. Then $P(c) = 0$, so $-P(c) = Q(c) = 0$. Thus, every zero of $P(x)$ is a zero of $Q(x)$. Now let r be a zero of $Q(x)$. Then $Q(r) = 0$, so $-P(r) = 0$ and $P(r) = 0$. Every zero of $Q(x)$ is also a zero of $P(x)$. Thus, $P(x)$ and $Q(x)$ have the same zeros.

59. $f(x) = x^2 - 8x + 10$

a) $-\frac{b}{2a} = -\frac{-8}{2 \cdot 1} = -(-4) = 4$

$f(4) = 4^2 - 8 \cdot 4 + 10 = -6$

The vertex is $(4, -6)$.

b) The line of symmetry is $x = 4$.

c) Since the coefficient of x^2 is positive, there is a minimum function value. It is the second coordinate of the vertex, -6. It occurs when $x = 4$.

60. $f(x) = 3x^2 - 6x - 1$

a) $-\frac{b}{2a} = -\frac{-6}{2 \cdot 3} = 1$

$f(1) = 3 \cdot 1^2 - 6 \cdot 1 - 1 = -4$

The vertex is $(1, -4)$.

b) $x = 1$

c) Minimum: -4 at $x = 1$

61.

$$-\frac{4}{5}x + 8 = 0$$

$$-\frac{4}{5}x = -8 \quad \text{Subtracting 8}$$

$$-\frac{5}{4}\left(-\frac{4}{5}x\right) = -\frac{5}{4}(-8) \quad \text{Multiplying by } -\frac{5}{4}$$

$$x = 10$$

We can also graph $y = -\frac{4}{5}x + 8$ and use the Zero feature.

The solution is 10.

62.

$$x^2 - 8x - 33 = 0$$

$$(x - 11)(x + 3) = 0$$

$$x = 11 \text{ or } x = -3$$

We can also graph $y = x^2 - 8x - 33$ and use the Zero feature twice.

The solutions are -3 and 11.

63. $f(x) = 2x^3 - 5x^2 - 4x + 3$

a) $2x^3 - 5x^2 - 4x + 3 = 0$

$\dfrac{\text{Possibilities for } p}{\text{Possibilities for } q}: \dfrac{\pm 1, \pm 3}{\pm 1, \pm 2}$

Possibilities for p/q: $1, -1, 3, -3, \frac{1}{2}, -\frac{1}{2}, \frac{3}{2}, -\frac{3}{2}$

The first possibility that is a solution of $f(x) = 0$ is -1:

$$\begin{array}{r|rrrr} -1 & 2 & -5 & -4 & 3 \\ & & -2 & 7 & -3 \\ \hline & 2 & -7 & 3 & 0 \end{array}$$

Thus, -1 is a solution.

Then we have:

$$(x+1)(2x^2 - 7x + 3) = 0$$
$$(x+1)(2x-1)(x-3) = 0$$

The other solutions are $\frac{1}{2}$ and 3.

b) The graph of $y = f(x-1)$ is the graph of $y = f(x)$ shifted 1 unit right. Thus, we add 1 to each solution of $f(x) = 0$ to find the solutions of $f(x-1) = 0$. The solutions are $-1+1$, or 0; $\frac{1}{2}+1$, or $\frac{3}{2}$; and $3+1$, or 4.

c) The graph of $y = f(x+2)$ is the graph of $y = f(x)$ shifted 2 units left. Thus, we subtract 2 from each solution of $f(x) = 0$ to find the solutions of $f(x+2) = 0$. The solutions are $-1-2$, or -3; $\frac{1}{2} - 2$, or $-\frac{3}{2}$; and $3-2$, or 1.

d) The graph of $y = f(2x)$ is a horizontal shrinking of the graph of $y = f(x)$ by a factor of 2. We divide each solution of $f(x) = 0$ by 2 to find the solutions of $f(2x) = 0$. The solutions are $\frac{-1}{2}$ or $-\frac{1}{2}$; $\frac{1/2}{2}$, or $\frac{1}{4}$; and $\frac{3}{2}$.

64. By the rational zeros theorem, only ± 1, ± 2, ± 3, ± 4, ± 6, and ± 12 can be rational solutions of $x^4 - 12 = 0$. Since none of them is a solution, the equation has no rational solutions. But $\sqrt[4]{12}$ is a solution of the equation, so $\sqrt[4]{12}$ must be irrational.

65. See the answer section in the text.

66. $P(x) = x^6 - x^5 - 72x^4 - 81x^2 + 486x + 5832$

a) $x^6 - 6x^5 - 72x^4 - 81x^2 + 486x + 5832 = 0$

Synthetic division shows that we can factor as follows:

$$\begin{aligned} P(x) &= (x-3)(x+3)(x+6)(x^3 - 12x^2 + 9x - 108) \\ &= (x-3)(x+3)(x+6)[x^2(x-12) + 9(x-12)] \\ &= (x-3)(x+3)(x+6)(x-12)(x^2+9) \end{aligned}$$

The rational zeros are 3, -3, -6, and 12.

67. $P(x) = 2x^5 - 33x^4 - 84x^3 + 2203x^2 - 3348x - 10,080$

a) $2x^5 - 33x^4 - 84x^3 + 2203x^2 - 3348x - 10,080 = 0$

Trying some of the many possibilities for p/q, we find that 4 is a zero.

$$\begin{array}{r|rrrrrr} 4 & 2 & -33 & -84 & 2203 & -3348 & -10,080 \\ & & 8 & -100 & -736 & 5868 & 10,080 \\ \hline & 2 & -25 & -184 & 1467 & 2520 & 0 \end{array}$$

Then we have:

$$(x-4)(2x^4 - 25x^3 - 184x^2 + 1467x + 2520) = 0$$

We now use the fourth degree polynomial above to find another zero. Synthetic division shows that 4 is not a double zero, but 7 is a zero.

$$\begin{array}{r|rrrrr} 7 & 2 & -25 & -184 & 1467 & 2520 \\ & & 14 & -77 & -1827 & -2520 \\ \hline & 2 & -11 & -261 & -360 & 0 \end{array}$$

Now we have:

$$(x-4)(x-7)(2x^3 - 11x^2 - 261x - 360) = 0$$

Use the third degree polynomial above to find a third zero. Synthetic division shows that 7 is not a double zero, but 15 is a zero.

$$\begin{array}{r|rrrr} 15 & 2 & -11 & -261 & -360 \\ & & 30 & 285 & 360 \\ \hline & 2 & 19 & 24 & 0 \end{array}$$

We have:

$$\begin{aligned} P(x) &= (x-4)(x-7)(x-15)(2x^2 + 19x + 24) \\ &= (x-4)(x-7)(x-15)(2x+3)(x+8) \end{aligned}$$

The rational zeros are 4, 7, 15, $-\frac{3}{2}$, and -8.

Exercise Set 3.4

1. Graph (d) is the graph of $f(x) = \dfrac{8}{x^2 - 4}$.

$x^2 - 4 = 0$ when $x = \pm 2$, so $x = -2$ and $x = 2$ are vertical asymptotes.

The x-axis, $y = 0$, is the horizontal asymptote because the degree of the numerator is less than the degree of the denominator.

There is no oblique asymptote.

2. Graph (f) is the graph of $f(x) = \dfrac{8}{x^2 + 4}$.

$x^2 + 4 = 0$ has no real solutions, so there is no vertical asymptote.

The x-axis, $y = 0$, is the horizontal asymptote because the degree of the numerator is less than the degree of the denominator.

There is no oblique asymptote.

3. Graph (e) is the graph of $f(x) = \dfrac{8x}{x^2 - 4}$.

As in Exercise 1, $x = -2$ and $x = 2$ are vertical asymptotes.

The x-axis, $y = 0$, is the horizontal asymptote because the degree of the numerator is less than the degree of the denominator.

There is no oblique asymptote.

4. Graph (a) is the graph of $f(x) = \dfrac{8x^2}{x^2-4}$.

As in Exercise 1, $x = 2$ and $x = -2$ are vertical asymptotes.

The numerator and denominator have the same degree, so $y = 8/1$, or $y = 8$, is the horizontal asymptote.

There is no oblique asymptote.

5. Graph (c) is the graph of $f(x) = \dfrac{8x^3}{x^2-4}$.

As in Exercise 1, $x = -2$ and $x = 2$ are vertical asymptotes.

The degree of the numerator is greater than the degree of the denominator, so there is no horizontal asymptote but there is a vertical asymptote. To find it we first divide to find an equivalent expression.

$$\begin{array}{r} 8x \\ x^2-4\overline{)\,8x^3\qquad} \\ 8x^3-32x \\ \hline 32x \end{array}$$

$$\frac{8x^3}{x^2-4} = 8x + \frac{32x}{x^2-4}$$

Now we multiply by 1, using $(1/x^2)/(1/x^2)$.

$$\frac{32x}{x^2-4}\cdot\frac{\frac{1}{x^2}}{\frac{1}{x^2}} = \frac{\frac{32}{x}}{1-\frac{4}{x^2}}$$

As $|x|$ becomes very large, each expression with x in the denominator tends toward zero.

Then, as $|x| \to \infty$, we have

$$\frac{\frac{32}{x}}{1-\frac{4}{x^2}} \to \frac{0}{1-0}, \text{ or } 0.$$

Thus, as $|x|$ becomes very large, the graph of $f(x)$ gets very close to the graph of $y = 8x$, so $y = 8x$ is the oblique asymptote.

6. Graph (b) is the graph of $f(x) = \dfrac{8x^3}{x^2+4}$.

As in Exercise 2, there is no vertical asymptote.

The degree of the numerator is greater than the degree of the denominator, so there is no horizontal asymptote but there is a vertical asymptote. To find it we first divide to find an equivalent expression.

$$\frac{8x^3}{x^2+4} = 8x - \frac{32x}{x^2+4}$$

Now $\dfrac{32x}{x^2+4} = \dfrac{\frac{32}{x}}{1+\frac{4}{x^2}}$ and, as $|x| \to \infty$,

$$\frac{\frac{32}{x}}{1+\frac{4}{x^2}} \to \frac{0}{1+0}, \text{ or } 0.$$

Thus, as $y = 8x$ is the oblique asymptote.

7. $f(x) = \dfrac{1}{x+3}$

1. -3 is the zero of the denominator, so the domain excludes -3. It is $(-\infty,-3)\cup(-3,\infty)$. The line $x = -3$ is the vertical asymptote.
2. Because the degree of the numerator is less than the degree of the denominator, the x-axis is the horizontal asymptote. There is no oblique asymptote.
3. The numerator has no zeros, so there is no x-intercept.
4. $f(0) = \dfrac{1}{0+3} = \dfrac{1}{3}$, so $\left(0,\dfrac{1}{3}\right)$ is the y-intercept.
5. Find other function values to determine the shape of the graph and then draw it.

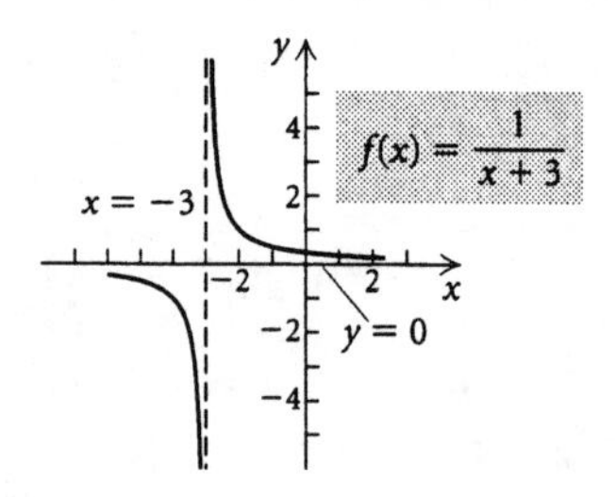

8. $f(x) = \dfrac{1}{x-5}$

1. 5 is the zero of the denominator, so the domain is $(-\infty,5)\cup(5,\infty)$ and $x = 5$ is the vertical asymptote.
2. Because the degree of the numerator is less than the degree of the denominator, the x-axis is the horizontal asymptote. There is no oblique asymptote.
3. The numerator has no zeros, so there is no x-intercept.
4. $f(0) = \dfrac{1}{0-5} = -\dfrac{1}{5}$, so $\left(0,-\dfrac{1}{5}\right)$ is the y-intercept.
5. Find other function values to determine the shape of the graph and then draw it.

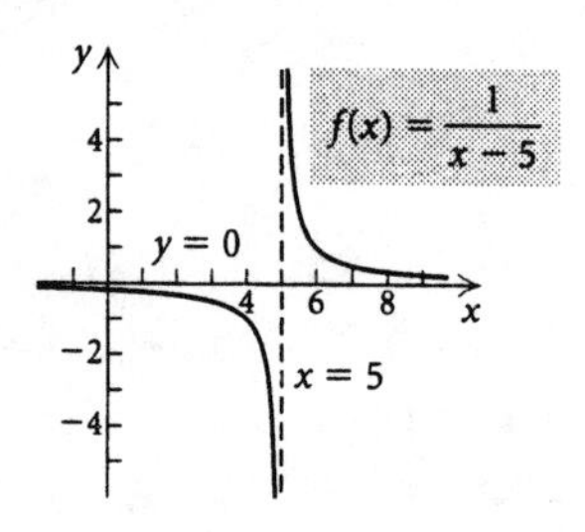

9. $f(x) = \dfrac{-2}{x-5}$

1. 5 is the zero of the denominator, so the domain excludes 5. It is $(-\infty, 5) \cup (5, \infty)$. The line $x = 5$ is the vertical asymptote.
2. Because the degree of the numerator is less than the degree of the denominator, the x-axis is the horizontal asymptote. There is no oblique asymptote.
3. The numerator has no zeros, so there is no x-intercept.
4. $f(0) = \dfrac{-2}{0-5} = \dfrac{2}{5}$, so $\left(0, \dfrac{2}{5}\right)$ is the y-intercept.
5. Find other function values to determine the shape of the graph and then draw it.

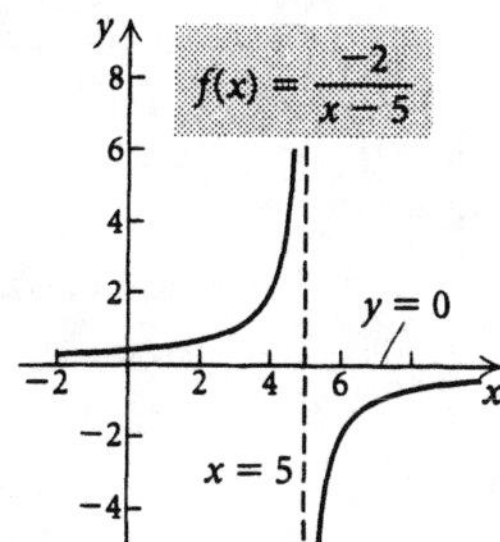

10. $f(x) = \dfrac{3}{3-x}$

1. 3 is the zero of the denominator, so the domain is $(-\infty, 3) \cup (3, \infty)$ and $x = 3$ is the vertical asymptote.
2. Because the degree of the numerator is less than the degree of the denominator, the x-axis is the horizontal asymptote. There is no oblique asymptote.
3. The numerator has no zeros, so there is no x-intercept.
4. $f(0) = \dfrac{3}{3-0} = 1$, so $(0, 1)$ is the y-intercept.
5. Find other function values to determine the shape of the graph and then draw it.

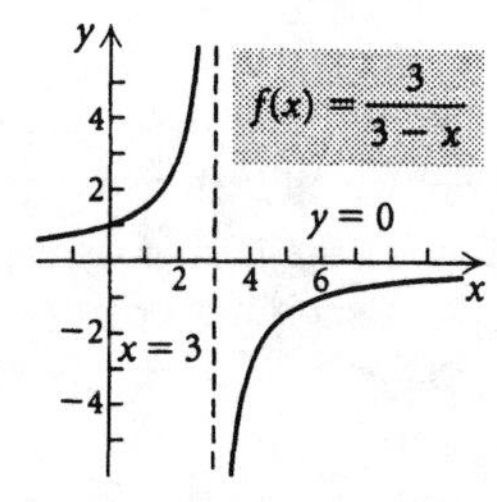

11. $f(x) = \dfrac{2x+1}{x}$

1. 0 is the zero of the denominator, so the domain excludes 0. It is $(-\infty, 0) \cup (0, \infty)$. The line $x = 0$, or the y-axis, is the vertical asymptote.
2. The numerator and denominator have the same degree, so the horizontal asymptote is determined by the ratio of the leading coefficients, 2/1, or 2. Thus, $y = 2$ is the horizontal asymptote. There is no oblique asymptote.
3. The zero of the numerator is the solution of $2x + 1 = 0$, or $-\dfrac{1}{2}$. The x-intercept is $\left(-\dfrac{1}{2}, 0\right)$.
4. Since 0 is not in the domain of the function, there is no y-intercept.
5. Find other function values to determine the shape of the graph and then draw it.

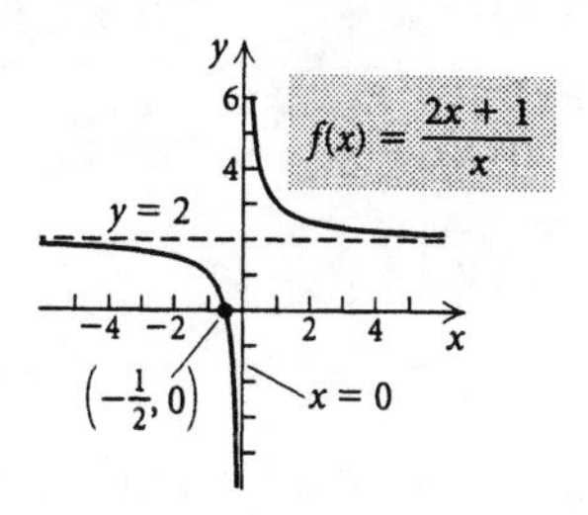

12. $f(x) = \dfrac{3x-1}{x}$

1. 0 is the zero of the denominator, so the domain is $(-\infty, 0) \cup (0, \infty)$. The line $x = 0$, or the y-axis, is the vertical asymptote.
2. The numerator and denominator have the same degree, so the horizontal asymptote is determined by the ratio of the leading coefficients, 3/1, or 3. Thus, $y = 3$ is the horizontal asymptote. There is no oblique asymptote.
3. The zero of the numerator is $\dfrac{1}{3}$, so the x-intercept is $\left(\dfrac{1}{3}, 0\right)$.
4. Since 0 is not in the domain of the function, there is no y-intercept.
5. Find other function values to determine the shape of the graph and then draw it.

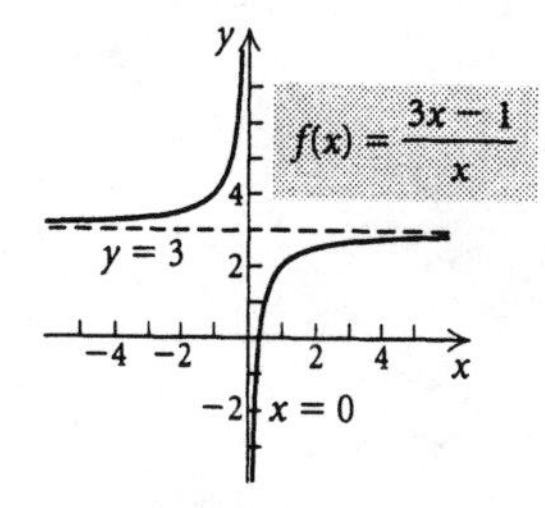

13. $f(x) = \dfrac{1}{(x-2)^2}$

1. 2 is the zero of the denominator, so the domain excludes 2. It is $(-\infty, 2) \cup (2, \infty)$. The line $x = 2$ is the vertical asymptote.
2. Because the degree of the numerator is less than the degree of the denominator, the x-axis is the horizontal asymptote. There is no oblique asymptote.

3. The numerator has no zeros, so there is no x-intercept.
4. $f(0) = \dfrac{1}{(0-2)^2} = \dfrac{1}{4}$, so $\left(0, \dfrac{1}{4}\right)$ is the y-intercept.
5. Find other function values to determine the shape of the graph and then draw it.

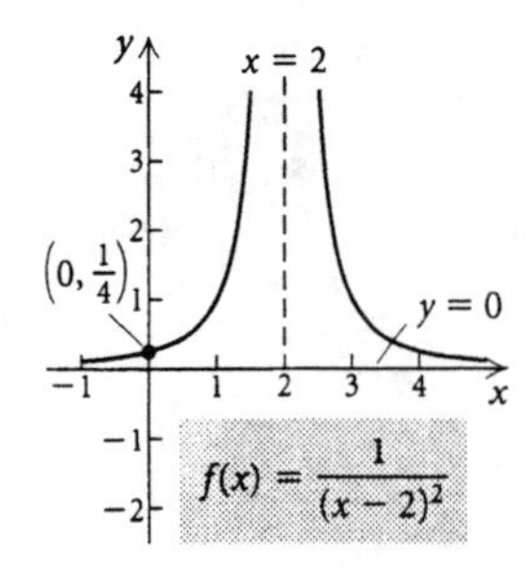

14. $f(x) = \dfrac{-2}{(x-3)^2}$

1. 3 is the zero of the denominator, so the domain is $(-\infty, 3) \cup (3, \infty)$ and $x = 3$ is the vertical asymptote.
2. Because the degree of the numerator is less than the degree of the denominator, the x-axis is the horizontal asymptote. There is no oblique asymptote.
3. The numerator has no zeros, so there is no x-intercept.
4. $f(0) = \dfrac{-2}{(0-3)^2} = -\dfrac{2}{9}$, so $\left(0, -\dfrac{2}{9}\right)$ is the y-intercept.
5. Find other function values to determine the shape of the graph and then draw it.

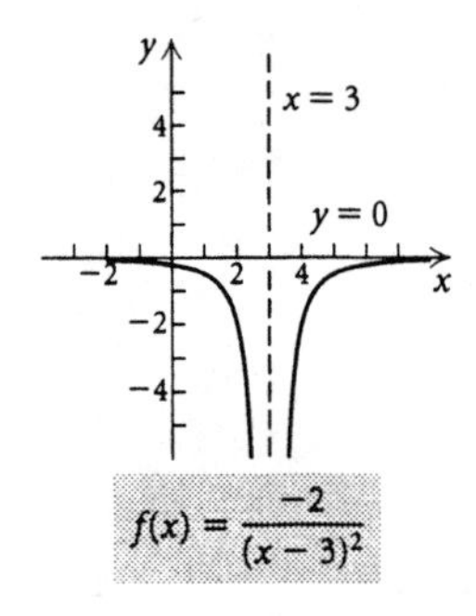

15. $f(x) = \dfrac{1}{x^2}$

1. 0 is the zero of the denominator, so the domain excludes 0. It is $(-\infty, 0)\cup(0, \infty)$. The line $x = 0$, or the y-axis, is the vertical asymptote.
2. Because the degree of the numerator is less than the degree of the denominator, the x-axis is the horizontal asymptote. There is no oblique asymptote.
3. The numerator has no zeros, so there is no x-intercept.
4. Since 0 is not in the domain of the function, there is no y-intercept.
5. Find other function values to determine the shape of the graph and then draw it.

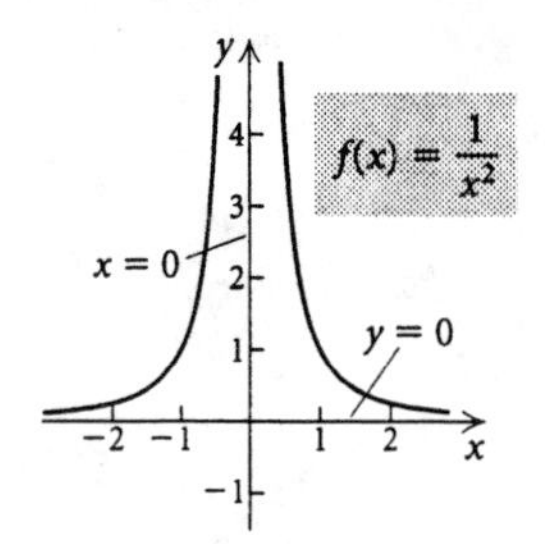

16. $f(x) = \dfrac{1}{3x^2}$

1. 0 is the zero of the denominator, so the domain is $(-\infty, 0)\cup(0, \infty)$. The line $x = 0$, or the y-axis, is the vertical asymptote.
2. Because the degree of the numerator is less than the degree of the denominator, the x-axis is the horizontal asymptote. There is no oblique asymptote.
3. The numerator has no zeros, so there is no x-intercept.
4. Since 0 is not in the domain of the function, there is no y-intercept.
5. Find other function values to determine the shape of the graph and then draw it.

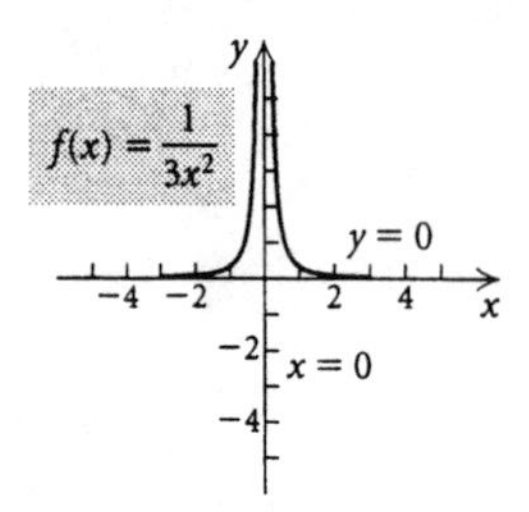

17. $f(x) = \dfrac{1}{x^2+3}$

1. The denominator has no real-number zeros, so the domain is the set of all real numbers and there is no vertical asymptote.
2. Because the degree of the numerator is less than the degree of the denominator, the x-axis is the horizontal asymptote. There is no oblique asymptote.
3. The numerator has no zeros, so there is no x-intercept.
4. $f(0) = \dfrac{1}{0^2+3} = \dfrac{1}{3}$, so $\left(0, \dfrac{1}{3}\right)$ is the y-intercept.
5. Find other function values to determine the shape of the graph and then draw it.

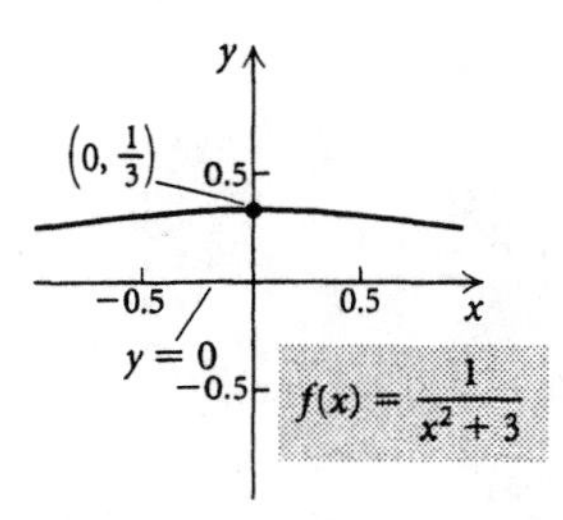

18. $f(x) = \dfrac{-1}{x^2+2}$

1. The denominator has no real-number zeros, so the domain is the set of all real numbers and there is no vertical asymptote.
2. Because the degree of the numerator is less than the degree of the denominator, the x-axis is the horizontal asymptote. There is no oblique asymptote.
3. The numerator has no zeros, so there is no x-intercept.
4. $f(0) = \dfrac{-1}{0^2+2} = -\dfrac{1}{2}$, so $\left(0, -\dfrac{1}{2}\right)$ is the y-intercept.
5. Find other function values to determine the shape of the graph and then draw it.

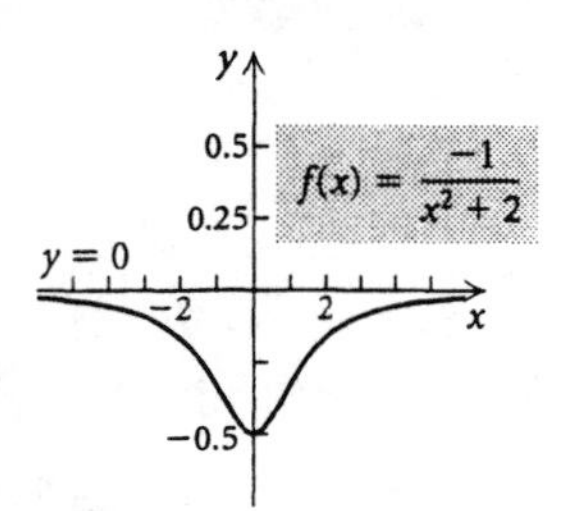

19. $f(x) = \dfrac{x^2-4}{x-2} = \dfrac{(x+2)(x-2)}{x-2} = x+2,\ x \neq 2$

The graph is the same as the graph of $f(x) = x+2$ except at $x = 2$, where there is a hole. The zero of $f(x) = x+2$ is -2, so the x-intercept is $(-2, 0)$; $f(0) = 2$, so the y-intercept is $(0, 2)$.

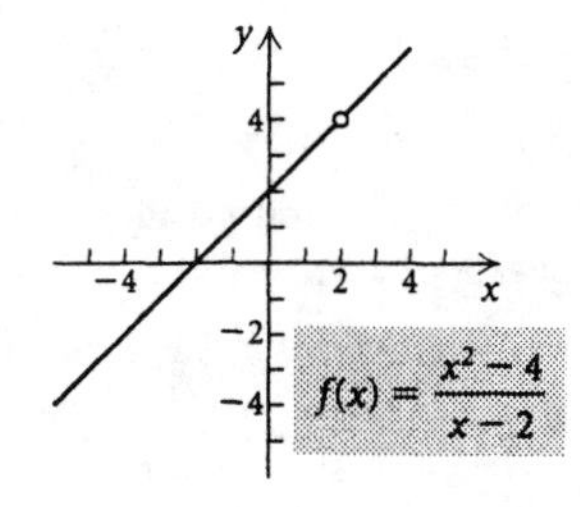

20. $f(x) = \dfrac{x^2-9}{x+3} = \dfrac{(x+3)(x-3)}{x+3} = x-3,\ x \neq -3$

The zero of $f(x) = x-3$ is 3, so the x-intercept is $(3, 0)$; $f(0) = -3$, so the y-intercept is $(0, -3)$.

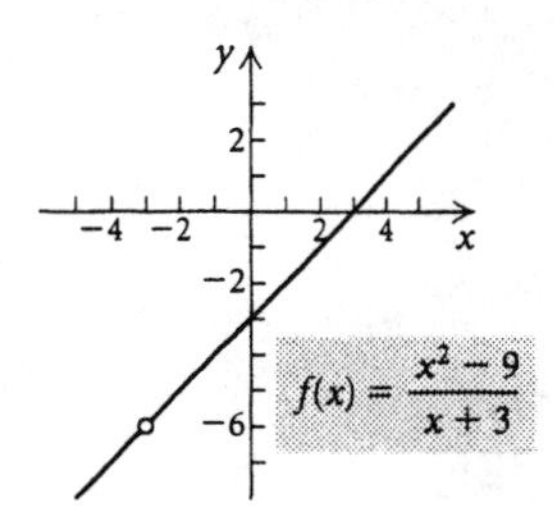

21. $f(x) = \dfrac{x-1}{x+2}$

1. -2 is the zero of the denominator, so the domain excludes -2. It is $(-\infty, -2) \cup (-2, \infty)$. The line $x = -2$ is the vertical asymptote.
2. The numerator and denominator have the same degree, so the horizontal asymptote is determined by the ratio of the leading coefficients, $1/1$, or 1. Thus, $y = 1$ is the horizontal asymptote. There is no oblique asymptote.
3. The zero of the numerator is 1, so the x-intercept is $(1, 0)$.
4. $f(0) = \dfrac{0-1}{0+2} = -\dfrac{1}{2}$, so $\left(0, -\dfrac{1}{2}\right)$ is the y-intercept.
5. Find other function values to determine the shape of the graph and then draw it.

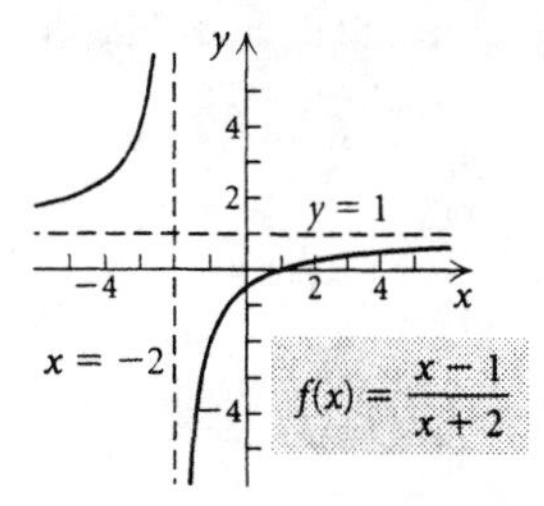

22. $f(x) = \dfrac{x-2}{x+1}$

1. -1 is the zero of the denominator, so the domain is $(-\infty, -1) \cup (-1, \infty)$ and $x = -1$ is the vertical asymptote.
2. The numerator and denominator have the same degree, so the horizontal asymptote is determined by the ratio of the leading coefficients, $1/1$, or 1. Thus, $y = 1$ is the horizontal asymptote. There is no oblique asymptote.
3. The zero of the numerator is 2, so the x-intercept is $(2, 0)$.
4. $f(0) = \dfrac{0-2}{0+1} = -2$, so $(0, -2)$ is the y-intercept.
5. Find other function values to determine the shape of the graph and then draw it.

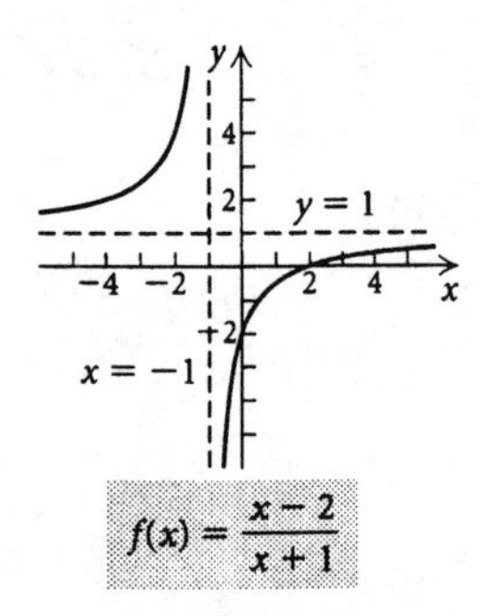

$f(x) = \dfrac{x-2}{x+1}$

23. $f(x) = \dfrac{x+3}{2x^2-5x-3}$

1. The zeros of the denominator are the solutions of $2x^2 - 5x - 3 = 0$. Since $2x^2 - 5x - 3 = (2x+1)(x-3)$, the zeros are $-\frac{1}{2}$ and 3. Thus, the domain is $\left(-\infty, -\frac{1}{2}\right) \cup \left(-\frac{1}{2}, 3\right) \cup (3, \infty)$ and the lines $x = -\frac{1}{2}$ and $x = 3$ are vertical asymptotes.
2. Because the degree of the numerator is less than the degree of the denominator, the x-axis is the horizontal asymptote. There is no oblique asymptote.
3. -3 is the zero of the numerator, so $(-3, 0)$ is the x-intercept.
4. $f(0) = \dfrac{0+3}{2 \cdot 0^2 - 5 \cdot 0 - 3} = -1$, so $(0, -1)$ is the y-intercept.
5. Find other function values to determine the shape of the graph and then draw it.

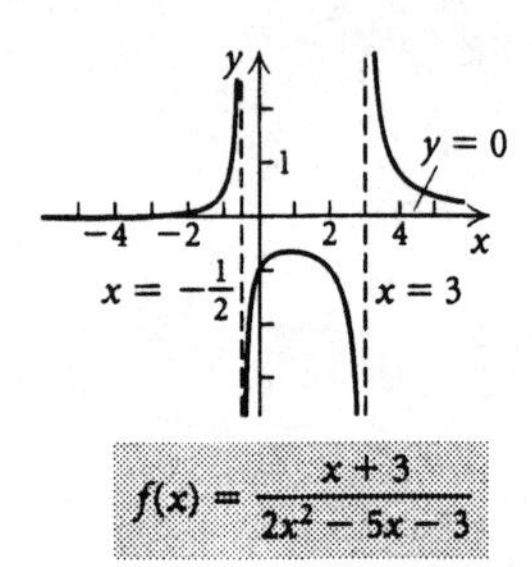

$f(x) = \dfrac{x+3}{2x^2-5x-3}$

24. $f(x) = \dfrac{3x}{x^2+5x+4}$

1. $x^2 + 5x + 4 = (x+4)(x+1)$, so the domain excludes -4 and -1. It is $(-\infty, -4) \cup (-4, -1) \cup (-1, \infty)$ and the lines $x = -4$ and $x = -1$ are vertical asymptotes.
2. Because the degree of the numerator is less than the degree of the denominator, the x-axis is the horizontal asymptote. There is no oblique asymptote.
3. 0 is the zero of the numerator, so $(0, 0)$ is the x-intercept.
4. From part (3) we see that $(0, 0)$ is the y-intercept.
5. Find other function values to determine the shape of the graph and then draw it.

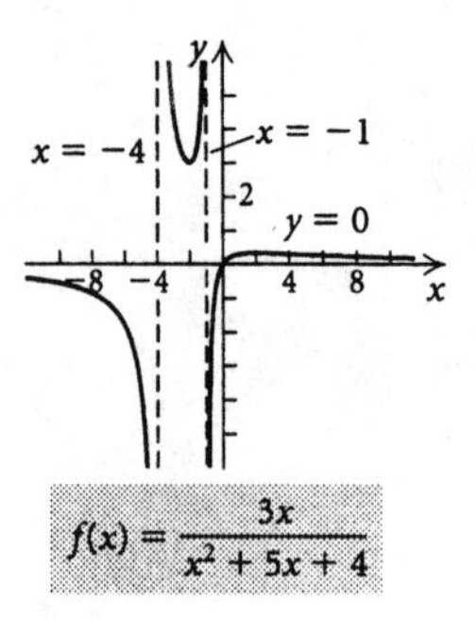

$f(x) = \dfrac{3x}{x^2+5x+4}$

25. $f(x) = \dfrac{x^2-9}{x+1}$

1. -1 is the zero of the denominator, so the domain excludes -1. It is $(-\infty, -1) \cup (-1, \infty)$. The line $x = -1$ is the vertical asymptote.
2. Because the degree of the numerator is one greater than the degree of the denominator, there is an oblique asymptote. Using division, we find that $\dfrac{x^2-9}{x+1} = x - 1 + \dfrac{-8}{x+1}$. As $|x|$ becomes very large, the graph of $f(x)$ gets close to the graph of $y = x - 1$. Thus, the line $y = x - 1$ is the oblique asymptote.
3. Since $x^2 - 9 = (x+3)(x-3)$, the zeros of the numerator are -3 and 3. Thus, the x-intercepts are $(-3, 0)$ and $(3, 0)$.
4. $f(0) = \dfrac{0^2-9}{0+1} = -9$, so $(0, -9)$ is the y-intercept.
5. Find other function values to determine the shape of the graph and then draw it.

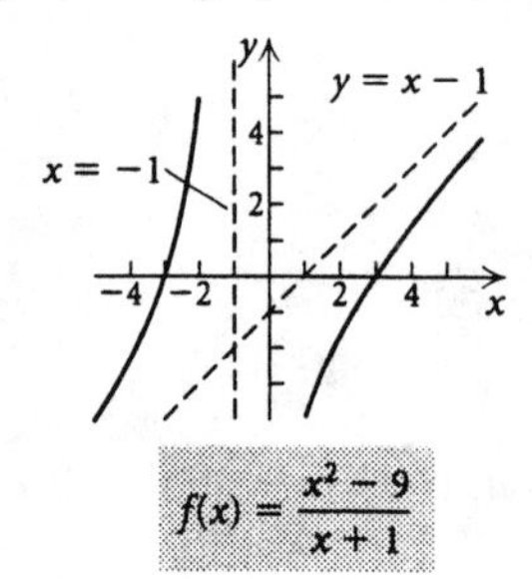

$f(x) = \dfrac{x^2-9}{x+1}$

26. $f(x) = \dfrac{x^2-4}{x-1}$

1. 1 is the zero of the denominator, so the domain is $(-\infty, 1) \cup (1, \infty)$ and $x = 1$ is the vertical asymptote.
2. $\dfrac{x^2-4}{x-1} = x + 1 + \dfrac{-3}{x-1}$, so $y = x + 1$ is the oblique asymptote.
3. $x^2 - 4 = (x+2)(x-2)$, so the x-intercepts are $(-2, 0)$ and $(2, 0)$.
4. $f(0) = \dfrac{0^2-4}{0-1} = 4$, so $(0, 4)$ is the y-intercept.
5. Find other function values to determine the shape of the graph and then draw it.

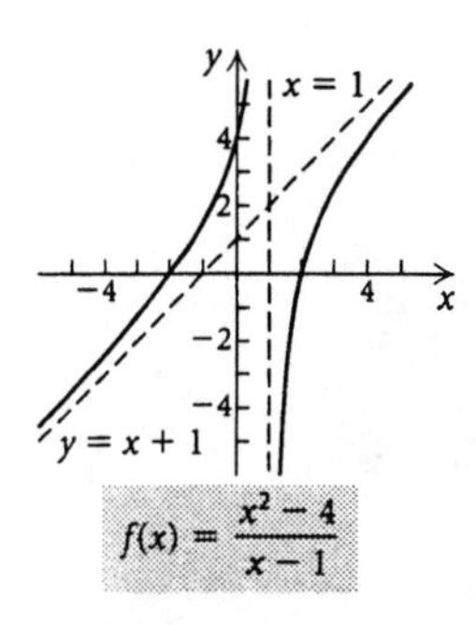

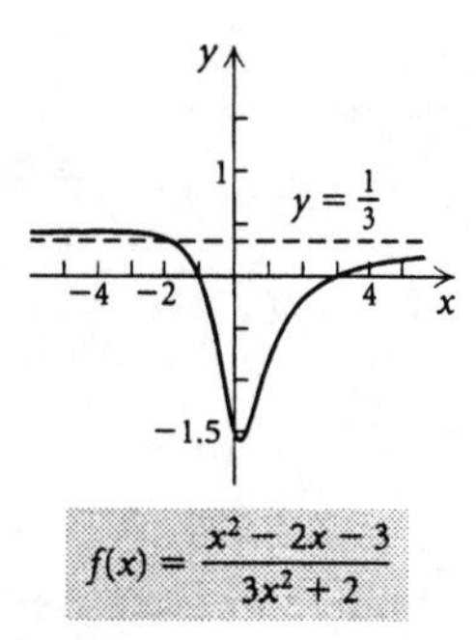

27. $f(x) = \dfrac{x^2 + x - 2}{2x^2 + 1}$

1. The denominator has no real-number zeros, so the domain is the set of all real numbers and there is no vertical asymptote.
2. The numerator and denominator have the same degree, so the horizontal asymptote is determined by the ratio of the leading coefficients, $1/2$. Thus, $y = 1/2$ is the horizontal asymptote. There is no oblique asymptote.
3. Since $x^2 + x - 2 = (x+2)(x-1)$, the zeros of the numerator are -2 and 1. Thus, the x-intercepts are $(-2, 0)$ and $(1, 0)$.
4. $f(0) = \dfrac{0^2 + 0 - 2}{2 \cdot 0^2 + 1} = -2$, so $(0, -2)$ is the y-intercept.
5. Find other function values to determine the shape of the graph and then draw it.

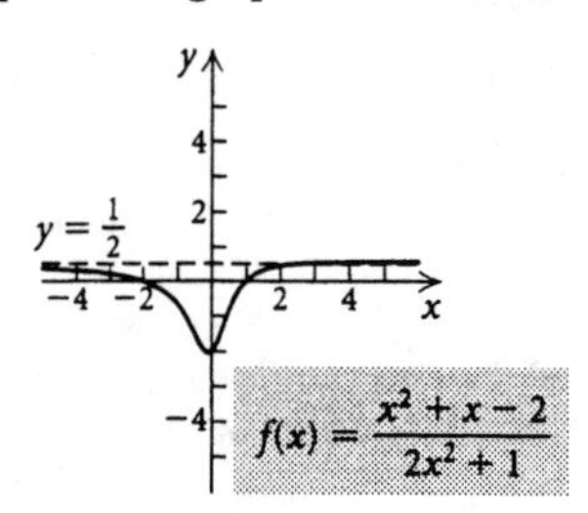

28. $f(x) = \dfrac{x^2 - 2x - 3}{3x^2 + 2}$

1. The denominator has no real-number zeros, so the domain is the set of all real numbers and there is no vertical asymptote.
2. The numerator and denominator have the same degree, so the horizontal asymptote is determined by the ratio of the leading coefficients, $1/3$. Thus, $y = 1/3$ is the horizontal asymptote. There is no oblique asymptote.
3. $x^2 - 2x - 3 = (x+1)(x-3)$, so the x-intercepts are $(-1, 0)$ and $(3, 0)$.
4. $f(0) = \dfrac{0^2 - 2 \cdot 0 - 3}{3 \cdot 0^2 + 2} = -\dfrac{3}{2}$, so $\left(0, -\dfrac{3}{2}\right)$ is the y-intercept.
5. Find other function values to determine the shape of the graph and then draw it.

29. $g(x) = \dfrac{3x^2 - x - 2}{x - 1} = \dfrac{(3x+2)(x-1)}{x-1} = 3x + 2$, $x \neq 1$

The graph is the same as the graph of $g(x) = 3x + 2$ except at $x = 1$, where there is a hole.

The zero of $g(x) = 3x + 2$ is $-\dfrac{2}{3}$, so the x-intercept is $\left(-\dfrac{2}{3}, 0\right)$; $g(0) = 2$, so the y-intercept is $(0, 2)$.

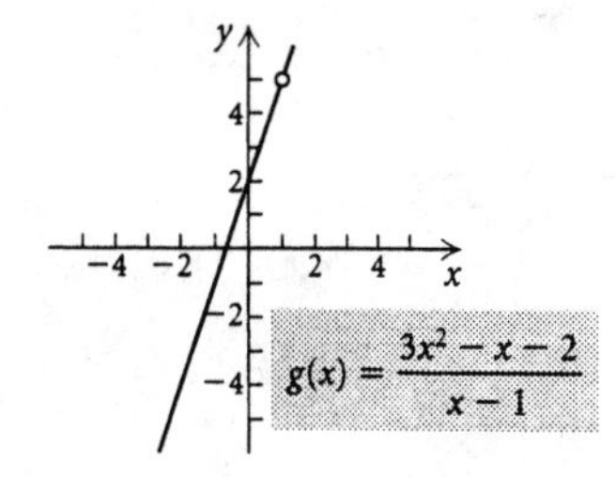

30. $f(x) = \dfrac{2x + 1}{2x^2 - 5x - 3} = \dfrac{2x + 1}{(2x+1)(x-3)} = \dfrac{1}{x - 3}$, $x \neq -\dfrac{1}{2}$

$f(x) = \dfrac{1}{x - 3}$ has no zero, so there is no x-intercept;

$f(0) = -\dfrac{1}{3}$, so the y-intercept is $\left(0, -\dfrac{1}{3}\right)$.

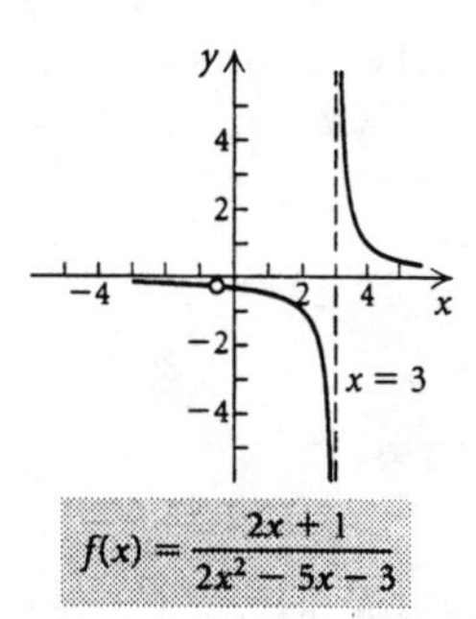

31. $f(x) = \dfrac{x - 1}{x^2 - 2x - 3}$

1. The zeros of the denominator are the solutions of
$x^2 - 2x - 3 = 0$. Since $x^2 - 2x - 3 = (x+1)(x-3)$, the zeros are -1 and 3. Thus, the domain is $(-\infty, -1) \cup (-1, 3) \cup (3, \infty)$ and the lines $x = -1$ and $y = 3$ are the vertical asymptotes.

2. Because the degree of the numerator is less than the degree of the denominator, the x-axis is the horizontal asymptote. There is no oblique asymptote.
3. 1 is the zero of the numerator, so $(1, 0)$ is the x-intercept.
4. $f(0) = \dfrac{0-1}{0^2 - 2\cdot 0 - 3} = \dfrac{1}{3}$, so $\left(0, \dfrac{1}{3}\right)$ is the y-intercept.
5. Find other function values to determine the shape of the graph and then draw it.

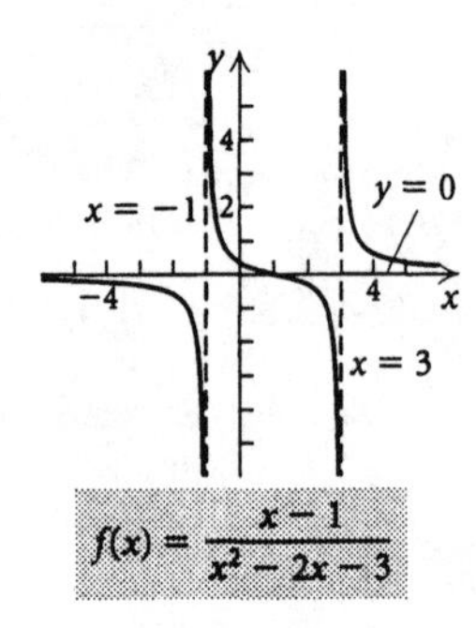

32. $f(x) = \dfrac{x+2}{x^2+2x-15}$

1. $x^2+2x-15 = (x+5)(x-3)$, so the domain excludes -5 and 3. It is $(-\infty, -5)\cup(-5, 3)\cup(3, \infty)$ and the lines $x = -5$ and $x = 3$ are vertical asymptotes.
2. Because the degree of the numerator is less than the degree of the denominator, the x-axis is the horizontal asymptote. There is no oblique asymptote.
3. -2 is the zero of the numerator, so $(-2, 0)$ is the x-intercept.
4. $f(0) = \dfrac{0+2}{0^2+2\cdot 0-15} = -\dfrac{2}{15}$, so $\left(0, -\dfrac{2}{15}\right)$ is the y-intercept.
5. Find other function values to determine the shape of the graph and then draw it.

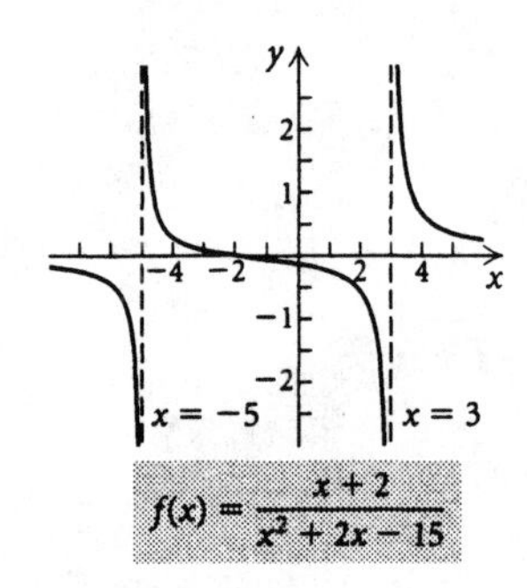

33. $f(x) = \dfrac{x-3}{(x+1)^3}$

1. -1 is the zero of the denominator, so the domain excludes -1. It is $(-\infty, -1)\cup(-1, \infty)$. The line $x = 1$ is the vertical asymptote.
2. Because the degree of the numerator is less than the degree of the denominator, the x-axis is the horizontal asymptote. There is no oblique asymptote.
3. 3 is the zero of the numerator, so $(3, 0)$ is the x-intercept.
4. $f(0) = \dfrac{0-3}{(0+1)^3} = -3$, so $(0, -3)$ is the y-intercept.
5. Find other function values to determine the shape of the graph and then draw it.

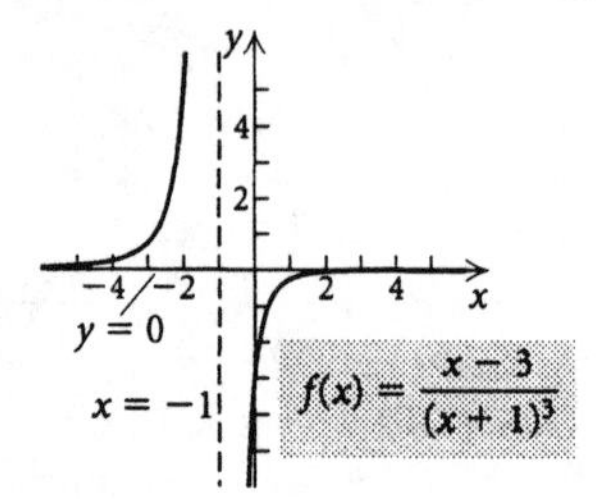

34. $f(x) = \dfrac{x+2}{(x-1)^3}$

1. 1 is the zero of the denominator, so the domain is $(-\infty, 1)\cup(1, \infty)$ and $x = 1$ is the vertical asymptote.
2. Because the degree of the numerator is less than the degree of the denominator, the x-axis is the horizontal asymptote. There is no oblique asymptote.
3. -2 is the zero of the numerator, so $(-2, 0)$ is the x-intercept.
4. $f(0) = \dfrac{0+2}{(0-1)^3} = -2$, so $(0, -2)$, is the y-intercept.
5. Find other function values to determine the shape of the graph and then draw it.

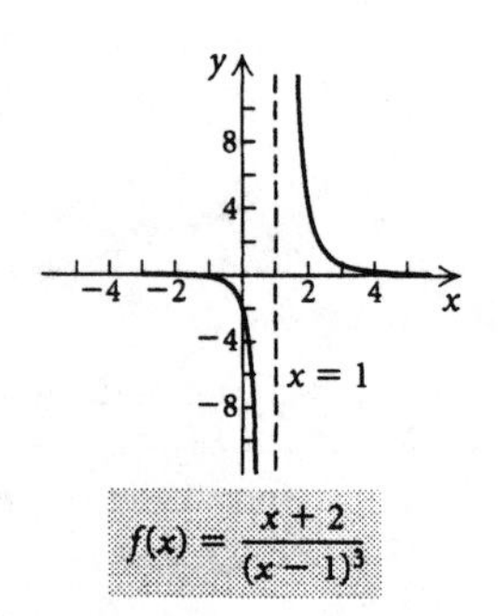

35. $f(x) = \dfrac{x^3+1}{x}$

1. 0 is the zero of the denominator, so the domain excludes 0. It is $(-\infty, 0)\cup(0, \infty)$. The line $x = 0$, or the y-axis, is the vertical asymptote.
2. Because the degree of the numerator is more than one greater than the degree of the denominator, there is no horizontal or oblique asymptote.

3. The real-number zero of the numerator is -1, so the x-intercept is $(-1, 0)$.
4. Since 0 is not in the domain of the function, there is no y-intercept.
5. Find other function values to determine the shape of the graph and then draw it.

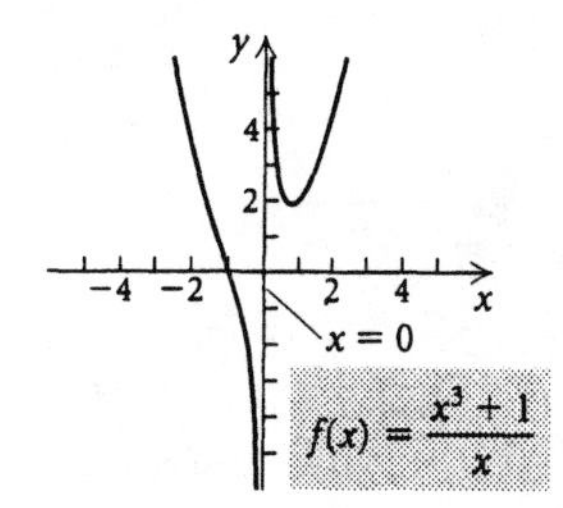

36. $f(x) = \dfrac{x^3 - 1}{x}$

1. 0 is the zero of the denominator, so the domain excludes 0. It is $(-\infty, 0) \cup (0, \infty)$. The line $x = 0$, or the y-axis, is the vertical asymptote.
2. Because the degree of the numerator is more than one greater than the degree of the denominator, there is no horizontal or oblique asymptote.
3. The real-number zero of the numerator is 1, so the x-intercept is $(1, 0)$.
4. Since 0 is not in the domain of the function, there is no y-intercept.
5. Find other function values to determine the shape of the graph and then draw it.

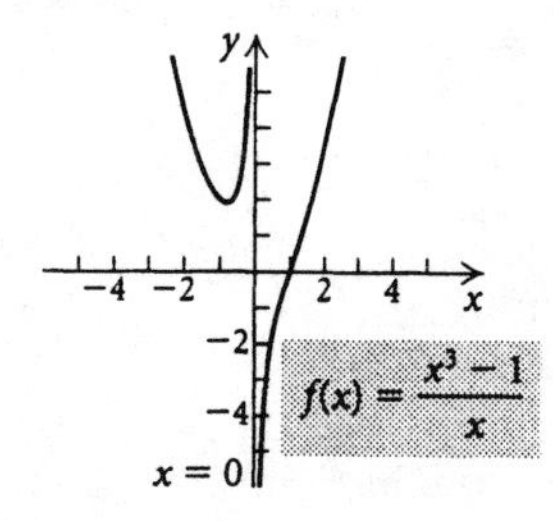

37. $f(x) = \dfrac{x^3 + 2x^2 - 15x}{x^2 - 5x - 14}$

1. The zeros of the denominator are the solutions of $x^2 - 5x - 14 = 0$. Since $x^2 - 5x - 14 = (x+2)(x-7)$, the zeros are -2 and 7. Thus, the domain is $(-\infty, -2) \cup (-2, 7) \cup (7, \infty)$ and the lines $x = -2$ and $x = 7$ are the vertical asymptotes.
2. Because the degree of the numerator is one greater than the degree of the denominator, there is an oblique asymptote. Using division, we find that $\dfrac{x^3 + 2x^2 - 15x}{x^2 - 5x - 14} = x + 7 + \dfrac{34x + 98}{x^2 - 5x - 14}$. As $|x|$ becomes very large, the graph of $f(x)$ gets close to the graph of $y = x + 7$. Thus, the line $y = x + 7$ is the oblique asymptote.
3. The zeros of the numerator are the solutions of $x^3 + 2x^2 - 15x = 0$. Since $x^3 + 2x^2 - 15x = x(x+5)(x-3)$, the zeros are 0, -5, and 3. Thus, the x-intercepts are $(-5, 0)$, $(0, 0)$, and $(3, 0)$.
4. From part (3) we see that $(0, 0)$ is the y-intercept.
5. Find other function values to determine the shape of the graph and then draw it.

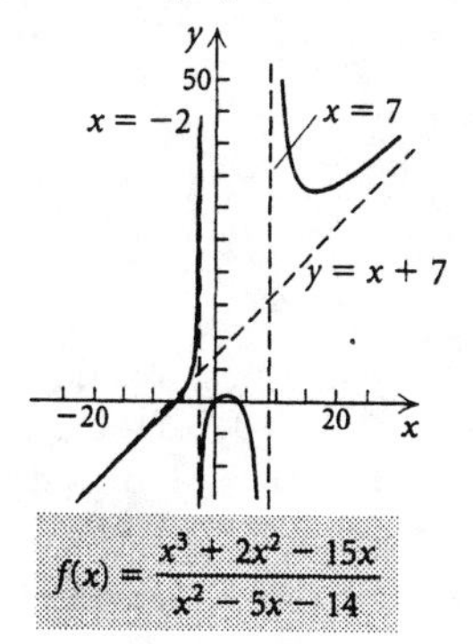

38. $f(x) = \dfrac{x^3 + 2x^2 - 3x}{x^2 - 25}$

1. The zeros of the denominator are -5 and 5, so the domain excludes -5 and 5. It is $(-\infty, -5) \cup (-5, 5) \cup (5, \infty)$ and the lines $x = -5$ and $x = 5$ are the vertical asymptotes.
2. $\dfrac{x^3 + 2x^2 - 3x}{x^2 - 25} = x + 2 + \dfrac{-28x + 50}{x^2 - 25}$, so $y = x + 2$ is the oblique asymptote.
3. $x^3 + 2x^2 - 3x = x(x+3)(x-1)$, so the x-intercepts are $(0, 0)$, $(-3, 0)$, and $(1, 0)$.
4. From part (3) we see that $(0, 0)$ is the y-intercept.
5. Find other function values to determine the shape of the graph and then draw it.

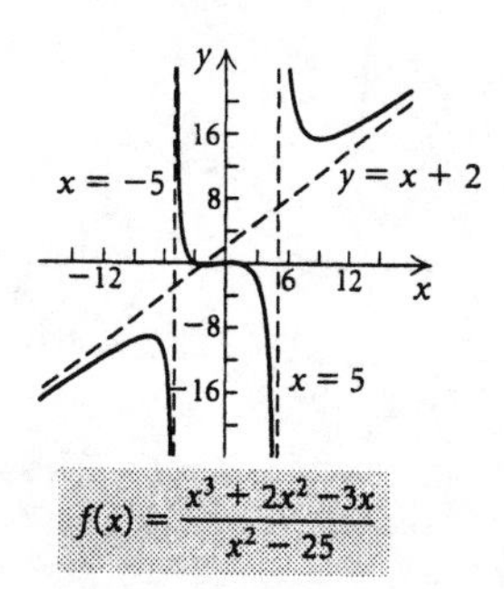

39. $f(x) = \dfrac{5x^4}{x^4 + 1}$

1. The denominator has no real-number zeros, so the domain is the set of all real numbers and there is no vertical asymptote.
2. The numerator and denominator have the same degree, so the horizontal asymptote is determined by the ratio of the leading coefficients, 5/1, or 5. Thus, $y = 5$ is the horizontal asymptote. There is no oblique asymptote.
3. The zero of the numerator is 0, so $(0, 0)$ is the x-intercept.

4. From part (3) we see that $(0, 0)$ is the y-intercept.
5. Find other function values to determine the shape of the graph and then draw it.

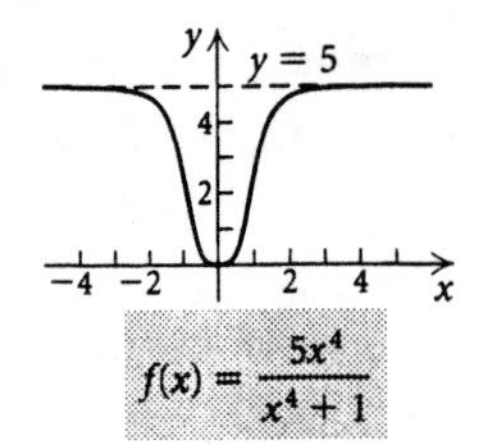

40. $f(x) = \dfrac{x+1}{x^2+x-6}$

1. $x^2+x-6 = (x+3)(x-2)$, so the domain excludes -3 and 2. It is $(-\infty, -3) \cup (-3, 2) \cup (2, \infty)$ and the lines $x = -3$ and $x = 2$ are vertical asymptotes.
2. Because the degree of the numerator is less than the degree of the denominator, the x-axis is the horizontal asymptote. There is no oblique asymptote.
3. The zero of the numerator is -1, so the x-intercept is $(-1, 0)$.
4. $f(0) = \dfrac{0+1}{0^2+0-6} = -\dfrac{1}{6}$, so $\left(0, -\dfrac{1}{6}\right)$ is the y-intercept.
5. Find other function values to determine the shape of the graph and then draw it.

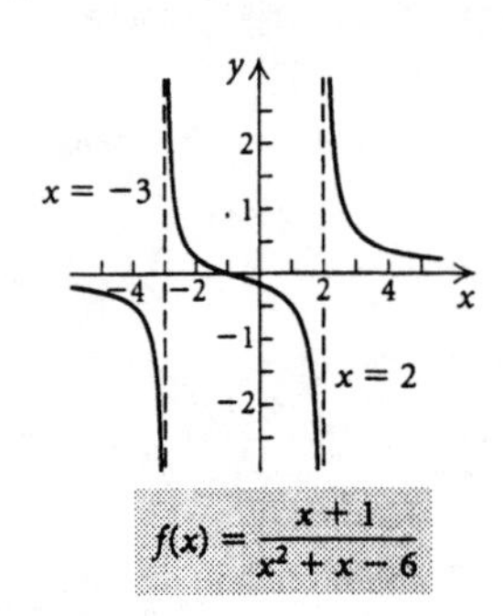

41. $f(x) = \dfrac{x^2}{x^2-x-2}$

1. The zeros of the denominator are the solutions of $x^2-x-2=0$. Since $x^2-x-2=(x+1)(x-2)$, the zeros are -1 and 2. Thus, the domain is $(-\infty, -1) \cup (-1, 2) \cup (2, \infty)$ and the lines $x = -1$ and $x = 2$ are the vertical asymptotes.
2. The numerator and denominator have the same degree, so the horizontal asymptote is determined by the ratio of the leading coefficients, $1/1$, or 1. Thus, $y = 1$ is the horizontal asymptote. There is no oblique asymptote.
3. The zero of the numerator is 0, so the x-intercept is $(0, 0)$.
4. From part (3) we see that $(0, 0)$ is the y-intercept.
5. Find other function values to determine the shape of the graph and then draw it.

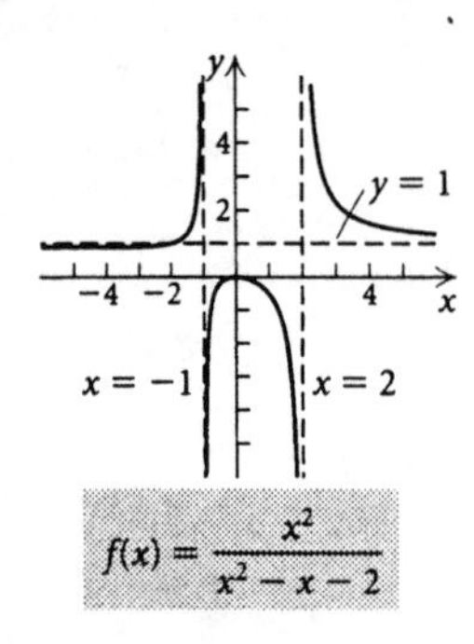

42. $f(x) = \dfrac{x^2-x-2}{x+2}$

1. -2 is the zero of the denominator, so the domain is $(-\infty, -2) \cup (-2, \infty)$ and the line $x = -2$ is the vertical asymptote.
2. $\dfrac{x^2-x-2}{x+2} = x - 3 + \dfrac{4}{x+2}$, so $y = x - 3$ is the oblique asymptote.
3. $x^2 - x - 2 = (x+1)(x-2)$, so the x-intercepts are $(-1, 0)$ and $(2, 0)$.
4. $f(0) = \dfrac{0^2-0-2}{0+2} = -1$, so $(0, -1)$ is the y-intercept.
5. Find other function values to determine the shape of the graph and then draw it.

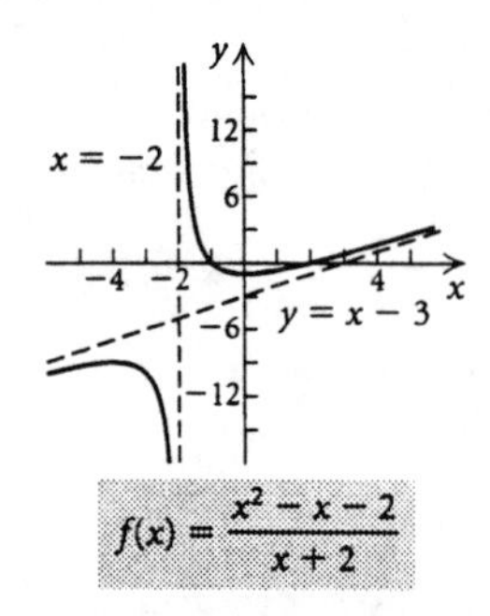

43. Answers may vary. The numbers -4 and 5 must be zeros of the denominator. A function that satisfies these conditions is

$f(x) = \dfrac{1}{(x+4)(x-5)}$, or $f(x) = \dfrac{1}{x^2-x-20}$.

44. Answers may vary. The numbers -4 and 5 must be zeros of the denominator and -2 must be a zero of the numerator.

$$f(x) = \frac{x+2}{(x+4)(x-5)}, \text{ or } f(x) = \frac{x+2}{x^2-x-20}.$$

45. Answers may vary. The numbers -4 and 5 must be zeros of the denominator and -2 must be a zero of the numerator. In addition, the numerator and denominator must have the same degree and the ratio of their leading coefficients must be $3/2$. A function that satisfies these conditions is

$f(x) = \dfrac{3x(x+2)}{2(x+4)(x-5)}$, or $f(x) = \dfrac{3x^2+6x}{2x^2-2x-40}$.

Another function that satisfies these conditions is $g(x) = \frac{3(x+2)^2}{2(x+4)(x-5)}$, or $g(x) = \frac{3x^2+12x+12}{2x^2-2x-40}$.

46. Answers may vary. The degree of the numerator must be 1 greater than the degree of the denominator and the quotient, when long division is performed, must be $x-1$. If we let the remainder be 1, a function that satisfies these conditions is $f(x) = x - 1 + \frac{1}{x}$, or $f(x) = \frac{x^2-x+1}{x}$.

47. a)

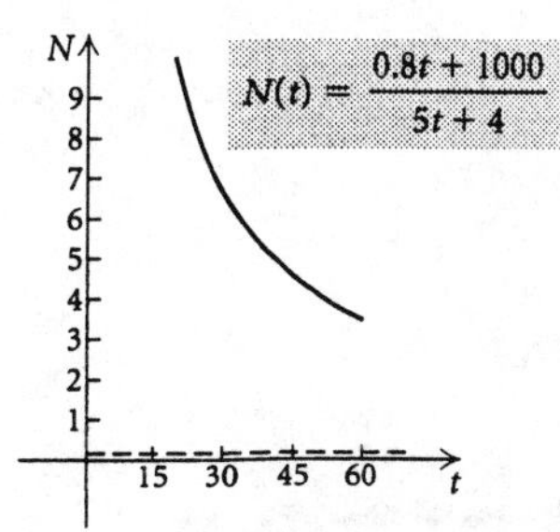

The horizontal asymptote of $N(t)$ is the ratio of the leading coefficients of the numerator and denominator, 0.8/5, or 0.16. Thus, $N(t) \to 0.16$ as $t \to \infty$.

b) The medication never completely disappears from the body; a trace amount remains.

48. a)

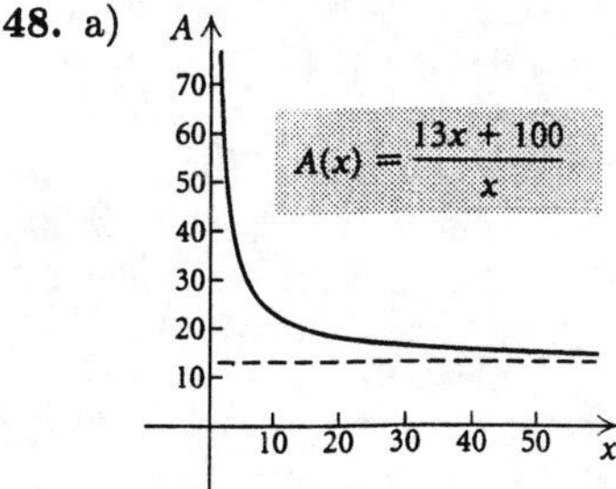

$A(x) \to 13/1$, or 13 as $x \to \infty$.

b) As more videotapes are produced, the average cost approaches \$13.

49. a)

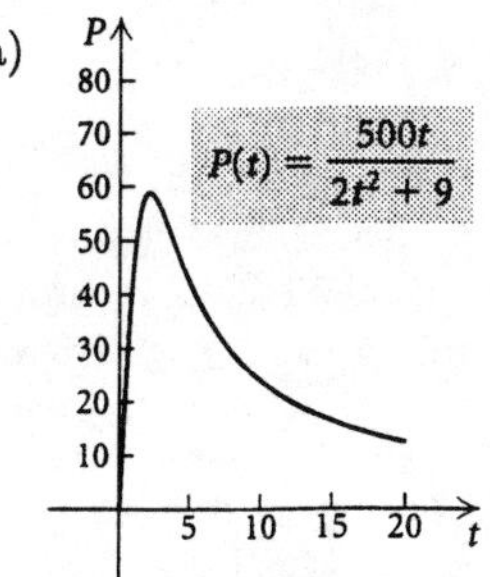

b) Use the table feature of a grapher (in ASK mode) or substitution.

$P(0) = 0$; $P(1) = 45.455$ thousand, or 45,455;

$P(3) = 55.556$ thousand, or 55,556;

$P(8) = 29.197$ thousand, or 29,197

c) The degree of the numerator is less than the degree of the denominator, so the x-axis is the horizontal asymptote. Thus, $P(t) \to 0$ as $t \to \infty$.

d) Eventually, no one will live in Lordsburg.

e) Using the Maximum feature of a grapher, we find that the maximum value is $f(2.1213) = 58.926$ thousand, or 58,926.

50. a) $S = $ area of base $+ 4 \cdot$ area of a side

$= x^2 + 4xy$

Now we express y in terms of x:

Volume $= 108 = x^2y$

$\frac{108}{x^2} = y$

Thus, $S = x^2 + 4x\left(\frac{108}{x^2}\right)$, or

$S = x^2 + \frac{432}{x}$.

b)

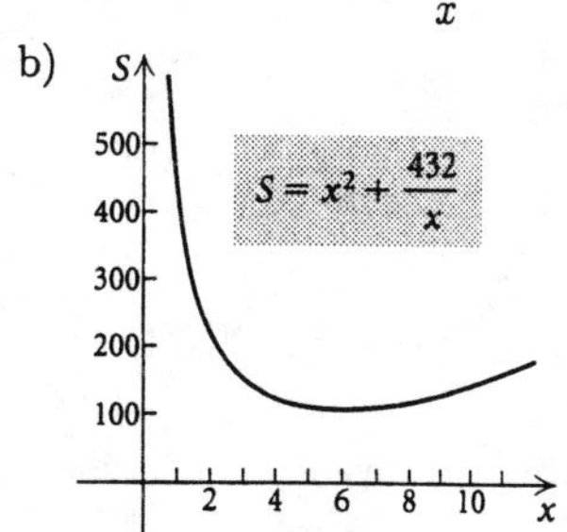

c) Using the Minimum feature on a grapher, we estimate that the minimum surface area is 108 cm^2. This occurs when $x = 6$ cm.

51. The denominator has no real-number zeros.

52. A horizontal asymptote occurs when the degree of the numerator of a rational function is less than or equal to the degree of the denominator. An oblique asymptote occurs when the degree of the numerator is 1 greater than the degree of the denominator. Thus, a rational function cannot have both a horizontal asymptote and an oblique asymptote.

53. $y_1 = \frac{x^3+4}{x} = x^2 + \frac{4}{x}$

As $|x| \to \infty$, $\frac{4}{x} \to 0$ and the value of $y_1 \to x^2$. Thus, the parabola $y_2 = x^2$ can be thought of as a nonlinear asymptote for y_1. The graph confirms this.

54.

$y = kx$

$22 = k \cdot \frac{1}{2}$

$44 = k$

$y = 44x$ Equation of variation

55. $y = \frac{k}{x}$

$\frac{2}{5} = \frac{k}{10}$ Substituting

$4 = k$ Variation constant

$y = \frac{4}{x}$ Equation of variation

56. $y = \frac{kxz^2}{w}$

$0.1 = \frac{k \cdot 10,000 \cdot 10^2}{5}$

$0.0000005 = k$

$y = \frac{0.0000005xz^2}{w}$ Equation of variation

57. $T = \frac{k}{P}$

$6 = \frac{k}{8}$ Substituting

$48 = k$ Variation constant

$T = \frac{48}{P}$ Equation of variation

$T = \frac{48}{12}$ Substituting

$T = 4$

It will take 4 hr for 12 employees to make the calls.

58. $f(x) = \frac{x^4 + 3x^2}{x^2 + 1} = x^2 + 2 + \frac{-2}{x^2 + 1}$

As $|x| \to \infty$, $\frac{-2}{x^2+1} \to 0$ and the value of $f(x) \to x^2 + 2$. Thus, the nonlinear asymptote is $y = x^2 + 2$.

59. $f(x) = \frac{x^5 + 2x^3 + 4x^2}{x^2 + 2} = x^3 + 4 + \frac{-8}{x^2 + 2}$

As $|x| \to \infty$, $\frac{-8}{x^2+2} \to 0$ and the value of $f(x) \to x^3 + 4$. Thus, the nonlinear asymptote is $y = x^3 + 4$.

60.

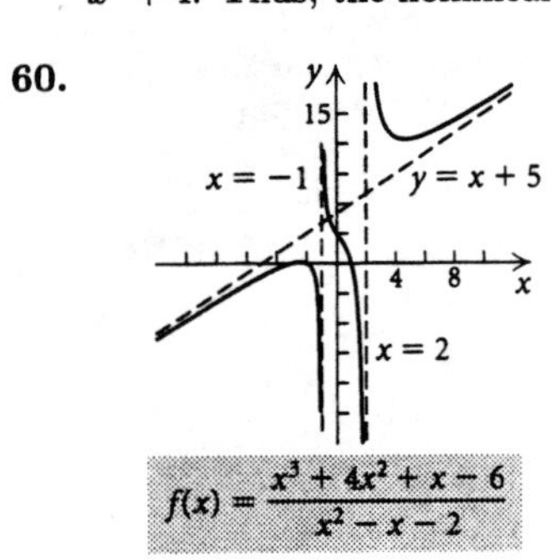

61.

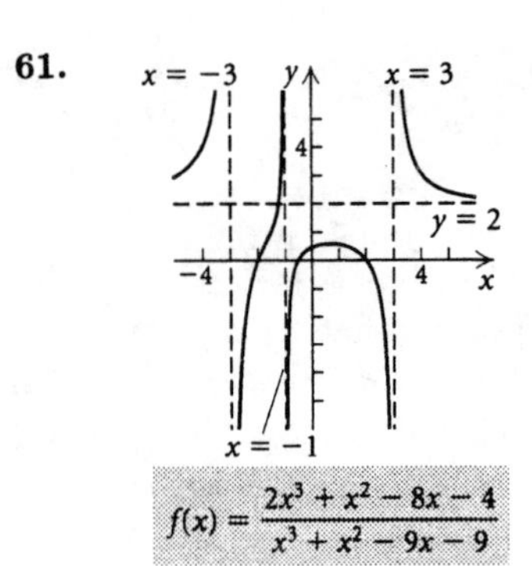

62.

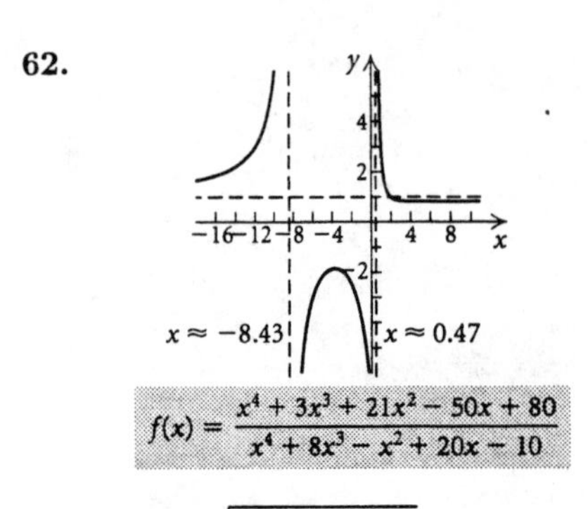

63. $f(x) = \sqrt{\frac{72}{x^2 - 4x - 21}}$

The radicand must be nonnegative and the denominator must be nonzero. Thus, the values of x for which $x^2 - 4x - 21 > 0$ comprise the domain. By inspecting the graph of $y = x^2 - 4x - 21$ we see that the domain is $\{x|x < -3 \text{ or } x > 7\}$, or $(-\infty, -3) \cup (7, \infty)$.

64. $f(x) = \sqrt{x^2 - 4x - 21}$

The radicand must be nonnegative. By inspecting the graph of $y = x^2 - 4x - 21$ we see that the domain is $\{x|x \le -3 \text{ or } x \ge 7\}$, or $(-\infty, -3] \cup [7, \infty)$.

Exercise Set 3.5

1. First we find an equivalent inequality with 0 on one side.

$$x^3 + 6x^2 < x + 30$$

$$x^3 + 6x^2 - x - 30 < 30$$

From the graph we see that the x-intercepts of the related function occur at $x = -5$, $x = -3$, and $x = 2$. They divide the x-axis into the intervals $(-\infty, -5)$, $(-5, -3)$, $(-3, 2)$, and $(2, \infty)$. From the graph we see that the function has negative values only on $(-\infty, -5)$ and $(-3, 2)$. Thus, the solution set is $(-\infty, -5) \cup (-3, 2)$.

2. From the graph we see that the x-intercepts of the related function occur at $x = -4$, $x = -3$, $x = 2$, and $x = 5$. They divide the x-axis into the intervals $(-\infty, -4)$, $(-4, -3)$, $(-3, 2)$, $(2, 5)$, and $(5, \infty)$. The function has positive values only on $(-\infty, -4)$, $(-3, 2)$, and $(5, \infty)$. Since the inequality symbol is $\ge$, the endpoints of the intervals are included in the solution set. It is $(-\infty, -4] \cup [-3, 2] \cup [5, \infty)$.

3. By observing the graph or the denomination of the function, we see that the function is not defined for $x = -2$ or $x = 2$. We also see that 0 is a zero of the function. These numbers divide the x-axis into the intervals $(-\infty, -2)$, $(-2, 0)$, $(0, 2)$, and $(2, \infty)$. From the graph we see that the function has positive values only on $(-2, 0)$ and $(2, \infty)$. Since the inequality symbol is $\geq$, 0 must be included in the solution set. It is $(-2, 0] \cup (2, \infty)$.

4. By observing the graph or the denomination of the function, we see that the function is not defined for $x = -2$ or $x = 2$. We also see that the function has no zeros. Thus, the number line is divided into the intervals $(-\infty, -2)$, $(-2, 2)$, and $(2, \infty)$. From the graph we see that the function has negative values only on $(-2, 2)$. Thus, the solution set is $(-2, 2)$.

5. $(x-1)(x+4) < 0$

   The related equation is $(x-1)(x+4) = 0$. Using the principle of zero products or by observing the graph of $y = (x-1)(x+4)$, we see that the solutions of the related equation are 1 and -4. These numbers divide the x-axis into the intervals $(-\infty, -4)$, $(-4, 1)$, and $(1, \infty)$. We let $f(x) = (x-1)(x+4)$ and test a value in each interval.

   $(-\infty, -4)$: $f(-5) = 6 > 0$

   $(-4, 1)$: $f(0) = -4 < 0$

   $(1, \infty)$: $f(2) = 6 > 0$

   Function values are negative only in the interval $(-4, 1)$. The graph of $y = (x-1)(x+4)$ can also be used to determine this. The solution set is $(-4, 1)$.

6. $(x+3)(x-5) < 0$

   The related equation is $(x+3)(x-5) = 0$. Its solutions are -3 and 5. These numbers divide the x-axis into the intervals $(-\infty, -3)$, $(-3, 5)$, and $(5, \infty)$. Let $f(x) =$
   $(x+3)(x-5)$ and test a value in each interval.

   $(-\infty, -3)$: $f(-4) = 9 > 0$

   $(-3, 5)$: $f(0) = -15 < 0$

   $(5, \infty)$: $f(6) = 9 > 0$

   Function values are negative only in the interval $(-3, 5)$. The graph of $y = (x+3)(x-5)$ can also be used to determine this. The solution set is $(-3, 5)$.

7. $(x-4)(x+2) \geq 0$

   The related equation is $(x-4)(x+2) = 0$. Using the principle of zero products or by observing the graph of $y = (x-4)(x+2)$, we see that the solutions of the related equation are 4 and -2. These numbers divide the x-axis into the intervals $(-\infty, -2)$, $(-2, 4)$, and $(4, \infty)$. We let $f(x) = (x-4)(x+2)$ and test a value in each interval.

   $(-\infty, -2)$: $f(-3) = 7 > 0$

   $(-2, 4)$: $f(0) = -8 < 0$

   $(4, \infty)$: $f(5) = 7 > 0$

   Function values are positive on $(-\infty, -2)$ and $(4, \infty)$. The graph of $y = (x-4)(x+2)$ can also be used to determine this. Since the inequality symbol is $\geq$, the endpoints of the intervals must be included in the solution set. It is $(-\infty, -2] \cup [4, \infty)$.

8. $(x-2)(x+1) \geq 0$

   The related equation is $(x-2)(x+1) = 0$. Its solutions are 2 and -1. These numbers divide the x-axis into the intervals $(-\infty, -1)$, $(-1, 2)$, and $(2, \infty)$. We let $f(x) = (x-2)(x+1)$ and test a value in each interval.

   $(-\infty, -1)$: $f(-2) = 4 > 0$

   $(-1, 2)$: $f(0) = -2 < 0$

   $(2, \infty)$: $f(3) = 4 > 0$

   Function values are positive on $(-\infty, -1)$ and $(2, \infty)$. The graph of $y = (x-2)(x+1)$ can also be used to determine this. Since the inequality symbol is $\geq$, the endpoints of the intervals must be included in the solution set. It is $(-\infty, -1] \cup [2, \infty)$.

9. $x^2 + x - 2 > 0$ Polynomial inequality

   $x^2 + x - 2 = 0$ Related equation

   $(x+2)(x-1) = 0$ Factoring

   Using the principle of zero products or by observing the graph of $y = x^2 + x - 2$, we see that the solutions of the related equation are -2 and 1. These numbers divide the x-axis into the intervals $(-\infty, -2)$, $(-2, 1)$, and $(1, \infty)$. We let $f(x) = x^2 + x - 2$ and test a value in each interval.

   $(-\infty, -2)$: $f(-3) = 4 > 0$

   $(-2, 1)$: $f(0) = -2 < 0$

   $(1, \infty)$: $f(2) = 4 > 0$

   Function values are positive on $(-\infty, -2)$ and $(1, \infty)$. The graph of $y = x^2 + x - 2$ can also be used to determine this. The solution set is $(-\infty, -2) \cup (1, \infty)$.

10. $x^2 - x - 6 > 0$ Polynomial inequality

    $x^2 - x - 6 = 0$ Related equation

    $(x-3)(x+2) = 0$ Factoring

    The solutions of the related equation are 3 and -2. These numbers divide the x-axis into the intervals $(-\infty, -2)$, $(-2, 3)$, and $(3, \infty)$. We let $f(x) = x^2 - x - 6$ and test a value in each interval.

    $(-\infty, -2)$: $f(-3) = 6 > 0$

    $(-2, 3)$: $f(0) = -6 < 0$

    $(3, \infty)$: $f(4) = 6 > 0$

    Function values are positive on $(-\infty, -2)$ and $(3, \infty)$. The graph of $y = x^2 - x - 6$ can also be used to determine this. The solution set is $(-\infty, -2) \cup (3, \infty)$.

11. $x^2 > 25$ Polynomial inequality

    $x^2 - 25 > 0$ Equivalent inequality with 0 on one side

    $x^2 - 25 = 0$ Related equation

    $(x+5)(x-5) = 0$ Factoring

Using the principle of zero products or by observing the graph of $y = x^2 - 25$, we see that the solutions of the related equation are -5 and 5. These numbers divide the x-axis into the intervals $(-\infty, -5)$, $(-5, 5)$, and $(5, \infty)$. We let $f(x) = x^2 - 25$ and test a value in each interval.

$(-\infty, -5)$: $f(-6) = 11 > 0$

$(-5, 5)$: $f(0) = -25 < 0$

$(5, \infty)$: $f(6) = 11 > 0$

Function values are positive on $(-\infty, -5)$ and $(5, \infty)$. The graph of $y = x^2 - 25$ can also be used to determine this. The solution set is $(-\infty, -5) \cup (5, \infty)$.

12.
$x^2 \le 1$ Polynomial inequality
$x^2 - 1 \le 0$ Equivalent inequality with 0 on one side
$x^2 - 1 = 0$ Related equation
$(x+1)(x-1) = 0$ Factoring

The solutions of the related equation are -1 and 1. These numbers divide the x-axis into the intervals $(-\infty, -1)$, $(-1, 1)$, and $(1, \infty)$. We let $f(x) = x^2 - 1$ and test a value in each interval.

$(-\infty, -1)$: $f(-2) = 3 > 0$

$(-1, 1)$: $f(0) = -1 < 0$

$(1, \infty)$: $f(2) = 3 > 0$

Function values are negative only on $(-1, 1)$. The graph of $y = x^2 - 1$ can also be used to determine this. Since the inequality symbol is $\le$, the endpoints of the interval must be included in the solution set. It is $[-1, 1]$.

13.
$4 - x^2 \le 0$ Polynomial inequality
$4 - x^2 = 0$ Related equation
$(2+x)(2-x) = 0$ Factoring

Using the principle of zero products or by observing the graph of $y = 4 - x^2$, we see that the solutions of the related equation are -2 and 2. These numbers divide the x-axis into the intervals $(-\infty, -2)$, $(-2, 2)$, and $(2, \infty)$. We let $f(x) = 4 - x^2$ and test a value in each interval.

$(-\infty, -2)$: $f(-3) = -5 < 0$

$(-2, 2)$: $f(0) = 4 > 0$

$(2, \infty)$: $f(3) = -5 < 0$

Function values are negative on $(-\infty, -2)$ and $(2, \infty)$. The graph of $y = 4 - x^2$ can also be used to determine this. Since the inequality symbol is $\le$, the endpoints of the intervals must be included in the solution set. It is $(-\infty, -2] \cup [2, \infty)$.

14. $11 - x^2 \ge 0$ Polynomial inequality
$11 - x^2 = 0$ Related equation

The solutions of the related equation are $\pm\sqrt{11}$. These numbers divide the x-axis into the intervals $(-\infty, -\sqrt{11})$, $(-\sqrt{11}, \sqrt{11})$, and $(\sqrt{11}, \infty)$. We let $f(x) = 11 - x^2$ and test a value in each interval.

$(-\infty, -\sqrt{11})$: $f(-4) = -5 < 0$

$(-\sqrt{11}, \sqrt{11})$: $f(0) = 11 > 0$

$(\sqrt{11}, \infty)$: $f(4) = -5 < 0$

Function values are positive only on $(-\sqrt{11}, \sqrt{11})$. The graph of $y = 11 - x^2$ can also be used to determine this. Since the inequality symbol is $\ge$, the endpoints of the interval must be included in the solution set. It is $[-\sqrt{11}, \sqrt{11}]$.

15.
$6x - 9 - x^2 < 0$ Polynomial inequality
$6x - 9 - x^2 = 0$ Related equation
$-(x^2 - 6x + 9) = 0$ Factoring out -1 and rearranging
$-(x-3)(x-3) = 0$ Factoring

Using the principle of zero products or by observing the graph of $y = 6x - 9 - x^2$, we see that the solution of the related equation is 3. This number divides the x-axis into the intervals $(-\infty, 3)$ and $(3, \infty)$. We let $f(x) = 6x - 9 - x^2$ and test a value in each interval.

$(-\infty, 3)$: $f(-4) = -49 < 0$

$(3, \infty)$: $f(4) = -49 < 0$

Function values are negative on both intervals. The solution set is $(-\infty, 3) \cup (3, \infty)$.

16.
$x^2 + 2x + 1 \le 0$ Polynomial inequality
$x^2 + 2x + 1 = 0$ Related equation
$(x+1)(x+1) = 0$ Factoring

The solution of the related equation is -1. This number divides the x-axis into the intervals $(-\infty, -1)$, and $(-1, \infty)$. We let $f(x) = x^2 + 2x + 1$ and test a value in each interval.

$(-\infty, -1)$: $f(-2) = 1 > 0$

$(-1, \infty)$: $f(0) = 1 > 0$

Function values are negative in neither interval. The function is equal to 0 when $x = -1$. Thus, the solution set is $\{-1\}$.

17.
$x^2 + 12 < 4x$ Polynomial inequality
$x^2 - 4x + 12 < 0$ Equivalent inequality with 0 on one side
$x^2 - 4x + 12 = 0$ Related equation

Using the quadratic formula or by observing the graph of $y = x^2 - 4x + 12$ we see that the related equation has no real-number solutions. The graph lies entirely above the x-axis, so the inequality has no solution. We could determine this algebraically by letting $f(x) = x^2 - 4x + 12$ and testing any real number (since there are no real-number solutions of $f(x) = 0$ to divide the x-axis into intervals). For example, $f(0) = 12 > 0$, so we see algebraically that the inequality has no solution. The solution set is $\emptyset$.

18.
$x^2 - 8 > 6x$ Polynomial inequality
$x^2 - 6x - 8 > 0$ Equivalent inequality with 0 on one side
$x^2 - 6x - 8 = 0$ Related equation

Using the quadratic formula, we find that the solutions of the related equation are $3-\sqrt{17}$ and $3+\sqrt{17}$. These numbers divide the x-axis into the intervals $(-\infty, 3-\sqrt{17})$, $(3-\sqrt{17}, 3+\sqrt{17})$, and $(3+\sqrt{17}, \infty)$. We let $f(x) = x^2 - 6x - 8$ and test a value in each interval.

$(-\infty, 3-\sqrt{17})$: $f(-2) = 8 > 0$

$(3-\sqrt{17}, 3+\sqrt{17})$: $f(0) = -8 < 0$

$(3+\sqrt{17}, \infty)$: $f(8) = 8 > 0$

Function values are positive on $(-\infty, 3-\sqrt{17})$ and $(3+\sqrt{17}, \infty)$. The graph of $y = x^2-6x-8$ can also be used to determine that function values are positive on $(-\infty, -1.123)$ and $(7.123, \infty)$. The solution set is $(-\infty, 3-\sqrt{17}) \cup (3+\sqrt{17}, \infty)$ or approximately $(-\infty, -1.123) \cup (7.123, \infty)$.

19.

$4x^3 - 7x^2 \le 15x$	Polynomial inequality
$4x^3 - 7x^2 - 15x \le 0$	Equivalent inequality with 0 on one side
$4x^3 - 7x^2 - 15x = 0$	Related equation
$x(4x+5)(x-3) = 0$	Factoring

Using the principle of zero products or by observing the graph of $y = 4x^3 - 7x^2 - 15x$, we see that the solutions of the related equation are 0, $-\frac{5}{4}$, and 3. These numbers divide the x-axis into the intervals $\left(-\infty, -\frac{5}{4}\right)$, $\left(-\frac{5}{4}, 0\right)$, $(0, 3)$, and $(3, \infty)$. We let $f(x) = 4x^3 - 7x^2 - 15x$ and test a value in each interval.

$\left(-\infty, -\frac{5}{4}\right)$: $f(-2) = -30 < 0$

$\left(-\frac{5}{4}, 0\right)$: $f(-1) = 4 > 0$

$(0, 3)$ $f(1) = -18 < 0$

$(3, \infty)$: $f(4) = 84 > 0$

Function values are negative on $\left(-\infty, -\frac{5}{4}\right)$ and $(0, 3)$. The graph of $y = 4x^3 - 7x^2 - 15x$ can also be used to determine this. Since the inequality symbol is $\le$, the endpoints of the intervals must be included in the solution set. It is $\left(-\infty, -\frac{5}{4}\right] \cup [0, 3]$.

20.

$2x^3 - x^2 < 5$	Polynomial inequality
$2x^3 - x^2 - 5 < 0$	Equivalent inequality with 0 on one side
$2x^3 - x^2 - 5 = 0$	Related equation

After trying all the possibilities, we find that the related equation has no rational zeros. Using the graph of $y = 2x^3 - x^2 - 5$, we find that the only real-number solutions of the related equation is approximately 1.546. This number divides the x-axis into the intervals $(-\infty, 1.546)$ and $(1.546, \infty)$. We let $f(x) = 2x^3 - x^2 - 5$ and test a value in each interval.

$(-\infty, 1.546)$: $f(0) = -5 < 0$

$(1.546, \infty)$: $f(2) = 7 > 0$

Function values are negative on $(-\infty, 1.546)$. This is the solution set. The graph can also be used to determine this.

21.

$x^3 + 3x^2 - x - 3 \ge 0$	Polynomial inequality
$x^3 + 3x^2 - x - 3 = 0$	Related equation
$x^2(x+3) - (x+3) = 0$	Factoring
$(x^2-1)(x+3) = 0$	
$(x+1)(x-1)(x+3) = 0$	

Using the principle of zero products or by observing the graph of $y = x^3 + 3x^2 - x - 3$, we see that the solutions of the related equation are -1, 1, and -3. These numbers divide the x-axis into the intervals $(-\infty, -3)$, $(-3, -1)$, $(-1, 1)$, and $(1, \infty)$. We let $f(x) = x^3 + 3x^2 - x - 3$ and test a value in each interval.

$(-\infty, -3)$: $f(-4) = -15 < 0$

$(-3, -1)$: $f(-2) = 3 > 0$

$(-1, 1)$: $f(0) = -3 < 0$

$(1, \infty)$: $f(2) = 15 > 0$

Function values are positive on $(-3, -1)$ and $(1, \infty)$. The graph of $y = x^3 + 3x^2 - x - 3$ can also be used to determine this. Since the inequality symbol is $\ge$, the endpoints of the intervals must be included in the solution set. It is $[-3, -1] \cup [1, \infty)$.

22.

$x^3 + x^2 - 4x - 4 \ge 0$	Polynomial inequality
$x^3 + x^2 - 4x - 4 = 0$	Related equation
$(x+2)(x-2)(x+1) = 0$	Factoring

The solutions of the related equation are -2, 2, and -1. These numbers divide the x-axis into the intervals $(-\infty, -2)$, $(-2, -1)$, $(-1, 2)$, and $(2, \infty)$. We let $f(x) = x^3 + x^2 - 4x - 4$ and test a value in each interval.

$(-\infty, -2)$: $f(-3) = -10 < 0$

$(-2, -1)$: $f(-1.5) = 0.875 > 0$

$(-1, 2)$: $f(0) = -4 < 0$

$(2, \infty)$: $f(3) = 20 > 0$

Function values are positive only on $(-2, -1)$ and $(2, \infty)$. The graph of $y = x^3 + x^2 - 4x - 4$ can also be used to determine this. Since the inequality symbol is $\ge$, the endpoints of the interval must be included in the solution set. It is $[-2, -1] \cup [2, \infty)$.

23.

$x^3 - 2x^2 < 5x - 6$	Polynomial inequality
$x^3 - 2x^2 - 5x + 6 < 0$	Equivalent inequality with 0 on one side
$x^3 - 2x^2 - 5x + 6 = 0$	Related equation

Using the techniques of Section 2.5, we find that the solutions of the related equation are -2, 1, and 3. We can also use the graph of $y = x^3 - 2x^2 - 5x + 6$ to find these solutions. They divide the x-axis into the intervals $(-\infty, -2)$, $(-2, 1)$, $(1, 3)$, and $(3, \infty)$. Let $f(x) = x^3 - 2x^2 - 5x + 6$ and test a value in each interval.

$(-\infty, -2)$: $f(-3) = -24 < 0$

$(-2, 1)$: $f(0) = 6 > 0$

$(1, 3)$: $f(2) = -4 < 0$

$(3, \infty)$: $f(4) = 18 > 0$

Function values are negative on $(-\infty, -2)$ and $(1, 3)$. This can also be determined graphically. The solution set is $(-\infty, -2) \cup (1, 3)$.

24.

$x^3 + x \leq 6 - 4x^2$	Polynomial inequality
$x^3 + 4x^2 + x - 6 \leq 0$	Equivalent inequality with 0 on one side
$x^3 + 4x^2 + x - 6 = 0$	Related equation

Using the techniques of Section 2.5, we find that the solutions of the related equation are -3, -2, and 1. We can also use the graph of $y = x^3 + 4x^2 + x - 6$ to find these solutions. They divide the x-axis into the intervals $(-\infty, -3)$, $(-3, -2)$, $(-2, 1)$, and $(1, \infty)$. Let $f(x) = x^3 + 4x^2 + x - 6$ and test a value in each interval.

$(-\infty, -3)$: $f(-4) = -10 < 0$

$(-3, -2)$: $f(-2.5) = 0.875 > 0$

$(-2, 1)$: $f(0) = -6 < 0$

$(1, \infty)$: $f(2) = 20 > 0$

Function values are negative on $(-\infty, -3)$ and $(-2, 1)$. This can also be determined graphically. Since the inequality symbol is $\leq$, the endpoints of the intervals must be included in the solution set. It is $(-\infty, -3] \cup [-2, 1]$.

25.

$x^5 + x^2 \geq 2x^3 + 2$	Polynomial inequality
$x^5 - 2x^3 + x^2 - 2 \geq 0$	Related inequality with 0 on one side
$x^5 - 2x^3 + x^2 - 2 = 0$	Related equation
$x^3(x^2 - 2) + x^2 - 2 = 0$	Factoring
$(x^3 + 1)(x^2 - 2) = 0$	

Using the principle of zero products or by observing the graph of $y = x^5 - 2x^3 + x^2 - 2$, we see that the real-number solutions of the related equation are -1, $-\sqrt{2}$, and $\sqrt{2}$. These numbers divide the x-axis into the intervals $(-\infty, -\sqrt{2})$, $(-\sqrt{2}, -1)$, $(-1, \sqrt{2})$, and $(\sqrt{2}, \infty)$. We let $f(x) = x^5 - 2x^3 + x^2 - 2$ and test a value in each interval.

$(-\infty, -\sqrt{2})$: $f(-2) = -14 < 0$

$(-\sqrt{2}, -1)$: $f(-1.3) \approx 0.37107 > 0$

$(-1, \sqrt{2})$: $f(0) = -2 < 0$

$(\sqrt{2}, \infty)$: $f(2) = 18 > 0$

Function values are positive on $(-\sqrt{2}, -1)$ and $(\sqrt{2}, \infty)$. This can also be determined graphically. Since the inequality symbol is $\geq$, the endpoints of the intervals must be included in the solution set. It is $[-\sqrt{2}, -1] \cup [\sqrt{2}, \infty)$.

26.

$x^5 + 24 > 3x^3 + 8x^2$	Polynomial inequality
$x^5 - 3x^3 - 8x^2 + 24 > 0$	Equivalent inequality with 0 on one side
$x^5 - 3x^3 - 8x^2 + 24 = 0$	Related equation
$x^3(x^2 - 3) - 8(x^2 - 3) = 0$	Factoring
$(x^3 - 8)(x^2 - 3) = 0$	

Using the principle of zero products or by observing the graph of $y = x^5 - 3x^3 - 8x^2 + 24$, we see that the real-number solutions of the related equation are 2, $-\sqrt{3}$, and $\sqrt{3}$. These numbers divide the x-axis into the intervals $(-\infty, -\sqrt{3})$, $(-\sqrt{3}, \sqrt{3})$, $(\sqrt{3}, 2)$, and $(2, \infty)$. We let $f(x) = x^5 - 3x^3 - 8x^2 + 24$ and test a value in each interval.

$(-\infty, -\sqrt{3})$: $f(-2) = -16 < 0$

$(-\sqrt{3}, \sqrt{3})$: $f(0) = 24 > 0$

$(\sqrt{3}, 2)$: $f(1.8) \approx -0.5203 < 0$

$(2, \infty)$: $f(3) = 114 > 0$

Function values are positive on $(-\sqrt{3}, \sqrt{3})$ and $(2, \infty)$. The graph can also be used to determine this. The solution set is $(-\sqrt{3}, \sqrt{3}) \cup (2, \infty)$.

27.

$2x^3 + 6 \leq 5x^2 + x$	Polynomial inequality
$2x^3 - 5x^2 - x + 6 \leq 0$	Equivalent inequality with 0 on one side
$2x^3 - 5x^2 - x + 6 = 0$	Related equation

Using the techniques of Section 2.5, we find that the solutions of the related equation are -1, $\frac{3}{2}$, and 2. We can also use the graph of $y = 2x^3 - 5x^2 - x + 6$ to find these solutions. They divide the x-axis into the intervals $(-\infty, -1)$, $\left(-1, \frac{3}{2}\right)$, $\left(\frac{3}{2}, 2\right)$, and $(2, \infty)$. Let $f(x) = 2x^3 - 5x^2 - x + 6$ and test a value in each interval.

$(-\infty, -1)$: $f(-2) = -28 < 0$

$\left(-1, \frac{3}{2}\right)$: $f(0) = 6 > 0$

$\left(\frac{3}{2}, 2\right)$: $f(1.6) = -0.208 < 0$

$(2, \infty)$: $f(3) = 12 > 0$

Function values are negative in $(-\infty, -1)$ and $\left(\frac{3}{2}, 2\right)$. This can also be determined graphically.

Since the inequality symbol is $\leq$, the endpoints of the intervals must be included in the solution set. The solution set is $\left(-\infty, -1\right] \cup \left[\frac{3}{2}, 2\right]$.

28.

$2x^3 + x^2 < 10 + 11x$	Polynomial inequality
$2x^3 + x^2 - 11x - 10 < 0$	Equivalent inequality with 0 on one side
$2x^3 + x^2 - 11x - 10 = 0$	Related equation

Using the techniques of Section 2.5 or by observing the graph of $y = 2x^3 + x^2 - 11x - 10$, we see that the real-number solutions of the related equation are -2, -1, and $\frac{5}{2}$. These numbers divide the x-axis into the intervals $(-\infty, -2)$, $(-2, -1)$, $\left(-1, \frac{5}{2}\right)$, and $\left(\frac{5}{2}, \infty\right)$. We let $f(x) = 2x^3 + x^2 - 11x - 10$ and test a value in each interval.

$(-\infty, -2)$: $f(-3) = -22 < 0$

$(-2, -1)$: $f(-1.5) = 2 > 0$

$\left(-1, \frac{5}{2}\right)$: $f(0) = -10 < 0$

$\left(\frac{5}{2}, \infty\right)$: $f(3) = 20 > 0$

Function values are negative on $(-\infty, -2)$ and $\left(-1, \frac{5}{2}\right)$. The graph can also be used to determine this. The solution set is $(-\infty, -2) \cup \left(-1, \frac{5}{2}\right)$.

29.

$x^3 + 5x^2 - 25x \leq 125$	Polynomial inequality
$x^3 + 5x^2 - 25x - 125 \leq 0$	Equivalent inequality with 0 on one side
$x^3 + 5x^2 - 25x - 125 = 0$	Related equation
$x^2(x + 5) - 25(x + 5) = 0$	Factoring
$(x^2 - 25)(x + 5) = 0$	
$(x + 5)(x - 5)(x + 5) = 0$	

Using the principle of zero products or by observing the graph of $y = x^3 + 5x^2 - 25x - 125$, we see that the solutions of the related equation are -5 and 5. These numbers divide the x-axis into the intervals $(-\infty, -5)$, $(-5, 5)$, and $(5, \infty)$. We let $f(x) = x^3 + 5x^2 - 25x - 125$ and test a value in each interval.

$(-\infty, -5)$: $f(-6) = -11 < 0$

$(-5, 5)$: $f(0) = -125 < 0$

$(5, \infty)$: $f(6) = 121 > 0$

Function values are negative on $(-\infty, -5)$ and $(-5, 5)$. This can also be determined graphically. Since the inequality symbol is $\leq$, the endpoints of the intervals must be included in the solution set. It is $(-\infty, -5] \cup [-5, 5]$ or $(-\infty, 5]$.

30.

$x^3 - 9x + 27 \geq 3x^2$	Polynomial inequality
$x^3 - 3x^2 - 9x + 27 \geq 0$	Equivalent inequality with 0 on one side
$x^3 - 3x^2 - 9x + 27 = 0$	Related equation
$x^2(x - 3) - 9(x - 3) = 0$	Factoring
$(x^2 - 9)(x - 3) = 0$	
$(x + 3)(x - 3)(x - 3) = 0$	

The solutions of the related equation are -3 and 3. These numbers divide the x-axis into the intervals $(-\infty, -3)$, $(-3, 3)$, and $(3, \infty)$. We let $f(x) = x^3 - 3x^2 - 9x + 27$ and test a value in each interval.

$(-\infty, -3)$: $f(-4) = -49 < 0$

$(-3, 3)$: $f(0) = 27 > 0$

$(3, \infty)$: $f(4) = 7 > 0$

Function values are positive only on $(-3, 3)$ and $(3, \infty)$. The graph of $y = x^3 - 3x^2 - 9x + 27$ can also be used to determine this. Since the inequality symbol is $\geq$, the endpoints of the intervals must be included in the solution set. It is $[-3, 3] \cup [3, \infty)$, or $[-3, \infty)$.

31.

$0.1x^3 - 0.6x^2 - 0.1x + 2 < 0$	Polynomial inequality
$0.1x^3 - 0.6x^2 - 0.1x + 2 = 0$	Related equation

After trying all the possibilities, we find that the related equation has no rational zeros. Using the graph of $y = 0.1x^3 - 0.6x^2 - 0.1x + 2$, we find that the only real-number solutions of the related equation are approximately -1.680, 2.154, and 5.526. These numbers divide the x-axis into the intervals $(-\infty, -1.680)$, $(-1.680, 2.154)$, $(2.154, 5.526)$, and $(5.526, \infty)$. We let $f(x) = 0.1x^3 - 0.6x^2 - 0.1x + 2$ and test a value in each interval.

$(-\infty, -1.680)$: $f(-2) = -1 < 0$

$(-1.680, 2.154)$: $f(0) = 2 > 0$

$(2.154, 5.526)$: $f(3) = -1 < 0$

$(5.526, \infty)$: $f(6) = 1.4 > 0$

Function values are negative on $(-\infty, -1.680)$ and $(2.154, 5.526)$. The graph can also be used to determine this. The solution set is $(-\infty, -1.680) \cup (2.154, 5.526)$.

32.

$$19.2x^3 + 12.8x^2 + 144 \geq 172.8x + 3.2x^4$$

$$-3.2x^4 + 19.2x^3 + 12.8x^2 - 172.8x + 144 \geq 0$$

$$-3.2x^4 + 19.2x^3 + 12.8x^2 - 172.8x + 144 = 0$$

Related equation

The solutions of the related equation are -3, 1, 3, and 5. These numbers divide the x-axis into the intervals $(-\infty, -3)$, $(-3, 1)$, $(1, 3)$, $(3, 5)$, and $(5, \infty)$. We let $f(x) = -3.2x^4 + 19.2x^3 + 12.8x^2 - 172.8x + 144$ and test a value in each interval.

$(-\infty, -3)$: $f(-4) = -1008 < 0$

$(-3, 1)$: $f(0) = 144 > 0$

$(1, 3)$: $f(2) = -48 < 0$

$(3, 5)$: $f(4) = 672 > 0$

$(5, \infty)$: $f(6) = -432 < 0$

Function values are positive only on $(-3, 1)$ and $(3, 5)$. The graph of $y = -3.2x^4 + 19.2x^3 + 12.8x^2 - 172.8x + 144$ can also be used to determine this. Since the inequality symbol is $\geq$, the endpoints of the intervals must be included in the solution set. It is $[-3, 1] \cup [3, 5]$.

33. $\frac{1}{x+4} > 0$ Rational inequality

$\frac{1}{x+4} = 0$ Related equation

The denominator of $f(x) = \frac{1}{x+4}$ is 0 when $x = -4$, so the function is not defined for $x = -4$. The related equation has no solution. Thus, the only critical value is -4. It divides the x-axis into the intervals $(-\infty, -4)$ and $(-4, \infty)$. We test a value in each interval.

$(-\infty, -4)$: $f(-5) = -1 < 0$

$(-4, \infty)$: $f(0) = \frac{1}{4} > 0$

Function values are positive on $(-4, \infty)$. This can also be determined from the graph of $y = \frac{1}{x+4}$. The solution set is $(-4, \infty)$.

34. $\frac{1}{x-3} \leq 0$ Rational inequality

$\frac{1}{x-3} = 0$ Related equation

The denominator of $f(x) = \frac{1}{x-3}$ is 0 when $x = 3$, so the function is not defined for $x = 3$. The related equation has no solution. Thus, the only critical value is 3. It divides the x-axis into the intervals $(-\infty, 3)$ and $(3, \infty)$. We test a value in each interval.

$(-\infty, 3)$: $f(0) = -\frac{1}{3} < 0$

$(3, \infty)$: $f(4) = 1 > 0$

Function values are negative on $(-\infty, 3)$. This can also be determined from the graph of $y = \frac{1}{x-3}$. Note that since 3 is not in the domain of $f(x)$, it cannot be included in the solution set. It is $(-\infty, 3)$.

35. $\frac{-4}{2x+5} < 0$ Rational inequality

$\frac{-4}{2x+5} = 0$ Related equation

The denominator of $f(x) = \frac{-4}{2x+5}$ is 0 when $x = -\frac{5}{2}$, so the function is not defined for $x = -\frac{5}{2}$. The related equation has no solution. Thus, the only critical value is $-\frac{5}{2}$. It divides the x-axis into the intervals $\left(-\infty, -\frac{5}{2}\right)$ and $\left(-\frac{5}{2}, \infty\right)$. We test a value in each interval.

$\left(-\infty, -\frac{5}{2}\right)$: $f(-3) = 4 > 0$

$\left(-\frac{5}{2}, \infty\right)$: $f(0) = -\frac{4}{5} < 0$

Function values are negative on $\left(-\frac{5}{2}, \infty\right)$. This can also be determined from the graph of $y = \frac{-4}{2x+5}$. The solution set is $\left(-\frac{5}{2}, \infty\right)$.

36. $\frac{-2}{5-x} \geq 0$ Rational inequality

$\frac{-2}{5-x} = 0$ Related equation

The denominator of $f(x) = \frac{-2}{5-x}$ is 0 when $x = 5$, so the function is not defined for $x = 5$. The related equation has no solution. Thus, the only critical value is 5. It divides the x-axis into the intervals $(-\infty, 5)$ and $(5, \infty)$. We test a value in each interval.

$(-\infty, 5)$: $f(0) = -\frac{2}{5} < 0$

$(5, \infty)$: $f(6) = 2 > 0$

Function values are positive on $(5, \infty)$. This can also be determined from the graph of $y = \frac{-2}{5-x}$. Note that since 5 is not in the domain of $f(x)$, it cannot be included in the solution set. It is $(5, \infty)$.

37. $\frac{x-4}{x+3} - \frac{x+2}{x-1} \leq 0$

The denominator of $f(x) = \frac{x-4}{x+3} - \frac{x+2}{x-1}$ is 0 when $x = -3$ or $x = 1$, so the function is not defined for these values of x. We solve the related equation $f(x) = 0$.

$$\frac{x-4}{x+3} - \frac{x+2}{x-1} = 0$$

$$(x+3)(x-1)\left(\frac{x-4}{x+3} - \frac{x+2}{x-1}\right) = (x+3)(x-1)\cdot 0$$

$$(x-1)(x-4) - (x+3)(x+2) = 0$$

$$x^2 - 5x + 4 - (x^2 + 5x + 6) = 0$$

$$-10x - 2 = 0$$

$$-10x = 2$$

$$x = -\frac{1}{5}$$

The critical values are -3, $-\frac{1}{5}$, and 1. They divide the x-axis into the intervals $(-\infty, -3)$, $\left(3, -\frac{1}{5}\right)$, $\left(-\frac{1}{5}, 1\right)$, and $(1, \infty)$. We test a value in each interval.

$(-\infty, -3)$: $f(-4) = 7.6 > 0$

$\left(-3, -\frac{1}{5}\right)$: $f(-1) = -2 < 0$

$\left(-\frac{1}{5}, 1\right)$: $f(0) = \frac{2}{3} > 0$

$(1, \infty)$: $f(2) = -4.4 < 0$

Function values are negative on $\left(-3, -\frac{1}{5}\right)$ and $(1, \infty)$. This can also be determined from the graph of $y = \frac{x-4}{x+3} - \frac{x+2}{x-1}$. Note that since the inequality symbol is $\leq$ and $f\left(-\frac{1}{5}\right) = 0$, then $-\frac{1}{5}$ must be included in the solution set. Note also that since neither -3 nor 1 is in the domain of $f(x)$, they are not included in the solution set. It is $\left(-3, -\frac{1}{5}\right] \cup (1, \infty)$.

38. $\frac{x+1}{x-2} + \frac{x-3}{x-1} < 0$

The denominator of $f(x) = \frac{x+1}{x-2} + \frac{x-3}{x-1}$ is 0 when $x = 2$ or $x = 1$, so the function is not defined for these values of x. We solve the related equation $f(x) = 0$.

$$\frac{x+1}{x-2} + \frac{x-3}{x-1} = 0$$

$$(x-1)(x+1) + (x-2)(x-3) = 0$$

Multiplying by $(x-2)(x-1)$

$$x^2 - 1 + x^2 - 5x + 6 = 0$$

$$2x^2 - 5x + 5 = 0$$

This equation has no real-number solutions, so the critical values are 1 and 2. They divide the x-axis into the intervals $(-\infty, 1)$, $(1, 2)$, and $(2, \infty)$. We test a value in each interval.

$(-\infty, 1)$: $f(0) = 2.5 > 0$

$(1, 2)$: $f(1.5) = -8 < 0$

$(2, \infty)$: $f(3) = 4 > 0$

Function values are negative on $(1, 2)$. This can also be determined from the graph of $y = \frac{x+1}{x-2} + \frac{x-3}{x-1}$. The solution set is $(1, 2)$.

39. $\frac{2x-1}{x+3} \geq \frac{x+1}{3x+1}$ Rational inequality

$\frac{2x-1}{x+3} - \frac{x+1}{3x+1} \geq 0$ Equivalent inequality with 0 on one side

The denominator of $f(x) = \frac{2x-1}{x+3} - \frac{x+1}{3x+1}$ is 0 when $x = -3$ or $x = -\frac{1}{3}$, so the function is not defined for these values of x. We solve the related equation $f(x) = 0$.

$$\frac{2x-1}{x+3} - \frac{x+1}{3x+1} = 0$$

$$(x+3)(3x+1)\left(\frac{2x-1}{x+3} - \frac{x+1}{3x+1}\right) = (x+3)(3x+1) \cdot 0$$

$$(3x+1)(2x-1) - (x+3)(x+1) = 0$$

$$6x^2 - x - 1 - (x^2 + 4x + 3) = 0$$

$$5x^2 - 5x - 4 = 0$$

Using the quadratic formula we find that $x = \frac{5 \pm \sqrt{105}}{10}$. Then the critical values are -3, $\frac{5 - \sqrt{105}}{10}$, $-\frac{1}{3}$, and $\frac{5 + \sqrt{105}}{10}$. They divide the x-axis into the intervals $(-\infty, -3)$, $\left(-3, \frac{5-\sqrt{105}}{10}\right)$, $\left(\frac{5-\sqrt{105}}{10}, -\frac{1}{3}\right)$, $\left(-\frac{1}{3}, \frac{5+\sqrt{105}}{10}\right)$, and $\left(\frac{5+\sqrt{105}}{10}, \infty\right)$. We test a value in each interval.

$(-\infty, -3)$: $f(-4) \approx 8.7273 > 0$

$\left(-3, \frac{5-\sqrt{105}}{10}\right)$: $f(-2) = -5.2 < 0$

$\left(\frac{5-\sqrt{105}}{10}, -\frac{1}{3}\right)$: $f(-0.4) \approx 2.3077 > 0$

$\left(-\frac{1}{3}, \frac{5+\sqrt{105}}{10}\right)$: $f(0) \approx -1.333 < 0$

$\left(\frac{5+\sqrt{105}}{10}, \infty\right)$: $f(2) \approx 0.1714 > 0$

Function values are positive on $(-\infty, -3)$, $\left(\frac{5-\sqrt{105}}{10}, -\frac{1}{3}\right)$ and $\left(\frac{5+\sqrt{105}}{10}, \infty\right)$. This can also be determined from the graph of $y = \frac{2x-1}{x+3} - \frac{x+1}{3x+1}$. Note that since the inequality symbol is $\geq$ and $f\left(\frac{5 \pm \sqrt{105}}{10}\right) = 0$, then $\frac{5 \pm \sqrt{105}}{10}$ must be included in the solution set. Note also that since neither -3 nor $-\frac{1}{3}$ is in the domain of $f(x)$, they are not included in the solution set. It is

$$(-\infty, -3) \cup \left[\frac{5-\sqrt{105}}{10}, -\frac{1}{3}\right) \cup \left[\frac{5+\sqrt{105}}{10}, \infty\right).$$

40. $\frac{x+5}{x-4} > \frac{3x+2}{2x+1}$ Rational inequality

$\frac{x+5}{x-4} - \frac{3x+2}{2x+1} > 0$ Equivalent inequality with 0 on one side

The denominator of $f(x) = \frac{x+5}{x-4} - \frac{3x+2}{2x+1}$ is 0 when $x = 4$ or $x = -\frac{1}{2}$, so the function is not defined for these values of x. We solve the related equation $f(x) = 0$.

$$\frac{x+5}{x-4} - \frac{3x+2}{2x+1} = 0$$

$$(2x+1)(x+5) - (x-4)(3x+2) = 0$$

Multiplying by $(x-4)(2x+1)$

$$2x^2 + 11x + 5 - (3x^2 - 10x - 8) = 0$$

$$-x^2 + 21x + 13 = 0$$

Using the quadratic formula we find that $x = \frac{21 \pm \sqrt{493}}{2}$. Then the critical values are $\frac{21-\sqrt{493}}{2}$, $-\frac{1}{2}$, 4, and $\frac{21+\sqrt{493}}{2}$. They divide the x-axis into the intervals $\left(-\infty, \frac{21-\sqrt{493}}{2}\right)$, $\left(\frac{21-\sqrt{493}}{2}, -\frac{1}{2}\right)$, $\left(-\frac{1}{2}, 4\right)$, $\left(4, \frac{21+\sqrt{493}}{2}\right)$, and $\left(\frac{21+\sqrt{493}}{2}, \infty\right)$. We test a value in each interval.

$\left(-\infty, \frac{21-\sqrt{493}}{2}\right)$: $f(-1) = -1.8 < 0$

$\left(\frac{21-\sqrt{493}}{2}, -\frac{1}{2}\right)$: $f(-0.55) = 2.522 > 0$

$\left(-\frac{1}{2}, 4\right)$: $f(0) = -3.25 < 0$

$\left(4, \frac{21+\sqrt{493}}{2}\right)$: $f(5) \approx 8.4545 > 0$

$\left(\frac{21+\sqrt{493}}{2}, \infty\right)$: $f(22) \approx -0.0111 < 0$

Function values are positive on $\left(\frac{21-\sqrt{493}}{2}, -\frac{1}{2}\right)$ and $\left(4, \frac{21+\sqrt{493}}{2}\right)$. This can also be determined from the graph of $y = \frac{x+5}{x-4} - \frac{3x+2}{2x+1}$. The solution set is $\left(\frac{21-\sqrt{493}}{2}, -\frac{1}{2}\right) \cup \left(4, \frac{21+\sqrt{493}}{2}\right)$.

41. $\frac{x+1}{x-2} \geq 3$ Rational inequality

$\frac{x+1}{x-2} - 3 \geq 0$ Equivalent inequality with 0 on one side

The denominator of $f(x) = \frac{x+1}{x-2} - 3$ is 0 when $x = 2$, so the function is not defined for this value of x. We solve the related equation $f(x) = 0$.

$$\frac{x+1}{x-2} - 3 = 0$$

$$(x-2)\left(\frac{x+1}{x-2} - 3\right) = (x-2) \cdot 0$$

$$x + 1 - 3(x-2) = 0$$

$$x + 1 - 3x + 6 = 0$$

$$-2x + 7 = 0$$

$$-2x = -7$$

$$x = \frac{7}{2}$$

The critical values are 2 and $\frac{7}{2}$. They divide the x-axis into the intervals $(-\infty, 2)$, $\left(2, \frac{7}{2}\right)$, and $\left(\frac{7}{2}, \infty\right)$. We test a value in each interval.

$(-\infty, 2)$: $f(0) = -3.5 < 0$

$\left(2, \frac{7}{2}\right)$: $f(3) = 1 > 0$

$\left(\frac{7}{2}, \infty\right)$: $f(4) = -0.5 < 0$

Function values are positive on $\left(2, \frac{7}{2}\right)$. This can also be determined from the graph of $y = \frac{x+1}{x-2} - 3$. Note that since the inequality symbol is $\geq$ and $f\left(\frac{7}{2}\right) = 0$, then $\frac{7}{2}$ must be included in the solution set. Note also that since 2 is not in the domain of $f(x)$, it is not included in the solution set. It is $\left(2, \frac{7}{2}\right]$.

42. $\frac{x}{x-5} < 2$ Rational inequality

$\frac{x}{x-5} - 2 < 0$ Equivalent inequality with 0 on one side

The denominator of $f(x) = \frac{x}{x-5} - 2$ is 0 when $x = 5$, so the function is not defined for this value of x. We solve the related equation $f(x) = 0$.

$$\frac{x}{x-5} - 2 = 0$$

$x - 2(x-5) = 0$ Multiplying by $x - 5$

$$x - 2x + 10 = 0$$

$$-x + 10 = 0$$

$$x = 10$$

The critical values are 5 and 10. They divide the x-axis into the intervals $(-\infty, 5)$, $(5, 10)$, and $(10, \infty)$. We test a value in each interval.

$(-\infty, 5)$: $f(0) = -2 < 0$

$(5, 10)$: $f(6) = 4 > 0$

$(10, \infty)$: $f(11) = -\frac{1}{6} < 0$

Function values are negative on $(-\infty, 5)$ and $(10, \infty)$. This can also be determined from the graph of $y = \frac{x}{x-5} - 2$. The solution set is $(-\infty, 5) \cup (10, \infty)$.

43. $x - 2 > \frac{1}{x}$ Rational inequality

$x - 2 - \frac{1}{x} > 0$ Equivalent inequality with 0 on one side

The denominator of $f(x) = x - 2 - \frac{1}{x}$ is 0 when $x = 0$, so the function is not defined for this value of x. We solve the related equation $f(x) = 0$.

$$x - 2 - \frac{1}{x} = 0$$
$$x\left(x - 2 - \frac{1}{x}\right) = x \cdot 0$$
$$x^2 - 2x - x \cdot \frac{1}{x} = 0$$
$$x^2 - 2x - 1 = 0$$

Using the quadratic formula we find that $x = 1 \pm \sqrt{2}$. The critical values are $1 - \sqrt{2}$, 0, and $1 + \sqrt{2}$. They divide the x-axis into the intervals $(-\infty, 1 - \sqrt{2})$, $(1 - \sqrt{2}, 0)$, $(0, 1 + \sqrt{2})$, and $(1 + \sqrt{2}, \infty)$. We test a value in each interval.

$(-\infty, 1 - \sqrt{2})$: $f(-1) = -2 < 0$

$(1 - \sqrt{2}, 0)$: $f(-0.1) = 7.9 > 0$

$(0, 1 + \sqrt{2})$: $f(1) = -2 < 0$

$(1 + \sqrt{2}, \infty)$: $f(3) = \frac{2}{3} > 0$

Function values are positive on $(1 - \sqrt{2}, 0)$ and $(1 + \sqrt{2}, \infty)$. This can also be determined from the graph of $y = x - 2 - \frac{1}{x}$. The solution set is $(1 - \sqrt{2}, 0) \cup (1 + \sqrt{2}, \infty)$.

44. $4 \geq \frac{4}{x} + x$ Rational inequality

$4 - \frac{4}{x} - x \geq 0$ Equivalent inequality with 0 on one side

The denominator of $f(x) = 4 - \frac{4}{x} - x$ is 0 when $x = 0$, so the function is not defined for this value of x. We solve the related equation $f(x) = 0$.

$$4 - \frac{4}{x} - x = 0$$
$$4x - 4 - x^2 = 0 \quad \text{Multiplying by } x$$
$$-(x - 2)^2 = 0$$
$$x = 2$$

The critical values are 0 and 2. They divide the x-axis into the intervals $(-\infty, 0)$, $(0, 2)$, and $(2, \infty)$. We test a value in each interval.

$(-\infty, 0)$: $f(-1) = 9 > 0$

$(0, 2)$: $f(1) = -1 < 0$

$(2, \infty)$: $f(3) = -\frac{1}{3} < 0$

Function values are positive on $(-\infty, 0)$. This can also be determined from the graph of $y = 4 - \frac{4}{x} - x$. Note that since the inequality symbol is $\geq$ and $f(2) = 0$, then 2 must be included in the solution set. Note also that since 0 is not in the domain of $f(x)$, it is not included in the solution set. It is $(-\infty, 0) \cup \{2\}$.

45.
$$\frac{2}{x^2 - 4x + 3} \leq \frac{5}{x^2 - 9}$$
$$\frac{2}{x^2 - 4x + 3} - \frac{5}{x^2 - 9} \leq 0$$
$$\frac{2}{(x-1)(x-3)} - \frac{5}{(x+3)(x-3)} \leq 0$$

The denominator of $f(x) = \frac{2}{(x-1)(x-3)} - \frac{5}{(x+3)(x-3)}$ is 0 when $x = 1$, 3, or -3, so the function is not defined for these values of x. We solve the related equation $f(x) = 0$.

$$\frac{2}{(x-1)(x-3)} - \frac{5}{(x+3)(x-3)} = 0$$
$$(x-1)(x-3)(x+3)\left(\frac{2}{(x-1)(x-3)} - \frac{5}{(x+3)(x-3)}\right) = (x-1)(x-3)(x+3) \cdot 0$$
$$2(x+3) - 5(x-1) = 0$$
$$2x + 6 - 5x + 5 = 0$$
$$-3x + 11 = 0$$
$$-3x = -11$$
$$x = \frac{11}{3}$$

The critical values are -3, 1, 3, and $\frac{11}{3}$. They divide the x-axis into the intervals $(-\infty, -3)$, $(-3, 1)$, $(1, 3)$, $\left(3, \frac{11}{3}\right)$, and $\left(\frac{11}{3}, \infty\right)$. We test a value in each interval.

$(-\infty, -3)$: $f(-4) \approx -0.6571 < 0$

$(-3, 1)$: $f(0) \approx 1.2222 > 0$

$(1, 3)$: $f(2) = -1 < 0$

$\left(3, \frac{11}{3}\right)$: $f(3.5) \approx 0.6154 > 0$

$\left(\frac{11}{3}, \infty\right)$: $f(4) \approx -0.0476 < 0$

Function values are negative on $(-\infty, -3)$, $(1, 3)$, and $\left(\frac{11}{3}, \infty\right)$. They can also be determined from the graph of $y = \frac{2}{(x-1)(x-3)} - \frac{5}{(x+3)(x-3)}$. Note that since the inequality symbol is $\leq$ and $f\left(\frac{11}{3}\right) = 0$, then $\frac{11}{3}$ must be included in the solution set. Note also that since -3, 1, and 3 are not in the domain of $f(x)$, they are not included in the solution set. It is $(-\infty, -3) \cup (1, 3) \cup \left[\frac{11}{3}, \infty\right)$.

46.
$$\frac{3}{x^2 - 4} \leq \frac{5}{x^2 + 7x + 10}$$
$$\frac{3}{(x+2)(x-2)} - \frac{5}{(x+2)(x+5)} \leq 0$$

The denominator of $f(x) = \frac{3}{(x+2)(x-2)} - \frac{5}{(x+2)(x+5)}$ is 0 when $x = -2$, 2, or -5, so the function is not defined for these values of x. We solve the related equation $f(x) = 0$.

$$\frac{3}{(x+2)(x-2)} - \frac{5}{(x+2)(x+5)} = 0$$

$$3(x+5) - 5(x-2) = 0 \quad \text{Multiplying by } (x+2)(x-2)(x+5)$$

$$3x + 15 - 5x + 10 = 0$$

$$-2x + 25 = 0$$

$$x = \frac{25}{2}$$

The critical values are -5, -2, 2, and $\frac{25}{2}$. They divide the x-axis into the intervals $(-\infty, -5)$, $(-5, -2)$, $(-2, 2)$, $\left(2, \frac{25}{2}\right)$, and $\left(\frac{25}{2}, \infty\right)$. We test a value in each interval.

$(-\infty, -5)$: $f(-6) \approx -1.156 < 0$

$(-5, -2)$: $f(-3) = 3.1 > 0$

$(-2, 2)$: $f(0) = -1.25 < 0$

$\left(2, \frac{25}{2}\right)$: $f(3) = 0.475 > 0$

$\left(\frac{25}{2}, \infty\right)$: $f(13) \approx -0.0003 < 0$

Function values are negative on $(-\infty, -5)$, $(-2, 2)$, and $\left(\frac{25}{2}, \infty\right)$. They can also be determined from the graph of $y = \frac{3}{(x+2)(x-2)} - \frac{5}{(x+2)(x+5)}$. Note that since the inequality symbol is $\leq$ and $f\left(\frac{25}{2}\right) = 0$, then $\frac{25}{2}$ must be included in the solution set. Note also that since -5, -2, and 2 are not in the domain of $f(x)$, they are not included in the solution set. It is $(-\infty, -5) \cup (-2, 2) \cup \left[\frac{25}{2}, \infty\right)$.

47.

$$\frac{3}{x^2+1} \geq \frac{6}{5x^2+2}$$

$$\frac{3}{x^2+1} - \frac{6}{5x^2+2} \geq 0$$

The denominator of $f(x) = \frac{3}{x^2+1} - \frac{6}{5x^2+2}$ has no real-number zeros. We solve the related equation $f(x) = 0$.

$$\frac{3}{x^2+1} - \frac{6}{5x^2+2} = 0$$

$$(x^2+1)(5x^2+2)\left(\frac{3}{x^2+1} - \frac{6}{5x^2+2}\right) = (x^2+1)(5x^2+2) \cdot 0$$

$$3(5x^2+2) - 6(x^2+1) = 0$$

$$15x^2 + 6 - 6x^2 - 6 = 0$$

$$9x^2 = 0$$

$$x^2 = 0$$

$$x = 0$$

The only critical value is 0. It divides the x-axis into the intervals $(-\infty, 0)$ and $(0, \infty)$. We test a value in each interval.

$(-\infty, 0)$: $f(-1) \approx 0.64286 > 0$

$(0, 0)$: $f(1) \approx 0.64286 > 0$

Function values are positive on both intervals. Note that since the inequality symbol is $\geq$ and $f(0) = 0$, then 0 must be included in the solution set. It is $(-\infty, 0] \cup [0, \infty)$, or $(-\infty, \infty)$.

48.

$$\frac{4}{x^2-9} < \frac{3}{x^2-25}$$

$$\frac{4}{x^2-9} - \frac{3}{x^2-25} < 0$$

$$\frac{4}{(x+3)(x-3)} - \frac{3}{(x+5)(x-5)} < 0$$

The denominator of $f(x) = \frac{4}{(x+3)(x-3)} - \frac{3}{(x+5)(x-5)}$ is 0 when $x = -3$, 3, -5, or 5, so the function is not defined for these values of x. We solve the related equation $f(x) = 0$.

$$\frac{4}{(x+3)(x-3)} - \frac{3}{(x+5)(x-5)} = 0$$

$$4(x+5)(x-5) - 3(x+3)(x-3) = 0$$

Multiplying by $(x+3)(x-3)(x+5)(x-5)$

$$4x^2 - 100 - 3x^2 + 27 = 0$$

$$x^2 - 73 = 0$$

$$x = \pm\sqrt{73}$$

The critical values are $-\sqrt{73}$, -5, -3, 3, 5, and $\sqrt{73}$. They divide the x-axis into the intervals $(-\infty, -\sqrt{73})$, $(-\sqrt{73}, -5)$, $(-5, -3)$, $(-3, 3)$, $(3, 5)$, $(5, \sqrt{73})$, and $(\sqrt{73}, \infty)$. We test a value in each interval.

$(-\infty, -\sqrt{73})$: $f(-9) \approx 0.00198 > 0$

$(-\sqrt{73}, -5)$: $f(-6) \approx -0.1246 < 0$

$(-5, -3)$: $f(-4) \approx 0.90476 > 0$

$(-3, 3)$: $f(0) \approx -0.3244 < 0$

$(3, 5)$: $f(4) \approx 0.90476 > 0$

$(5, \sqrt{73})$: $f(6) \approx -0.1246 < 0$

$(\sqrt{73}, \infty)$: $f(9) \approx 0.00198 > 0$

Function values are negative on $(-\sqrt{73}, -5)$, $(-3, 3)$, and $(5, \sqrt{73})$. This can also be determined from the graph of $y = \frac{4}{(x+3)(x-3)} - \frac{3}{(x+5)(x-5)}$. The solution set is $(-\sqrt{73}, -5) \cup (-3, 3) \cup (5, \sqrt{73})$.

49.

$$\frac{5}{x^2+3x} < \frac{3}{2x+1}$$

$$\frac{5}{x^2+3x} - \frac{3}{2x+1} < 0$$

$$\frac{5}{x(x+3)} - \frac{3}{2x+1} < 0$$

The denominator of $f(x) = \dfrac{5}{x(x+3)} - \dfrac{3}{2x+1}$ is 0 when $x = 0, -3$, or $-\dfrac{1}{2}$, so the function is not defined for these values of x. We solve the related equation $f(x) = 0$.

$$\frac{5}{x(x+3)} - \frac{3}{2x+1} = 0$$

$$x(x+3)(2x+1)\left(\frac{5}{x(x+3)} - \frac{3}{2x+1}\right) = x(x+3)(2x+1)\cdot 0$$

$$5(2x+1) - 3x(x+3) = 0$$

$$10x + 5 - 3x^2 - 9x = 0$$

$$-3x^2 + x + 5 = 0$$

Using the quadratic formula we find that $x = \dfrac{1 \pm \sqrt{61}}{6}$. The critical values are -3, $\dfrac{1-\sqrt{61}}{6}$, $-\dfrac{1}{2}$, 0, and $\dfrac{1+\sqrt{61}}{6}$. They divide the x-axis into the intervals $(-\infty, -3)$, $\left(-3, \dfrac{1-\sqrt{61}}{6}\right)$, $\left(\dfrac{1-\sqrt{61}}{6}, -\dfrac{1}{2}\right)$, $\left(-\dfrac{1}{2}, 0\right)$, $\left(0, \dfrac{1+\sqrt{61}}{6}\right)$, and $\left(\dfrac{1+\sqrt{61}}{6}, \infty\right)$.

We test a value in each interval.

$(-\infty, -3)$: $f(-4) \approx 1.6786 > 0$

$\left(-3, \dfrac{1-\sqrt{61}}{6}\right)$: $f(-2) = -1.5 < 0$

$\left(\dfrac{1-\sqrt{61}}{6}, -\dfrac{1}{2}\right)$: $f(-1) = 0.5 > 0$

$\left(-\dfrac{1}{2}, 0\right)$: $f(-0.1) \approx -20.99 < 0$

$\left(0, \dfrac{1+\sqrt{61}}{6}\right)$: $f(1) = 0.25 > 0$

$\left(\dfrac{1+\sqrt{61}}{6}, \infty\right)$: $f(2) = -0.1 < 0$

Function values are negative on $\left(-3, \dfrac{1-\sqrt{61}}{6}\right)$, $\left(-\dfrac{1}{2}, 0\right)$ and $\left(\dfrac{1+\sqrt{61}}{6}, \infty\right)$. This can also be determined from the graph of $y = \dfrac{5}{x(x+3)} - \dfrac{3}{2x+1}$.

The solution set is

$$\left(-3, \frac{1-\sqrt{61}}{6}\right) \cup \left(-\frac{1}{2}, 0\right) \cup \left(\frac{1+\sqrt{61}}{6}, \infty\right).$$

50.

$$\frac{2}{x^2+3} > \frac{3}{5+4x^2}$$

$$\frac{2}{x^2+3} - \frac{3}{5+4x^2} > 0$$

The denominator of $f(x) = \dfrac{2}{x^2+3} - \dfrac{3}{5+4x^2}$ has no real-number zeros. We solve the related equation $f(x) = 0$.

$$\frac{2}{x^2+3} - \frac{3}{5+4x^2} = 0$$

$$2(5+4x^2) - 3(x^2+3) = 0 \quad \text{Multiplying by } (x^2+3)(5+4x^2)$$

$$10 + 8x^2 - 3x^2 - 9 = 0$$

$$5x^2 + 1 = 0$$

This equation has no real-number solutions. Thus, there are no critical values. We test a value in $(-\infty, \infty)$: $f(0) = \dfrac{1}{15} > 0$. The function is positive on $(-\infty, \infty)$. This is the solution set. This can also be determined from the graph of $y = \dfrac{2}{x^2+3} - \dfrac{3}{5+4x^2}$.

51.

$$\frac{5x}{7x-2} > \frac{x}{x+1}$$

$$\frac{5x}{7x-2} - \frac{x}{x+1} > 0$$

The denominator of $f(x) = \dfrac{5x}{7x-2} - \dfrac{x}{x+1}$ is 0 when $x = \dfrac{2}{7}$ or $x = -1$, so the function is not defined for these values of x. We solve the related equation $f(x) = 0$.

$$\frac{5x}{7x-2} - \frac{x}{x+1} = 0$$

$$(7x-2)(x+1)\left(\frac{5x}{7x-2} - \frac{x}{x+1}\right) = (7x-2)(x+1)\cdot 0$$

$$5x(x+1) - x(7x-2) = 0$$

$$5x^2 + 5x - 7x^2 + 2x = 0$$

$$-2x^2 + 7x = 0$$

$$-x(2x-7) = 0$$

$$x = 0 \quad or \quad x = \frac{7}{2}$$

The critical values are -1, 0, $\dfrac{2}{7}$, and $\dfrac{7}{2}$. They divide the x-axis into the intervals $(-\infty, -1)$, $(-1, 0)$, $\left(0, \dfrac{2}{7}\right)$, $\left(\dfrac{2}{7}, \dfrac{7}{2}\right)$, and $\left(\dfrac{7}{2}, \infty\right)$. We test a value in each interval.

$(-\infty, -1)$: $f(-2) = -1.375 < 0$

$(-1, 0)$: $f(-0.5) \approx 1.4545 > 0$

$\left(0, \dfrac{2}{7}\right)$: $f(0.1) \approx -0.4755 < 0$

$\left(\dfrac{2}{7}, \dfrac{7}{2}\right)$: $f(1) = 0.5 > 0$

$\left(\dfrac{7}{2}, \infty\right)$: $f(4) \approx -0.0308 < 0$

Function values are positive on $(-1, 0)$ and $\left(\dfrac{2}{7}, \dfrac{7}{2}\right)$. This can also be determined from the graph of

$y = \dfrac{5x}{7x-2} - \dfrac{x}{x+1}$. The solution set is $(-1,0) \cup \left(\dfrac{2}{7}, \dfrac{7}{2}\right)$.

52. $\dfrac{x^2-x-2}{x^2+5x+6} < 0$

$\dfrac{x^2-x-2}{(x+3)(x+2)} < 0$

The denominator of $f(x) = \dfrac{x^2-x-2}{(x+3)(x+2)}$ is 0 when $x = -3$ or $x = -2$, so the function is not defined for these values of x. We solve the related equation $f(x) = 0$.

$$\frac{x^2-x-2}{(x+3)(x+2)} = 0$$

$$(x-2)(x+1) = 0$$

$$x = 2 \quad or \quad x = -1$$

The critical values are -3, -2, -1, and 2. They divide the x-axis into the intervals $(-\infty,-3)$, $(-3,-2)$, $(-2,-1)$, $(-1,2)$, and $(2,\infty)$. We test a value in each interval.

$(-\infty,-3)$: $f(-4) = 9 > 0$

$(-3,-2)$: $f(-2.5) = -27 < 0$

$(-2,-1)$: $f(-1.5) \approx 2.3333 > 0$

$(-1,2)$: $f(0) \approx -0.3333 < 0$

$(2,\infty)$: $f(3) \approx 0.13333 > 0$

Function values are negative on $(-3,-2)$ and $(-1,2)$. This can also be determined from the graph of $y = \dfrac{x^2-x-2}{(x+3)(x+2)}$. The solution set is $(-3,-2) \cup (-1,2)$.

53. $\dfrac{x}{x^2+4x-5} + \dfrac{3}{x^2-25} \le \dfrac{2x}{x^2-6x+5}$

$\dfrac{x}{x^2+4x-5} + \dfrac{3}{x^2-25} - \dfrac{2x}{x^2-6x+5} \le 0$

$\dfrac{x}{(x+5)(x-1)} + \dfrac{3}{(x+5)(x-5)} - \dfrac{2x}{(x-5)(x-1)} \le 0$

The denominator of

$f(x) = \dfrac{x}{(x+5)(x-1)} + \dfrac{3}{(x+5)(x-5)} - \dfrac{2x}{(x-5)(x-1)}$

is 0 when $x = -5$, 1, or 5, so the function is not defined for these values of x. We solve the related equation $f(x) = 0$.

$$\frac{x}{(x+5)(x-1)} + \frac{3}{(x+5)(x-5)} - \frac{2x}{(x-5)(x-1)} = 0$$

$$x(x-5) + 3(x-1) - 2x(x+5) = 0$$

Multiplying by $(x+5)(x-1)(x-5)$

$$x^2 - 5x + 3x - 3 - 2x^2 - 10x = 0$$

$$-x^2 - 12x - 3 = 0$$

$$x^2 + 12x + 3 = 0$$

Using the quadratic formula, we find that $x = -6 \pm \sqrt{33}$. The critical values are $-6-\sqrt{33}$, -5, $-6+\sqrt{33}$, 1, and 5. They divide the x-axis into the intervals $(-\infty, -6-\sqrt{33})$, $(-6-\sqrt{33}, -5)$, $(-5, -6+\sqrt{33})$, $(-6+\sqrt{33}, 1)$, $(1,5)$, and $(5,\infty)$. We test a value in each interval.

$(-\infty, -6-\sqrt{33})$: $f(-12) \approx 0.00194 > 0$

$(-6-\sqrt{33}, -5)$: $f(-6) \approx -0.4286 < 0$

$(-5, -6+\sqrt{33})$: $f(-1) \approx 0.16667 > 0$

$(-6+\sqrt{33}, 1)$: $f(0) = -0.12 < 0$

$(1,5)$: $f(2) \approx 1.4762 > 0$

$(5,\infty)$: $f(6) \approx -2.018 < 0$

Function values are negative on $(-6-\sqrt{33}, -5)$, $(-6+\sqrt{33}, 1)$, and $(5,\infty)$. They can also be determined from the graph of $y = \dfrac{x}{(x+5)(x-1)} + \dfrac{3}{(x+5)(x-5)} - \dfrac{2x}{(x-5)(x-1)}$. Note that since the inequality symbol is $\le$ and $f(-6 \pm \sqrt{33}) = 0$, then $-6-\sqrt{33}$ and $-6+\sqrt{33}$ must be included in the solution set. Note also that since -5, 1, and 5 are not in the domain of $f(x)$, they are not included in the solution set. It is $[-6-\sqrt{33}, -5) \cup [-6+\sqrt{33}, 1) \cup (5,\infty)$.

54. $\dfrac{2x}{x^2-9} + \dfrac{x}{x^2+x-12} \ge \dfrac{3x}{x^2+7x+12}$

$\dfrac{2x}{(x+3)(x-3)} + \dfrac{x}{(x+4)(x-3)} - \dfrac{3x}{(x+4)(x+3)} \ge 0$

The denominator of $f(x) = \dfrac{2x}{(x+3)(x-3)} + \dfrac{x}{(x+4)(x-3)} - \dfrac{3x}{(x+4)(x+3)}$ is 0 when $x = -3$, 3, or -4, so the function is not defined for these values of x. We solve the related equation $f(x) = 0$.

$$\frac{2x}{(x+3)(x-3)} + \frac{x}{(x+4)(x-3)} - \frac{3x}{(x+4)(x+3)} = 0$$

$$2x(x+4) + x(x+3) - 3x(x-3) = 0$$

$$2x^2 + 8x + x^2 + 3x - 3x^2 + 9x = 0$$

$$20x = 0$$

$$x = 0$$

The critical values are -4, -3, 0, and 3. They divide the x-axis into the intervals $(-\infty,-4)$, $(-4,-3)$, $(-3,0)$, $(0,3)$, and $(3,\infty)$. We test a value in each interval.

$(-\infty,-4)$: $f(-5) = 6.25 > 0$

$(-4,-3)$: $f(-3.5) \approx -43.08 < 0$

$(-3,0)$: $f(-1) \approx 0.83333 > 0$

$(0,3)$: $f(1) = -0.5 < 0$

$(3,\infty)$: $f(4) \approx 1.4286 > 0$

Function values are positive on $(-\infty,-4)$, $(-3,0)$, and $(3,\infty)$. This can also be determined from the

graph of
$y = \frac{2x}{(x+3)(x-3)} + \frac{x}{(x+4)(x-3)} - \frac{3x}{(x+4)(x+3)}$.
Note that since the inequality symbol is $\geq$ and $f(0) = 0$, then 0 must be included in the solution set. Note also that since -4, -3, and 3 are not in the domain of $f(x)$, they are not included in the solution set. It is $(-\infty, -4) \cup (-3, 0] \cup (3, \infty)$.

55. We write and solve a rational inequality.

$$\frac{4t}{t^2+1} + 98.6 > 100$$

$$\frac{4t}{t^2+1} - 1.4 > 0$$

The denominator of $f(t) = \frac{4t}{t^2+1} - 1.4$ has no real-number zeros. We solve the related equation $f(t) = 0$.

$$\frac{4t}{t^2+1} - 1.4 = 0$$

$$4t - 1.4(t^2+1) = 0 \quad \text{Multiplying by } t^2+1$$

$$4t - 1.4t^2 - 1.4 = 0$$

Using the quadratic formula, we find that $t = \frac{4 \pm \sqrt{8.16}}{2.8}$; that is, $t \approx 0.408$ or $t \approx 2.449$. This can also be determined from the graph of $y = \frac{4x}{x^2+1} - 1.4$. These numbers divide the t-axis into the intervals $(-\infty, 0.408)$, $(0.408, 2.449)$, and $(2.449, \infty)$. We test a value in each interval.

$(-\infty, 0.408)$: $f(0) = -1.4 < 0$

$(0.408, 2.449)$: $f(1) = 0.6 > 0$

$(2.449, \infty)$: $f(3) = -0.2 < 0$

Function values are positive on $(0.408, 2.449)$. This can also be determined from the graph. The solution set is $(0.408, 2.449)$.

56. We write and solve a rational inequality.

$$\frac{500t}{2t^2+9} \geq 40$$

$$\frac{500t}{2t^2+9} - 40 \geq 0$$

The denominator of $f(t) = \frac{500t}{2t^2+9} - 40$ has no real-number zeros. We solve the related equation $f(t) = 0$.

$$\frac{500t}{2t^2+9} - 40 = 0$$

$$500t - 80t^2 - 360 = 0 \quad \text{Multiplying by } 2t^2+9$$

Using the quadratic formula, we find that $t = \frac{25 \pm \sqrt{337}}{8}$; that is, $t \approx 0.830$ or $t \approx 5.420$. This can also be determined from the graph of $y = \frac{500x}{2x^2+9} - 40$. These numbers divide the t-axis into the intervals $(-\infty, 0.830)$, $(0.830, 5.420)$, and $(5.420, \infty)$. We test a value in each interval.

$(-\infty, 0.830)$: $f(0) = -40 < 0$

$(0.830, 5.420)$: $f(1) \approx 5.4545 > 0$

$(5.420, \infty)$: $f(6) \approx -2.963 < 0$

Function values are positive on $(0.830, 5.420)$. This can also be determined from the graph. The solution set is $(0.830, 5.420)$.

57. a) We write and solve a polynomial inequality.

$$-3x^2 + 630x - 6000 > 0 \quad (x \geq 0)$$

We first solve the related equation.

$$-3x^2 + 630x - 6000 = 0$$

$$x^2 - 210x + 2000 = 0 \quad \text{Dividing by } -3$$

$$(x-10)(x-200) = 0 \quad \text{Factoring}$$

Using the principle of zero products or by observing the graph of $y = -3x^2 + 630 - 6000$, we see that the solutions of the related equation are 10 and 200. These numbers divide the x-axis into the intervals $(-\infty, 10)$, $(10, 200)$, and $(200, \infty)$. Since we are restricting our discussion to nonnegative values of x, we consider the intervals $[0, 10)$, $(10, 200)$, and $(200, \infty)$.

We let $f(x) = -3x^2 + 630x - 6000$ and test a value in each interval.

$[0, 10)$: $f(0) = -6000 < 0$

$(10, 200)$: $f(11) = 567 > 0$

$(200, \infty)$: $f(201) = -573 < 0$

Function values are positive only on $(10, 200)$. This can also be determined graphically. The solution set is $\{x|10 < x < 200\}$, or $(10, 200)$.

b) From part (a), we see that function values are negative on $[0, 10)$ and $(200, \infty)$. Thus, the solution set is $\{x|0 < x < 10 \textit{ or } x > 200\}$, or $(0, 10) \cup (200, \infty)$.

58. a) We write and solve a polynomial inequality.

$$-16t^2 + 32t + 1920 > 1920$$

$$-16t^2 + 32t > 0$$

$$-16t^2 + 32t = 0 \quad \text{Related equation}$$

$$-16t(t-2) = 0 \quad \text{Factoring}$$

The solutions of the related equation are 0 and 2. This could also be determined from the graph of $y = -16x^2 + 32x$. These numbers divide the t-axis into the intervals $(-\infty, 0)$, $(0, 2)$, and $(2, \infty)$. Since only nonnegative values of t have meaning in this application, we restrict our discussion to the intervals $(0, 2)$ and $(2, \infty)$. We let $f(x) = -16t^2 + 32t$ and test a value in each interval.

$(0, 2)$: $f(1) = 16 > 0$

$(2, \infty)$: $f(3) = -48 < 0$

Function values are positive on $(0, 2)$. This can also be determined graphically. The solution set is $\{t|0 < t < 2\}$, or $(0, 2)$.

b) We write and solve a polynomial inequality.

$$-16t^2 + 32t + 1920 < 640$$
$$-16t^2 + 32t + 1280 < 0$$
$$-16t^2 + 32t^2 + 1280 = 0 \quad \text{Related equation}$$
$$t^2 - 2t - 80 = 0$$
$$(t - 10)(t + 8) = 0$$

The solutions of the related equation are 10 and -8. This could also be determined from the graph of $y = -16t^2 + 32t + 1280$. These numbers divide the t-axis into the intervals $(-\infty, -8)$, $(-8, 10)$, and $(10, \infty)$. As in part (a), we will not consider negative values of t. In addition, note that the nonnegative solution of $S(t) = 0$ is 12. This means that the object reaches the ground in 12 sec. Thus, we also restrict our discussion to values of t such that $t \leq 12$. We consider the intervals $[0, 10)$ and $(10, 12]$. We let $f(x) = -16t^2 + 32t + 1280$ and test a value in each interval.

$[0, 10)$: $f(0) = 1280 > 0$

$(10, 12]$: $f(11) = -304 < 0$

Function values are negative on $(10, 12]$. This can also be determined graphically. The solution set is $\{x|10 < x \leq 12\}$, or $(10, 12]$.

59. We write an inequality.

$$27 \leq \frac{n(n-3)}{2} \leq 230$$
$$54 \leq n(n-3) \leq 460 \quad \text{Multiplying by 2}$$
$$54 \leq n^2 - 3n \leq 460$$

We write this as two inequalities.

$$54 \leq n^2 - 3n \quad \textit{and} \quad n^2 - 3n \leq 460$$

Solve each inequality.

$$n^2 - 3n \geq 54$$
$$n^2 - 3n - 54 \geq 0$$
$$n^2 - 3n - 54 = 0 \quad \text{Related equation}$$
$$(n + 6)(n - 9) = 0$$
$$n = -6 \quad or \quad n = 9$$

Since only positive values of n have meaning in this application, we consider the intervals $(0, 9)$ and $(9, \infty)$. Let $f(n) = n^2 - 3n - 54$ and test a value in each interval.

$(0, 9)$: $f(1) = -56 < 0$

$(9, \infty)$: $f(10) = 16 > 0$

Function values are positive on $(9, \infty)$. Since the inequality symbol is $\geq$, 9 must also be included in the solution set for this portion of the inequality. It is $\{n|n \geq 9\}$.

Now solve the second inequality.

$$n^2 - 3n \leq 460$$
$$n^2 - 3n - 460 \leq 0$$
$$n^2 - 3n - 460 = 0 \quad \text{Related equation}$$
$$(n + 20)(n - 23) = 0$$
$$n = -20 \quad or \quad n = 23$$

We consider only positive values of n as above. Thus, we consider the intervals $(0, 23)$ and $(23, \infty)$. Let $f(n) = n^2 - 3n - 460$ and test a value in each interval.

$(0, 23)$: $f(1) = -462 < 0$

$(23, \infty)$: $f(24) = 44 > 0$

Function values are negative on $(0, 23)$. Since the inequality symbol is $\leq$, 23 must also be included in the solution set for this portion of the inequality. It is $\{n|0 < n \leq 23\}$.

The solution set of the original inequality is $\{n|n \geq 9 \text{ and } 0 < n \leq 23\}$, or $\{n|9 \leq n \leq 23\}$.

60. We write an inequality.

$$66 \leq \frac{n(n-1)}{2} \leq 300$$
$$132 \leq n^2 - n \leq 600$$

We write this as two inequalities.

$$132 \leq n^2 - n \quad \textit{and} \quad n^2 - n \leq 600$$

Solve each inequality.

$$n^2 - n \geq 132$$
$$n^2 - n - 132 \geq 0$$
$$n^2 - n - 132 = 0 \quad \text{Related equation}$$
$$(n + 11)(n - 12) = 0$$
$$n = -11 \quad or \quad n = 12$$

Since only positive values of n have meaning in this application, we consider the intervals $(0, 12)$ and $(12, \infty)$. Let $f(n) = n^2 - n - 132$ and test a value in each interval.

$(0, 12)$: $f(1) = -132 < 0$

$(12, \infty)$: $f(13) = 24 > 0$

Function values are positive on $(12, \infty)$. Since the inequality symbol is $\geq$, 12 must also be included in the solution set for this portion of the inequality. It is $\{n|n \geq 12\}$.

Now solve the second inequality.

$$n^2 - n \leq 600$$
$$n^2 - n - 600 \leq 0$$
$$n^2 - n - 600 = 0 \quad \text{Related equation}$$
$$(n + 24)(n - 25) = 0$$
$$n = -24 \quad or \quad n = 25$$

We consider only positive values of n as above. Thus, we consider the intervals $(0, 25)$ and $(25, \infty)$. Let $f(n) = n^2 - n - 600$ and test a value in each interval.

$(0, 25)$: $f(1) = -600 < 0$

$(25, \infty)$: $f(26) = 50 > 0$

Function values are negative on $(0, 25)$. Since the inequality symbol is $\leq$, 25 must also be included in the solution set for this portion of the inequality. It is $\{n|0 < n \leq 25\}$.

The solution set of the original inequality is $\{n|n \geq 12 \text{ and } 0 < n \leq 25\}$, or $\{n|12 \leq n \leq 25\}$.

61. A quadratic inequality $ax^2 + bx + c \le 0$, $a > 0$, or $ax^2 + bx + c \ge 0$, $a < 0$, has a solution set that is a closed interval.

62. We need to know these values because they cannot be included in the solution set. The sign of the function often changes at these values as well.

63.
$$(x-h)^2 + (y-k)^2 = r^2$$
$$[x-(-2)]^2 + (y-4)^2 = 3^2$$
$$(x+2)^2 + (y-4)^2 = 9$$

64. $r = \frac{7/2}{2} = \frac{7}{4}$
$$(x-0)^2 + [y-(-3)]^2 = \left(\frac{7}{4}\right)^2$$
$$x^2 + (y+3)^2 = \frac{49}{16}$$

65. $h(x) = -2x^2 + 3x - 8$

a) $-\frac{b}{2a} = -\frac{3}{2(-2)} = \frac{3}{4}$

$$h\left(\frac{3}{4}\right) = -2\left(\frac{3}{4}\right)^2 + 3\cdot\frac{3}{4} - 8 = -\frac{55}{8}$$

The vertex is $\left(\frac{3}{4}, -\frac{55}{8}\right)$.

b) The coefficient of x^2 is negative, so there is a maximum value. It is the second coordinate of the vertex, $-\frac{55}{8}$. It occurs at $x = \frac{3}{4}$.

c) The range is $\left(-\infty, -\frac{55}{8}\right]$.

66. $g(x) = x^2 - 10x + 2$

a) $-\frac{b}{2a} = -\frac{-10}{2\cdot 1} = 5$

$g(5) = 5^2 - 10\cdot 5 + 2 = -23$

The vertex is $(5, -23)$.

b) Minimum: -23 at $x = 5$

c) $[-23, \infty)$

67.
$$x^2 + 9 \le 6x$$
$$x^2 - 6x + 9 \le 0$$
$$(x-3)^2 \le 0 \quad \text{Factoring}$$

Note that $(x-3)^2 \ge 0$ for all values of x and that $(x-3)^2 = 0$ when $x = 3$. Thus, the solution set is $\{3\}$.

68.
$$x^4 - 6x^2 + 5 > 0$$
$$x^4 - 6x^2 + 5 = 0 \quad \text{Related equation}$$
$$(x^2-1)(x^2-5) = 0$$
$$x = \pm 1 \ \textit{or} \ x = \pm\sqrt{5}$$

Let $f(x) = x^4 - 6x^2 + 5$ and test a value in each of the intervals determined by the solutions of the related equation.

$(-\infty, -\sqrt{5})$: $f(-3) = 32 > 0$

$(-\sqrt{5}, -1)$: $f(-2) = -3 < 0$

$(-1, 1)$: $f(0) = 5 > 0$

$(1, \sqrt{5})$: $f(2) = -3 < 0$

$(\sqrt{5}, \infty)$: $f(3) = 32 > 0$

The solution set is $(-\infty, -\sqrt{5}) \cup (-1, 1) \cup (\sqrt{5}, \infty)$.

69.
$$x^4 + 3x^2 > 4x - 15$$
$$x^4 + 3x^2 - 4x + 15 > 0$$

The graph of $y = x^4 + 3x^2 - 4x + 15$ lies entirely above the x-axis. Thus, the solution set is the set of all real numbers, or $(-\infty, \infty)$.

70.
$$\left|\frac{x+3}{x-4}\right| < 2$$
$$-2 < \frac{x+3}{x-4} < 2$$
$$-2 < \frac{x+3}{x-4} \ \textit{and} \ \frac{x+3}{x-4} < 2$$

First solve $-2 < \frac{x+3}{x-4}$.
$$\frac{x+3}{x-4} + 2 > 0$$

The denominator of $f(x) = \frac{x+3}{x-4} + 2$ is 0 when $x = 4$, so the function is not defined for this value of x. Now solve the related equation.
$$\frac{x+3}{x-4} + 2 = 0$$
$$x + 3 + 2(x-4) = 0 \quad \text{Multiplying by } x-4$$
$$x + 3 + 2x - 8 = 0$$
$$3x - 5 = 0$$
$$x = \frac{5}{3}$$

The critical values are $\frac{5}{3}$ and 4. Test a value in each of the intervals determined by them.

$\left(-\infty, \frac{5}{3}\right)$: $f(0) = 1.25 > 0$

$\left(\frac{5}{3}, 4\right)$: $f(2) = -0.5 < 0$

$(4, \infty)$: $f(5) = 10 > 0$

The solution set for this portion of the inequality is $\left(-\infty, \frac{5}{3}\right) \cup (4, \infty)$.

Next solve $\frac{x+3}{x-4} < 2$, or $\frac{x+3}{x-4} - 2 < 0$.

The denominator of $f(x) = \frac{x+3}{x-4} - 2$ is 0 when $x = 4$, so the function is not defined for this value of x. Now solve the related equation.

$$\frac{x+3}{x-4} - 2 = 0$$
$$x + 3 - 2(x-4) = 0 \quad \text{Multiplying by } x-4$$
$$x + 3 - 2x + 8 = 0$$
$$-x + 11 = 0$$
$$x = 11$$

The critical values are 4 and 11. Test a value in each of the intervals determined by them.

$(-\infty, 4)$: $f(0) = -2.75 < 0$

$(4, 11)$: $f(5) = 6 > 0$

$(11, \infty)$: $f(12) = -0.125 < 0$

The solution set for this portion of the inequality is $(-\infty, 4) \cup (11, \infty)$.

The solution set of the original inequality is

$\left(\left(-\infty, \frac{5}{3}\right) \cup (4, \infty)\right)$ and $((-\infty, 4) \cup (11, \infty))$, or $\left(-\infty, \frac{5}{3}\right) \cup (11, \infty)$.

71. $|x^2 - 5| = |5 - x^2| = 5 - x^2$ when $5 - x^2 \geq 0$. Thus we solve $5 - x^2 \geq 0$.

$$5 - x^2 \geq 0$$
$$5 - x^2 = 0 \quad \text{Related equation}$$
$$5 = x^2$$
$$\pm\sqrt{5} = x$$

Let $f(x) = 5 - x^2$ and test a value in each of the intervals determined by the solutions of the related equation.

$(-\infty, -\sqrt{5})$: $f(-3) = -4 < 0$

$(-\sqrt{5}, \sqrt{5})$: $f(0) = 5 > 0$

$(\sqrt{5}, \infty)$: $f(3) = -4 < 0$

Function values are positive on $(-\sqrt{5}, \sqrt{5})$. Since the inequality symbol is $\geq$, the endpoints of the interval must be included in the solution set. It is $[-\sqrt{5}, \sqrt{5}]$.

72. $(7-x)^{-2} < 0$

$$\frac{1}{(7-x)^2} < 0$$

Since $(7-x)^2 \geq 0$ for all real numbers x, then $\frac{1}{(7-x)^2} > 0$ for all values of x in the domain of $f(x) = \frac{1}{(7-x)^2}$. Thus, the solution set is $\emptyset$.

73.
$$2|x|^2 - |x| + 2 \leq 5$$
$$2|x|^2 - |x| - 3 \leq 0$$
$$2|x|^2 - |x| - 3 = 0 \quad \text{Related equation}$$
$$(2|x| - 3)(|x| + 1) = 0 \quad \text{Factoring}$$
$$2|x| - 3 = 0 \quad or \quad |x| + 1 = 0$$
$$|x| = \frac{3}{2} \quad or \quad |x| = -1$$

The solution of the first equation is $x = -\frac{3}{2}$ or $x = \frac{3}{2}$. The second equation has no solution. Let $f(x) = 2|x|^2 - |x| - 3$ and test a value in each interval determined by the solutions of the related equation.

$\left(-\infty, -\frac{3}{2}\right)$: $f(-2) = 3 > 0$

$\left(-\frac{3}{2}, \frac{3}{2}\right)$: $f(0) = -3 < 0$

$\left(\frac{3}{2}, \infty\right)$: $f(2) = 3 > 0$

Function values are negative on $\left(-\frac{3}{2}, \frac{3}{2}\right)$. Since the inequality symbol is $\leq$, the endpoints of the interval must also be included in the solution set. It is $\left[-\frac{3}{2}, \frac{3}{2}\right]$.

74.
$$|x|^2 - 4|x| + 4 \geq 9$$
$$|x|^2 - 4|x| - 5 \geq 0$$
$$|x|^2 - 4|x| - 5 = 0 \quad \text{Related equation}$$
$$(|x| + 1)(|x| - 5) = 0$$
$$|x| + 1 = 0 \quad or \quad |x| - 5 = 0$$
$$|x| = 0 \quad or \quad |x| = 5$$

The first equation has no solution. The solution of the second equation is $x = -5$ or $x = 5$. Let $f(x) = |x|^2 - 4|x| - 5$ and test a value in each of the intervals determined by the solutions of the related equation.

$(-\infty, -5)$: $f(-6) = 7 > 0$

$(-5, 5)$: $f(0) = -5 < 0$

$(5, \infty)$: $f(6) = 7 > 0$

Function values are positive on $(-\infty, -5)$ and $(5, \infty)$. Since the inequality symbol is $\geq$, the endpoints of the intervals must be included in the solution set. It is $(-\infty, -5] \cup [5, \infty)$.

75.
$$\left|1 + \frac{1}{x}\right| < 3$$
$$-3 < 1 + \frac{1}{x} < 3$$
$$-3 < 1 + \frac{1}{x} \quad and \quad 1 + \frac{1}{x} < 3$$

First solve $-3 < 1 + \frac{1}{x}$.

$$0 < 4 + \frac{1}{x}, \quad \text{or} \quad \frac{1}{x} + 4 > 0$$

The denominator of $f(x) = \frac{1}{x} + 4$ is 0 when $x = 0$, so the

function is not defined for this value of x. Now solve the related equation.

$$\frac{1}{x} + 4 = 0$$
$$1 + 4x = 0 \quad \text{Multiplying by } x$$
$$x = -\frac{1}{4}$$

The critical values are $-\frac{1}{4}$ and 0. Test a value in each of the intervals determined by them.

$\left(\infty,-\frac{1}{4}\right)$: $f(-1)=3>0$

$\left(-\frac{1}{4},0\right)$: $f(-0.1)=-6<0$

$(0,\infty)$: $f(1)=5>0$

The solution set for this portion of the inequality is $\left(-\infty,-\frac{1}{4}\right)\cup(0,\infty)$.

Next solve $1+\frac{1}{x}<3$, or $\frac{1}{x}-2<0$. The denominator of $f(x)=\frac{1}{x}-2$ is 0 when $x=0$, so the function is not defined for this value of x. Now solve the related equation.

$$\frac{1}{x}-2=0$$
$$1-2x=0 \quad \text{Multiplying by } x$$
$$x=\frac{1}{2}$$

The critical values are 0 and $\frac{1}{2}$. Test a value in each of the intervals determined by them.

$(-\infty,0)$: $f(-1)=-3<0$

$\left(0,\frac{1}{2}\right)$: $f(0.1)=8>0$

$\left(\frac{1}{2},\infty\right)$: $f(1)=-1<0$

The solution set for this portion of the inequality is $(-\infty,0)\cup\left(\frac{1}{2},\infty\right)$.

The solution set of the original inequality is

$$\left(\left(-\infty,-\frac{1}{4}\right)\cup(0,\infty)\right)\textit{and}\left((-\infty,0)\cup\left(\frac{1}{2},\infty\right)\right),$$

or $\left(-\infty,-\frac{1}{4}\right)\cup\left(\frac{1}{2},\infty\right)$.

76. $$\left|2-\frac{1}{x}\right|\le 2+\left|\frac{1}{x}\right|$$
$$\left|2-\frac{1}{x}\right|-2-\left|\frac{1}{x}\right|\le 0$$

By examining the graph or the table for

$y=\left|2-\frac{1}{x}\right|-2-\left|\frac{1}{x}\right|$, we see that $y=0$ for $x<0$, $y<0$ for

$x>0$, and 0 is not in the domain. Thus, the solution set is $(-\infty,0)\cup(0,\infty)$.

77. $$|x^2+3x-1|<3$$
$$-3<x^2+3x-1<3$$
$$-3<x^2+3x-1 \quad \textit{and} \quad x^2+3x-1<3$$

First solve $-3<x^2+3x-1$, or $x^2+3x-1>-3$.

$$x^2+3x-1>-3$$
$$x^2+3x+2>0$$
$$x^2+3x+2=0 \quad \text{Related equation}$$
$$(x+2)(x+1)=0$$
$$x=-2 \quad \textit{or} \quad x=-1$$

Let $f(x)=x^2+3x+2$ and test a value in each of the intervals determined by the solution of the related equation.

$(-\infty,-2)$: $f(-3)=2>0$

$(-2,-1)$: $f(-1.5)=-0.25<0$

$(-1,\infty)$: $f(0)=2>0$

The solution set for this portion of the inequality is $(-\infty,-2)\cup(-1,\infty)$.

Next solve $x^2+3x-1<3$.

$$x^2+3x-1<3$$
$$x^2+3x-4<0$$
$$x^2+3x-4=0 \quad \text{Related equation}$$
$$(x+4)(x-1)=0$$
$$x=-4 \quad \textit{or} \quad x=1$$

Let $f(x)=x^2+3x-4$ and test a value in each of the intervals determined by the solution of the related equation.

$(-\infty,-4)$: $f(-5)=6>0$

$(-4,1)$: $f(0)=-4<0$

$(1,\infty)$: $f(2)=6>0$

The solution set for this portion of the inequality is $(-4,1)$.

The solution set of the original inequality is $((-\infty,-2)\cup(-1,\infty))$ *and* $(-4,1)$, or $(-4,-2)\cup(-1,1)$.

78. $|1+5x-x^2|\ge 5$

$1+5x-x^2\le -5$ *or* $1+5x-x^2\ge 5$

First solve $1+5x-x^2\le -5$.

$$1+5x-x^2\le -5$$
$$6+5x-x^2\le 0$$
$$6+5x-x^2=0 \quad \text{Related equation}$$
$$(6-x)(1+x)=0$$
$$x=6 \quad \textit{or} \quad x=-1$$

Let $f(x)=6+5x-x^2$ and test a value in each of the intervals determined by the solution of the related equation.

$(-\infty,-1)$: $f(-2)=-8<0$

$(-1,6)$: $f(0)=6>0$

$(6,\infty)$: $f(7)=-8<0$

The solution set for this portion of the inequality is $(-\infty,-1]\cup[6,\infty)$.

Next solve $1+5x-x^2\ge 5$.

$$1 + 5x - x^2 \geq 5$$
$$-4 + 5x - x^2 \geq 0$$
$$x^2 - 5x + 4 \leq 0$$
$$x^2 - 5x + 4 = 0 \quad \text{Related equation}$$
$$(x - 1)(x - 4) = 0$$
$$x = 1 \quad or \quad x = 4$$

Let $f(x) = x^2 - 5x + 4$ and test a value in each of the intervals determined by the solution of the related equation.

$(-\infty, 1)$: $f(0) = 4 > 0$

$(1, 4)$: $f(2) = -2 < 0$

$(4, \infty)$: $f(5) = 4 > 0$

The solution set for this portion of the inequality is $[1, 4]$.

The solution set of the original inequality is $\left((-\infty, -1] \cup [6, \infty)\right)$ *or* $[1, 4]$, or $(-\infty, -1] \cup [1, 4] \cup [6, \infty)$.

79. First find a quadratic equation with solutions -4 and 3.

$$(x + 4)(x - 3) = 0$$
$$x^2 + x - 12 = 0$$

Observe that the graph of $y = x^2 + x - 12$ lies below the x-axis on $(-4, 3)$. Thus, a quadratic inequality for which the solution set is $(-4, 3)$ is $x^2 + x - 12 < 0$. Answers may vary.

80. First find the polynomial with solutions -4, 3, and 7.

$$(x + 4)(x - 3)(x - 7) = 0$$
$$x^3 - 6x^2 - 19x + 84 = 0$$

Observe that the graph of $y = x^3 - 6x^2 - 19x + 84$ lies on or above the x-axis on $[-4, 3] \cup [7, \infty)$. Thus, a polynomial inequality for which the solution set is $[-4, 3] \cup [7, \infty)$ is $x^3 - 6x^2 - 19x + 84 \geq 0$. Answers may vary.

Chapter 4

Exponential and Logarithmic Functions

Exercise Set 4.1

1. $(f \circ g)(x) = f(g(x)) = f(x-3) = x-3+3 = x$
$(g \circ f)(x) = g(f(x)) = g(x+3) = x+3-3 = x$

2. $(f \circ g)(x) = f\left(\frac{5}{4}x\right) = \frac{4}{5} \cdot \frac{5}{4}x = x$

$(g \circ f)(x) = g\left(\frac{4}{5}x\right) = \frac{5}{4} \cdot \frac{4}{5}x = x$

3. $(f \circ g)(x) = f(g(x)) = f\left(\frac{x+7}{3}\right) =$

$3\left(\frac{x+7}{3}\right) - 7 = x+7-7 = x$

$(g \circ f)(x) = g(f(x)) = g(3x-7) = \frac{(3x-7)+7}{3} =$

$\frac{3x}{3} = x$

4. $(f \circ g)(x) = f(1.5x+1.2) = \frac{2}{3}(1.5x+1.2) - \frac{4}{5}$

$x + 0.8 - \frac{4}{5} = x$

$(g \circ f)(x) = g\left(\frac{2}{3}x - \frac{4}{5}\right) = 1.5\left(\frac{2}{3}x - \frac{4}{5}\right) + 1.2 =$
$x - 1.2 + 1.2 = x$

5. $(f \circ g)(x) = f(g(x)) = f(0.05) = 20$
$(g \circ f)(x) = g(f(x)) = g(20) = 0.05$

6. $(f \circ g)(x) = (\sqrt[4]{x})^4 = x$
$(g \circ f)(x) = \sqrt[4]{x^4} = |x|$

7. $(f \circ g)(x) = f(g(x)) = f(x^2-5) =$
$\sqrt{x^2-5+5} = \sqrt{x^2} = |x|$
$(g \circ f)(x) = g(f(x)) = g(\sqrt{x+5}) =$
$(\sqrt{x+5})^2 - 5 = x+5-5 = x$

8. $(f \circ g)(x) = (\sqrt[5]{x+2})^5 - 2 = x+2-2 = x$
$(g \circ f)(x) = \sqrt[5]{x^5-2+2} = \sqrt[5]{x^5} = x$

9. $(f \circ g)(x) = f(g(x)) = f\left(\frac{1}{1+x}\right) =$

$\frac{1-\left(\frac{1}{1+x}\right)}{\frac{1}{1+x}} \; \frac{\frac{1+x-1}{1+x}}{\frac{1}{1+x}} =$

$\frac{x}{1+x} \cdot \frac{1+x}{1} = x$

$(g \circ f)(x) = g(f(x)) = g\left(\frac{1-x}{x}\right) =$

$\frac{1}{1+\left(\frac{1-x}{x}\right)} = \frac{1}{\frac{x+1-x}{x}} =$

$\frac{1}{\frac{1}{x}} = 1 \cdot \frac{x}{1} = x$

10. $(f \circ g)(x) = \frac{\left(\frac{3x-4}{5x-2}\right)^2 - 1}{\left(\frac{3x-4}{5x-2}\right)^2 + 1} = \frac{-8x^2-2x+6}{17x^2-22x+10}$

$(g \circ f)(x) = \frac{3\left(\frac{x^2-1}{x^2+1}\right) - 4}{5\left(\frac{x^2-1}{x^2+1}\right) - 2} = \frac{-x^2-7}{3x^2-7}$

11. $(f \circ g)(x) = f(g(x)) = f(x+1) =$
$(x+1)^3 - 5(x+1)^2 + 3(x+1) + 7 =$
$x^3 + 3x^2 + 3x + 1 - 5x^2 - 10x - 5 + 3x + 3 + 7 =$
$x^3 - 2x^2 - 4x + 6$

$(g \circ f)(x) = g(f(x)) = g(x^3 - 5x^2 + 3x + 7) =$
$x^3 - 5x^2 + 3x + 7 + 1 = x^3 - 5x^2 + 3x + 8$

12. $(g \circ f)(x) = x^3 + 2x^2 - 3x - 9 - 1 =$
$x^3 + 2x^2 - 3x - 10$
$(g \circ f)(x) = (x-1)^3 + 2(x-1)^2 - 3(x-1) - 9 =$
$x^3 - 3x^2 + 3x - 1 + 2x^2 - 4x + 2 - 3x + 3 - 9 =$
$x^3 - x^2 - 4x - 5$

13. $h(x) = (4+3x)^5$
This is $4+3x$ to the 5th power. The most obvious answer is $f(x) = x^5$ and $g(x) = 4+3x$.

14. $f(x) = \sqrt[3]{x}$, $g(x) = x^2 - 8$

15. $h(x) = \frac{1}{(x-2)^4}$
This is 1 divided by $(x-2)$ to the 4th power. One obvious answer is $f(x) = \frac{1}{x^4}$ and $g(x) = (x-2)$. Another possibility is $f(x) = \frac{1}{x}$ and $g(x) = (x-2)^4$.

16. $f(x) = \frac{1}{\sqrt{x}}$, $g(x) = 3x + 7$

17. $f(x) = \frac{x-1}{x+1}$, $g(x) = x^3$

18. $f(x) = |x|$, $g(x) = 9x^2 - 4$

19. $f(x) = x^6$, $g(x) = \dfrac{2 + x^3}{2 - x^3}$

20. $f(x) = x^4$, $g(x) = \sqrt{x} - 3$

21. $f(x) = \sqrt{x}$, $g(x) = \dfrac{x - 5}{x + 2}$

22. $f(x) = \sqrt{1 + x}$, $g(x) = \sqrt{1 + x}$

23. $f(x) = x^3 - 5x^2 + 3x - 1$, $g(x) = x + 2$

24. $f(x) = 2x^{5/3} + 5x^{2/3}$, $g(x) = x - 1$, or
$f(x) = 2x^5 + 5x^2$, $g(x) = (x - 1)^{1/3}$

25. $f(x) = y(x) \circ s(x) = y(s(x))$
$f(x) = 2(x - 32 + 12) = 2(x - 20)$

26. a) Use the distance formula, distance = rate × time. Substitute 3 for the rate and t for time.

$r(t) = 3t$

b) Use the formula for the area of a circle.

$A(r) = \pi r^2$

c) $(A \circ r)(t) = A(r(t)) = A(3t) = \pi(3t)^2 = 9\pi t^2$

This function gives the area of the ripple in terms of time t.

27. We interchange the first and second coordinates of each ordered pair to find the inverse of the relation. It is

$$\{(8,7), (8,-2), (-4,3), (-8,8)\}.$$

28. $\{(1,0), (6,5), (-4,-2)\}$

29. We interchange the first and second coordinates of each ordered pair to find the inverse of the relation. It is

$$\{(-1,-1), (4,-3)\}.$$

30. $\{(3,-1), (5,2), (5,-3), (0,2)\}$

31. Interchange x and y.

$y = 4x - 5$
$\downarrow \quad \downarrow$
$x = 4y - 5$

32. $2y^2 + 5x^2 = 4$

33. Interchange x and y.

$x^3 y = -5$
$\downarrow \downarrow$
$y^3 x = -5$

34. $x = 3y^2 - 5y + 9$

35. Graph $x = y^2 - 3$. Some points on the graph are $(-3,0)$, $(-2,-1)$, $(-2,1)$, $(1,-2)$, and $(1,2)$. Plot these points and draw the curves. Then reflect the graph across the line $y = x$.

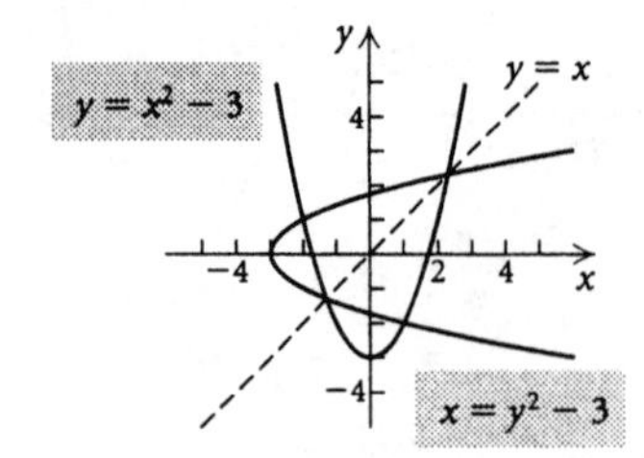

36.

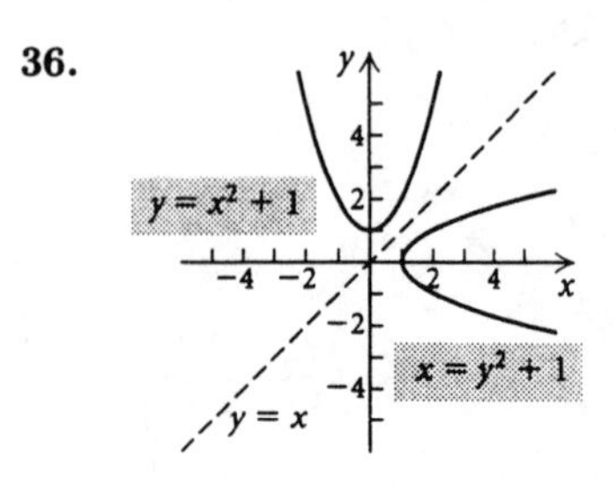

37. Graph $y = |x|$. Some points on the graph are $(0,0)$, $(-2,2)$, $(2,2)$, $(-5,5)$, and $(5,5)$. Plot these points and draw the graph. Then reflect the graph across the line $y = x$.

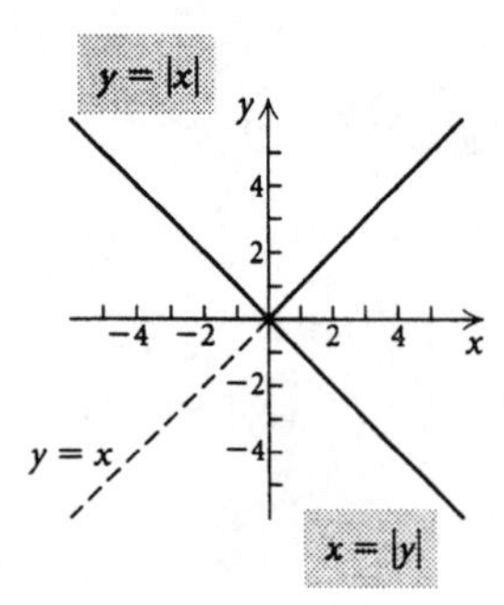

38. See the graph in Exercise 37.

39. The function is one-to-one, because no horizontal line crosses the graph more than once.

40. The function is one-to-one, because no horizontal line crosses the graph more than once.

41. The function is not one-to-one, because there are many horizontal lines that cross the graph more than once.

42. The function is not one-to-one, because there are many horizontal lines that cross the graph more than once.

43. The function is not one-to-one, because there are many horizontal lines that cross the graph more than once.

44. The function is one-to-one, because no horizontal line crosses the graph more than once.

45. The function is one-to-one, because no horizontal line crosses the graph more than once.

46. The function is one-to-one, because no horizontal line crosses the graph more than once.

47. The graph of $f(x) = 5x - 8$ is shown below.

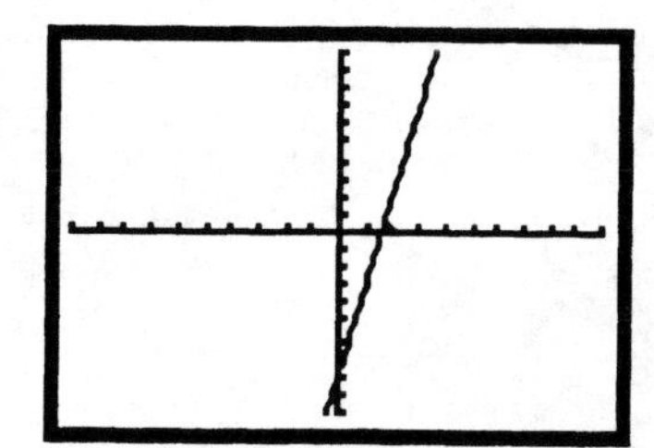

Since there is no horizontal line that crosses the graph more than once, the function is one-to-one.

48. The graph of $f(x) = 3 + 4x$ is shown below.

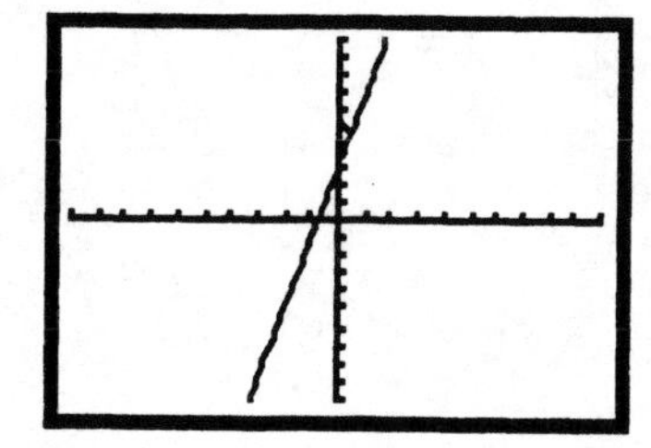

Since there is no horizontal line that crosses the graph more than once, the function is one-to-one.

49. The graph of $f(x) = 1 - x^2$ is shown below.

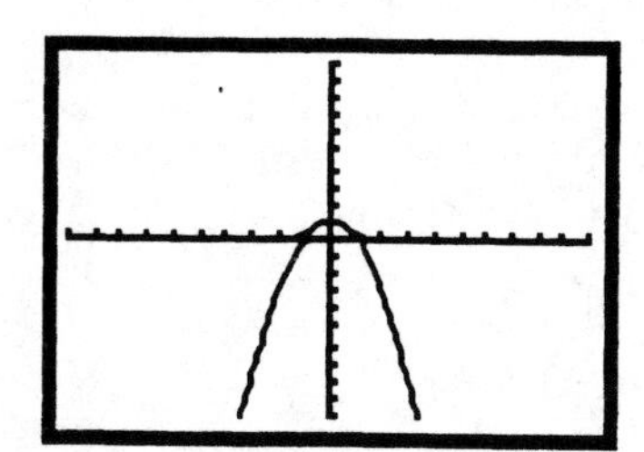

Since there are many horizontal lines that cross the graph more than once, the function is not one-to-one.

50. The graph of $f(x) = |x| - 2$ is shown below.

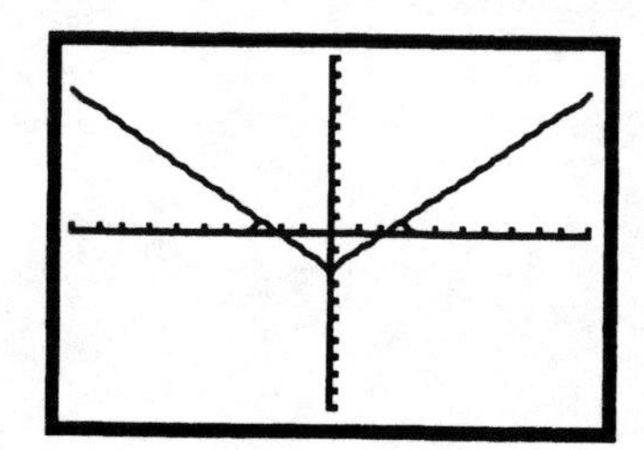

Since there are many horizontal lines that cross the graph more than once, the function is not one-to-one.

51. The graph of $f(x) = |x + 2|$ is shown below.

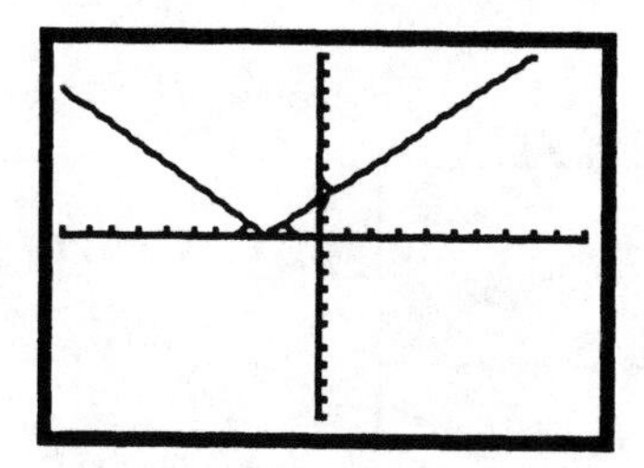

Since there are many horizontal lines that cross the graph more than once, the function is not one-to-one.

52. The graph of $f(x) = -0.8$ is shown below.

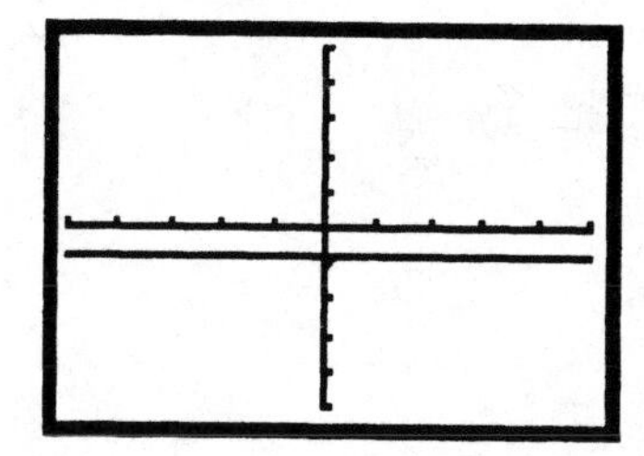

Since the horizontal line $y = -0.8$ crosses the graph more than once, the function is not one-to-one.

53. The graph of $f(x) = -\frac{4}{x}$ is shown below.

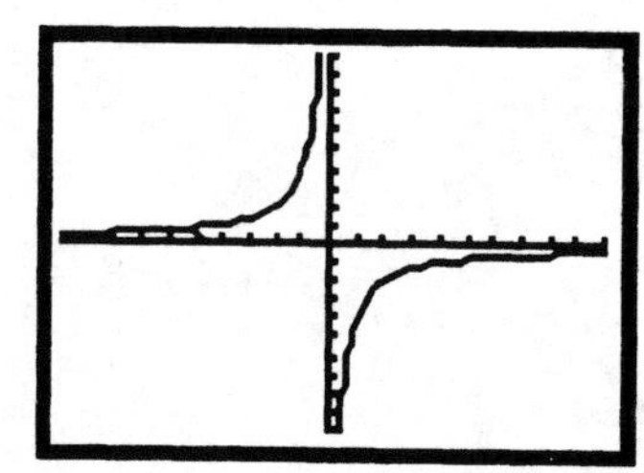

Since there is no horizontal line that crosses the graph more than once, the function is one-to-one.

54. The graph of $f(x) = \frac{2}{x+3}$ is shown below.

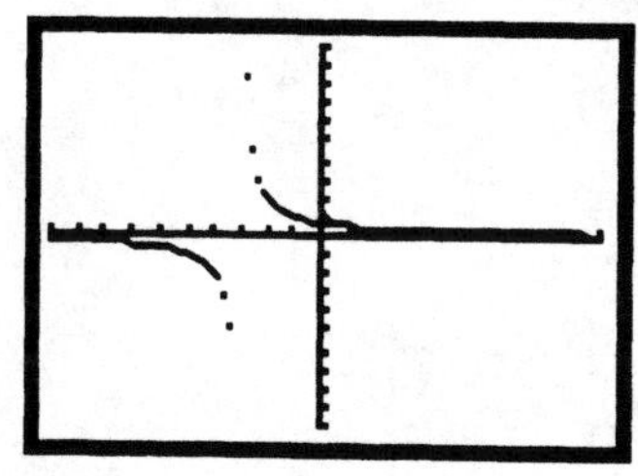

Since there is no horizontal line that crosses the graph more than once, the function is one-to-one.

55. $y_1 = 0.8x + 1.7,$
$y_2 = \dfrac{x - 1.7}{0.8}$

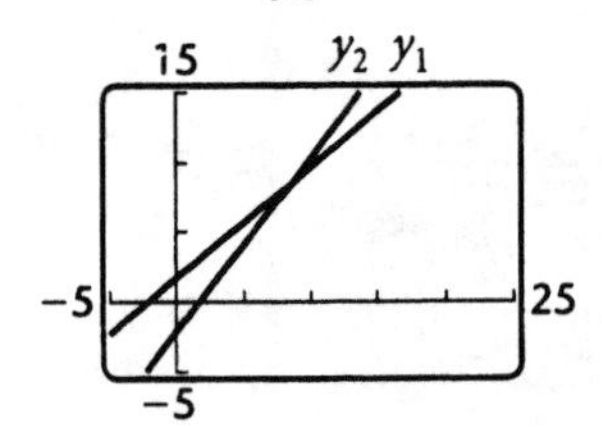

Both the domain and the range of f are the set of all real numbers. Then both the domain and the range of f^{-1} are also the set of all real numbers.

56. $y_1 = 2.7 - 1.08x,$
$y_2 = \dfrac{2.7 - x}{1.08}$

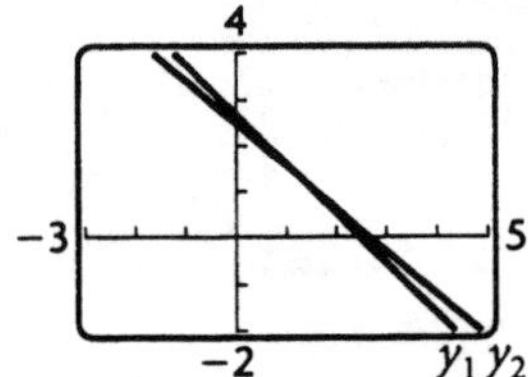

Both the domain and the range of f are the set of all real numbers. Then both the domain and the range of f^{-1} are also the set of all real numbers.

57. $y_1 = \frac{1}{2}x - 4,$
$y_2 = 2x + 8$

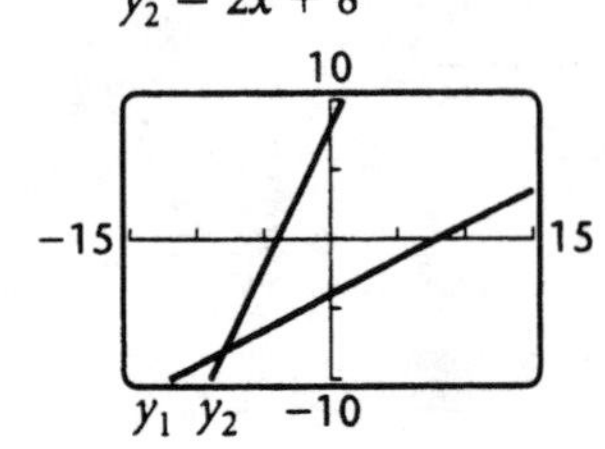

Both the domain and the range of f are the set of all real numbers. Then both the domain and the range of f^{-1} are also the set of all real numbers.

58. $y_1 = x^3 - 1,$
$y_2 = \sqrt[3]{x + 1}$

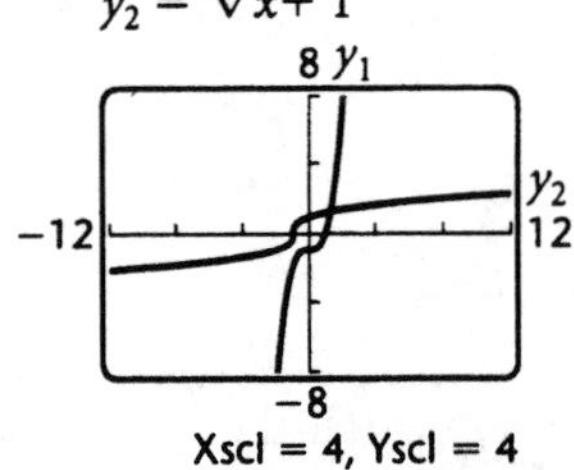

Both the domain and the range of f are the set of all real numbers. Then both the domain and the range of f^{-1} are also the set of all real numbers.

59. $y_1 = \sqrt{x - 3},$
$y_2 = x^2 + 3, x \geqslant 0$

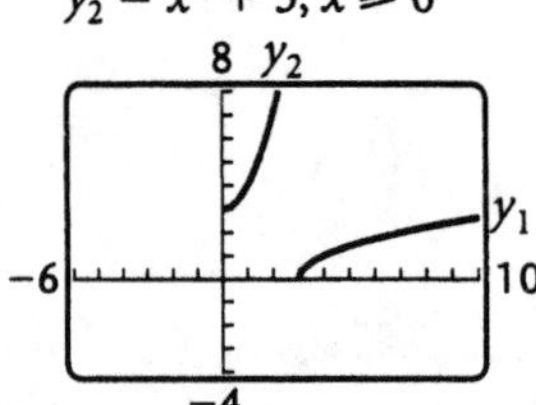

The domain of f is $[3, \infty)$ and the range of f is $[0, \infty)$. Then the domain of f^{-1} is $[0, \infty)$ and the range of f^{-1} is $[3, \infty)$.

60. $y_1 = -\dfrac{2}{x}, \; y_2 = -\dfrac{2}{x}$

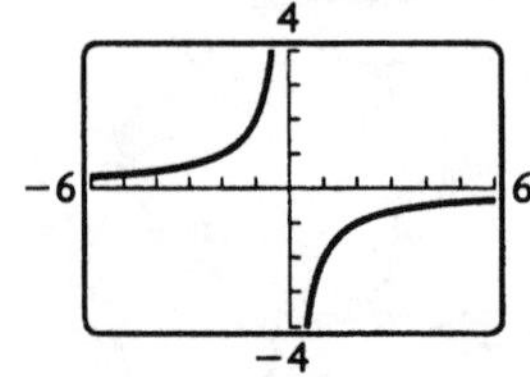

Both the domain and the range of f are $(-\infty, 0) \cup (0, \infty)$. Since $f^{-1} = f$, f^{-1} has the same domain and range.

61. 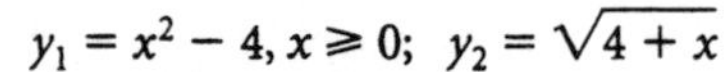
$y_1 = x^2 - 4, x \geqslant 0; \; y_2 = \sqrt{4 + x}$

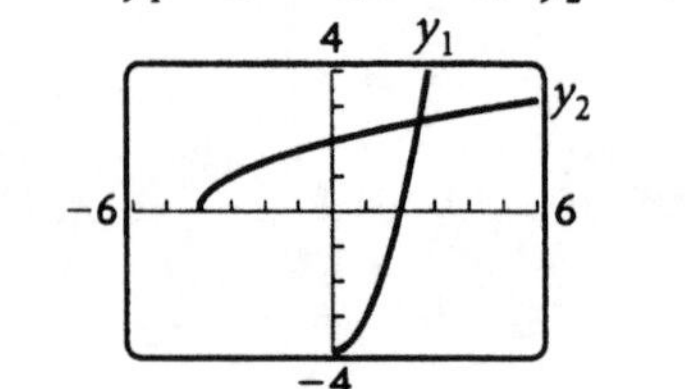

Since it is specified that $x \geq 0$, the domain of f is $[0, \infty)$. The range of f is $[-4, \infty)$. Then the domain of f^{-1} is $[-4, \infty)$ and the range of f^{-1} is $[0, \infty)$.

62. $y_1 = 3 - x^2, x \geqslant 0;$
$y_2 = \sqrt{3 - x}$

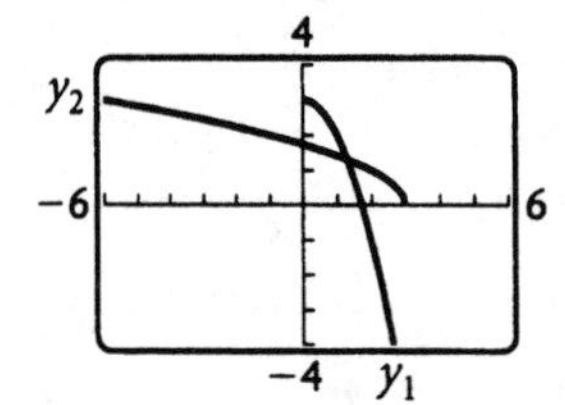

Since it is specified that $x \geq 0$, the domain of f is $[0, \infty)$. The range of f is $(-\infty, 3]$. Then the domain of f^{-1} is $(-\infty, 3]$ and the range of f^{-1} is $[0, \infty)$.

63. $y_1 = (3x - 9)^3, \; y_2 = \dfrac{\sqrt[3]{x} + 9}{3}$

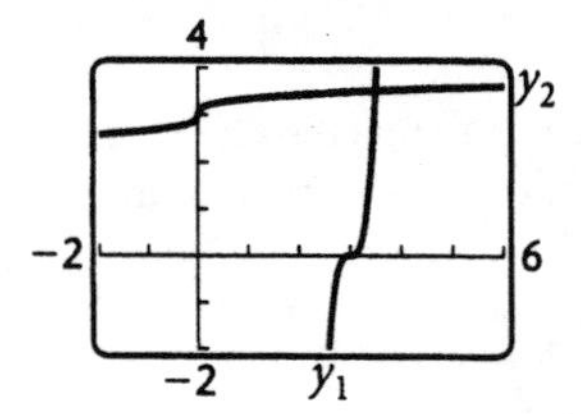

Both the domain and the range of f are the set of all real numbers. Then both the domain and the range of f^{-1} are also the set of all real numbers.

64.

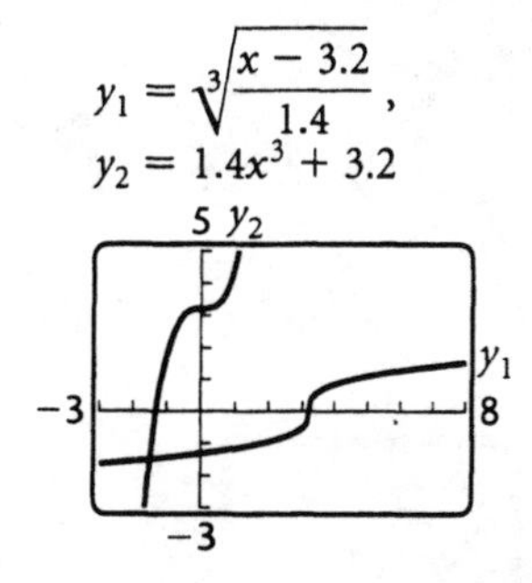

Both the domain and the range of f are the set of all real numbers. Then both the domain and the range of f^{-1} are also the set of all real numbers.

65. a) The graph of $f(x) = x + 4$ is shown below. It passes the horizontal line test, so it is one-to-one.

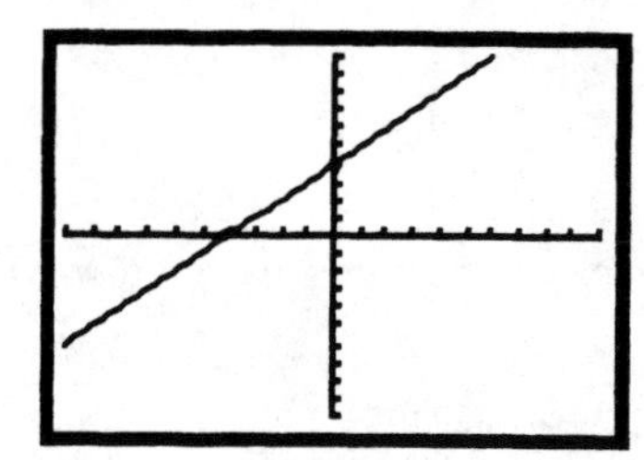

b) Replace $f(x)$ with y: $y = x + 4$

Interchange x and y: $x = y + 4$

Solve for y: $x - 4 = y$

Replace y with $f^{-1}(x)$: $f^{-1}(x) = x - 4$

66. a) The graph of $f(x) = 7 - x$ is shown below. It passes the horizontal line test, so it is one-to-one.

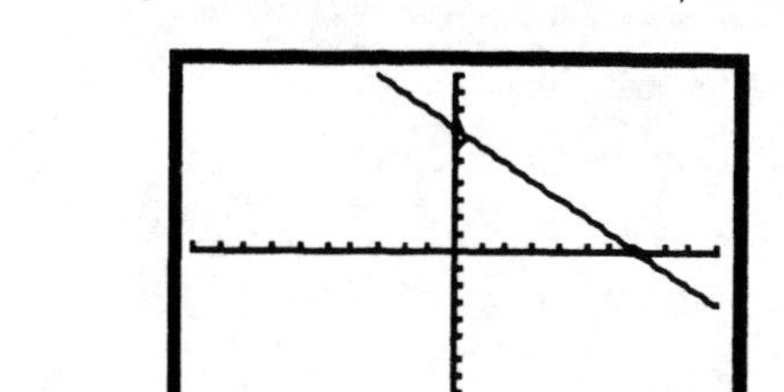

b) Replace $f(x)$ with y: $y = 7 - x$

Interchange x and y: $x = 7 - y$

Solve for y: $y = 7 - x$

Replace y with $f^{-1}(x)$: $f^{-1}(x) = 7 - x$

67. a) The graph of $f(x) = 2x - 1$ is shown below. It passes the horizontal line test, so it is one-to-one.

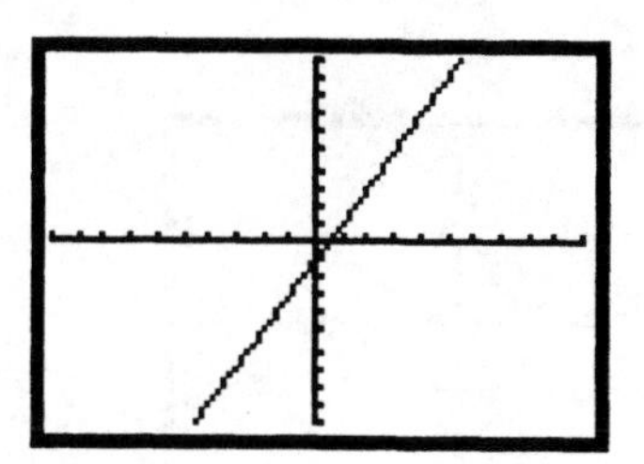

b) Replace $f(x)$ with y: $y = 2x - 1$

Interchange x and y: $x = 2y - 1$

Solve for y: $\dfrac{x+1}{2} = y$

Replace y with $f^{-1}(x)$: $f^{-1}(x) = \dfrac{x+1}{2}$

68. a) The graph of $f(x) = 5x + 8$ is shown below. It passes the horizontal line test, so it is one-to-one.

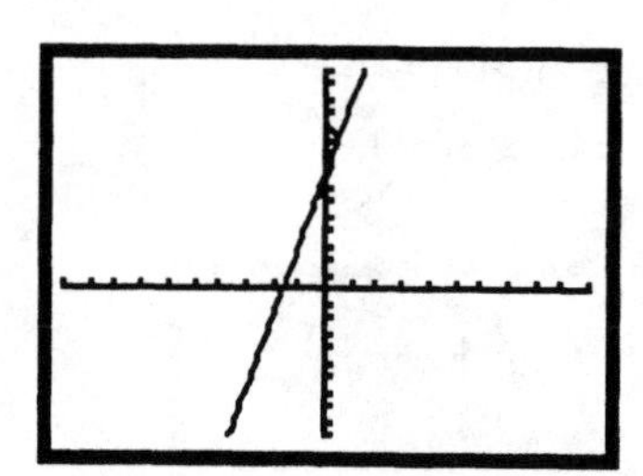

b) Replace $f(x)$ with y: $y = 5x + 8$

Interchange x and y: $x = 5y + 8$

Solve for y: $\dfrac{x-8}{5} = y$

Replace y with $f^{-1}(x)$: $f^{-1}(x) = \dfrac{x-8}{5}$

69. a) The graph of $f(x) = \dfrac{4}{x+7}$ is shown below. It passes the horizontal line test, so the function is one-to-one.

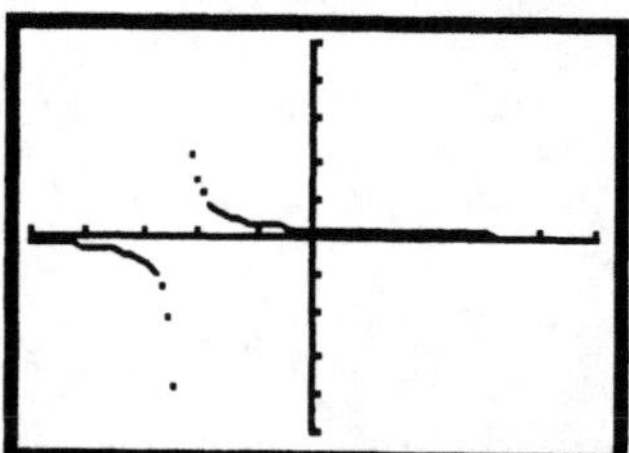

b) Replace $f(x)$ with y: $y = \dfrac{4}{x+7}$

Interchange x and y: $x = \dfrac{4}{y+7}$

Solve for y: $x(y+7) = 4$

$$y + 7 = \frac{4}{x}$$

$$y = \frac{4}{x} - 7$$

Replace y with $f^{-1}(x)$: $f^{-1}(x) = \dfrac{4}{x} - 7$

70. a) The graph of $f(x) = -\frac{3}{x}$ is shown below. It passes the horizontal line test, so it is one-to-one.

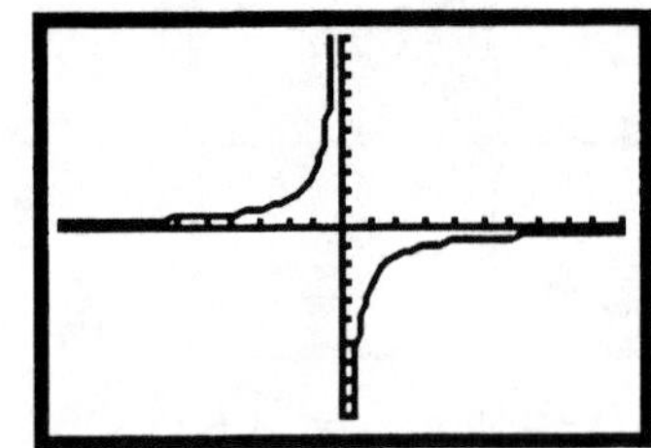

b) Replace $f(x)$ with y: $y = -\frac{3}{x}$

Interchange x and y: $x = -\frac{3}{y}$

Solve for y: $y = -\frac{3}{x}$

Replace y with $f^{-1}(x)$: $f^{-1}(x) = -\frac{3}{x}$

71. a) The graph of $f(x) = \frac{x+4}{x-3}$ is shown below. It passes the horizontal line test, so the function is one-to-one.

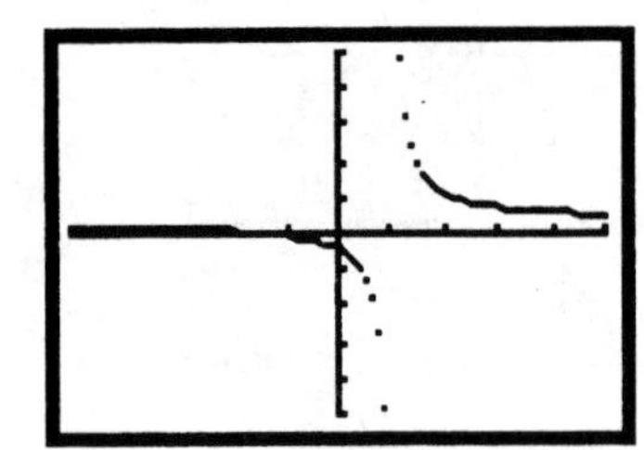

b) Replace $f(x)$ with y: $y = \frac{x+4}{x-3}$

Interchange x and y: $x = \frac{y+4}{y-3}$

Solve for y:
$$(y-3)x = y+4$$
$$xy - 3x = y+4$$
$$xy - y = 3x+4$$
$$y(x-1) = 3x+4$$
$$y = \frac{3x+4}{x-1}$$

Replace y with $f^{-1}(x)$: $f^{-1}(x) = \frac{3x+4}{x-1}$

72. a) The graph of $f(x) = \frac{5x-3}{2x+1}$ is shown below. It passes the horizontal line test, so it is one-to-one.

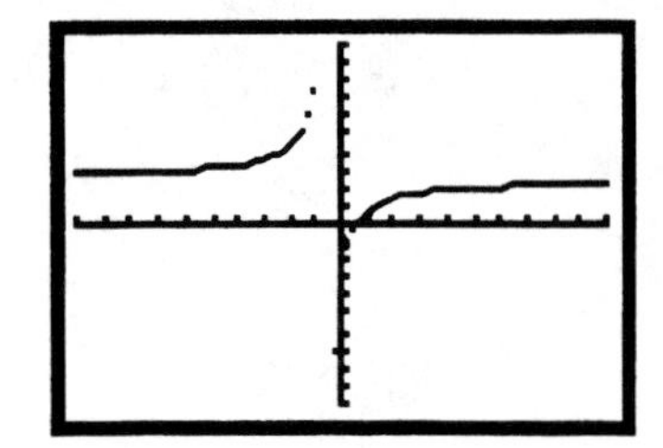

b) Replace $f(x)$ with y: $y = \frac{5x-3}{2x+1}$

Interchange x and y: $x = \frac{5y-3}{2y+1}$

Solve for y: $\frac{x+3}{5-2x} = y$

Replace y with $f^{-1}(x)$: $f^{-1}(x) = \frac{x+3}{5-2x}$

73. a) The graph of $f(x) = x^3 - 1$ is shown below. It passes the horizontal line test, so the function is one-to-one.

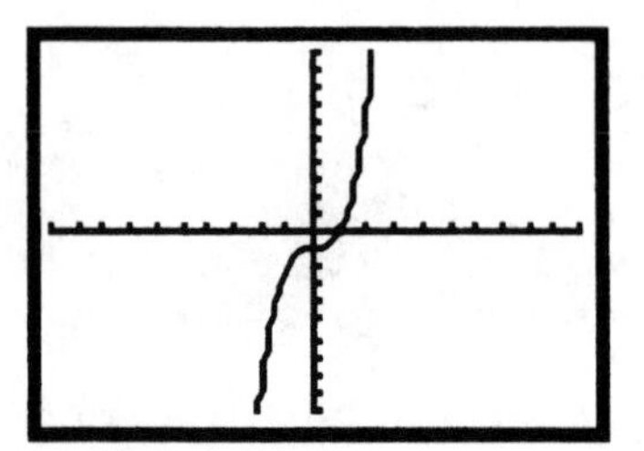

b) Replace $f(x)$ with y: $y = x^3 - 1$

Interchange x and y: $x = y^3 - 1$

Solve for y:
$$x + 1 = y^3$$
$$\sqrt[3]{x+1} = y$$

Replace y with $f^{-1}(x)$: $f^{-1}(x) = \sqrt[3]{x+1}$

74. a) The graph of $f(x) = (x+5)^3$ is shown below. It passes the horizontal line test, so it is one-to-one.

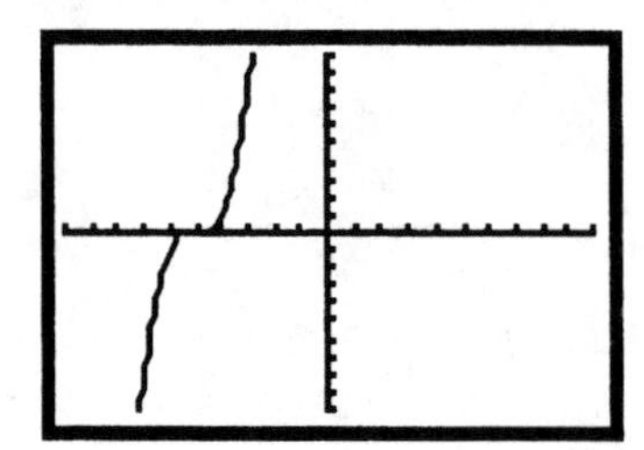

b) Replace $f(x)$ with y: $y = (x+5)^3$

Interchange x and y: $x = (y+5)^3$

Solve for y: $\sqrt[3]{x} - 5 = y$

Replace y with $f^{-1}(x)$: $f^{-1}(x) = \sqrt[3]{x} - 5$

75. a) The graph of $f(x) = x\sqrt{4-x^2}$ is shown below. Since there are many horizontal lines that cross the graph more than once, the function is not one-to-one and thus does not have an inverse that is a function.

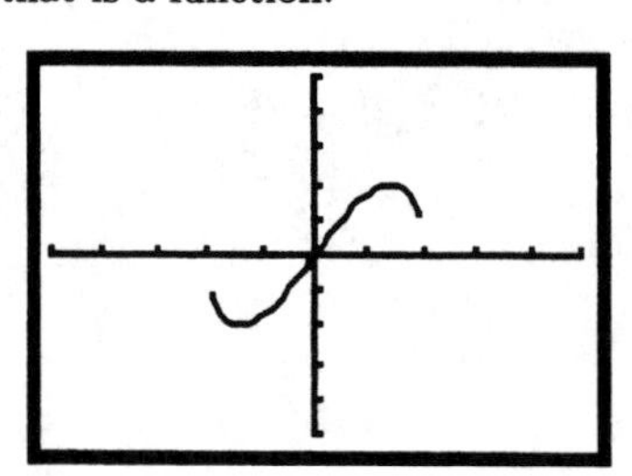

76. a) The graph of $f(x) = 4x^5 - 20x^3 + 2x^2 - 5x + 1$ is shown below. Since there are many horizontal lines that cross the graph more than once, the function is not one-to-one and thus does not have an inverse that is a function.

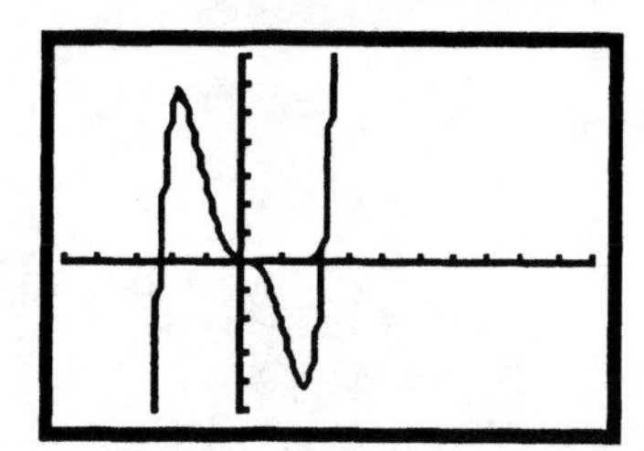

77. a) The graph of $f(x) = 5x^2 - 2$, $x \geq 0$ is shown below. It passes the horizontal line test, so it is one-to-one.

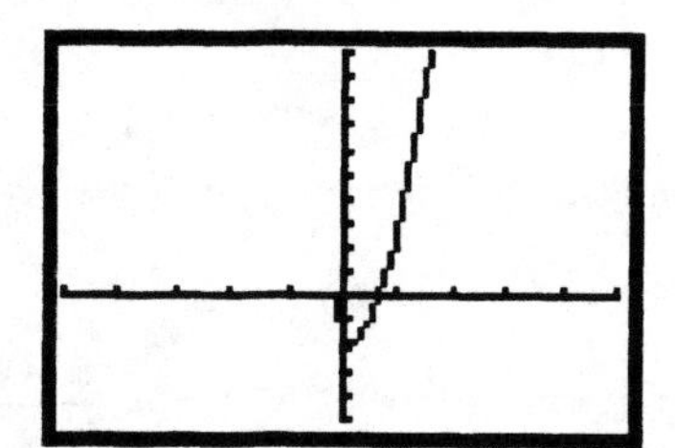

b) Replace $f(x)$ with y: $y = 5x^2 - 2$

Interchange x and y: $x = 5y^2 - 2$

Solve for y: $x + 2 = 5y^2$

$$\frac{x+2}{5} = y^2$$

$$\sqrt{\frac{x+2}{5}} = y$$

(We take the principal square root, because $x \geq 0$ in the original equation.)

Replace y with $f^{-1}(x)$: $f^{-1}(x) = \sqrt{\frac{x+2}{5}}$ for all x in the range of $f(x)$, or $f^{-1}(x) = \sqrt{\frac{x+2}{5}}$, $x \geq -2$

78. a) The graph of $f(x) = 4x^2 + 3$, $x \geq 0$ is shown below. It passes the horizontal line test, so the function is one-to-one.

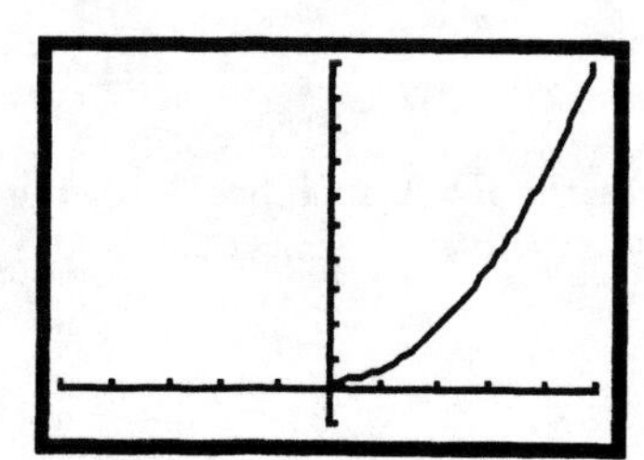

b) Replace $f(x)$ with y: $y = 4x^2 + 3$

Interchange x and y: $x = 4y^2 + 3$

Solve for y: $x - 3 = 4y^2$

$$\frac{x-3}{4} = y^2$$

$$\frac{\sqrt{x-3}}{2} = y$$

(We take the principal square root since $x \geq 0$ in the original function.)

Replace y with $f^{-1}(x)$: $f^{-1}(x) = \frac{\sqrt{x-3}}{2}$ for all x in the range of $f(x)$, or $f^{-1}(x) = \frac{\sqrt{x-3}}{2}$, $x \geq 3$

79. a) The graph of $f(x) = \sqrt{x+1}$ is shown below. It passes the horizontal line test, so the function is one-to-one.

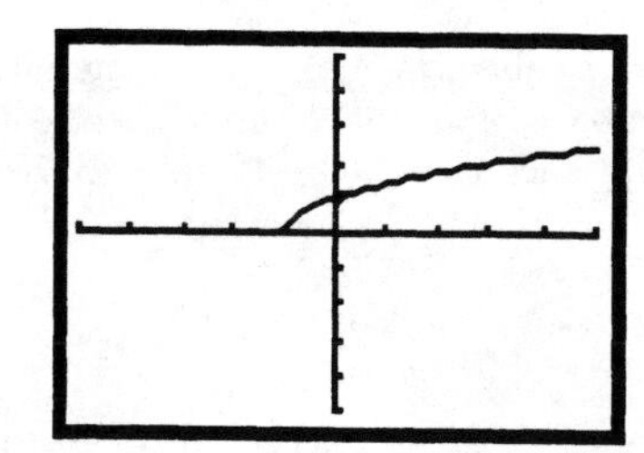

b) Replace $f(x)$ with y: $y = \sqrt{x+1}$

Interchange x and y: $x = \sqrt{y+1}$

Solve for y: $x^2 = y + 1$

$$x^2 - 1 = y$$

Replace y with $f^{-1}(x)$: $f^{-1}(x) = x^2 - 1$ for all x in the range of $f(x)$, or $f^{-1}(x) = x^2 - 1$, $x \geq 0$.

80. a) The graph of $f(x) = \sqrt[3]{x-8}$ is shown below. It passes the horizontal line test, so the function is one-to-one.

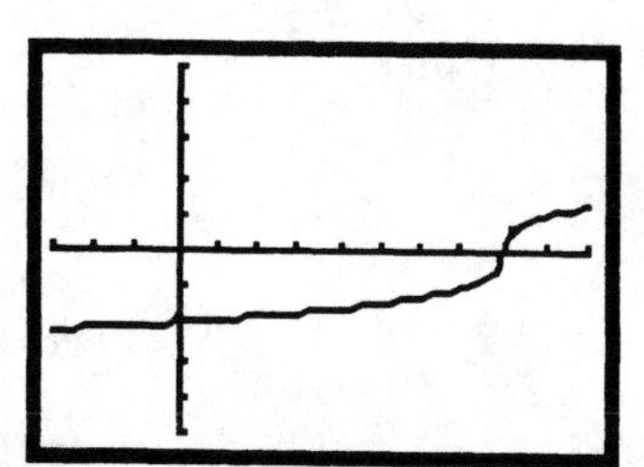

b) Replace $f(x)$ with y: $y = \sqrt[3]{x-8}$

Interchange x and y: $x = \sqrt[3]{y-8}$

Solve for y: $x^3 + 8 = y$

Replace y with $f^{-1}(x)$: $f^{-1}(x) = x^3 + 8$

81. $f(x) = 3x$

The function f multiplies an input by 3. Then to reverse this procedure, f^{-1} would divide each of its inputs by 3. Thus, $f^{-1}(x) = \frac{x}{3}$, or $f^{-1}(x) = \frac{1}{3}x$.

82. $f(x) = \frac{1}{4}x + 7$

The function f multiplies an input by $\frac{1}{4}$ and then adds 7. To reverse this procedure, f^{-1} would subtract 7 from each of its inputs and then multiply by 4. Thus, $f^{-1}(x) = 4(x - 7)$.

83. $f(x) = -x$

The outputs of f are the opposites, or additive inverses, of the inputs. Then the outputs of f^{-1} are the opposites of its inputs. Thus, $f^{-1}(x) = -x$.

84. $f(x) = \sqrt[3]{x} - 5$

The function f takes the cube root of an input and then subtracts 5. To reverse this procedure, f^{-1} would add 5 to each of its inputs and then raise the result to the third power. Thus, $f^{-1}(x) = (x + 5)^3$.

85. $f(x) = \sqrt[3]{x - 5}$

The function f subtracts 5 from each input and then takes the cube root of the result. To reverse this procedure, f^{-1} would raise each input to the third power and then add 5 to the result. Thus, $f^{-1}(x) = x^3 + 5$.

86. $f(x) = x^{-1}$

The outputs of f are the reciprocals of the inputs. Then the outputs of f^{-1} are the reciprocals of its inputs. Thus, $f^{-1}(x) = x^{-1}$.

87. We find $(f^{-1} \circ f)(x)$ and $(f \circ f^{-1})(x)$ and check to see that each is x.

$(f^{-1} \circ f)(x) = f^{-1}(f(x)) = f^{-1}\left(\frac{7}{8}x\right) = \frac{8}{7}\left(\frac{7}{8}x\right) = x$

$(f \circ f^{-1})(x) = f(f^{-1}(x)) = f\left(\frac{8}{7}x\right) = \frac{7}{8}\left(\frac{8}{7}x\right) = x$

88. $(f^{-1} \circ f)(x) = 4\left(\frac{x+5}{4}\right) - 5 = x + 5 - 5 = x$

$(f \circ f^{-1})(x) = \frac{4x - 5 + 5}{4} = \frac{4x}{4} = x$

89. We find $(f^{-1} \circ f)(x)$ and $(f \circ f^{-1})(x)$ and check to see that each is x.

$(f^{-1} \circ f)(x) = f^{-1}(f(x)) = f^{-1}\left(\frac{1-x}{x}\right) =$

$\frac{1}{\frac{1-x}{x} + 1} = \frac{1}{\frac{1-x+x}{x}} = \frac{1}{\frac{1}{x}} = x$

$(f \circ f^{-1})(x) = f(f^{-1}(x)) = f\left(\frac{1}{x+1}\right) =$

$\frac{1 - \frac{1}{x+1}}{\frac{1}{x+1}} = \frac{\frac{x+1-1}{x+1}}{\frac{1}{x+1}} = \frac{\frac{x}{x+1}}{\frac{1}{x+1}} = x$

90. $(f^{-1} \circ f)(x) = (\sqrt[3]{x+4})^3 - 4 = x + 4 - 4 = x$

$(f \circ f^{-1})(x) = \sqrt[3]{x^3 - 4 + 4} = \sqrt[3]{x^3} = x$

91. Since $f(f^{-1}(x)) = f^{-1}(f(x)) = x$, then $f(f^{-1}(5)) = 5$ and $f^{-1}(f(a)) = a$.

92. Since $f^{-1}(f(x)) = f(f^{-1}(x)) = x$, then $f^{-1}(f(p)) = p$ and $f(f^{-1}(1253)) = 1253$.

93. a) $g(x) = 2(x + 12)$

$g(6) = 2(6 + 12) = 2 \cdot 18 = 36$

$g(8) = 2(8 + 12) = 2 \cdot 20 = 40$

$g(10) = 2(10 + 12) = 2 \cdot 22 = 44$

$g(14) = 2(14 + 12) = 2 \cdot 26 = 52$

$g(18) = 2(18 + 12) = 2 \cdot 30 = 60$

b) The graph passes the horizontal line test and thus has an inverse that is a function.

Replace $g(x)$ with y: $y = 2(x + 12)$

Interchange x and y: $x = 2(y + 12)$

Solve for y: $\frac{x - 24}{2} = y$

Replace y with $g^{-1}(x)$: $g^{-1}(x) = \frac{x - 24}{2}$, or $\frac{x}{2} - 12$

c) $g^{-1}(36) = \frac{36 - 24}{2} = \frac{12}{2} = 6$

$g^{-1}(40) = \frac{40 - 24}{2} = \frac{16}{2} = 8$

$g^{-1}(44) = \frac{44 - 24}{2} = \frac{20}{2} = 10$

$g^{-1}(52) = \frac{52 - 24}{2} = \frac{28}{2} = 14$

$g^{-1}(60) = \frac{60 - 24}{2} = \frac{36}{2} = 18$

94. $C(x) = \frac{100 + 5x}{x}$

Replace $C(x)$ with y: $y = \frac{100 + 5x}{x}$

Interchange x and y: $x = \frac{100 + 5y}{y}$

Solve for y: $y = \frac{100}{x - 5}$

Replace y with $C^{-1}(x)$: $C^{-1}(x) = \frac{100}{x - 5}$

$C^{-1}(x)$ gives the number of people in the group, where x is the cost per person, in dollars.

95. a) $D(r) = \frac{11r + 5}{10}$

$D(0) = \frac{11 \cdot 0 + 5}{10} = \frac{5}{10} = 0.5$

$D(10) = \frac{11 \cdot 10 + 5}{10} = \frac{115}{10} = 11.5$

$$D(20) = \frac{11 \cdot 20 + 5}{10} = \frac{225}{10} = 22.5$$

$$D(50) = \frac{11 \cdot 50 + 5}{10} = \frac{555}{10} = 55.5$$

$$D(65) = \frac{11 \cdot 65 + 5}{10} = \frac{720}{10} = 72$$

b), d) $y_1 = \dfrac{11x + 5}{10}$, $y_2 = \dfrac{10x - 5}{11}$

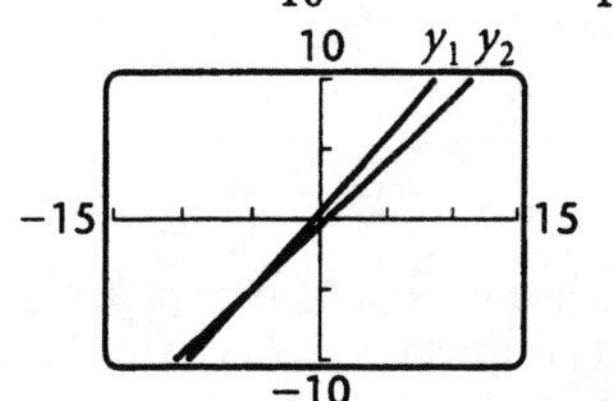

c) Replace $D(r)$ with y: $y = \dfrac{11r + 5}{10}$

Interchange r and y: $r = \dfrac{11y + 5}{10}$

Solve for y: $10r = 11y + 5$

$$10r - 5 = 11y$$

$$\frac{10r - 5}{11} = y$$

Replace y with $D^{-1}(r)$: $D^{-1}(r) = \dfrac{10r - 5}{11}$

$D^{-1}(r)$ represents the speed, in miles per hour, that the car is traveling when the reaction distance is r feet.

d) See part (b).

96. a) In 1998, $t = 1998 - 1995$, or 3.

$N(3) = 0.6514(3) + 53.1599 = 55.1141$

In 2000, $t = 2000 - 1995$, or 5.

$N(5) = 0.6514(5) + 53.1599 = 56.4169$

b) $y_1 = 0.6514x + 53.1599$,

$y_2 = \dfrac{x - 53.1599}{0.6514}$

100 y_1 y_2 0 100 0

c) $N^{-1}(t)$ represents the number of years after 1995 when t loaves of bread are consumed per person per year.

97. Use the TRACE feature to find the coordinates of some points on the graph of the function. Then interchange the coordinates of each ordered pair to find some points on the graph of the inverse. Plot these points and draw the graph.

98. C and F are inverses.

99. Graph $y = \dfrac{3}{2}x - 4$.

When $x = 0$, $y = \dfrac{3}{2} \cdot 0 - 4 = 0 - 4 = -4$.

When $x = 2$, $y = \dfrac{3}{2} \cdot 2 - 4 = 3 - 4 = -1$.

When $x = 4$, $y = \dfrac{3}{2} \cdot 4 - 4 = 6 - 4 = 2$.

Plot the points $(0, -4)$, $(2, -1)$, and $(4, 2)$ and connect them with a straight line.

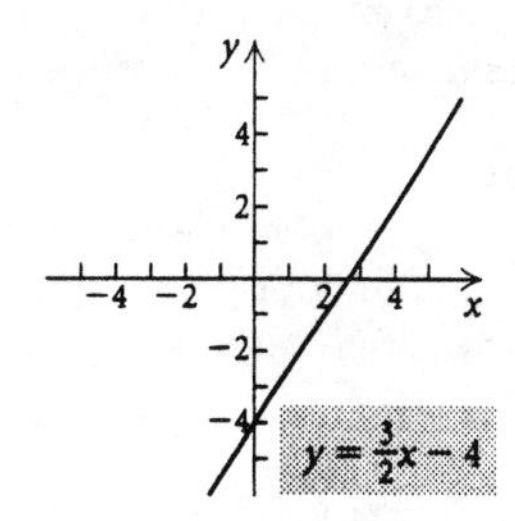

100.

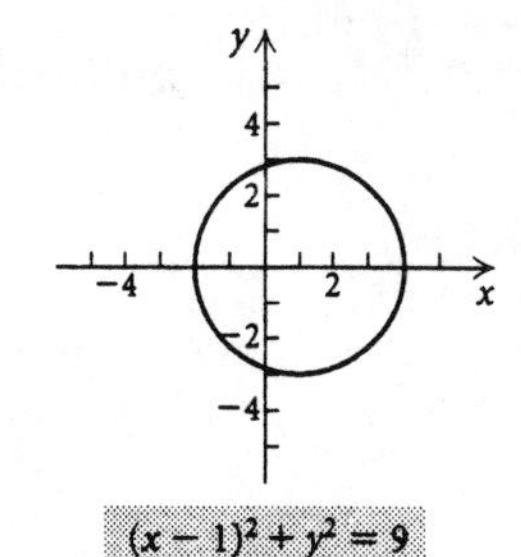

101. Graph $y = \dfrac{x + 3}{x^2 - x - 2}$.

1. The zeros of the denominator are -1 and 2, so the domain is $(-\infty, -1) \cup (-1, 2) \cup (2, \infty)$ and the lines $x = -1$ and $x = 2$ are vertical asymptotes.

2. Because the degree of the numerator is less than the degree of the denominator, the x-axis is the horizontal asymptote. There is no oblique asymptote.

3. The zero of the numerator is -3, so the x-intercept is $(-3, 0)$.

4. When $x = 0$, $y = \dfrac{0 + 3}{0^2 - 0 - 2} = -\dfrac{3}{2}$, so the y-intercept is $\left(0, -\dfrac{3}{2}\right)$.

5. Find other function values to determine the shape of the graph and then draw it.

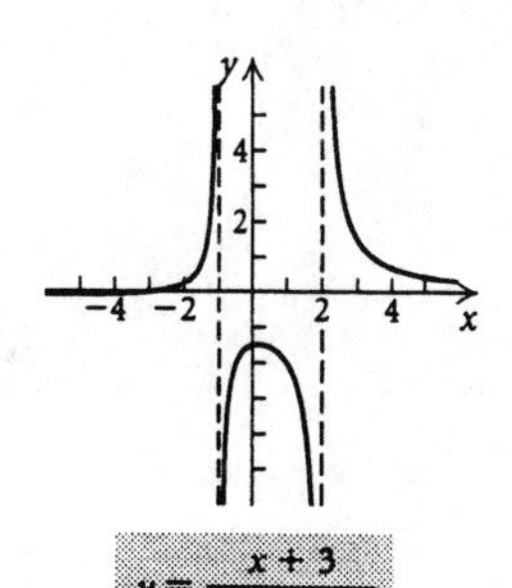

102.

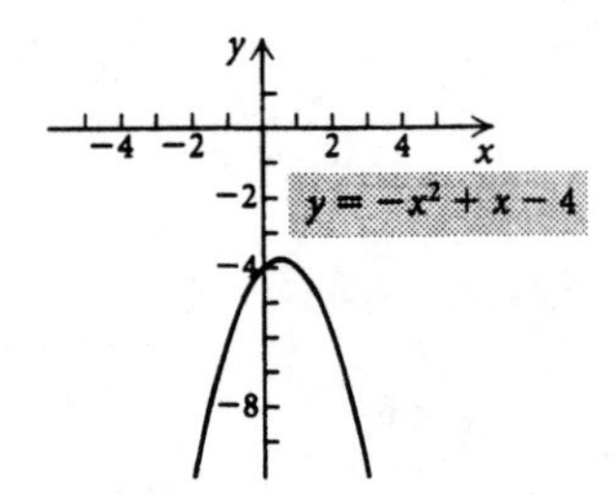

103. Graph $y_1 = f(x)$ and $y_2 = g(x)$ and observe that the graphs are reflections of each other across the line $y = x$. Thus, the functions are inverses of each other.

104. Graph $y_1 = f(x)$ and $y_2 = g(x)$ and observe that the graphs are not reflections of each other across the line $y = x$. Thus, the functions are not inverses of each other.

105. Graph $y_1 = f(x)$ and $y_2 = g(x)$ and observe that the graphs are not reflections of each other across the line $y = x$. Thus, the functions are not inverses of each other.

106. Graph $y_1 = f(x)$ and $y_2 = g(x)$ and observe that the graphs are reflections of each other across the line $y = x$. Thus, the functions are inverses of each other.

107. The graph of $f(x) = x^2 - 3$ is a parabola with vertex $(0, -3)$. If we consider x-values such that $x \geq 0$, then the graph is the right-hand side of the parabola and it passes the horizontal line test. We find the inverse of $f(x) = x^2 - 3$, $x \geq 0$.

Replace $f(x)$ with y: $y = x^2 - 3$

Interchange x and y: $x = y^2 - 3$

Solve for y: $x + 3 = y^2$

$\sqrt{x+3} = y$

(We take the principal square root, because $x \geq 0$ in the original equation.)

Replace y with $f^{-1}(x)$: $f^{-1}(x) = \sqrt{x+3}$ for all x in the range of $f(x)$, or $f^{-1}(x) = \sqrt{x+3}$, $x \geq -3$.

Answers may vary. There are other restrictions that also make $f(x)$ one-to-one.

108. No; the graph of f does not pass the horizontal line test.

109. Answers may vary. $f(x) = \dfrac{3}{x}$, $f(x) = 1 - x$, $f(x) = x$.

Exercise Set 4.2

1. $e^4 \approx 54.5982$

2. $e^{10} \approx 22,026.4658$

3. $e^{-2.458} \approx 0.0856$

4. $\left(\dfrac{1}{e^3}\right)^2 \approx 0.0025$

5. Graph $f(x) = 3^x$.

Compute some function values, plot the corresponding points, and connect them with a smooth curve.

x	$y = f(x)$	(x, y)
-3	$\frac{1}{27}$	$\left(-3, \frac{1}{27}\right)$
-2	$\frac{1}{9}$	$\left(-2, \frac{1}{9}\right)$
-1	$\frac{1}{3}$	$\left(-1, \frac{1}{3}\right)$
0	1	$(0, 1)$
1	3	$(1, 3)$
2	9	$(2, 9)$
3	27	$(3, 27)$

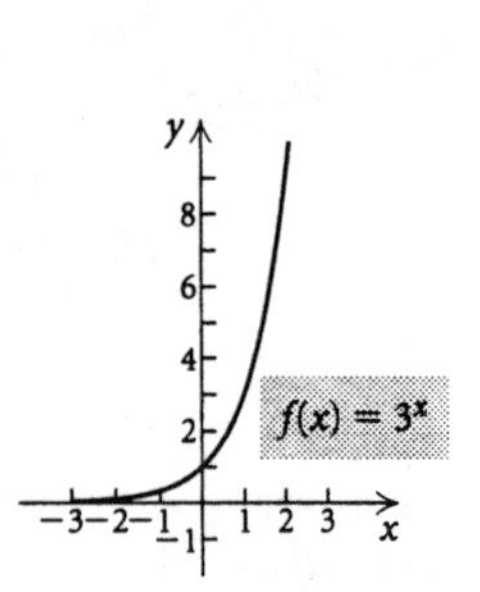

6. Graph $f(x) = 5^x$.

x	$y = f(x)$	(x, y)
-3	$\frac{1}{125}$	$\left(-3, \frac{1}{125}\right)$
-2	$\frac{1}{25}$	$\left(-2, \frac{1}{25}\right)$
-1	$\frac{1}{5}$	$\left(-1, \frac{1}{5}\right)$
0	1	$(0, 1)$
1	5	$(1, 5)$
2	25	$(2, 25)$
3	125	$(3, 125)$

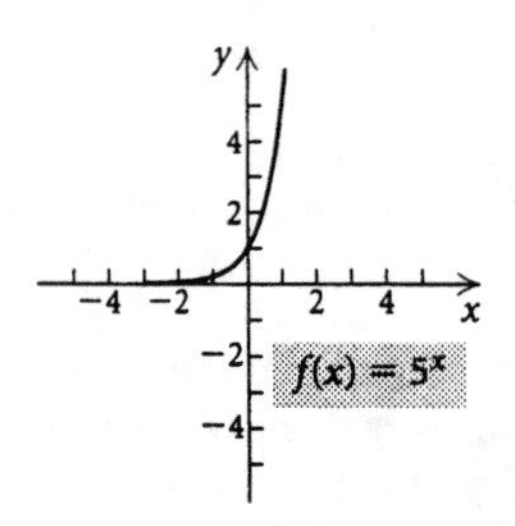

7. Graph $f(x) = 6^x$.

Compute some function values, plot the corresponding points, and connect them with a smooth curve.

x	$y = f(x)$	(x, y)
-3	$\frac{1}{216}$	$\left(-3, \frac{1}{216}\right)$
-2	$\frac{1}{36}$	$\left(-2, \frac{1}{36}\right)$
-1	$\frac{1}{6}$	$\left(-1, \frac{1}{6}\right)$
0	1	$(0, 1)$
1	6	$(1, 6)$
2	36	$(2, 36)$
3	216	$(3, 216)$

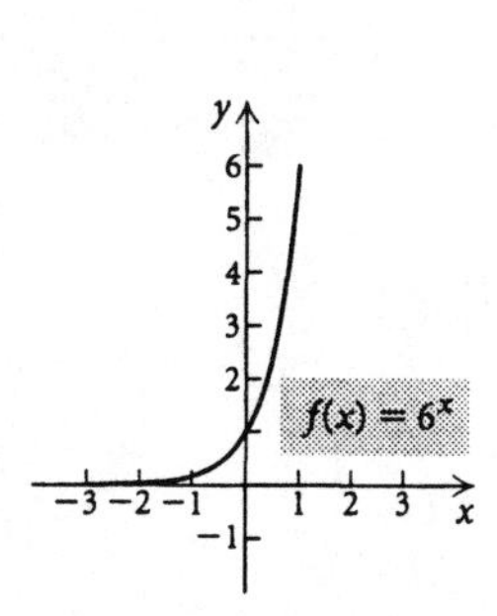

8. Graph $f(x) = 3^{-x}$.

x	$y = f(x)$	(x, y)
-3	27	$(-3, 27)$
-2	9	$(-2, 9)$
-1	3	$(-1, 3)$
0	1	$(0, 1)$
1	$\frac{1}{3}$	$\left(1, \frac{1}{3}\right)$
2	$\frac{1}{9}$	$\left(2, \frac{1}{9}\right)$
3	$\frac{1}{27}$	$\left(3, \frac{1}{27}\right)$

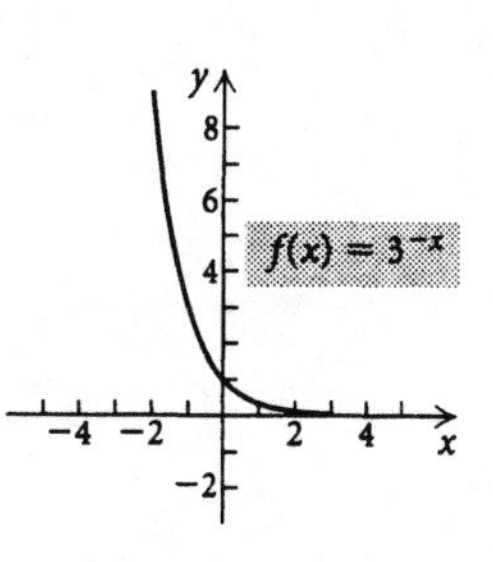

9. Graph $f(x) = \left(\frac{1}{4}\right)^x$.

Compute some function values, plot the corresponding points, and connect them with a smooth curve.

x	$y = f(x)$	(x, y)
-3	64	$(-3, 64)$
-2	16	$(-2, 16)$
-1	4	$(-1, 4)$
0	1	$(0, 1)$
1	$\frac{1}{4}$	$\left(1, \frac{1}{4}\right)$
2	$\frac{1}{16}$	$\left(2, \frac{1}{16}\right)$
3	$\frac{1}{64}$	$\left(3, \frac{1}{64}\right)$

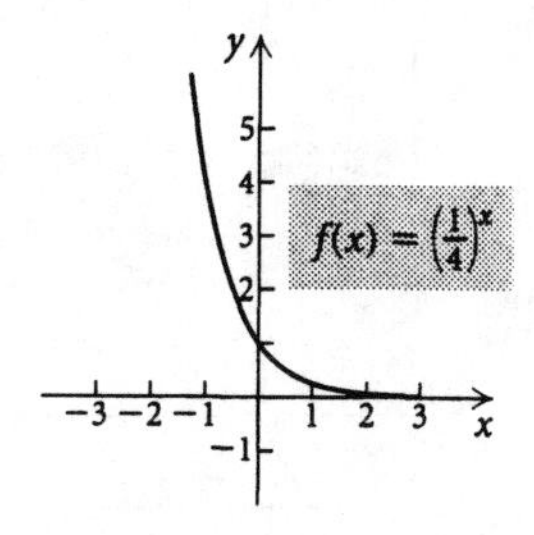

10. Graph $f(x) = \left(\frac{2}{3}\right)^x$.

x	$y = f(x)$	(x, y)
-3	$\frac{27}{8}$	$\left(-3, \frac{27}{8}\right)$
-2	$\frac{9}{4}$	$\left(-2, \frac{9}{4}\right)$
-1	$\frac{3}{2}$	$\left(-1, \frac{3}{2}\right)$
0	1	$(0, 1)$
1	$\frac{2}{3}$	$\left(1, \frac{2}{3}\right)$
2	$\frac{4}{9}$	$\left(2, \frac{4}{9}\right)$
3	$\frac{8}{27}$	$\left(3, \frac{8}{27}\right)$

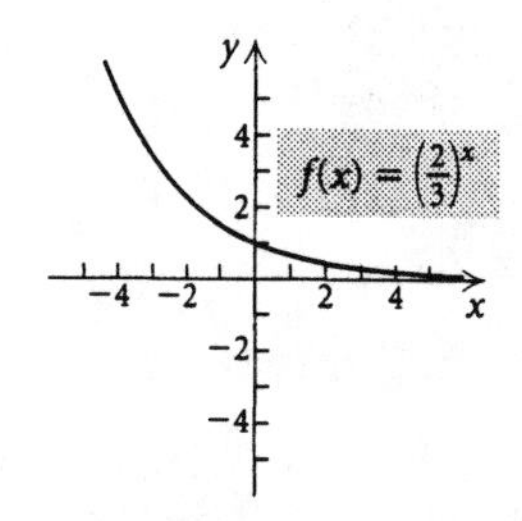

11. Graph $x = 3^y$.

Choose values for y and compute the corresponding x-values. Plot the points (x, y) and connect them with a smooth curve.

x	y	(x, y)
$\frac{1}{27}$	-3	$\left(\frac{1}{27}, -3\right)$
$\frac{1}{9}$	-2	$\left(\frac{1}{9}, -2\right)$
$\frac{1}{3}$	-1	$\left(\frac{1}{3}, -1\right)$
1	0	$(1, 0)$
3	1	$(3, 1)$
9	2	$(9, 2)$
27	3	$(27, 3)$

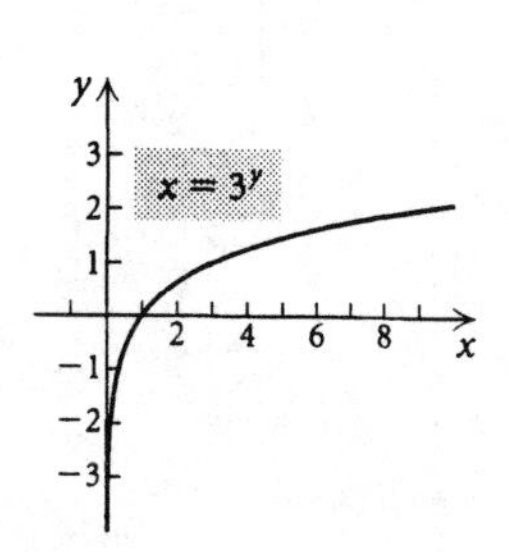

12. Graph $x = 4^y$.

x	y	(x, y)
$\frac{1}{64}$	-3	$\left(\frac{1}{64}, -3\right)$
$\frac{1}{16}$	-2	$\left(\frac{1}{16}, -2\right)$
$\frac{1}{4}$	-1	$\left(\frac{1}{4}, -1\right)$
1	0	$(1, 0)$
4	1	$(4, 1)$
16	2	$(16, 2)$
64	3	$(64, 3)$

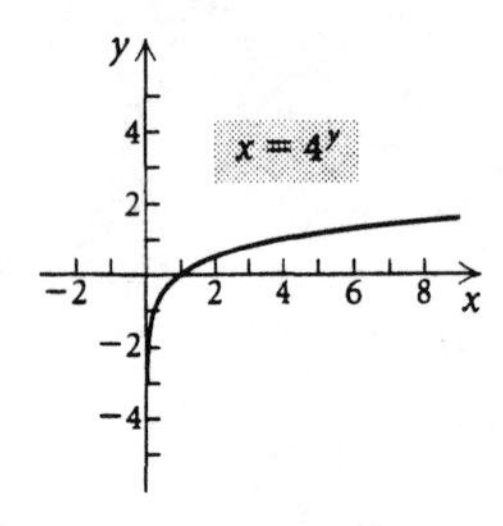

13. Graph $x = \left(\frac{1}{2}\right)^y$.

Choose values for y and compute the corresponding x-values. Plot the points (x, y) and connect them with a smooth curve.

x	y	(x, y)
8	-3	$(8, -3)$
4	-2	$(4, -2)$
2	-1	$(2, -1)$
1	0	$(1, 0)$
$\frac{1}{2}$	1	$\left(\frac{1}{2}, 1\right)$
$\frac{1}{4}$	2	$\left(\frac{1}{4}, 2\right)$
$\frac{1}{8}$	3	$\left(\frac{1}{8}, 3\right)$

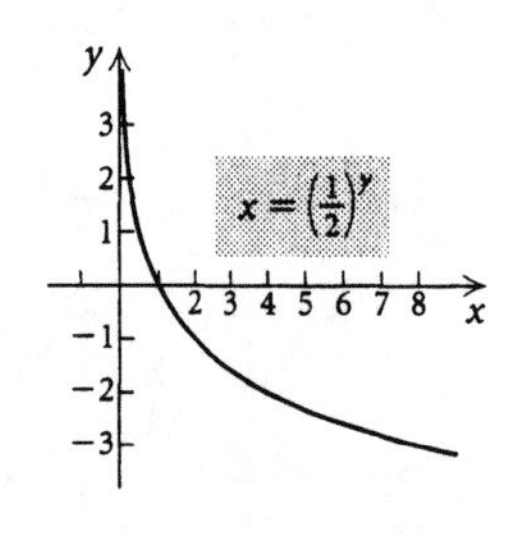

14. Graph $x = \left(\frac{4}{3}\right)^y$.

x	y	(x, y)
$\frac{27}{64}$	-3	$\left(\frac{27}{64}, -3\right)$
$\frac{9}{16}$	-2	$\left(\frac{9}{16}, -2\right)$
$\frac{3}{4}$	-1	$\left(\frac{3}{4}, -1\right)$
1	0	$(1, 0)$
$\frac{4}{3}$	1	$\left(\frac{4}{3}, 1\right)$
$\frac{16}{9}$	2	$\left(\frac{16}{9}, 2\right)$
$\frac{64}{27}$	3	$\left(\frac{64}{27}, 3\right)$

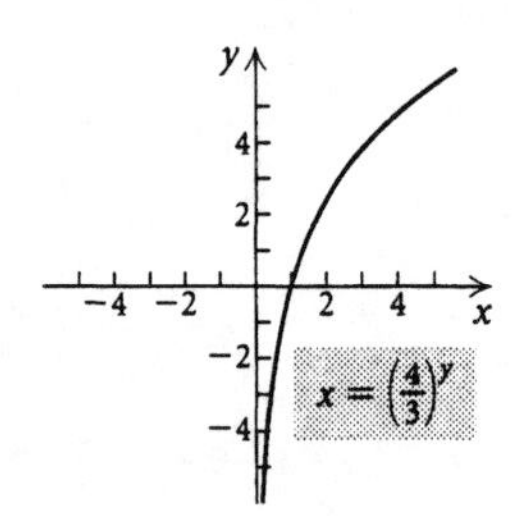

15. Graph $y = \frac{1}{4}e^x$.

Choose values for x and compute the corresponding y-values. Plot the points (x, y) and connect them with a smooth curve.

x	y	(x, y)
-3	0.0124	$(-3, 0.0124)$
-2	0.0338	$(-2, 0.0338)$
-1	0.0920	$(-1, 0.0920)$
0	0.25	$(0, 0.25)$
1	0.6796	$(1, 0.6796)$
2	1.8473	$(2, 1.8473)$
3	5.0214	$(3, 5.0214)$

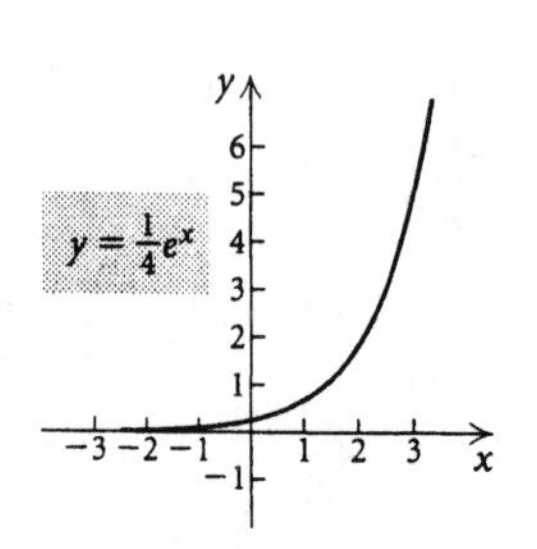

16. Graph $y = 2e^{-x}$.

x	y	(x, y)
-3	40.1711	$(-3, 40.1711)$
-2	14.7781	$(-2, 14.7781)$
-1	5.4366	$(-1, 5.4366)$
0	2	$(0, 2)$
1	0.7358	$(1, 0.7358)$
2	0.2707	$(2, 0.2707)$
3	0.0996	$(3, 0.0996)$

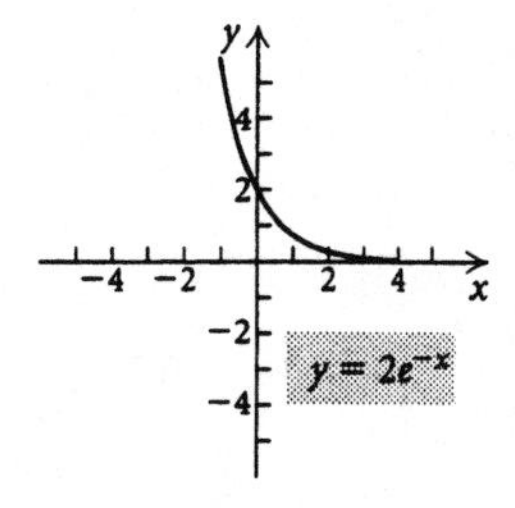

17. Graph $f(x) = 1 - e^{-x}$.

Compute some function values, plot the corresponding points, and connect them with a smooth curve.

x	y	(x, y)
−3	−19.0855	(−3, −19.0855)
−2	−6.3891	(−2, −6.3891)
−1	−1.7183	(−1, −1.7183)
0	0	(0, 0)
1	0.6321	(1, 0.6321)
2	0.8647	(2, 0.8647)
3	0.9502	(3, 0.9502)

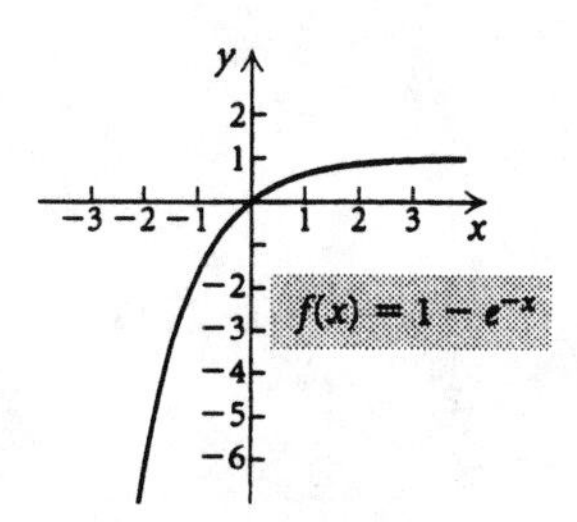

18. Graph $f(x) = e^x - 2$.

x	y	(x, y)
−3	−1.9502	(−3, −1.9502)
−2	−1.8647	(−2, −1.8647)
−1	−1.6321	(−1, −1.6321)
0	−1	(0, −1)
1	0.7183	(1, 0.7183)
2	5.3891	(2, 5.3891)
3	18.0855	(3, 18.0855)

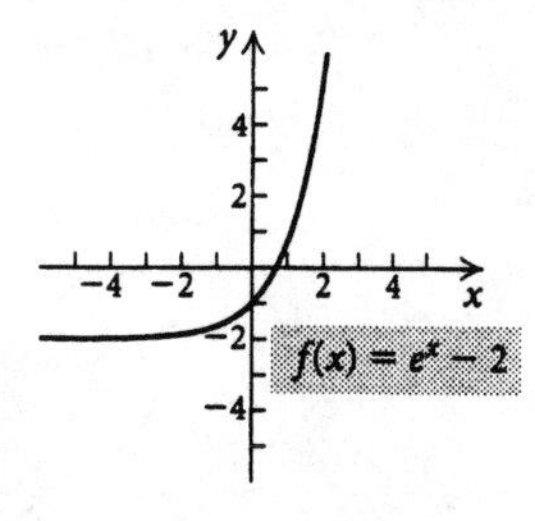

19. Shift the graph of $y = 2^x$ left 1 unit.

$y = 2^{x+1}$

20. Shift the graph of $y = 2^x$ right 1 unit.

$y = 2^{x-1}$

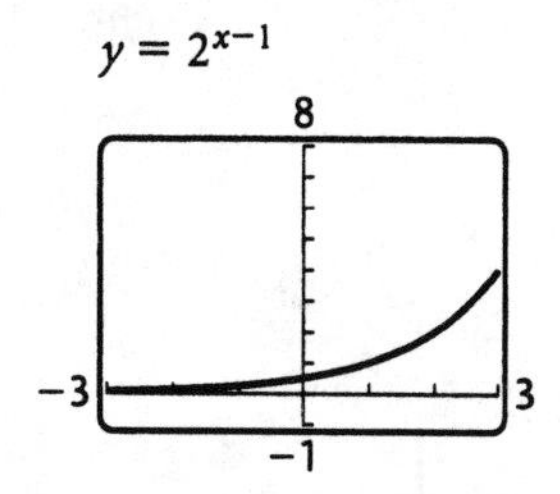

21. Shift the graph of $y = 2^x$ down 3 units.

$y = 2^x - 3$

22. Shift the graph of $y = 2^x$ up 1 unit.

$y = 2^x + 1$

23. Reflect the graph of $y = 3^x$ across the y-axis, then across the x-axis, and then shift it up 4 units.

$y = 4 - 3^{-x}$

24. Shift the graph of $y = 2^x$ right 1 unit and down 3 units.

$y = 2^{x-1} - 3$

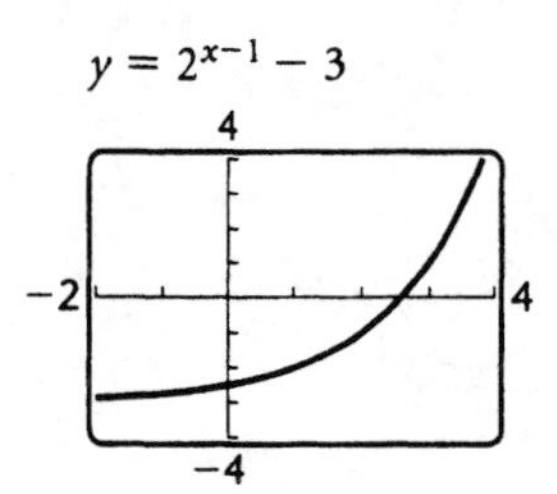

25. Shift the graph of $y = \left(\frac{3}{2}\right)^x$ right 1 unit.

$y = \left(\frac{3}{2}\right)^{x-1}$

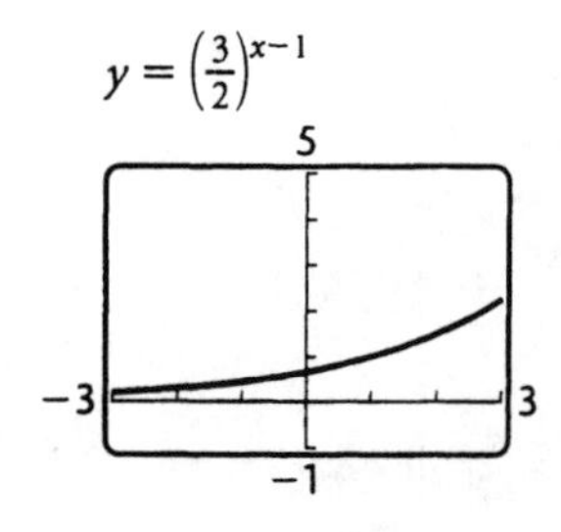

26. Shift the graph of $y = 3^x$ right 4 units and reflect it across the y-axis.

$y = 3^{4-x}$

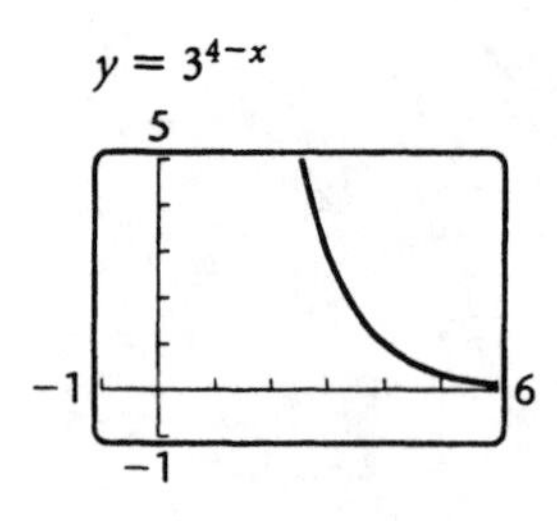

27. Shift the graph of $y = 2^x$ left 3 units and down 5 units.

$y = 2^{x+3} - 5$

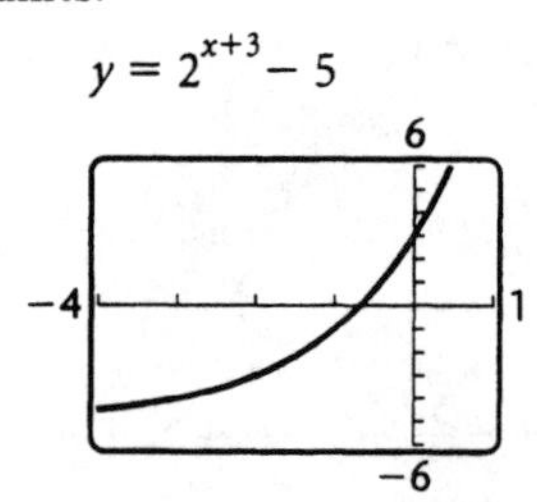

28. Shift the graph of $y = 3^x$ right 2 units and reflect it across the x-axis.

$y = -3^{x-2}$

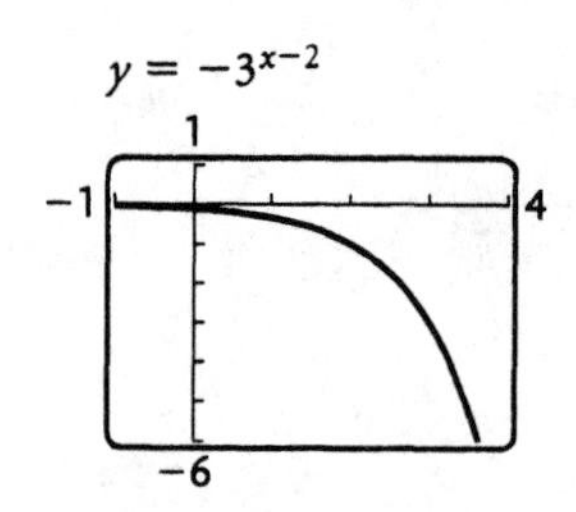

29. Shrink the graph of $y = e^x$ horizontally.

$y = e^{2x}$

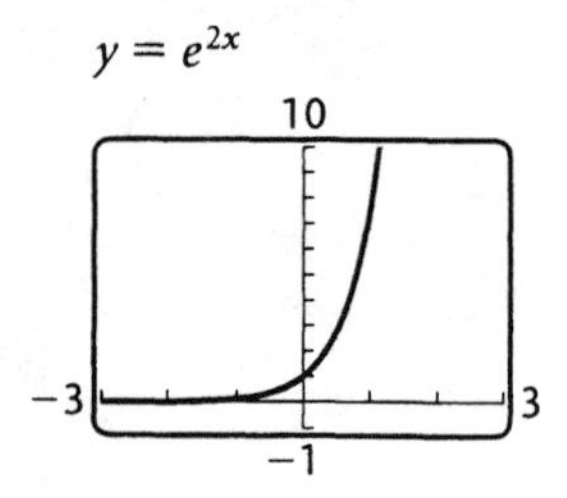

30. Stretch the graph of $y = e^x$ horizontally and reflect it across the y-axis.

$y = e^{-0.2x}$

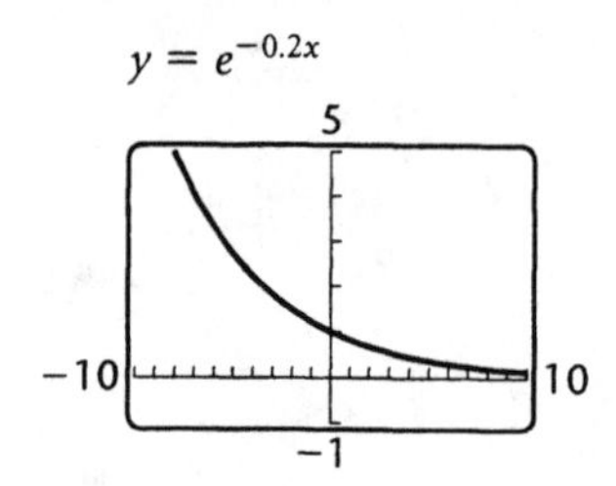

31. Shift the graph of $y = e^x$ left 1 unit and reflect it across the y-axis.

$y = e^{-x+1}$

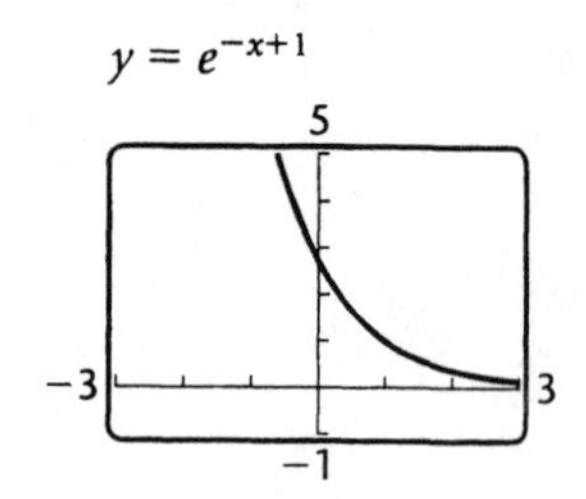

32. Shrink the graph of $y = e^x$ horizontally and shift it up 1 unit.

$y = e^{2x} + 1$

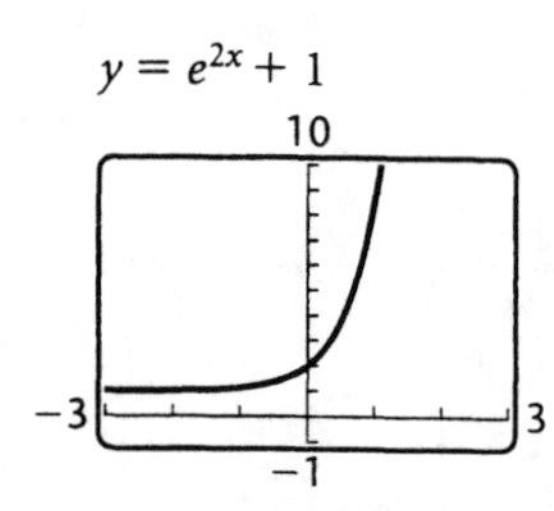

33. Reflect the graph of $y = e^x$ across the y-axis and then across the x-axis; shift it up 1 unit and then stretch it vertically.

$y = 2(1 - e^{-x}), x > 0$

34. Stretch the graph of $y = e^x$ horizontally and reflect it across the y-axis; then reflect it down across the x-axis and shift it up 1 unit.

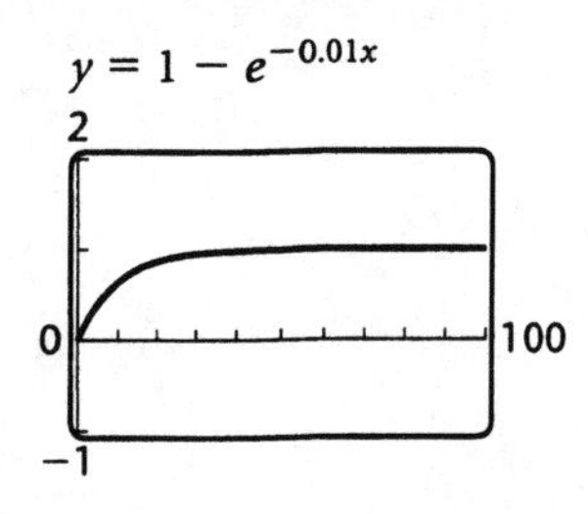

35. a) In 1998, $t = 1998 - 1992 = 6$.
$N(6) = 274,390(1.195)^6 \approx 799,053$

b) In 2001, $t = 2001 - 1992 = 9$.
$N(9) = 274,390(1.195)^9 \approx 1,363,576$

c) $y = 274{,}390(1.195)^t$

d) Using Intersect feature, we find that the first coordinate of the point of intersection of $y_1 = 274,390(1.195)^x$ and $y_2 = 3,000,000$ is about 13. Thus, it will take about 13 years for the number of AIDS cases reported in the United States to reach 3 million.

We could also graph $y = 274,390(1.195)^x - 3,000,000$ and use the Zero method to find the result.

36. a) $N(10) = 3000(2)^{10/20} \approx 4243;$
$N(20) = 3000(2)^{20/20} = 6000;$
$N(30) = 3000(2)^{30/20} \approx 8485;$
$N(40) = 3000(2)^{40/20} = 12,000;$
$N(60) = 3000(2)^{60/20} = 24,000$

b) $y = 3000(2)^{t/20}$

c) Using the Intersect feature, we find that the first coordinate of the point of intersection of $y_1 = 3000(2)^{x/20}$ and $y_2 = 100,000,000$ is about 300. Thus it takes about 300 min, or 5 hr, for a bladder infection to be possible.

We could also graph $y = 3000(2)^{x/20} - 100,000,000$ and use the Zero method to find this result.

37. a) $N(0) = 350,000\left(\frac{2}{3}\right)^0 = 350,000$

$N(1) = 350,000\left(\frac{2}{3}\right)^1 \approx 233,333$

$N(4) = 350,000\left(\frac{2}{3}\right)^4 \approx 69,136$

$N(10) = 350,000\left(\frac{2}{3}\right)^{10} \approx 6070$

b)

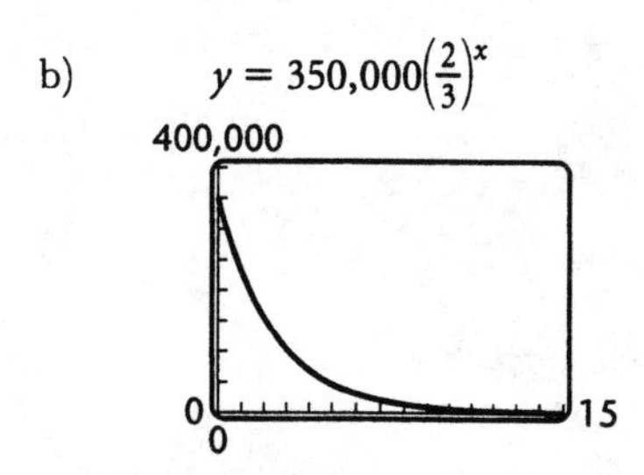

c) Using the Intersect feature, we find that the first coordinate of the point of intersection of $y_1 = 350,000\left(\frac{2}{3}\right)^x$ and $y_2 = 2000$ is about 13. Thus, 2000 cans will still be in use after about 13 years.

38. a) $A(t) = 10,000\left(1 + \frac{0.064}{2}\right)^{2t} = 10,000(1.032)^{2t}$

b) $A(0) = 10,000(1.032)^{2\cdot 0} = \$10,000$
$A(4) = 10,000(1.032)^{2\cdot 4} \approx \$12,865.82$
$A(8) = 10,000(1.032)^{2\cdot 8} \approx \$16,552.94$
$A(10) = 10,000(1.032)^{2\cdot 10} \approx \$18,775.61$
$A(18) = 10,000(1.032)^{2\cdot 18} \approx \$31,079.15$

c) $y = 10{,}000(1.032)^{2x}$

d) Using the Intersect feature, we find that the first coordinate of the point of intersection of $y_1 = 10,000(1.032)^{2x}$ and $y_2 = 100,000$ is about 36.6. Thus, the account will contain \$100,000 after about 37 years.

39. a) $y = 5800(0.8)^x$

b) $V(0) = 5800(0.8)^0 = \$5800$

$V(1) = 5800(0.8)^1 = \$4640$

$V(2) = 5800(0.8)^2 = \$3712$

$V(5) = 5800(0.8)^5 \approx \$1900.54$

$V(10) = 5800(0.8)^{10} \approx \$622.77$

c) Using the Intersect feature, we find that the first coordinate of the point of intersection of $y_1 = 5800(0.8)^x$ and $y_2 = 500$ is about 11. Thus, the machine will be replaced after about 11 years.

40. a) In 1995, $t = 1995 - 1985 = 10$. Similarly, in 1999, $t = 14$; in 2000, $t = 15$; and in 2004, $t = 19$.

$C(10) = 0.4703477987(1.523963441)^{10} \approx 31.8$ million

$C(14) = 0.4703477987(1.523963441)^{14} \approx 171.4$ million

$C(15) = 0.4703477987(1.523963441)^{15} \approx 261.2$ million

$C(19) = 0.4703477987(1.523963441)^{19} \approx 1409.1$ million

b) $y = 0.4703477987(1.523963441)^x$

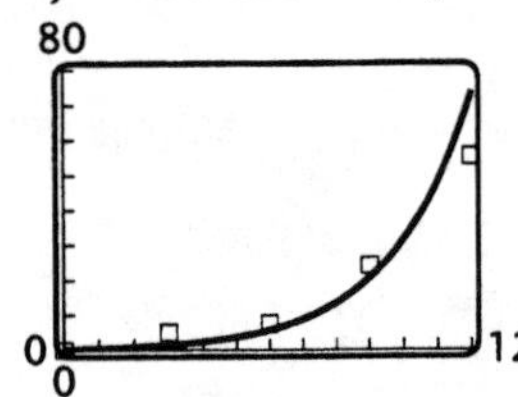

c) Using the Intersect feature, we find that the first coordinate of the point of intersection of $y_1 = 0.4703477987(1.523963441)^x$ and $y_2 = 500$ is approximately 16.54. Thus, there will be 500,000,000 cellular phone subscribers about 16.54 years after 1985.

41. a) $y = 46.6(1.018)^x$

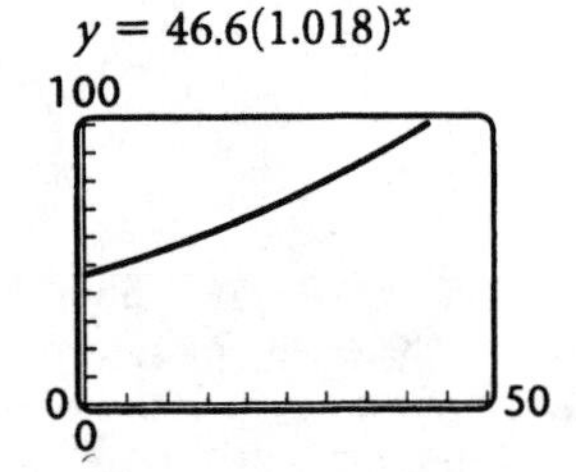

b) In 2000, $t = 2000 - 1981 = 19$.

$N(19) = 46.6(1.018)^{19} \approx 65.4$ billion cubic feet

In 2010, $t = 2010 - 1981 = 29$.

$N(29) = 46.6(1.018)^{29} \approx 78.2$ billion cubic feet

c) Using the Intersect feature, we find that the first coordinate of the point of intersection of $y_1 = 46.6(1.018)^x$ and $y_2 = 93.4$ is about 39. Then the demand for timber will be 93.4 billion cubic feet after about 39 years.

42. a) $y = 200[1 - (0.86)^x]$

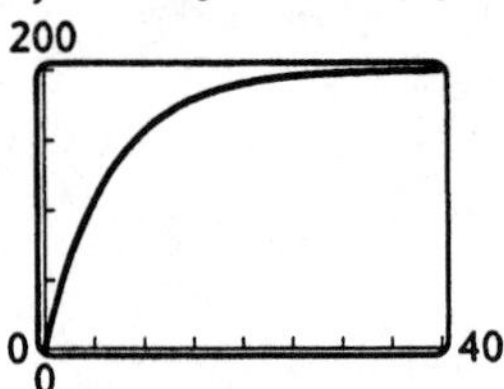

b) $S(10) = 200[1 - (0.86)^{10}] \approx 155.7$ words per minute

$S(20) = 200[1 - (0.86)^{20}] \approx 190.2$ words per minute

$S(40) = 200[1 - (0.86)^{40}] \approx 199.5$ words per minute

$S(100) \approx 200[1 - (0.86)^{100}] \approx 199.9999$ words per minute

c) Using the Intersect feature, we find that the first coordinate of the point of intersection of $y_1 = 200[1 - (0.86)^x]$ and $y_2 = 100$ is about 4.6. Thus, Sarah's speed is 100 words per minute after she practices for about 4.6 hr.

d) Using the Table feature we find that the graph has a horizontal asymptote, $y = 200$. This means that Sarah could eventually type close to 200 words per minute but never 200 words per minute or more.

43. a) $y = 100(1 - e^{-0.04x})$

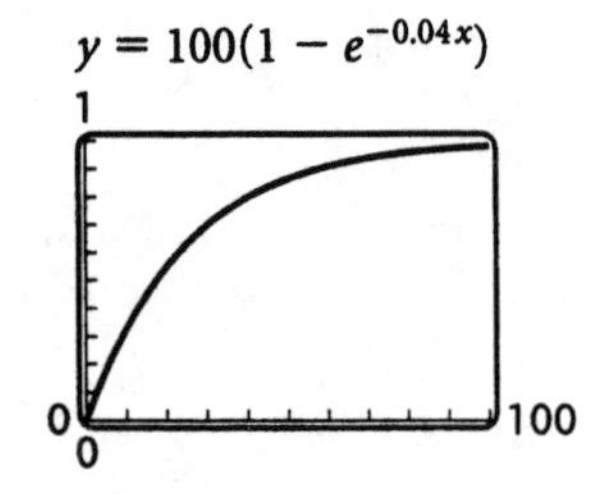

b) $f(25) = 100(1 - e^{-0.04(25)}) \approx 63\%$.

c)

$$90 = 100(1 - e^{-0.04t})$$
$$0.9 = 1 - e^{-0.04t}$$
$$-0.1 = -e^{-0.04t}$$
$$0.1 = e^{-0.04t}$$
$$\ln 0.1 = \ln e^{-0.04t}$$
$$\ln 0.1 = -0.04t$$
$$\frac{\ln 0.01}{-0.04} = t$$
$$58 \approx t$$

After about 58 days, 90% of the target market will have bought the product.

44. a) $y = 58(1 - e^{-1.1x}) + 20$

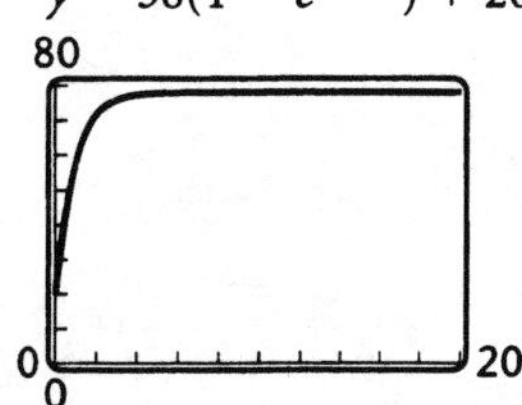

b) $V(1) = \$58(1 - e^{-1.1(1)}) + \$20 \approx \$58.69$

$V(2) = \$58(1 - e^{-1.1(2)}) + \$20 \approx \$71.57$

$V(4) = \$58(1 - e^{-1.1(4)}) + \$20 \approx \$77.29$

$V(6) = \$58(1 - e^{-1.1(6)}) + \$20 \approx \$77.92$

$V(12) = \$58(1 - e^{-1.1(12)}) + \$20 \approx \$78.00$

c) Using the Intersect feature, we find that the first coordinate of the point of intersection of $y_1 = 58(1 - e^{-1.1x}) + 20$ and $y_2 = 75$ is about 2.7. Thus, the value of the stock will be \$75 after about 2.7 months.

45. Graph (c) is the graph of $y = 3^x - 3^{-x}$.

46. Graph (j) is the graph of $y = 3^{-(x+1)^2}$.

47. Graph (a) is the graph of $f(x) = -2.3^x$.

48. Graph (d) is the graph of $f(x) = 30,000(1.4)^x$.

49. Graph (l) is the graph of $y = 2^{-|x|}$.

50. Graph (n) is the graph of $y = 2^{-(x-1)}$.

51. Graph (g) is the graph of $f(x) = (0.58)^x - 1$.

52. Graph (b) is the graph of $y = 2^x + 2^{-x}$.

53. Graph (i) is the graph of $g(x) = e^{|x|}$.

54. Graph (h) is the graph of $f(x) = |2^x - 1|$.

55. Graph (k) is the graph of $y = 2^{-x^2}$.

56. Graph (e) is the graph of $y = |2^{x^2} - 8|$.

57. Graph (m) is the graph of $g(x) = \dfrac{e^x - e^{-x}}{2}$.

58. Graph (f) is the graph of $f(x) = \dfrac{e^x + e^{-x}}{2}$.

59. Graph $y_1 = |1 - 3^x|$ and $y_2 = 4 + 3^{-x^2}$. Use the Intersect feature to find their point of intersection, $(1.481, 4.090)$.

60. Graph $y_1 = 4^x + 4^{-x}$ and $y_2 = 8 - 2x - x^2$. Use the Intersect feature to find their points of intersection, $(-1.551, 8.697)$ and $(1.078, 4.682)$.

61. Graph $y_1 = 2e^x - 3$ and $y_2 = \dfrac{e^x}{x}$. Use the Intersect feature to find their points of intersection, $(-0.402, -1.662)$ and $(1.051, 2.722)$.

62. Graph $y_1 = \dfrac{1}{e^x + 1}$ and $y_2 = 0.3x + \dfrac{7}{9}$. Use the Intersect feature to find their point of intersection, $(-0.510, 0.625)$.

63. Graph $y_1 = 5.3^x - 4.2^x$ and $y_2 = 1073$. Use the Intersect feature to find the first coordinate of their point of intersection. The solution is 4.448.

64. Graph $y_1 = e^x$ and $y_2 = x^3$. Use the Intersect feature to find the first coordinates of their points of intersection. The solutions are 1.857 and 4.536.

65. Graph $y_1 = 2^x$ and $y_2 = 1$. Use the Intersect feature to find their point of intersection, $(0, 1)$. Note that the graph of y_1 lies above the graph of y_2 for all points to the right of this point. Thus the solution set is $(0, \infty)$.

66. Graph $y_1 = 3^x$ and $y_2 = 1$. Use the Intersect feature to find their point of intersection, $(0, 1)$. Note that the graph of y_1 lies on or below the graph of y_2 for all points to the left of this point. Thus the solution set is $(-\infty, 0]$.

67. Graph $y_1 = 2^x + 3^x$ and $y_2 = x^2 + x^3$. Use the Intersect feature to find the first coordinates of their points of intersection. The solutions are 2.294 and 3.228.

68. Graph $y_1 = 31,245e^{-3x}$ and $y_2 = 523,467$. Use the Intersect feature to find the first coordinate of their point of intersection. The solution is -0.940.

69. Some differences are as follows: The range of f is $(-\infty, \infty)$ whereas the range of g is $(0, \infty)$; f has no asymptotes but g has a horizontal asymptote, the x-axis; the y-intercept of f is $(0, 0)$ and the y-intercept of g is $(0, 1)$.

70. The most interest will be earned the eighth year, because the principle is greatest during that year.

71. Both pairs of equations are inverses of each other.

72. Graph $y = x^3 - 2x + 4$ and use the Zero feature. The real-number zero is -2.

73. Graph $y = x^4 - 5x + 3$ and use the Zero feature twice. The real-number zeros are approximately 0.63188461 and 1.4252375.

74. Graph $y_1 = 3x^2 - 6$ and $y_2 = 5x$ and use the Intersect feature twice to find the first coordinate of each point of intersection of the graphs. The solutions are approximately -0.808143 and 2.4748096.

75. Graph $y = x^3 + 6x^2 - 16x$ and use the Zero feature three times. The solutions are -8, 0, and 2.

76. $7^\pi \approx 451.8078726$ and $\pi^7 \approx 3020.293228$, so π^7 is larger.

$70^{80} \approx 4.054 \times 10^{147}$ and $80^{70} \approx 1.646 \times 10^{133}$, so 70^{80} is larger.

77. Scrolling through the table set in AUTO mode, we see that function values approach 1 as x increases. Similarly, we could set the table in ASK mode and enter increasing values of x. In either case, we see that the horizontal asymptote is $y = 1$.

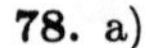

78. a)

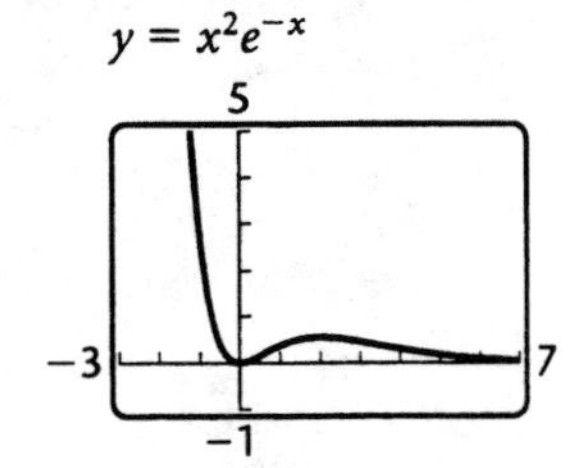

b) Use the Zero feature to find the first coordinate of the x-intercept. The only zero of the function is 0.

c) Relative minimum: 0 at $x = 0$;

relative maximum: 0.541 at $x = 2$

79. a)

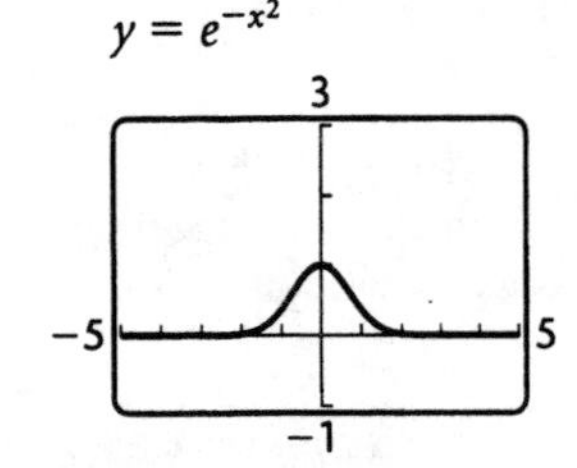

b) There are no x-intercepts, so the function has no zeros.

c) Relative minimum: none;

relative maximum: 1 at $x = 0$

Exercise Set 4.3

1. Graph $y = \log_3 x$.

The equation $y = \log_3 x$ is equivalent to $x = 3^y$. We can find ordered pairs that are solutions by choosing values for y and computing the corresponding x-values.

For $y = -2$, $x = 3^{-2} = \frac{1}{9}$.

For $y = -1$, $x = 3^{-1} = \frac{1}{3}$.

For $y = 0$, $x = 3^0 = 1$.

For $y = 1$, $x = 3^1 = 3$.

For $y = 2$, $x = 3^2 = 9$.

x, or 3^y	y
$\frac{1}{9}$	-2
$\frac{1}{3}$	-1
1	0
3	1
9	2

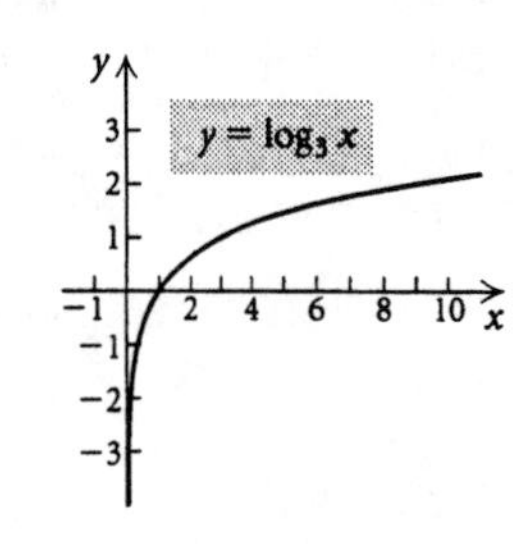

2. $y = \log_4 x$ is equivalent to $x = 4^y$.

x, or 4^y	y
$\frac{1}{16}$	-2
$\frac{1}{4}$	-1
1	0
4	1
16	2

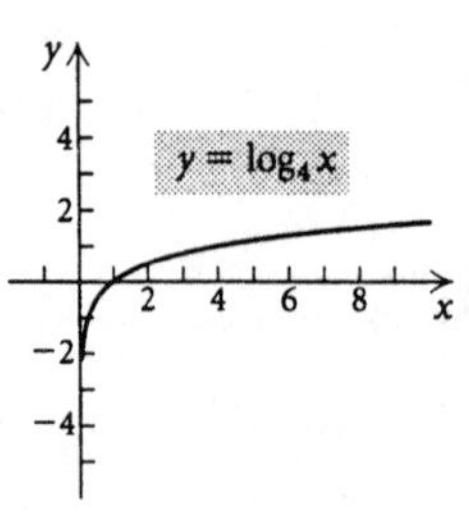

3. Graph $f(x) = \log x$.

Think of $f(x)$ as y. The equation $y = \log x$ is equivalent to $x = 10^y$. We can find ordered pairs that are solutions by choosing values for y and computing the corresponding x-values.

For $y = -2$, $x = 10^{-2} = 0.01$.

For $y = -1$, $x = 10^{-1} = 0.1$.

For $y = 0$, $x = 10^0 = 1$.

For $y = 1$, $x = 10^1 = 10$.

For $y = 2$, $x = 10^2 = 100$.

x, or 10^y	y
0.01	-2
0.1	-1
1	0
10	1
100	2

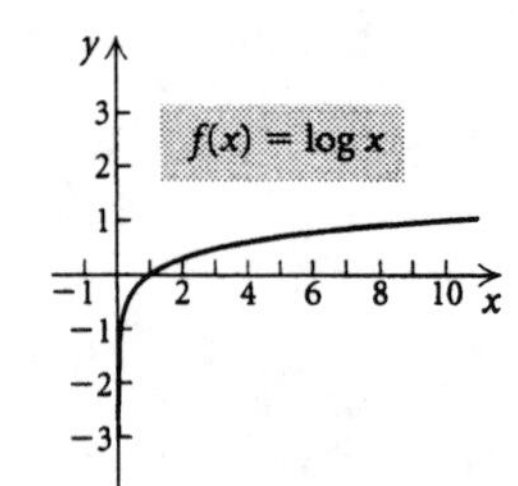

4. See Example 9.

5. $\log_2 16 = 4$ because the exponent to which we raise 2 to get 16 is 4.

6. $\log_3 9 = 2$, because the exponent to which we raise 3 to get 9 is 2.

7. $\log_5 125 = 3$, because the exponent to which we raise 5 to get 125 is 3.

8. $\log_2 64 = 6$, because the exponent to which we raise 2 to get 64 is 6.

9. $\log 0.001 = -3$, because the exponent to which we raise 10 to get 0.001 is -3.

10. $\log 100 = 2$, because the exponent to which we raise 10 to get 100 is 2.

11. $\log_2 \frac{1}{4} = -2$, because the exponent to which we raise 2 to

get $\frac{1}{4}$ is -2.

12. $\log_8 2 = \frac{1}{3}$, because the exponent to which we raise 8 to

get 2 is $\frac{1}{3}$.

13. $\ln 1 = 0$, because the exponent to which we raise e to get 1 is 0.

14. $\ln e = 1$, because the exponent to which we raise e to get e is 1.

15. $\log 10 = 1$, because the exponent to which we raise 10 to get 10 is 1.

16. $\log 1 = 0$, because the exponent to which we raise 10 to get 1 is 0.

17.

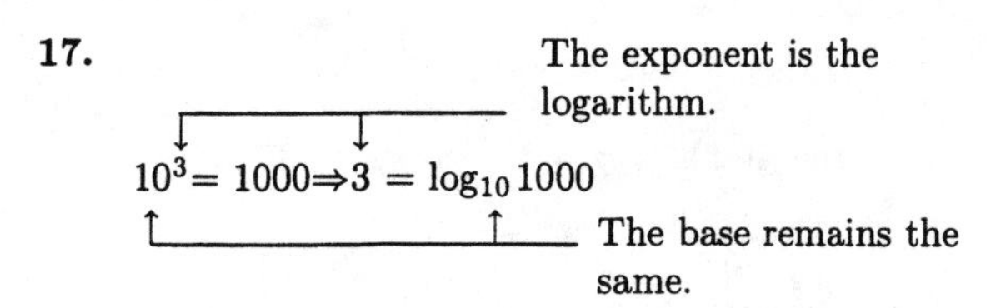

18. $5^{-3} = \frac{1}{125} \Rightarrow \log_5 \frac{1}{125} = -3$

19.

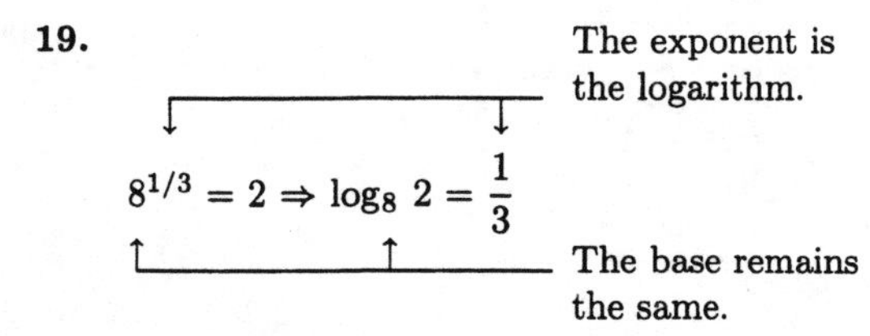

20. $10^{0.3010} = 2 \Rightarrow \log_{10} 2 = 0.0310$

21. $e^3 = t \Rightarrow \log_e t = 3$

22. $Q^t = x \Rightarrow \log_Q x = t$

23. $e^2 = 7.3891 \Rightarrow \log_e 7.3891 = 2$

24. $e^{-1} = 0.3679 \Rightarrow \log_e 0.3679 = -1$

25. $p^k = 3 \Rightarrow \log_p 3 = k$

26. $e^{-t} = 4000 \Rightarrow \log_e 4000 = -t$

27. The logarithm is the exponent.

$\log_5 5 = 1 \Rightarrow 5^1 = 5$

The base remains the same.

28. $t = \log_4 7 \Rightarrow 7 = 4^t$

29. $\log 0.01 = -2$ is equivalent to $\log_{10} 0.01 = -2$.

The logarithm is the exponent.

$\log_{10} 0.01 = -2 \Rightarrow 10^{-2} = 0.01$

The base remains the same.

30. $\log 7 = 0.845 \Rightarrow 10^{0.845} = 7$

31. $\ln 30 = 3.4012 \Rightarrow e^{3.4012} = 30$

32. $\ln 0.38 = -0.9676 \Rightarrow e^{-0.9676} = 0.38$

33. $\log_a M = -x \Rightarrow a^{-x} = M$

34. $\log_t Q = k \Rightarrow t^k = Q$

35. $\log_a T^3 = x \Rightarrow a^x = T^3$

36. $\ln W^5 = t \Rightarrow e^t = W^5$

37. $\log 3 \approx 0.4771$

38. $\log 8 \approx 0.9031$

39. $\log 532 \approx 2.7259$

40. $\log 93,100 \approx 4.9689$

41. $\log 0.57 \approx -0.2441$

42. $\log 0.082 \approx -1.0862$

43. $\log(-2)$ does not exist. (The grapher gives an error message.)

44. $\ln 50 \approx 3.9120$

45. $\ln 2 \approx 0.6931$

46. $\ln(-4)$ does not exist. (The grapher gives an error message.)

47. $\ln 809.3 \approx 6.6962$

48. $\ln 0.00037 \approx -7.9020$

49. $\ln(-1.32)$ does not exist. (The grapher gives an error message.)

50. $\ln 0$ does not exist. (The grapher gives an error message.)

51. Let $a = 10$, $b = 4$, and $M = 100$ and substitute in the change-of-base formula.

$$\log_4 100 = \frac{\log_{10} 100}{\log_{10} 4} \approx \frac{2}{0.602060} \approx 3.3219$$

52. $\log_3 20 = \frac{\log 20}{\log 3} \approx 2.7268$

53. Let $a = e$, $b = 100$, and $M = 0.3$ and substitute in the change-of-base formula.

$$\log_{100} 0.3 = \frac{\ln 0.3}{\ln 100} \approx \frac{-1.203973}{4.605170} \approx -0.2614$$

54. $\log_\pi 100 = \frac{\ln 100}{\ln \pi} = 4.0229$

55. Let $a = 10$, $b = 200$, and $M = 50$ and substitute in the change-of-base formula.

$$\log_{200} 50 = \frac{\log_{10} 50}{\log_{10} 200} \approx \frac{1.698970}{2.301030} \approx 0.7384$$

56. $\log_{5.3} 1700 = \dfrac{\log 1700}{\log 5.3} \approx 4.4602$

57. Shift the graph of $y = \log_2 x$ left 3 units. To use the grapher, we must first change the base. Here we change from base 2 to base 10. We get $y = \dfrac{\log(x+3)}{\log 2}$, using $\log_b M = \dfrac{\log_a M}{\log_a b}$ with $b = 2$, $M = x + 3$, and $a = 10$.

$$y = \frac{\log(x+3)}{\log 2}$$

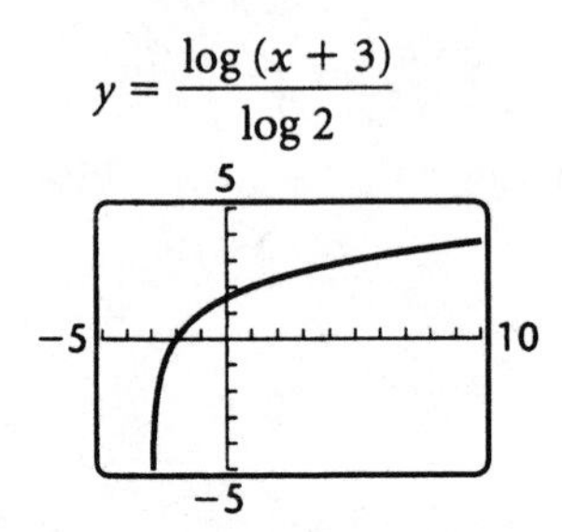

Domain: $(-3, \infty)$

Vertical asymptote: $x = -3$

58. Shift the graph of $y = \log_3 x$ right 2 units. Change from base 3 to base 10: $y = \dfrac{\log(x-2)}{\log 3}$.

$$y = \frac{\log(x-2)}{\log 3}$$

Domain: $(2, \infty)$

Vertical asymptote: $x = 2$

59. Shift the graph of $y = \log_3 x$ down 1 unit. To use the grapher, we must first change the base. Here we change from base 3 to base 10. We get $y = \dfrac{\log x}{\log 3} - 1$, using $\log_b M = \dfrac{\log_a M}{\log_a b}$ with $b = 3$, $M = x$, and $a = 10$.

$$y = \frac{\log x}{\log 3} - 1$$

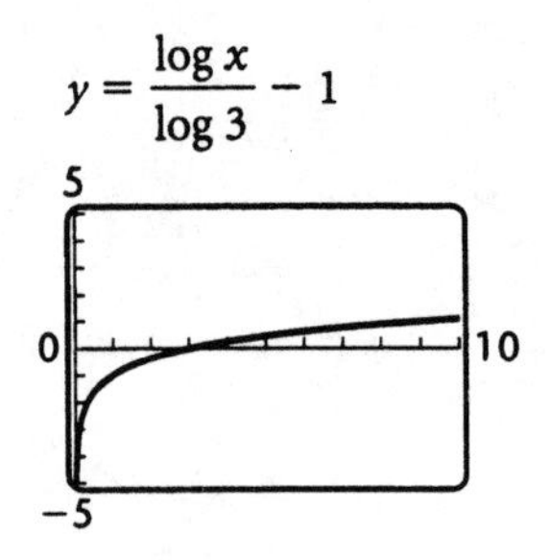

Domain: $(0, \infty)$

Vertical asymptote: $x = 0$

60. Shift the graph of $y = \log_2 x$ up 3 units. Change from base 2 to base 10: $y = 3 + \dfrac{\log x}{\log 2}$.

$$y = 3 + \frac{\log x}{\log 2}$$

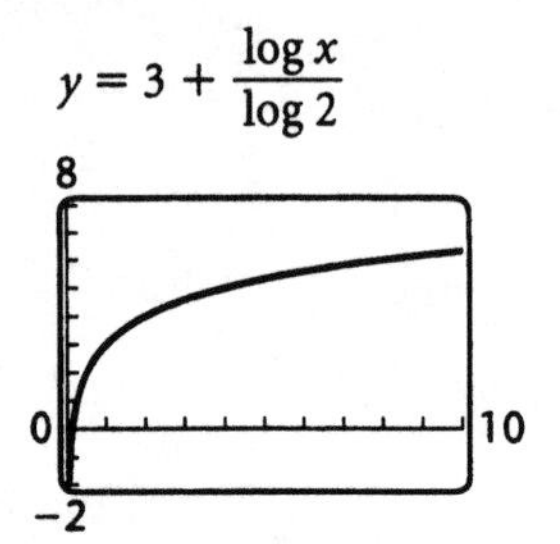

Domain: $(0, \infty)$

Vertical asymptote: $x = 0$

61. Stretch the graph of $y = \ln x$ vertically.

$$y = 4 \ln x$$

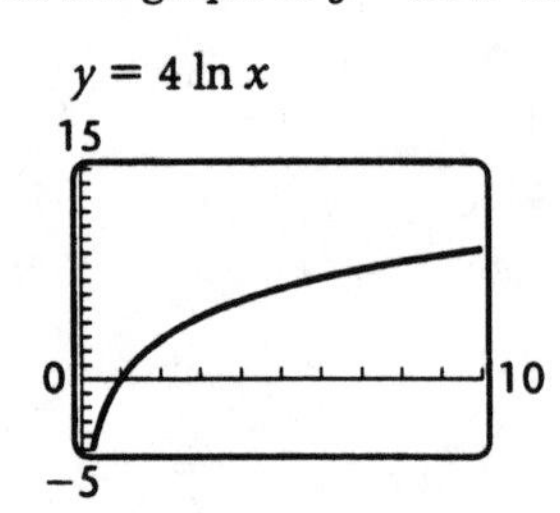

Domain: $(0, \infty)$

Vertical asymptote: $x = 0$

62. Shrink the graph of $y = \ln x$ vertically.

$$y = \frac{1}{2} \ln x$$

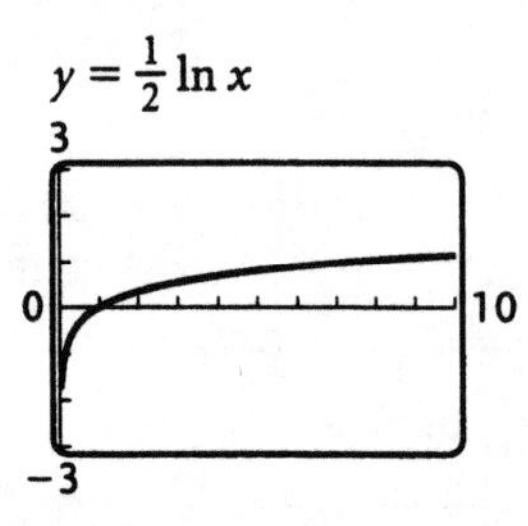

Domain: $(0, \infty)$

Vertical asymptote: $x = 0$

63. Reflect the graph of $y = \ln x$ across the x-axis and then shift it up 2 units.

$y = 2 - \ln x$

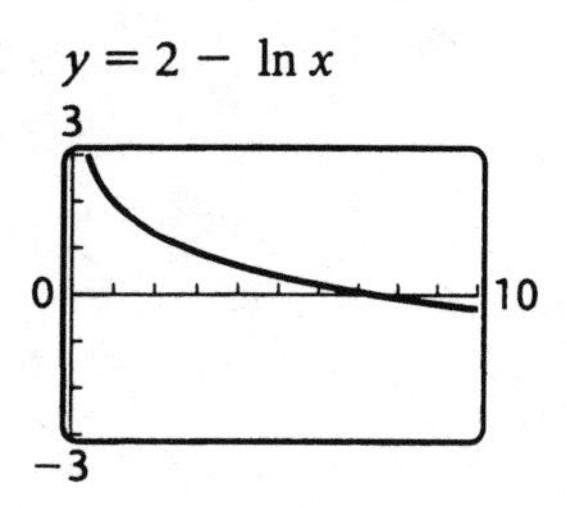

Domain: $(0, \infty)$

Vertical asymptote: $x = 0$

64. Shift the graph of $y = \ln x$ left 1 unit.

$y = \ln(x + 1)$

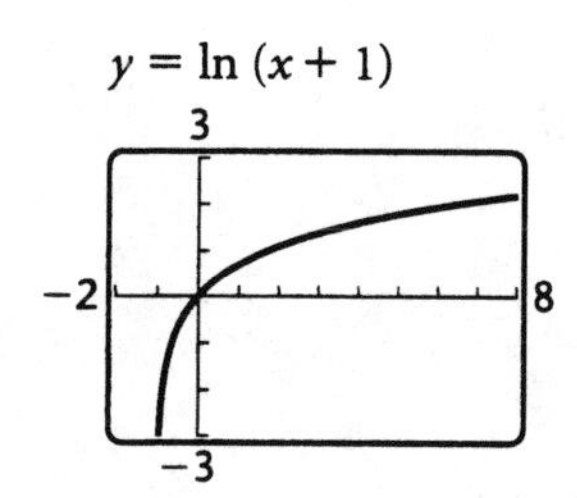

Domain: $(-1, \infty)$

Vertical asymptote: $x = -1$

65. Graph $y_1 = 3^x$ and $y_2 = \log_3 x = \dfrac{\log x}{\log 3}$ in the same window, or graph $y = 3^x$ and use the grapher feature that automatically graphs inverses to graph $y = \log_3 x$. On a hand-drawn graph, we can graph $y = 3^x$ and then reflect this graph across the line $y = x$ to get the graph of $y = \log_3 x$.

$y_1 = 3^x$,
$y_2 = \dfrac{\log x}{\log 3}$

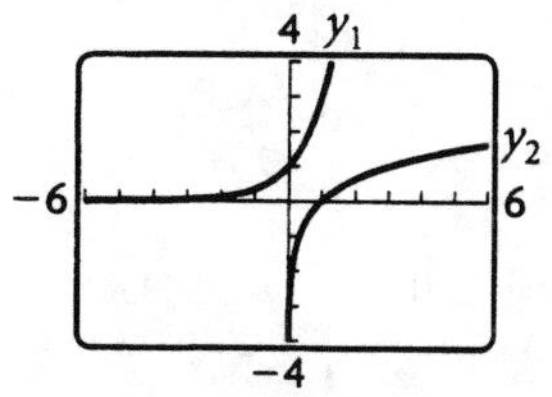

4 y_1, y_2, −6, 6, −4

66. Graph $y_1 = \log_4 x = \dfrac{\log x}{\log 4}$ and $y_2 = 4^x$ in the same window, or graph $y = 4^x$ and use the grapher feature that automatically graphs inverses to graph $y = \log_4 x$. (In the second alternative, we choose to graph $y = 4^x$ first since this can be done without using the change-of-base formula.) On a hand-drawn graph, we can graph $y = \log_4 x$ and then reflect this graph across the line $y = x$ to get the graph of $y = 4^x$.

$y_1 = \dfrac{\log x}{\log 4}$,
$y_2 = 4^x$

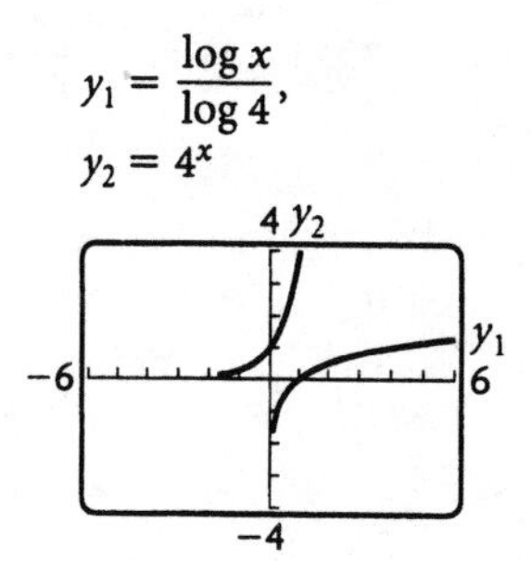

67. Graph $y_1 = \log x$ and $y_2 = 10^x$ in the same window, or graph $y = \log x$ and use the grapher feature that automatically graphs inverses to graph $y = 10^x$. On a hand-drawn graph, we can graph $y = \log x$ and then reflect this graph across the line $y = x$ to get the graph of $y = 10^x$.

$y_1 = \log x$,
$y_2 = 10^x$

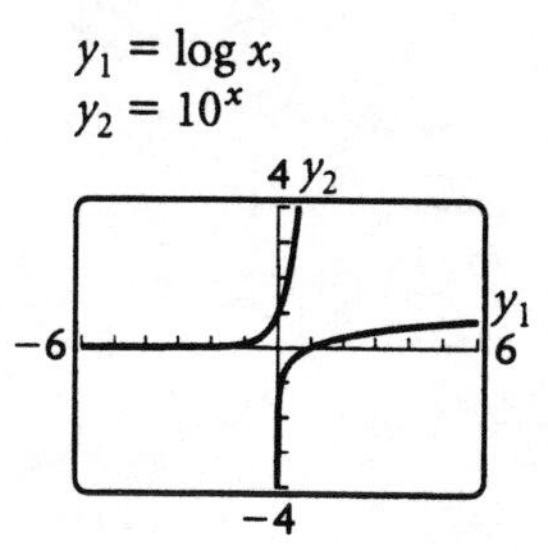

68. Graph $y_1 = e^x$ and $y_2 = \ln x$ in the same window, or graph $y = e^x$ and use the grapher feature that automatically graphs inverses to graph $y = \ln x$. On a hand-drawn graph, we can graph $y = e^x$ and then reflect this graph across the line $y = x$ to get the graph of $y = \ln x$.

$y_1 = e^x$, $y_2 = \ln x$

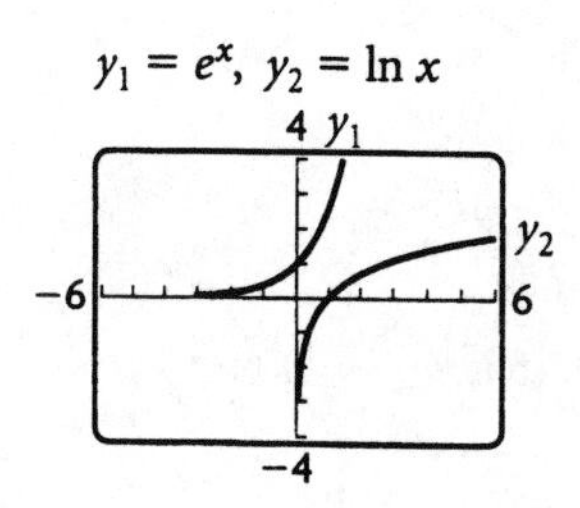

69. a) We substitute 419.681 for P, since P is in thousands.

$$w(419.681) = 0.37 \ln 419.681 + 0.05 \approx 2.3 \text{ ft/sec}$$

b) We substitute 2721.547 for P, since P is in thousands.

$$w(2721.547) = 0.37 \ln 2721.547 + 0.05 \approx 3.0 \text{ ft/sec}$$

c) We substitute 350.363 for P, since P is in thousands.

$$w(350.363) = 0.37 \ln 350.363 + 0.05 \approx 2.2 \text{ ft/sec}$$

d) We substitute 149.799 for P, since P is in thousands.

$$w(149.799) = 0.37 \ln 149.799.4 + 0.05$$
$$\approx 1.9 \text{ ft/sec}$$

e) We substitute 94.466 for P, since P is in thousands.

$$w(94.466) = 0.37 \ln 94.466 + 0.05$$
$$\approx 1.7 \text{ ft/sec}$$

70. a) $R = \log \dfrac{10^{7.85} \cdot I_0}{I_0} = \log 10^{7.85} = 7.85$

b) $R = \log \dfrac{10^{8.25} \cdot I_0}{I_0} = \log 10^{8.25} = 8.25$

c) $R = \log \dfrac{10^{9.6} \cdot I_0}{I_0} = \log 10^{9.6} = 9.6$

d) $R = \log \dfrac{10^{7.85} \cdot I_0}{I_0} = \log 10^{7.85} = 7.85$

e) $R = \log \dfrac{10^{6.9} \cdot I_0}{I_0} = \log 10^{6.9} = 6.9$

71. a)
$$S(0) = 78 - 15\log(0+1)$$
$$= 78 - 15\log 1$$
$$= 78 - 15 \cdot 0$$
$$= 78\%$$

b)
$$S(4) = 78 - 15\log(4+1)$$
$$= 78 - 15\log 5$$
$$\approx 78 - 15(0.698970)$$
$$\approx 67.5\%$$

$$S(24) = 78 - 15\log(24+1)$$
$$= 78 - 15\log 25$$
$$\approx 78 - 15(1.397940)$$
$$\approx 57\%$$

c) $y = 78 - 15 \log (x + 1), x \geqslant 0$

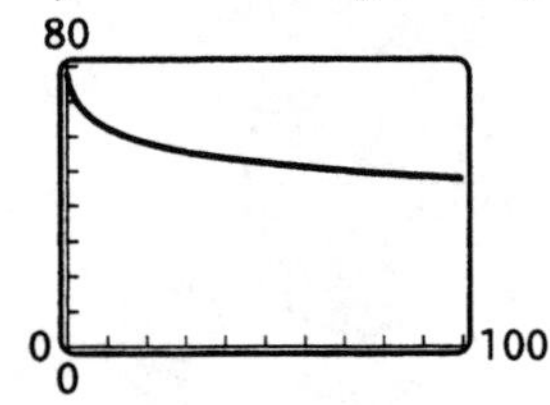

d)
$$50 = 78 - 15\log(x+1)$$
$$-28 = -15\log(x+1) \quad \text{Subtracting 78}$$
$$\frac{28}{15} = \log(x+1) \quad \text{Dividing by } -15$$
$$x + 1 = 10^{28/15} \quad \text{Using the definition of logarithm}$$
$$x = 10^{28/15} - 1$$
$$x \approx 73 \text{ months}$$

72. a) $pH = -\log[1.6 \times 10^{-4}] \approx -(-3.8) = 3.8$

b) $pH = -\log[0.0013] \approx -(-2.9) = 2.9$

c) $pH = -\log[6.3 \times 10^{-7}] \approx -(-6.2) = 6.2$

d) $pH = -\log[1.6 \times 10^{-8}] \approx -(-7.8) = 7.8$

e) $pH = -\log[6.3 \times 10^{-5}] \approx -(-4.2) = 4.2$

73. a)
$$7 = -\log[H^+]$$
$$-7 = \log[H^+]$$
$$H^+ = 10^{-7} \quad \text{Using the definition of logarithm}$$

b)
$$5.4 = -\log[H^+]$$
$$-5.4 = \log[H^+]$$
$$H^+ = 10^{-5.4} \quad \text{Using the definition of logarithm}$$
$$H^+ \approx 4.0 \times 10^{-6}$$

c)
$$3.2 = -\log[H^+]$$
$$-3.2 = \log[H^+]$$
$$H^+ = 10^{-3.2} \quad \text{Using the definition of logarithm}$$
$$H^+ \approx 6.3 \times 10^{-4}$$

d)
$$4.8 = -\log[H^+]$$
$$-4.8 = \log[H^+]$$
$$H^+ = 10^{-4.8} \quad \text{Using the definition of logarithm}$$
$$H^+ \approx 1.6 \times 10^{-5}$$

74. a) $N(1) = 1000 + 200 \ln 1 = 1000$

b) $N(5) = 1000 + 200 \ln 5 \approx 1332$

c) $y = 1000 + 200 \ln x$

5000

0 0 10

d)
$$2000 = 1000 + 200 \ln a$$
$$1000 = 200 \ln a$$
$$5 = \ln a$$
$$a = e^5$$
$$a \approx \$148 \text{ thousand}$$

75. a)
$$L = 10\log \frac{2510 \cdot I_0}{I_0}$$
$$= 10 \log 2510$$
$$\approx 34 \text{ decibels}$$

b)
$$L = 10\log \frac{2,500,000 \cdot I_0}{I_0}$$
$$= 10\log 2,500,000$$
$$\approx 64 \text{ decibels}$$

c) $L = 10\log\dfrac{10^6 \cdot I_0}{I_0}$
$= 10\log 10^6$
$= 60$ decibels

d) $L = 10\log\dfrac{10^9 \cdot I_0}{I_0}$
$= 10\log 10^9$
$= 90$ decibels

76. Reflect the graph of $f(x) = \ln x$ across the line $y = x$ to obtain the graph of $h(x) = e^x$. Then shift this graph 2 units right to obtain the graph of $g(x) = e^{x-2}$.

77. Reflect the graph of $f(x) = e^x$ across the line $y = x$ to obtain the graph of $h(x) = \ln x$. Then shift this graph up 3 units to obtain the graph of $g(x) = 3 + \ln x$.

78.
$$\begin{array}{r|rrrrr} -1 & 1 & -2 & 0 & 1 & -6 \\ & & -1 & 3 & -3 & 2 \\ \hline & 1 & -3 & 3 & -2 & -4 \end{array}$$

$f(-1) = -4$

79.
$$\begin{array}{r|rrrr} -5 & 1 & -6 & 3 & 10 \\ & & -5 & 55 & -290 \\ \hline & 1 & -11 & 58 & -280 \end{array}$$

The remainder is -280, so $f(-5) = -280$.

80. $f(x) = (x - 4i)(x + 4i)(x - 1)$
$= (x^2 + 16)(x - 1)$
$= x^3 - x^2 + 16x - 16$

81. $f(x) = (x - \sqrt{7})(x + \sqrt{7})(x - 0)$
$= (x^2 - 7)(x)$
$= x^3 - 7x$

82. Using the change-of-base formula, we get

$\dfrac{\log_3 64}{\log_3 16} = \log_{16} 64.$

Let $\log_{16} 64 = x$. Then we have

$16^x = 64$ Using the definition of logarithm
$(2^4)^x = 2^6$
$2^{4x} = 2^6$, so
$4x = 6$
$x = \dfrac{6}{4} = \dfrac{3}{2}$

Thus, $\dfrac{\log_3 64}{\log_3 16} = \dfrac{3}{2}$.

83. Using the change-of-base formula, we get

$\dfrac{\log_5 8}{\log_2 8} = \log_2 8 = 3.$

84. $f(x) = \log_4 x^2$

x^2 must be positive, so the domain is $(-\infty 0)\cup(0, \infty)$.

85. $f(x) = \log_5 x^3$

x^3 must be positive. Since $x^3 > 0$ for $x > 0$, the domain is $(0, \infty)$.

86. $f(x) = \log(3x - 4)$

$3x - 4$ must be positive. We have

$3x - 4 > 0$
$x > \dfrac{4}{3}$

The domain is $\left(\dfrac{4}{3}, \infty\right)$.

87. $f(x) = \ln|x|$

$|x|$ must be positive. Since $|x| > 0$ for $x \neq 0$, the domain is $(-\infty, 0) \cup (0, \infty)$.

88. Graph $y_1 = \log_2(x - 3) = \dfrac{\log(x - 3)}{\log 2}$ and $y_2 = 4$. Observe that the graph of y_1 lies on or above the graph of y_2 for all inputs greater than or equal to 19. Thus, the solution set is $[19, \infty)$.

89. Graph $y = \log_2(2x + 5) = \dfrac{\log(2x + 5)}{\log 2}$. Observe that outputs are negative for inputs between $-\dfrac{5}{2}$ and -2. Thus, the solution set is $\left(-\dfrac{5}{2}, -2\right)$.

90. Graph $y_1 = 4$ and $y_2 = \dfrac{4}{e^x + 1}$ and use the Intersect feature to find the point(s) of intersection of the graphs. The point of intersection is $(1.250, 0.891)$.

91. Graph (d) is the graph of $f(x) = \ln|x|$.

92. Graph (c) is the graph of $f(x) = |\ln x|$.

93. Graph (b) is the graph of $f(x) = \ln x^2$.

94. Graph (a) is the graph of $g(x) = |\ln(x - 1)|$.

95. a) $y = x \ln x$

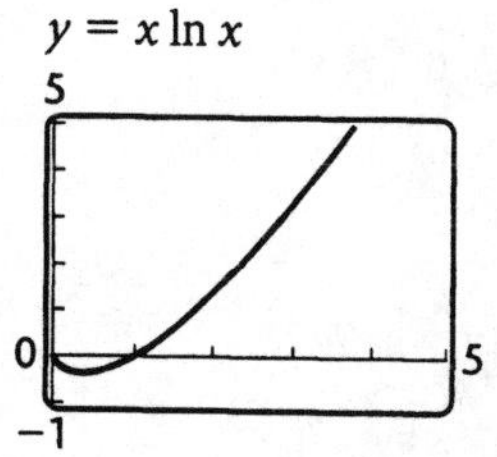

b) Use the Zero feature. The zero is 1.

c) Use the Minimum feature. The relative minimum is -0.368 at $x = 0.368$. There is no relative maximum.

96. a) $y = x^2 \ln x$

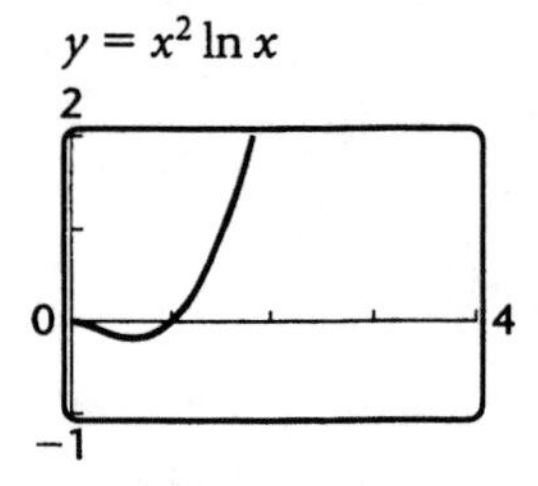

b) 1

c) Relative minimum: -0.184 at $x = 0.607$

Relative maximum: none

97. a) $y = \dfrac{\ln x}{x^2}$

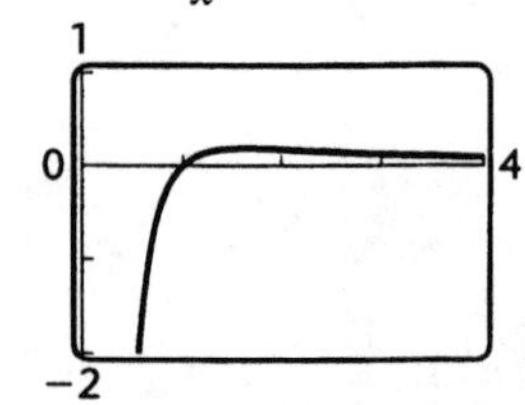

b) Use the Zero feature. The zero is 1.

c) Use the Maximum feature. There is no relative minimum. The relative maximum is 0.184 at $x = 1.649$.

98. a) $y = e^{-x} \ln x$

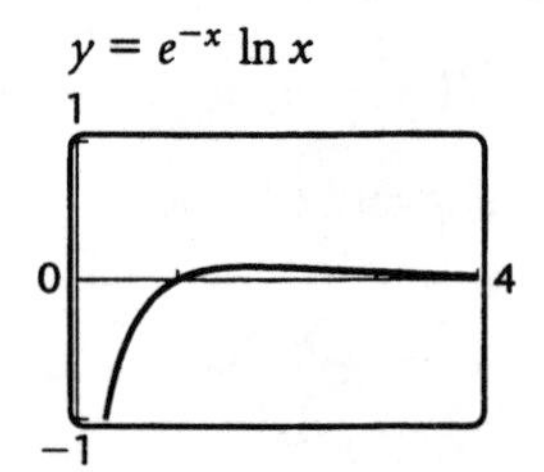

b) 1

c) Relative minimum: none

Relative maximum: 0.097 at $x = 1.763$

Exercise Set 4.4

1. Use the product rule.

$\log_3(81 \cdot 27) = \log_3 81 + \log_3 27$

2. $\log_2(8 \cdot 64) = \log_2 8 + \log_2 64$

3. Use the product rule.

$\log_5(5 \cdot 125) = \log_5 5 + \log_5 125$

4. $\log_4(64 \cdot 32) = \log_4 64 + \log_4 32$

5. Use the product rule.

$\log_t 8Y = \log_t 8 + \log_t Y$

6. $\log_e Qx = \log_e Q + \log_e x$

7. Use the power rule.

$\log_b t^3 = 3\log_b t$

8. $\log_a x^4 = 4\log_a x$

9. Use the power rule.

$\log y^8 = 8\log y$

10. $\ln y^5 = 5\ \ln y$

11. Use the power rule.

$\log_c K^{-6} = -6\log_c K$

12. $\log_b Q^{-8} = -8\log_b Q$

13. Use the difference rule.

$\log_t \dfrac{M}{8} = \log_t M - \log_t 8$

14. $\log_a \dfrac{76}{13} = \log_a 76 - \log_a 13$

15. Use the difference rule.

$\log_a \dfrac{x}{y} = \log_a x - \log_a y$

16. $\log_b \dfrac{3}{w} = \log_b 3 - \log_b w$

17. $\log_a 6xy^5z^4$

$= \log_a 6 + \log_a x + \log_a y^5 + \log_a z^4$ Product rule

$= \log_a 6 + \log_a x + 5\log_a y + 4\log_a z$ Power rule

18. $\log_a x^3y^2z$

$= \log_a x^3 + \log_a y^2 + \log_a z$

$= 3\log_a x + 2\log_a y + \log_a z$

19. $\log_b \dfrac{p^2q^5}{m^4b^9}$

$= \log_b p^2q^5 - \log_b m^4b^9$ Quotient rule

$= \log_b p^2 + \log_b q^5 - (\log_b m^4 + \log_b b^9)$ Product rule

$= \log_b p^2 + \log_b q^5 - \log_b m^4 - \log_b b^9$

$= \log_b p^2 + \log_b q^5 - \log_b m^4 - 9 \quad (\log_b b^9 = 9)$

$= 2\log_b p + 5\log_b q - 4\log_b m - 9$ Power rule

20. $\log_b \dfrac{x^2y}{b^3} = \log_b x^2y - \log_b b^3$

$= \log_b x^2 + \log_b y - \log_b b^3$

$= \log_b x^2 + \log_b y - 3$

$= 2\log_b x + \log_b y - 3$

21. $\frac{1}{2}\log_a \frac{x^6}{p^5q^8}$

$= \frac{1}{2}[\log_a x^6 - \log_a(p^5q^8)]$ Quotient rule

$= \frac{1}{2}[\log_a x^6 - (\log_a p^5 + \log_a q^8)]$ Product rule

$= \frac{1}{2}(\log_a x^6 - \log_a p^5 - \log_a q^8)$

$= \frac{1}{2}(6\log_a x - 5\log_a p - 8\log_a q)$ Power rule

$= 3\log_a x - \frac{5}{2}\log_a p - 4\log_a q$

22. $\log_c \sqrt[3]{\frac{y^3z^2}{x^4}}$

$= \frac{1}{3}\log_c \frac{y^3z^2}{x^4}$

$= \frac{1}{3}(\log_c y^3z^2 - \log_c x^4)$

$= \frac{1}{3}(\log_c y^3 + \log_c z^2 - \log_c x^4)$

$= \frac{1}{3}(3\log_c y + 2\log_c z - 4\log_c x)$

$= \log_c y + \frac{2}{3}\log_c z - \frac{4}{3}\log_c x$

23. $\log_a \sqrt[4]{\frac{m^8n^{12}}{a^3b^5}}$

$= \frac{1}{4}\log_a \frac{m^8n^{12}}{a^3b^5}$ Power rule

$= \frac{1}{4}(\log_a m^8n^{12} - \log_a a^3b^5)$ Quotient rule

$= \frac{1}{4}[\log_a m^8 + \log_a n^{12} - (\log_a a^3 + \log_a b^5)]$ Product rule

$= \frac{1}{4}(\log_a m^8 + \log_a n^{12} - \log_a a^3 - \log_a b^5)$

$= \frac{1}{4}(\log_a m^8 + \log_a n^{12} - 3 - \log_a b^5)$ $(\log_a a^3 = 3)$

$= \frac{1}{4}(8\log_a m + 12\log_a n - 3 - 5\log_a b)$ Power rule

$= 2\log_a m + 3\log_a n - \frac{3}{4} - \frac{5}{4}\log_a b$

24. $\log_a \sqrt{\frac{a^6b^8}{a^2b^5}} = \log_a \sqrt{a^4b^3}$

$= \frac{1}{2}(\log_a a^4 + \log_a b^3)$

$= \frac{1}{2}(4 + 3\log_a b)$

$= 2 + \frac{3}{2}\log_a b$

25. $\log_a 75 + \log_a 2$

$= \log_a(75 \cdot 2)$ Product rule

$= \log_a 150$

26. $\log 0.01 + \log 1000 = \log(0.01 \cdot 1000) = \log 10 = 1$

27. $\log 10{,}000 - \log 100$

$= \log \frac{10{,}000}{100}$ Quotient rule

$= \log 100$

$= 2$

28. $\ln 54 - \ln 6 = \ln \frac{54}{6} = \ln 9$

29. $\frac{1}{2}\log_a x + 4\log_a y - 3\log_a x$

$= \log_a x^{1/2} + \log_a y^4 - \log_a x^3$ Power rule

$= \log_a x^{1/2}y^4 - \log_a x^3$ Product rule

$= \log_a \frac{x^{1/2}y^4}{x^3}$ Quotient rule

$= \log_a x^{-5/2}y^4$, or $\log_a \frac{y^4}{x^{5/2}}$ Simplifying

30. $\frac{2}{5}\log_a x - \frac{1}{3}\log_a y = \log_a x^{2/5} - \log_a y^{1/3} =$

$\log_a \frac{x^{2/5}}{y^{1/3}}$

31. $\ln x^2 - 2\ln\sqrt{x}$

$= \ln x^2 - \ln(\sqrt{x})^2$ Power rule

$= \ln x^2 - \ln x$ $[(\sqrt{x})^2 = x]$

$= \ln \frac{x^2}{x}$ Quotient rule

$= \ln x$

32. $\ln 2x + 3(\ln x - \ln y) = \ln 2x + 3\ln \frac{x}{y}$

$= \ln 2x + \ln \left(\frac{x}{y}\right)^3$

$= \ln 2x\left(\frac{x}{y}\right)^3$

$= \ln \frac{2x^4}{y^3}$

33. $\ln(x^2 - 4) - \ln(x + 2)$

$= \ln \frac{x^2 - 4}{x + 2}$ Quotient rule

$= \ln \frac{(x+2)(x-2)}{x+2}$ Factoring

$= \ln(x - 2)$ Removing a factor of 1

34. $\log_a \frac{a}{\sqrt{x}} - \log_a \sqrt{ax} = \log_a \frac{a}{\sqrt{x}\sqrt{ax}}$

$= \log_a \frac{\sqrt{a}}{x}$

$= \log_a \sqrt{a} - \log_a x$

$= \frac{1}{2}\log_a a - \log_a x$

$= \frac{1}{2} - \log_a x$

35. $\ln x - 3[\ln(x-5) + \ln(x+5)]$

$= \ln x - 3\ln[(x-5)(x+5)]$ Product rule

$= \ln x - 3\ln(x^2-25)$

$= \ln x - \ln(x^2-25)^3$ Power rule

$= \ln \frac{x}{(x^2-25)^3}$ Quotient rule

36. $\frac{2}{3}[\ln(x^2-9) - \ln(x+3)] + \ln(x+y)$

$= \frac{2}{3}\ln\frac{x^2-9}{x+3} + \ln(x+y)$

$= \frac{2}{3}\ln\frac{(x+3)(x-3)}{x+3} + \ln(x+y)$

$= \frac{2}{3}\ln(x-3) + \ln(x+y)$

$= \ln(x-3)^{2/3} + \ln(x+y)$

$= \ln[(x-3)^{2/3}(x+y)]$

37. $\frac{3}{2}\ln 4x^6 - \frac{4}{5}\ln 2y^{10}$

$= \frac{3}{2}\ln 2^2x^6 - \frac{4}{5}\ln 2y^{10}$ Writing 4 as 2^2

$= \ln(2^2x^6)^{3/2} - \ln(2y^{10})^{4/5}$ Power rule

$= \ln(2^3x^9) - \ln(2^{4/5}y^8)$

$= \ln\frac{2^3x^9}{2^{4/5}y^8}$ Quotient rule

$= \ln\frac{2^{11/5}x^9}{y^8}$

38. $120(\ln\sqrt[5]{x^3} + \ln\sqrt[3]{y^2} - \ln\sqrt[4]{16z^5})$

$= 120\left(\ln\frac{\sqrt[5]{x^3}\sqrt[3]{y^2}}{\sqrt[4]{16z^5}}\right)$

$= 120\left(\frac{x^{3/5}y^{2/3}}{2z^{5/4}}\right)$

$= \ln\left(\frac{x^{3/5}y^{2/3}}{2z^{5/4}}\right)^{120}$

$= \ln\frac{x^{72}y^{80}}{2^{120}z^{150}}$

39. $\log_b \frac{3}{5} = \log_b 3 - \log_b 5$ Quotient rule

$= 1.0986 - 1.6094$

$= -0.5108$

40. $\log_b 15 = \log_b(3\cdot 5)$

$= \log_b 3 + \log_b 5$

$= 1.0986 + 1.6094$

$= 2.708$

41. $\log_b \frac{1}{5} = \log_b 1 - \log_b 5$ Quotient rule

$= 0 - 1.6094$ $(\log_b 1 = 0)$

$= -1.6094$

42. $\log_b \frac{5}{3} = \log_b 5 - \log_b 3$

$= 1.6094 - 1.0986$

$= 0.5108$

43. $\log_b \sqrt{b} = \log_b b^{1/2}$

$= \frac{1}{2}\log_b b$ Power rule

$= \frac{1}{2}\cdot 1$ $(\log_b b = 1)$

$= \frac{1}{2}$

44. $\log_b \sqrt{b^3} = \log_b b^{3/2}$

$= \frac{3}{2}\log_b b$

$= \frac{3}{2}\cdot 1$

$= \frac{3}{2}$

45. $\log_b 5b = \log_b 5 + \log_b b$ Product rule

$= 1.6094 + 1$ $(\log_b b = 1)$

$= 2.6094$

46. $\log_b 9 = \log_b 3^2$

$= 2\log_b 3$ Power rule

$= 2(1.0986)$

$= 2.1972$

47. $\log_b 75 = \log_b(3\cdot 5^2)$

$= \log_b 3 + \log_b 5^2$ Product rule

$= \log_b 3 + 2\log_b 5$ Power rule

$= 1.0986 + 2(1.6094)$

$= 4.3174$

48. $\log_b \frac{1}{b} = \log_b 1 - \log_b b$

$= 0 - 1$

$= -1$

49. $\log_p p^3 = 3$ $(\log_a a^x = x)$

50. $\log_t t^{2713} = 2713$

51. $\log_e e^{|x-4|} = |x-4|$ $(\log_a a^x = x)$

52. $\log_q q^{\sqrt{3}} = \sqrt{3}$

53. $3^{\log_3 4x} = 4x \quad (a^{\log_a x} = x)$

54. $5^{\log_5 (4x-3)} = 4x - 3$

55. $10^{\log w} = w \quad (a^{\log_a x} = x)$

56. $e^{\ln x^3} = x^3$

57. $\ln e^{8t} = 8t \quad (\log_a a^x = x)$

58. $\log 10^{-k} = -k$

59. $f(x) = a^x$, $g(x) = \log_a x$

Since f and g are inverses, we know that $(f \circ g)(x) = x$ and $(g \circ f)(x) = x$. Now $(f \circ g)(x) = f(g(x)) = f(\log_a x) = a^{\log_a x}$, so we know that $a^{\log_a x} = x$. Also $(g \circ f)(x) = g(f(x)) = g(a^x) = \log_a a^x$, so we know that $\log_a a^x = x$. These results are alternate proofs of the Logarithm of a Base to a Power property and the Base to a Logarithmic Power property.

60. $\log_a ab^3 \neq (\log_a a)(\log_a b^3)$. If the first step had been correct, then so would the second step. The correct procedure follows.

$\log_a ab^3 = \log_a a + \log_a b^3 = 1 + 3\log_a b$

61.
$$\begin{aligned}(1-4i)(7+6i) &= 7 + 6i - 28i - 24i^2\\ &= 7 + 6i - 28i + 24\\ &= 31 - 22i\end{aligned}$$

62.
$$\begin{aligned}\frac{2-i}{3+1} &= \frac{2-i}{3+i}\cdot\frac{3-i}{3-i}\\ &= \frac{6-5i+i^2}{9-i^2}\\ &= \frac{6-5i-1}{9+1}\\ &= \frac{5-5i}{10}\\ &= \frac{1}{2}-\frac{1}{2}i\end{aligned}$$

63. Graph $y = 2x^2 - 13x - 7$ and use the Zero feature twice. The x-intercepts are $(-0.5, 0)$ and $(7, 0)$. The zeros are -0.5 and 7.

64. Graph $y = x^3 - 3x^2 + 3x - 1$ and use the Zero feature. The x-intercept is $(1, 0)$ and the zero is 1.

65.
$$\begin{aligned}5^{\log_5 8} &= 2x\\ 8 &= 2x \quad (a^{\log_a x} = x)\\ 4 &= x\end{aligned}$$

The solution is 4.

66.
$$\begin{aligned}\ln e^{3x-5} &= -8\\ 3x - 5 &= -8\\ 3x &= -3\\ x &= -1\end{aligned}$$

The solution is -1.

67.
$$\begin{aligned}&\log_a(x^2+xy+y^2) + \log_a(x-y)\\ &= \log_a[(x^2+xy+y^2)(x-y)] \quad \text{Product rule}\\ &= \log_a(x^3-y^3) \quad \text{Multiplying}\end{aligned}$$

68.
$$\begin{aligned}&\log_a(a^{10}-b^{10}) - \log_a(a+b)\\ &= \log_a\frac{a^{10}-b^{10}}{a+b}, \text{ or}\\ &\log_a(a^9 - a^8b + a^7b^2 - a^6b^3 + a^5b^4 - a^4b^5 + \\ &a^3b^6 - a^2b^7 + ab^8 - b^9)\end{aligned}$$

69.
$$\begin{aligned}&\log_a\frac{x-y}{\sqrt{x^2-y^2}}\\ &= \log_a\frac{x-y}{(x^2-y^2)^{1/2}}\\ &= \log_a(x-y) - \log_a(x^2-y^2)^{1/2} \quad \text{Quotient rule}\\ &= \log_a(x-y) - \frac{1}{2}\log_a(x^2-y^2) \quad \text{Power rule}\\ &= \log_a(x-y) - \frac{1}{2}\log_a[(x+y)(x-y)]\\ &= \log_a(x-y) - \frac{1}{2}[\log_a(x+y) + \log_a(x-y)]\\ &\qquad \text{Product rule}\\ &= \log_a(x-y) - \frac{1}{2}\log_a(x+y) - \frac{1}{2}\log_a(x-y)\\ &= \frac{1}{2}\log_a(x-y) - \frac{1}{2}\log_a(x+y)\end{aligned}$$

70.
$$\begin{aligned}&\log_a\sqrt{9-x^2}\\ &= \log_a(9-x^2)^{1/2}\\ &= \frac{1}{2}\log_a(9-x^2)\\ &= \frac{1}{2}\log_a[(3+x)(3-x)]\\ &= \frac{1}{2}[\log_a(3+x) + \log_a(3-x)]\\ &= \frac{1}{2}\log_a(3+x) + \frac{1}{2}\log_a(3-x)\end{aligned}$$

71. $\log_a \frac{\sqrt[4]{y^2z^5}}{\sqrt[4]{x^3z^{-2}}}$

$= \log_a \sqrt[4]{\frac{y^2z^5}{x^3z^{-2}}}$

$= \log_a \sqrt[4]{\frac{y^2z^7}{x^3}}$

$= \log_a \left(\frac{y^2z^7}{x^3}\right)^{1/4}$

$= \frac{1}{4}\log_a \left(\frac{y^2z^7}{x^3}\right)$ Power rule

$= \frac{1}{4}(\log_a y^2z^7 - \log_a x^3)$ Quotient rule

$= \frac{1}{4}(\log_a y^2 + \log_a z^7 - \log_a x^3)$ Product rule

$= \frac{1}{4}(2\log_a y + 7\log_a z - 3\log_a x)$ Power rule

$= \frac{1}{4}(2\cdot 3 + 7\cdot 4 - 3\cdot 2)$

$= \frac{1}{4}\cdot 28$

$= 7$

72. $\log_a M + \log_a N = \log_a(M+N)$
Let $a = 10$, $M = 1$, and $N = 10$. Then $\log_{10} 1 + \log_{10} 10 = 0 + 1 = 1$, but $\log_{10}(1+10) = \log_{10} 11 \approx 1.0414$. Thus, the statement is false.

73. $\log_a M - \log_a N = \log_a \frac{M}{N}$
This is the quotient rule, so it is true.

74. $\frac{\log_a M}{\log_a N} = \log_a M - \log_a N$

Let $M = a^2$ and $N = a$. Then $\frac{\log_a a^2}{\log_a a} = \frac{2}{1} = 2$, but $\log_a a^2 - \log_a a = 2 - 1 = 1$. Thus, the statement is false.

75. $\frac{\log_a M}{x} = \frac{1}{x}\log_a M = \log_a M^{1/x}$. The statement is true by the power rule.

76. $\log_a x^3 = 3\log_a x$ is true by the power rule.

77. $\log_a 8x = \log_a 8 + \log_a x = \log_a x + \log_a 8$. The statement is true by the product rule and the commutative property of addition.

78. $\log_N(M\cdot N)^x = x\log_N(M\cdot N)$

$= x(\log_N M + \log_N N)$

$= x(\log_N M + 1)$

$= x\log_N M + x$

The statement is true.

79. $\log_a \left(\frac{1}{x}\right) = \log_a x^{-1} = -1\cdot\log_a x = -1\cdot 2 = -2$

80. $\log_a x = 2$

$a^2 = x$

Let $\log_{1/a} x = n$ and solve for n.

$\log_{1/a} a^2 = n$ Substituting a^2 for x

$\left(\frac{1}{a}\right)^n = a^2$

$(a^{-1})^n = a^2$

$a^{-n} = a^2$

$-n = 2$

$n = -2$

Thus, $\log_{1/a} x = -2$ when $\log_a x = 2$.

81. We use the change-of-base formula.

$\log_{10} 11\cdot\log_{11} 12\cdot\log_{12} 13\cdots$
$\log_{998} 999\cdot\log_{999} 1000$

$= \log_{10} 11\cdot\frac{\log_{10} 12}{\log_{10} 11}\cdot\frac{\log_{10} 13}{\log_{10} 12}\cdots$
$\frac{\log_{10} 999}{\log_{10} 998}\cdot\frac{\log_{10} 1000}{\log_{10} 999}$

$= \frac{\log_{10} 11}{\log_{10} 11}\cdot\frac{\log_{10} 12}{\log_{10} 12}\cdots\frac{\log_{10} 999}{\log_{10} 999}\cdot\log_{10} 1000$

$= \log_{10} 1000$

$= 3$

82. $\log_a \frac{1}{x} = \log_a 1 - \log_a x = -\log_a x$.
Let $-\log_a x = y$. Then $\log_a x = -y$ and $x = a^{-y} = a^{-1\cdot y} = \left(\frac{1}{a}\right)^y$, so $\log_{1/a} x = y$. Thus, $\log_a\left(\frac{1}{x}\right) = -\log_a x = \log_{1/a} x$.

83. $\log_a \left(\frac{x+\sqrt{x^2-5}}{5}\right)$

$= \log_a \left(\frac{x+\sqrt{x^2-5}}{5}\cdot\frac{x-\sqrt{x^2-5}}{x-\sqrt{x^2-5}}\right)$

$= \log_a \left(\frac{5}{5(x-\sqrt{x^2-5})}\right) = \log_a \left(\frac{1}{x-\sqrt{x^2-5}}\right)$

$= \log_a 1 - \log_a(x-\sqrt{x^2-5})$

$= -\log_a(x-\sqrt{x^2-5})$

Exercise Set 4.5

1. $3^x = 81$

$3^x = 3^4$

$x = 4$ The exponents are the same.

The solution is 4.

2. $2^x = 32$

$2^x = 2^5$

$x = 5$

The solution is 5.

3. $2^{2x} = 8$

$2^{2x} = 2^3$

$2x = 3$ The exponents are the same.

$x = \frac{3}{2}$

The solution is $\frac{3}{2}$.

4. $3^{7x} = 27$

$3^{7x} = 3^3$

$7x = 3$

$x = \frac{3}{7}$

The solution is $\frac{3}{7}$.

5. $2^x = 33$

$\log 2^x = \log 33$ Taking the common logarithm on both sides

$x \log 2 = \log 33$ Power rule

$x = \frac{\log 33}{\log 2}$

$x \approx \frac{1.5185}{0.3010}$

$x \approx 5.044$

The solution is 5.044.

6. $2^x = 40$

$\log 2^x = \log 40$

$x \log 2 = \log 40$

$x = \frac{\log 40}{\log 2}$

$x \approx \frac{1.6021}{0.3010}$

$x \approx 5.322$

The solution is 5.322.

7. $5^{4x-7} = 125$

$5^{4x-7} = 5^3$

$4x - 7 = 3$

$4x = 10$

$x = \frac{10}{4} = \frac{5}{2}$

The solution is $\frac{5}{2}$.

8. $4^{3x-5} = 16$

$4^{3x-5} = 4^2$

$3x - 5 = 2$

$3x = 7$

$x = \frac{7}{3}$

The solution is $\frac{7}{3}$.

9. $27 = 3^{5x} \cdot 9^{x^2}$

$3^3 = 3^{5x} \cdot (3^2)^{x^2}$

$3^3 = 3^{5x} \cdot 3^{2x^2}$

$3^3 = 3^{5x+2x^2}$

$3 = 5x + 2x^2$

$0 = 2x^2 + 5x - 3$

$0 = (2x - 1)(x + 3)$

$x = \frac{1}{2}$ *or* $x = -3$

The solutions are -3 and $\frac{1}{2}$.

10. $3^{x^2+4x} = \frac{1}{27}$

$3^{x^2+4x} = 3^{-3}$

$x^2 + 4x = -3$

$x^2 + 4x + 3 = 0$

$(x + 3)(x + 1) = 0$

$x = -3$ *or* $x = -1$

The solutions are -3 and -1.

11. $84^x = 70$

$\log 84^x = \log 70$

$x \log 84 = \log 70$

$x = \frac{\log 70}{\log 84}$

$x \approx \frac{1.8451}{1.9243}$

$x \approx 0.959$

The solution is 0.959.

12. $28^x = 10^{-3x}$

$\log 28^x = \log 10^{-3x}$

$x \log 28 = -3x$

$x \log 28 + 3x = 0$

$x(\log 28 + 3) = 0$

$x = 0$

The solution is 0.

13. $e^t = 1000$

$\ln e^t = \ln 1000$

$t = \ln 1000$ Using $\log_a a^x = x$

$t \approx 6.908$

The solution is 6.908.

14. $e^{-t} = 0.04$

$\ln e^{-t} = \ln 0.04$

$-t = \ln 0.04$

$t = -\ln 0.04 \approx 3.219$

The solution is 3.219.

15.
$$\begin{aligned} e^{-0.03t} &= 0.08 \\ \ln e^{-0.03t} &= \ln 0.08 \\ -0.03t &= \ln 0.08 \\ t &= \frac{\ln 0.08}{-0.03} \\ t &\approx \frac{-2.5257}{-0.03} \\ t &\approx 84.191 \end{aligned}$$
The solution is 84.191.

16.
$$\begin{aligned} 1000e^{0.09t} &= 5000 \\ e^{0.09t} &= 5 \\ \ln e^{0.09t} &= \ln 5 \\ 0.09t &= \ln 5 \\ t &= \frac{\ln 5}{0.09} \\ t &\approx 17.883 \end{aligned}$$
The solution is 17.883.

17.
$$\begin{aligned} 3^x &= 2^{x-1} \\ \ln 3^x &= \ln 2^{x-1} \\ x\ln 3 &= (x-1)\ln 2 \\ x\ln 3 &= x\ln 2 - \ln 2 \\ \ln 2 &= x\ln 2 - x\ln 3 \\ \ln 2 &= x(\ln 2 - \ln 3) \\ \frac{\ln 2}{\ln 2 - \ln 3} &= x \\ \frac{0.6931}{0.6931 - 1.0986} &\approx x \\ -1.710 &\approx x \end{aligned}$$
The solution is -1.710.

18.
$$\begin{aligned} 5^{x+2} &= 4^{1-x} \\ \log 5^{x+2} &= \log 4^{1-x} \\ (x+2)\log 5 &= (1-x)\log 4 \\ x\log 5 + 2\log 5 &= \log 4 - x\log 4 \\ x\log 5 + x\log 4 &= \log 4 - 2\log 5 \\ x(\log 5 + \log 4) &= \log 4 - 2\log 5 \\ x &= \frac{\log 4 - 2\log 5}{\log 5 + \log 4} \\ x &\approx -0.612 \end{aligned}$$
The solution is -0.612.

19.
$$\begin{aligned} (3.9)^x &= 48 \\ \log(3.9)^x &= \log 48 \\ x\log 3.9 &= \log 48 \\ x &= \frac{\log 48}{\log 3.9} \\ x &\approx \frac{1.6812}{0.5911} \\ x &\approx 2.844 \end{aligned}$$
The solution is 2.844.

20.
$$\begin{aligned} 250 - (1.87)^x &= 0 \\ 250 &= (1.87)^x \\ \log 250 &= \log(1.87)^x \\ \log 250 &= x\log 1.87 \\ \frac{\log 250}{\log 1.87} &= x \\ 8.821 &\approx x \end{aligned}$$
The solution is 8.821.

21.
$$\begin{aligned} e^x + e^{-x} &= 5 \\ e^{2x} + 1 &= 5e^x && \text{Multiplying by } e^x \\ e^{2x} - 5e^x + 1 &= 0 && \text{This equation is quadratic in } e^x. \\ e^x &= \frac{5 \pm \sqrt{21}}{2} \\ x &= \ln\left(\frac{5 \pm \sqrt{21}}{2}\right) \approx \pm 1.567 \end{aligned}$$
The solutions are -1.567 and 1.567.

22.
$$\begin{aligned} e^x - 6e^{-x} &= 1 \\ e^{2x} - 6 &= e^x \\ e^{2x} - e^x - 6 &= 0 \\ (e^x - 3)(e^x + 2) &= 0 \\ e^x &= 3 \quad \textit{or} \quad e^x = -2 \\ \ln e^x &= \ln 3 \qquad\quad \text{No solution} \\ x &= \ln 3 \\ x &\approx 1.099 \end{aligned}$$
The solution is 1.099.

23.
$$\begin{aligned} \frac{e^x + e^{-x}}{e^x - e^{-x}} &= 3 \\ e^x + e^{-x} &= 3e^x - 3e^{-x} && \text{Multiplying by } e^x - e^{-x} \\ 4e^{-x} &= 2e^x && \text{Subtracting } e^x \text{ and adding } e^{-x} \\ 2e^{-x} &= e^x \\ 2 &= e^{2x} && \text{Multiplying by } e^x \\ \ln 2 &= \ln e^{2x} \\ \ln 2 &= 2x \\ \frac{\ln 2}{2} &= x \\ 0.347 &\approx x \end{aligned}$$
The solution is 0.347.

24.
$$\begin{aligned} \frac{5^x - 5^{-x}}{5^x + 5^{-x}} &= 8 \\ 5^{-x} - 5^{-x} &= 8\cdot 5^x + 8\cdot 5^{-x} \\ -9\cdot 5^{-x} &= 7\cdot 5^x \\ -9 &= 7\cdot 5^{2x} && \text{Multiplying by } 5^x \\ -\frac{9}{7} &= 5^{2x} \end{aligned}$$
The number 5 raised to any power is non-negative. Thus, the equation has no solution.

25. $\log_5 x = 4$

$x = 5^4$ Writing an equivalent exponential equation

$x = 625$

The solution is 625.

26. $\log_2 x = -3$

$x = 2^{-3}$

$x = \frac{1}{8}$

The solution is $\frac{1}{8}$.

27. $\log x = -4$ The base is 10.

$x = 10^{-4}$, or 0.0001

The solution is 0.0001.

28. $\log x = 1$

$x = 10^1 = 10$

The solution is 10.

29. $\ln x = 1$ The base is e.

$x = e^1 = e$

The solution is e.

30. $\ln x = -2$

$x = e^{-2}$, or $\frac{1}{e^2}$

The solution is e^{-2}, or $\frac{1}{e^2}$.

31. $\log_2(10 + 3x) = 5$

$2^5 = 10 + 3x$

$32 = 10 + 3x$

$22 = 3x$

$\frac{22}{3} = x$

The answer checks. The solution is $\frac{22}{3}$.

32. $\log_5(8 - 7x) = 3$

$5^3 = 8 - 7x$

$125 = 8 - 7x$

$117 = -7x$

$-\frac{117}{7} = x$

The answer checks. The solution is $-\frac{117}{7}$.

33. $\log x + \log(x - 9) = 1$ The base is 10.

$\log_{10}[x(x - 9)] = 1$

$x(x - 9) = 10^1$

$x^2 - 9x = 10$

$x^2 - 9x - 10 = 0$

$(x - 10)(x + 1) = 0$

$x = 10$ or $x = -1$

Check: For 10:

$$\begin{array}{r|l} \log x + \log(x-9) & = 1 \\ \hline \log 10 + \log(10-9) & ?\ 1 \\ \log 10 + \log 1 & \\ 1 + 0 & \\ 1 & 1 \quad \text{TRUE} \end{array}$$

For -1:

$$\begin{array}{r|l} \log x + \log(x-9) & = 1 \\ \hline \log(-1) + \log(-1-9) & ?\ 1 \end{array}$$

The number -1 does not check, because negative numbers do not have logarithms. The solution is 10.

34. $\log_2(x + 1) + \log_2(x - 1) = 3$

$\log_2[(x + 1)(x - 1)] = 3$

$(x + 1)(x - 1) = 2^3$

$x^2 - 1 = 8$

$x^2 = 9$

$x = \pm 3$

The number 3 checks, but -3 does not. The solution is 3.

35. $\log_8(x + 1) - \log_8 x = 2$

$\log_8\left(\frac{x+1}{x}\right) = 2$ Quotient rule

$\frac{x+1}{x} = 8^2$

$\frac{x+1}{x} = 64$

$x + 1 = 64x$

$1 = 63x$

$\frac{1}{63} = x$

The answer checks. The solution is $\frac{1}{63}$.

36. $\log x - \log(x + 3) = -1$

$\log_{10} \frac{x}{x+3} = -1$

$\frac{x}{x+3} = 10^{-1}$

$\frac{x}{x+3} = \frac{1}{10}$

$10x = x + 3$

$9x = 3$

$x = \frac{1}{3}$

The answer checks. The solution is $\frac{1}{3}$.

37. $\log_4(x+3) + \log_4(x-3) = 2$

$$\log_4[(x+3)(x-3)] = 2 \quad \text{Product rule}$$
$$(x+3)(x-3) = 4^2$$
$$x^2 - 9 = 16$$
$$x^2 = 25$$
$$x = \pm 5$$

The number 5 checks, but -5 does not. The solution is 5.

38. $\ln(x+1) - \ln x = \ln 4$

$$\ln \frac{x+1}{x} = \ln 4$$
$$\frac{x+1}{x} = 4$$
$$x + 1 = 4x$$
$$1 = 3x$$
$$\frac{1}{3} = x$$

The answer checks. The solution is $\frac{1}{3}$.

39. $\log(2x+1) - \log(x-2) = 1$

$$\log\left(\frac{2x+1}{x-2}\right) = 1 \quad \text{Quotient rule}$$
$$\frac{2x+1}{x-2} = 10^1 = 10$$
$$2x + 1 = 10x - 20$$

Multiplying by $x - 2$

$$21 = 8x$$
$$\frac{21}{8} = x$$

The answer checks. The solution is $\frac{21}{8}$.

40. $\log_5(x+4) + \log_5(x-4) = 2$

$$\log_5[(x+4)(x-4)] = 2$$
$$x^2 - 16 = 25$$
$$x^2 = 41$$
$$x = \pm\sqrt{41}$$

Only $\sqrt{41}$ checks. The solution is $\sqrt{41}$.

41. $e^{7.2x} = 14.009$

Graph $y_1 = e^{7.2x}$ and $y_2 = 14.009$ and find the first coordinate of the point of intersection using the Intersect feature. The solution is 0.367.

42. $0.082e^{0.05x} = 0.034$

Graph $y_1 = 0.082e^{0.05x}$ and $y_2 = 0.034$ and find the first coordinate of the point of intersection using the Intersect feature. The solution is -17.607.

43. $xe^{3x} - 1 = 3$

Graph $y_1 = xe^{3x} - 1$ and $y_2 = 3$ and find the first coordinate of the point of intersection using the Intersect feature. The solution is 0.621.

44. $5e^{5x} + 10 = 3x + 40$

Graph $y_1 = 5e^{5x} + 10$ and $y_2 = 3x + 40$ and find the first coordinates of the points of intersection using the Intersect feature. The solutions are -10 and 0.366.

45. $4\ln(x+3.4) = 2.5$

Graph $y_1 = 4\ln(x+3.4)$ and $y_2 = 2.5$ and find the first coordinate of the point of intersection using the Intersect feature. The solution is -1.532.

46. $\ln x^2 = -x^2$

Graph $y_1 = \ln x^2$ and $y_2 = -x^2$ and find the first coordinates of the points of intersection using the Intersect feature. The solutions are -0.753 and 0.753.

47. $\log_8 x + \log_8(x+2) = 2$

Graph $y_1 = \dfrac{\log x}{\log 8} + \dfrac{\log(x+2)}{\log 8}$ and $y_2 = 2$ and find the first coordinate of the point of intersection using the intersect feature. The solution is 7.062.

48. $\log_3 x + 7 = 4 - \log_5 x$

Graph $y_1 = \dfrac{\log x}{\log 3} + 7$ and $y_2 = 4 - \dfrac{\log x}{\log 5}$ and find the first coordinate of the point of intersection using the Intersect feature. The solution is 0.141.

49. $\log_5(x+7) - \log_5(2x-3) = 1$

Graph $y_1 = \dfrac{\log(x+7)}{\log 5} - \dfrac{\log(2x-3)}{\log 5}$ and $y_2 = 1$ and find the first coordinate of the point of intersection using the Intersect feature. The solution is 2.444.

50. Graph $y_1 = \ln 3x$ and $y_2 = 3x - 8$ and use the Intersect feature to find the points of intersection. They are $(0.0001, -7.9997)$ and $(3.445, 2.336)$.

51. Solving the first equation for y, we get $y = \dfrac{12.4 - 2.3x}{3.8}$. Graph $y_1 = \dfrac{12.4 - 2.3x}{3.8}$ and $y_2 = 1.1\ln(x - 2.05)$ and use the Intersect feature to find the point of intersection. It is $(4.093, 0.786)$.

52. Graph $y_1 = 2.3\ln(x+10.7)$ and $y_2 = 10e^{-0.07x^2}$ and use the Intersect feature to find the points of intersection. They are $(-9.694, 0.014)$, $(-3.334, 4.593)$, and $(2.714, 5.971)$.

53. Graph $y_1 = 2.3\ln(x+10.7)$ and $y_2 = 10e^{-0.007x^2}$ and use the Intersect feature to find the point of intersection. It is $(7.586, 6.684)$.

54. The final result would have been the same, but to find t we would have computed $\dfrac{\log 2500}{0.08 \log e}$.

It seems best to take the natural logarithm on both sides since the final computation for t is simpler.

55. Trace along the graph of $y = \ln x$ to find the x-value that corresponds to the value on the right-hand side of the equation.

56. $f(x) = -x^2 + 6x - 8$

a) $-\frac{b}{2a} = -\frac{6}{2(-1)} = 3$

$f(3) = -3^2 + 6 \cdot 3 - 8 = 1$

The vertex is $(3, 1)$.

b) $x = 3$

c) Maximum: 1 at $x = 3$

57. $g(x) = x^2 - 6$

a) $-\frac{b}{2a} = -\frac{0}{2 \cdot 1} = 0$

$g(0) = 0^2 - 6 = -6$

The vertex is $(0, -6)$.

b) The line of symmetry is $x = 0$.

c) Since the coefficient of the x^2-term is positive, the function has a minimum value. It is the second coordinate of the vertex, -6, and it occurs when $x = 0$.

58. $H(x) = 3x^2 - 12x + 16$

a) $-\frac{b}{2a} = -\frac{-12}{2 \cdot 3} = 2$

$H(2) = 3 \cdot 2^2 - 12 \cdot 2 + 16 = 4$

The vertex is $(2, 4)$.

b) $x = 2$

c) Minimum: 4 at $x = 2$

59. $G(x) = -2x^2 - 4x - 7$

a) $-\frac{b}{2a} = -\frac{-4}{2(-2)} = -1$

$G(-1) = -2(-1)^2 - 4(-1) - 7 = -5$

The vertex is $(-1, -5)$.

b) The line of symmetry is $x = -1$.

c) Since the coefficient of the x^2-term is negative, the function has a maximum value. It is the second coordinate of the vertex, -5, and it occurs when $x = -1$.

60. $\ln(\ln x) = 2$

$\ln x = e^2$

$x = e^{e^2} \approx 1618.178$

The answer checks. The solution is e^{e^2}, or 1618.178.

61. $\ln(\log x) = 0$

$\log x = e^0$

$\log x = 1$

$x = 10^1 = 10$

The answer checks. The solution is 10.

62. $\ln \sqrt[4]{x} = \sqrt{\ln x}$

$\frac{1}{4}\ln x = \sqrt{\ln x}$

$\frac{1}{16}(\ln x)^2 = \ln x$ Squaring both sides

$\frac{1}{16}(\ln x)^2 - \ln x = 0$

Let $u = \ln x$ and substitute.

$\frac{1}{16}u^2 - u = 0$

$u\left(\frac{1}{16}u - 1\right) = 0$

$u = 0$ *or* $\frac{1}{16}u - 1 = 0$

$u = 0$ *or* $u = 16$

$\ln x = 0$ *or* $\ln x = 16$

$x = e^0$ *or* $x = e^{16}$

$x = 1$ *or* $x = e^{16} \approx 8,886,110.521$

Both answers check. The solutions are 1 and e^{16}, or 1 and 8,886,110.521.

63. $\sqrt{\ln x} = \ln \sqrt{x}$

$\sqrt{\ln x} = \frac{1}{2}\ln x$ Power rule

$\ln x = \frac{1}{4}(\ln x)^2$ Squaring both sides

$0 = \frac{1}{4}(\ln x)^2 - \ln x$

Let $u = \ln x$ and substitute.

$\frac{1}{4}u^2 - u = 0$

$u\left(\frac{1}{4}u - 1\right) = 0$

$u = 0$ *or* $\frac{1}{4}u - 1 = 0$

$u = 0$ *or* $\frac{1}{4}u = 1$

$u = 0$ *or* $u = 4$

$\ln x = 0$ *or* $\ln x = 4$

$x = e^0 = 1$ *or* $x = e^4 \approx 54.598$

Both answers check. The solutions are 1 and e^4, or 1 and 54.598.

64. $\log_3(\log_4 x) = 0$

$\log_4 x = 3^0$

$\log_4 x = 1$

$x = 4^1$

$x = 4$

The answer checks. The solution is 4.

65. $(\log_3 x)^2 - \log_3 x^2 = 3$

$(\log_3 x)^2 - 2\log_3 x - 3 = 0$

Let $u = \log_3 x$, substitute:

$u^2 - 2u - 3 = 0$
$(u-3)(u+1) = 0$
$u = 3$ *or* $u = -1$
$\log_3 x = 3$ *or* $\log_3 x = -1$
$x = 3^3$ *or* $x = 3^{-1}$
$x = 27$ *or* $x = \frac{1}{3}$

Both answers check. The solutions are $\frac{1}{3}$ and 27.

66. $(\log x)^2 - \log x^2 = 3$
$(\log x)^2 - 2\log x - 3 = 0$
Let $u = \log x$ and substitute.
$u^2 - 2u - 3 = 0$
$(u+1)(u-3) = 0$
$u = -1$ *or* $u = 3$
$\log x = -1$ *or* $\log x = 3$
$x = \frac{1}{10}$ *or* $x = 1000$

Both answers check. The solutions are $\frac{1}{10}$ and 1000.

67. $\ln x^2 = (\ln x)^2$
$2\ln x = (\ln x)^2$
$0 = (\ln x)^2 - 2\ln x$
Let $u = \ln x$ and substitute.
$0 = u^2 - 2u$
$0 = u(u-2)$
$u = 0$ *or* $u = 2$
$\ln x = 0$ *or* $\ln x = 2$
$x = 1$ *or* $x = e^2 \approx 7.389$

Both answers check. The solutions are 1 and e^2, or 1 and 7.389.

68. $e^{2x} - 9 \cdot e^x + 14 = 0$
$(e^x - 2)(e^x - 7) = 0$
$e^x = 2$ *or* $e^x = 7$
$\ln e^x = \ln 2$ *or* $\ln e^x = \ln 7$
$x = \ln 2$ *or* $x = \ln 7$
$x \approx 0.693$ *or* $x \approx 1.946$

The solutions are 0.693 and 1.946.

69. $5^{2x} - 3 \cdot 5^x + 2 = 0$
$(5^x - 1)(5^x - 2) = 0$ This equation is quadratic in 5^x.
$5^x = 1$ *or* $5^x = 2$
$\log 5^x = \log 1$ *or* $\log 5^x = \log 2$
$x \log 5 = 0$ *or* $x \log 5 = \log 2$
$x = 0$ *or* $x = \frac{\log 2}{\log 5} \approx 0.431$

The solutions are 0 and 0.431.

70. $x\left(\ln \frac{1}{6}\right) = \ln 6$
$x(\ln 1 - \ln 6) = \ln 6$
$-x\ln 6 = \ln 6$ $(\ln 1 = 0)$
$x = -1$

The solution is -1.

71. $\log_3 |x| = 2$
$|x| = 3^2$
$|x| = 9$
$x = -9$ *or* $x = 9$

Both answers check. The solutions are -9 and 9.

72. $x^{\log x} = \frac{x^3}{100}$
$\log x^{\log x} = \log \frac{x^3}{100}$
$\log x \cdot \log x = \log x^3 - \log 100$
$(\log x)^2 = 3\log x - 2$
$(\log x)^2 - 3\log x + 2 = 0$
Let $u = \log x$ and substitute.
$u^2 - 3u + 2 = 0$
$(u-1)(u-2) = 0$
$u = 1$ *or* $u = 2$
$\log x = 1$ *or* $\log x = 2$
$x = 10$ *or* $x = 10^2 = 100$

Both answers check. The solutions are 10 and 100.

73. $\ln x^{\ln x} = 4$
$\ln x \cdot \ln x = 4$
$(\ln x)^2 = 4$
$\ln x = \pm 2$
$\ln x = -2$ *or* $\ln x = 2$
$x = e^{-2}$ *or* $x = e^2$
$x \approx 0.135$ *or* $x \approx 7.389$

Both answers check. The solutions are e^{-2} and e^2, or 0.135 and 7.389.

74. $\frac{(e^{3x+1})^2}{e^4} = e^{10x}$
$\frac{e^{6x+2}}{e^4} = e^{10x}$
$e^{6x-2} = e^{10x}$
$6x - 2 = 10x$
$-2 = 4x$
$-\frac{1}{2} = x$

The solution is $-\frac{1}{2}$.

75. $$\frac{\sqrt{(e^{2x}\cdot e^{-5x})^{-4}}}{e^x \div e^{-x}} = e^7$$
$$\frac{\sqrt{e^{12x}}}{e^{x-(-x)}} = e^7$$
$$\frac{e^{6x}}{e^{2x}} = e^7$$
$$e^{4x} = e^7$$
$$4x = 7$$
$$x = \frac{7}{4}$$

The solution is $\frac{7}{4}$.

76. $|\log_a x| = \log_a |x|$, $a > 1$

Graph $y_1 = |\log x|$ and $y_2 = \log |x|$. (Here we let $a = 10$.) Observe that the graphs intersect at $x = 1$ and that they coincide for $x > 1$. This will be the case for all $a > 1$. Thus, the solution set is $[1, \infty)$.

77. $\ln(x-2) > 4$

Graph $y_1 = \ln(x-2)$ and $y_2 = 4$. Using the Intersect feature, we find that the first coordinate of the point of intersection of the graphs is 56.598. Observe that the graph of y_1 lies above the graph of y_2 for all x-values greater than 56.598. Thus, the solution set is $(56.598, \infty)$.

78. $$e^x < \frac{4}{5}$$
$$\ln e^x < \ln 0.8$$
$$x < -0.223$$

The solution set is $(-\infty, -0.223)$.

This exercise could also be done graphically.

79. $|\log_5 x| + 3\log_5 |x| = 4$

Note that we must have $x > 0$. First consider the case when $0 < x < 1$. When $0 < x < 1$, then $\log_5 x < 0$, so $|\log_5 x| = -\log_5 x$ and $|x| = x$. Thus we have:

$$-\log_5 x + 3\log_5 x = 4$$
$$2\log_5 x = 4$$
$$\log_5 x^2 = 4$$
$$x^2 = 5^4$$
$$x = 5^2$$
$$x = 25 \quad \text{(Recall that } x > 0.)$$

25 cannot be a solution since we assumed $0 < x < 1$.

Now consider the case when $x > 1$. In this case $\log_5 x > 0$, so $|\log_5 x| = \log_5 x$ and $|x| = x$. Thus we have:

$$\log_5 x + 3\log_5 x = 4$$
$$4\log_5 x = 4$$
$$\log_5 x = 1$$
$$x = 5$$

This answer checks. The solution is 5.

80. $|2^{x^2} - 8| = 3$

$$2^{x^2} - 8 = -3 \quad or \quad 2^{x^2} - 8 = 3$$
$$2^{x^2} = 5 \quad or \quad 2^{x^2} = 11$$
$$\log 2^{x^2} = \log 5 \quad or \quad \log 2^{x^2} = \log 11$$
$$x^2 \log 2 = \log 5 \quad or \quad x^2 \log 2 = \log 11$$
$$x^2 = \frac{\log 5}{\log 2} \quad or \quad x^2 = \frac{\log 11}{\log 2}$$
$$x = \pm 1.524 \quad or \quad x = \pm 1.860$$

The solutions are -1.860, -1.524, 1.524, and 1.860.

81. $a = \log_8 225$, so $8^a = 225 = 15^2$.

$b = \log_2 15$, so $2^b = 15$.

Then
$$8^a = (2^b)^2$$
$$(2^3)^a = 2^{2b}$$
$$2^{3a} = 2^{2b}$$
$$3a = 2b$$
$$a = \frac{2}{3}b.$$

82. $\log_5 125 = 3$ and $\log_{125} 5 = \frac{1}{3}$, so $a = (\log_{125} 5)^{\log_5 125}$ is equivalent to $a = \left(\frac{1}{3}\right)^3 = \frac{1}{27}$.

Then $\log_3 a = \log_3 \frac{1}{27} = -3$.

83. $\log_2[\log_3(\log_4 x)] = 0$ yields $x = 64$.

$\log_3[\log_2(\log_4 y)] = 0$ yields $y = 16$.

$\log_4[\log_3(\log_2 z)] = 0$ yields $z = 8$.

Then $x + y + z = 64 + 16 + 8 = 88$.

84. $f(x) = e^x - e^{-x}$

Replace $f(x)$ with y: $y = e^x - e^{-x}$

Interchange x and y: $x = e^y - e^{-y}$

Solve for y: $xe^y = e^{2y} - 1$ Multiplying by e^y

$$0 = e^{2y} - xe^y - 1$$

Using the quadratic formula with $a = 1$, $b = -x$, and $c = -1$ and taking the positive square root (since $e^y > 0$), we get $e^y = \frac{x + \sqrt{x^2+4}}{2}$. Then we have

$$\ln e^y = \ln\left(\frac{x+\sqrt{x^2+4}}{2}\right)$$
$$y = \ln\left(\frac{x+\sqrt{x^2+4}}{2}\right)$$

Replace y with $f^{-1}(x)$:

$$f^{-1}(x) = \ln\left(\frac{x+\sqrt{x^2+4}}{2}\right).$$

Exercise Set 4.6

1. a) Substitute 6.0 for P_0 and 0.013 for k in $P(t) = P_0e^{kt}$. We have:

   $P(t) = 6.0e^{0.013t}$, where $P(t)$ is in billions and t is the number of years after 1999.

   b) In 2005, $t = 2005 - 1999 = 6$.

   $P(6) = 6.0e^{0.013(6)} \approx 6.5$ billion

   In 2010, $t = 2010 - 1999 = 11$.

   $P(11) = 6.0e^{0.013(11)} \approx 6.9$ billion

   c) Substitute 8 for $P(t)$ and solve for t.

   $$8 = 6.0e^{0.013t}$$
   $$\frac{4}{3} = e^{0.013t}$$
   $$\ln\frac{4}{3} = \ln e^{0.013t}$$
   $$\ln\frac{4}{3} = 0.013t$$
   $$\frac{\ln\frac{4}{3}}{0.013} = t$$
   $$22.1 \approx t$$

   The world population will be 8 billion about 22.1 yr after 1999.

   d) $T = \dfrac{\ln 2}{0.013} \approx 53.3$ yr

2. a) $P(t) = 100e^{0.117t}$

   b) $y = 100e^{0.117x}$

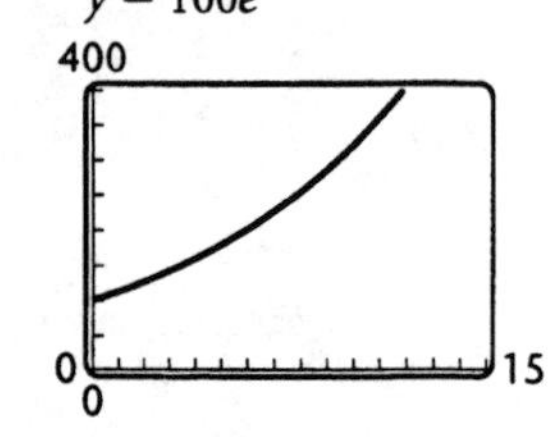

   c) $P(7) = 100e^{0.117(7)} \approx 227$

   d) $t = \dfrac{\ln 2}{0.117} \approx 5.9$ days

3. a) $T = \dfrac{\ln 2}{0.019} \approx 36.5$ yr

   b) $k = \dfrac{\ln 2}{346} \approx 0.2\%$ per yr

   c) $T = \dfrac{\ln 2}{0.033} \approx 21.0$ yr

   d) $T = \dfrac{\ln 2}{0.005} \approx 138.6$ yr

   e) $k = \dfrac{\ln 2}{20.4} \approx 3.4\%$ per yr

   f) $k = \dfrac{\ln 2}{31.5} \approx 2.2\%$ per yr

4. a) In 1996, $t = 1996 - 1985 = 11$.

   $$3600 = 500e^{11k}$$
   $$7.2 = e^{11k}$$
   $$\ln 7.2 = \ln e^{11k}$$
   $$\ln 7.2 = 11k$$
   $$\frac{\ln 7.2}{11} = k$$
   $$0.179 \approx k$$

   $P(t) = 500e^{0.179t}$

   b) In 2000, $t = 2000 - 1985 = 15$.

   $P(15) = 500e^{0.179(15)} \approx 7300$

   In 2004, $t = 2004 - 1985 = 19$.

   $P(19) = 500e^{0.179(19)} \approx 15{,}000$

5. $$P(t) = P_0e^{kt}$$
   $$24{,}313{,}062{,}400 = 5{,}644{,}000e^{0.026t}$$
   $$\frac{24{,}313{,}062{,}400}{5{,}644{,}000} = e^{0.026t}$$
   $$\ln\left(\frac{24{,}313{,}062{,}400}{5{,}644{,}000}\right) = \ln e^{0.026t}$$
   $$\ln\left(\frac{24{,}313{,}062{,}400}{5{,}644{,}000}\right) = 0.026t$$
   $$\frac{\ln\left(\frac{24{,}313{,}062{,}400}{5{,}644{,}000}\right)}{0.026} = t$$
   $$322 \approx t$$

   There will be one person for every square yard of land about 322 yr after 1998.

6. In 2003, $t = 2003 - 1626 = 377$.

   $P(377) = 24e^{0.08(377)} \approx \$301{,}000{,}000{,}000{,}000$

7. a) Substitute 10,000 for P_0 and 5.4%, or 0.054 for k.

   $P(t) = 10{,}000e^{0.054t}$

   b) We can use the TABLE feature of a grapher, set in ASK mode, to evaluate $P(t) = 10{,}000e^{0.054t}$ for the desired values of t. We enter $y = 10{,}000e^{0.054x}$.

   $P(1) \approx \$10{,}555$

   $P(2) \approx \$11{,}140$

   $P(5) \approx \$13{,}100$

   $P(10) \approx \$17{,}160$

   c) $T = \dfrac{\ln 2}{0.054} \approx 12.8$ yr

8. a) $T = \dfrac{\ln 2}{0.062} \approx 11.2$ yr

   $P(5) = 35{,}000e^{0.062(5)} \approx \$47{,}719.88$

b)
$$\begin{aligned} 7130.90 &= 5000e^{5k} \\ 1.4618 &= e^{5k} \\ \ln 1.4618 &= \ln e^{5k} \\ \ln 1.4618 &= 5k \\ \frac{\ln 1.4618}{5} &= k \\ 0.071 &\approx k \\ 7.1\% &\approx k \end{aligned}$$

$$T = \frac{\ln 2}{0.071} \approx 9.8 \text{ yr}$$

c)
$$\begin{aligned} 11,414.71 &= P_0 e^{0.084(5)} \\ \frac{11,414.71}{e^{0.084(5)}} &= P_0 \\ \$7500 &\approx P_0 \end{aligned}$$

$$T = \frac{\ln 2}{0.084} \approx 8.3 \text{ yr}$$

d) $k = \frac{\ln 2}{11} \approx 0.063$, or 6.3%

$$\begin{aligned} 17,539.32 &= P_0 e^{0.063(5)} \\ \frac{17,539.32}{e^{0.063(5)}} &= P_0 \\ \$12,800 &\approx P_0 \end{aligned}$$

9. We use the function found in Example 5. If the mummy has lost 46% of its carbon-14 from an initial amount P_0, then $54\%P_0$, or $0.54P_0$ remains. We substitute in the function.

$$\begin{aligned} 0.54P_0 &= P_0 e^{-0.00012t} \\ 0.54 &= e^{-0.00012t} \\ \ln 0.54 &= \ln e^{-0.00012t} \\ \ln 0.54 &= -0.00012t \\ \frac{\ln 0.54}{-0.00012} &= t \\ 5135 &\approx t \end{aligned}$$

The mummy is about 5135 years old.

10. $35\%P_0$ of the carbon-14 has been lost, so $65\%P_0$, $0.65P_0$ remains.

$$\begin{aligned} 0.65P_0 &= P_0 e^{-0.00012t} \\ 0.65 &= e^{-0.00012t} \\ \ln 0.65 &= \ln e^{-0.00012t} \\ \ln 0.65 &= -0.00012t \\ \frac{\ln 0.65}{-0.00012} &= t \\ 3590 &\approx t \end{aligned}$$

The statue is about 3590 years old.

11. a) $K = \frac{\ln 2}{3} \approx 0.231$, or 23.1% per min

b) $k = \frac{\ln 2}{22} \approx 0.0315$, or 3.15% per yr

c) $T = \frac{\ln 2}{0.096} \approx 7.2$ days

d) $T = \frac{\ln 2}{0.063} \approx 11$ yr

e) $k = \frac{\ln 2}{25} \approx 0.028$, or 2.8% per yr

f) $k = \frac{\ln 2}{4560} \approx 0.00015$, or 0.015% per yr

g) $k = \frac{\ln 2}{23,105} \approx 0.00003$, or 0.003% per yr

12. a)
$$\begin{aligned} t &= 1995 - 1950 = 45 \\ N(t) &= N_0 e^{-kt} \\ 2,071,520 &= 5,647,800e^{-k(45)} \\ \frac{2,071,520}{5,647,800} &= e^{-45k} \\ \ln\left(\frac{2,071,520}{5,647,800}\right) &= \ln e^{-45k} \\ \ln\left(\frac{2,071,520}{5,647,800}\right) &= -45k \\ \frac{\ln\left(\frac{2,071,520}{5,647,800}\right)}{-45} &= k \\ 0.022 &\approx k \end{aligned}$$

$$N(t) = 5,647,800e^{-0.022t}$$

b) In 2000, $t = 2000 - 1950 = 50$.

$N(50) = 5,647,800e^{-0.022(50)} \approx 1,879,989$

In 2005, $t = 2005 - 1950 = 55$.

$N(55) = 5,647,800e^{-0.022(55)} \approx 1,684,159$

In 2010, $t = 2010 - 1950 = 60$.

$N(60) = 5,647,800e^{-0.022(60)} \approx 1,508,727$

c)
$$\begin{aligned} 100,000 &= 5,647,800e^{-0.022t} \\ \frac{100,000}{5,647,800} &= e^{-0.022t} \\ \ln\left(\frac{100,000}{5,647,800}\right) &= \ln e^{-0.022t} \\ \ln\left(\frac{100,000}{5,647,800}\right) &= -0.022t \\ \frac{\ln\left(\frac{100,000}{5,647,800}\right)}{-0.022} &= t \\ 183 &\approx t \end{aligned}$$

Only 100,000 farms will remain about 183 years after 1950, or in 2133.

13. a) Substitute 1996 − 1985, or 11, for t; 80 for P_0; and 67 for $P(11)$ and solve for k.

$$67 = 80e^{-11k}$$
$$0.8375 = e^{-11k}$$
$$\ln 0.8375 = \ln e^{-11k}$$
$$\ln 0.8375 = -11k$$
$$\frac{\ln 0.8375}{-11} = k$$
$$0.016 \approx k$$

The desired equation is $P(t) = 80e^{-0.016t}$.

b) In 2002, $t = 2002 - 1985 = 17$.

$P(17) = 80e^{-0.016(17)} \approx 61$ lb per person

c)
$$20 = 80e^{-0.016t}$$
$$0.25 = e^{-0.016t}$$
$$\ln 0.25 = \ln e^{-0.016t}$$
$$\ln 0.25 = -0.016t$$
$$\frac{\ln 0.25}{-0.016} = t$$
$$86.6 \approx t$$

The average annual consumption of beef will be 20 lb per person after about 86.6 years.

14. a) $t = 1997 - 1987 = 10$

$$20 = 8e^{k(10)}$$
$$2.5 = e^{10k}$$
$$\ln 2.5 = \ln e^{10k}$$
$$\ln 2.5 = 10k$$
$$\frac{\ln 2.5}{10} = k$$
$$0.0916 \approx k$$

$V(t) = 8e^{0.0916t}$, where t is the number of years after 1987.

b) In 2000, $t = 2000 - 1987 = 13$.

$V(13) = 8e^{0.0916(13)} \approx \$26$

c) $T = \dfrac{\ln 2}{0.0916} \approx 7.6\text{yr}$

d)
$$2000 = 8e^{0.0916t}$$
$$250 = e^{0.0916t}$$
$$\ln 250 = \ln e^{0.0916t}$$
$$\ln 250 = 0.0916t$$
$$\frac{\ln 250}{0.0916} = t$$
$$60 \approx t$$

The value of the card will be \$2000 about 60 yr after 1987.

e) The \$200 value in 2000 is nearly 8 times the estimated value found in part (b). The value increased dramatically after McGwire's 70-homerun season in 1998.

15. a) $y = \dfrac{2000}{1 + 19.9e^{-0.6x}}$

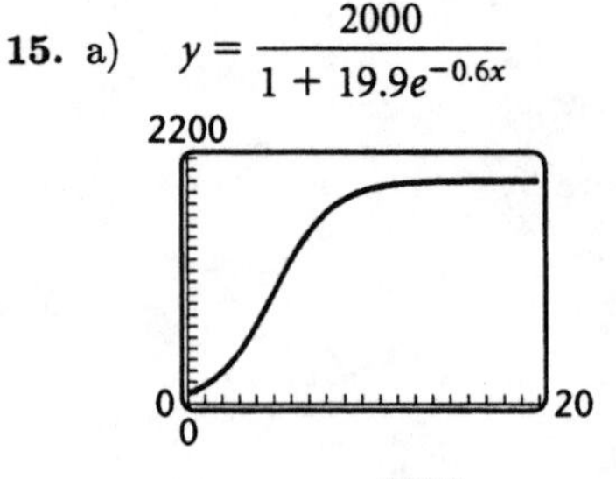

b) $N(0) = \dfrac{2000}{1 + 19.9e^{-0.6(0)}} \approx 96$

c) $N(2) = \dfrac{2000}{1 + 19.9e^{-0.6(2)}} \approx 286$

$N(5) = \dfrac{2000}{1 + 19.9e^{-0.6(5)}} \approx 1005$

$N(8) = \dfrac{2000}{1 + 19.9e^{-0.6(8)}} \approx 1719$

$N(12) = \dfrac{2000}{1 + 19.9e^{-0.6(12)}} \approx 1971$

$N(16) = \dfrac{2000}{1 + 19.9e^{-0.6(16)}} \approx 1997$

16. a) $y = \dfrac{50}{1 + 22e^{-0.6x}}$

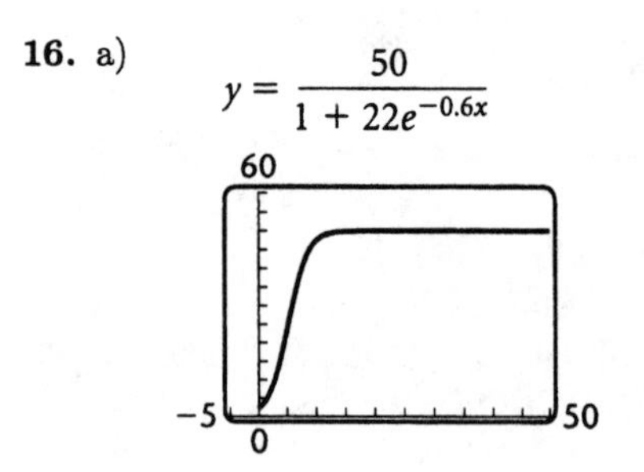

b) $N(0) = \dfrac{50}{1 + 22e^{-0.6(0)}} \approx 2$

c) In 1996, $t = 1996 - 1984 = 12$.

$N(12) = \dfrac{50}{1 + 22e^{-0.6(12)}} \approx 49$

In 2002, $t = 2002 - 1984 = 18$.

$N(18) = \dfrac{50}{1 + 22e^{-0.6(18)}} \approx 49.98$

d) No; the function increases toward a limiting value of 50 as $t \to \infty$, but it never actually reaches that value.

17. a) $y = \dfrac{2500}{1 + 5.25e^{-0.32x}}$

b) $P(0) = \dfrac{2500}{1 + 5.25e^{-0.32(0)}} = 400$

$P(1) = \dfrac{2500}{1 + 5.25e^{-0.32(1)}} \approx 520$

$P(5) = \dfrac{2500}{1 + 5.25e^{-0.32(5)}} \approx 1214$

$$P(10) = \frac{2500}{1 + 5.25e^{-0.32(10)}} \approx 2059$$

$$P(15) = \frac{2500}{1 + 5.25e^{-0.32(15)}} \approx 2396$$

$$P(20) = \frac{2500}{1 + 5.25e^{-0.32(20)}} \approx 2478$$

18. The data have the pattern of a decreasing exponential function, so function (d) might be used as a model.

19. The data have an S-shaped pattern, so function (f) might be used as a model.

20. The data have a parabolic pattern, so function (a) might be used as a model.

21. The data fit the pattern of a polynomial function with degree greater than two. Thus, function (b) might be used as a model.

22. The data have a logarithmic pattern, so function (e) might be used as a model.

23. The data have the pattern of an increasing exponential function, so function (c) might be used as a model.

24. a)

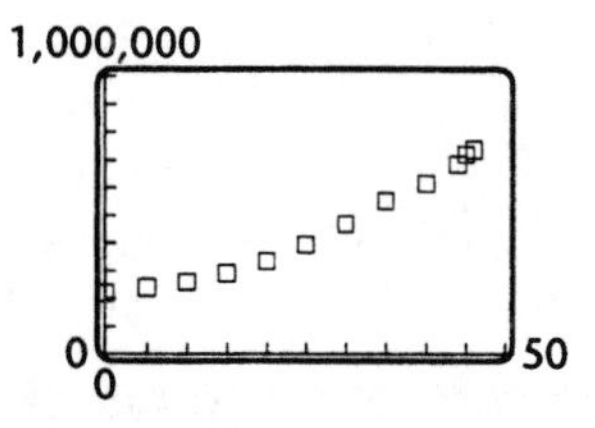

An exponential function appears to fit the data.

b) $y = 204,297.6056(1.027905316)^x$

We can convert this equation to an equation with base e, if desired.

$$y = 204,297.6056e^{x(\ln 1.027905316)}$$
$$= 204,297.6056e^{0.0275230577x}$$

In each case x is the number of years after 1950.

The coefficient of correlation r is approximately 0.9962. Since this is close to 1, the function is a good fit.

c) 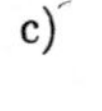

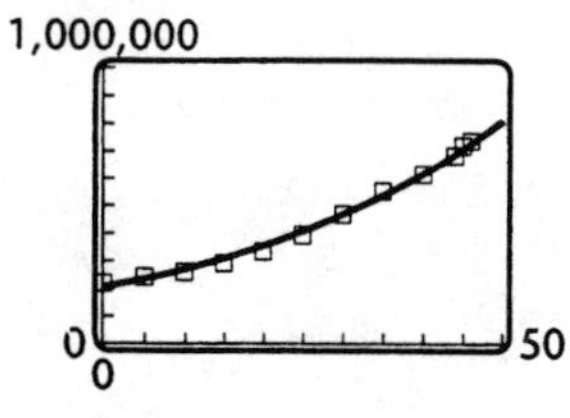

d) In 2000, $x = 2000 - 1950 = 50$.

$y = 204,297.6056(1.027905316)^{50} \approx 808,945$

In 2025, $x = 2025 - 1950 = 75$.

$y = 204,297.6056(1.027905316)^{75} \approx 1,609,706$

25. a) $y = 4.195491964(1.025306189)^x$

We can convert this equation to an equation with base e, if desired.

$$y = 4.195491964e^{x(\ln 1.025306189)}$$
$$= 4.195491964e^{0.024991289x}$$

In each case x is the number of years after 2000 and y is in millions.

The coefficient of correlation r is approximately 0.9954. Since this is close to 1, the function is a good fit.

b)

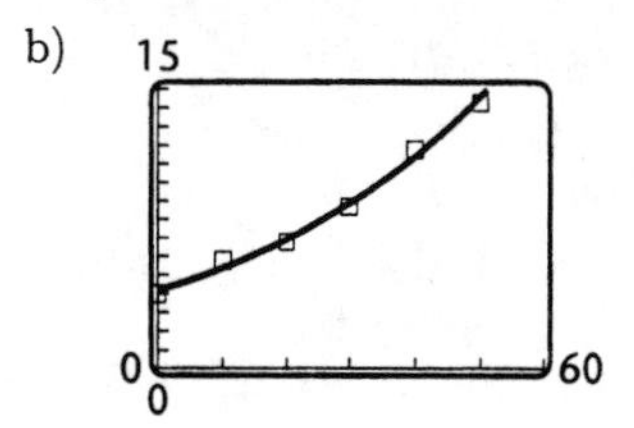

c) In 2005, $t = 2005 - 2000 = 5$.

$y = 4.195491964(1.025306189)^5 \approx 4.8$ million

In 2025, $t = 2025 - 2000 = 25$.

$y = 4.195491964(1.025306189)^{25} \approx 7.8$ million

In 2100, $t = 2100 - 2000 = 100$.

$y = 4.195491964(1.025306189)^{100} \approx 51.1$ million

26. a) Using the logarithmic regression feature on a grapher, we get $y = 84.94353992 - 0.5412834098 \ln x$.

b) For $x = 8$, $y \approx 83.8\%$.

For $x = 10$, $y \approx 83.7\%$.

For $x = 24$, $y \approx 83.2\%$.

For $x = 36$, $y \approx 83.0\%$.

c) $82 = 84.94353992 - 0.5412834098 \ln x$

Graph $y_1 = 82$ and $y_2 = 84.94353992 - 0.5412834098 \ln x$ and find the first coordinate of the point of intersection of the graphs. It is approximately 230, so test scores will fall below 82% after about 230 months.

27. a)

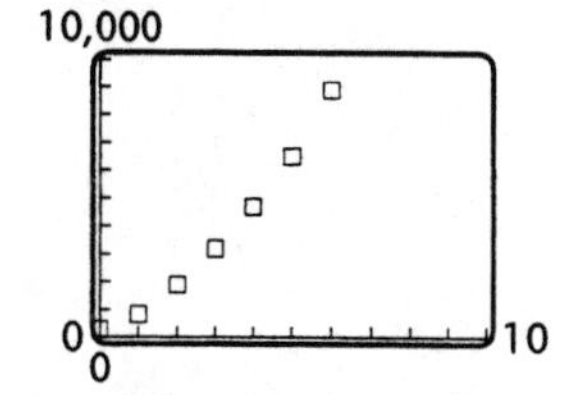

b) Linear: $y = 1429.214286x - 530.0714286$; $r^2 \approx 0.9641$

Quadratic: $y = 158.0952381x^2 + 480.6428571x + 260.4047619$; $R^2 \approx 0.9995$

Exponential: $y = 445.8787388(1.736315606)^x$; $r^2 \approx 0.9361$

The value of R^2 is highest for the quadratic function, os we determine that this function fits the data best.

c) 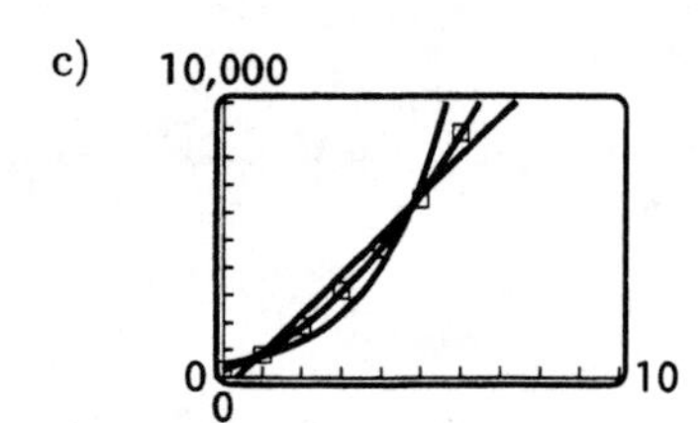

d) In 2010, $x = 2010 - 1996 = 14$.

For the linear function, when $x = 14$, $y \approx \$19,479$ million, or \$19.479 billion.

For the quadratic function, when $x = 14$, $y \approx \$37,976$ million, or \$37.976 billion.

For the exponential function, when $x = 14$, $y \approx \$1,009,295$ million, or \$1009.295 billion.

Given the rate at which the revenues in the table are increasing, it appears that the quadratic function provides the most realistic prediction. Answers may vary.

28. a) Using the logistic regression feature on a grapher, we get $y = \dfrac{99.98884912}{1 + 489.2438401e^{-0.1299899024x}}$.

b) For $x = 55$, $y \approx 72.2\%$.

For $x = 100$, $y \approx 99.9\%$.

c) $y = 100$ is an asymptote; as more and more ads are run, the percent of people who bought the product approaches 100%.

29. Answers will vary.

30. Measure the atmospheric pressure P at the top of the building. Substitute that value in the equation $P = 14.7e^{-0.00005a}$, and solve for the height, or altitude, a, of the top of the building. Also measure the atmospheric pressure at the base of the building and solve for the altitude of the base. Then subtract to find the height of the building.

31. $y = 6 = 0 \cdot x + 6$

Slope: 0; y-intercept $(0, 6)$

32.
$$3x - 10y = 14$$
$$3x - 14 = 10y$$
$$\frac{3}{10}x - \frac{7}{5} = y$$

Slope: $\frac{3}{10}$; y-intercept: $\left(0, -\frac{7}{5}\right)$

33. $y = 2x - \dfrac{3}{13}$

Slope: 2, y-intercept: $\left(0, -\frac{3}{13}\right)$

34. $x = -4$

Slope: undefined; y-intercept: none

35.
$$P(t) = P_0e^{kt}$$
$$50,000 = P_0e^{0.07(18)}$$
$$\frac{50,000}{e^{0.07(18)}} = P_0$$
$$\$14,182.70 \approx P_0$$

36. a)
$$P = P_0e^{kt}$$
$$\frac{P}{e^{kt}} = P_0, \text{ or}$$
$$Pe^{-kt} = P_0$$

b) $P_0 = 50,000e^{-0.064(18)} \approx \$15,800.21$

37.
$$480e^{-0.003p} = 150e^{0.004p}$$
$$\frac{480}{150} = \frac{e^{0.004p}}{e^{-0.003p}}$$
$$3.2 = e^{0.007p}$$
$$\ln 3.2 = \ln e^{0.007p}$$
$$\ln 3.2 = 0.007p$$
$$\frac{\ln 3.2}{0.007} = p$$
$$\$166.16 \approx p$$

38.
$$P(4000) = P_0e^{-0.00012(4000)}$$
$$= 0.619P_0, \text{ or } 61.9\%P_0$$

Thus, about 61.9% of the carbon-14 remains, so about 38.1% has been lost.

39. To find k we substitute 105 for T_1, 0 for T_0, 5 for t, and 70 for $T(t)$ and solve for k.
$$70 = 0 + |105 - 0|e^{-5k}$$
$$70 = 105e^{-5k}$$
$$\frac{70}{105} = e^{-5k}$$
$$\ln\frac{70}{105} = \ln e^{-5k}$$
$$\ln\frac{70}{105} = -5k$$
$$\frac{\ln\frac{70}{105}}{-5} = k$$
$$0.081 \approx k$$

The function is $T(t) = 105e^{-0.081t}$.

Now we find $T(10)$.

$T(10) = 105e^{-0.081(10)} \approx 46.7$ °F

40. To find k we substitute 94.6 for T_1, 70 for T_0, 60 for t (1 hr = 60 min), and 93.4 for $T(t)$.

$$93.4 = 70 + |94.6 - 70|e^{-k(60)}$$
$$23.4 = 24.6e^{-60k}$$
$$\frac{23.4}{24.6} = e^{-60k}$$
$$\ln\frac{23.4}{24.6} = \ln e^{-60k}$$
$$\ln\frac{23.4}{24.6} = -60k$$
$$k = \frac{\ln\frac{23.4}{24.6}}{-60} \approx 0.0008$$

The function is $T(t) = 70 + 24.6e^{-0.0008t}$.

We substitute 98.6 for $T(t)$ and solve for t.

$$98.6 = 70 + 24.6e^{-0.0008t}$$
$$28.6 = 24.6e^{-0.0008t}$$
$$\frac{28.6}{24.6} = e^{-0.0008t}$$
$$\ln\frac{28.6}{24.6} = \ln e^{-0.0008t}$$
$$\ln\frac{28.6}{24.6} = -0.0008t$$
$$t = \frac{\ln\frac{28.6}{24.6}}{-0.0008} \approx -188$$

The murder was committed at approximately 188 minutes, or about 3 hours, before 12:00 PM, or at about 9:00 AM. (Answers may vary slightly due to rounding differences.)

41.

$$i = \frac{V}{R}\left[1 - e^{-(R/L)t}\right]$$
$$\frac{iR}{V} = 1 - e^{-(R/L)t}$$
$$e^{-(R/L)t} = 1 - \frac{iR}{V}$$
$$\ln e^{-(R/L)t} = \ln\left(1 - \frac{iR}{V}\right)$$
$$-\frac{R}{L}t = \ln\left(1 - \frac{iR}{V}\right)$$
$$t = -\frac{L}{R}\left[\ln\left(1 - \frac{iR}{V}\right)\right]$$

42. a) At 1 m: $I = I_0e^{-1.4(1)} \approx 0.247I_0$

24.7% of I_0 remains.

At 3 m: $I = I_0e^{-1.4(3)} \approx 0.015I_0$

1.5% of I_0 remains.

At 5 m: $I = I_0e^{-1.4(5)} \approx 0.0009I_0$

0.09% of I_0 remains.

At 50 m: $I = I_0e^{-1.4(50)} \approx (3.98 \times 10^{-31})I_0$

Now, $3.98 \times 10^{-31} = (3.98 \times 10^{-29}) \times 10^{-2}$, so

(3.98×10^{-29})% remains.

b) $I = I_0e^{-1.4(10)} \approx 0.0000008I_0$

Thus, 0.00008% remains.

43.

$$y = ae^x$$
$$\ln y = \ln(ae^x)$$
$$\ln y = \ln a + \ln e^x$$
$$\ln y = \ln a + x$$
$$Y = x + \ln a$$

This function is of the form $y = mx+b$, so it is linear.

44.

$$y = ax^b$$
$$\ln y = \ln(ax^b)$$
$$\ln y = \ln a + b\ln x$$
$$Y = \ln a + bX$$

This function is of the form $y = mx+b$, so it is linear.

Chapter 5

Systems of Equations and Matrices

Exercise Set 5.1

1. Graph (c) is the graph of this system.

2. Graph (e) is the graph of this system.

3. Graph (f) is the graph of this system.

4. Graph (a) is the graph of this system.

5. Graph (b) is the graph of this system.

6. Graph (d) is the graph of this system.

7. Graph $y_1 = 2 - x$ and $y_2 = -3x$ in the same window and find the point of intersection.

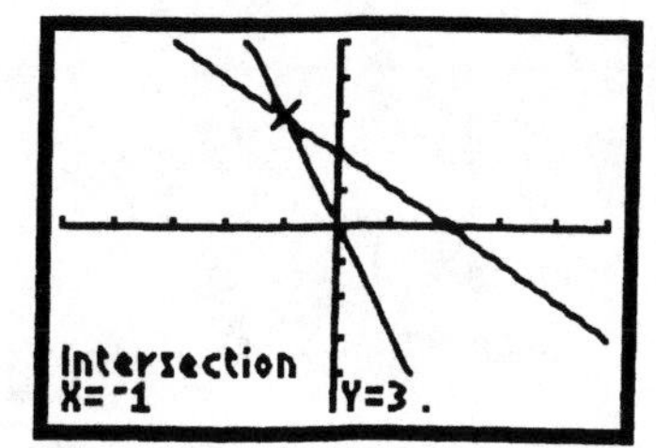

The solution is $(-1, 3)$.

8. Graph $y_1 = 1 - x$ and $y_2 = 7 - 3x$ in the same window and find the point of intersection.

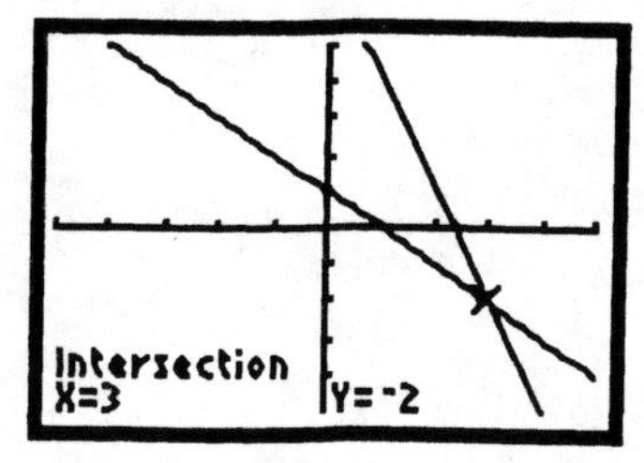

The solution is $(3, -2)$.

9. Graph $y_1 = -\frac{1}{2}x + \frac{3}{2}$ and $y_2 = -\frac{1}{4}x + \frac{13}{8}$ in the same window and find the point of intersection.

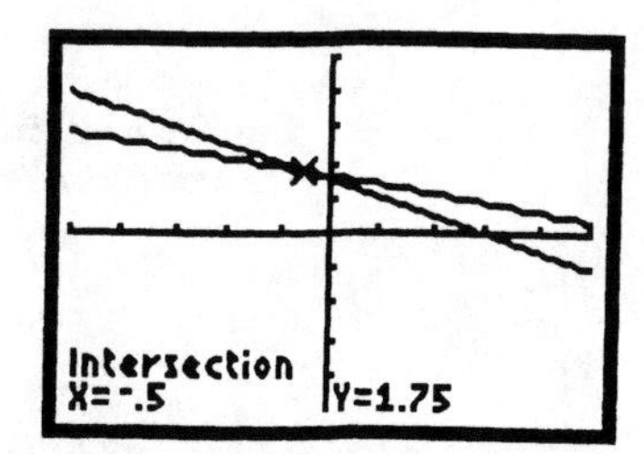

The solution is $(-0.5, 1.75)$.

10. Graph $y_1 = -\frac{1}{2}x - \frac{1}{3}$ and $y_2 = \frac{1}{2}x - 2$ in the same window and find the point of intersection.

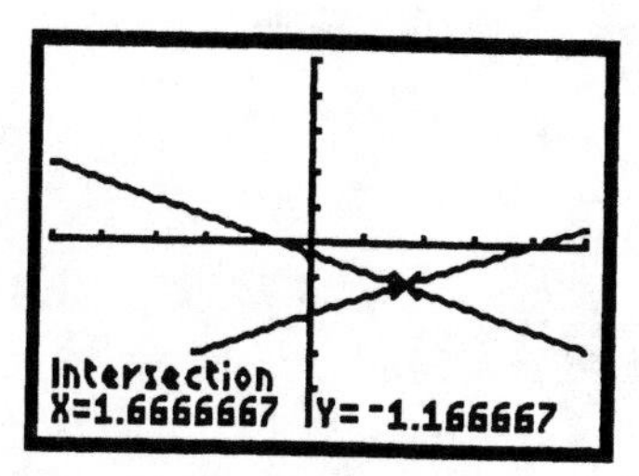

The solution is about $(1.667, -1.167)$.

11. Graph $y_1 = 2x - 1$ and $y_2 = 2x + 1$ in the same window and find the point of intersection.

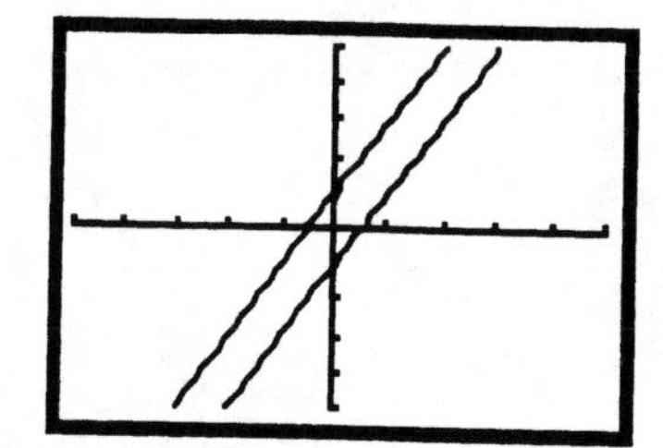

The lines do not intersect, so there is no solution.

12. When we write both equations in $y =$ form we find that they are identical: $y = 2x - 1$. Thus, the system of equations has infinitely many solutions.

13.
$$x + y = 9, \quad (1)$$
$$2x - 3y = -2 \quad (2)$$

Solve equation (1) for either x or y. We choose to solve for y.

$$y = 9 - x$$

Then substitute $9 - x$ for y in equation (2) and solve the resulting equation.

$$2x - 3(9 - x) = -2$$
$$2x - 27 + 3x = -2$$
$$5x - 27 = -2$$
$$5x = 25$$
$$x = 5$$

Now substitute 5 for x in either equation (1) or (2) and solve for y.

$5 + y = 9$ Using equation (1)

$y = 4$

The solution is $(5, 4)$. The grapher confirms that this is correct.

14. $3x - y = 5,\quad (1)$

$x + y = \frac{1}{2}\quad (2)$

Solve equation (2) for y.

$$y = \frac{1}{2} - x$$

Substitute in equation (1) and solve for x.

$$3x - \left(\frac{1}{2} - x\right) = 5$$
$$3x - \frac{1}{2} + x = 5$$
$$4x = \frac{11}{2}$$
$$x = \frac{11}{8}$$

Back-substitute to find y.

$$\frac{11}{8} + y = \frac{1}{2}\quad \text{Using equation (2)}$$
$$y = -\frac{7}{8}$$

The solution is $\left(\frac{11}{8}, -\frac{7}{8}\right)$.

15. $x - 2y = 7,\quad (1)$

$x = y + 4\quad (2)$

Use equation (2) and substitute $y+4$ for x in equation (1). Then solve for y.

$$y + 4 - 2y = 7$$
$$-y + 4 = 7$$
$$-y = 3$$
$$y = -3$$

Substitute -3 for y in equation (2) to find x.

$$x = -3 + 4 = 1$$

The solution is $(1, -3)$. The grapher confirms that this is correct.

16. $x + 4y = 6,\quad (1)$

$x = -3y + 3\quad (2)$

Substitute $-3y + 3$ for x in equation (1) and solve for y.

$$-3y + 3 + 4y = 6$$
$$y = 3$$

Back-substitute to find x.

$$x = -3 \cdot 3 + 3 = -6$$

The solution is $(-6, 3)$.

17. $y = 2x - 6,\quad (1)$

$5x - 3y = 16\quad (2)$

Use equation (1) and substitute $2x - 6$ for y in equation (2). Then solve for x.

$$5x - 3(2x - 6) = 16$$
$$5x - 6x + 18 = 16$$
$$-x + 18 = 16$$
$$-x = -2$$
$$x = 2$$

Substitute 2 for x in equation (1) to find y.

$$y = 2 \cdot 2 - 6 = 4 - 6 = -2$$

The solution is $(2, -2)$. The grapher confirms that this is correct.

18. $3x + 5y = 2,\quad (1)$

$2x - y = -3\quad (2)$

Solve equation (2) for y.

$$y = 2x + 3$$

Substitute $2x + 3$ for y in equation (1) and solve for x.

$$3x + 5(2x + 3) = 2$$
$$3x + 10x + 15 = 2$$
$$13x = -13$$
$$x = -1$$

Back-substitute to find y.

$$2(-1) - y = -3\quad \text{Using equation (2)}$$
$$-2 - y = -3$$
$$1 = y$$

The solution is $(-1, 1)$.

19. $x - 5y = 4,\quad (1)$

$y = 7 - 2x\quad (2)$

Use equation (2) and substitute $7 - 2x$ for y in equation (1). Then solve for x.

$$x - 5(7 - 2x) = 4$$
$$x - 35 + 10x = 4$$
$$11x - 35 = 4$$
$$11x = 39$$
$$x = \frac{39}{11}$$

Substitute $\frac{39}{11}$ for x in equation (2) to find y.

$$y = 7 - 2 \cdot \frac{39}{11} = 7 - \frac{78}{11} = -\frac{1}{11}$$

The solution is $\left(\frac{39}{11}, -\frac{1}{11}\right)$. The grapher confirms that this is correct.

20. $5x + 3y = -1,\quad (1)$

$x + y = 1\quad (2)$

Solve equation (2) for either x or y. We choose to solve for x.

$$x = 1 - y$$

Substitute $1 - y$ for x in equation (1) and solve for y.

$$5(1-y)+3y=-1$$
$$5-5y+3y=-1$$
$$-2y=-6$$
$$y=3$$

Back-substitute to find x.

$x+3=1$ Using equation (2)

$$x=-2$$

The solution is $(-2,3)$.

21. $2x-3y=5,$ (1)

$5x+4y=1$ (2)

Solve one equation for either x or y. We choose to solve equation (1) for x.

$$2x-3y=5$$
$$2x=3y+5$$
$$x=\frac{3}{2}y+\frac{5}{2}$$

Substitute $\frac{3}{2}y+\frac{5}{2}$ for x in equation (2) and solve for y.

$$5\left(\frac{3}{2}y+\frac{5}{2}\right)+4y=1$$
$$\frac{15}{2}y+\frac{25}{2}+4y=1$$
$$\frac{23}{2}y+\frac{25}{2}=1$$
$$\frac{23}{2}y=-\frac{23}{2}$$
$$y=-1$$

Substitute -1 for y in either equation (1) or (2) and solve for x.

$2x-3(-1)=5$ Using equation (1)

$$2x+3=5$$
$$2x=2$$
$$x=1$$

The solution is $(1,-1)$. The grapher confirms that this is correct.

22. $3x+4y=6,$ (1)

$2x+3y=5$ (2)

Solve one equation for either x or y. We choose to solve equation (2) for x.

$$x=-\frac{3}{2}y+\frac{5}{2}$$

Substitute $-\frac{3}{2}y+\frac{5}{2}$ for x in equation (1) and solve for y.

$$3\left(-\frac{3}{2}y+\frac{5}{2}\right)+4y=6$$
$$-\frac{9}{2}y+\frac{15}{2}+4y=6$$
$$-\frac{1}{2}y=-\frac{3}{2}$$
$$y=3$$

Back-substitute to find x.

$3x+4\cdot 3=6$ Using equation (1)

$$3x=-6$$
$$x=-2$$

The solution is $(-2,3)$.

23. $x+2y=7,$ (1)

$x-2y=-5$ (2)

We add the equations to eliminate y.

$$x+2y=7$$
$$x-2y=-5$$
$$2x\quad=2 \quad \text{Adding}$$
$$x=1$$

Back-substitute in either equation and solve for y.

$1+2y=7$ Using equation (1)

$$2y=6$$
$$y=3$$

The solution is $(1,3)$. The grapher confirms that this is correct.

Since the system of equations has exactly one solution it is consistent and independent.

24. $3x+4y=-2$ (1)

$-3x-5y=1$ (2)

$-y=-1$ Adding

$$y=1$$

Back-substitute to find x.

$3x+4\cdot 1=-2$ Using equation (1)

$$3x=-6$$
$$x=-2$$

The solution is $(-2,1)$. Since the system of equations has exactly one solution it is consistent and independent.

25. $x-3y=2,$ (1)

$6x+5y=-34$ (2)

Multiply equation (1) by -6 and add it to equation (2) to eliminate x.

$$-6x+18y=-12$$
$$6x+5y=-34$$
$$23y=-46$$
$$y=-2$$

Back-substitute to find x.

$x-3(-2)=2$ Using equation (1)

$$x+6=2$$
$$x=-4$$

The solution is $(-4,-2)$. The grapher confirms that this is correct.

Since the system of equations has exactly one solution it is consistent and independent.

26. $x + 3y = 0$, (1)

$20x - 15y = 75$ (2)

Multiply equation (1) by 5 and add.

$$\begin{aligned} 5x + 15y &= 0 \\ 20x - 15y &= 75 \\ \hline 25x \quad &= 75 \\ x &= 3 \end{aligned}$$

Back-substitute to find y.

$$\begin{aligned} 3 + 3y &= 0 \quad \text{Using equation (1)} \\ 3y &= -3 \\ y &= -1 \end{aligned}$$

The solution is $(3, -1)$. Since the system of equations has exactly one solution it is consistent and independent.

27. $0.3x - 0.2y = -0.9$,

$0.2x - 0.3y = -0.6$

First, multiply each equation by 10 to clear the decimals.

$3x - 2y = -9$ (1)

$2x - 3y = -6$ (2)

Now multiply equation (1) by 3 and equation (2) by -2 and add to eliminate y.

$$\begin{aligned} 9x - 6y &= -27 \\ -4x + 6y &= 12 \\ \hline 5x \quad &= -15 \\ x &= -3 \end{aligned}$$

Back-substitute to find y.

$$\begin{aligned} 3(-3) - 2y &= -9 \quad \text{Using equation (1)} \\ -9 - 2y &= -9 \\ -2y &= 0 \\ y &= 0 \end{aligned}$$

The solution is $(-3, 0)$. The grapher confirms that this is correct. Since the system of equations has exactly one solution it is consistent and independent.

28. $0.2x - 0.3y = 0.3$,

$0.4x + 0.6y = -0.2$

First, multiply each equation by 10 to clear the decimals.

$2x - 3y = 3$ (1)

$4x + 6y = -2$ (2)

Now multiply equation (1) by 2 and add.

$$\begin{aligned} 4x - 6y &= 6 \\ 4x + 6y &= -2 \\ \hline 8x \quad &= 4 \\ x &= \frac{1}{2} \end{aligned}$$

Back-substitute to find y.

$$\begin{aligned} 4 \cdot \frac{1}{2} + 6y &= -2 \quad \text{Using equation (2)} \\ 2 + 6y &= -2 \\ 6y &= -4 \\ y &= -\frac{2}{3} \end{aligned}$$

The solution is $\left(\frac{1}{2}, -\frac{2}{3}\right)$. Since the system of equations has exactly one solution it is consistent and independent.

29. $3x - 12y = 6$, (1)

$2x - 8y = 4$ (2)

Multiply equation (1) by 2 and equation (2) by -3 and add.

$$\begin{aligned} 6x - 24y &= 12 \\ -6x + 24y &= -12 \\ \hline 0 &= 0 \end{aligned}$$

The equation $0 = 0$ is true for all values of x and y. Thus, the system of equations has infinitely many solutions. Solving either equation for y, we can write $y = \frac{1}{4}x - \frac{1}{2}$ so the solutions are ordered pairs of the form $\left(x, \frac{1}{4}x - \frac{1}{2}\right)$. Equivalently, if we solve either equation for x we get $x = 4y + 2$ so the solutions can also be expressed as $(4y + 2, y)$. The grapher confirms that the graphs of the equations coincide.

Since there are infinitely many solutions, the system of equations is consistent and dependent.

30. $2x + 6y = 7$, (1)

$3x + 9y = 10$ (2)

Multiply equation (1) by 3 and equation (2) by -2 and add.

$$\begin{aligned} 6x + 18y &= 21 \\ -6x - 18y &= -20 \\ \hline 0 &= 1 \end{aligned}$$

We get a false equation so there is no solution. Since there is no solution the system of equations is inconsistent and independent.

31. $\frac{1}{5}x + \frac{1}{2}y = 6$, (1)

$\frac{3}{5}x - \frac{1}{2}y = 2$ (2)

We could multiply both equations by 10 to clear fractions, but since the y-coefficients differ only by sign we will just add to eliminate y.

$$\begin{aligned} \frac{1}{5}x + \frac{1}{2}y &= 6 \\ \frac{3}{5}x - \frac{1}{2}y &= 2 \\ \hline \frac{4}{5}x \quad &= 8 \\ x &= 10 \end{aligned}$$

Back-substitute to find y.

$\frac{1}{5}\cdot 10+\frac{1}{2}y=6$ Using equation (1)

$2+\frac{1}{2}y=6$

$\frac{1}{2}y=4$

$y=8$

The solution is $(10,8)$. The grapher confirms that this is correct.

Since the system of equations has exactly one solution it is consistent and independent.

32. $\frac{2}{3}x+\frac{3}{5}y=-17,$

$\frac{1}{2}x-\frac{1}{3}y=-1$

Multiply the first equation by 15 and the second by 6 to clear fractions.

$10x+9y=-255$ (1)

$3x-2y=-6$ (2)

Now multiply equation (1) by 2 and equation (2) by 9 and add.

$$\begin{array}{r} 20x+18y=-510 \\ 27x-18y=-54 \\ \hline 47x\quad\quad\; =-564 \end{array}$$

$x=-12$

Back-substitute to find y.

$3(-12)-2y=-6$ Using equation (2)

$-36-2y=-6$

$-2y=30$

$y=-15$

The solution is $(-12,-15)$. Since the system of equations has exactly one solution it is consistent and independent.

33. $2x=5-3y,$ (1)

$4x=11-7y$ (2)

We rewrite the equations.

$2x+3y=5,$ (1a)

$4x+7y=11$ (2a)

Multiply equation (2a) by -2 and add to eliminate x.

$$\begin{array}{r} -4x-6y=-10 \\ 4x+7y=11 \\ \hline y=1 \end{array}$$

Back-substitute to find x.

$2x=5-3\cdot 1$ Using equation (1)

$2x=2$

$x=1$

The solution is $(1,1)$. The grapher confirms that this is correct.

Since the system of equations has exactly one solution it is consistent and independent.

34. $7(x-y)=14,$ (1)

$2x=y+5$ (2)

$x-y=2,$ (1a) Dividing equation (1) by 7

$2x-y=5$ (2a) Rewriting equation (2)

Multiply equation (1a) by -1 and add.

$$\begin{array}{r} -x+y=-2 \\ 2x-y=5 \\ \hline x\quad\; =3 \end{array}$$

Back-substitute to find y.

$3-y=2$ Using equation (1a)

$1=y$

The solution is $(3,1)$. Since the system of equations has exactly one solution it is consistent and independent.

35. ***Familiarize***. Let $x=$ the number of video rentals and $y=$ the number of boxes of popcorn given away. Then the total cost of the rentals is $1\cdot x$, or x, and the total cost of the popcorn is $2y$.

Translate. The number of new members is the same as the number of incentives so we have one equation:

$x+y=48.$

The total cost of the incentive was \$86. This gives us another equation:

$x+2y=86.$

Carry out. We solve the system

$x+\;y=48,$ (1)

$x+2y=86.$ (2)

Multiply equation (1) by -1 and add.

$$\begin{array}{r} -x-\;y=-48 \\ x+2y=86 \\ \hline y=38 \end{array}$$

Back-substitute to find x.

$x+38=48$ Using equation (1)

$x=10$

Check. When 10 rentals and 38 boxes of popcorn are given away, a total of 48 new members have signed up. Ten rentals cost the store \$10 and 38 boxes of popcorn cost $2\cdot 38$, or \$76, so the total cost of the incentives was \$10 + \$76, or \$86. The solution checks.

State. Ten rentals and 38 boxes of popcorn were given away.

36. Let $x=$ the number of tickets sold for pavilion seats and $y=$ the number sold for lawn seats.

Solve: $x+y=1500,$

$25x+15y=28,500.$

$x=600,\ y=900$

37. ***Familiarize***. Let $x =$ the cost of an adult's admission and $y =$ the cost of a child's admission. Then the total cost for 4 adults was $4x$ and for 16 children was $16y$.

Translate. Each adult's admission costs twice as much as each child's admission, so we have one equation:

$x = 2y$.

The total admission cost was \$72, so we have:

$4x + 16y = 72$

Solve. We solve the system

$x = 2y$, (1)

$4x + 16y = 72$. (2)

Substitute $2y$ for x in equation (2) and solve for y.

$$4 \cdot 2y + 16y = 72$$
$$24y = 72$$
$$y = 3$$

Back-substitute in equation (1) to find x.

$x = 2 \cdot 3 = 6$

Check. The cost of an adult's admission, \$6, is twice the cost of a child's admission, \$3. The total admission cost was $4 \cdot \$6 + 16 \cdot \$3 = \$24 + \$48 = \$72$. The solution checks.

State. Each adult's admission is \$6 and each child's admission is \$3.

38. Let $x =$ the number of standard-delivery packages and $y =$ the number of express-delivery packages.

Solve: $x + y = 120$,

$3.5x + 7.5y = 596$.

$x = 76$, $y = 44$

39. ***Familiarize and Translate***. We use the system of equations given in the problem.

$y = 70 + 2x$ (1)

$y = 175 - 5x$, (2)

Carry out. Substitute $175 - 5x$ for y in equation (1) and solve for x.

$$175 - 5x = 70 + 2x$$
$$105 = 7x \quad \text{Adding } 5x \text{ and subtracting } 70$$
$$15 = x$$

Back-substitute in either equation to find y. We choose equation (1).

$y = 70 + 2 \cdot 15 = 70 + 30 = 100$

Check. Substituting 15 for x and 100 for y in both of the original equations yields true equations, so the solution checks. We could also check graphically by finding the point of intersection of equations (1) and (2).

State. The equilibrium point is $(15, \$100)$.

40. Solve: $y = 240 + 40x$,

$y = 500 - 25x$.

$x = 4$, $y = 400$, so the equilibrium point is $(4, \$400)$.

41. ***Familiarize and Translate***. We find the value of x for which $C = R$, where

$C = 14x + 350$,

$R = 16.5x$.

Carry out. When $C = R$ we have:

$$14x + 350 = 16.5x$$
$$350 = 2.5x$$
$$140 = x$$

Check. When $x = 140$, $C = 14 \cdot 140 + 350$, or 2310 and $R = 16.5(140)$, or 2310. Since $C = R$, the solution checks.

State. 140 units must be produced and sold in order to break even.

42. Solve $C = R$, where

$C = 8.5x + 75$,

$R = 10x$.

$x = 50$

43. ***Familiarize and Translate***. We find the value of x for which $C = R$, where

$C = 15x + 12,000$,

$R = 18x - 6000$.

Carry out. When $C = R$ we have:

$$15x + 12,000 = 18x - 6000$$
$$18,000 = 3x \quad \text{Subtracting } 15x \text{ and adding } 6000$$
$$6000 = x$$

Check. When $x = 6000$, $C = 15 \cdot 6000 + 12,000$, or 102,000 and $R = 18 \cdot 6000 - 6000$, or 102,000. Since $C = R$, the solution checks.

State. 6000 units must be produced and sold in order to break even.

44. Solve $C = R$, where

$C = 3x + 400$,

$R = 7x - 600$.

$x = 250$

45. ***Familiarize***. Let $x =$ the number of servings of spaghetti and meatballs required and $y =$ the number of servings of iceberg lettuce required. Then x servings of spaghetti contain $260x$ Cal and $32x$ g of carbohydrates; y servings of lettuce contain $5y$ Cal and $1 \cdot y$ or y, g of carbohydrates.

Translate. One equation comes from the fact that 400 Cal are desired:

$260x + 5y = 400$.

A second equation comes from the fact that 50g of carbohydrates are required:

$32x + y = 50.$

Solve. We solve the system

$260x + 5y = 400, \quad (1)$

$32x + y = 50. \quad (2)$

Multiply equation (2) by -5 and add.

$$\begin{array}{r} 260x + 5y = 400 \\ -160x - 5y = -250 \\ \hline 100x \qquad = 150 \\ x = 1.5 \end{array}$$

Back-substitute to find y.

$32(1.5) + y = 50$ Using equation (2)

$48 + y = 50$

$y = 2$

Check. 1.5 servings of spaghetti contain 260(1.5), or 390 Cal and 32(1.5), or 48 g of carbohydrates; 2 servings of lettuce contain $5 \cdot 2$, or 10 Cal and $1 \cdot 2$, or 2 g of carbohydrates. Together they contain 390+10, or 400 Cal and 48 + 2, or 50 g of carbohydrates. The solution checks.

State. 1.5 servings of spaghetti and meatballs and 2 servings of iceberg lettuce are required.

46. Let x = the number of servings of tomato soup and y = the number of slices of whole wheat bread required.

Solve: $100x + 70y = 230,$

$18x + 13y = 42.$

$x = 1.25, y = 1.5$

47. ***Familiarize.*** It helps to make a drawing. Then organize the information in a table. Let x = the speed of the boat and y = the speed of the stream. The speed upstream is $x - y$. The speed downstream is $x + y$.

46 km 2 hr $(x + y)$ km/h

Downstream

51 km 3 hr $(x - y)$ km/h

Upstream

	Distance	Speed	Time
Downstream	46	$x+y$	2
Upstream	51	$x-y$	3

Translate. Using $d = rt$ in each row of the table, we get a system of equations.

$46 = (x + y)2 \qquad x + y = 23, \quad (1)$

or

$51 = (x - y)3 \qquad x - y = 17 \quad (2)$

Carry out. We begin by adding equations (1) and (2).

$$\begin{array}{r} x + y = 23 \\ x - y = 17 \\ \hline 2x \qquad = 40 \\ x = 20 \end{array}$$

Back-substitute to find y.

$20 + y = 23$ Using equation (1)

$y = 23$

Check. The speed downstream is 20+3, or 23 km/h. The distance traveled downstream in 2 hr is $23 \cdot 2$, or 46 km. The speed upstream is $20 - 3$, or 17 km/h. The distance traveled upstream in 3 hr is $17 \cdot 3$, or 51 km. The solution checks.

State. The speed of the boat is 20 km/h. The speed of the stream is 3 km/h.

48. Let x = the speed of the plane and y = the speed of the wind.

	Speed	Time	Distance
Downwind	$x+y$	3	3000
Upwind	$x-y$	4	3000

Solve: $(x + y)3 = 3000,$

$(x - y)4 = 3000.$

$x = 875$ km/h, $y = 125$ km/h

49. ***Familiarize.*** Let x = the amount invested at 7% and y = the amount invested at 9%. Then the interest from the investments is $7\%x$ and $9\%y$, or $0.07x$ and $0.09y$.

Translate.

The total investment is $15,000.

$x + y = 15,000$

The total interest is $1230.

$0.07x + 0.09y = 1230$

We have a system of equations:

$x + y = 15,000,$

$0.07x + 0.09y = 1230$

Multiplying the second equation by 100 to clear the decimals, we have:

$x + y = 15,000, \quad (1)$

$7x + 9y = 123,000. \quad (2)$

Carry out. We begin by multiplying equation (1) by -7 and adding.

$$\begin{array}{r} -7x - 7y = -105,000 \\ 7x + 9y = 123,000 \\ \hline 2y = 18,000 \\ y = 9000 \end{array}$$

Back-substitute to find x.

$x + 9000 = 15,000$ Using equation (1)

$x = 6000$

Check. The total investment is $6000 + $9000, or $15,000. The total interest is 0.07($6000) +

0.09($9000), or $420 + $810, or $1230. The solution checks.

State. $6000 was invested at 7% and $9000 was invested at 9%.

50. Let $x =$ the number of short-sleeved shirts sold and $y =$ the number of long-sleeved shirts sold.

Solve: $x + y = 36,$

$12x + 18y = 522.$

$x = 21$, $y = 15$

51. ***Familiarize.*** Let $x =$ the number of pounds of French roast coffee used and $y =$ the number of pounds of Kenyan coffee. We organize the information in a table.

	French roast	Kenyan	Mixture
Amount	x	y	10 lb
Price per pound	\$9.00	\$7.50	\$8.40
Total cost	$\$9x$	$\$7.50y$	\$8.40(10), or \$84

Translate. The first and third rows of the table give us a system of equations.

$x + y = 10,$

$9x + 7.5y = 84$

Multiply the second equation by 10 to clear the decimals.

$x + y = 10, \quad (1)$

$90x + 75y = 840 \quad (2)$

Carry out. Begin by multiplying equation (1) by -75 and adding.

$$\begin{array}{r} -75x - 75y = -750 \\ 90x + 75y = 840 \\ \hline 15x \qquad = 90 \\ x = 6 \end{array}$$

Back-substitute to find y.

$6 + y = 10$ Using equation (1)

$y = 4$

Check. The total amount of coffee in the mixture is $6 + 4$, or 10 lb. The total value of the mixture is 6($9) + 4($7.50), or $54 + $30, or $84. The solution checks.

State. 6 lb of French roast coffee and 4 lb of Kenyan coffee should be used.

52. Let $x =$ the speed of the plane and $y =$ the speed of the wind.

	Distance	Speed	Time
LA to NY	3000	$x + y$	5
NY to LA	3000	$x - y$	6

Solve: $3000 = 5(x + y),$

$3000 = 6(x - y).$

$x = 550$ mph, $y = 50$ mph

53. ***Familiarize.*** Let $x =$ the monthly sales, $C =$ the earnings with the straight-commission plan, and $S =$ the earnings with the salary-plus-commission plan. Then 8% of sales is represented by $8\%x$, or $0.08x$, and 1% of sales is represented by $1\%x$, or $0.01x$.

Translate.

Straight commission pays 8% of sales.

$C = 0.08x$

Salary plus commission pays $1500 plus 1% of sales.

$S = 1500 + 0.01x$

Carry out. We find the value of x for which $C = S$. We have:

$0.08x = 1500 + 0.01x$

$0.07x = 1500$

$x \approx 21,428.57$

Check. When $x \approx 21,428.57$, $C \approx 0.08(21,428.57) \approx 1714.29$ and $S \approx 1500 + 0.01(21,428.57) \approx 1714.29$. Since $C = S$, the solution checks.

State. The two plans pay the same amount for monthly sales of about $21,428.57.

54. Let $d =$ the distance traveled by the slower plane and $t =$ the time the planes travel.

	Distance	Speed	Time
Slower plane	d	190	t
Faster plane	$780 - d$	200	t

Solve: $d = 190t,$

$780 - d = 200t.$

$t = 2$ hr

55. a) $t(x) = 1.69541779x + 44.15363881,$

$r(x) = -0.4528301887x + 68.83018868$

b) Graph $y_1 = t(x)$ and $y_2 = r(x)$ and find the first coordinate of the point of intersection of the graphs. It is approximately 11.5, so the number of take-out meals purchased was equal to the number of restaurant meals purchased about 11.5 years after 1984.

56. a) $m(x) = -0.1293948127x + 76.54841499,$

$w(x) = 0.4250720461x + 54.80576369$

b) Graph $y_1 = m(x)$ and $y_2 = w(x)$ and find the first coordinate or the point of intersection of the graphs. It is approximately 39.2, so the percentages of men and women in the labor force will be equal about 39.2 years after 1985.

57. When a variable is not alone on one side of an equation or when solving for a variable is difficult or produces an expression containing fractions, the elimination method is preferable to the substitution method.

58. The solution of the equation $2x + 5 = 3x - 7$ is the first coordinate of the point of intersection of the graphs of $y_1 = 2x + 5$ and $y_2 = 3x - 7$. The solution of the system of equations $y = 2x + 5$, $y = 3x - 7$ is the ordered pair that is the point of intersection of y_1 and y_2.

59. $2x - 5 = 0$

$2x = 5$

$x = \frac{5}{2}$

We could also graph $y = 2x - 5$ and use the Zero feature.

The solution is $\frac{5}{2}$, or 2.5.

60. $6 - 3x = 0$

$6 = 3x$

$2 = x$

We could also graph $y = 6 - 3x$ and use the Zero feature.

The solution is 2.

61. $x^2 - 4x + 3 = 0$

$(x - 1)(x - 3) = 0$

$x - 1 = 0$ *or* $x - 3 = 0$

$x = 1$ *or* $x = 3$

We could also graph $y = x^2 - 4x + 3$ and use the Zero feature twice.

The solutions are 1 and 3.

62. $f(x) = 3x^2 - x - 5$

We must use the quadratic formula to find exact values for the zeros.

$$x = \frac{-(-1) \pm \sqrt{(-1)^2 - 4 \cdot 3(-5)}}{2 \cdot 3}$$

$$= \frac{1 \pm \sqrt{61}}{6}$$

63. ***Familiarize***. Let x = the time spent jogging and y = the time spent walking. Then Nancy jogs $8x$ km and walks $4y$ km. We organize the information in a table.

Translate.

The total time is 1 hr.

$x + y = 1$

The total distance is 6 km.

$8x + 4y = 6$

Carry out. Solve the system

$x + y = 1$, (1)

$8x + 4y = 6$. (2)

Multiply equation (1) by -4 and add.

$-4x - 4y = -4$

$8x + 4y = 6$

$4x \quad = 2$

$x = \frac{1}{2}$

This is the time we need to find the distance spent jogging, so we could stop here. However, we will not be able to check the solution unless we find y also so we continue. We back-substitute.

$\frac{1}{2} + y = 1$ Using equation (1)

$y = \frac{1}{2}$

Then the distance jogged is $8 \cdot \frac{1}{2}$, or 4 km.

Check. The total time is $\frac{1}{2}$ hr + $\frac{1}{2}$ hr, or 1 hr. The total distance is 4 km + 2 km, or 6 km. The solution checks.

State. Nancy jogged 4 km on each trip.

64. Let x = the number of one-turtleneck orders and y = the number of two-turtleneck orders.

Solve: $x + 2y = 1250$,

$15x + 25y = 16,750$.

$y = 400$

65. ***Familiarize and Translate***. We let x and y represent the speeds of the trains. Organize the information in a table. Using $d = rt$, we let $3x$, $2y$, $1.5x$, and $3y$ represent the distances the trains travel.

First situation:

3 hours x km/h y km/h 2 hours
Union Central
216 km

Second situation:

1.5 hours x km/h y km/h 3 hours
Union Central
216 km

	Distance traveled in first situation	Distance traveled in second situation
Train_1 (from Union to Central)	$3x$	$1.5x$
Train_2 (from Central to Union)	$2y$	$3y$
Total	216	216

The total distance in each situation is 216 km. Thus, we have a system of equations.

$3x + 2y = 216,$ (1)
$1.5x + 3y = 216$ (2)

Carry out. Multiply equation (2) by -2 and add.

$$\begin{array}{r} 3x + 2y = 216 \\ -3x - 6y = -432 \\ \hline -4y = -216 \\ y = 54 \end{array}$$

Back-substitute to find x.

$3x + 2 \cdot 54 = 216$ Using equation (1)
$3x + 108 = 216$
$3x = 108$
$x = 36$

Check. If $x = 36$ and $y = 54$, the total distance the trains travel in the first situation is $3 \cdot 36 + 2 \cdot 54$, or 216 km. The total distance they travel in the second situation is $1.5 \cdot 36 + 3 \cdot 54$, or 216 km. The solution checks.

State. The speed of the first train is 36 km/h. The speed of the second train is 54 km/h.

66. Let x = the amount of mixture replaced by 100% antifreeze and y = the amount of 30% mixture retained.

	Replaced	Retained	Total
Amount	x	y	16 L
Percent of antifreeze	100%	30%	50%
Amount of antifreeze	$100\%x$	$30\%y$	$50\% \times 16$, or 8 L

Solve: $x + y = 16,$
$x + 0.3y = 8.$

$x = 4\frac{4}{7}$ L

67. Substitute the given solutions in the equation $Ax + By = 1$ to get a system of equations.

$3A - B = 1,$ (1)
$-4A - 2B = 1$ (2)

Multiply equation (1) by -2 and add.

$$\begin{array}{r} -6A + 2B = -2 \\ -4A - 2B = 1 \\ \hline -10A = -1 \\ A = \frac{1}{10} \end{array}$$

Back-substitute to find B.

$3\left(\frac{1}{10}\right) - B = 1$ Using equation (1)

$\frac{3}{10} - B = 1$

$-B = \frac{7}{10}$

$B = -\frac{7}{10}$

We have $A = \frac{1}{10}$ and $B = -\frac{7}{10}$.

68. Let x = the number of people ahead of you and y = the number of people behind you.

Solve: $x = 2 + y,$
$x + 1 + y = 3y.$

$x = 5$

69. ***Familiarize.*** Let x = the number of city miles driven and y = the number of highway miles driven. The $x/19$ and $y/28$ represent the number of gallons of gasoline used driving x mi and y mi, respectively.

Translate.

The car was driven a total of 405 mi.

$x + y = 405$

A total of 23 gal of gasoline was used.

$\frac{x}{19} + \frac{y}{28} = 18$

Carry out. We solve the system

$x + y = 405,$
$\frac{x}{19} + \frac{y}{28} = 18.$

Multiply the second equation by 532 to clear fractions.

$x + y = 405,$ (1)
$28x + 19y = 9576$ (2)

Now multiply equation (1) by -19 and add.

$$\begin{array}{r} -19x - 19y = -7695 \\ 28x + 19y = 9576 \\ \hline 9x = 1881 \\ x = 209 \end{array}$$

Back-substitute to find y.

$209 + y = 405$ Using equation (1)
$y = 196$

Check. A total of $209 + 196$, or 405 mi was driven. City driving used $209/19$, or 11 gal and highway driving used $196/28$, or 7 gal. Thus, a total of $11 + 7$, or 18 gal of gasoline was used. The solution checks.

State. 209 mi were driven in the city and 196 mi were driven on the highway.

70. First we convert the given distances to miles:

$300 \text{ ft} = \frac{300}{5280} \text{ mi} = \frac{5}{88} \text{ mi},$

$500 \text{ ft} = \frac{500}{5280} \text{ mi} = \frac{25}{264} \text{ mi}$

Then at 10 mph, Heather can run to point P in $\frac{\frac{5}{88}}{10}$, or $\frac{1}{176}$ hr, and she can run to point Q in $\frac{\frac{25}{264}}{10}$, or $\frac{5}{528}$ hr (using $d = rt$, or $\frac{d}{r} = t$).

Let d = the distance, in miles, from the train to point P in the drawing in the text, and let r the speed of the train, in miles per hour.

	Distance	Speed	Time
Going to P	d	r	$\frac{1}{176}$
Going to Q	$d+\frac{5}{88}+\frac{25}{264}$	r	$\frac{5}{528}$

Solve: $d = r\left(\frac{1}{176}\right)$,

$$d+\frac{5}{88}+\frac{25}{264} = r\left(\frac{5}{528}\right).$$

$r = 40$ mph

Exercise Set 5.2

1. $x + y + z = 2,$ (1)
$6x - 4y + 5z = 31,$ (2)
$5x + 2y + 2z = 13$ (3)

Multiply equation (1) by -6 and add it to equation (2). We also multiply equation (1) by -5 and add it to equation (3).

$x + y + z = 2$ (1)
$-10y - z = 19$ (4)
$-3y - 3z = 3$ (5)

Multiply the last equation by 10 to make the y-coefficient a multiple of the y-coefficient in equation (4).

$x + y + z = 2$ (1)
$-10y - z = 19$ (4)
$-30y - 30z = 30$ (6)

Multiply equation (4) by -3 and add it to equation (6).

$x + y + z = 2$ (1)
$-10y - z = 19$ (4)
$-27z = -27$ (7)

Solve equation (7) for z.

$-27z = -27$
$z = 1$

Back-substitute 1 for z in equation (4) and solve for y.

$-10y - 1 = 19$
$-10y = 20$
$y = -2$

Back-substitute 1 for z for -2 and y in equation (1) and solve for x.

$x + (-2) + 1 = 2$
$x - 1 = 2$
$x = 3$

The solution is $(3, -2, 1)$.

2. $x + 6y + 3z = 4,$ (1)
$2x + y + 2z = 3,$ (2)
$3x - 2y + z = 0$ (3)

Multiply equation (1) by -2 and add it to equation (2). We also multiply equation (1) by -3 and add it to equation (3).

$x + 6y + 3z = 4,$ (1)
$-11y - 4z = -5,$ (4)
$-20y - 8z = -12$ (5)

Multiply equation (5) by 11.

$x + 6y + 3z = 4,$ (1)
$-11y - 4z = -5,$ (4)
$-220y - 88z = -132$ (6)

Multiply equation (4) by -20 and add it to equation (6).

$x + 6y + 3z = 4,$ (1)
$-11y - 4z = -5,$ (4)
$-8z = -32$ (7)

Complete the solution.

$-8z = -32$
$z = 4$
$-11y - 4 \cdot 4 = -5$
$-11y = 11$
$y = -1$
$x + 6(-1) + 3 \cdot 4 = 4$
$x = -2$

The solution is $(-2, -1, 4)$.

3. $x - y + 2z = -3$ (1)
$x + 2y + 3z = 4$ (2)
$2x + y + z = -3$ (3)

Multiply equation (1) by -1 and add it to equation (2). We also multiply equation (1) by -2 and add it to equation (3).

$x - y + 2z = -3$ (1)
$3y + z = 7$ (4)
$3y - 3z = 3$ (5)

Multiply equation (4) by -1 and add it to equation (5).

$x - y + 2z = -3$ (1)
$3y + z = 7$ (4)
$-4z = -4$ (6)

Solve equation (6) for z.

$$-4z = -4$$
$$z = 1$$

Back-substitute 1 for z in equation (4) and solve for y.

$$3y + 1 = 7$$
$$3y = 6$$
$$y = 2$$

Back-substitute 1 for z and 2 for y in equation (1) and solve for x.

$$x - 2 + 2 \cdot 1 = -3$$
$$x = -3$$

The solution is $(-3, 2, 1)$.

4. $x + y + z = 6,$ (1)

$2x - y - z = -3,$ (2)

$x - 2y + 3z = 6$ (3)

Multiply equation (1) by -2 and add it to equation (2). Also, multiply equation (1) by -1 and add it to equation (3).

$x + y + z = 6,$ (1)

$-3y - 3z = -15,$ (4)

$-3y + 2z = 0$ (5)

Multiply equation (4) by -1 and add it to equation (5).

$x + y + z = 6,$ (1)

$-3y - 3z = -15,$ (4)

$5z = 15$ (6)

Complete the solution.

$$5z = 15$$
$$z = 3$$
$$-3y - 3 \cdot 3 = -15$$
$$-3y = -6$$
$$y = 2$$
$$x + 2 + 3 = 6$$
$$x = 1$$

The solution is $(1, 2, 3)$.

5. $x + 2y - z = 5,$ (1)

$2x - 4y + z = 0,$ (2)

$3x + 2y + 2z = 3$ (3)

Multiply equation (1) by -2 and add it to equation (2). Also, multiply equation (1) by -3 and add it to equation (3).

$x + 2y - z = 5,$ (1)

$-8y + 3z = -10,$ (4)

$-4y + 5z = -12$ (5)

Multiply equation (5) by 2 to make the y-coefficient a multiple of the y-coefficient of equation (4).

$x + 2y - z = 5,$ (1)

$-8y + 3z = -10,$ (4)

$-8y + 10z = -24$ (6)

Multiply equation (4) by -1 and add it to equation (6).

$x + 2y - z = 5,$ (1)

$-8y + 3z = -10,$ (4)

$7z = -14$ (7)

Solve equation (7) for z.

$$7z = -14$$
$$z = -2$$

Back-substitute -2 for z in equation (4) and solve for y.

$$-8y + 3(-2) = -10$$
$$-8y - 6 = -10$$
$$-8y = -4$$
$$y = \frac{1}{2}$$

Back-substitute $\frac{1}{2}$ for y and -2 for z in equation (1) and solve for x.

$$x + 2 \cdot \frac{1}{2} - (-2) = 5$$
$$x + 1 + 2 = 5$$
$$x = 2$$

The solution is $\left(2, \frac{1}{2}, -2\right)$.

6. $2x + 3y - z = 1,$ (1)

$x + 2y + 5z = 4,$ (2)

$3x - y - 8z = -7$ (3)

Interchange equations (1) and (2).

$x + 2y + 5z = 4,$ (2)

$2x + 3y - z = 1,$ (1)

$3x - y - 8z = -7$ (3)

Multiply equation (2) by -2 and add it to equation (1). Multiply equation (2) by -3 and add it to equation (3).

$x + 2y + 5z = 4,$ (2)

$-y - 11z = -7,$ (4)

$-7y - 23z = -19$ (5)

Multiply equation (4) by -7 and add it to equation (5).

$x + 2y + 5z = 4,$ (2)

$-y - 11z = -7$ (4)

$54z = 30$

Complete the solution.

$$54z = 30$$
$$z = \frac{5}{9}$$

$$-y - 11 \cdot \frac{5}{9} = -7$$
$$-y = -\frac{8}{9}$$
$$y = \frac{8}{9}$$
$$x + 2 \cdot \frac{8}{9} + 5 \cdot \frac{5}{9} = 4$$
$$x = -\frac{5}{9}$$

The solution is $\left(-\frac{5}{9}, \frac{8}{9}, \frac{5}{9}\right)$.

7. $x + 2y - z = -8,$ (1)
$2x - y + z = 4,$ (2)
$8x + y + z = 2$ (3)

Multiply equation (1) by -2 and add it to equation (2). Also, multiply equation (1) by -8 and add it to equation (3).

$x + 2y - z = -8,$ (1)
$-5y + 3z = 20,$ (4)
$-15y + 9z = 66$ (5)

Multiply equation (4) by -3 and add it to equation (5).

$x + 2y - z = -8,$ (1)
$-5y + 3z = 20,$ (4)
$0 = 6$ (6)

Equation (6) is false, so the system of equations has no solution.

8. $x + 2y - z = 4,$ (1)
$4x - 3y + z = 8,$ (2)
$5x - y = 12$ (3)

Multiply equation (1) by -4 and add it to equation (2). Also, multiply equation (1) by -5 and add it to equation (3).

$x + 2y - z = 4,$ (1)
$-11y + 5z = -8,$ (4)
$-11y + 5z = -8$ (5)

Multiply equation (4) by -1 and add it to equation (5).

$x + 2y - z = 4,$ (1)
$-11y + 5z = -8,$ (4)
$0 = 0$ (6)

The equation $0 = 0$ tells us that equation (3) of the original system is dependent on the first two equations. The system of equations has infinitely many solutions and is equivalent to

$x + 2y - z = 4,$ (1)
$4x - 3y + z = 8.$ (2)

To find an expression for the solutions, we first solve equation (4) for either y or z. We choose to solve for y.

$$-11y + 5z = -8$$
$$-11y = -5z - 8$$
$$y = \frac{5z + 8}{11}$$

Back-substitute in equation (1) to find an expression for x in terms of z.

$$x + 2\left(\frac{5z + 8}{11}\right) - z = 4$$
$$x + \frac{10z}{11} + \frac{16}{11} - z = 4$$
$$x = \frac{z}{11} + \frac{28}{11} = \frac{z + 28}{11}$$

The solutions are given by $\left(\frac{z + 28}{11}, \frac{5z + 8}{11}, z\right)$, where z is any real number.

9. $2x + y - 3z = 1,$ (1)
$x - 4y + z = 6,$ (2)
$4x - 7y - z = 13$ (3)

Interchange equations (1) and (2).

$x - 4y + z = 6,$ (2)
$2x + y - 3z = 1,$ (1)
$4x - 7y - z = 13$ (3)

Multiply equation (2) by -2 and add it to equation (1). Also, multiply equation (2) by -4 and add it to equation (3).

$x - 4y + z = 6,$ (2)
$9y - 5z = -11,$ (4)
$9y - 5z = -11$ (5)

Multiply equation (4) by -1 and add it to equation (5).

$x - 4y + z = 6,$ (1)
$9y - 5z = -11,$ (4)
$0 = 0$ (6)

The equation $0 = 0$ tells us that equation (3) of the original system is dependent on the first two equations. The system of equations has infinitely many solutions and is equivalent to

$2x + y - 3z = 1,$ (1)
$x - 4y + z = 6.$ (2)

To find an expression for the solutions, we first solve equation (4) for either y or z. We choose to solve for z.

$$9y - 5z = -11$$
$$-5z = -9y - 11$$
$$z = \frac{9y + 11}{5}$$

Back-substitute in equation (2) to find an expression for x in terms of y.

$$x - 4y + \frac{9y+11}{5} = 6$$
$$x - 4y + \frac{9}{5}y + \frac{11}{5} = 6$$
$$x = \frac{11}{5}y + \frac{19}{5} = \frac{11y+19}{5}$$

The solutions are given by $\left(\frac{11y+19}{5}, y, \frac{9y+11}{5}\right)$, where y is any real number.

10. $x + 3y + 4z = 1, \quad (1)$
$3x + 4y + 5z = 3, \quad (2)$
$x + 8y + 11z = 2 \quad (3)$

Multiply equation (1) by -3 and add it to equation (2). Also, multiply equation (1) by -1 and add it to equation (3).

$x + 3y + 4z = 1, \quad (1)$
$-5y - 7z = 0, \quad (4)$
$5y + 7z = 1 \quad (5)$

Add equation (4) to equation (5).

$x + 3y + 4z = 1, \quad (1)$
$-5y - 7z = 0, \quad (4)$
$0 = 1 \quad (6)$

Equation (6) is false, so the system of equations has no solution.

11. $4a + 9b = 8, \quad (1)$
$8a + 6c = -1, \quad (2)$
$6b + 6c = -1 \quad (3)$

Multiply equation (1) by -2 and add it to equation (2).

$4a + 9b = 8, \quad (1)$
$-18b + 6c = -17, \quad (4)$
$6b + 6c = -1 \quad (3)$

Multiply equation (3) by 3 to make the b-coefficient a multiple of the b-coefficient in equation (4).

$4a + 9b = 8, \quad (1)$
$-18b + 6c = -17, \quad (4)$
$18b + 18c = -3 \quad (5)$

Add equation (4) to equation (5).

$4a + 9b = 8, \quad (1)$
$-18b + 6c = -17, \quad (4)$
$24c = -20 \quad (6)$

Solve equation (6) for c.

$$24c = -20$$
$$c = -\frac{20}{24} = -\frac{5}{6}$$

Back-substitute $-\frac{5}{6}$ for c in equation (4) and solve for b.

$$-18b + 6c = -17$$
$$-18b + 6\left(-\frac{5}{6}\right) = -17$$
$$-18b - 5 = -17$$
$$-18b = -12$$
$$b = \frac{12}{18} = \frac{2}{3}$$

Back-substitute $\frac{2}{3}$ for b in equation (1) and solve for a.

$$4a + 9b = 8$$
$$4a + 9 \cdot \frac{2}{3} = 8$$
$$4a + 6 = 8$$
$$4a = 2$$
$$a = \frac{1}{2}$$

The solution is $\left(\frac{1}{2}, \frac{2}{3}, -\frac{5}{6}\right)$.

12. $3p + 2r = 11, \quad (1)$
$q - 7r = 4, \quad (2)$
$p - 6q = 1 \quad (3)$

Interchange equations (1) and (3).

$p - 6q = 1, \quad (3)$
$q - 7r = 4, \quad (2)$
$3p + 2r = 11 \quad (1)$

Multiply equation (3) by -3 and add it to equation (1).

$p - 6q = 1, \quad (3)$
$q - 7r = 4, \quad (2)$
$18q + 2r = 8 \quad (4)$

Multiply equation (2) by -18 and add it to equation (4).

$p - 6q = 1, \quad (3)$
$q - 7r = 4, \quad (2)$
$128r = -64 \quad (5)$

Complete the solution.

$$128r = -64$$
$$r = -\frac{1}{2}$$
$$q - 7\left(-\frac{1}{2}\right) = 4$$
$$q = \frac{1}{2}$$
$$p - 6 \cdot \frac{1}{2} = 1$$
$$p = 4$$

The solution is $\left(4, \frac{1}{2}, -\frac{1}{2}\right)$.

13.
$$\begin{aligned} w + x + y + z &= 2 \quad (1)\\ w + 2x + 2y + 4z &= 1 \quad (2)\\ -w + x - y - z &= -6 \quad (3)\\ -w + 3x + y - z &= -2 \quad (4) \end{aligned}$$

Multiply equation (1) by -1 and add to equation (2). Add equation (1) to equation (3) and to equation (4).

$$\begin{aligned} w + x + y + z &= 2 \quad (1)\\ x + y + 3z &= -1 \quad (5)\\ 2x &= -4 \quad (6)\\ 4x + 2y &= 0 \quad (7) \end{aligned}$$

Solve equation (6) for x.

$$2x = -4$$
$$x = -2$$

Back-substitute -2 for x in equation (7) and solve for y.

$$4(-2) + 2y = 0$$
$$-8 + 2y = 0$$
$$2y = 8$$
$$y = 4$$

Back-substitute -2 for x and 4 for y in equation (5) and solve for z.

$$-2 + 4 + 3z = -1$$
$$3z = -3$$
$$z = -1$$

Back-substitute -2 for x, 4 for y, and -1 for z in equation (1) and solve for w.

$$w - 2 + 4 - 1 = 2$$
$$w = 1$$

The solution is $(1, -2, 4, -1)$.

14.
$$\begin{aligned} w + x - y + z &= 0, \quad (1)\\ -w + 2x + 2y + z &= 5, \quad (2)\\ -w + 3x + y - z &= -4, \quad (3)\\ -2w + x + y - 3z &= -7 \quad (4) \end{aligned}$$

Add equation (1) to equation (2) and equation (3). Also, multiply equation (1) by 2 and add it to equation (4).

$$\begin{aligned} w + x - y + z &= 0, \quad (1)\\ 3x + y + 2z &= 5, \quad (5)\\ 4x &= -4, \quad (6)\\ 3x - y - z &= -7 \quad (7) \end{aligned}$$

Multiply equation (6) by 3.

$$\begin{aligned} w + x - y + z &= 0, \quad (1)\\ 3x + y + 2z &= 5, \quad (5)\\ 12x &= -12, \quad (8)\\ 3x - y - z &= -7 \quad (7) \end{aligned}$$

Multiply equation (5) by -4 and add it to equation (8). Also, multiply equation (5) by -1 and add it to equation (7).

$$\begin{aligned} w + x - y + z &= 0, \quad (1)\\ 3x + y + 2z &= 5, \quad (5)\\ -4y - 8z &= -32, \quad (9)\\ -2y - 3z &= -12 \quad (10) \end{aligned}$$

Multiply equation (10) by 2.

$$\begin{aligned} w + x - y + z &= 0, \quad (1)\\ 3x + y + 2z &= 5, \quad (5)\\ -4y - 8z &= -32, \quad (9)\\ -4y - 6z &= -24 \quad (11) \end{aligned}$$

Multiply equation (9) by -1 and add it to equation (11).

$$\begin{aligned} w + x - y + z &= 0, \quad (1)\\ 3x + y + 2z &= 5, \quad (5)\\ -4y - 8z &= -32, \quad (9)\\ 2z &= 8 \quad (12) \end{aligned}$$

Complete the solution.

$$2z = 8$$
$$z = 4$$
$$-4y - 8 \cdot 4 = -32$$
$$-4y = 0$$
$$y = 0$$
$$3x + 0 + 2 \cdot 4 = 5$$
$$3x = -3$$
$$x = -1$$
$$w - 1 - 0 + 4 = 0$$
$$w = -3$$

The solution is $(-3, -1, 0, 4)$.

15. ***Familiarize***. Let x = the number of orders under 10 lb, y = the number of orders from 10 lb up to 15 lb, and z = the number of orders of 15 lb or more. Then the total shipping charges for each category of order are $\$3x$, $\$5y$, and $\$7.50z$.

Translate.

The total number of orders was 150.

$$x + y + z = 150$$

Total shipping charges were \$680.

$$3x + 5y + 7.5z = 680$$

The number of orders under 10 lb was three times the number of orders weighing 15 lb or more.

$$x = 3z$$

We have a system of equations

$$\begin{aligned} x + y + z &= 150, & x + y + z &= 150,\\ 3x + 5y + 7.5z &= 680, \quad \text{or} & 30x + 50y + 75z &= 6800,\\ x &= 3z & x - 3z &= 0 \end{aligned}$$

Carry out. Solving the system of equations, we get $(60, 70, 20)$.

Check. The total number of orders is $60 + 70 + 20$, or 150. The total shipping charges are $\$3 \cdot 60 +$

$\$5 \cdot 70 + \$7.50(20)$, or \$680. The number of orders under 10 lb, 60, is three times the number of orders weighing over 15 lb, 20. The solution checks.

State. There were 60 packages under 10 lb, 70 packages from 10 lb up to 15 lb, and 20 packages weighing 15 lb or more.

16. Let x, y, and z represent the number of orders of \$30 or less, from \$30 to \$70, and over \$70, respectively.

Solve: $x + y + z = 600,$

$4x + 6y + 7z = 3340,$

$x = z + 80.$

$x = 180$, $y = 320$, $z = 100$

17. ***Familiarize.*** Let x, y, and z represent the number of servings of ground beef, baked potato, and strawberries required, respectively. One serving of ground beef contains $245x$ Cal, $0x$ or 0 g of carbohydrates, and $9x$ mg of calcium. One baked potato contains $145y$ Cal, $34y$ g of carbohydrates, and $8y$ mg of calcium. One serving of strawberries contains $45z$ Cal, $10z$ g of carbohydrates, and $21z$ mg of calcium.

Translate.

The total number of calories is 485.

$$245x + 145y + 45z = 485$$

A total of 41.5 g of carbohydrates is required.

$$34y + 10z = 41.5$$

A total of 35 mg of calcium is required.

$$9x + 8y + 21z = 35$$

We have a system of equations.

$$245x + 145y + 45z = 485,$$
$$34y + 10z = 41.5,$$
$$9x + 8y + 21z = 35$$

Carry out. Solving the system of equations, we get $(1.25, 1, 0.75)$.

Check. 1.25 servings of ground beef contains 306.25 Cal, no carbohydrates, and 11.25 mg of calcium; 1 baked potato contains 145 Cal, 34 g of carbohydrates, and 8 mg of calcium; 0.75 servings of strawberries contains 33.75 Cal, 7.5 g of carbohydrates, and 15.75 mg of calcium. Thus, there are a total of $306.25 + 145 + 33.75$, or 485 Cal, $34 + 7.5$, or 41.5 g of carbohydrates, and $11.25 + 8 + 15.75$, or 35 mg of calcium. The solution checks.

State. 1.25 servings of ground beef, 1 baked potato, and 0.75 serving of strawberries are required.

18. Let x, y, and z represent the number of servings of chicken, mashed potatoes, and peas to be used, respectively.

Solve: $140x + 160y + 125z = 415,$

$27x + 4y + 8z = 50.5,$

$64x + 636y + 139z = 553.$

$x = 1.5$, $y = 0.5$, $z = 1$

19. ***Familiarize.*** Let x, y, and z represent the amounts invested at 4%, 6%, and 7%, respectively. Then the annual interest from the investments is $4\%x$, $6\%y$, and $7\%z$, or $0.04x$, $0.06y$, and $0.07z$.

Translate.

A total of \$5000 was invested.

$$x + y + z = 5000$$

The total interest is \$302.

$$0.04x + 0.06y + 0.07z = 302$$

The amount invested at 7% is \$1500 more than the amount invested at 4%.

$$z = x + 1500$$

We have a system of equations.

$$x + y + z = 5000,$$
$$0.04x + 0.06y + 0.07z = 302,$$
$$z = x + 1500$$

or

$$x + y + z = 5000,$$
$$4x + 6y + 7z = 30,200,$$
$$-x + z = 1500$$

Carry out. Solving the system of equations, we get $(1300, 900, 2800)$.

Check. The total investment was $\$1300 + \$900 + \$2800$, or \$5000. The total interest was $0.04(\$1300) + 0.06(\$900) + 0.07(\$2800) = \$52 + \$54 + \$196$, or \$302. The amount invested at 7%, \$2800, is \$1500 more than the amount invested at 4%, \$1300. The solution checks.

State. \$1300 was invested at 4%, \$900 at 6%, and \$2800 at 7%.

20. Let x, y, and z represent the amounts invested at 8%, 9%, and 10%, respectively.

Solve: $0.08x + 0.09y + 0.1z = 336,$

$y = x + 500,$

$z = 3y.$

$x = \$300$, $y = \$800$, $z = \$2400$

21. ***Familiarize.*** Let x, y, and z represent the prices of orange juice, a raisin bagel, and a cup of coffee, respectively. The new price for orange juice is $x + 50\%x$, or $x + 0.5x$, or $1.5x$; the new price of a bagel is $y + 20\%y$, or $y + 0.2y$, or $1.2y$.

Translate.

Orange juice, a raisin bagel, and a cup of coffee cost \$3.

$$x + y + x = 3$$

After the price increase, orange juice, a raisin bagel, and a cup of coffee will cost \$3.75.

$$1.5x + 1.2y + z = 3.75$$

After the price increases, orange juice will cost twice as much as coffee.

$$1.5x = 2z$$

We have a system of equations.

$$x+y+z=3, \qquad x+y+z=3,$$
$$1.5x+1.2y+z=3.75, \text{ or } 150x+120y+100z=375,$$
$$1.5x=2z \qquad 15x-20z=0$$

Carry out. Solving the system of equations, we get $(1, 1.25, 0.75)$.

Check. If orange juice costs \$1, a bagel costs \$1.25, and a cup of coffee costs \$0.75, then together they cost \$1+\$1.25+\$0.75, or \$3. After the price increases orange juice will cost 1.5(\$1), or \$1.50 and a bagel will cost 1.2(\$1.25) or \$1.50. Then orange juice, a bagel, and coffee will cost \$1.50 + \$1.50 + \$0.75, or \$3.75. After the price increase the price of orange juice, \$1.50, will be twice the price of coffee, \$0.75. The solution checks.

State. Before the increase orange juice costs \$1, a raisin bagel cost \$1.25, and a cup of coffee cost \$0.75.

22. Let x, y, and z represent the prices of a carton of milk, a donut, and a cup of coffee, respectively.

Solve: $x+2y+z=1.85,$

$3y+2z=2.30,$

$x+y+2z=1.75.$

$x=\$0.45$, $y=\$0.50$, $z=\$0.40$

Then 2 cartons of milk and 2 donuts will cost 2(\$0.45) + 2(\$0.50), or \$1.90. They will not have enough money.

23. ***Familiarize.*** Let x, y, and z represent the volume of passenger traffic on domestic airways, by bus, and by railroads, respectively, in billions of passenger-miles.

Translate.

The total volume of passenger traffic was 405 billion passenger-miles.

$x+y+z=405$

The volume of bus traffic was 10 billion passenger-miles more than the volume of railroad traffic.

$y=z+10$

The volume of railroad traffic was 329 billion passenger-miles less than the volume of traffic on domestic airways.

$z=x-329$

We have a system of equations.

$$x+y+z=405, \qquad x+y+z=405,$$
$$y=z+10, \quad \text{or} \quad y-z=10,$$
$$z=x-329 \qquad -x+z=-329$$

Carry out. Solving the system of equations, we get $(351, 32, 22)$.

Check. The total volume is 351 + 32 + 22, or 405 billion passenger-miles. The volume of bus traffic, 32 billion passenger-miles, is 10 billion passenger-miles more than the volume of railroad traffic, 22 billion passenger-miles. The volume of railroad traffic, 22 billion passenger-miles, is 329 billion passenger-miles less than the volume of traffic on domestic airways, 351 billion passenger-miles. The solution checks.

State. The volume of traffic on domestic airways was 351 billion passenger-miles, by bus was 32 billion passenger-miles, and by railroad was 22 billion passenger-miles.

24. Let x, y, and z represent the per capita consumption of cheddar, mozzarella, and Swiss cheese, in pounds.

Solve: $x+y+z=18.1,$

$y+z=x-0.3,$

$y=z+6.5.$

$x=9.2$ lb, $y=7.7$ lb, $z=1.2$ lb

25. ***Familiarize.*** Let x, y, and z represent the number of par-3, par-4, and par-5 holes, respectively. A golfer who shoots par on every hole has a score of $3x$ from the par-3 holes, $4y$ from the par-4 holes, and $5z$ from the par-5 holes.

Translate.

The total number of holes is 18.

$x+y+z=18$

A golfer who shoots par on every hole has a score of 72.

$3x+4y+5z=72$

The sum of the number of par-3 holes and the number of par-5 holes is 8.

$x+z=8$

We have a system of equations.

$$x+y+z=18,$$
$$3x+4y+5z=72,$$
$$x+z=8$$

Carry out. We solve the system. The solution is $(4, 10, 4)$.

Check. The total number of holes is 4 + 10 + 4, or 18. A golfer who shoots par on every hole has a score of $3\cdot4+4\cdot10+5\cdot4$, or 72. The sum of the number of par-3 holes and the number of par-5 holes is 4 + 4, or 8. The solution checks.

State. There are 4 par-3 holes, 10 par-4 holes, and 4 par-5 holes.

26. Let x, y, and z represent the number of par-3, par-4, and par-5 holes, respectively.

Solve: $x+y+z=18,$

$3x+4y+5z=70,$

$y=2z.$

$x=6$, $y=8$, $z=4$

27. a) Substitute the data points $(0, 211)$, $(10, 237)$, and $(16, 203)$ in the function $f(x) = ax^2 + bx + c$.

$211 = a \cdot 0^2 + b \cdot 0 + c$

$237 = a \cdot 10^2 + b \cdot 10 + c$

$203 = a \cdot 16^2 + b \cdot 16 + c$

We have a system of equations.

$$c = 211,$$
$$100a + 10b + c = 237,$$
$$256a + 16b + c = 203$$

Solving the system of equations, we get

$\left(-\frac{31}{60}, \frac{233}{30}, 211\right)$, so

$f(x) = -\frac{31}{60}x^2 + \frac{233}{30}x + 211.$

b) In 2005, $x = 2005 - 1980$, or 25.

$f(25) = -\frac{31}{60}(25)^2 + \frac{233}{30}(25) + 211 \approx 82$

There will be about 82 thousand marriages in California in 2005.

28. a) Solve: $26 = a \cdot 0^2 + b \cdot 0 + c,$

$21 = a \cdot 2^2 + b \cdot 2 + c,$

$23 = a \cdot 4^2 + b \cdot 4 + c,$

or

$$c = 26,$$
$$4a + 2b + c = 21,$$
$$16a + 4b + c = 23.$$

$a = \frac{7}{8}$, $b = -\frac{17}{4}$, $c = 26$, so

$f(x) = \frac{7}{8}x^2 - \frac{17}{4}x + 26.$

b) In 2003, $x = 2003 - 1992$, or 11.

$f(11) = \frac{7}{8} \cdot 11^2 - \frac{17}{4} \cdot 11 + 26 \approx 85.$

The per capita consumption of coffee in 2003 will be about 85 gal.

29. a) Substitute the data points $(0, 594)$, $(5, 567)$, and $(11, 576)$ in the function $f(x) = ax^2 + bx + c$.

$594 = a \cdot 0^2 + b \cdot 0 + c$

$567 = a \cdot 5^2 + b \cdot 5 + c$

$576 = a \cdot 11^2 + b \cdot 11 + c$

We have a system of equations.

$$c = 594,$$
$$25a + 5b + c = 567,$$
$$121a + 11b + c = 576$$

Solving the system of equations, we get

$\left(\frac{69}{110}, -\frac{939}{110}, 594\right)$, so

$f(x) = \frac{69}{110}x^2 - \frac{939}{110}x + 594.$

b) In 2005, $x = 2005 - 1985$, or 20.

$f(20) = \frac{69}{110}(20)^2 - \frac{939}{110}(20) + 594 \approx 674.$

The per capita milk consumption in 2005 will be about 674 lb.

30. a) Solve: $771 = a \cdot 0^2 + b \cdot 0 + c,$

$755 = a \cdot 5^2 + b \cdot 5 + c,$

$773 = a \cdot 7^2 + b \cdot 7 + c,$

or

$$c = 771,$$
$$25a + 5b + c = 775,$$
$$49a + 7b + c = 773.$$

$a = \frac{61}{35}$, $b = -\frac{417}{35}$, $c = 771$, so

$f(x) = \frac{61}{35}x^2 - \frac{417}{35}x + 771.$

b) In 2007, $x = 2007 - 1990$, or 17.

$f(17) = \frac{61}{35}(17)^2 - \frac{417}{35}(17) + 771 \approx 1072$ million metric tons.

31. Add a non-zero multiple of one equation to a non-zero multiple of the other equation, where the multiples are not opposites.

32. Answers will vary.

33. $(3 - 4i) - (-2 - i) = 3 - 4i + 2 + i =$

$(3 + 2) + (-4 + 1)i = 5 - 3i$

34. $(5 + 2i) + (1 - 4i) = 6 - 2i$

35. $(1 - 2i)(6 + 2i) = 6 + 2i - 12i - 4i^2 =$

$6 + 2i - 12i + 4 = 10 - 10i$

36. $\frac{3 + i}{4 - 3i} = \frac{3 + i}{4 - 3i} \cdot \frac{4 + 3i}{4 + 3i} = \frac{12 + 9i + 4i + 3i^2}{16 - 9i^2} =$

$\frac{12 + 9i + 4i - 3}{16 + 9} = \frac{9 + 13i}{25} = \frac{9}{25} + \frac{13}{25}i$

37. $\frac{2}{x} - \frac{1}{y} - \frac{3}{z} = -1,$

$\frac{2}{x} - \frac{1}{y} + \frac{1}{z} = -9,$

$\frac{1}{x} + \frac{2}{y} - \frac{4}{z} = 17$

First substitute u for $\frac{1}{x}$, v for $\frac{1}{y}$, and w for $\frac{1}{z}$ and solve for u, v, and w.

$$2u - v - 3w = -1,$$
$$2u - v + w = -9,$$
$$u + 2v - 4w = 17$$

Solving this system we get $(-1, 5, -2)$.

If $u = -1$, and $u = \frac{1}{x}$, then $-1 = \frac{1}{x}$, or $x = -1$.

If $v = 5$ and $v = \frac{1}{y}$, then $5 = \frac{1}{y}$, or $y = \frac{1}{5}$.

If $w = -2$ and $w = \frac{1}{z}$, then $-2 = \frac{1}{z}$, or $z = -\frac{1}{2}$.

The solution of the original system is $\left(-1, \frac{1}{5} - \frac{1}{2}\right)$.

38. $\frac{2}{x} + \frac{2}{y} - \frac{3}{z} = 3,$

$\frac{1}{x} - \frac{2}{y} - \frac{3}{z} = 9,$

$\frac{7}{x} - \frac{2}{y} + \frac{9}{z} = -39$

Substitute u for $\frac{1}{x}$, v for $\frac{1}{y}$, and w for $\frac{1}{z}$ and solve for u, v, and w.

$2u + 2v - 3w = 3,$

$u - 2v - 3w = 9,$

$7u - 2v + 9w = -39$

Solving this system we get $(-2, -1, -3)$.

Then $\frac{1}{x} = -2$, or $x = -\frac{1}{2}$; $\frac{1}{y} = -1$, or $y = -1$; and $\frac{1}{z} = -3$, or $z = -\frac{1}{3}$. The solution of the original system is

$\left(-\frac{1}{2}, -1, -\frac{1}{3}\right)$.

39. Label the angle measures at the tips of the stars a, b, c, d, and e. Also label the angles of the pentagon p, q, r, s, and t.

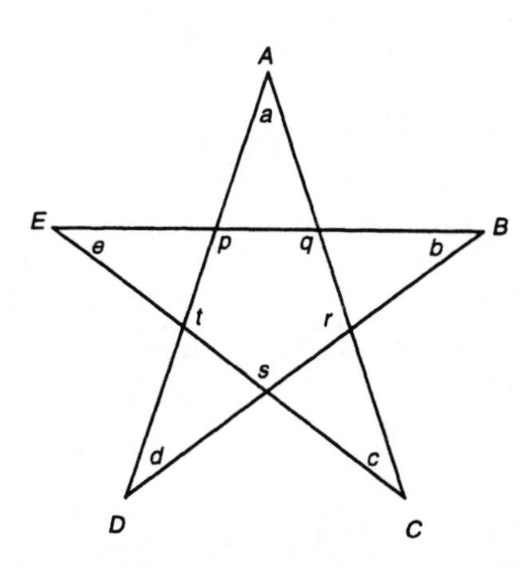

Using the geometric fact that the sum of the angle measures of a triangle is 180°, we get 5 equations.

$p + b + d = 180$

$q + c + e = 180$

$r + a + d = 180$

$s + b + e = 180$

$t + a + c = 180$

Adding these equations, we get

$(p + q + r + s + t) + 2a + 2b + 2c + 2d + 2e = 5(180).$

The sum of the angle measures of any convex polygon with n sides is given by the formula $S = (n - 2)180$.

Thus $p + q + r + s + t = (5 - 2)180$, or 540. We substitute and solve for $a + b + c + d + e$.

$$540 + 2(a + b + c + d + e) = 900$$
$$2(a + b + c + d + e) = 360$$
$$a + b + c + d + e = 180$$

The sum of the angle measures at the tips of the star is 180°.

40. Let h, t, and u represent the hundred's, ten's, and unit's digit of the year, respectively. (We know the thousand's digit is 1.) Using the given information we know the following:

$1 + h + t + u = 24,$

$u = 1 + h,$

$t = k \cdot 3$ and $u = m \cdot 3$

where k, m are positive integers.

We know $h > 5$ (there was no transcontinental railroad before 1600); and, since

$u = 1 + h = m \cdot 3$, $h \neq 6$, $h \neq 7$, $h \neq 9$.

Thus $h = 8$ and $u = 9$. Then $1 + 8 + t + 9 = 24$, or $t = 6$.

The year is 1869.

41. Substituting, we get

$A + \frac{3}{4}B + 3C = 12,$

$\frac{4}{3}A + B + 2C = 12,$

$2A + B + C = 12,$ or

$4A + 3B + 12C = 48,$

$4A + 3B + 6C = 36,$ Clearing fractions

$2A + B + C = 12.$

Solving the system of equations, we get $(3, 4, 2)$. The equation is $3x + 4y + 2z = 12$.

42. Solve: $1 = B - M - 2N,$

$2 = B - 3M + 6N,$

$1 = B - \frac{3}{2}M - N$

$B = 2$, $M = \frac{1}{2}$, $N = \frac{1}{4}$, so $y = 2 - \frac{1}{2}x - \frac{1}{4}z$.

43. Substituting, we get

$59 = a(-2)^3 + b(-2)^2 + c(-2) + d,$

$13 = a(-1)^3 + b(-1)^2 + c(-1) + d,$

$-1 = a \cdot 1^3 + b \cdot 1^2 + c \cdot 1 + d,$

$-17 = a \cdot 2^3 + b \cdot 2^2 + c \cdot 2 + d,$ or

$-8a + 4b - 2c + d = 59,$

$-a + b - c + d = 13,$

$a + b + c + d = -1,$

$8a + 4b + 2c + d = -17.$

Solving the system of equations, we get

$(-4, 5, -3, 1)$, so $y = -4x^3 + 5x^2 - 3x + 1$.

44. Solve: $-39 = a(-2)^3 + b(-2)^2 + c(-2) + d,$

$-12 = a(-1)^3 + b(-1)^2 + c(-1) + d,$

$-6 = a \cdot 1^3 + b \cdot 1^2 + c \cdot 1 + d,$

$16 = a \cdot 3^3 + b \cdot 3^2 + c \cdot 3 + d,$ or

$-8a + 4b - 2c + d = -39,$

$-a + b - c + d = -12,$

$a + b + c + d = -6,$

$27a + 9b + 3c + d = 16.$

$a = 2, b = -4, c = 1, d = -5$, so $y = 2x^3 - 4x^2 + x - 5$.

45. ***Familiarize and Translate***. Let a, s, and c represent the number of adults, students, and children in attendance, respectively.

The total attendance was 100.

$a + s + c = 100$

The total amount of money taken in was \$100.

(Express 50 cents as $\frac{1}{2}$ dollar.)

$10a + 3s + \frac{1}{2}c = 100$

The resulting system is

$a + s + c = 100,$

$10a + 3s + \frac{1}{2}c = 100.$

Carry out. Multiply the first equation by -3 and add it to the second equation to obtain $7a - \frac{5}{2}c = -200$ or $a = \frac{5}{14}(c - 80)$ where $(c-80)$ is a positive multiple of 14 (because a must be a positive integer). That is $(c - 80) = k \cdot 14$ or $c = 80 + k \cdot 14$, where k is a positive integer. If $k > 1$, then $c > 100$. This is impossible since the total attendance is 100. Thus $k = 1$, so $c = 80 + 1 \cdot 14 = 94$. Then $a = \frac{5}{14}(94 - 80) = \frac{5}{14} \cdot 14 = 5$, and $5 + s + 94 = 10$, or $s = 1$.

Check. The total attendance is $5 + 1 + 94$, or 100.

The total amount of money taken in was $\$10 \cdot 5 + \$3 \cdot 1 + \$\frac{1}{2} \cdot 94 = \$100$. The result checks.

State. There were 5 adults, 1 student, and 94 children in attendance.

Exercise Set 5.3

1. The matrix has 3 rows and 2 columns, so its order is 3×2.

2. The matrix has 4 rows and 1 column, so its order is 4×1.

3. The matrix has 1 row and 4 columns, so its order is 1×4.

4. The matrix has 1 row and 1 column, so its order is 1×1.

5. The matrix has 3 rows and 3 columns, so its order is 3×3.

6. The matrix has 2 rows and 4 columns, so its order is 2×4.

7. We omit the variables and replace the equals signs with a vertical line.

$$\left[\begin{array}{cc|c} 2 & -1 & 7 \\ 1 & 4 & -5 \end{array}\right]$$

8. $$\left[\begin{array}{cc|c} 3 & 2 & 8 \\ 2 & -3 & 15 \end{array}\right]$$

9. We omit the variables, writing zeros for the missing terms, and replace the equals signs with a vertical line.

$$\left[\begin{array}{ccc|c} 1 & -2 & 3 & 12 \\ 2 & 0 & -4 & 8 \\ 0 & 3 & 1 & 7 \end{array}\right]$$

10. $$\left[\begin{array}{ccc|c} 1 & 1 & -1 & 7 \\ 0 & 3 & 2 & 1 \\ -2 & -5 & 0 & 6 \end{array}\right]$$

11. Insert variables and replace the vertical line with equals signs.

$3x - 5y = 1,$

$x + 4y = -2$

12. $x + 2y = -6,$

$4x + y = -3$

13. Insert variables and replace the vertical line with equals signs.

$2x + y - 4z = 12,$

$3x \quad + 5z = -1,$

$x - y + z = 2$

14. $-x - 2y + 3z = 6,$

$4y + z = 2,$

$2x - y \quad = 9$

15. $4x + 2y = 11,$

$3x - y = 2$

Write the augmented matrix. We will use Gaussian elimination.

$$\left[\begin{array}{cc|c} 4 & 2 & 11 \\ 3 & -1 & 2 \end{array}\right]$$

Multiply row 2 by 4 to make the first number in row 2 a multiple of 4.

$$\left[\begin{array}{cc|c} 4 & 2 & 11 \\ 12 & -4 & 8 \end{array}\right]$$

Multiply row 1 by -3 and add it to row 2.

$$\left[\begin{array}{cc|c} 4 & 2 & 11 \\ 0 & -10 & -25 \end{array}\right]$$

Multiply row 1 by $\frac{1}{4}$ and row 2 by $-\frac{1}{10}$.

$$\left[\begin{array}{cc|c} 1 & \frac{1}{2} & \frac{11}{4} \\ 0 & 1 & \frac{5}{2} \end{array}\right]$$

Write the system of equations that corresponds to the last matrix.

$$x + \frac{1}{2}y = \frac{11}{4}, \quad (1)$$

$$y = \frac{5}{2} \quad (2)$$

Back-substitute in equation (1) and solve for x.

$$x + \frac{1}{2} \cdot \frac{5}{2} = \frac{11}{4}$$

$$x + \frac{5}{4} = \frac{11}{4}$$

$$x = \frac{6}{4} = \frac{3}{2}$$

The solution is $\left(\frac{3}{2}, \frac{5}{2}\right)$.

16. $2x + y = 1,$

$3x + 2y = -2$

Write the augmented matrix. We will use Gauss-Jordan elimination.

$$\left[\begin{array}{cc|c} 2 & 1 & 1 \\ 3 & 2 & -2 \end{array}\right]$$

Multiply row 2 by 2.

$$\left[\begin{array}{cc|c} 2 & 1 & 1 \\ 6 & 4 & -4 \end{array}\right]$$

Multiply row 1 by -3 and add it to row 2.

$$\left[\begin{array}{cc|c} 2 & 1 & 1 \\ 0 & 1 & -7 \end{array}\right]$$

Multiply row 2 by -1 and add it to row 1.

$$\left[\begin{array}{cc|c} 2 & 0 & 8 \\ 0 & 1 & -7 \end{array}\right]$$

Multiply row 1 by $\frac{1}{2}$.

$$\left[\begin{array}{cc|c} 1 & 0 & 4 \\ 0 & 1 & -7 \end{array}\right]$$

The solution is $(4, -7)$.

17. $5x - 2y = -3,$

$2x + 5y = -24$

Write the augmented matrix. We will use Gaussian elimination.

$$\left[\begin{array}{cc|c} 5 & -2 & -3 \\ 2 & 5 & -24 \end{array}\right]$$

Multiply row 2 by 5 to make the first number in row 2 a multiple of 5.

$$\left[\begin{array}{cc|c} 5 & -2 & -3 \\ 10 & 25 & -120 \end{array}\right]$$

Multiply row 1 by -2 and add it to row 2.

$$\left[\begin{array}{cc|c} 5 & -2 & -3 \\ 0 & 29 & -114 \end{array}\right]$$

Multiply row 1 by $\frac{1}{5}$ and row 2 by $\frac{1}{29}$.

$$\left[\begin{array}{cc|c} 1 & -\frac{2}{5} & -\frac{3}{5} \\ 0 & 1 & -\frac{114}{29} \end{array}\right]$$

Write the system of equations that corresponds to the last matrix.

$$x - \frac{2}{5}y = -\frac{3}{5}, \quad (1)$$

$$y = -\frac{114}{29} \quad (2)$$

Back-substitute in equation (1) and solve for x.

$$x - \frac{2}{5}\left(-\frac{114}{29}\right) = -\frac{3}{5}$$

$$x + \frac{228}{145} = -\frac{3}{5}$$

$$x = -\frac{315}{145} = -\frac{63}{29}$$

The solution is $\left(-\frac{63}{29}, -\frac{114}{29}\right)$.

18. $2x + y = 1,$

$3x - 6y = 4$

Write the augmented matrix. We will use Gaussian elimination.

$$\left[\begin{array}{cc|c} 2 & 1 & 1 \\ 3 & -6 & 4 \end{array}\right]$$

Multiply row 2 by 2.

$$\left[\begin{array}{cc|c} 2 & 1 & 1 \\ 6 & -12 & 8 \end{array}\right]$$

Multiply row 1 by -3 and add it to row 2.

$$\left[\begin{array}{cc|c} 2 & 1 & 1 \\ 0 & -15 & 5 \end{array}\right]$$

Multiply row 1 by $\frac{1}{2}$ and row 2 by $-\frac{1}{15}$.

$$\left[\begin{array}{cc|c} 1 & \frac{1}{2} & \frac{1}{2} \\ 0 & 1 & -\frac{1}{3} \end{array}\right]$$

We have:

$$x + \frac{1}{2}y = 1, \quad (1)$$
$$y = -\frac{1}{3} \quad (2)$$

Back-substitute in equation (1) and solve for x.

$$x + \frac{1}{2}\left(-\frac{1}{3}\right) = \frac{1}{2}$$
$$x = \frac{2}{3}$$

The solution is $\left(\frac{2}{3}, -\frac{1}{3}\right)$.

19. $3x + 4y = 7,$
$-5x + 2y = 10$

Write the augmented matrix. We will use Gaussian elimination.

$$\left[\begin{array}{cc|c} 3 & 4 & 7 \\ -5 & 2 & 10 \end{array}\right]$$

Multiply row 2 by 3 to make the first number in row 2 a multiple of 3.

$$\left[\begin{array}{cc|c} 3 & 4 & 7 \\ -15 & 6 & 30 \end{array}\right]$$

Multiply row 1 by 5 and add it to row 2.

$$\left[\begin{array}{cc|c} 3 & 4 & 7 \\ 0 & 26 & 65 \end{array}\right]$$

Multiply row 1 by $\frac{1}{3}$ and row 2 by $\frac{1}{26}$.

$$\left[\begin{array}{cc|c} 1 & \frac{4}{3} & \frac{7}{3} \\ 0 & 1 & \frac{5}{2} \end{array}\right]$$

Write the system of equations that corresponds to the last matrix.

$$x + \frac{4}{3}y = \frac{7}{3}, \quad (1)$$
$$y = \frac{5}{2} \quad (2)$$

Back-substitute in equation (1) and solve for x.

$$x + \frac{4}{3} \cdot \frac{5}{2} = \frac{7}{3}$$
$$x + \frac{10}{3} = \frac{7}{3}$$
$$x = -\frac{3}{3} = -1$$

The solution is $\left(-1, \frac{5}{2}\right)$.

20. $5x - 3y = -2,$
$4x + 2y = 5$

Write the augmented matrix. We will use Gaussian elimination.

$$\left[\begin{array}{cc|c} 5 & -3 & -2 \\ 4 & 2 & 5 \end{array}\right]$$

Multiply row 2 by 5.

$$\left[\begin{array}{cc|c} 5 & -3 & -2 \\ 20 & 10 & 25 \end{array}\right]$$

Multiply row 1 by -4 and add it to row 2.

$$\left[\begin{array}{cc|c} 5 & -3 & -2 \\ 0 & 22 & 33 \end{array}\right]$$

Multiply row 1 by $\frac{1}{5}$ and row 2 by $\frac{1}{22}$.

$$\left[\begin{array}{cc|c} 1 & -\frac{3}{5} & -\frac{2}{5} \\ 0 & 1 & \frac{3}{2} \end{array}\right]$$

We have:

$$x - \frac{3}{5}y = -\frac{2}{5}, \quad (1)$$
$$y = \frac{3}{2} \quad (2)$$

Back-substitute in (1) and solve for x.

$$x - \frac{3}{5} \cdot \frac{3}{2} = -\frac{2}{5}$$
$$x - \frac{9}{10} = -\frac{2}{5}$$
$$x = \frac{1}{2}$$

The solution is $\left(\frac{1}{2}, \frac{3}{2}\right)$.

21. $3x + 2y = 6,$
$2x - 3y = -9$

Write the augmented matrix. We will use Gauss-Jordan elimination.

$$\left[\begin{array}{cc|c} 3 & 2 & 6 \\ 2 & -3 & -9 \end{array}\right]$$

Multiply row 2 by 3 to make the first number in row 2 a multiple of 3.

$$\left[\begin{array}{cc|c} 3 & 2 & 6 \\ 6 & -9 & -27 \end{array}\right]$$

Multiply row 1 by -2 and add it to row 2.

$$\left[\begin{array}{cc|c} 3 & 2 & 6 \\ 0 & -13 & -39 \end{array}\right]$$

Multiply row 2 by $-\frac{1}{13}$.

$$\left[\begin{array}{cc|c} 3 & 2 & 6 \\ 0 & 1 & 3 \end{array}\right]$$

Multiply row 2 by -2 and add it to row 1.

$$\left[\begin{array}{cc|c} 3 & 0 & 0 \\ 0 & 1 & 3 \end{array}\right]$$

Multiply row 1 by $\frac{1}{3}$.

$$\left[\begin{array}{cc|c} 1 & 0 & 0 \\ 0 & 1 & 3 \end{array}\right]$$

We have $x = 0$, $y = 3$. The solution is $(0, 3)$.

22. $x - 4y = 9,$
$2x + 5y = 5$

Write the augmented matrix. We will use Gauss-Jordan elimination.

$$\left[\begin{array}{cc|c} 1 & -4 & 9 \\ 2 & 5 & 5 \end{array}\right]$$

Multiply row 1 by -2 and add it to row 2.

$$\left[\begin{array}{cc|c} 1 & -4 & 9 \\ 0 & 13 & -13 \end{array}\right]$$

Multiply row 2 by $\frac{1}{13}$.

$$\left[\begin{array}{cc|c} 1 & -4 & 9 \\ 0 & 1 & -1 \end{array}\right]$$

Multiply row 2 by 4 and add it to row 1.

$$\left[\begin{array}{cc|c} 1 & 0 & 5 \\ 0 & 1 & -1 \end{array}\right]$$

The solution is $(5, -1)$.

23. $x - 3y = 8,$
$2x - 6y = 3$

Write the augmented matrix.

$$\left[\begin{array}{cc|c} 1 & -3 & 8 \\ 2 & -6 & 3 \end{array}\right]$$

Multiply row 1 by -2 and add it to row 2.

$$\left[\begin{array}{cc|c} 1 & -3 & 8 \\ 0 & 0 & -13 \end{array}\right]$$

The last row corresponds to the false equation $0 = -13$, so there is no solution.

24. $4x - 8y = 12,$
$-x + 2y = -3$

Write the augmented matrix.

$$\left[\begin{array}{cc|c} 4 & -8 & 12 \\ -1 & 2 & -3 \end{array}\right]$$

Interchange the rows.

$$\left[\begin{array}{cc|c} -1 & 2 & -3 \\ 4 & -8 & 12 \end{array}\right]$$

Multiply row 1 by 4 and add it to row 2.

$$\left[\begin{array}{cc|c} -1 & 2 & -3 \\ 0 & 0 & 0 \end{array}\right]$$

The last row corresponds to $0 = 0$, so the system is dependent and equivalent to $-x + 2y = -3$. Solving this system for x gives us $x = 2y + 3$. Then the solutions are of the form $(2y + 3, y)$, where y is any real number.

25. $-2x + 6y = 4,$
$3x - 9y = -6$

Write the augmented matrix.

$$\left[\begin{array}{cc|c} -2 & 6 & 4 \\ 3 & -9 & -6 \end{array}\right]$$

Multiply row 1 by $-\frac{1}{2}$.

$$\left[\begin{array}{rr|r} 1 & -3 & -2 \\ 3 & -9 & -6 \end{array}\right]$$

Multiply row 1 by -3 and add it to row 2.

$$\left[\begin{array}{rr|r} 1 & -3 & -2 \\ 0 & 0 & 0 \end{array}\right]$$

The last row corresponds to the equation $0 = 0$ which is true for all values of x and y. Thus, the system of equations is dependent and is equivalent to the first equation $-2x + 6y = 4$, or $x - 3y = -2$. Solving for x, we get $x = 3y - 2$. Then the solutions are of the form $(3y - 2, y)$, where y is any real number.

26. $6x + 2y = -10,$
$-3x - y = 6$

Write the augmented matrix.

$$\left[\begin{array}{rr|r} 6 & 2 & -10 \\ -3 & -1 & 6 \end{array}\right]$$

Interchange the rows.

$$\left[\begin{array}{rr|r} -3 & -1 & 6 \\ 6 & 2 & -10 \end{array}\right]$$

Multiply row 1 by 2 and add it to row 2.

$$\left[\begin{array}{rr|r} 3 & -1 & 6 \\ 0 & 0 & 2 \end{array}\right]$$

The last row corresponds to the false equation $0 = 2$, so there is no solution.

27. $x + 2y - 3z = 9,$
$2x - y + 2z = -8,$
$3x - y - 4z = 3$

Write the augmented matrix. We will use Gauss-Jordan elimination.

$$\left[\begin{array}{rrr|r} 1 & 2 & -3 & 9 \\ 2 & -1 & 2 & -8 \\ 3 & -1 & -4 & 3 \end{array}\right]$$

Multiply row 1 by -2 and add it to row 2. Also, multiply row 1 by -3 and add it to row 3.

$$\left[\begin{array}{rrr|r} 1 & 2 & -3 & 9 \\ 0 & -5 & 8 & -26 \\ 0 & -7 & 5 & -24 \end{array}\right]$$

Multiply row 2 by $-\frac{1}{5}$ to get a 1 in the second row, second column.

$$\left[\begin{array}{rrr|r} 1 & 2 & -3 & 9 \\ 0 & 1 & -\frac{8}{5} & \frac{26}{5} \\ 0 & -7 & 5 & -24 \end{array}\right]$$

Multiply row 2 by -2 and add it to row 1. Also, multiply row 2 by 7 and add it to row 3.

$$\left[\begin{array}{rrr|r} 1 & 0 & \frac{1}{5} & -\frac{7}{5} \\ 0 & 1 & -\frac{8}{5} & \frac{26}{5} \\ 0 & 0 & -\frac{31}{5} & \frac{62}{5} \end{array}\right]$$

Multiply row 3 by $-\frac{5}{31}$ to get a 1 in the third row, third column.

$$\left[\begin{array}{rrr|r} 1 & 0 & \frac{1}{5} & -\frac{7}{5} \\ 0 & 1 & -\frac{8}{5} & \frac{26}{5} \\ 0 & 0 & 1 & -2 \end{array}\right]$$

Multiply row 3 by $-\frac{1}{5}$ and add it to row 1. Also, multiply row 3 by $\frac{8}{5}$ and add it to row 2.

$$\left[\begin{array}{rrr|r} 1 & 0 & 0 & -1 \\ 0 & 1 & 0 & 2 \\ 0 & 0 & 1 & -2 \end{array}\right]$$

We have $x = -1$, $y = 2$, $z = -2$. The solution is $(-1, 2, -2)$.

28. $x - y + 2z = 0,$
$x - 2y + 3z = -1,$
$2x - 2y + z = -3$

Write the augmented matrix. We will use Gauss-Jordan elimination.

$$\left[\begin{array}{rrr|r} 1 & -1 & 2 & 0 \\ 1 & -2 & 3 & -1 \\ 2 & -2 & 1 & -3 \end{array}\right]$$

Multiply row 1 by -1 and add it to row 2. Also, multiply row 1 by -2 and add it to row 3.

$$\left[\begin{array}{rrr|r} 1 & -1 & 2 & 0 \\ 0 & -1 & 1 & -1 \\ 0 & 0 & -3 & -3 \end{array}\right]$$

Multiply row 2 by -1.

$$\left[\begin{array}{rrr|r} 1 & -1 & 2 & 0 \\ 0 & 1 & -1 & 1 \\ 0 & 0 & -3 & -3 \end{array}\right]$$

Add row 2 to row 1.

$$\left[\begin{array}{ccc|c} 1 & 0 & 1 & 1 \\ 0 & 1 & -1 & 1 \\ 0 & 0 & -3 & -3 \end{array}\right]$$

Multiply row 3 by $-\frac{1}{3}$.

$$\left[\begin{array}{ccc|c} 1 & 0 & 1 & 1 \\ 0 & 1 & -1 & 1 \\ 0 & 0 & 1 & 1 \end{array}\right]$$

Multiply row 3 by -1 and add it to row 1. Also, add row 3 to row 2.

$$\left[\begin{array}{ccc|c} 1 & 0 & 0 & 0 \\ 0 & 1 & 0 & 2 \\ 0 & 0 & 1 & 1 \end{array}\right]$$

The solution is $(0, 2, 1)$.

29. $4x - y - 3z = 1,$
$8x + y - z = 5,$
$2x + y + 2z = 5$

Write the augmented matrix. We will use Gauss-Jordan elimination.

$$\left[\begin{array}{ccc|c} 4 & -1 & -3 & 1 \\ 8 & 1 & -1 & 5 \\ 2 & 1 & 2 & 5 \end{array}\right]$$

First interchange rows 1 and 3 so that each number below the first number in the first row is a multiple of that number.

$$\left[\begin{array}{ccc|c} 2 & 1 & 2 & 5 \\ 8 & 1 & -1 & 5 \\ 4 & -1 & -3 & 1 \end{array}\right]$$

Multiply row 1 by -4 and add it to row 2. Also, multiply row 1 by -2 and add it to row 3.

$$\left[\begin{array}{ccc|c} 2 & 1 & 2 & 5 \\ 0 & -3 & -9 & -15 \\ 0 & -3 & -7 & -9 \end{array}\right]$$

Multiply row 2 by -1 and add it to row 3.

$$\left[\begin{array}{ccc|c} 2 & 1 & 2 & 5 \\ 0 & -3 & -9 & -15 \\ 0 & 0 & 2 & 6 \end{array}\right]$$

Multiply row 2 by $-\frac{1}{3}$ to get a 1 in the second row, second column.

$$\left[\begin{array}{ccc|c} 2 & 1 & 2 & 5 \\ 0 & 1 & 3 & 5 \\ 0 & 0 & 2 & 6 \end{array}\right]$$

Multiply row 2 by -1 and add it to row 1.

$$\left[\begin{array}{ccc|c} 2 & 0 & -1 & 0 \\ 0 & 1 & 3 & 5 \\ 0 & 0 & 2 & 6 \end{array}\right]$$

Multiply row 3 by $\frac{1}{2}$ to get a 1 in the third row, third column.

$$\left[\begin{array}{ccc|c} 2 & 0 & -1 & 0 \\ 0 & 1 & 3 & 5 \\ 0 & 0 & 1 & 3 \end{array}\right]$$

Add row 3 to row 1. Also multiply row 3 by -3 and add it to row 2.

$$\left[\begin{array}{ccc|c} 2 & 0 & 0 & 3 \\ 0 & 1 & 0 & -4 \\ 0 & 0 & 1 & 3 \end{array}\right]$$

Finally, multiply row 1 by $\frac{1}{2}$.

$$\left[\begin{array}{ccc|c} 1 & 0 & 0 & \frac{3}{2} \\ 0 & 1 & 0 & -4 \\ 0 & 0 & 1 & 3 \end{array}\right]$$

We have $x = \frac{3}{2}$, $y = -4$, $z = 3$. The solution is $\left(\frac{3}{2}, -4, 3\right)$.

30. $3x + 2y + 2z = 3,$
$x + 2y - z = 5,$
$2x - 4y + z = 0$

Write the augmented matrix. We will use Gauss-Jordan elimination.

$$\left[\begin{array}{ccc|c} 3 & 2 & 2 & 3 \\ 1 & 2 & -1 & 5 \\ 2 & -4 & 1 & 0 \end{array}\right]$$

Interchange the first two rows.

$$\left[\begin{array}{ccc|c} 1 & 2 & -1 & 5 \\ 3 & 2 & 2 & 3 \\ 2 & -4 & 1 & 0 \end{array}\right]$$

Multiply row 1 by -3 and add it to row 2. Also, multiply row 1 by -2 and add it to row 3.

$$\left[\begin{array}{ccc|c} 1 & 2 & -1 & 5 \\ 0 & -4 & 5 & -12 \\ 0 & -8 & 3 & -10 \end{array}\right]$$

Multiply row 2 by $-\frac{1}{4}$.

$$\left[\begin{array}{ccc|c} 1 & 2 & -1 & 5 \\ 0 & 1 & -\frac{5}{4} & 3 \\ 0 & -8 & 3 & -10 \end{array}\right]$$

Multiply row 2 by -2 and add it to row 1. Also, multiply row 2 by 8 and add it to row 3.

$$\left[\begin{array}{ccc|c} 1 & 0 & \frac{3}{2} & -1 \\ 0 & 1 & -\frac{5}{4} & 3 \\ 0 & 0 & -7 & 14 \end{array}\right]$$

Multiply row 3 by $-\frac{1}{7}$.

$$\left[\begin{array}{ccc|c} 1 & 0 & \frac{3}{2} & -1 \\ 0 & 1 & -\frac{5}{4} & 3 \\ 0 & 0 & 1 & -2 \end{array}\right]$$

Multiply row 3 by $-\frac{3}{2}$ and add it to row 1. Also, multiply row 3 by $\frac{5}{4}$ and add it to row 2.

$$\left[\begin{array}{ccc|c} 1 & 0 & 0 & 2 \\ 0 & 1 & 0 & \frac{1}{2} \\ 0 & 0 & 1 & -2 \end{array}\right]$$

The solution is $\left(2, \frac{1}{2}, -2\right)$.

31. $x - 2y + 3z = 4,$
$3x + y - z = 0,$
$2x + 3y - 5z = 1$

Write the augmented matrix. We will use Gaussian elimination.

$$\left[\begin{array}{ccc|c} 1 & -2 & 3 & -4 \\ 3 & 1 & -1 & 0 \\ 2 & 3 & -5 & 1 \end{array}\right]$$

Multiply row 1 by -3 and add it to row 2. Also, multiply row 1 by -2 and add it to row 3.

$$\left[\begin{array}{ccc|c} 1 & -2 & 3 & -4 \\ 0 & 7 & -10 & 12 \\ 0 & 7 & -11 & 9 \end{array}\right]$$

Multiply row 2 by -1 and add it to row 3.

$$\left[\begin{array}{ccc|c} 1 & -2 & 3 & -4 \\ 0 & 7 & -10 & 12 \\ 0 & 0 & -1 & -3 \end{array}\right]$$

Multiply row 2 by $\frac{1}{7}$ and multiply row 3 by -1.

$$\left[\begin{array}{ccc|c} 1 & -2 & 3 & -4 \\ 0 & 1 & -\frac{10}{7} & \frac{12}{7} \\ 0 & 0 & 1 & 3 \end{array}\right]$$

Now write the system of equations that corresponds to the last matrix.

$$x - 2y + 3z = -4, \quad (1)$$
$$y - \frac{10}{7}z = \frac{12}{7}, \quad (2)$$
$$z = 3 \quad (3)$$

Back-substitute 3 for z in equation (2) and solve for y.

$$y - \frac{10}{7} \cdot 3 = \frac{12}{7}$$
$$y - \frac{30}{7} = \frac{12}{7}$$
$$y = \frac{42}{7} = 6$$

Back-substitute 6 for y and 3 for z in equation (1) and solve for x.

$$x - 2 \cdot 6 + 3 \cdot 3 = -4$$
$$x - 3 = -4$$
$$x = -1$$

The solution is $(-1, 6, 3)$.

32. $2x - 3y + 2z = 2,$
$x + 4y - z = 9,$
$-3x + y - 5z = 5$

Write the augmented matrix. We will use Gaussian elimination.

$$\left[\begin{array}{ccc|c} 2 & -3 & 2 & 2 \\ 1 & 4 & -1 & 9 \\ -3 & 1 & -5 & 5 \end{array}\right]$$

Interchange the first two rows.

$$\left[\begin{array}{ccc|c} 1 & 4 & -1 & 9 \\ 2 & -3 & 2 & 2 \\ -3 & 1 & -5 & 5 \end{array}\right]$$

Multiply row 1 by -2 and add it to row 2. Also, multiply row 1 by 3 and add it to row 3.

$$\left[\begin{array}{ccc|c} 1 & 4 & -1 & 9 \\ 0 & -11 & 4 & -16 \\ 0 & 13 & -8 & 32 \end{array}\right]$$

Multiply row 3 by 11.

$$\left[\begin{array}{ccc|c} 1 & 4 & -1 & 9 \\ 0 & -11 & 4 & -16 \\ 0 & 143 & -88 & 352 \end{array}\right]$$

Multiply row 2 by 13 and add it to row 3.

$$\left[\begin{array}{ccc|c} 1 & 4 & -1 & 9 \\ 0 & -11 & 4 & -16 \\ 0 & 0 & -36 & 144 \end{array}\right]$$

Multiply row 2 by $-\frac{1}{11}$ and multiply row 3 by $-\frac{1}{36}$.

$$\left[\begin{array}{ccc|c} 1 & 4 & -1 & 9 \\ 0 & 1 & -\frac{4}{11} & \frac{16}{11} \\ 0 & 0 & 1 & -4 \end{array}\right]$$

We have

$$x + 4y - z = 9, \quad (1)$$
$$y - \frac{4}{11}z = \frac{16}{11}, \quad (2)$$
$$z = -4. \quad (3)$$

Back-substitute in (2) to find y.

$$y - \frac{4}{11}(-4) = \frac{16}{11}$$
$$y + \frac{16}{11} = \frac{16}{11}$$
$$y = 0$$

Back-substitute in (1) to find x.

$$x - 4 \cdot 0 - (-4) = 9$$
$$x = 5$$

The solution is $(5, 0, -4)$.

33. $2x - 4y - 3z = 3,$
$x + 3y + z = -1,$
$5x + y - 2z = 2$

Write the augmented matrix.

$$\left[\begin{array}{ccc|c} 2 & -4 & -3 & 3 \\ 1 & 3 & 1 & -1 \\ 5 & 1 & -2 & 2 \end{array}\right]$$

Interchange the first two rows to get a 1 in the first row, first column.

$$\left[\begin{array}{ccc|c} 1 & 3 & 1 & -1 \\ 2 & -4 & -3 & 3 \\ 5 & 1 & -2 & 2 \end{array}\right]$$

Multiply row 1 by -2 and add it to row 2. Also, multiply row 1 by -5 and add it to row 3.

$$\left[\begin{array}{ccc|c} 1 & 3 & 1 & -1 \\ 0 & -10 & -5 & 5 \\ 0 & -14 & -7 & 7 \end{array}\right]$$

Multiply row 2 by $-\frac{1}{10}$ to get a 1 in the second row, second column.

$$\left[\begin{array}{ccc|c} 1 & 3 & 1 & -1 \\ 0 & 1 & \frac{1}{2} & -\frac{1}{2} \\ 0 & -14 & -7 & 7 \end{array}\right]$$

Multiply row 2 by 14 and add it to row 3.

$$\left[\begin{array}{ccc|c} 1 & 3 & 1 & -1 \\ 0 & 1 & \frac{1}{2} & -\frac{1}{2} \\ 0 & 0 & 0 & 0 \end{array}\right]$$

The last row corresponds to the equation $0 = 0$. This indicates that the system of equations is dependent. It is equivalent to

$$x + 3y + z = -1,$$
$$y + \frac{1}{2}z = -\frac{1}{2}$$

We solve the second equation for y.

$$y = -\frac{1}{2}z - \frac{1}{2}$$

Substitute for y in the first equation and solve for x.

$$x + 3\left(-\frac{1}{2}z - \frac{1}{2}\right) + z = -1$$
$$x - \frac{3}{2}z - \frac{3}{2} + z = -1$$
$$x = \frac{1}{2}z + \frac{1}{2}$$

The solution is $\left(\frac{1}{2}z + \frac{1}{2}, -\frac{1}{2}z - \frac{1}{2}, z\right)$, where z is any real number.

34. $x + y - 3z = 4,$
$4x + 5y + z = 1,$
$2x + 3y + 7z = -7$

Write the augmented matrix.

$$\left[\begin{array}{ccc|c} 1 & 1 & -3 & 4 \\ 4 & 5 & 1 & 1 \\ 2 & 3 & 7 & -7 \end{array}\right]$$

Multiply row 1 by -4 and add it to row 2. Also, multiply row 1 by -2 and add it to row 3.

$$\left[\begin{array}{ccc|c} 1 & 1 & -3 & 4 \\ 0 & 1 & 13 & -15 \\ 0 & 1 & 13 & -15 \end{array}\right]$$

Multiply row 2 by -1 and add it to row 3.

$$\left[\begin{array}{ccc|c} 1 & 1 & -3 & 4 \\ 0 & 1 & 13 & -15 \\ 0 & 0 & 0 & 0 \end{array}\right]$$

We have a dependent system of equations that is equivalent to

$$x + y - 3z = 4,$$
$$y + 13z = -15.$$

Solve the second equation for y.

$$y = -13z - 15$$

Substitute for y in the first equation and solve for x.

$$x - 13z - 15 - 3z = 4$$
$$x = 16z + 19$$

The solution is $(16z + 19, -13z - 15, z)$, where z is any real number.

35. $p + q + r = 1,$
$p + 2q + 3r = 4,$
$4p + 5q + 6r = 7$

Write the augmented matrix.

$$\left[\begin{array}{ccc|c} 1 & 1 & 1 & 1 \\ 1 & 2 & 3 & 4 \\ 4 & 5 & 6 & 7 \end{array}\right]$$

Multiply row 1 by -1 and add it to row 2. Also, multiply row 1 by -4 and add it to row 3.

$$\left[\begin{array}{ccc|c} 1 & 1 & 1 & 1 \\ 0 & 1 & 2 & 3 \\ 0 & 1 & 2 & 3 \end{array}\right]$$

Multiply row 2 by -1 and add it to row 3.

$$\left[\begin{array}{ccc|c} 1 & 1 & 1 & 1 \\ 0 & 1 & 2 & 3 \\ 0 & 0 & 0 & 0 \end{array}\right]$$

The last row corresponds to the equation $0 = 0$. This indicates that the system of equations is dependent. It is equivalent to

$p + q + r = 1,$
$q + 2r = 3.$

We solve the second equation for q.

$q = -2r + 3$

Substitute for y in the first equation and solve for p.

$p - 2r + 3 + r = 1$
$p - r + 3 = 1$
$p = r - 2$

The solution is $(r-2, -2r+3, r)$, where r is any real number.

36. $m + n + t = 9,$
$m - n - t = -15,$
$3m + n + t = 2$

Write the augmented matrix.

$$\left[\begin{array}{ccc|c} 1 & 1 & 1 & 9 \\ 1 & -1 & -1 & -15 \\ 3 & 1 & 1 & 2 \end{array}\right]$$

Multiply row 1 by -1 and add it to row 2. Also, multiply row 1 by -3 and add it to row 3.

$$\left[\begin{array}{ccc|c} 1 & 1 & 1 & 9 \\ 0 & -2 & -2 & -24 \\ 0 & -2 & -2 & -25 \end{array}\right]$$

Multiply row 2 by -1 and add it to row 3.

$$\left[\begin{array}{ccc|c} 1 & 1 & 1 & 9 \\ 0 & -2 & -2 & -24 \\ 0 & 0 & 0 & -1 \end{array}\right]$$

The last row corresponds to the false equation $0 = -1$. Thus, the system of equations has no solution.

37. $a + b - c = 7,$
$a - b + c = 5,$
$3a + b - c = -1$

Write the augmented matrix.

$$\left[\begin{array}{ccc|c} 1 & 1 & -1 & 7 \\ 1 & -1 & 1 & 5 \\ 3 & 1 & -1 & -1 \end{array}\right]$$

Multiply row 1 by -1 and add it to row 2. Also, multiply row 1 by -3 and add it to row 3.

$$\left[\begin{array}{ccc|c} 1 & 1 & -1 & 7 \\ 0 & -2 & 2 & -2 \\ 0 & -2 & 2 & -22 \end{array}\right]$$

Multiply row 2 by -1 and add it to row 3.

$$\left[\begin{array}{ccc|c} 1 & 1 & -1 & 7 \\ 0 & -2 & 2 & -2 \\ 0 & 0 & 0 & -20 \end{array}\right]$$

The last row corresponds to the false equation $0 = -20$. Thus, the system of equations has no solution.

38. $a - b + c = 3,$
$2a + b - 3c = 5,$
$4a + b - c = 11$

Write the augmented matrix. We will use Gaussian elimination.

$$\left[\begin{array}{ccc|c} 1 & -1 & 1 & 3 \\ 2 & 1 & -3 & 5 \\ 4 & 1 & -1 & 11 \end{array}\right]$$

Multiply row 1 by -2 and add it to row 2. Also, multiply row 1 by -4 and add it to row 3.

$$\left[\begin{array}{ccc|c} 1 & -1 & 1 & 3 \\ 0 & 3 & -5 & -1 \\ 0 & 5 & -5 & -1 \end{array}\right]$$

Multiply row 3 by 3.

$$\left[\begin{array}{ccc|c} 1 & -1 & 1 & 3 \\ 0 & 3 & -5 & -1 \\ 0 & 15 & -15 & -3 \end{array}\right]$$

Multiply row 2 by -5 and add it to row 3.

$$\left[\begin{array}{ccc|c} 1 & -1 & 1 & 3 \\ 0 & 3 & -5 & -1 \\ 0 & 0 & 10 & 2 \end{array}\right]$$

Multiply row 2 by $\frac{1}{3}$ and multiply row 3 by $\frac{1}{10}$.

$$\left[\begin{array}{ccc|c} 1 & -1 & 1 & 3 \\ 0 & 1 & -\frac{5}{3} & -\frac{1}{3} \\ 0 & 0 & 1 & \frac{1}{5} \end{array}\right]$$

We have

$$x - y + z = 3, \quad (1)$$

$$y - \frac{5}{3}z = -\frac{1}{3}, \quad (2)$$

$$z = \frac{1}{5}. \quad (3)$$

Back-substitute in equation (2) to find y.

$$y - \frac{5}{3}\cdot\frac{1}{5} = -\frac{1}{3}$$

$$y = 0$$

Back-substitute in equation (1) to find x.

$$x - 0 + \frac{1}{5} = 3$$

$$x = \frac{14}{5}$$

The solution is $\left(\frac{14}{5}, 0, \frac{1}{5}\right)$.

39. $-2w + 2x + 2y - 2z = -10,$

$w + x + y + z = -5,$

$3w + x - y + 4z = -2,$

$w + 3x - 2y + 2z = -6$

Write the augmented matrix. We will use Gaussian elimination.

$$\left[\begin{array}{cccc|c} -2 & 2 & 2 & -2 & -10 \\ 1 & 1 & 1 & 1 & -5 \\ 3 & 1 & -1 & 4 & -2 \\ 1 & 3 & -2 & 2 & -6 \end{array}\right]$$

Interchange rows 1 and 2.

$$\left[\begin{array}{cccc|c} 1 & 1 & 1 & 1 & -5 \\ -2 & 2 & 2 & -2 & -10 \\ 3 & 1 & -1 & 4 & -2 \\ 1 & 3 & -2 & 2 & -6 \end{array}\right]$$

Multiply row 1 by 2 and add it to row 2. Multiply row 1 by -3 and add it to row 3. Multiply row 1 by -1 and add it to row 4.

$$\left[\begin{array}{cccc|c} 1 & 1 & 1 & 1 & -5 \\ 0 & 4 & 4 & 0 & -20 \\ 0 & -2 & -4 & 1 & 13 \\ 0 & 2 & -3 & 1 & -1 \end{array}\right]$$

Interchange rows 2 and 3.

$$\left[\begin{array}{cccc|c} 1 & 1 & 1 & 1 & -5 \\ 0 & -2 & -4 & 1 & 13 \\ 0 & 4 & 4 & 0 & -20 \\ 0 & 2 & -3 & 1 & -1 \end{array}\right]$$

Multiply row 2 by 2 and add it to row 3. Add row 2 to row 4.

$$\left[\begin{array}{cccc|c} 1 & 1 & 1 & 1 & -5 \\ 0 & -2 & -4 & 1 & 13 \\ 0 & 0 & -4 & 2 & 6 \\ 0 & 0 & -7 & 2 & 12 \end{array}\right]$$

Multiply row 4 by 4.

$$\left[\begin{array}{cccc|c} 1 & 1 & 1 & 1 & -5 \\ 0 & -2 & -4 & 1 & 13 \\ 0 & 0 & -4 & 2 & 6 \\ 0 & 0 & -28 & 8 & 48 \end{array}\right]$$

Multiply row 3 by -7 and add it to row 4.

$$\left[\begin{array}{cccc|c} 1 & 1 & 1 & 1 & -5 \\ 0 & -2 & -4 & 1 & 13 \\ 0 & 0 & -4 & 2 & 6 \\ 0 & 0 & 0 & -6 & 6 \end{array}\right]$$

Multiply row 2 by $-\frac{1}{2}$, row 3 by $-\frac{1}{4}$, and row 6 by $-\frac{1}{6}$.

$$\left[\begin{array}{cccc|c} 1 & 1 & 1 & 1 & -5 \\ 0 & 1 & 2 & -\frac{1}{2} & -\frac{13}{2} \\ 0 & 0 & 1 & -\frac{1}{2} & -\frac{3}{2} \\ 0 & 0 & 0 & 1 & -1 \end{array}\right]$$

Write the system of equations that corresponds to the last matrix.

$$w + x + y + z = -5, \quad (1)$$

$$x + 2y - \frac{1}{2}z = -\frac{13}{2}, \quad (2)$$

$$y - \frac{1}{2}z = -\frac{3}{2}, \quad (3)$$

$$z = -1 \quad (4)$$

Back-substitute in equation (3) and solve for y.

$$y - \frac{1}{2}(-1) = -\frac{3}{2}$$

$$y + \frac{1}{2} = -\frac{3}{2}$$

$$y = -2$$

Back-substitute in equation (2) and solve for x.

$$x + 2(-2) - \frac{1}{2}(-1) = -\frac{13}{2}$$

$$x - 4 + \frac{1}{2} = -\frac{13}{2}$$

$$x = -3$$

Back-substitute in equation (1) and solve for w.

$$w - 3 - 2 - 1 = -5$$

$$w = 1$$

The solution is $(1, -3, -2, -1)$.

40. $-w + 2x - 3y + z = -8,$
$-w + x + y - z = -4,$
$w + x + y + z = 22,$
$-w + x - y - z = -14$

Write the augmented matrix. We will use Gauss-Jordan elimination.

$$\left[\begin{array}{rrrr|r} -1 & 2 & -3 & 1 & -8 \\ -1 & 1 & 1 & -1 & -4 \\ 1 & 1 & 1 & 1 & 22 \\ -1 & 1 & -1 & -1 & -14 \end{array}\right]$$

Multiply row 1 by -1.

$$\left[\begin{array}{rrrr|r} 1 & -2 & 3 & -1 & 8 \\ -1 & 1 & 1 & -1 & -4 \\ 1 & 1 & 1 & 1 & 22 \\ -1 & 1 & -1 & -1 & -14 \end{array}\right]$$

Add row 1 to row 2 and to row 4. Multiply row 1 by -1 and add it to row 3.

$$\left[\begin{array}{rrrr|r} 1 & -2 & 3 & -1 & 8 \\ 0 & -1 & 4 & -2 & 4 \\ 0 & 3 & -2 & 2 & 14 \\ 0 & -1 & 2 & -2 & -6 \end{array}\right]$$

Multiply row 2 by -1.

$$\left[\begin{array}{rrrr|r} 1 & -2 & 3 & -1 & 8 \\ 0 & 1 & -4 & 2 & -4 \\ 0 & 3 & -2 & 2 & 14 \\ 0 & -1 & 2 & -2 & -6 \end{array}\right]$$

Multiply row 2 by -3 and add it to row 3. Also, add row 2 to row 4.

$$\left[\begin{array}{rrrr|r} 1 & -2 & 3 & -1 & 8 \\ 0 & 1 & -4 & 2 & -4 \\ 0 & 0 & 10 & -4 & 26 \\ 0 & 0 & -2 & 0 & -10 \end{array}\right]$$

Interchange row 3 and row 4.

$$\left[\begin{array}{rrrr|r} 1 & -2 & 3 & -1 & 8 \\ 0 & 1 & -4 & 2 & -4 \\ 0 & 0 & -2 & 0 & -10 \\ 0 & 0 & 10 & -4 & 26 \end{array}\right]$$

Multiply row 3 by $-\frac{1}{2}$.

$$\left[\begin{array}{rrrr|r} 1 & -2 & 3 & -1 & 8 \\ 0 & 1 & -4 & 2 & -4 \\ 0 & 0 & 1 & 0 & 5 \\ 0 & 0 & 10 & -4 & 26 \end{array}\right]$$

Multiply row 3 by -10 and add it to row 4.

$$\left[\begin{array}{rrrr|r} 1 & -2 & 3 & -1 & 8 \\ 0 & 1 & -4 & 2 & -4 \\ 0 & 0 & 1 & 0 & 5 \\ 0 & 0 & 0 & -4 & -24 \end{array}\right]$$

Multiply row 4 by $-\frac{1}{4}$.

$$\left[\begin{array}{rrrr|r} 1 & -2 & 3 & -1 & 8 \\ 0 & 1 & -4 & 2 & -4 \\ 0 & 0 & 1 & 0 & 5 \\ 0 & 0 & 0 & 1 & 6 \end{array}\right]$$

Add row 4 to row 1. Also, multiply row 4 by -2 and add it to row 2.

$$\left[\begin{array}{rrrr|r} 1 & -2 & 3 & 0 & 14 \\ 0 & 1 & -4 & 0 & -16 \\ 0 & 0 & 1 & 0 & 5 \\ 0 & 0 & 0 & 1 & 6 \end{array}\right]$$

Multiply row 3 by -3 and add it to row 1. Also, multiply row 3 by 4 and add it to row 2.

$$\left[\begin{array}{rrrr|r} 1 & -2 & 0 & 0 & -1 \\ 0 & 1 & 0 & 0 & 4 \\ 0 & 0 & 1 & 0 & 5 \\ 0 & 0 & 0 & 1 & 6 \end{array}\right]$$

Multiply row 2 by 2 and add it to row 1.

$$\left[\begin{array}{rrrr|r} 1 & 0 & 0 & 0 & 7 \\ 0 & 1 & 0 & 0 & 4 \\ 0 & 0 & 1 & 0 & 5 \\ 0 & 0 & 0 & 1 & 6 \end{array}\right]$$

The solution is $(7, 4, 5, 6)$.

41. ***Familiarize.*** Let x = the number of hours the Houlihans were out before 11 P.M. and y = the number of hours after 11 P.M. Then they pay the babysitter \$$5x$ before 11 P.M. and \$$7.50y$ after 11 P.M.

Translate.

The Houlihans were out for a total of 5 hr.

$x + y = 5$

They paid the sitter a total of \$30.

$5x + 7.5y = 30$

Carry out. Use Gaussian elimination or Gauss-Jordan elimination to solve the system of equations.

$x + y = 5,$
$5x + 7.5y = 30.$

The solution is $(3, 2)$. The coordinate $y = 2$ indicates that the Houlihans were out 2 hr after 11 P.M., so they came home at 1 A.M.

Check. The total time is $3+2$, or 5 hr. The total pay is $\$5 \cdot 3 + \$7.50(2)$, or $\$15 + \$15$, or \$30. The solution checks.

State. The Houlihans came home at 1 A.M.

42. Let x, y, and z represent the amount spent on advertising in fiscal years 1998, 1999, and 2000, respectively, in millions of dollars.

Solve: $x+y+z=11$,
$z=3x$,
$y=z-3$

$x = \$2$ million, $y = \$3$ million, $z = \$6$ million

43. ***Familiarize.*** Let x, y, and z represent the amounts borrowed at 8%, 10%, and 12%, respectively. Then the annual interest is $8\%x$, $10\%y$, and $12\%z$, or $0.08x$, $0.1y$, and $0.12z$.

Translate.

The total amount borrowed was \$30,000.

$$x+y+z=30,000$$

The total annual interest was \$3040.

$$0.08x+0.1y+0.12z=3040$$

The total amount borrowed at 8% and 10% was twice the amount borrowed at 12%.

$$x+y=2z$$

We have a system of equations.

$$x+y+z=30,000,$$
$$0.08x+0.1y+0.12z=3040,$$
$$x+y=2z, \text{ or}$$

$$x+y+z=30,000,$$
$$0.08x+0.1y+0.12z=3040,$$
$$x+y-2z=0$$

Carry out. Using Gaussian elimination or Gauss-Jordan elimination, we find that the solution is $(8000, 12,000, 10,000)$.

Check. The total amount borrowed was $\$8000 + \$12,000 + \$10,000$, or \$30,000. The total annual interest was $0.08(\$8000) + 0.1(\$12,000) + 0.12(\$10,000)$, or $\$640 + \$1200 + \$1200$, or \$3040. The total amount borrowed at 8% and 10%, $\$8000 + \$12,000$ or \$20,000, was twice the amount borrowed at 12%, \$10,000. The solution checks.

State. The amounts borrowed at 8%, 10%, and 12% were \$8000, \$12,000 and \$10,000, respectively.

44. Let x and y represent the number of 33¢ and 22¢ stamps purchased, respectively.

Solve: $x+y=60$,
$0.33x+0.22y=15.84$

$x=24$, $y=36$

45. Answers will vary.

46. See page 385 of the text.

47.
$$2x^2+x=7$$
$$2x^2+x-7=0$$
$$a=2,\ b=1,\ c=-7$$
$$x=\frac{-b\pm\sqrt{b^2-4ac}}{2a}$$
$$=\frac{-1\pm\sqrt{1^2-4\cdot2\cdot(-7)}}{2\cdot2}=\frac{-1\pm\sqrt{1+56}}{4}$$
$$=\frac{-1\pm\sqrt{57}}{4}$$

The solutions are $\frac{-1+\sqrt{57}}{4}$ and $\frac{-1-\sqrt{57}}{4}$, or $\frac{-1\pm\sqrt{57}}{4}$.

We could find approximate solutions by graphing $y_1=2x^2+x$ and $y_2=7$ and using the Intersect feature twice to find the first coordinates of the points of intersection of the graphs.

48. $\frac{1}{x+1}-\frac{6}{x-1}=1$, LCD is $(x+1)(x-1)$

$$(x+1)(x-1)\left(\frac{1}{x+1}-\frac{6}{x-1}\right)=(x+1)(x-1)\cdot1$$
$$x-1-6(x+1)=x^2-1$$
$$x-1-6x-6=x^2-1$$
$$0=x^2+5x+6$$
$$0=(x+3)(x+2)$$

$x=-3$ *or* $x=-2$

Both numbers check.

49.
$$\sqrt{2x+1}-1=\sqrt{2x-4}$$
$$(\sqrt{2x+1}-1)^2=(\sqrt{2x-4})^2 \quad \text{Squaring both sides}$$
$$2x+1-2\sqrt{2x+1}+1=2x-4$$
$$2x+2-2\sqrt{2x+1}=2x-4$$
$$2-2\sqrt{2x+1}=-4 \quad \text{Subtracting } 2x$$
$$-2\sqrt{2x+1}=-6 \quad \text{Subtracting 2}$$
$$\sqrt{2x+1}=3 \quad \text{Dividing by } -2$$
$$(\sqrt{2x+1})^2=3^2 \quad \text{Squaring both sides}$$
$$2x+1=9$$
$$2x=8$$
$$x=4$$

The number 4 checks. It is the solution.

50. $x-\sqrt{x}-6=0$

Let $u=\sqrt{x}$.

$$u^2-u-6=0$$
$$(u-3)(u+2)=0$$

$u=3$ *or* $u=-2$

$\sqrt{x}=3$ *or* $\sqrt{x}=-2$

$x=9$ No solution

The number 9 checks. It is the solution.

51. Substitute to find three equations.

$$12 = a(-3)^2 + b(-3) + c$$
$$-7 = a(-1)^2 + b(-1) + c$$
$$-2 = a \cdot 1^2 + b \cdot 1 + c$$

We have a system of equations.

$$9a - 3b + c = 12,$$
$$a - b + c = -7,$$
$$a + b + c = -2$$

Write the augmented matrix. We will use Gaussian elimination.

$$\left[\begin{array}{ccc|c} 9 & -3 & 1 & 12 \\ 1 & -1 & 1 & -7 \\ 1 & 1 & 1 & -2 \end{array}\right]$$

Interchange the first two rows.

$$\left[\begin{array}{ccc|c} 1 & -1 & 1 & -7 \\ 9 & -3 & 1 & 12 \\ 1 & 1 & 1 & -2 \end{array}\right]$$

Multiply row 1 by -9 and add it to row 2. Also, multiply row 1 by -1 and add it to row 3.

$$\left[\begin{array}{ccc|c} 1 & -1 & 1 & -7 \\ 0 & 6 & -8 & 75 \\ 0 & 2 & 0 & 5 \end{array}\right]$$

Interchange row 2 and row 3.

$$\left[\begin{array}{ccc|c} 1 & -1 & 1 & -7 \\ 0 & 2 & 0 & 5 \\ 0 & 6 & -8 & 75 \end{array}\right]$$

Multiply row 2 by -3 and add it to row 3.

$$\left[\begin{array}{ccc|c} 1 & -1 & 1 & -7 \\ 0 & 2 & 0 & 5 \\ 0 & 0 & -8 & 60 \end{array}\right]$$

Multiply row 2 by $\frac{1}{2}$ and row 3 by $-\frac{1}{8}$.

$$\left[\begin{array}{ccc|c} 1 & -1 & 1 & -7 \\ 0 & 1 & 0 & \frac{5}{2} \\ 0 & 0 & 1 & -\frac{15}{2} \end{array}\right]$$

Write the system of equations that corresponds to the last matrix.

$$x - y + z = -7,$$
$$y = \frac{5}{2},$$
$$z = -\frac{15}{2}$$

Back-substitute $\frac{5}{2}$ for y and $-\frac{15}{2}$ for z in the first equation and solve for x.

$$x - \frac{5}{2} - \frac{15}{2} = -7$$
$$x - 10 = -7$$
$$x = 3$$

The solution is $\left(3, \frac{5}{2}, -\frac{15}{2}\right)$, so the equation is $y = 3x^2 + \frac{5}{2}x - \frac{15}{2}$.

52. Solve: $0 = a(-1)^2 + b(-1) + c,$

$$-3 = a \cdot 1^2 + b \cdot 1 + c,$$
$$-22 = a \cdot 3^2 + b \cdot 3 + c, \text{ or}$$
$$a - b + c = 0,$$
$$a + b + c = -3,$$
$$9a + 3b + c = -22$$

$a = -2$, $b = -\frac{3}{2}$, $c = \frac{1}{2}$, so $y = -2x^2 - \frac{3}{2}x + \frac{1}{2}$.

53. $\begin{bmatrix} 1 & 5 \\ 3 & 2 \end{bmatrix}$

Multiply row 1 by -3 and add it to row 2.

$$\begin{bmatrix} 1 & 5 \\ 0 & -13 \end{bmatrix}$$

Multiply row 2 by $-\frac{1}{13}$.

$$\begin{bmatrix} 1 & 5 \\ 0 & 1 \end{bmatrix}$$ Row-echelon form

Multiply row 2 by -5 and add it to row 1.

$$\begin{bmatrix} 1 & 0 \\ 0 & 1 \end{bmatrix}$$ Reduced row-echelon form

54. $x - y + 3z = -8,$

$$2x + 3y - z = 5,$$
$$3x + 2y + 2kz = -3k$$

Write the augmented matrix.

$$\left[\begin{array}{ccc|c} 1 & -1 & 3 & -8 \\ 2 & 3 & -1 & 5 \\ 3 & 2 & 2k & -3k \end{array}\right]$$

Multiply row 1 by -2 and add it to row 2. Also, multiply row 1 by -3 and add it to row 3.

$$\left[\begin{array}{ccc|c} 1 & -1 & 3 & -8 \\ 0 & 5 & -7 & 21 \\ 0 & 5 & 2k-9 & -3k+24 \end{array}\right]$$

Multiply row 2 by -1 and add it to row 3.

$$\left[\begin{array}{ccc|c} 1 & -1 & 3 & -8 \\ 0 & 5 & -7 & 21 \\ 0 & 0 & 2k-2 & -3k+3 \end{array}\right]$$

a) If the system has no solution we have:

$2k - 2 = 0$ and $-3k + 3 \neq 0$

$k = 1$ and $k \neq 1$

This is impossible, so there is no value of k for which the system has no solution.

b) If the system has exactly one solution, we have:

$2k - 2 \neq -3k + 3$

$5k \neq 5$

$k \neq 1$

c) If the system has infinitely many solutions, we have:

$2k - 2 = 0$ and $-3k + 3 = 0$

$k = 1$ and $k = 1$, or

$k = 1$

55. $y = x + z$,

$3y + 5z = 4$,

$x + 4 = y + 3z$, or

$x - y + z = 0$,

$3y + 5z = 4$,

$x - y - 3z = -4$

Write the augmented matrix. We will use Gauss-Jordan elimination.

$$\left[\begin{array}{ccc|c} 1 & -1 & 1 & 0 \\ 0 & 3 & 5 & 4 \\ 1 & -1 & -3 & -4 \end{array}\right]$$

Multiply row 1 by -1 and add it to row 3.

$$\left[\begin{array}{ccc|c} 1 & -1 & 1 & 0 \\ 0 & 3 & 5 & 4 \\ 0 & 0 & -4 & -4 \end{array}\right]$$

Multiply row 3 by $-\frac{1}{4}$.

$$\left[\begin{array}{ccc|c} 1 & -1 & 1 & 0 \\ 0 & 3 & 5 & 4 \\ 0 & 0 & 1 & 1 \end{array}\right]$$

Multiply row 3 by -1 and add it to row 1. Also, multiply row 3 by -5 and add it to row 2.

$$\left[\begin{array}{ccc|c} 1 & -1 & 0 & -1 \\ 0 & 3 & 0 & -1 \\ 0 & 0 & 1 & 1 \end{array}\right]$$

Multiply row 2 by $\frac{1}{3}$.

$$\left[\begin{array}{ccc|c} 1 & -1 & 0 & -1 \\ 0 & 1 & 0 & -\frac{1}{3} \\ 0 & 0 & 1 & 1 \end{array}\right]$$

Add row 2 to row 1.

$$\left[\begin{array}{ccc|c} 1 & 0 & 0 & -\frac{4}{3} \\ 0 & 1 & 0 & -\frac{1}{3} \\ 0 & 0 & 1 & 1 \end{array}\right]$$

Read the solution from the last matrix. It is $\left(-\frac{4}{3}, -\frac{1}{3}, 1\right)$.

56. $x + y = 2z$,

$2x - 5z = 4$,

$x - z = y + 8$, or

$x + y - 2z = 0$,

$2x - 5z = 4$,

$x - y - z = 8$

Write the augmented matrix. We will use Gauss-Jordan elimination.

$$\left[\begin{array}{ccc|c} 1 & 1 & -2 & 0 \\ 2 & 0 & -5 & 4 \\ 1 & -1 & -1 & 8 \end{array}\right]$$

Multiply row 1 by -2 and add it to row 2. Also, multiply row 1 by -1 and add it to row 3.

$$\left[\begin{array}{ccc|c} 1 & 1 & -2 & 0 \\ 0 & -2 & -1 & 4 \\ 0 & -2 & 1 & 8 \end{array}\right]$$

Multiply row 2 by $-\frac{1}{2}$.

$$\left[\begin{array}{ccc|c} 1 & 1 & -2 & 0 \\ 0 & 1 & \frac{1}{2} & -2 \\ 0 & -2 & 1 & 8 \end{array}\right]$$

Multiply row 2 by -1 and add it to row 1. Also, multiply row 2 by 2 and add it to row 3.

$$\left[\begin{array}{ccc|c} 1 & 0 & -\frac{5}{2} & 2 \\ 0 & 1 & \frac{1}{2} & -2 \\ 0 & 0 & 2 & 4 \end{array}\right]$$

Multiply row 3 by $\frac{1}{2}$.

$$\left[\begin{array}{ccc|c} 1 & 0 & -\frac{5}{2} & 2 \\ 0 & 1 & \frac{1}{2} & -2 \\ 0 & 0 & 1 & 2 \end{array}\right]$$

Multiply row 3 by $\frac{5}{2}$ and add it to row 1. Also, multiply row 3 by $-\frac{1}{2}$ and add it to row 2.

$$\left[\begin{array}{ccc|c} 1 & 0 & 0 & 7 \\ 0 & 1 & 0 & -3 \\ 0 & 0 & 1 & 2 \end{array}\right]$$

The solution is $(7, -3, 2)$.

57. $x - 4y + 2z = 7,$

$3x + y + 3z = -5$

Write the augmented matrix.

$$\left[\begin{array}{ccc|c} 1 & -4 & 2 & 7 \\ 3 & 1 & 3 & -5 \end{array}\right]$$

Multiply row 1 by -3 and add it to row 2.

$$\left[\begin{array}{ccc|c} 1 & -4 & 2 & 7 \\ 0 & 13 & -3 & -26 \end{array}\right]$$

Multiply row 2 by $\frac{1}{13}$.

$$\left[\begin{array}{ccc|c} 1 & -4 & 2 & 7 \\ 0 & 1 & -\frac{3}{13} & -2 \end{array}\right]$$

Write the system of equations that corresponds to the last matrix.

$$x - 4y + 2z = 7,$$
$$y - \frac{3}{13}z = -2$$

Solve the second equation for y.

$$y = \frac{3}{13}z - 2$$

Substitute in the first equation and solve for x.

$$x - 4\left(\frac{3}{13}z - 2\right) + 2z = 7$$
$$x - \frac{12}{13}z + 8 + 2z = 7$$
$$x = -\frac{14}{13}z - 1$$

The solution is $\left(-\frac{14}{13}z - 1, \frac{3}{13}z - 2, z\right)$, where z is any real number.

58. $x - y - 3z = 3,$

$-x + 3y + z = -7$

Write the augmented matrix.

$$\left[\begin{array}{ccc|c} 1 & -1 & -3 & 3 \\ -1 & 3 & 1 & -7 \end{array}\right]$$

Add row 1 to row 2.

$$\left[\begin{array}{ccc|c} 1 & -1 & -3 & 3 \\ 0 & 2 & -2 & -4 \end{array}\right]$$

Multiply row 2 by $\frac{1}{2}$.

$$\left[\begin{array}{ccc|c} 1 & -1 & -3 & 3 \\ 0 & 1 & -1 & -2 \end{array}\right]$$

We have

$$x - y - 3z = 3,$$
$$y - z = -2.$$

Then $y = z - 2$. Substitute in the first equation and solve for x.

$$x - (z - 2) - 3z = 3$$
$$x - z + 2 - 3z = 3$$
$$x = 4z + 1$$

The solution is $(4z + 1, z - 2, z)$, where z is any real number.

59. $4x + 5y = 3,$

$-2x + y = 9,$

$3x - 2y = -15$

Write the augmented matrix.

$$\left[\begin{array}{cc|c} 4 & 5 & 3 \\ -2 & 1 & 9 \\ 3 & -2 & -15 \end{array}\right]$$

Multiply row 2 by 2 and row 3 by 4.

$$\left[\begin{array}{cc|c} 4 & 5 & 3 \\ -4 & 2 & 18 \\ 12 & -8 & -60 \end{array}\right]$$

Add row 1 to row 2. Also, multiply row 1 by -3 and add it to row 3.

$$\left[\begin{array}{cc|c} 4 & 5 & 3 \\ 0 & 7 & 21 \\ 0 & -23 & -69 \end{array}\right]$$

Multiply row 2 by $\frac{1}{7}$ and row 3 by $-\frac{1}{23}$.

$$\left[\begin{array}{cc|c} 4 & 5 & 3 \\ 0 & 1 & 3 \\ 0 & 1 & 3 \end{array}\right]$$

Multiply row 2 by -1 and add it to row 3.

$$\left[\begin{array}{cc|c} 4 & 5 & 3 \\ 0 & 1 & 3 \\ 0 & 0 & 0 \end{array}\right]$$

The last row corresponds to the equation $0 = 0$. Thus we have a dependent system that is equivalent to

$$4x + 5y = 3, \quad (1)$$
$$y = 3. \quad (2)$$

Back-substitute in equation (1) to find x.

$$4x + 5 \cdot 3 = 3$$
$$4x + 15 = 3$$
$$4x = -12$$
$$x = -3$$

The solution is $(-3, 3)$.

60. $2x - 3y = -1,$
$-x + 2y = -2,$
$3x - 5y = 1$

Write the augmented matrix.

$$\left[\begin{array}{rr|r} 2 & -3 & -1 \\ -1 & 2 & -2 \\ 3 & -5 & 1 \end{array}\right]$$

Interchange the first two rows.

$$\left[\begin{array}{rr|r} -1 & 2 & -2 \\ 2 & -3 & -1 \\ 3 & -5 & 1 \end{array}\right]$$

Multiply row 1 by 2 and add it to row 2. Also, multiply row 1 by 3 and add it to row 3.

$$\left[\begin{array}{rr|r} -1 & 2 & -2 \\ 0 & 1 & -5 \\ 0 & 1 & -5 \end{array}\right]$$

Multiply row 2 by -1 and add it to row 3.

$$\left[\begin{array}{rr|r} -1 & 2 & -2 \\ 0 & 1 & -5 \\ 0 & 0 & 0 \end{array}\right]$$

We have a dependent system that is equivalent to

$$-x + 2y = -2, \quad (1)$$
$$y = -5. \quad (2)$$

Back-substitute in equation (1) to find x.

$$-x + 2(-5) = -2$$
$$-x - 10 = -2$$
$$-x = 8$$
$$x = -8$$

The solution is $(-8, -5)$.

Exercise Set 5.4

1. $\begin{bmatrix} 5 & x \end{bmatrix} = \begin{bmatrix} y & -3 \end{bmatrix}$

Corresponding entries of the two matrices must be equal. Thus we have $5 = y$ and $x = -3$.

2. $\begin{bmatrix} 6x \\ 25 \end{bmatrix} = \begin{bmatrix} -9 \\ 5y \end{bmatrix}$

$$6x = -9 \quad \text{and} \quad 25 = 5y$$
$$x = -\frac{3}{2} \quad \text{and} \quad 5 = y$$

3. $\begin{bmatrix} 3 & 2x \\ y & -8 \end{bmatrix} = \begin{bmatrix} 3 & -2 \\ 1 & -8 \end{bmatrix}$

Corresponding entries of the two matrices must be equal. Thus, we have:

$$2x = -2 \quad \text{and} \quad y = 1$$
$$x = -1 \quad \text{and} \quad y = 1$$

4. $\begin{bmatrix} x-1 & 4 \\ y+3 & -7 \end{bmatrix} = \begin{bmatrix} 0 & 4 \\ -2 & -7 \end{bmatrix}$

$$x - 1 = 0 \quad \text{and} \quad y + 3 = -2$$
$$x = 1 \quad \text{and} \quad y = -5$$

5. $\mathbf{A} + \mathbf{B} = \begin{bmatrix} 1 & 2 \\ 4 & 3 \end{bmatrix} + \begin{bmatrix} -3 & 5 \\ 2 & -1 \end{bmatrix}$

$$= \begin{bmatrix} 1+(-3) & 2+5 \\ 4+2 & 3+(-1) \end{bmatrix}$$

$$= \begin{bmatrix} -2 & 7 \\ 6 & 2 \end{bmatrix}$$

6. $\mathbf{B} + \mathbf{A} = \mathbf{A} + \mathbf{B} = \begin{bmatrix} -2 & 7 \\ 6 & 2 \end{bmatrix}$ (See Exercise 5.)

7. $\mathbf{E} + \mathbf{O} = \begin{bmatrix} 1 & 3 \\ 2 & 6 \end{bmatrix} + \begin{bmatrix} 0 & 0 \\ 0 & 0 \end{bmatrix}$

$$= \begin{bmatrix} 1+0 & 3+0 \\ 2+0 & 6+0 \end{bmatrix}$$

$$= \begin{bmatrix} 1 & 3 \\ 2 & 6 \end{bmatrix}$$

8. $2\mathbf{A} = \begin{bmatrix} 2 \cdot 1 & 2 \cdot 2 \\ 2 \cdot 4 & 2 \cdot 3 \end{bmatrix} = \begin{bmatrix} 2 & 4 \\ 8 & 6 \end{bmatrix}$

9. $3\mathbf{F} = 3\begin{bmatrix} 3 & 3 \\ -1 & -1 \end{bmatrix}$

$$= \begin{bmatrix} 3 \cdot 3 & 3 \cdot 3 \\ 3 \cdot (-1) & 3 \cdot (-1) \end{bmatrix}$$

$$= \begin{bmatrix} 9 & 9 \\ -3 & -3 \end{bmatrix}$$

10. $(-1)\mathbf{D} = \begin{bmatrix} -1 \cdot 1 & -1 \cdot 1 \\ -1 \cdot 1 & -1 \cdot 1 \end{bmatrix} = \begin{bmatrix} -1 & -1 \\ -1 & -1 \end{bmatrix}$

11. $3\mathbf{F} = 3\begin{bmatrix} 3 & 3 \\ -1 & -1 \end{bmatrix} = \begin{bmatrix} 9 & 9 \\ -3 & -3 \end{bmatrix}$,

$2\mathbf{A} = 2\begin{bmatrix} 1 & 2 \\ 4 & 3 \end{bmatrix} = \begin{bmatrix} 2 & 4 \\ 8 & 6 \end{bmatrix}$

$$3\mathbf{F} + 2\mathbf{A} = \begin{bmatrix} 9 & 9 \\ -3 & -3 \end{bmatrix} + \begin{bmatrix} 2 & 4 \\ 8 & 6 \end{bmatrix}$$
$$= \begin{bmatrix} 9+2 & 9+4 \\ -3+8 & -3+6 \end{bmatrix}$$
$$= \begin{bmatrix} 11 & 13 \\ 5 & 3 \end{bmatrix}$$

12. $\mathbf{A} - \mathbf{B} = \begin{bmatrix} 1 & 2 \\ 4 & 3 \end{bmatrix} - \begin{bmatrix} -3 & 5 \\ 2 & -1 \end{bmatrix} = \begin{bmatrix} 4 & -3 \\ 2 & 4 \end{bmatrix}$

13. $\mathbf{B} - \mathbf{A} = \begin{bmatrix} -3 & 5 \\ 2 & -1 \end{bmatrix} - \begin{bmatrix} 1 & 2 \\ 4 & 3 \end{bmatrix}$

$$= \begin{bmatrix} -3 & 5 \\ 2 & -1 \end{bmatrix} + \begin{bmatrix} -1 & -2 \\ -4 & -3 \end{bmatrix}$$

$[\mathbf{B} - \mathbf{A} = \mathbf{B} + (-\mathbf{A})]$

$$= \begin{bmatrix} -3+(-1) & 5+(-2) \\ 2+(-4) & -1+(-3) \end{bmatrix}$$
$$= \begin{bmatrix} -4 & 3 \\ -2 & -4 \end{bmatrix}$$

14. $\mathbf{AB} = \begin{bmatrix} 1 & 2 \\ 4 & 3 \end{bmatrix}\begin{bmatrix} -3 & 5 \\ 2 & -1 \end{bmatrix}$

$$= \begin{bmatrix} 1(-3)+2\cdot 2 & 1\cdot 5+2(-1) \\ 4(-3)+3\cdot 2 & 4\cdot 5+3(-1) \end{bmatrix}$$
$$= \begin{bmatrix} 1 & 3 \\ -6 & 17 \end{bmatrix}$$

15. $\mathbf{BA} = \begin{bmatrix} -3 & 5 \\ 2 & -1 \end{bmatrix}\begin{bmatrix} 1 & 2 \\ 4 & 3 \end{bmatrix}$

$$= \begin{bmatrix} -3\cdot 1+5\cdot 4 & -3\cdot 2+5\cdot 3 \\ 2\cdot 1+(-1)4 & 2\cdot 2+(-1)3 \end{bmatrix}$$
$$= \begin{bmatrix} 17 & 9 \\ -2 & 1 \end{bmatrix}$$

16. $\mathbf{OF} = \mathbf{O} = \begin{bmatrix} 0 & 0 \\ 0 & 0 \end{bmatrix}$

17. $\mathbf{CD} = \begin{bmatrix} 1 & -1 \\ -1 & 1 \end{bmatrix}\begin{bmatrix} 1 & 1 \\ 1 & 1 \end{bmatrix}$

$$= \begin{bmatrix} 1\cdot 1+(-1)\cdot 1 & 1\cdot 1+(-1)\cdot 1 \\ -1\cdot 1+1\cdot 1 & -1\cdot 1+1\cdot 1 \end{bmatrix}$$
$$= \begin{bmatrix} 0 & 0 \\ 0 & 0 \end{bmatrix}$$

18. $\mathbf{EF} = \begin{bmatrix} 1 & 3 \\ 2 & 6 \end{bmatrix}\begin{bmatrix} 3 & 3 \\ -1 & -1 \end{bmatrix}$

$$= \begin{bmatrix} 1\cdot 3+3(-1) & 1\cdot 3+3(-1) \\ 2\cdot 3+6(-1) & 2\cdot 3+6(-1) \end{bmatrix}$$
$$= \begin{bmatrix} 0 & 0 \\ 0 & 0 \end{bmatrix}$$

19. $\mathbf{AI} = \begin{bmatrix} 1 & 2 \\ 4 & 3 \end{bmatrix}\begin{bmatrix} 1 & 0 \\ 0 & 1 \end{bmatrix}$

$$= \begin{bmatrix} 1\cdot 1+2\cdot 0 & 1\cdot 0+2\cdot 1 \\ 4\cdot 1+3\cdot 0 & 4\cdot 0+3\cdot 1 \end{bmatrix}$$
$$= \begin{bmatrix} 1 & 2 \\ 4 & 3 \end{bmatrix}$$

20. $\mathbf{IA} = \mathbf{A} = \begin{bmatrix} 1 & 2 \\ 4 & 3 \end{bmatrix}$

21. a) $\mathbf{B} = \begin{bmatrix} 150 & 80 & 40 \end{bmatrix}$

b) $\$150 + 5\% \cdot \$150 = 1.05(\$150) = \$157.50$

$\$80 + 5\% \cdot \$80 = 1.05(\$80) = \$84$

$\$40 + 5\% \cdot \$40 = 1.05(\$40) = \$42$

We write the matrix that corresponds to these amounts.

$\mathbf{R} = \begin{bmatrix} 157.5 & 84 & 42 \end{bmatrix}$

c) $\mathbf{B}+\mathbf{R} = \begin{bmatrix} 150 & 80 & 40 \end{bmatrix} + \begin{bmatrix} 157.5 & 84 & 42 \end{bmatrix}$
$= \begin{bmatrix} 307.5 & 164 & 82 \end{bmatrix}$

The entries represent the total budget in each type of expenditure for June and July.

22. a) $\mathbf{A} = \begin{bmatrix} 40 & 20 & 30 \end{bmatrix}$

b) $1.1(40) = 44$, $1.1(20) = 22$, $1.1(30) = 33$

$\mathbf{B} = \begin{bmatrix} 44 & 22 & 33 \end{bmatrix}$

c) $\mathbf{A}+\mathbf{B} = \begin{bmatrix} 40 & 20 & 30 \end{bmatrix} + \begin{bmatrix} 44 & 22 & 33 \end{bmatrix}$
$= \begin{bmatrix} 84 & 42 & 63 \end{bmatrix}$

The entries represent the total amount of each type of produce ordered for both weeks.

23. a) $\mathbf{C} = \begin{bmatrix} 140 & 27 & 3 & 13 & 64 \end{bmatrix}$

$\mathbf{P} = \begin{bmatrix} 180 & 4 & 11 & 24 & 662 \end{bmatrix}$

$\mathbf{B} = \begin{bmatrix} 50 & 5 & 1 & 82 & 20 \end{bmatrix}$

b) $\mathbf{C} + 2\mathbf{P} + 3\mathbf{B}$

$= \begin{bmatrix} 140 & 27 & 3 & 13 & 64 \end{bmatrix} +$
$\begin{bmatrix} 360 & 8 & 22 & 48 & 1324 \end{bmatrix} +$
$\begin{bmatrix} 150 & 15 & 3 & 246 & 60 \end{bmatrix}$
$= \begin{bmatrix} 650 & 50 & 28 & 307 & 1448 \end{bmatrix}$

The entries represent the total nutritional value of one serving of chicken, 1 cup of potato salad, and 3 broccoli spears.

24. a) $\mathbf{P} = \begin{bmatrix} 290 & 15 & 9 & 39 \end{bmatrix}$

$\mathbf{G} = \begin{bmatrix} 70 & 2 & 0 & 17 \end{bmatrix}$

$\mathbf{M} = \begin{bmatrix} 150 & 8 & 8 & 11 \end{bmatrix}$

b) $3\mathbf{P} + 2\mathbf{G} + 2\mathbf{M}$

$= \begin{bmatrix} 870 & 45 & 27 & 117 \end{bmatrix} +$
$\begin{bmatrix} 140 & 4 & 0 & 34 \end{bmatrix} +$
$\begin{bmatrix} 300 & 16 & 16 & 22 \end{bmatrix}$
$= \begin{bmatrix} 1310 & 65 & 43 & 173 \end{bmatrix}$

The entries represent the total nutritional value of 3 slices of pizza, 1 cup of gelatin, and 2 cups of whole milk.

25. Use a grapher.

$$\begin{bmatrix} -1 & 0 & 7 \\ 3 & -5 & 2 \end{bmatrix}\begin{bmatrix} 6 \\ -4 \\ 1 \end{bmatrix} = \begin{bmatrix} 1 \\ 40 \end{bmatrix}$$

26. Use a grapher.

$$\begin{bmatrix} 6 & -1 & 2 \end{bmatrix}\begin{bmatrix} 1 & 4 \\ -2 & 0 \\ 5 & -3 \end{bmatrix} = \begin{bmatrix} 18 & 18 \end{bmatrix}$$

27. Use a grapher.

$$\begin{bmatrix} -2 & 4 \\ 5 & 1 \\ -1 & -3 \end{bmatrix}\begin{bmatrix} 3 & -6 \\ -1 & 4 \end{bmatrix} = \begin{bmatrix} -10 & 28 \\ 14 & -26 \\ 0 & -6 \end{bmatrix}$$

28. Use a grapher.

$$\begin{bmatrix} 2 & -1 & 0 \\ 0 & 5 & 4 \end{bmatrix}\begin{bmatrix} -3 & 1 & 0 \\ 0 & 2 & -1 \\ 5 & 0 & 4 \end{bmatrix} = \begin{bmatrix} -6 & 0 & 1 \\ 20 & 10 & 11 \end{bmatrix}$$

29. $\begin{bmatrix} 1 \\ -5 \\ 3 \end{bmatrix}\begin{bmatrix} -6 & 5 & 8 \\ 0 & 4 & -1 \end{bmatrix}$

The grapher produces an error message when this multiplication is attempted. This product is not defined because the number of columns of the first matrix, 1, is not equal to the number of rows of the second matrix, 2.

30. Use a grapher.

$$\begin{bmatrix} 2 & 0 & 0 \\ 0 & -1 & 0 \\ 0 & 0 & 3 \end{bmatrix}\begin{bmatrix} 0 & -4 & 3 \\ 2 & 1 & 0 \\ -1 & 0 & 6 \end{bmatrix} = \begin{bmatrix} 0 & -8 & 6 \\ -2 & -1 & 0 \\ -3 & 0 & 18 \end{bmatrix}$$

31. Use a grapher.

$$\begin{bmatrix} 1 & -4 & 3 \\ 0 & 8 & 0 \\ -2 & -1 & 5 \end{bmatrix}\begin{bmatrix} 3 & 0 & 0 \\ 0 & -4 & 0 \\ 0 & 0 & 1 \end{bmatrix} = \begin{bmatrix} 3 & 16 & 3 \\ 0 & -32 & 0 \\ -6 & 4 & 5 \end{bmatrix}$$

32. $\begin{bmatrix} 4 \\ -5 \end{bmatrix}\begin{bmatrix} 2 & 0 \\ 6 & -7 \\ 0 & -3 \end{bmatrix}$

The grapher produces an error message when this multiplication is attempted. This product is not defined because the number of columns of the first matrix, 1, is not equal to the number of rows of the second matrix, 3.

33. a) $\mathbf{M} = \begin{bmatrix} 45.29 & 6.63 & 10.94 & 7.42 & 8.01 \\ 53.78 & 4.95 & 9.83 & 6.16 & 12.56 \\ 47.13 & 8.47 & 12.66 & 8.29 & 9.43 \\ 51.64 & 7.12 & 11.57 & 9.35 & 10.72 \end{bmatrix}$

b) $\mathbf{N} = \begin{bmatrix} 65 & 48 & 93 & 57 \end{bmatrix}$

c) Use a grapher.

$\mathbf{NM} =$

$\begin{bmatrix} 12{,}851.86 & 1862.1 & 3019.81 & 2081.9 & 2611.56 \end{bmatrix}$

d) The entries of **NM** represent the total cost, in cents, of each item for the day's meals.

34. a) $\mathbf{M} = \begin{bmatrix} 1 & 2.5 & 0.75 & 0.5 \\ 0 & 0.5 & 0.25 & 0 \\ 0.75 & 0.25 & 0.5 & 0.5 \\ 0.5 & 0 & 0.5 & 1 \end{bmatrix}$

b) $\mathbf{C} = \begin{bmatrix} 15 & 28 & 54 & 83 \end{bmatrix}$

c) $\mathbf{CM} = \begin{bmatrix} 97 & 65 & 86.75 & 117.5 \end{bmatrix}$

d) The entries of **CM** represent the total cost, in cents, of each menu item.

35. a) $\mathbf{S} = \begin{bmatrix} 8 & 15 \\ 6 & 10 \\ 4 & 3 \end{bmatrix}$

b) $\mathbf{C} = \begin{bmatrix} 3 & 1.5 & 2 \end{bmatrix}$

c) $\mathbf{CS} = \begin{bmatrix} 41 & 66 \end{bmatrix}$

d) The entries of **CS** represent the total cost, in dollars, of ingredients for each coffee shop.

36. a) $\mathbf{M} = \begin{bmatrix} 900 & 500 \\ 450 & 1000 \\ 600 & 700 \end{bmatrix}$

b) $\mathbf{P} = \begin{bmatrix} 5 & 8 & 4 \end{bmatrix}$

c) $\mathbf{PM} = \begin{bmatrix} 10{,}500 & 13{,}300 \end{bmatrix}$

d) The entries of **PM** represent the total profit from each distributor.

37. a) $\mathbf{P} = \begin{bmatrix} 6 & 4.5 & 5.2 \end{bmatrix}$

b) $\mathbf{PS} = \begin{bmatrix} 6 & 4.5 & 5.2 \end{bmatrix}\begin{bmatrix} 8 & 15 \\ 6 & 10 \\ 4 & 3 \end{bmatrix}$

$= \begin{bmatrix} 95.8 & 150.6 \end{bmatrix}$

The profit from Mugsey's Coffee Shop is \$95.80, and the profit from The Coffee Club is \$150.60.

38. a) $\mathbf{C} = \begin{bmatrix} 20 & 25 & 15 \end{bmatrix}$

b) $\mathbf{CM} = \begin{bmatrix} 20 & 25 & 15 \end{bmatrix}\begin{bmatrix} 900 & 500 \\ 450 & 1000 \\ 600 & 700 \end{bmatrix}$

$= \begin{bmatrix} 38{,}250 & 45{,}500 \end{bmatrix}$

The total production costs for the products shipped to Distributors 1 and 2 are \$38,250 and \$45,500, respectively.

39. $2x - 3y = 7,$

$x + 5y = -6$

Write the coefficients on the left in a matrix. Then write the product of that matrix and the column matrix containing the variables, and set the result equal to the column matrix containing the constants on the right.

$$\begin{bmatrix} 2 & -3 \\ 1 & 5 \end{bmatrix}\begin{bmatrix} x \\ y \end{bmatrix} = \begin{bmatrix} 7 \\ -6 \end{bmatrix}$$

40. $\begin{bmatrix} -1 & 1 \\ 5 & -4 \end{bmatrix}\begin{bmatrix} x \\ y \end{bmatrix} = \begin{bmatrix} 3 \\ 16 \end{bmatrix}$

41. $x + y - 2z = 6,$
$3x - y + z = 7,$
$2x + 5y - 3z = 8$

Write the coefficients on the left in a matrix. Then write the product of that matrix and the column matrix containing the variables, and set the result equal to the column matrix containing the constants on the right.

$$\begin{bmatrix} 1 & 1 & -2 \\ 3 & -1 & 1 \\ 2 & 5 & -3 \end{bmatrix}\begin{bmatrix} x \\ y \\ z \end{bmatrix} = \begin{bmatrix} 6 \\ 7 \\ 8 \end{bmatrix}$$

42. $\begin{bmatrix} 3 & -1 & 1 \\ 1 & 2 & -1 \\ 4 & 3 & -2 \end{bmatrix}\begin{bmatrix} x \\ y \\ z \end{bmatrix} = \begin{bmatrix} 1 \\ 3 \\ 11 \end{bmatrix}$

43. $3x - 2y + 4z = 17,$
$2x + y - 5z = 13$

Write the coefficients on the left in a matrix. Then write the product of that matrix and the column matrix containing the variables, and set the result equal to the column matrix containing the constants on the right.

$$\begin{bmatrix} 3 & -2 & 4 \\ 2 & 1 & -5 \end{bmatrix}\begin{bmatrix} x \\ y \\ z \end{bmatrix} = \begin{bmatrix} 17 \\ 13 \end{bmatrix}$$

44. $\begin{bmatrix} 3 & 2 & 5 \\ 4 & -3 & 2 \end{bmatrix}\begin{bmatrix} x \\ y \\ z \end{bmatrix} = \begin{bmatrix} 9 \\ 10 \end{bmatrix}$

45. $-4w + x - y + 2z = 12,$
$w + 2x - y - z = 0,$
$-w + x + 4y - 3z = 1,$
$2w + 3x + 5y - 7z = 9$

Write the coefficients on the left in a matrix. Then write the product of that matrix and the column matrix containing the variables, and set the result equal to the column matrix containing the constants on the right.

$$\begin{bmatrix} -4 & 1 & -1 & 2 \\ 1 & 2 & -1 & -1 \\ -1 & 1 & 4 & -3 \\ 2 & 3 & 5 & -7 \end{bmatrix}\begin{bmatrix} w \\ x \\ y \\ z \end{bmatrix} = \begin{bmatrix} 12 \\ 0 \\ 1 \\ 9 \end{bmatrix}$$

46. $\begin{bmatrix} 12 & 2 & 4 & -5 \\ -1 & 4 & -1 & 12 \\ 2 & -1 & 4 & 0 \\ 0 & 2 & 10 & 1 \end{bmatrix}\begin{bmatrix} w \\ x \\ y \\ z \end{bmatrix} = \begin{bmatrix} 2 \\ 5 \\ 13 \\ 5 \end{bmatrix}$

47. No; see Exercise 17, for example.

48. She could make decisions regarding cost and profit, including how many of each product to prepare and distribute to each shop.

49. $f(x) = x^2 - 3x - 10$

a) $-\dfrac{b}{2a} = -\dfrac{-3}{2\cdot 1} = \dfrac{3}{2}$

$f\left(\dfrac{3}{2}\right) = \left(\dfrac{3}{2}\right)^2 - 3\left(\dfrac{3}{2}\right) - 10 = -\dfrac{49}{4}$

The vertex is $\left(\dfrac{3}{2}, -\dfrac{49}{4}\right)$.

b) The line of symmetry is $x = \dfrac{3}{2}$.

c) Since the coefficient of x^2 is positive, the function has a minimum value. It is the second coordinate of the vertex, $-\dfrac{49}{4}$.

50. $f(x) = 2x^2 - 5x - 3$

a) $-\dfrac{b}{2a} = -\dfrac{-5}{2\cdot 2} = \dfrac{5}{4}$

$f\left(\dfrac{5}{4}\right) = 2\left(\dfrac{5}{4}\right)^2 - 5\left(\dfrac{5}{4}\right) - 3 = -\dfrac{49}{8}$

The vertex is $\left(\dfrac{5}{4}, -\dfrac{49}{8}\right)$.

b) $x = \dfrac{5}{4}$

c) Minimum: $-\dfrac{49}{8}$

51. $f(x) = -x^2 - 3x + 5$

a) $-\dfrac{b}{2a} = -\dfrac{-3}{2(-1)} = -\dfrac{3}{2}$

$f\left(-\dfrac{3}{2}\right) = -\left(-\dfrac{3}{2}\right)^2 - 3\left(-\dfrac{3}{2}\right) + 5 = \dfrac{29}{4}$

The vertex is $\left(-\dfrac{3}{2}, \dfrac{29}{4}\right)$.

b) The line of symmetry is $x = -\dfrac{3}{2}$.

c) Since the coefficient of x^2 is negative, the function has a maximum value. It is the second coordinate of the vertex, $\dfrac{29}{4}$.

52. $f(x) = -3x^2 + 4x + 1$

a) $-\dfrac{b}{2a} = -\dfrac{4}{2(-3)} = \dfrac{2}{3}$

$f\left(\dfrac{2}{3}\right) = -3\left(\dfrac{2}{3}\right)^2 + 4\left(\dfrac{2}{3}\right) + 1 = \dfrac{7}{3}$

The vertex is $\left(\dfrac{2}{3}, \dfrac{7}{3}\right)$.

b) $x = \dfrac{2}{3}$

c) Maximum: $\dfrac{7}{3}$

53. $\mathbf{A} = \begin{bmatrix} -1 & 0 \\ 2 & 1 \end{bmatrix}$, $\mathbf{B} = \begin{bmatrix} 1 & -1 \\ 0 & 2 \end{bmatrix}$

$$(\mathbf{A}+\mathbf{B})(\mathbf{A}-\mathbf{B}) = \begin{bmatrix} 0 & -1 \\ 2 & 3 \end{bmatrix}\begin{bmatrix} -2 & 1 \\ 2 & -1 \end{bmatrix} = \begin{bmatrix} -2 & 1 \\ 2 & -1 \end{bmatrix}$$

$\mathbf{A}^2 - \mathbf{B}^2$

$$= \begin{bmatrix} -1 & 0 \\ 2 & 1 \end{bmatrix}\begin{bmatrix} -1 & 0 \\ 2 & 1 \end{bmatrix} - \begin{bmatrix} 1 & -1 \\ 0 & 2 \end{bmatrix}\begin{bmatrix} 1 & -1 \\ 0 & 2 \end{bmatrix}$$

$$= \begin{bmatrix} 1 & 0 \\ 0 & 1 \end{bmatrix} - \begin{bmatrix} 1 & -3 \\ 0 & 4 \end{bmatrix}$$

$$= \begin{bmatrix} 0 & 3 \\ 0 & -3 \end{bmatrix}$$

Thus $(\mathbf{A}+\mathbf{B})(\mathbf{A}-\mathbf{B}) \neq \mathbf{A}^2 - \mathbf{B}^2$.

54. $\mathbf{A} = \begin{bmatrix} -1 & 0 \\ 2 & 1 \end{bmatrix}$, $\mathbf{B} = \begin{bmatrix} 1 & -1 \\ 0 & 2 \end{bmatrix}$

$$(\mathbf{A}+\mathbf{B})(\mathbf{A}+\mathbf{B}) = \begin{bmatrix} 0 & -1 \\ 2 & 3 \end{bmatrix}\begin{bmatrix} 0 & -1 \\ 2 & 3 \end{bmatrix} = \begin{bmatrix} -2 & -3 \\ 6 & 7 \end{bmatrix}$$

We found $\mathbf{A}^2$ and $\mathbf{B}^2$ in Exercise 53.

$$2\mathbf{AB} = 2\begin{bmatrix} -1 & 0 \\ 2 & 1 \end{bmatrix}\begin{bmatrix} 1 & -1 \\ 0 & 2 \end{bmatrix} = \begin{bmatrix} -2 & 2 \\ 4 & 0 \end{bmatrix}$$

$\mathbf{A}^2 + 2\mathbf{AB} + \mathbf{B}^2$

$$= \begin{bmatrix} 1 & 0 \\ 0 & 1 \end{bmatrix} + \begin{bmatrix} -2 & 2 \\ 4 & 0 \end{bmatrix} + \begin{bmatrix} 1 & -3 \\ 0 & 4 \end{bmatrix}$$

$$= \begin{bmatrix} 0 & -1 \\ 4 & 5 \end{bmatrix}$$

Thus $(\mathbf{A}+\mathbf{B})(\mathbf{A}+\mathbf{B}) \neq \mathbf{A}^2 + 2\mathbf{AB} + \mathbf{B}^2$.

55. In Exercise 53 we found that $(\mathbf{A}+\mathbf{B})(\mathbf{A}-\mathbf{B}) = \begin{bmatrix} -2 & 1 \\ 2 & -1 \end{bmatrix}$ and we also found $\mathbf{A}^2$ and $\mathbf{B}^2$.

$$\mathbf{BA} = \begin{bmatrix} 1 & -1 \\ 0 & 2 \end{bmatrix}\begin{bmatrix} -1 & 0 \\ 2 & 1 \end{bmatrix} = \begin{bmatrix} -3 & -1 \\ 4 & 2 \end{bmatrix}$$

$$\mathbf{AB} = \begin{bmatrix} -1 & 0 \\ 2 & 1 \end{bmatrix}\begin{bmatrix} 1 & -1 \\ 0 & 2 \end{bmatrix} = \begin{bmatrix} -1 & 1 \\ 2 & 0 \end{bmatrix}$$

$\mathbf{A}^2 + \mathbf{BA} - \mathbf{AB} - \mathbf{B}^2$

$$= \begin{bmatrix} 1 & 0 \\ 0 & 1 \end{bmatrix} + \begin{bmatrix} -3 & -1 \\ 4 & 2 \end{bmatrix} - \begin{bmatrix} -1 & 1 \\ 2 & 0 \end{bmatrix} - \begin{bmatrix} 1 & -3 \\ 0 & 4 \end{bmatrix}$$

$$= \begin{bmatrix} -2 & 1 \\ 2 & -1 \end{bmatrix}$$

Thus $(\mathbf{A}+\mathbf{B})(\mathbf{A}-\mathbf{B}) = \mathbf{A}^2 + \mathbf{BA} - \mathbf{AB} - \mathbf{B}^2$.

56. In Exercise 54 we found that

$$(\mathbf{A}+\mathbf{B})(\mathbf{A}+\mathbf{B}) = \begin{bmatrix} -2 & -3 \\ 6 & 7 \end{bmatrix}.$$

We found $\mathbf{A}^2$ and $\mathbf{B}^2$ in Exercise 53, and we found $\mathbf{BA}$ and $\mathbf{AB}$ in Exercise 55.

$\mathbf{A}^2 + \mathbf{BA} + \mathbf{AB} + \mathbf{B}^2$

$$= \begin{bmatrix} 1 & 0 \\ 0 & 1 \end{bmatrix} + \begin{bmatrix} -3 & -1 \\ 4 & 2 \end{bmatrix} + \begin{bmatrix} -1 & 1 \\ 2 & 0 \end{bmatrix} + \begin{bmatrix} 1 & -3 \\ 0 & 4 \end{bmatrix}$$

$$= \begin{bmatrix} -2 & -3 \\ 6 & 7 \end{bmatrix}$$

Thus $(\mathbf{A}+\mathbf{B})(\mathbf{A}+\mathbf{B}) = \mathbf{A}^2 + \mathbf{BA} + \mathbf{AB} + \mathbf{B}^2$.

57. See the answer section in the text.

58. $\mathbf{A} + (\mathbf{B} + \mathbf{C})$

$$= \begin{bmatrix} a_{11}+(b_{11}+c_{11}) & \cdots & a_{1n}+(b_{1n}+c_{1n}) \\ a_{21}+(b_{21}+c_{21}) & \cdots & a_{2n}+(b_{2n}+c_{2n}) \\ \vdots & \vdots & \vdots \\ a_{m1}+(b_{m1}+c_{m1}) & \cdots & a_{mn}+(b_{mn}+c_{mn}) \end{bmatrix}$$

$$= \begin{bmatrix} (a_{11}+b_{11})+c_{11} & \cdots & (a_{1n}+b_{1n})+c_{1n} \\ (a_{21}+b_{21})+c_{21} & \cdots & (a_{2n}+b_{2n})+c_{2n} \\ \vdots & \vdots & \vdots \\ (a_{m1}+b_{m1})+c_{m1} & \cdots & (a_{mn}+b_{mn})+c_{mn} \end{bmatrix}$$

$$= (\mathbf{A}+\mathbf{B})+\mathbf{C}$$

59. See the answer section in the text.

60. $k(\mathbf{A}+\mathbf{B})$

$$= \begin{bmatrix} k(a_{11}+b_{11}) & \cdots & k(a_{1n}+b_{1n}) \\ k(a_{21}+b_{21}) & \cdots & k(a_{2n}+b_{2n}) \\ \vdots & & \vdots \\ k(a_{m1}+b_{m1}) & \cdots & k(a_{mn}+b_{mn}) \end{bmatrix}$$

$$= \begin{bmatrix} ka_{11}+kb_{11} & \cdots & ka_{1n}+kb_{1n} \\ ka_{21}+kb_{21} & \cdots & ka_{2n}+kb_{2n} \\ \vdots & & \vdots \\ ka_{m1}+kb_{m1} & \cdots & ka_{mn}+kb_{mn} \end{bmatrix}$$

$$= k\mathbf{A} + k\mathbf{B}$$

61. See the answer section in the text.

Exercise Set 5.5

1. $\mathbf{BA} = \begin{bmatrix} 7 & 3 \\ 2 & 1 \end{bmatrix}\begin{bmatrix} 1 & -3 \\ -2 & 7 \end{bmatrix} = \begin{bmatrix} 1 & 0 \\ 0 & 1 \end{bmatrix}$

$\mathbf{AB} = \begin{bmatrix} 1 & -3 \\ -2 & 7 \end{bmatrix}\begin{bmatrix} 7 & 3 \\ 2 & 1 \end{bmatrix} = \begin{bmatrix} 1 & 0 \\ 0 & 1 \end{bmatrix}$

Since $\mathbf{BA} = \mathbf{I} = \mathbf{AB}$, $\mathbf{B}$ is the inverse of $\mathbf{A}$.

2. $\mathbf{BA} = \mathbf{I} = \mathbf{AB}$, so $\mathbf{B}$ is the inverse of $\mathbf{A}$.

3. $\mathbf{BA} = \begin{bmatrix} 2 & 3 & 2 \\ 3 & 3 & 4 \\ 1 & 1 & 1 \end{bmatrix} \begin{bmatrix} -1 & -1 & 6 \\ 1 & 0 & -2 \\ 1 & 0 & -3 \end{bmatrix} = \begin{bmatrix} 3 & -2 & 0 \\ 4 & -3 & 0 \\ 1 & -1 & 1 \end{bmatrix}$

Since $\mathbf{BA} \neq \mathbf{I}$, $\mathbf{B}$ is not the inverse of $\mathbf{A}$.

4. $\mathbf{BA} = \begin{bmatrix} 1 & 0 & -24 \\ 0 & 1 & 8 \\ 0 & 0 & 17 \end{bmatrix} \neq \mathbf{I}$, so $\mathbf{B}$ is not the inverse of $\mathbf{A}$.

5. $\mathbf{A} = \begin{bmatrix} 3 & 2 \\ 5 & 3 \end{bmatrix}$

Write the augmented matrix.

$$\left[\begin{array}{cc|cc} 3 & 2 & 1 & 0 \\ 5 & 3 & 0 & 1 \end{array}\right]$$

Multiply row 2 by 3.

$$\left[\begin{array}{cc|cc} 3 & 2 & 1 & 0 \\ 15 & 9 & 0 & 3 \end{array}\right]$$

Multiply row 1 by -5 and add it to row 2.

$$\left[\begin{array}{cc|cc} 3 & 2 & 1 & 0 \\ 0 & -1 & -5 & 3 \end{array}\right]$$

Multiply row 2 by 2 and add it to row 1.

$$\left[\begin{array}{cc|cc} 3 & 0 & -9 & 6 \\ 0 & -1 & -5 & 3 \end{array}\right]$$

Multiply row 1 by $\frac{1}{3}$ and row 2 by -1.

$$\left[\begin{array}{cc|cc} 1 & 0 & -3 & 2 \\ 0 & 1 & 5 & -3 \end{array}\right]$$

Then $\mathbf{A}^{-1} = \begin{bmatrix} -3 & 2 \\ 5 & -3 \end{bmatrix}$.

6. $\mathbf{A} = \begin{bmatrix} 3 & 5 \\ 1 & 2 \end{bmatrix}$

Write the augmented matrix.

$$\left[\begin{array}{cc|cc} 3 & 5 & 1 & 0 \\ 1 & 2 & 0 & 1 \end{array}\right]$$

Interchange the rows.

$$\left[\begin{array}{cc|cc} 1 & 2 & 0 & 1 \\ 3 & 5 & 1 & 0 \end{array}\right]$$

Multiply row 1 by -3 and add it to row 2.

$$\left[\begin{array}{cc|cc} 1 & 2 & 0 & 1 \\ 0 & -1 & 1 & -3 \end{array}\right]$$

Multiply row 2 by 2 and add it to row 1.

$$\left[\begin{array}{cc|cc} 1 & 0 & 2 & -5 \\ 0 & -1 & 1 & -3 \end{array}\right]$$

Multiply row 2 by -1.

$$\left[\begin{array}{cc|cc} 1 & 0 & 2 & -5 \\ 0 & 1 & -1 & 3 \end{array}\right]$$

Then $\mathbf{A}^{-1} = \begin{bmatrix} 2 & -5 \\ -1 & 3 \end{bmatrix}$.

7. $\mathbf{A} = \begin{bmatrix} 6 & 9 \\ 4 & 6 \end{bmatrix}$

Write the augmented matrix.

$$\left[\begin{array}{cc|cc} 6 & 9 & 1 & 0 \\ 4 & 6 & 0 & 1 \end{array}\right]$$

Multiply row 2 by 3.

$$\left[\begin{array}{cc|cc} 6 & 9 & 1 & 0 \\ 12 & 18 & 0 & 3 \end{array}\right]$$

Multiply row 1 by -2 and add it to row 2.

$$\left[\begin{array}{cc|cc} 6 & 9 & 1 & 0 \\ 0 & 0 & -2 & 3 \end{array}\right]$$

We cannot obtain the identity matrix on the left since the second row contains only zeros to the left of the vertical line. Thus, $\mathbf{A}^{-1}$ does not exist.

8. $\mathbf{A} = \begin{bmatrix} -4 & -6 \\ 2 & 3 \end{bmatrix}$

Write the augmented matrix.

$$\left[\begin{array}{cc|cc} -4 & -6 & 1 & 0 \\ 2 & 3 & 0 & 1 \end{array}\right]$$

Multiply row 2 by 2.

$$\left[\begin{array}{cc|cc} -4 & -6 & 1 & 0 \\ 4 & 6 & 0 & 2 \end{array}\right]$$

Add row 1 to row 2.

$$\left[\begin{array}{cc|cc} -4 & -6 & 1 & 0 \\ 0 & 0 & 1 & 2 \end{array}\right]$$

We cannot obtain the identity matrix on the left since the second row contains only zeros to the left of the vertical line. Thus, $\mathbf{A}^{-1}$ does not exist.

9. $\mathbf{A} = \begin{bmatrix} 3 & 1 & 0 \\ 1 & 1 & 1 \\ 1 & -1 & 2 \end{bmatrix}$

Write the augmented matrix.

$$\left[\begin{array}{ccc|ccc} 3 & 1 & 0 & 1 & 0 & 0 \\ 1 & 1 & 1 & 0 & 1 & 0 \\ 1 & -1 & 2 & 0 & 0 & 1 \end{array}\right]$$

Interchange the first two rows.

$$\left[\begin{array}{ccc|ccc} 1 & 1 & 1 & 0 & 1 & 0 \\ 3 & 1 & 0 & 1 & 0 & 0 \\ 1 & -1 & 2 & 0 & 0 & 1 \end{array}\right]$$

Multiply row 1 by -3 and add it to row 2. Also, multiply row 1 by -1 and add it to row 3.

$$\left[\begin{array}{ccc|ccc} 1 & 1 & 1 & 0 & 1 & 0 \\ 0 & -2 & -3 & 1 & -3 & 0 \\ 0 & -2 & 1 & 0 & -1 & 1 \end{array}\right]$$

Multiply row 2 by $-\frac{1}{2}$.

$$\left[\begin{array}{ccc|ccc} 1 & 1 & 1 & 0 & 1 & 0 \\ 0 & 1 & \frac{3}{2} & -\frac{1}{2} & \frac{3}{2} & 0 \\ 0 & -2 & 1 & 0 & -1 & 1 \end{array}\right]$$

Multiply row 2 by -1 and add it to row 1. Also, multiply row 2 by 2 and add it to row 3.

$$\left[\begin{array}{ccc|ccc} 1 & 0 & -\frac{1}{2} & \frac{1}{2} & -\frac{1}{2} & 0 \\ 0 & 1 & \frac{3}{2} & -\frac{1}{2} & \frac{3}{2} & 0 \\ 0 & 0 & 4 & -1 & 2 & 1 \end{array}\right]$$

Multiply row 3 by $\frac{1}{4}$.

$$\left[\begin{array}{ccc|ccc} 1 & 0 & -\frac{1}{2} & \frac{1}{2} & -\frac{1}{2} & 0 \\ 0 & 1 & \frac{3}{2} & -\frac{1}{2} & \frac{3}{2} & 0 \\ 0 & 0 & 1 & -\frac{1}{4} & \frac{1}{2} & \frac{1}{4} \end{array}\right]$$

Multiply row 3 by $\frac{1}{2}$ and add it to row 1. Also, multiply row 3 by $-\frac{3}{2}$ and add it to row 2.

$$\left[\begin{array}{ccc|ccc} 1 & 0 & 0 & \frac{3}{8} & -\frac{1}{4} & \frac{1}{8} \\ 0 & 1 & 0 & -\frac{1}{8} & \frac{3}{4} & -\frac{3}{8} \\ 0 & 0 & 1 & -\frac{1}{4} & \frac{1}{2} & \frac{1}{4} \end{array}\right]$$

Then $\mathbf{A}^{-1} = \begin{bmatrix} \frac{3}{8} & -\frac{1}{4} & \frac{1}{8} \\ -\frac{1}{8} & \frac{3}{4} & -\frac{3}{8} \\ -\frac{1}{4} & \frac{1}{2} & \frac{1}{4} \end{bmatrix}$.

10. $\mathbf{A} = \begin{bmatrix} 1 & 0 & 1 \\ 2 & 1 & 0 \\ 1 & -1 & 1 \end{bmatrix}$

Write the augmented matrix.

$$\left[\begin{array}{ccc|ccc} 1 & 0 & 1 & 1 & 0 & 0 \\ 2 & 1 & 0 & 0 & 1 & 0 \\ 1 & -1 & 1 & 0 & 0 & 1 \end{array}\right]$$

Multiply row 1 by -2 and add it to row 2. Also, multiply row 1 by -1 and add it to row 3.

$$\left[\begin{array}{ccc|ccc} 1 & 0 & 1 & 1 & 0 & 0 \\ 0 & 1 & -2 & -2 & 1 & 0 \\ 0 & -1 & 0 & -1 & 0 & 1 \end{array}\right]$$

Add row 2 to row 3.

$$\left[\begin{array}{ccc|ccc} 1 & 0 & 1 & 1 & 0 & 0 \\ 0 & 1 & -2 & -2 & 1 & 0 \\ 0 & 0 & -2 & -3 & 1 & 1 \end{array}\right]$$

Multiply row 3 by $-\frac{1}{2}$.

$$\left[\begin{array}{ccc|ccc} 1 & 0 & 1 & 1 & 0 & 0 \\ 0 & 1 & -2 & -2 & 1 & 0 \\ 0 & 0 & 1 & \frac{3}{2} & -\frac{1}{2} & -\frac{1}{2} \end{array}\right]$$

Multiply row 3 by -1 and add it to row 1. Also, multiply row 3 by 2 and add it to row 2.

$$\left[\begin{array}{ccc|ccc} 1 & 0 & 0 & -\frac{1}{2} & \frac{1}{2} & \frac{1}{2} \\ 0 & 1 & 0 & 1 & 0 & -1 \\ 0 & 0 & 1 & \frac{3}{2} & -\frac{1}{2} & -\frac{1}{2} \end{array}\right]$$

Then $\mathbf{A}^{-1} = \begin{bmatrix} -\frac{1}{2} & \frac{1}{2} & \frac{1}{2} \\ 1 & 0 & -1 \\ \frac{3}{2} & -\frac{1}{2} & -\frac{1}{2} \end{bmatrix}$.

11. $\mathbf{A} = \begin{bmatrix} 1 & -4 & 8 \\ 1 & -3 & 2 \\ 2 & -7 & 10 \end{bmatrix}$

Write the augmented matrix.

$$\left[\begin{array}{ccc|ccc} 1 & -4 & 8 & 1 & 0 & 0 \\ 1 & -3 & 2 & 0 & 1 & 0 \\ 2 & -7 & 10 & 0 & 0 & 1 \end{array}\right]$$

Multiply row 1 by -1 and add it to row 2. Also, multiply row 1 by -2 and add it to row 3.

$$\left[\begin{array}{ccc|ccc} 1 & -4 & 8 & 1 & 0 & 0 \\ 0 & 1 & -6 & -1 & 1 & 0 \\ 0 & 1 & -6 & -2 & 0 & 1 \end{array}\right]$$

Since the second and third rows are identical left of the vertical line, it will not be possible to obtain the identity matrix on the left side. Thus, $\mathbf{A}^{-1}$ does not exist.

12. $\mathbf{A} = \begin{bmatrix} -2 & 5 & 3 \\ 4 & -1 & 3 \\ 7 & -2 & 5 \end{bmatrix}$

Write the augmented matrix.

$$\left[\begin{array}{ccc|ccc} -2 & 5 & 3 & 1 & 0 & 0 \\ 4 & -1 & 3 & 0 & 1 & 0 \\ 7 & -2 & 5 & 0 & 0 & 1 \end{array}\right]$$

Multiply row 3 by 2.

$$\left[\begin{array}{ccc|ccc} -2 & 5 & 3 & 1 & 0 & 0 \\ 4 & -1 & 3 & 0 & 1 & 0 \\ 14 & -4 & 10 & 0 & 0 & 2 \end{array}\right]$$

Multiply row 1 by 2 and add it to row 2. Also, multiply row 1 by 7 and add it to row 3.

$$\left[\begin{array}{ccc|ccc} -2 & 5 & 3 & 1 & 0 & 0 \\ 0 & 9 & 9 & 2 & 1 & 0 \\ 0 & 31 & 31 & 7 & 0 & 2 \end{array}\right]$$

To the left of the vertical line, row 3 is a multiple of row 2 so it will not be possible to obtain the identity matrix on the left. Thus, $\mathbf{A}^{-1}$ does not exist.

13. Use a grapher.

$$\mathbf{A}^{-1} = \begin{bmatrix} 0.4 & -0.6 \\ 0.2 & -0.8 \end{bmatrix}$$

14. Use a grapher.

$$\mathbf{A}^{-1} = \begin{bmatrix} 0 & 1 \\ -1 & 0 \end{bmatrix}$$

15. Use a grapher.

$$\mathbf{A}^{-1} = \begin{bmatrix} -1 & -1 & -6 \\ 1 & 0 & 2 \\ 0 & 1 & 3 \end{bmatrix}$$

16. Use a grapher.

$$\mathbf{A}^{-1} = \begin{bmatrix} 0.3 & 0.3 & -0.1 \\ -0.4 & 0.6 & 0.8 \\ 0.5 & -0.5 & -0.5 \end{bmatrix}$$

17. Use a grapher.

$$\mathbf{A}^{-1} = \begin{bmatrix} 1 & 1 & 2 \\ 1 & 1 & 1 \\ 2 & 3 & 4 \end{bmatrix}$$

18. Use a grapher.

$$\mathbf{A}^{-1} = \begin{bmatrix} 0 & -1.5 & -1 \\ -1 & -1.5 & -2.5 \\ 0 & -1 & -0.5 \end{bmatrix}$$

19. The grapher produces an error message. Thus, $\mathbf{A}^{-1}$ does not exist.

20. The grapher produces an error message. Thus, $\mathbf{A}^{-1}$ does not exist.

21. Use a grapher.

$$\mathbf{A}^{-1} = \begin{bmatrix} 1 & -2 & 3 & 8 \\ 0 & 1 & -3 & 1 \\ 0 & 0 & 1 & -2 \\ 0 & 0 & 0 & -1 \end{bmatrix}$$

22. The grapher produces an error message. Thus, $\mathbf{A}^{-1}$ does not exist.

23. Use a grapher.

$$\mathbf{A}^{-1} = \begin{bmatrix} 0.25 & 0.25 & 1.25 & -0.25 \\ 0.5 & 1.25 & 1.75 & -1 \\ -0.25 & -0.25 & -0.75 & 0.75 \\ 0.25 & 0.5 & 0.75 & -0.5 \end{bmatrix}$$

24. Use a grapher.

$$\mathbf{A}^{-1} = \begin{bmatrix} 0.2 & 0.4 & 0.2 & -0.2 \\ -0.2 & -0.2 & 0.2 & 1 \\ 0.2 & 0.8 & 1.2 & 1 \\ 0.6 & 1.6 & 2 & 0.8 \end{bmatrix}$$

25. $4x + 3y = 2,$

$x - 2y = 6$

Write an equivalent matrix equation, $\mathbf{AX} = \mathbf{B}$.

$$\begin{bmatrix} 4 & 3 \\ 1 & -2 \end{bmatrix} \begin{bmatrix} x \\ y \end{bmatrix} = \begin{bmatrix} 2 \\ 6 \end{bmatrix}$$

Then $\mathbf{X} = \mathbf{A}^{-1}\mathbf{B} = \begin{bmatrix} 2 \\ -2 \end{bmatrix}$.

The solution is $(2, -2)$.

26. $2x - 3y = 7,$

$4x + y = -7$

Write an equivalent matrix equation, $\mathbf{AX} = \mathbf{B}$.

$$\begin{bmatrix} 2 & -3 \\ 4 & 1 \end{bmatrix} \begin{bmatrix} x \\ y \end{bmatrix} = \begin{bmatrix} 7 \\ -7 \end{bmatrix}$$

Then $\mathbf{X} = \mathbf{A}^{-1}\mathbf{B} = \begin{bmatrix} -1 \\ -3 \end{bmatrix}$.

The solution is $(-1, -3)$.

27. $5x + y = 2,$

$3x - 2y = -4$

Write an equivalent matrix equation, $\mathbf{AX} = \mathbf{B}$.

$$\begin{bmatrix} 5 & 1 \\ 3 & -2 \end{bmatrix} \begin{bmatrix} x \\ y \end{bmatrix} = \begin{bmatrix} 2 \\ -4 \end{bmatrix}$$

Then $\mathbf{X} = \mathbf{A}^{-1}\mathbf{B} = \begin{bmatrix} 0 \\ 2 \end{bmatrix}$.

The solution is $(0, 2)$.

28. $x - 6y = 5,$

$-x + 4y = -5$

Write an equivalent matrix equation, $\mathbf{AX} = \mathbf{B}$.

$$\begin{bmatrix} 1 & -6 \\ -1 & 4 \end{bmatrix} \begin{bmatrix} x \\ y \end{bmatrix} = \begin{bmatrix} 5 \\ -5 \end{bmatrix}$$

Then $\mathbf{X} = \mathbf{A}^{-1}\mathbf{B} = \begin{bmatrix} 5 \\ 0 \end{bmatrix}$.

The solution is $(5, 0)$.

29. $x + z = 1,$

$2x + y = 3,$

$x - y + z = 4$

Write an equivalent matrix equation, $\mathbf{AX} = \mathbf{B}$.

$$\begin{bmatrix} 1 & 0 & 1 \\ 2 & 1 & 0 \\ 1 & -1 & 1 \end{bmatrix} \begin{bmatrix} x \\ y \\ z \end{bmatrix} = \begin{bmatrix} 1 \\ 3 \\ 4 \end{bmatrix}$$

Then $\mathbf{X} = \mathbf{A}^{-1}\mathbf{B} = \begin{bmatrix} 3 \\ -3 \\ -2 \end{bmatrix}$.

The solution is $(3, -3, -2)$.

30. $x + 2y + 3z = -1,$

$2x - 3y + 4z = 2,$

$-3x + 5y - 6z = 4$

Write an equivalent matrix equation, $\mathbf{AX} = \mathbf{B}$.

$$\begin{bmatrix} 1 & 2 & 3 \\ 2 & -3 & 4 \\ -3 & 5 & -6 \end{bmatrix} \begin{bmatrix} x \\ y \\ z \end{bmatrix} = \begin{bmatrix} -1 \\ 2 \\ 4 \end{bmatrix}$$

Then $\mathbf{X} = \mathbf{A}^{-1}\mathbf{B} = \begin{bmatrix} 124 \\ 14 \\ -51 \end{bmatrix}$.

The solution is $(124, 14, -51)$.

31. $2x + 3y + 4z = 2,$

$x - 4y + 3z = 2,$

$5x + y + z = -4$

Write an equivalent matrix equation, $\mathbf{AX} = \mathbf{B}$.

$$\begin{bmatrix} 2 & 3 & 4 \\ 1 & -4 & 3 \\ 5 & 1 & 1 \end{bmatrix} \begin{bmatrix} x \\ y \\ z \end{bmatrix} = \begin{bmatrix} 2 \\ 2 \\ -4 \end{bmatrix}$$

Then $\mathbf{X} = \mathbf{A}^{-1}\mathbf{B} = \begin{bmatrix} -1 \\ 0 \\ 1 \end{bmatrix}$.

The solution is $(-1, 0, 1)$.

32. $x + y \quad = 2,$

$3x \quad + 2z = 5,$

$2x + 3y - 3z = 9$

Write an equivalent matrix equation, $\mathbf{AX} = \mathbf{B}$.

$$\begin{bmatrix} 1 & 1 & 0 \\ 3 & 0 & 2 \\ 2 & 3 & -3 \end{bmatrix}\begin{bmatrix} x \\ y \\ z \end{bmatrix} = \begin{bmatrix} 2 \\ 5 \\ 9 \end{bmatrix}$$

Then $\mathbf{X} = \mathbf{A}^{-1}\mathbf{B} = \begin{bmatrix} 3 \\ -1 \\ -2 \end{bmatrix}$.

The solution is $(3, -1, -2)$.

33. $2w - 3x + 4y - 5z = 0,$

$3w - 2x + 7y - 3z = 2,$

$w + x - y + z = 1,$

$-w - 3x - 6y + 4z = 6$

Write an equivalent matrix equation, $\mathbf{AX} = \mathbf{B}$.

$$\begin{bmatrix} 2 & -3 & 4 & -5 \\ 3 & -2 & 7 & -3 \\ 1 & 1 & -1 & 1 \\ -1 & -3 & -6 & 4 \end{bmatrix}\begin{bmatrix} w \\ x \\ y \\ z \end{bmatrix} = \begin{bmatrix} 0 \\ 2 \\ 1 \\ 6 \end{bmatrix}$$

Then $\mathbf{X} = \mathbf{A}^{-1}\mathbf{B} = \begin{bmatrix} 1 \\ -1 \\ 0 \\ 1 \end{bmatrix}$.

The solution is $(1, -1, 0, 1)$.

34. $5w - 4x + 3y - 2z = -6,$

$w + 4x - 2y + 3z = -5,$

$2w - 3x + 6y - 9z = 14,$

$3w - 5x + 2y - 4z = -3$

Write an equivalent matrix equation, $\mathbf{AX} = \mathbf{B}$.

$$\begin{bmatrix} 5 & -4 & 3 & -2 \\ 1 & 4 & -2 & 3 \\ 2 & -3 & 6 & -9 \\ 3 & -5 & 2 & -4 \end{bmatrix}\begin{bmatrix} w \\ x \\ y \\ z \end{bmatrix} = \begin{bmatrix} -6 \\ -5 \\ 14 \\ -3 \end{bmatrix}$$

Then $\mathbf{X} = \mathbf{A}^{-1}\mathbf{B} = \begin{bmatrix} -2 \\ 1 \\ 2 \\ -1 \end{bmatrix}$.

The solution is $(-2, 1, 2, -1)$.

35. ***Familiarize***. Let x = the number of hot dogs sold and y = the number of sausages.

Translate.

The total number of items sold was 145.

$x + y = 145$

The number of hot dogs sold is 45 more than the number of sausages.

$x = y + 45$

We have a system of equations:

$x + y = 145, \quad x + y = 145,$
or
$x = y + 45, \quad x - y = 45.$

Carry out. Write an equivalent matrix equation, $\mathbf{AX} = \mathbf{B}$.

$$\begin{bmatrix} 1 & 1 \\ 1 & -1 \end{bmatrix}\begin{bmatrix} x \\ y \end{bmatrix} = \begin{bmatrix} 145 \\ 45 \end{bmatrix}$$

Then $\mathbf{X} = \mathbf{A}^{-1}\mathbf{B} = \begin{bmatrix} 95 \\ 50 \end{bmatrix}$, so the solution is $(95, 50)$.

Check. The total number of items is $95 + 50$, or 145. The number of hot dogs, 95, is 45 more than the number of sausages. The solution checks.

State. Stefan sold 95 hot dogs and 50 Italian sausages.

36. Let x = the price of a lab record book and y = the price of a highlighter.

Solve: $4x + 3y = 13.93,$

$3x + 2y = 10.25$

Writing $\begin{bmatrix} 4 & 3 \\ 3 & 2 \end{bmatrix}\begin{bmatrix} x \\ y \end{bmatrix} = \begin{bmatrix} 13.93 \\ 10.25 \end{bmatrix}$, we find that $x = \$2.89$, $y = \$0.79$.

37. ***Familiarize***. Let x, y, and z represent the prices of one ton of topsoil, mulch, and pea gravel, respectively.

Translate.

Four tons of topsoil, 3 tons of mulch, and 6 tons of pea gravel costs \$2825.

$4x + 3y + 6z = 2825$

Five tons of topsoil, 2 tons of mulch, and 5 tons of pea gravel costs \$2663.

$5x + 2y + 5z = 2663$

Pea gravel costs \$17 less per ton than topsoil.

$z = x - 17$

We have a system of equations.

$4x + 3y + 6z = 2825,$

$5x + 2y + 5z = 2663,$

$z = x - 17$, or

$4x + 3y + 6z = 2825,$

$5x + 2y + 5z = 2663,$

$x \quad - z = 17$

Carry out. Write an equivalent matrix equation, $\mathbf{AX} = \mathbf{B}$.

$$\begin{bmatrix} 4 & 3 & 6 \\ 5 & 2 & 5 \\ 1 & 0 & -1 \end{bmatrix}\begin{bmatrix} x \\ y \\ z \end{bmatrix} = \begin{bmatrix} 2825 \\ 2663 \\ 17 \end{bmatrix}$$

Then $\mathbf{X} = \mathbf{A}^{-1}\mathbf{B} = \begin{bmatrix} 239 \\ 179 \\ 222 \end{bmatrix}$, so the solution is $(239, 179, 222)$.

Check. Four tons of topsoil, 3 tons of mulch, and 6 tons of pea gravel costs $4 \cdot \$239 + 3 \cdot \$179 + 6 \cdot \$222$, or $\$956 + \$537 + \$1332$, or \$2825. Five tons of topsoil, 2 tons of mulch, and 5 tons of pea gravel costs $5 \cdot$

\$239 + 2 · \$179 + 5 · \$222, or \$1195 + \$358 + \$1110, or \$2663. The price of pea gravel, \$222, is \$17 less than the price of topsoil, \$239. The solution checks.

State. The price of topsoil is \$239 per ton, of mulch is \$179 per ton, and of pea gravel is \$222 per ton.

38. Let x, y, and z represent the amounts invested at 5.15%, 6.05%, and 7.2%, respectively.

Solve: $x + y + z = 8500$,

$0.0515x + 0.0605y + 0.072z = 537.75$,

$z = x + 1500$

Writing

$$\begin{bmatrix} 1 & 1 & 1 \\ 0.0515 & 0.0605 & 0.072 \\ -1 & 0 & 1 \end{bmatrix}\begin{bmatrix} x \\ y \\ z \end{bmatrix} = \begin{bmatrix} 8500 \\ 537.75 \\ 1500 \end{bmatrix},$$

we find that $x = \$2500$, $y = \$2000$, $z = \$4000$.

39. No; for example, let $\mathbf{A} = \mathbf{B} = \begin{bmatrix} 1 & 0 \\ 0 & 1 \end{bmatrix}$.

Then $\mathbf{A} + \mathbf{B} = \begin{bmatrix} 2 & 0 \\ 0 & 2 \end{bmatrix}$ and $(\mathbf{A} + \mathbf{B})^{-1} = \begin{bmatrix} 0.5 & 0 \\ 0 & 0.5 \end{bmatrix}$,

but $\mathbf{A}^{-1} + \mathbf{B}^{-1} = \begin{bmatrix} 1 & 0 \\ 0 & 1 \end{bmatrix} + \begin{bmatrix} 1 & 0 \\ 0 & 1 \end{bmatrix} = \begin{bmatrix} 2 & 0 \\ 0 & 2 \end{bmatrix}$.

40. No; for example, let $\mathbf{A} = \begin{bmatrix} 3 & 2 \\ 5 & 3 \end{bmatrix}$ and $\mathbf{B} = \begin{bmatrix} 11 & 3 \\ 7 & 2 \end{bmatrix}$.

Then $\mathbf{AB} = \begin{bmatrix} 47 & 13 \\ 76 & 21 \end{bmatrix}$ and $(\mathbf{AB})^{-1} = \begin{bmatrix} -21 & 13 \\ 76 & -47 \end{bmatrix}$,

but

$\mathbf{A}^{-1}\mathbf{B}^{-1} = \begin{bmatrix} -3 & 2 \\ 5 & -3 \end{bmatrix}\begin{bmatrix} 2 & -3 \\ -7 & 11 \end{bmatrix} = \begin{bmatrix} -20 & 31 \\ 31 & -48 \end{bmatrix}$.

41.

$$\begin{array}{r|rrrr} -2 & 1 & -6 & 4 & -8 \\ & & -2 & 16 & -40 \\ \hline & 1 & -8 & 20 & -48 \end{array}$$

$f(-2) = -48$

42.

$$\begin{array}{r|rrrrr} 3 & 2 & -1 & 5 & 6 & -4 \\ & & 6 & 15 & 60 & 198 \\ \hline & 2 & 5 & 20 & 66 & 194 \end{array}$$

$f(3) = 194$

43. $f(x) = x^3 - 3x^2 - 6x + 8$

We use synthetic division to find one factor. We first try $x - 1$.

$$\begin{array}{r|rrrr} 1 & 1 & -3 & -6 & 8 \\ & & 1 & -2 & -8 \\ \hline & 1 & -2 & -8 & 0 \end{array}$$

Since $f(1) = 0$, $x - 1$ is a factor of $f(x)$. We have $f(x) = (x - 1)(x^2 - 2x - 8)$. Factoring the trinomial we get $f(x) = (x - 1)(x - 4)(x + 2)$.

44. $f(x) = x^4 + 2x^3 - 16x^2 - 2x + 15$

We try $x - 1$.

$$\begin{array}{r|rrrrr} 1 & 1 & 2 & -16 & -2 & 15 \\ & & 1 & 3 & -13 & -15 \\ \hline & 1 & 3 & -13 & -15 & 0 \end{array}$$

We have $f(x) = (x - 1)(x^3 + 3x^2 - 13x - 15)$. Now we factor the cubic polynomial. We try $x + 1$.

$$\begin{array}{r|rrrr} -1 & 1 & 3 & -13 & -15 \\ & & -1 & -2 & 15 \\ \hline & 1 & 2 & -15 & 0 \end{array}$$

Now we have $f(x) = (x - 1)(x + 1)(x^2 + 2x - 15) = (x - 1)(x + 1)(x + 5)(x - 3)$.

45. $\mathbf{A} = [x]$

Write the augmented matrix.

$\left[\begin{array}{c|c} x & 1 \end{array}\right]$

Multiply by $\dfrac{1}{x}$.

$\left[\begin{array}{c|c} 1 & \dfrac{1}{x} \end{array}\right]$

Then $\mathbf{A}^{-1}$ exists if and only if $x \neq 0$. $\mathbf{A}^{-1} = \left[\dfrac{1}{x}\right]$.

46. $\mathbf{A} = \begin{bmatrix} x & 0 \\ 0 & y \end{bmatrix}$

Write the augmented matrix.

$\left[\begin{array}{cc|cc} x & 0 & 1 & 0 \\ 0 & y & 0 & 1 \end{array}\right]$

Multiply row 1 by $\dfrac{1}{x}$ and row 2 by $\dfrac{1}{y}$.

$\left[\begin{array}{cc|cc} 1 & 0 & \dfrac{1}{x} & 0 \\ 0 & 1 & 0 & \dfrac{1}{y} \end{array}\right]$

Then $\mathbf{A}^{-1}$ exists if and only if $x \neq 0$ and $y \neq 0$, or if and only if $xy \neq 0$.

$\mathbf{A}^{-1} = \begin{bmatrix} \dfrac{1}{x} & 0 \\ 0 & \dfrac{1}{y} \end{bmatrix}$

47. $\mathbf{A} = \begin{bmatrix} 0 & 0 & x \\ 0 & y & 0 \\ z & 0 & 0 \end{bmatrix}$

Write the augmented matrix.

$\left[\begin{array}{ccc|ccc} 0 & 0 & x & 1 & 0 & 0 \\ 0 & y & 0 & 0 & 1 & 0 \\ z & 0 & 0 & 0 & 0 & 1 \end{array}\right]$

Interchange row 1 and row 3.

$\left[\begin{array}{ccc|ccc} z & 0 & 0 & 0 & 0 & 1 \\ 0 & y & 0 & 0 & 1 & 0 \\ 0 & 0 & x & 1 & 0 & 0 \end{array}\right]$

Multiply row 1 by $\frac{1}{z}$, row 2 by $\frac{1}{y}$, and row 3 by $\frac{1}{x}$.

$$\left[\begin{array}{ccc|ccc} 1 & 0 & 0 & 0 & 0 & \frac{1}{z} \\ 0 & 1 & 0 & 0 & \frac{1}{y} & 0 \\ 0 & 0 & 1 & \frac{1}{x} & 0 & 0 \end{array}\right]$$

Then $\mathbf{A}^{-1}$ exists if and only if $x \neq 0$ and $y \neq 0$ and $z \neq 0$, or if and only if $xyz \neq 0$.

$$\mathbf{A}^{-1} = \begin{bmatrix} 0 & 0 & \frac{1}{z} \\ 0 & \frac{1}{y} & 0 \\ \frac{1}{x} & 0 & 0 \end{bmatrix}$$

48. $\mathbf{A} = \begin{bmatrix} x & 1 & 1 & 1 \\ 0 & y & 0 & 0 \\ 0 & 0 & z & 0 \\ 0 & 0 & 0 & w \end{bmatrix}$

Write the augmented matrix.

$$\left[\begin{array}{cccc|cccc} x & 1 & 1 & 1 & 1 & 0 & 0 & 0 \\ 0 & y & 0 & 0 & 0 & 1 & 0 & 0 \\ 0 & 0 & z & 0 & 0 & 0 & 1 & 0 \\ 0 & 0 & 0 & w & 0 & 0 & 0 & 1 \end{array}\right]$$

Multiply row 4 by $-\frac{1}{w}$ and add it to row 1.

$$\left[\begin{array}{cccc|cccc} x & 1 & 1 & 0 & 1 & 0 & 0 & -\frac{1}{w} \\ 0 & y & 0 & 0 & 0 & 1 & 0 & 0 \\ 0 & 0 & z & 0 & 0 & 0 & 1 & 0 \\ 0 & 0 & 0 & w & 0 & 0 & 0 & 1 \end{array}\right]$$

Multiply row 3 by $-\frac{1}{z}$ and add it to row 1.

$$\left[\begin{array}{cccc|cccc} x & 1 & 0 & 0 & 1 & 0 & -\frac{1}{z} & -\frac{1}{w} \\ 0 & y & 0 & 0 & 0 & 1 & 0 & 0 \\ 0 & 0 & z & 0 & 0 & 0 & 1 & 0 \\ 0 & 0 & 0 & w & 0 & 0 & 0 & 1 \end{array}\right]$$

Multiply row 2 by $-\frac{1}{y}$ and add it to row 1.

$$\left[\begin{array}{cccc|cccc} x & 0 & 0 & 0 & 1 & -\frac{1}{y} & -\frac{1}{z} & -\frac{1}{w} \\ 0 & y & 0 & 0 & 0 & 1 & 0 & 0 \\ 0 & 0 & z & 0 & 0 & 0 & 1 & 0 \\ 0 & 0 & 0 & w & 0 & 0 & 0 & 1 \end{array}\right]$$

Multiply row 1 by $\frac{1}{x}$, row 2 by $\frac{1}{y}$, row 3 by $\frac{1}{z}$, and row 4 by $\frac{1}{w}$.

$$\left[\begin{array}{cccc|cccc} 1 & 0 & 0 & 0 & \frac{1}{x} & -\frac{1}{xy} & -\frac{1}{xz} & -\frac{1}{xw} \\ 0 & 1 & 0 & 0 & 0 & \frac{1}{y} & 0 & 0 \\ 0 & 0 & 1 & 0 & 0 & 0 & \frac{1}{z} & 0 \\ 0 & 0 & 0 & 1 & 0 & 0 & 0 & \frac{1}{w} \end{array}\right]$$

Then $\mathbf{A}^{-1}$ exists if and only if $w \neq 0$ and $x \neq 0$ and $y \neq 0$ and $z \neq 0$, or if and only if $wxyz \neq 0$.

$$\mathbf{A}^{-1} = \begin{bmatrix} \frac{1}{x} & -\frac{1}{xy} & -\frac{1}{xz} & -\frac{1}{xw} \\ 0 & \frac{1}{y} & 0 & 0 \\ 0 & 0 & \frac{1}{z} & 0 \\ 0 & 0 & 0 & \frac{1}{w} \end{bmatrix}$$

Exercise Set 5.6

1. Graph (f) is the graph of $y > x$.

2. Graph (c) is the graph of $y < -2x$.

3. Graph (h) is the graph of $y \leq x - 3$.

4. Graph (a) is the graph of $y \geq x + 5$.

5. Graph (g) is the graph of $2x + y < 4$.

6. Graph (d) is the graph of $3x + y < -12$.

7. Graph (b) is the graph of $2x - 5y > 10$.

8. Graph (e) is the graph of $3x - 9y < 18$.

9. Graph: $y > 2x$

   1. We first graph the related equation $y = 2x$. We draw the line dashed since the inequality symbol is $>$.

   2. To determine which half-plane to shade, test a point not on the line. We try $(1, 1)$ and substitute:

$y > 2x$		
1 ?	$2 \cdot 1$	
1	2	FALSE

   Since $1 > 2$ is false, $(1, 1)$ is not a solution, nor are any points in the half-plane containing $(1, 1)$. The points in the opposite half-plane are solutions, so we shade that half-plane and obtain the graph.

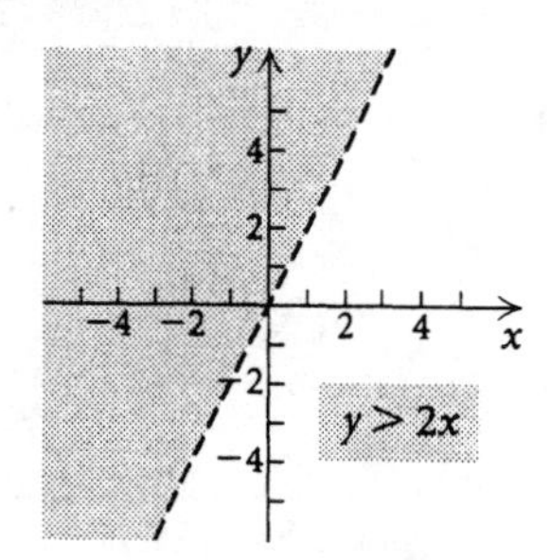

10.

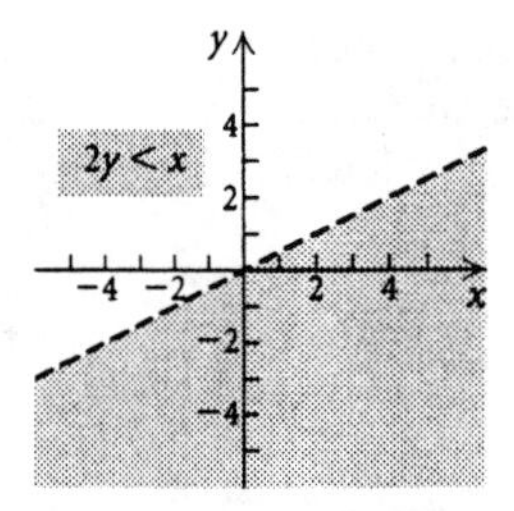

11. Graph: $y + x \geq 0$

1. First graph the related equation $y+x=0$. Draw the line solid since the inequality is $\geq$.
2. Next determine which half-plane to shade by testing a point not on the line. Here we use $(2,2)$ as a check.

$$\begin{array}{c|l} y + x \geq 0 & \\ \hline 2+2\ ?\ 0 & \\ 4 & 0 \quad \text{TRUE} \end{array}$$

Since $4 \geq 0$ is true, $(2,2)$ is a solution. Thus shade the half-plane containing $(2,2)$.

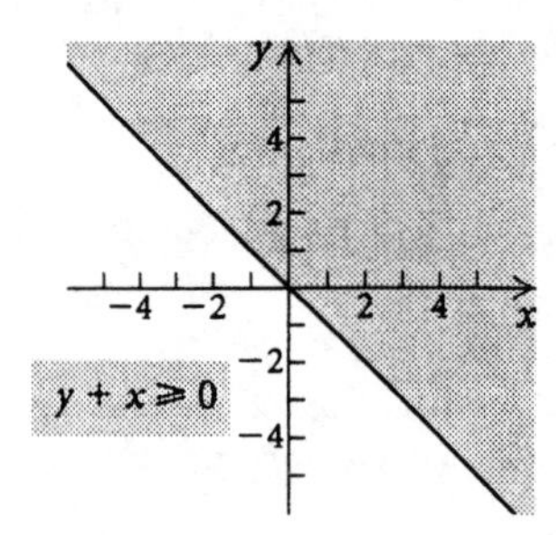

12.

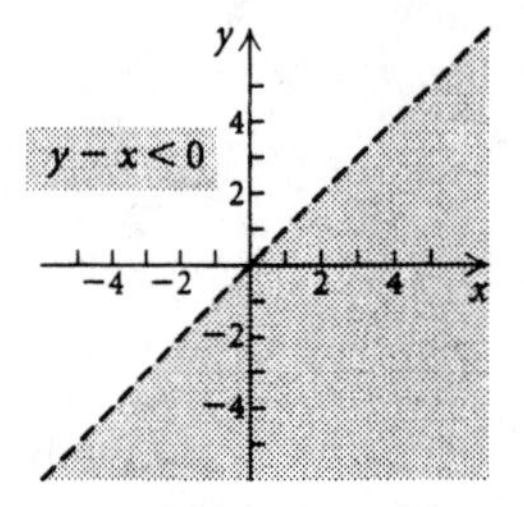

13. Graph: $y > x - 3$

1. We first graph the related equation $y = x - 3$. Draw the line dashed since the inequality symbol is $>$.
2. To determine which half-plane to shade, test a point not on the line. We try $(0,0)$.

$$\begin{array}{c|l} y > x - 3 & \\ \hline 0\ ?\ 0-3 & \\ 0 & -3 \quad \text{TRUE} \end{array}$$

Since $0 > -3$ is true, $(0,0)$ is a solution. Thus we shade the half-plane containing $(0,0)$.

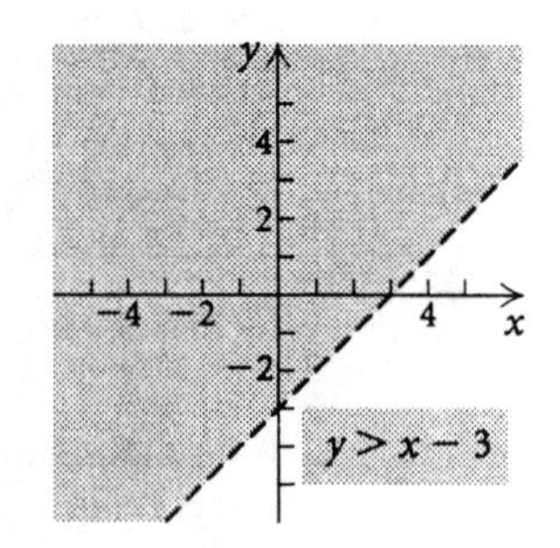

14.

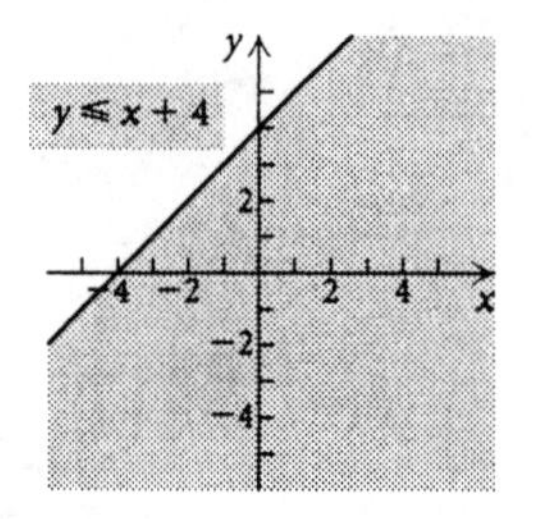

15. Graph: $x + y < 4$

1. First graph the related equation $x+y=4$. Draw the line dashed since the inequality is $<$.
2. To determine which half-plane to shade, test a point not on the line. We try $(0,0)$.

$$\begin{array}{c|l} x + y < 4 & \\ \hline 0+0\ ?\ 4 & \\ 0 & 4 \quad \text{TRUE} \end{array}$$

Since $0 < 4$ is true, $(0,0)$ is a solution. Thus shade the half-plane containing $(0,0)$.

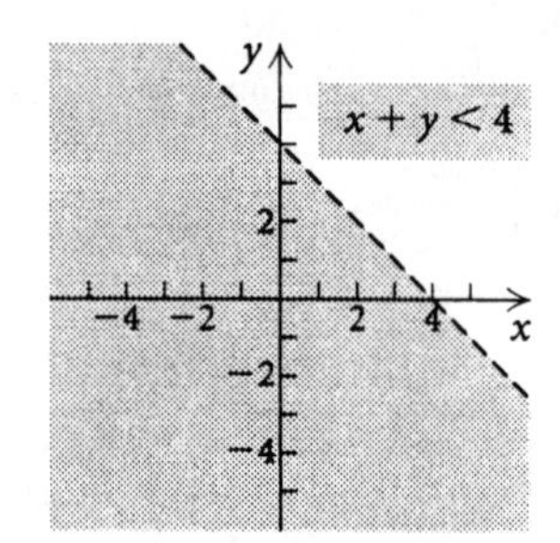

16.

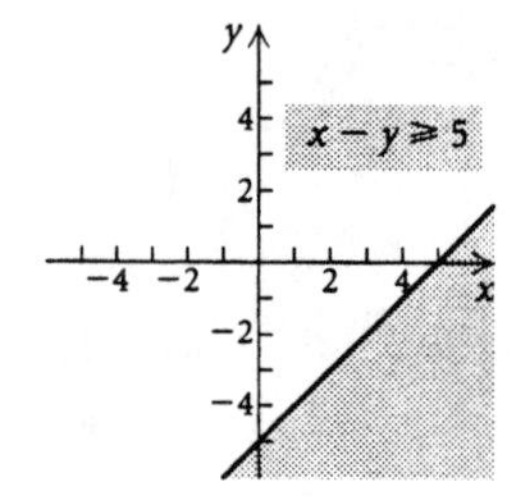

17. Graph: $3x - 2y \leq 6$

1. First graph the related equation $3x - 2y = 6$. Draw the line solid since the inequality is $\leq$.

2. To determine which half-plane to shade, test a point not on the line. We try $(0,0)$.

$$\begin{array}{c|c} \multicolumn{2}{c}{3x - 2y \le 6} \\ \hline 3(0) - 2(0) \ ? & 6 \\ 0 & 6 \quad \text{TRUE} \end{array}$$

Since $0 \le 6$ is true, $(0,0)$ is a solution. Thus shade the half-plane containing $(0,0)$.

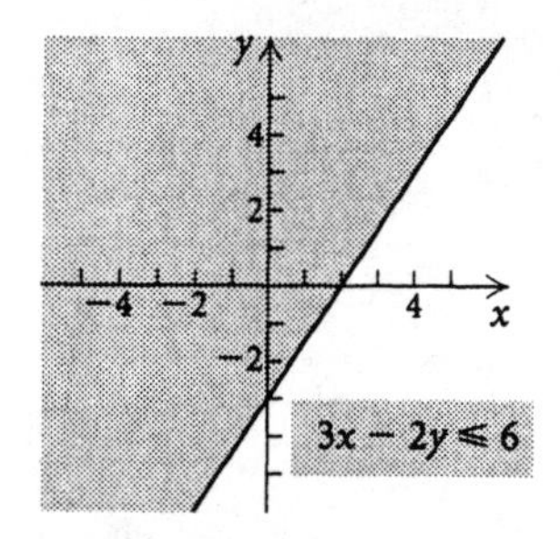

18.

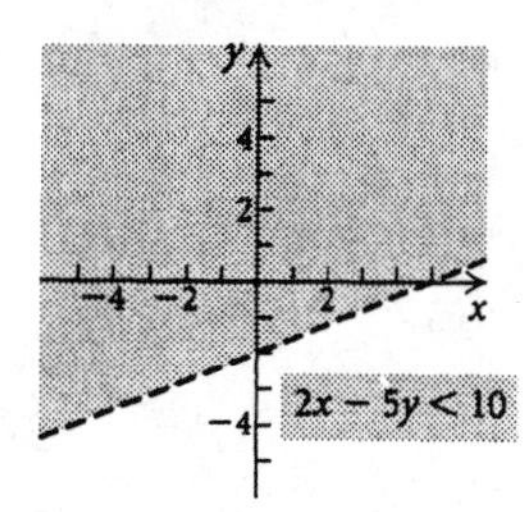

19. Graph: $3y + 2x \ge 6$

1. First graph the related equation $3y + 2x = 6$. Draw the line solid since the inequality is $\ge$.
2. To determine which half-plane to shade, test a point not on the line. We try $(0,0)$.

$$\begin{array}{c|c} \multicolumn{2}{c}{3y + 2x \ge 6} \\ \hline 3 \cdot 0 + 2 \cdot 0 \ ? & 6 \\ 0 & 6 \quad \text{FALSE} \end{array}$$

Since $0 \ge 6$ is false, $(0,0)$ is not a solution. We shade the half-plane which does not contain $(0,0)$.

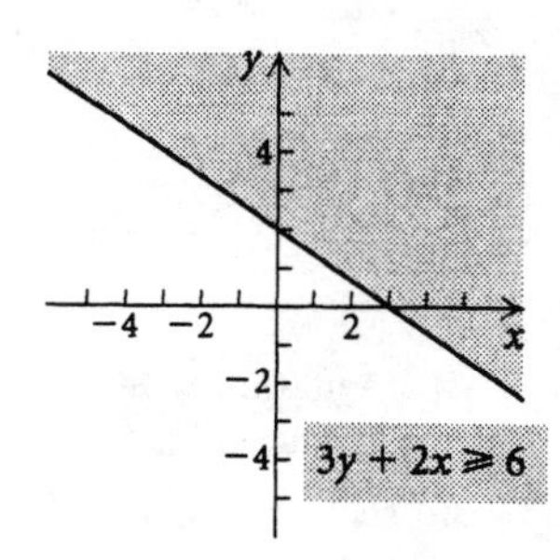

20.

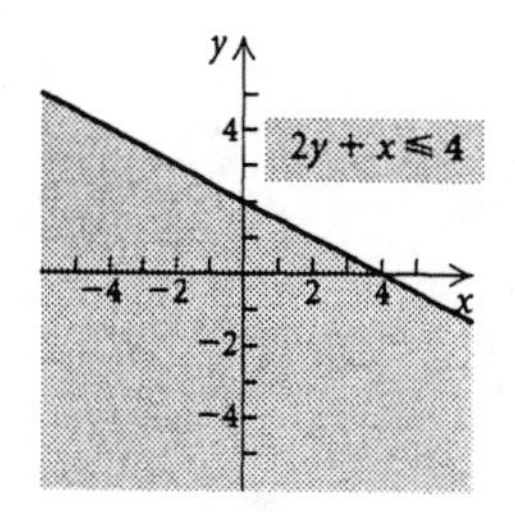

21. Graph: $3x - 2 \le 5x + y$

$-2 \le 2x + y$ Adding $-3x$

1. First graph the related equation $2x + y = -2$. Draw the line solid since the inequality is $\le$.
2. To determine which half-plane to shade, test a point not on the line. We try $(0,0)$.

$$\begin{array}{c|c} \multicolumn{2}{c}{2x + y \ge -2} \\ \hline 2(0) + 0 \ ? & -2 \\ 0 & -2 \quad \text{TRUE} \end{array}$$

Since $0 \ge -2$ is true, $(0,0)$ is a solution. Thus shade the half-plane containing the origin.

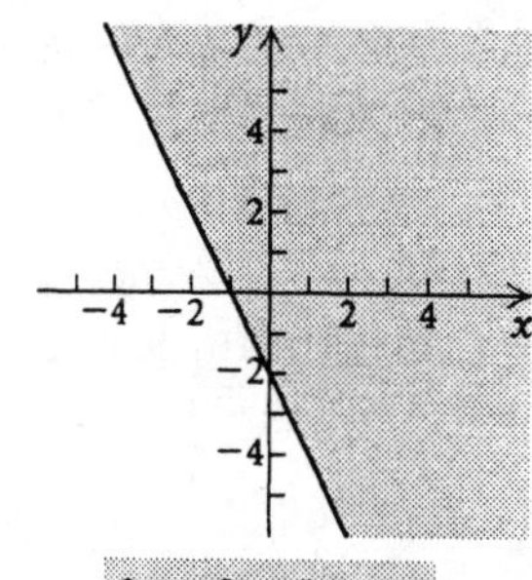

22.

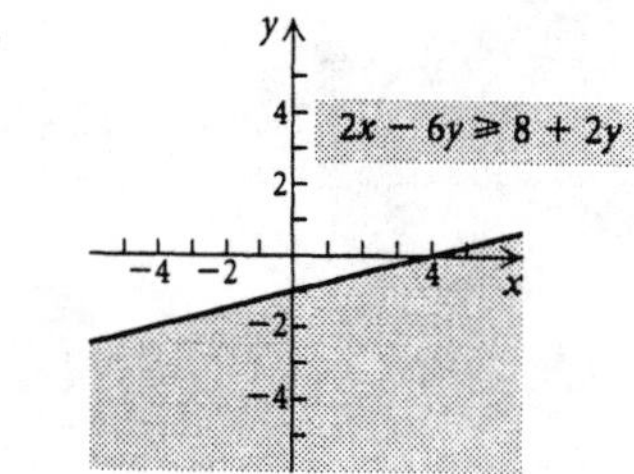

23. Graph: $x < -4$

1. We first graph the related equation $x = -4$. Draw the line dashed since the inequality is $<$.
2. To determine which half-plane to shade, test a point not on the line. We try $(0,0)$.

$$\begin{array}{c|c} \multicolumn{2}{c}{x < -4} \\ \hline 0 \ ? & -4 \quad \text{FALSE} \end{array}$$

Since $0 < -4$ is false, $(0,0)$ is not a solution. Thus, we shade the half-plane which does not contain the origin.

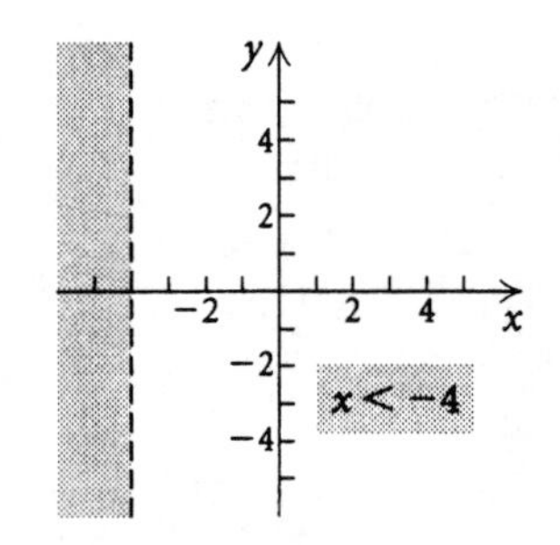

24.

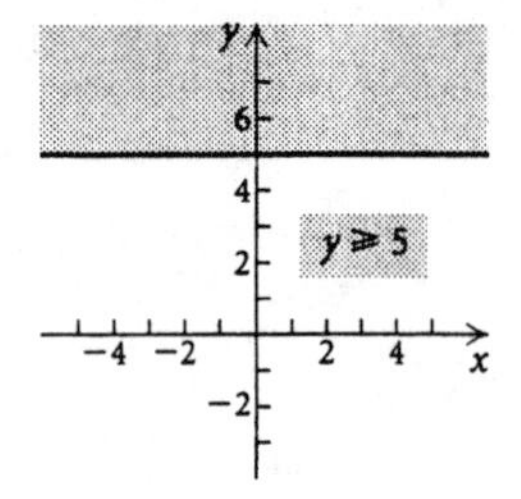

25. Graph: $y > -3$

1. First we graph the related equation $y = -3$. Draw the line dashed since the inequality is $>$.
2. To determine which half-plane to shade we test a point not on the line. We try $(0, 0)$.

$$\begin{array}{c|c} y > -3 & \\ \hline 0\ ?\ -3 & \text{TRUE} \end{array}$$

Since $0 > -3$ is true, $(0, 0)$ is a solution. We shade the half-plane containing $(0, 0)$.

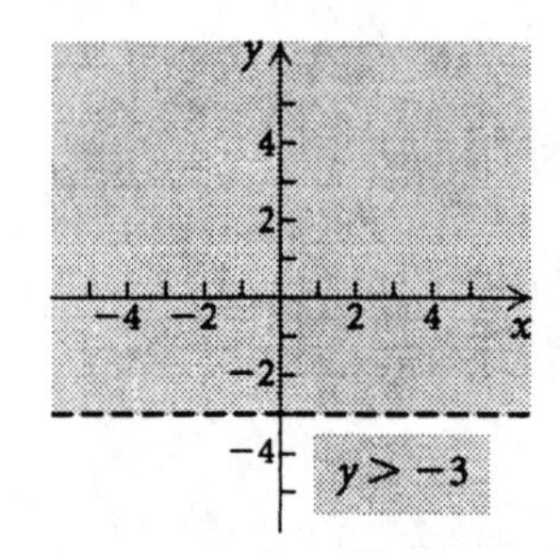

26.

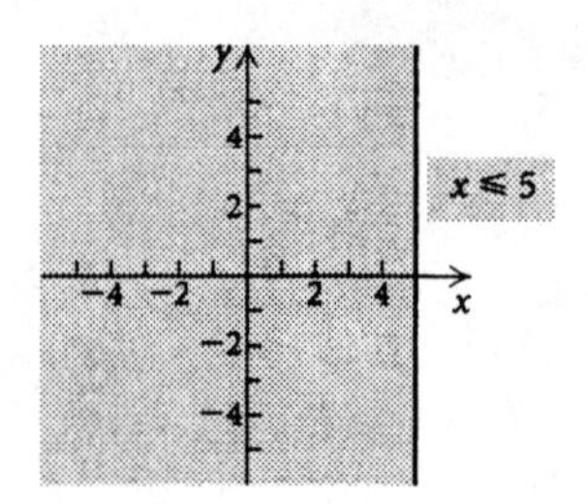

27. Graph: $-4 < y < -1$

This is a conjunction of two inequalities

$4 < y$ *and* $y < -1$.

We can graph $-4 < y$ and $y < -1$ separately and then graph the intersection, or region in both solution sets.

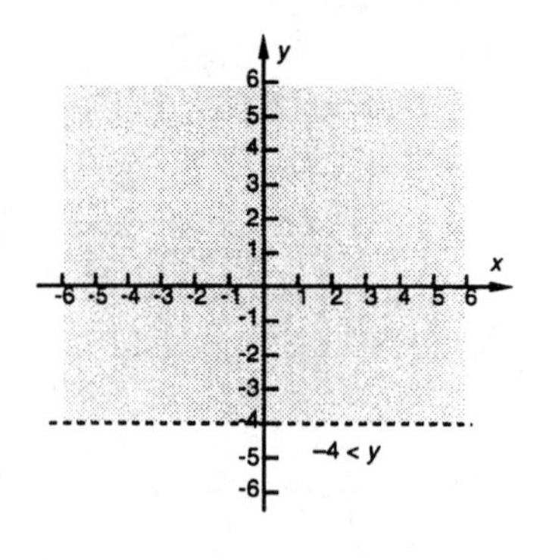

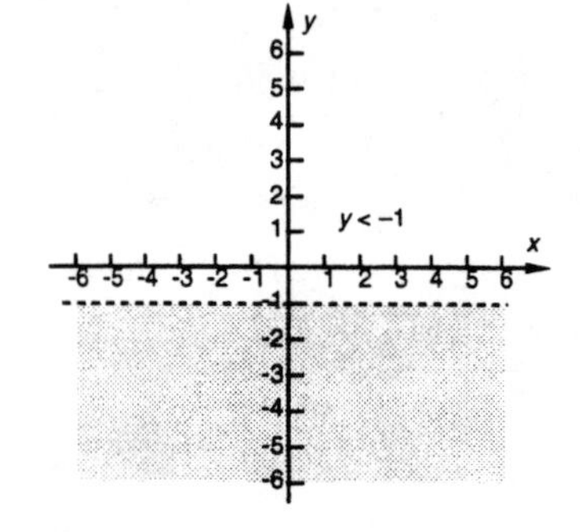

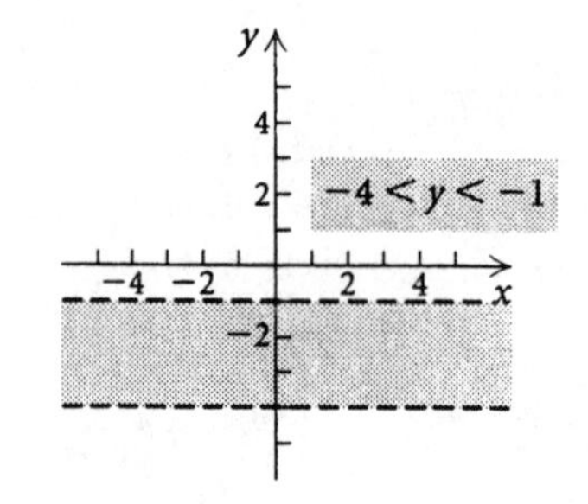

28.

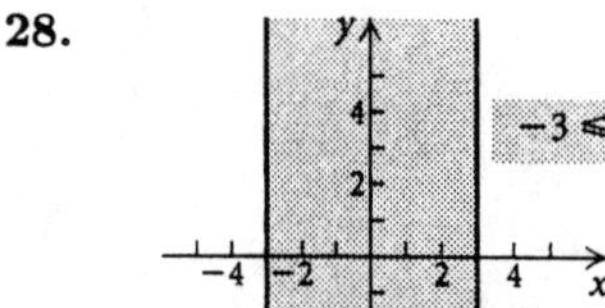

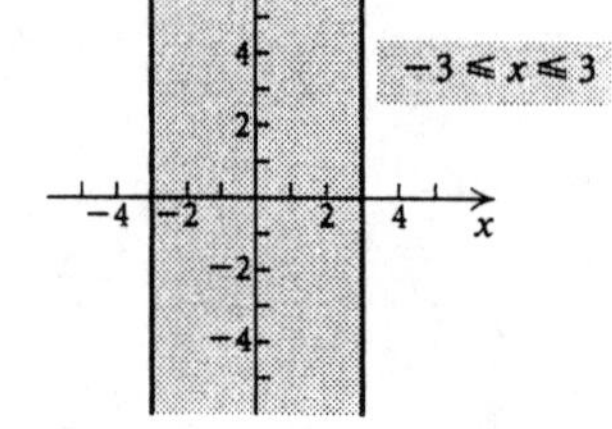

29. Graph: $y \geq |x|$

1. Graph the related equation $y = |x|$. Draw the line solid since the inequality symbol is $\geq$.
2. To determine the region to shade, observe that the solution set consists of all ordered pairs (x, y) where the second coordinate is greater than or equal to the absolute value of the first coordinate. We see that the solutions are the points on or above the graph of $y = |x|$.

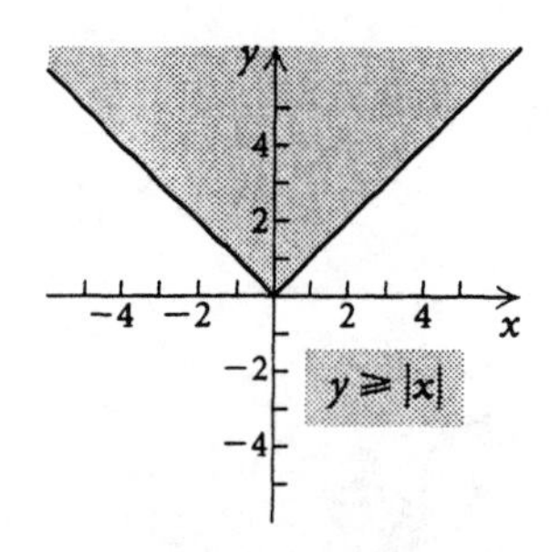

30.

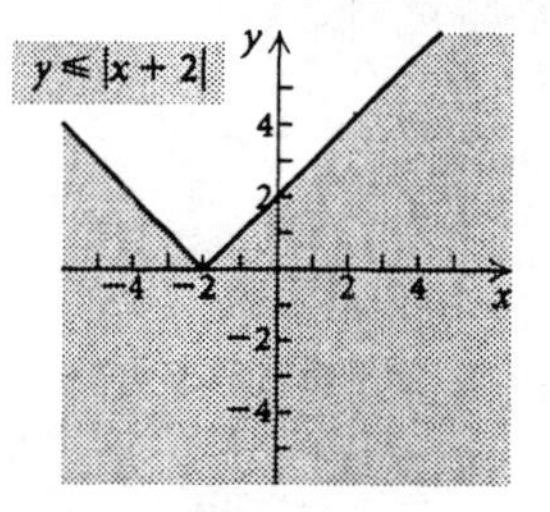

31. Graph (f) is the correct graph.

32. Graph (c) is the correct graph.

33. Graph (a) is the correct graph.

34. Graph (d) is the correct graph.

35. Graph (b) is the correct graph.

36. Graph (e) is the correct graph.

37. Graph: $y \leq x$,

$y \geq 3 - x$

We graph the related equations $y = x$ and $y = 3 - x$ using solid lines. The arrows at the ends of the lines indicate the half-plane containing the solution set for each inequality. Shade the region common to both solution sets.

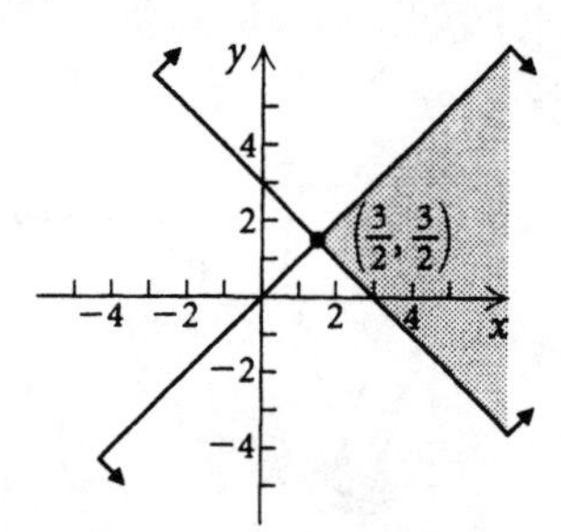

We find the vertex $\left(\dfrac{3}{2}, \dfrac{3}{2}\right)$ by solving the system

$y = x$,

$y = 3 - x$.

38. Graph: $y \geq x$,

$y \leq x - 5$

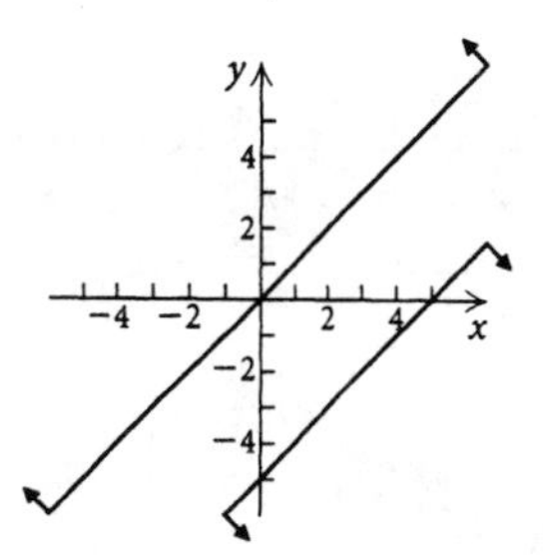

39. Graph: $y \geq x$,

$y \leq x - 4$

We graph the related equations $y = x$ and $y = x - 4$ using solid lines. The arrows at the ends of the lines indicate the half-plane containing the solution set for each inequality. There is not a region common to both solution sets, so there are no vertices.

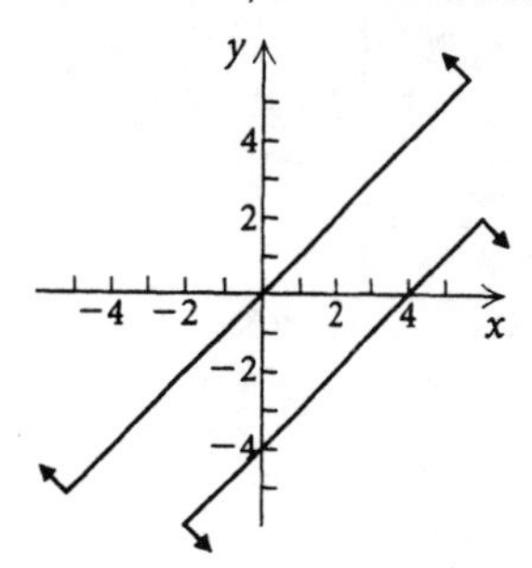

40. Graph: $y \geq x$,

$y \leq 2 - x$

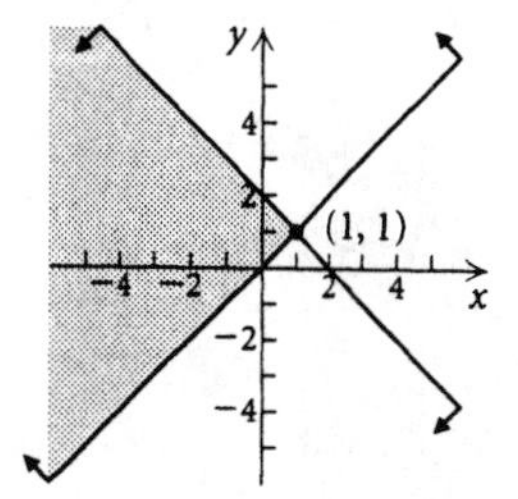

41. Graph: $y \geq -3$,

$x \geq 1$

We graph the related equations $y = -3$ and $x = 1$ using solid lines. The arrows at the ends of the lines indicate the half-plane containing the solution set for each inequality. Shade the region common to both solution sets.

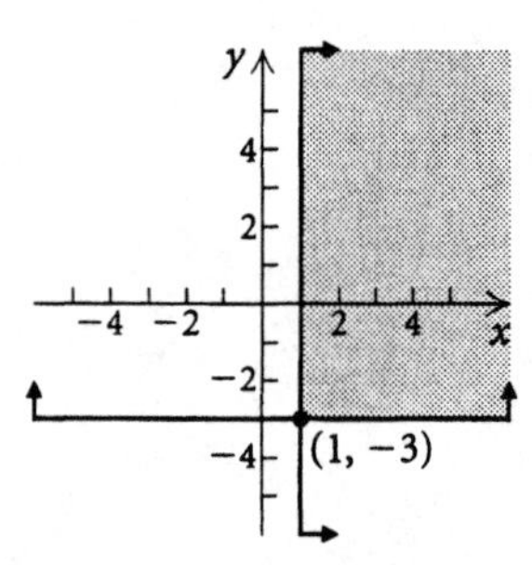

We find the vertex $(1, -3)$ by solving the system

$y = -3,$

$x = 1.$

42. Graph: $y \leq -2,$

$x \geq 2$

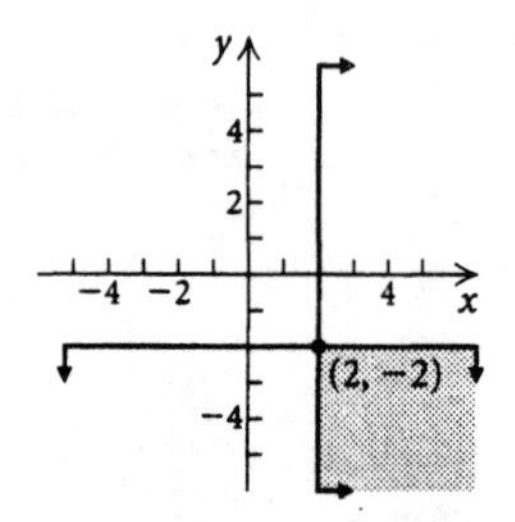

43. Graph: $x \leq 3,$

$y \geq 2 - 3x$

We graph the related equations $x = 3$ and $y = 2 - 3x$ using solid lines. The arrows at the ends of the lines indicate the half-plane containing the solution set for each inequality. Shade the region common to both solution sets.

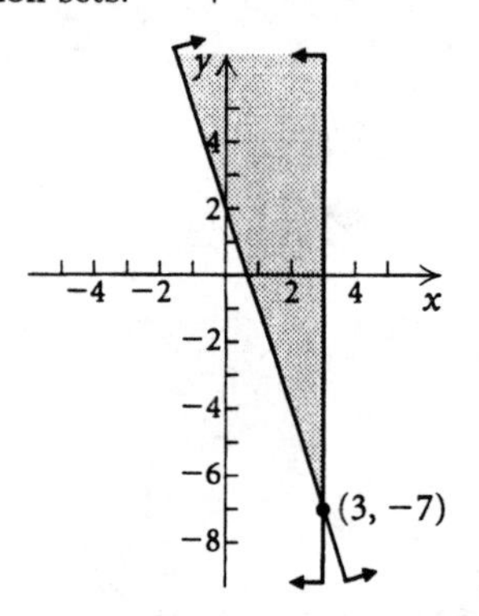

We find the vertex $(3, -7)$ by solving the system

$x = 3,$

$y = 2 - 3x.$

44. Graph: $x \geq -2,$

$y \leq 3 - 2x$

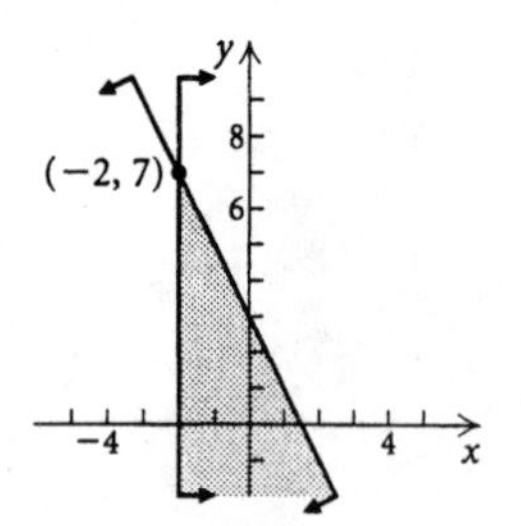

45. Graph: $x + y \leq 1,$

$x - y \leq 2$

We graph the related equations $x+y = 1$ and $x-y = 2$ using solid lines. The arrows at the ends of the lines indicate the half-plane containing the solution set for each inequality. Shade the region common to both solution sets.

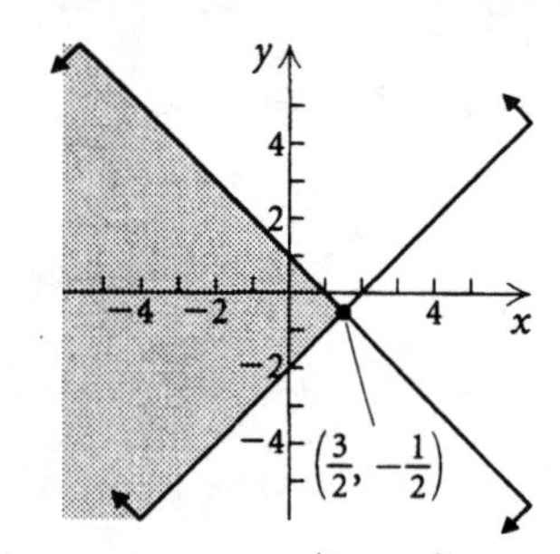

We find the vertex $\left(\frac{3}{2}, -\frac{1}{2}\right)$ by solving the system

$x + y = 1,$

$x - y = 2.$

46. Graph: $y + 3x \geq 0,$

$y + 3x \leq 2$

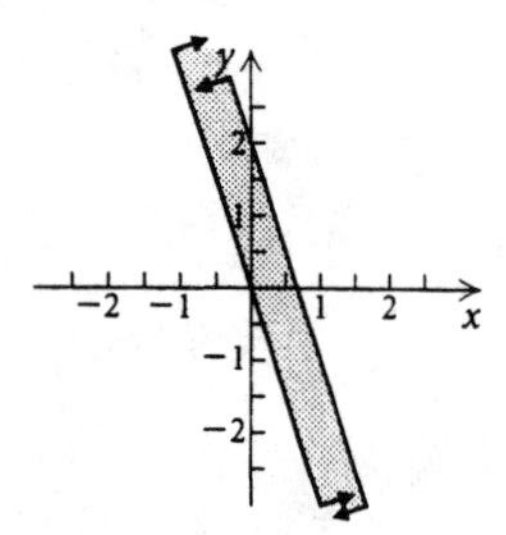

47. Graph: $2y - x \leq 2,$

$y + 3x \geq -1$

We graph the related equations $2y - x = 2$ and $y + 3x = -1$ using solid lines. The arrows at the ends of the lines indicate the half-plane containing the solution set for each inequality. Shade the region common to both solution sets.

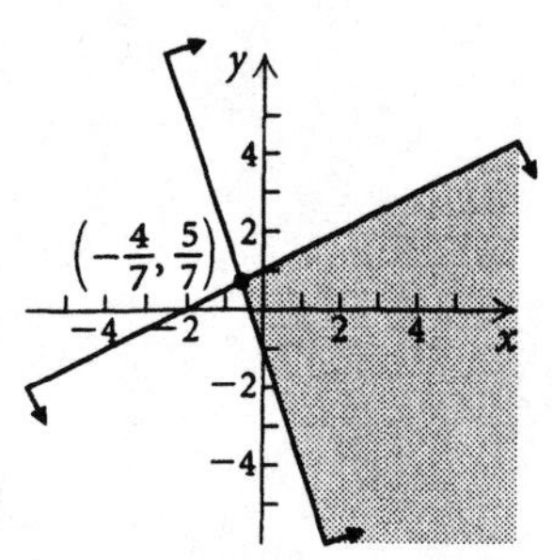

We find the vertex $\left(-\frac{4}{7}, \frac{5}{7}\right)$ by solving the system

$$2y - x = 2,$$
$$y + 3x = -1.$$

48. Graph: $y \leq 2x + 1,$

$x \geq -2x + 1,$

$x \leq 2$

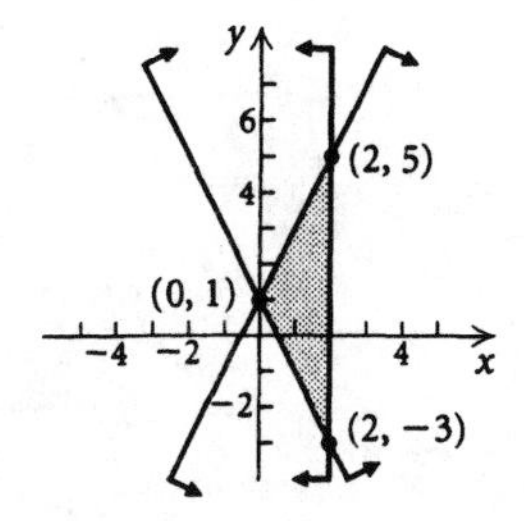

49. Graph: $x - y \leq 2,$

$x + 2y \geq 8,$

$y \leq 4$

We graph the related equations $x - y = 2$, $x + 2y = 8$, and $y = 4$ using solid lines. The arrows at the ends of the lines indicate the half-plane containing the solution set for each inequality. Shade the region common to all three solution sets.

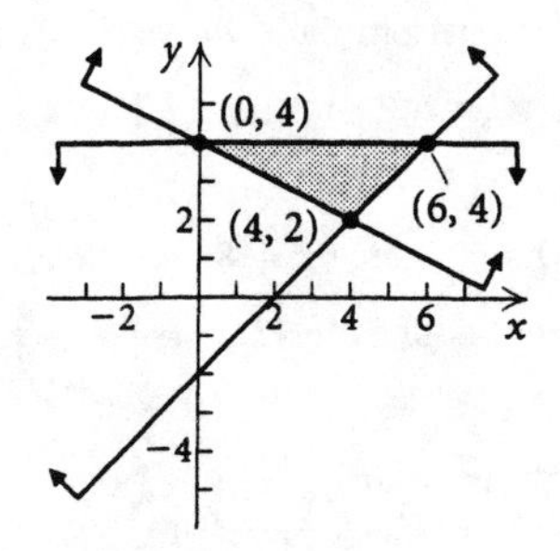

We find the vertex $(0, 4)$ by solving the system

$$x + 2y = 8,$$
$$y = 4.$$

We find the vertex $(6, 4)$ by solving the system

$$x - y = 2,$$
$$y = 4.$$

We find the vertex $(4, 2)$ by solving the system

$$x - y = 2,$$
$$x + 2y = 8.$$

50. Graph: $x + 2y \leq 12,$

$2x + y \leq 12,$

$x \geq 0,$

$y \geq 0$

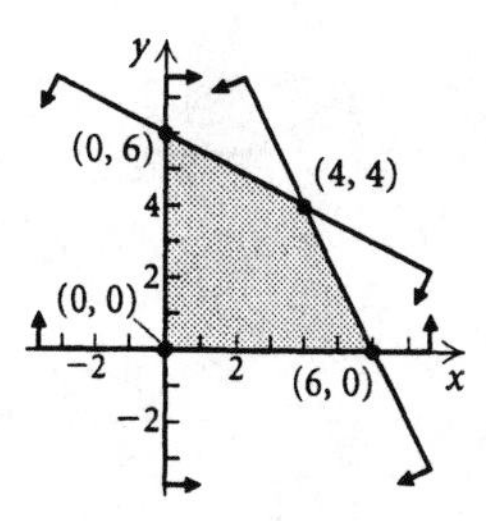

51. Graph: $4x - 3y \geq -12,$

$4x + 3y \geq -36,$

$y \leq 0,$

$x \leq 0$

Shade the intersection of the graphs of the four inequalities.

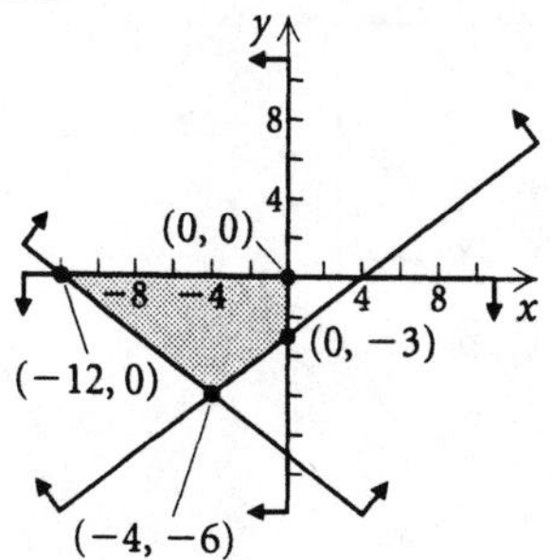

We find the vertex $(-12, 0)$ by solving the system

$$4y + 3x = -36,$$
$$y = 0.$$

We find the vertex $(0, 0)$ by solving the system

$$y = 0,$$
$$x = 0.$$

We find the vertex $(0, -3)$ by solving the system

$$4y - 3x = -12,$$
$$x = 0.$$

We find the vertex $(-4, -6)$ by solving the system

$$4y - 3x = -12,$$
$$4y + 3x = -36.$$

52. Graph: $8x + 5y \le 40$,
$x + 2y \le 8$,
$x \ge 0$,
$y \ge 0$

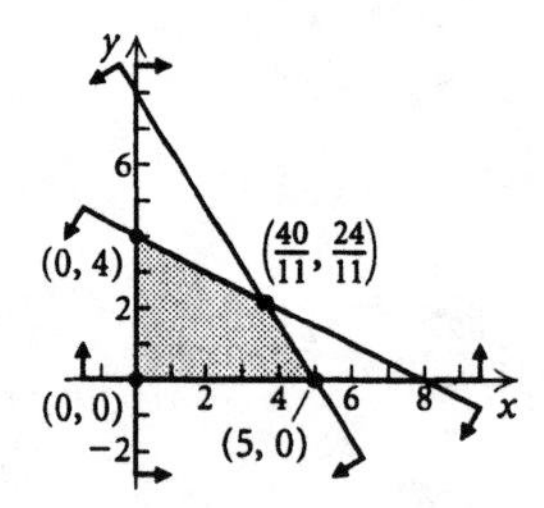

53. Graph: $3x + 4y \ge 12$,
$5x + 6y \le 30$,
$1 \le x \le 3$

Shade the intersection of the graphs of the given inequalities.

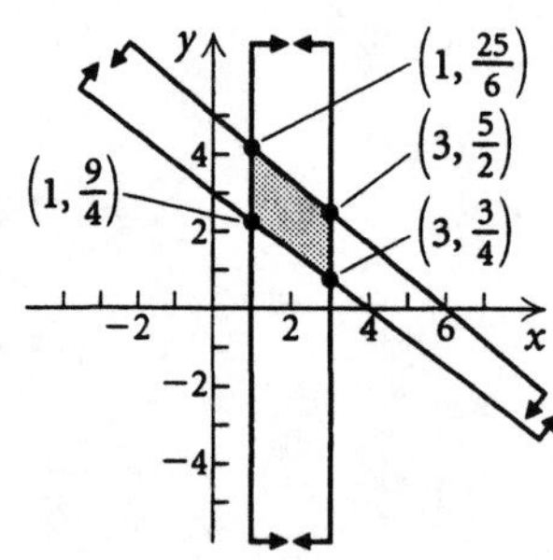

We find the vertex $\left(1, \dfrac{25}{6}\right)$ by solving the system

$$5x + 6y = 30,$$
$$x = 1.$$

We find the vertex $\left(3, \dfrac{5}{2}\right)$ by solving the system

$$5x + 6y = 30,$$
$$x = 3.$$

We find the vertex $\left(3, \dfrac{3}{4}\right)$ by solving the system

$$3x + 4y = 12,$$
$$x = 3.$$

We find the vertex $\left(1, \dfrac{9}{4}\right)$ by solving the system

$$3x + 4y = 12,$$
$$x = 1.$$

54. Graph: $y - x \ge 1$,
$y - x \le 3$,
$2 \le x \le 5$

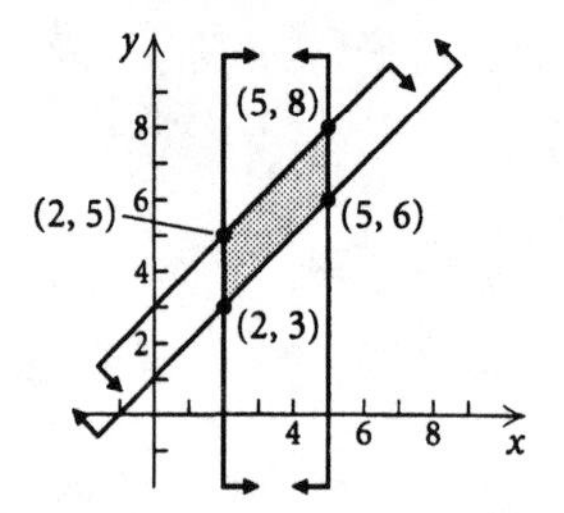

55. Find the maximum and minimum values of $P = 17x - 3y + 60$, subject to
$6x + 8y \le 48$,
$0 \le y \le 4$,
$0 \le x \le 3$.

Graph the system of inequalities and determine the vertices.

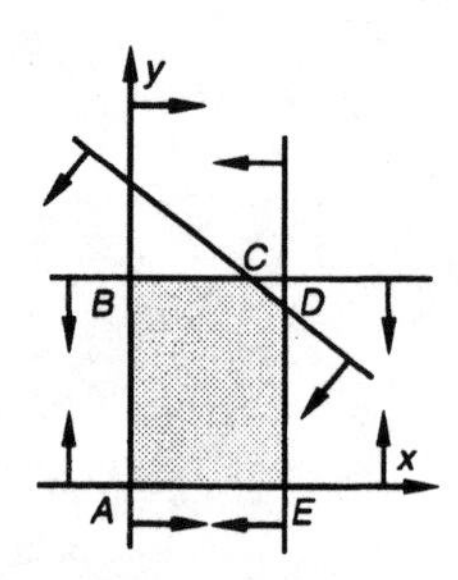

Vertex A: $(0, 0)$

Vertex B:
We solve the system $x = 0$ and $y = 4$. The coordinates of point B are $(0, 4)$.

Vertex C:
We solve the system $6x + 8y = 48$ and $y = 4$. The coordinates of point C are $\left(\dfrac{8}{3}, 4\right)$.

Vertex D:
We solve the system $6x + 8y = 48$ and $y = 7$. The coordinates of point D are $\left(7, \dfrac{3}{4}\right)$.

Vertex E:
We solve the system $x = 7$ and $y = 0$. The coordinates of point E are $(7, 0)$.

Evaluate the objective function P at each vertex.

Vertex	$P = 17x - 3y + 60$
$A(0,0)$	$17 \cdot 0 - 3 \cdot 0 + 60 = 60$
$B(0,4)$	$17 \cdot 0 - 3 \cdot 4 + 60 = 48$
$C\left(\frac{8}{3}, 3\right)$	$17 \cdot \frac{8}{3} - 3 \cdot 4 + 60 = 66\frac{2}{3}$
$D\left(7, \frac{3}{4}\right)$	$17 \cdot 7 - 3 \cdot \frac{3}{4} + 60 = 176\frac{3}{4}$
$E(7,0)$	$17 \cdot 7 - 3 \cdot 0 + 60 = 179$

The maximum value of P is 179 when $x = 7$ and $y = 0$.

The minimum value of P is 48 when $x = 0$ and $y = 4$.

56. We graph the system of inequalities and find the vertices:

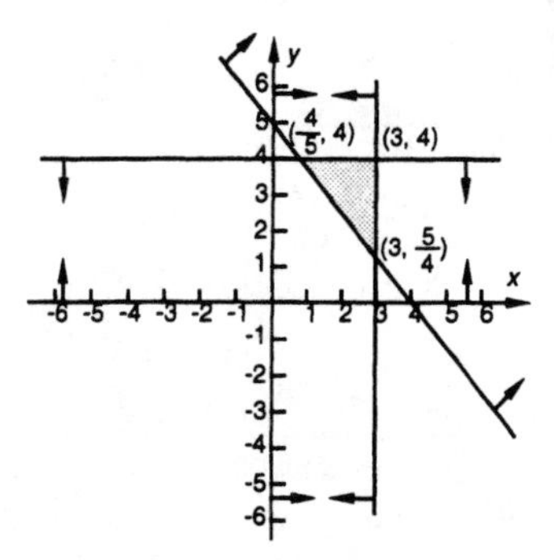

Vertex	$Q = 28x - 4y + 72$
$\left(\frac{4}{5}, 4\right)$	$78\frac{2}{5}$ ⟵ Minimum
$(3, 4)$	140
$\left(3, \frac{5}{4}\right)$	151 ⟵ Maximum

57. Find the maximum and minimum values of

$F = 5x + 36y$, subject to

$$5x + 3y \le 34,$$
$$3x + 5y \le 30,$$
$$x \ge 0,$$
$$y \ge 0.$$

Graph the system of inequalities and find the vertices.

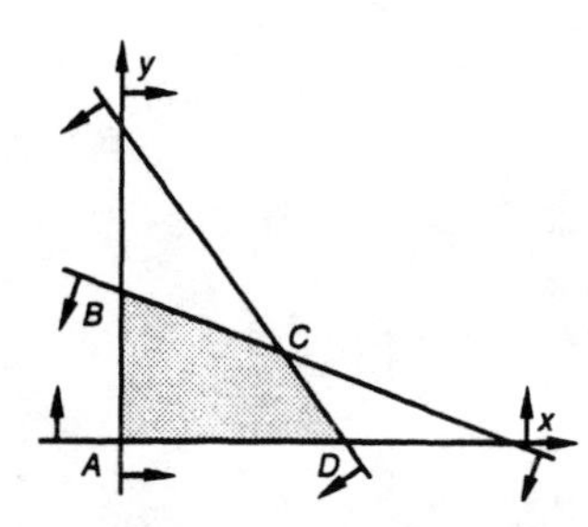

Vertex A: $(0, 0)$

Vertex B:

We solve the system $3x + 5y = 30$ and $x = 0$. The coordinates of point B are $(0, 6)$.

Vertex C:

We solve the system $5x + 3y = 34$ and $3x + 5y = 30$. The coordinates of point C are $(5, 3)$.

Vertex D:

We solve the system $5x + 3y = 34$ and $y = 0$. The coordinates of point D are $\left(\frac{34}{5}, 0\right)$.

Evaluate the objective function F at each vertex.

Vertex	$F = 5x + 36y$
$A(0, 0)$	$5 \cdot 0 + 36 \cdot 0 + = 0$
$B(0, 6)$	$5 \cdot 0 + 36 \cdot 6 = 216$
$C(5, 3)$	$5 \cdot 5 + 39 \cdot 3 = 133$
$D\left(\frac{34}{5}, 0\right)$	$5 \cdot \frac{34}{5} + 36 \cdot 0 = 34$

The maximum value of F is 216 when $x = 0$ and $y = 6$.

The minimum value of F is 0 when $x = 0$ and $y = 0$.

58. We graph the system of inequalities and find the vertices:

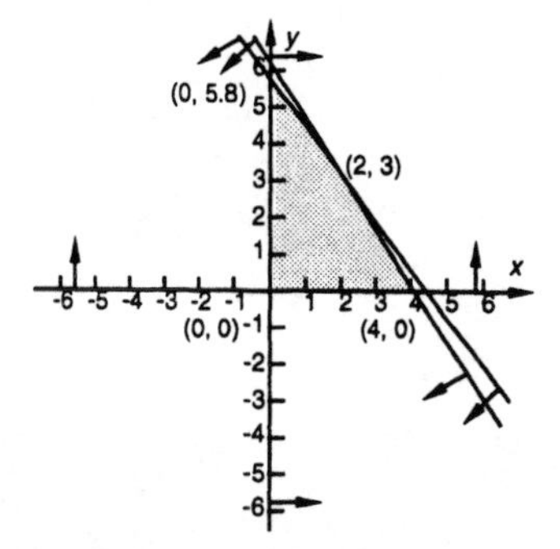

Vertex	$G = 16x + 14y$
$(0, 0)$	0 ⟵ Minimum
$(0, 5.8)$	81.2 ⟵ Maximum
$(2, 3)$	74
$(4, 0)$	64

59. Let $x =$ the number of jumbo biscuits and $y =$ the number of regular biscuits to be made per day. The income I is given by

$$I = 0.10x + 0.08y$$

subject to the constrains

$$x + y \le 200,$$
$$2x + y \le 300,$$
$$x \ge 0,$$
$$y \ge 0.$$

We graph the system of inequalities, determine the vertices, and find the value if I at each vertex.

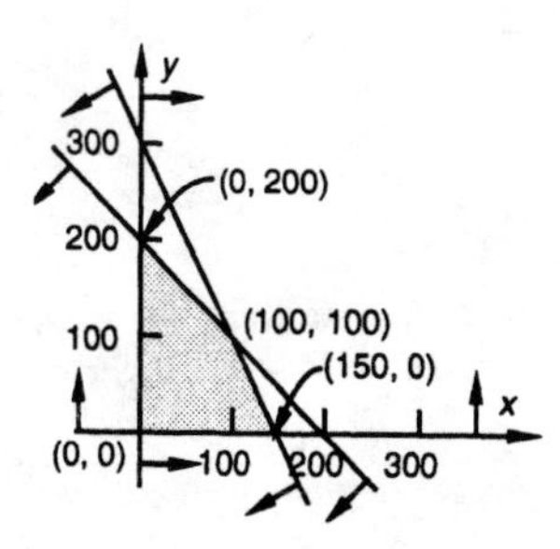

Vertex	$I = 0.10x + 0.08y$
$(0,0)$	$0.10(0) + 0.08(0) = 0$
$(0,200)$	$0.10(0) + 0.08(200) = 16$
$(100,100)$	$0.10(100) + 0.08(100) = 18$
$(150,0)$	$0.10(150) + 0.08(0) = 15$

The company will have a maximum income of \$18 when 100 of each type of biscuit is made.

60. Let x = the number of gallons the car uses and y = the number of gallons the moped uses. Find the maximum value of

$$M = 20x + 100y$$

subject to

$$x + y \leq 12,$$
$$0 \leq x \leq 10,$$
$$0 \leq y \leq 3.$$

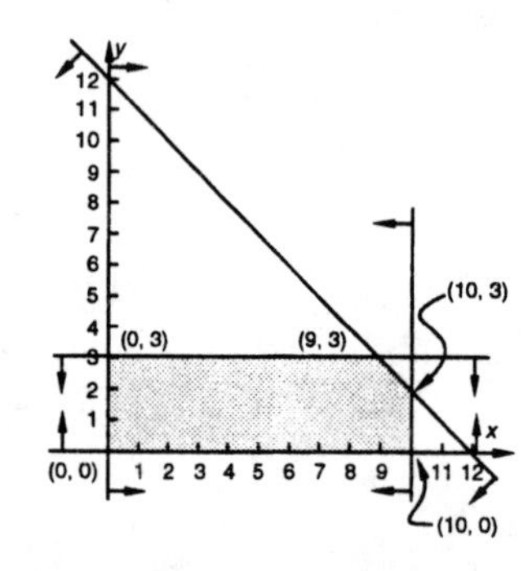

Vertex	$M = 20x + 100y$
$(0,0)$	0
$(0,3)$	300
$(9,3)$	480
$(10,2)$	400
$(10,0)$	200

The maximum number of miles is 480 when the car uses 9 gal and the moped uses 3 gal.

61. Let x = the number of units of lumber and y = the number of units of plywood produced per week. The profit P is given by

$$P = 20x + 30y$$

subject to the constraints

$$x + y \leq 400,$$
$$x \geq 100,$$
$$y \geq 150.$$

We graph the system of inequalities, determine the vertices and find the value of P at each vertex.

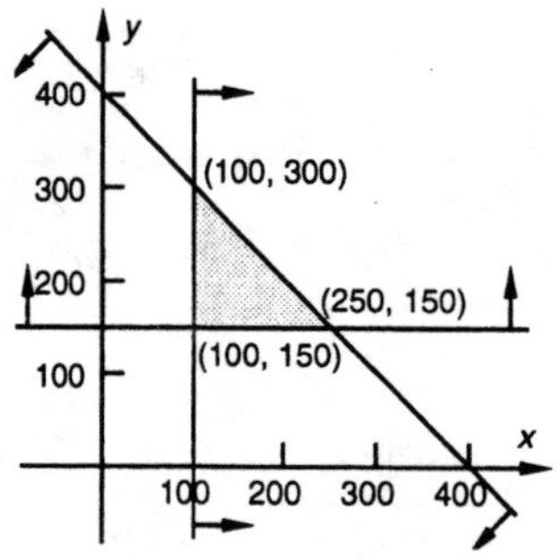

Vertex	$P = 20x + 30y$
$(100,150)$	$20 \cdot 100 + 30 \cdot 150 = 6500$
$(100,300)$	$20 \cdot 100 + 30 \cdot 300 = 11{,}000$
$(250,150)$	$20 \cdot 250 + 30 \cdot 150 = 9500$

The maximum profit of \$11,000 is achieved by producing 100 units of lumber and 300 units of plywood.

62. Let x = the corn acreage and y = the oats acreage. Find the maximum value of

$$P = 40x + 30y$$

subject to

$$x + y \leq 240,$$
$$2x + y \leq 320,$$
$$x \geq 0,$$
$$y \geq 0.$$

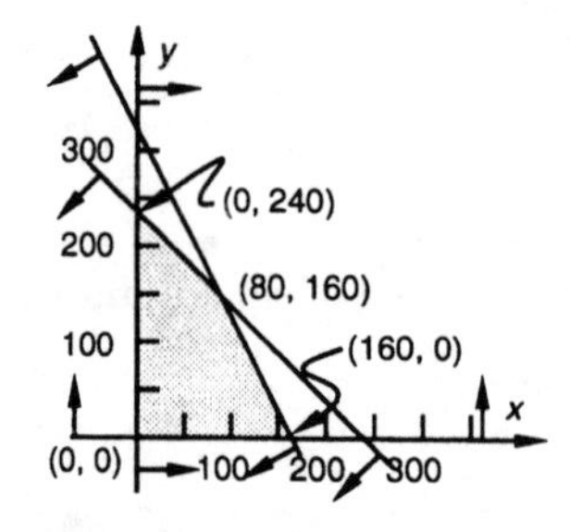

Vertex	$P = 40x + 30y$
$(0,0)$	0
$(0,240)$	7200
$(80,160)$	8000
$(160,0)$	6400

The maximum profit of \$8000 occurs when 80 acres of corn and 160 acres of oats are planted.

63. Let x = the number of sacks of soybean meal to be used and y = the number of sacks of oats. The minimum cost is given by

$$C = 15x + 5y$$

subject to the constraints

$$50x + 15y \geq 120,$$
$$8x + 5y \geq 24,$$
$$5x + y \geq 10,$$
$$x \geq 0,$$
$$y \geq 0.$$

Graph the system of inequalities, determine the vertices, and find the value of C at each vertex.

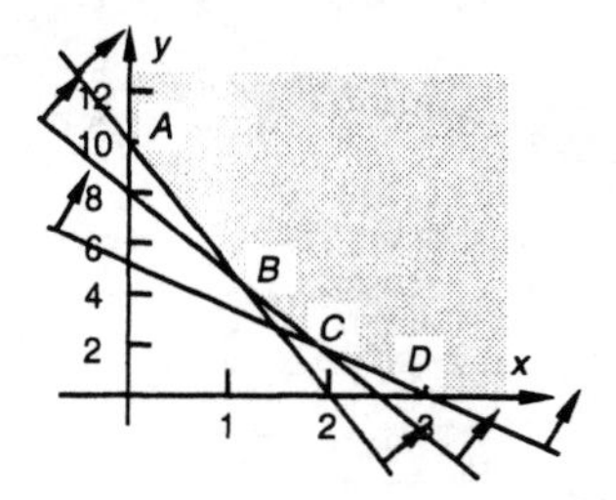

Vertex	$C = 15x + 5y$
$A(0, 10)$	$15 \cdot 0 + 5 \cdot 10 = 50$
$B\left(\frac{6}{5}, 4\right)$	$15 \cdot \frac{6}{5} + 5 \cdot 4 = 38$
$C\left(\frac{24}{13}, \frac{24}{13}\right)$	$15 \cdot \frac{24}{13} + 5 \cdot \frac{24}{13} = 36\frac{12}{13}$
$D(3, 0)$	$15 \cdot 3 + 5 \cdot 0 = 45$

The minimum cost of $\$36\frac{12}{13}$ is achieved by using $\frac{24}{13}$, or $1\frac{11}{13}$ sacks of soybean meal and $\frac{24}{13}$, or $1\frac{11}{13}$ sacks of oats.

64. Let x = the number of sacks of soybean meal to be used and y = the number of sacks of alfalfa. Find the minimum value of

$$C = 15x + 8y$$

subject to the constraints

$$50x + 20y \geq 120,$$
$$8x + 6y \geq 24,$$
$$5x + 8y \geq 10,$$
$$x \geq 0,$$
$$y \geq 0.$$

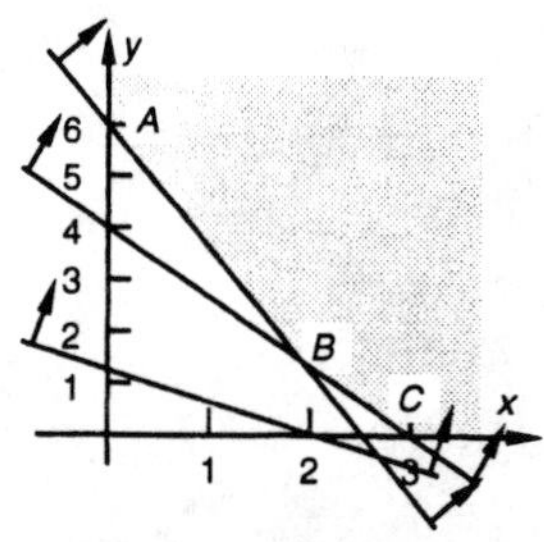

Vertex	$C = 15x + 8y$
$A(0, 6)$	$15 \cdot 0 + 8 \cdot 6 = 48$
$B\left(\frac{12}{7}, \frac{12}{7}\right)$	$15 \cdot \frac{12}{7} + 8 \cdot \frac{12}{7} = 39\frac{3}{7}$
$C(3, 0)$	$15 \cdot 3 + 8 \cdot 0 = 45$

The minimum cost of $\$39\frac{3}{7}$ is achieved by using $\frac{12}{7}$, or $1\frac{5}{7}$ sacks of soybean meal and $\frac{12}{7}$, or $1\frac{5}{7}$ sacks of alfalfa.

65. Let x = the amount invested in corporate bonds and y = the amount invested in municipal bonds. The income I is given by

$$I = 0.08x + 0.075y$$

subject to the constrains

$$x + y \leq 40,000,$$
$$6000 \leq x \leq 22,000,$$
$$y \leq 30,000.$$

We graph the system of inequalities, determine the vertices, and find the value of I at each vertex.

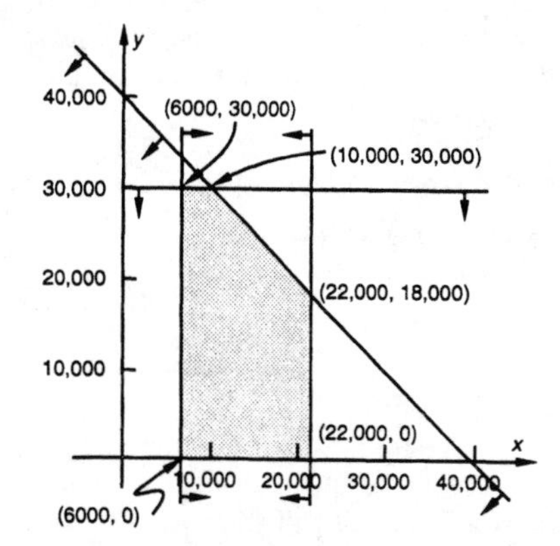

Vertex	$I = 0.08x + 0.075y$
$(6000, 0)$	480
$(6000, 30,000)$	2730
$(10,000, 30,000)$	3050
$(22,000, 18,000)$	3110
$(22,000, 0)$	1760

The maximum income of \$3110 occurs when \$22,000 is invested in corporate bonds and \$18,000 is invested in municipal bonds.

66. Let x = the amount invested in City Bank and y = the amount invested in People's Bank. Find the maximum value of

$$I = 0.06x + 0.065y$$

subject to

$$x + y \leq 22,000,$$
$$2000 \leq x \leq 14,000$$
$$0 \leq y \leq 15,000.$$

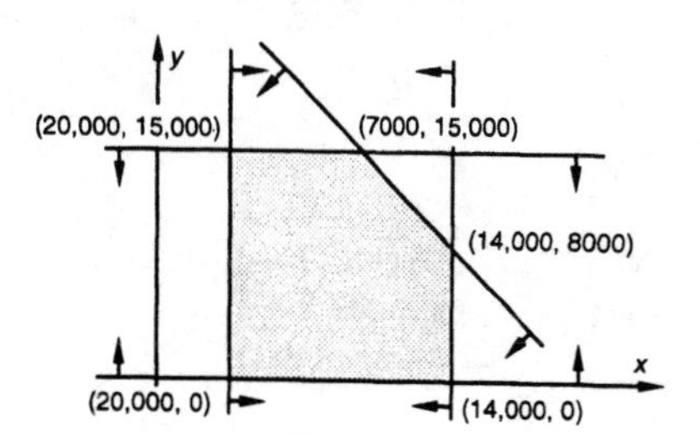

Vertex	$I = 0.06x + 0.065y$
$(2000, 0)$	120
$(2000, 15,000)$	1095
$(7000, 15,000)$	1395
$(14,000, 8,000)$	1360
$(14,000, 0)$	840

The maximum interest income is \$1395 when \$7000 is invested in City Bank and \$15,000 is invested in People's Bank.

67. Let x = the number of P_1 airplanes and y = the number of P_2 airplanes to be used. The operating cost C, in thousands of dollars, is given by

$$C = 12x + 10y$$

subject to the constraints

$$40x + 80y \geq 2000,$$
$$40x + 30y \geq 1500,$$
$$120x + 40y \geq 2400,$$
$$x \geq 0,$$
$$y \geq 0.$$

Graph the system of inequalities, determine the vertices, and find the value of C at each vertex.

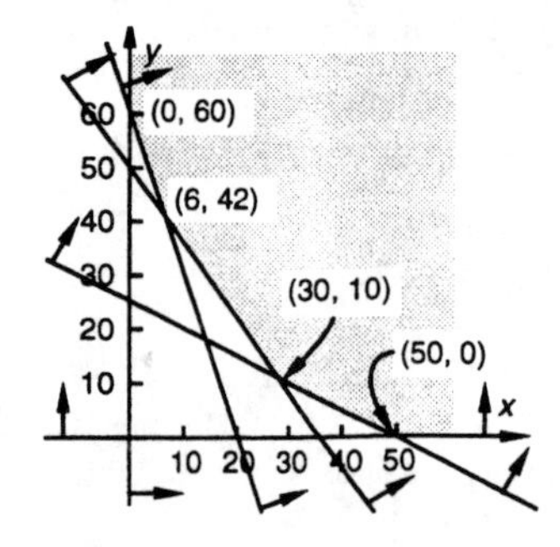

Vertex	$C = 12x + 10y$
$(0, 60)$	$12 \cdot 0 + 10 \cdot 60 = 600$
$(6, 42)$	$12 \cdot 6 + 10 \cdot 42 = 492$
$(30, 10)$	$12 \cdot 30 + 10 \cdot 10 = 460$
$(50, 0)$	$12 \cdot 50 + 10 \cdot 0 = 600$

The minimum cost of \$460 thousand is achieved using 30 P_1's and 10 P_2's.

68. Let x = the number of P_2 airplanes and y = the number of P_3 airplanes to be used. Find the minimum value of

$$C = 10x + 15y$$

subject to

$$80x + 40y \geq 2000,$$
$$30x + 40y \geq 1500,$$
$$40x + 80y \geq 2400,$$
$$x \geq 0,$$
$$y \geq 0.$$

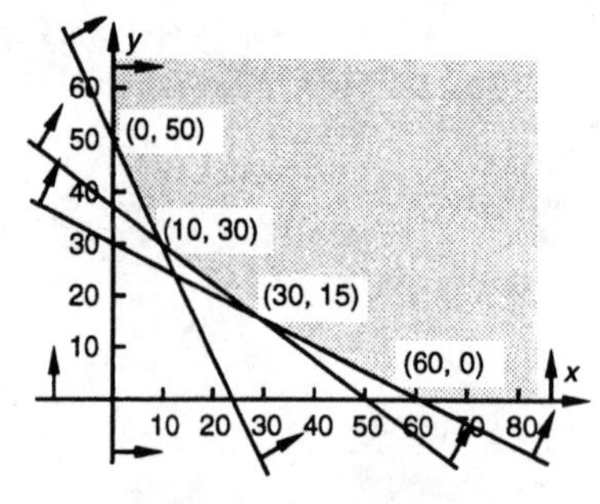

Vertex	$C = 10x + 15y$
$(0, 50)$	$10 \cdot 0 + 15 \cdot 50 = 750$
$(10, 30)$	$10 \cdot 10 + 15 \cdot 30 = 550$
$(30, 15)$	$10 \cdot 30 + 15 \cdot 15 = 525$
$(60, 0)$	$10 \cdot 60 + 15 \cdot 0 = 600$

The minimum cost of \$525 thousand is achieved using 30 P_2's and 15 P_3's.

69. Let x = the number of knit suits and y = the number of worsted suits made. The profit is given by

$$P = 34x + 31y$$

subject to

$$2x + 4y \leq 20,$$
$$4x + 2y \leq 16,$$
$$x \geq 0,$$
$$y \geq 0.$$

Graph the system of inequalities, determine the vertices, and find the value of P at each vertex.

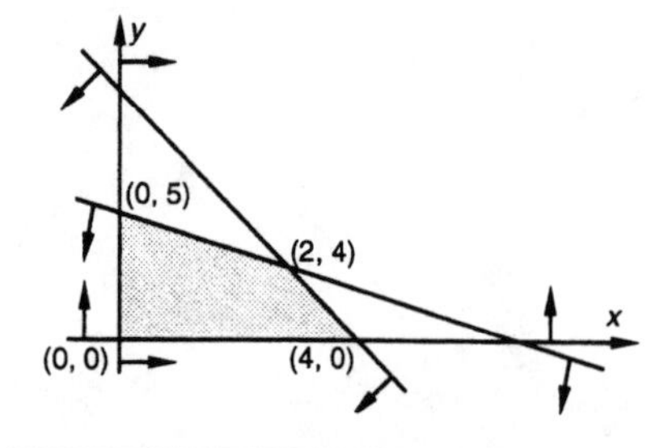

Vertex	$P = 34x + 31y$
$(0, 0)$	$34 \cdot 0 + 31 \cdot 0 = 0$
$(0, 5)$	$34 \cdot 0 + 31 \cdot 5 = 155$
$(2, 4)$	$34 \cdot 2 + 31 \cdot 4 = 192$
$(4, 0)$	$34 \cdot 4 + 31 \cdot 0 = 136$

The maximum profit per day is \$192 when 2 knit suits and 4 worsted suits are made.

70. Let x = the number of smaller gears and y = the number of larger gears produced each day. Find the maximum value of

$$P = 25x + 10y$$

subject to

$$4x + y \le 24,$$
$$x + y \le 9,$$
$$x \ge 0,$$
$$y \ge 0.$$

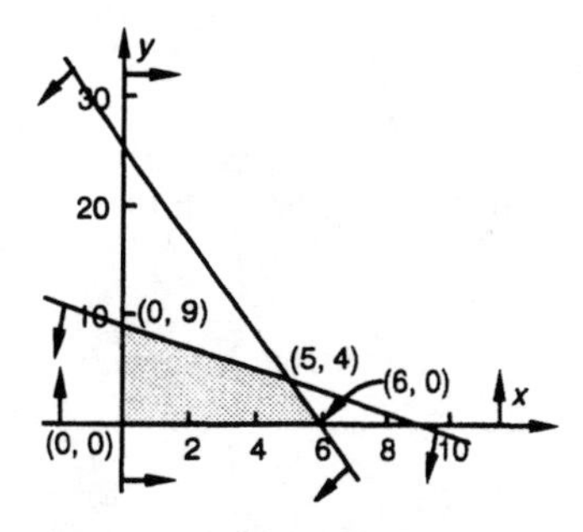

Vertex	$P = 25x + 10y$
$(0, 0)$	$25 \cdot 0 + 10 \cdot 0 = 0$
$(0, 9)$	$25 \cdot 0 + 10 \cdot 9 = 90$
$(5, 4)$	$25 \cdot 5 + 10 \cdot 4 = 165$
$(6, 0)$	$25 \cdot 6 + 10 \cdot 0 = 150$

The maximum profit of $165 per day is achieved when 5 smaller gears and 4 larger gears are produced.

71. Let x = the number of pounds of meat and y = the number of pounds of cheese in the diet in a week. The cost is given by

$$C = 3.50x + 4.60y$$

subject to

$$2x + 3y \ge 12,$$
$$2x + y \ge 6,$$
$$x \ge 0,$$
$$y \ge 0.$$

Graph the system of inequalities, determine the vertices, and find the value of C at each vertex.

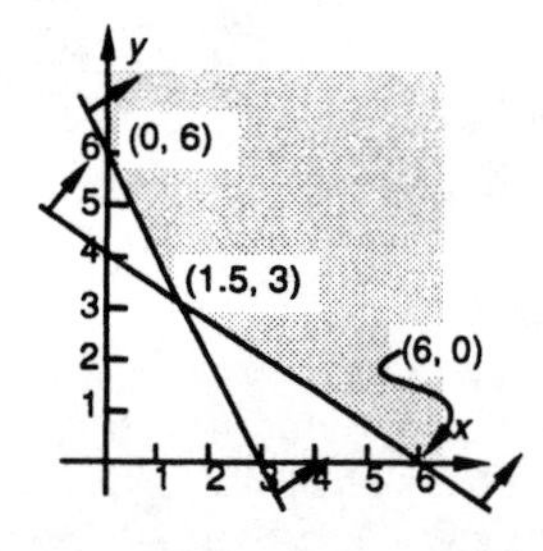

Vertex	$C = 3.50x + 4.60y$
$(0, 6)$	$3.50(0) + 4.60(6) = 27.60$
$(1.5, 3)$	$3.50(1.5) + 4.60(3) = 19.05$
$(6, 0)$	$3.50(6) + 4.60(0) = 21.00$

The minimum weekly cost of $19.05 is achieved when 1.5 lb of meat and 3 lb of cheese are used.

72. Let x = the number of teachers and y = the number of teacher's aides. Find the minimum value of

$$C = 35,000x + 18,000y$$

subject to

$$x + y \le 50,$$
$$x + y \ge 20,$$
$$y \ge 12,$$
$$x \ge 2y.$$

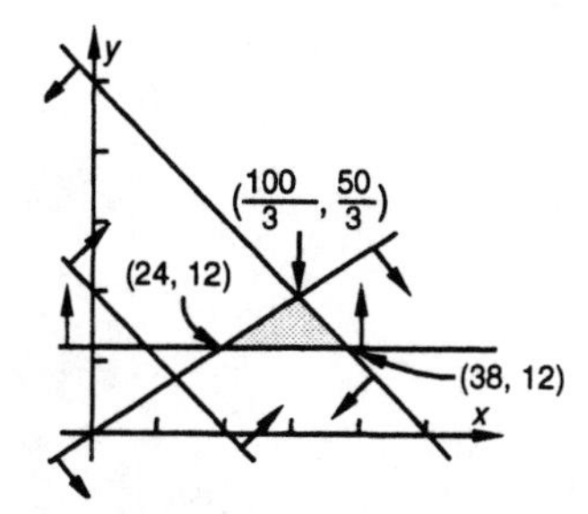

Vertex	$C = 35,000x + 18,000y$
$(24, 12)$	$35,000 \cdot 24 + 18,000 \cdot 12 = 1,056,000$
$\left(\frac{100}{3}, \frac{50}{3}\right)$	$35,000 \cdot \frac{100}{3} + 18,000 \cdot \frac{50}{3} \approx 1,466,667$
$(38, 12)$	$35,000 \cdot 38 + 18,000 \cdot 12 = 1,546,000$

The minimum cost of $1,056,000 is achieved when 24 teachers and 12 teacher's aides are hired.

73. Let x = the number of animal A and y = the number of animal B. The total number of animals is given by

$$T = x + y$$

subject to

$$x + 0.2y \le 600,$$
$$0.5x + y \le 525,$$
$$x \ge 0,$$
$$y \ge 0.$$

Graph the system of inequalities, determine the vertices, and find the value of T at each vertex.

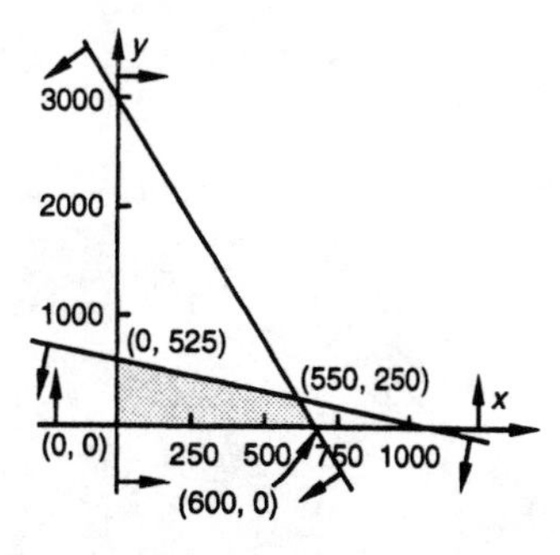

Vertex	$T = x + y$
$(0, 0)$	$0 + 0 = 0$
$(0, 525)$	$0 + 525 = 525$
$(550, 250)$	$550 + 250 = 800$
$(600, 0)$	$600 + 0 = 600$

The maximum total number of 800 is achieved when there are 550 of A and 250 of B.

74. Let x = the number of animal A and y = the number of animal B. Find the maximum value of

$$T = x + y$$

subject to

$$x + 0.2y \leq 1080,$$
$$0.5x + y \leq 810,$$
$$x \geq 0,$$
$$y \geq 0.$$

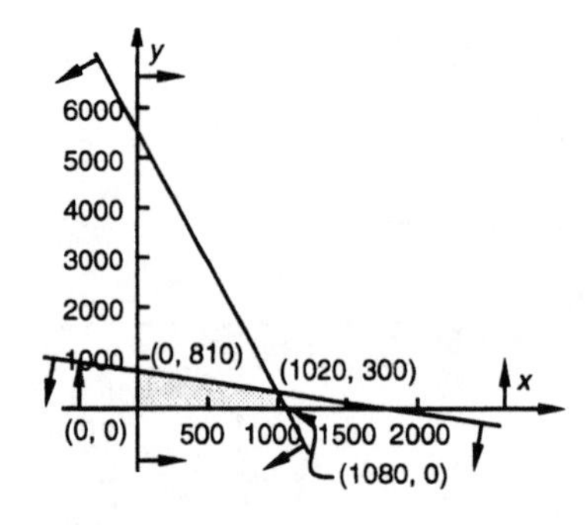

Vertex	$T = x + y$
$(0, 0)$	$0 + 0 = 0$
$(0, 810)$	$0 + 810 = 810$
$(1020, 300)$	$1020 + 300 = 1320$
$(1080, 0)$	$1080 + 0 = 1080$

The maximum total number of 1320 is achieved when there are 1020 of A and 300 of B.

75. Answers will vary. Exercise 59 can be used as a model, if desired.

76. The graph of a linear equation consists of a set of points on a line. The graph of a linear inequality consists of the set of points in a half-plane and might also include the points on the line that is the boundary of the half-plane.

77. $-5 \leq x + 2 < 4$

$-7 \leq x < 2$ Subtracting 2

The solution set is $\{x| -7 \leq x < 2\}$, or $[-7, 2)$.

78. $|x - 3| \geq 2$

$x - 3 \leq -2$ *or* $x - 3 \geq 2$

$x \leq 1$ *or* $x \geq 5$

The solution set is $\{x|x \leq 1 \text{ or } x \geq 5\}$, or $(-\infty, 1] \cup [5, \infty)$.

79. $x^2 - 2x \leq 3$ Polynomial inequality

$x^2 - 2x - 3 \leq 0$

$x^2 - 2x - 3 = 0$ Related equation

$(x + 1)(x - 3) = 0$ Factoring

Using the principle of zero products or by observing the graph of $y = x^2 - 2x - 3$, we see that the solutions of the related equation are -1 and 3. These numbers divide the x-axis into the intervals $(-\infty, -1)$, $(-1, 3)$, and $(3, \infty)$. We let $f(x) = x^2 - 2x - 3$ and test a value in each interval.

$(-\infty, -1)$: $f(-2) = 5 > 0$

$(-1, 3)$: $f(0) = -3 < 0$

$(3, \infty)$: $f(4) = 5 > 0$

Function values are negative on $(-1, 3)$. This can also be determined from the graph of $y = x^2 - 2x - 3$. Since the inequality symbol is $\leq$, the endpoints of the interval must be included in the solution set. It is $[-1, 3]$.

80. $\dfrac{x-1}{x+2} > 4$ Rational inequality

$\dfrac{x-1}{x+2} - 4 > 0$

$\dfrac{x-1}{x+2} - 4 = 0$ Related equation

The denominator of $f(x) = \dfrac{x-1}{x+2} - 4$ is 0 when $x = -2$, so the function is not defined for $x = -2$. We solve the related equation $f(x) = 0$.

$$\frac{x-1}{x+2} - 4 = 0$$

$x - 1 - 4(x + 2) = 0$ Multiplying by $x + 2$

$$x - 1 - 4x - 8 = 0$$
$$-3x - 9 = 0$$
$$-3x = 9$$
$$x = -3$$

Thus, the critical values are -3 and -2. They divide the x-axis into the intervals $(-\infty, -3)$, $(-3, -2)$, and $(-2, \infty)$. We test a value in each interval.

$(-\infty, -3)$: $f(-4) = -\dfrac{3}{2} < 0$

$(-3, -2)$: $f(-2.5) = 3 > 0$

$(-2, \infty)$: $f(0) = -\dfrac{9}{2} < 0$

Function values are positive on $(-3, -2)$. This can also be determined from the graph of $y = \dfrac{x-1}{x+2} - 4$. The solution set is $(-3, -2)$.

81. Graph: $y \geq x^2 - 2$,
$y \leq 2 - x^2$

First graph the related equations $y = x^2 - 2$ and $y = 2 - x^2$ using solid lines. The solution set consists of the region above the graph of $y = x^2 - 2$ and below the graph of $y = 2 - x^2$.

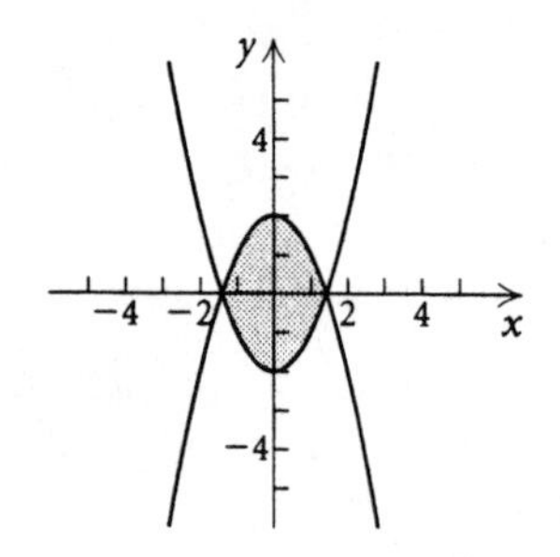

82. $y < x + 1$
$y \geq x^2$

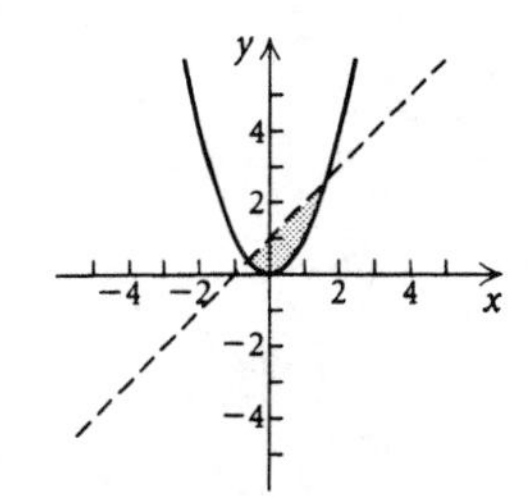

83. See the answer section in the text.

84.

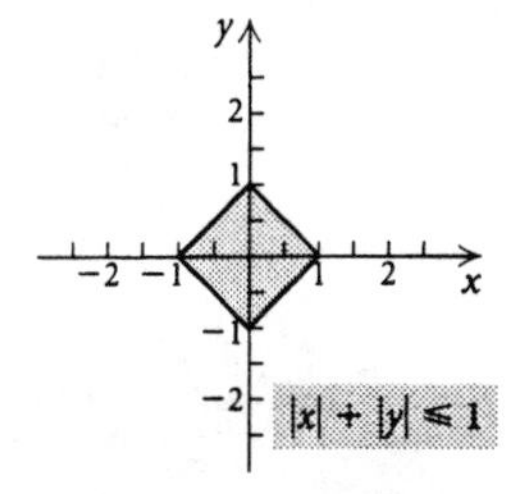

85. See the answer section in the text.

86.

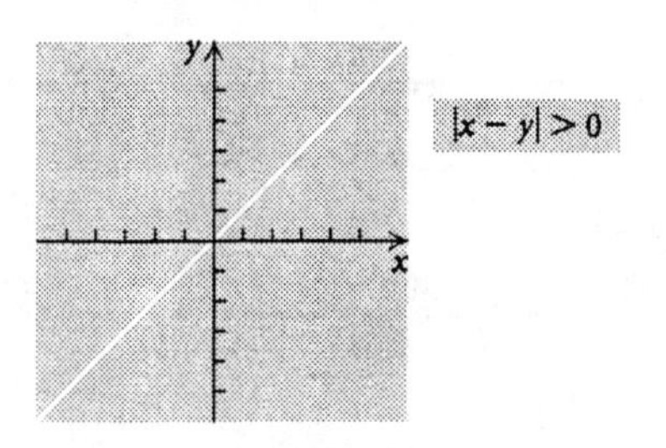

87. Let x = the number of less expensive speaker assemblies and y = the number of more expensive assemblies. The income is given by

$$I = 350x + 600y$$

subject to

$$y \leq 44$$
$$x + y \leq 60,$$
$$x + 2y \leq 90,$$
$$x \geq 0,$$
$$y \geq 0.$$

Graph the system of inequalities, determine the vertices, and find the value of I at each vertex.

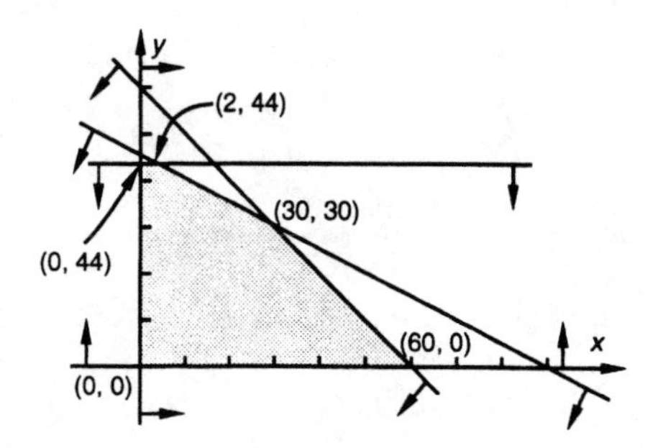

Vertex	$I = 350x + 600y$
$(0,0)$	$350 \cdot 0 + 600 \cdot 0 = 0$
$(0,44)$	$350 \cdot 0 + 600 \cdot 44 = 26,400$
$(2,44)$	$350 \cdot 2 + 600 \cdot 44 = 27,100$
$(30,30)$	$350 \cdot 30 + 600 \cdot 30 = 28,500$
$(60,0)$	$350 \cdot 60 + 600 \cdot 0 = 21,000$

The maximum income of $28,500 is achieved when 30 less expensive and 30 more expensive assemblies are made.

88. Let x = the number of chairs and y = the number of sofas produced. Find the maximum value of

$$I = 80x + 300y$$

subject to

$$20x + 100y \leq 1900,$$
$$x + 50y \leq 500,$$
$$2x + 20y \leq 240,$$
$$x \geq 0,$$
$$y \geq 0.$$

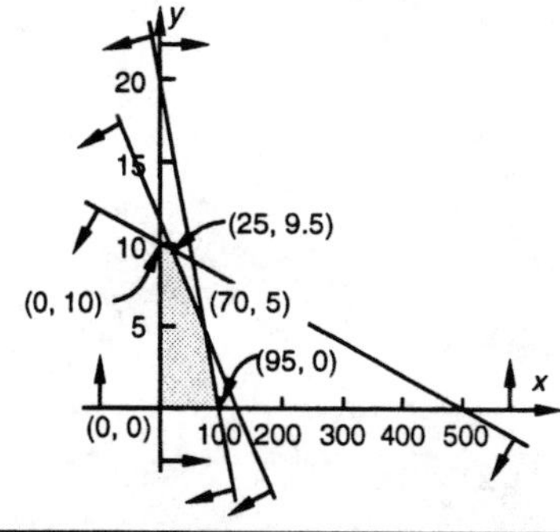

Vertex	$I = 80x + 300y$
$(0,0)$	$80 \cdot 0 + 300 \cdot 0 = 0$
$(0,10)$	$80 \cdot 0 + 300 \cdot 10 = 3000$
$(25,9.5)$	$80 \cdot 25 + 300(9.5) = 4850$
$(70,5)$	$80 \cdot 70 + 300 \cdot 5 = 7100$
$(95,0)$	$80 \cdot 95 + 300 \cdot 0 = 7600$

The maximum income of $7600 is achieved by making 95 chairs and 0 sofas.

Exercise Set 5.7

1. $$\frac{x+7}{(x-3)(x+2)} = \frac{A}{x-3} + \frac{B}{x+2}$$

$$\frac{x+7}{(x-3)(x+2)} = \frac{A(x+2)+B(x-3)}{(x-3)(x+2)} \quad \text{Adding}$$

Equate the numerators:

$x + 7 = A(x+2) + B(x-3)$

Let $x + 2 = 0$, or $x = -2$. Then we get

$-2 + 7 = 0 + B(-2-3)$

$5 = -5B$

$-1 = B$

Next let $x-3=0$, or $x=3$. Then we get

$$3+7=A(3+2)+0$$
$$10=5A$$
$$2=A$$

The decomposition is as follows:

$$\frac{2}{x-3}-\frac{1}{x+2}$$

2. $$\frac{2x}{(x+1)(x-1)}=\frac{A}{x+1}+\frac{B}{x-1}$$
$$\frac{2x}{(x+1)(x-1)}=\frac{A(x-1)+B(x+1)}{(x+1)(x-1)}$$
$$2x=A(x-1)+B(x+1)$$

Let $x=1$: $2\cdot 1=0+B(1+1)$
$$2=2B$$
$$1=B$$

Let $x=-1$: $2(-1)=A(-1-1)+0$
$$-2=-2A$$
$$1=A$$

The decomposition is $\frac{1}{x+1}+\frac{1}{x-1}$.

3. $$\frac{7x-1}{6x^2-5x+1}$$
$$=\frac{7x-1}{(3x-1)(2x-1)} \quad \text{Factoring the denominator}$$
$$=\frac{A}{3x-1}+\frac{B}{2x-1}$$
$$=\frac{A(2x-1)+B(3x-1)}{(3x-1)(2x-1)} \quad \text{Adding}$$

Equate the numerators:

$$7x-1=A(2x-1)+B(3x-1)$$

Let $2x-1=0$, or $x=\frac{1}{2}$. Then we get

$$7\left(\frac{1}{2}\right)-1=0+B\left(3\cdot\frac{1}{2}-1\right)$$
$$\frac{5}{2}=\frac{1}{2}B$$
$$5=B$$

Next let $3x-1=0$, or $x=\frac{1}{3}$. We get

$$7\left(\frac{1}{3}\right)-1=A\left(2\cdot\frac{1}{3}-1\right)$$
$$\frac{7}{3}-1=A\left(\frac{2}{3}-1\right)$$
$$\frac{4}{3}=-\frac{1}{3}A$$
$$-4=A$$

The decomposition is as follows:

$$-\frac{4}{3x-1}+\frac{5}{2x-1}$$

4. $$\frac{13x+46}{12x^2-11x-15}=\frac{13x+46}{(4x+3)(3x-5)}$$
$$=\frac{A}{4x+3}+\frac{B}{3x-5}$$
$$=\frac{A(3x-5)+B(4x+3)}{(4x+3)(3x-5)}$$

Let $x=\frac{5}{3}$: $13\cdot\frac{5}{3}+46=0+B\left(4\cdot\frac{5}{3}+3\right)$
$$\frac{203}{3}=\frac{29}{3}B$$
$$7=B$$

Let $x=-\frac{3}{4}$:

$$13\left(-\frac{3}{4}\right)+46=A\left[3\left(-\frac{3}{4}\right)-5\right]+0$$
$$\frac{145}{4}=-\frac{29}{4}A$$
$$-5=A$$

The decomposition is $-\frac{5}{4x+3}+\frac{7}{3x-5}$.

5. $$\frac{3x^2-11x-26}{(x^2-4)(x+1)}$$
$$=\frac{3x^2-11x-26}{(x+2)(x-2)(x+1)} \quad \text{Factoring the denominator}$$
$$=\frac{A}{x+2}+\frac{B}{x-2}+\frac{C}{x+1}$$
$$=\frac{A(x-2)(x+1)+B(x+2)(x+1)+C(x+2)(x-2)}{(x+2)(x-2)(x+1)} \quad \text{Adding}$$

Equate the numerators:

$$3x^2-11x-26=A(x-2)(x+1)+B(x+2)(x+1)+C(x+2)(x-2)$$

Let $x+2=0$ or $x=-2$. Then we get

$$3(-2)^2-11(-2)-26=A(-2-2)(-2+1)+0+0$$
$$12+22-26=A(-4)(-1)$$
$$8=4A$$
$$2=A$$

Next let $x-2=0$, or $x=2$. Then, we get

$$3\cdot 2^2-11\cdot 2-26=0+B(2+2)(2+1)+0$$
$$12-22-26=B\cdot 4\cdot 3$$
$$-36=12B$$
$$-3=B$$

Finally let $x+1=0$, or $x=-1$. We get

$$3(-1)^2-11(-1)-26=0+0+C(-1+2)(-1-2)$$
$$3+11-26=C(1)(-3)$$
$$-12=-3C$$
$$4=C$$

The decomposition is as follows:

$$\frac{2}{x+2}-\frac{3}{x-2}+\frac{4}{x+1}$$

6. $\dfrac{5x^2+9x-56}{(x-4)(x-2)(x+1)}$

$= \dfrac{A}{x-4} + \dfrac{B}{x-2} + \dfrac{C}{x+1}$

$= \dfrac{A(x-2)(x+1)+B(x-4)(x+1)+C(x-4)(x-2)}{(x-4)(x-2)(x+1)}$

$5x^2+9x-56 = A(x-2)(x+1)+ B(x-4)(x+1)+C(x-4)(x-2)$

Let $x=4$:

$5\cdot 4^2+9\cdot 4-56 = A(4-2)(4+1)+0+0$

$60 = 10A$

$6 = A$

Let $x=2$:

$5\cdot 2^2+9\cdot 2-56 = 0+B(2-4)(2+1)+0$

$-18 = -6B$

$3 = B$

Let $x=-1$:

$5(-1)^2+9(-1)-56 = 0+0+C(-1-4)(-1-2)$

$-60 = 15C$

$-4 = C$

The decomposition is $\dfrac{6}{x-4} + \dfrac{3}{x-2} - \dfrac{4}{x+1}$.

7. $\dfrac{9}{(x+2)^2(x-1)}$

$= \dfrac{A}{x+2} + \dfrac{B}{(x+2)^2} + \dfrac{C}{x-1}$

$= \dfrac{A(x+2)(x-1)+B(x-1)+C(x+2)^2}{(x+2)^2(x-1)}$ Adding

Equate the numerators:

$9 = A(x+2)(x-1)+B(x-1)+C(x+2)^2 \quad (1)$

Let $x-1=0$, or $x=1$. Then, we get

$9 = 0+0+C(1+2)^2$

$9 = 9C$

$1 = C$

Next let $x+2=0$, or $x=-2$. Then, we get

$9 = 0+B(-2-1)+0$

$9 = -3B$

$-3 = B$

To find A we first simplify equation (1).

$9 = A(x^2+x-2)+B(x-1)+C(x^2+4x+4)$

$= Ax^2+Ax-2A+Bx-B+Cx^2+4Cx+4C$

$= (A+C)x^2+(A+B+4C)x+(-2A-B+4C)$

Then we equate the coefficients of x^2.

$0 = A+C$

$0 = A+1$ Substituting 1 for C

$-1 = A$

The decomposition is as follows:

$-\dfrac{1}{x+2} - \dfrac{3}{(x+2)^2} + \dfrac{1}{x-1}$

8. $\dfrac{x^2-x-4}{(x-2)^3}$

$= \dfrac{A}{x-2} + \dfrac{B}{(x-2)^2} + \dfrac{C}{(x-2)^3}$

$= \dfrac{A(x-2)^2+B(x-2)+C}{(x-2)^3}$

$x^2-x-4 = A(x-2)^2+B(x-2)+C \quad (1)$

Let $x=2$: $2^2-2-4 = 0+0+C$

$-2 = C$

Simplify equation (1).

$x^2-x-4 = Ax^2-4Ax+4A+Bx-2B+C$

$= Ax^2+(-4A+B)x+(4A-2B+C)$

Then $1 = A$ and

$-1 = -4A+B$, so $B=3$.

The decomposition is $\dfrac{1}{x-2} + \dfrac{3}{(x-2)^2} - \dfrac{2}{(x-2)^3}$.

9. $\dfrac{2x^2+3x+1}{(x^2-1)(2x-1)}$

$= \dfrac{2x^2+3x+1}{(x+1)(x-1)(2x-1)}$ Factoring the denominator

$= \dfrac{A}{x+1} + \dfrac{B}{x-1} + \dfrac{C}{2x-1}$

$= \dfrac{A(x-1)(2x-1)+B(x+1)(2x-1)+C(x+1)(x-1)}{(x+1)(x-1)(2x-1)}$ Adding

Equate the numerators:

$2x^2+3x+1 = A(x-1)(2x-1)+ B(x+1)(2x-1)+C(x+1)(x-1)$

Let $x+1=0$, or $x=-1$. Then, we get

$2(-1)^2+3(-1)+1 = A(-1-1)[2(-1)-1]+0+0$

$2-3+1 = A(-2)(-3)$

$0 = 6A$

$0 = A$

Next let $x-1=0$, or $x=1$. Then, we get

$2\cdot 1^2+3\cdot 1+1 = 0+B(1+1)(2\cdot 1-1)+0$

$2+3+1 = B\cdot 2\cdot 1$

$6 = 2B$

$3 = B$

Finally we let $2x-1=0$, or $x=\frac{1}{2}$. We get

$2\left(\frac{1}{2}\right)^2+3\left(\frac{1}{2}\right)+1 = 0+0+C\left(\frac{1}{2}+1\right)\left(\frac{1}{2}-1\right)$

$\frac{1}{2}+\frac{3}{2}+1 = C\cdot\frac{3}{2}\cdot\left(-\frac{1}{2}\right)$

$3 = -\frac{3}{4}C$

$-4 = C$

The decomposition is as follows:

$\dfrac{3}{x-1} - \dfrac{4}{2x-1}$

10. $$\frac{x^2-10x+13}{(x^2-5x+6)(x-1)}$$

$$=\frac{x^2-10x+13}{(x-3)(x-2)(x-1)}$$

$$=\frac{A}{x-3}+\frac{B}{x-2}+\frac{C}{x-1}$$

$$=\frac{A(x-2)(x-1)+B(x-3)(x-1)+C(x-3)(x-2)}{(x-3)(x-2)(x-1)}$$

$$x^2-10x+13 = A(x-2)(x-1)+ B(x-3)(x-1)+C(x-3)(x-2)$$

Let $x=3$:

$$3^2-10\cdot 3+13 = A(3-2)(3-1)+0+0$$
$$-8 = 2A$$
$$-4 = A$$

Let $x=2$:

$$2^2-10\cdot 2+13 = 0+B(2-3)(2-1)+0$$
$$-3 = -B$$
$$3 = B$$

Let $x=1$:

$$1^2-10\cdot 1+13 = 0+0+C(1-3)(1-2)$$
$$4 = 2C$$
$$2 = C$$

The decomposition is $-\frac{4}{x-3}+\frac{3}{x-2}+\frac{2}{x-1}$.

11. $$\frac{x^4-3x^3-3x^2+10}{(x+1)^2(x-3)}$$

$$=\frac{x^4-3x^3-3x^2+10}{x^3-x^2-5x-3} \quad \text{Multiplying the denominator}$$

Since the degree of the numerator is greater than the degree of the denominator, we divide.

$$\begin{array}{r} x-\ \ 2 \\ x^3-x^2-5x-3\overline{\big)\,x^4-3x^3-3x^2+\ 0x+10} \\ x^4-\ x^3-5x^2-\ 3x \\ \hline -2x^3+2x^2+\ 3x+10 \\ -2x^3+2x^2+10x+\ 6 \\ \hline -\ 7x+\ 4 \end{array}$$

The original expression is thus equivalent to the following:

$$x-2+\frac{-7x+4}{x^3-x^2-5x-3}$$

We proceed to decompose the fraction.

$$\frac{-7x+4}{(x+1)^2(x-3)}$$

$$=\frac{A}{x+1}+\frac{B}{(x+1)^2}+\frac{C}{x-3}$$

$$=\frac{A(x+1)(x-3)+B(x-3)+C(x+1)^2}{(x+1)^2(x-3)} \quad \text{Adding}$$

Equate the numerators:

$$-7x+4 = A(x+1)(x-3)+B(x-3)+ C(x+1)^2 \qquad (1)$$

Let $x-3=0$, or $x=3$. Then, we get

$$-7\cdot 3+4 = 0+0+C(3+1)^2$$
$$-17 = 16C$$
$$-\frac{17}{16} = C$$

Let $x+1=0$, or $x=-1$. Then, we get

$$-7(-1)+4 = 0+B(-1-3)+0$$
$$11 = -4B$$
$$-\frac{11}{4} = B$$

To find A we first simplify equation (1).

$$-7x+4$$
$$= A(x^2-2x-3)+B(x-3)+C(x^2+2x+1)$$
$$= Ax^2-2Ax-3A+Bx-3B+Cx^2-2Cx+C$$
$$= (A+C)x^2+(-2A+B-2C)x+(-3A-3B+C)$$

Then equate the coefficients of x^2.

$$0 = A+C$$

Substituting $-\frac{17}{16}$ for C, we get $A=\frac{17}{16}$.

The decomposition is as follows:

$$\frac{17/16}{x+1}-\frac{11/4}{(x+1)^2}-\frac{17/16}{x-3}$$

The original expression is equivalent to the following:

$$x-2+\frac{17/16}{x+1}-\frac{11/4}{(x+1)^2}-\frac{17/16}{x-3}$$

12. $$\frac{10x^3-15x^2-35x}{x^2-x-6} = 10x-5+\frac{20x-30}{x^2-x-6} \quad \text{Dividing}$$

$$\frac{20x-30}{x^2-x-6} = \frac{20x-30}{(x-3)(x+2)}$$
$$=\frac{A}{x-3}+\frac{B}{x+2}$$
$$=\frac{A(x+2)+B(x-3)}{(x-3)(x+2)}$$

$$20x-30 = A(x+2)+B(x-3)$$

Let $x=3$: $20\cdot 3-30 = A(3+2)+0$

$$30 = 5A$$
$$6 = A$$

Let $x=-2$: $20(-2)-30 = 0+B(-2-3)$

$$-70 = -5B$$
$$14 = B$$

The decomposition is $10x-5+\frac{6}{x-3}+\frac{14}{x+2}$.

13. $$\frac{-x^2+2x-13}{(x^2+2)(x-1)}$$

$$=\frac{Ax+B}{x^2+2}+\frac{C}{x-1}$$

$$=\frac{(Ax+B)(x-1)+C(x^2+2)}{(x^2+2)(x-1)} \quad \text{Adding}$$

Equate the numerators:

$-x^2 + 2x - 13 = (Ax + B)(x - 1) + C(x^2 + 2) \quad (1)$

Let $x - 1 = 0$, or $x = 1$. Then we get

$$\begin{aligned} -1^2 + 2\cdot 1 - 13 &= 0 + C(1^2 + 2) \\ -1 + 2 - 13 &= C(1 + 2) \\ -12 &= 3C \\ -4 &= C \end{aligned}$$

To find A and B we first simplify equation (1).

$$\begin{aligned} &-x^2 + 2x - 13 \\ &= Ax^2 - Ax + Bx - B + Cx^2 + 2C \\ &= (A + C)x^2 + (-A + B)x + (-B + 2C) \end{aligned}$$

Equate the coefficients of x^2:

$-1 = A + C$

Substituting -4 for C, we get $A = 3$.

Equate the constant terms:

$-13 = -B + 2C$

Substituting -4 for C, we get $B = 5$.

The decomposition is as follows:

$$\frac{3x + 5}{x^2 + 2} - \frac{4}{x - 1}$$

14.

$$\begin{aligned} &\frac{26x^2 + 208x}{(x^2 + 1)(x + 5)} \\ &= \frac{Ax + B}{x^2 + 1} + \frac{C}{x + 5} \\ &= \frac{(Ax + B)(x + 5) + C(x^2 + 1)}{(x^2 + 1)(x + 5)} \end{aligned}$$

$26x^2 + 208x = (Ax + B)(x + 5) + C(x^2 + 1) \quad (1)$

Let $x = -5$:

$$\begin{aligned} 26(-5)^2 + 208(-5) &= 0 + C[(-5)^2 + 1] \\ -390 &= 26C \\ -15 &= C \end{aligned}$$

Simplify equation (1).

$$\begin{aligned} &26x^2 + 208x \\ &= Ax^2 + 5Ax + Bx + 5B + Cx^2 + C \\ &= (A + C)x^2 + (5A + B)x + (5B + C) \end{aligned}$$

$$\begin{aligned} 26 &= A + C \\ 26 &= A - 15 \\ 41 &= A \end{aligned}$$

$$\begin{aligned} 0 &= 5B + C \\ 0 &= 5B - 15 \\ 3 &= B \end{aligned}$$

The decomposition is $\dfrac{41x + 3}{x^2 + 1} - \dfrac{15}{x + 5}$.

15.

$$\begin{aligned} &\frac{6 + 26x - x^2}{(2x - 1)(x + 2)^2} \\ &= \frac{A}{2x - 1} + \frac{B}{x + 2} + \frac{C}{(x + 2)^2} \\ &= \frac{A(x + 2)^2 + B(2x - 1)(x + 2) + C(2x - 1)}{(2x - 1)(x + 2)^2} \end{aligned}$$

Adding

Equate the numerators:

$6 + 26x - x^2 = A(x + 2)^2 + B(2x - 1)(x + 2) + C(2x - 1) \quad (1)$

Let $2x - 1 = 0$, or $x = \frac{1}{2}$. Then, we get

$$\begin{aligned} 6 + 26\cdot\frac{1}{2} - \left(\frac{1}{2}\right)^2 &= A\left(\frac{1}{2} + 2\right)^2 + 0 + 0 \\ 6 + 13 - \frac{1}{4} &= A\left(\frac{5}{2}\right)^2 \\ \frac{75}{4} &= \frac{25}{4}A \\ 3 &= A \end{aligned}$$

Let $x + 2 = 0$, or $x = -2$. We get

$$\begin{aligned} 6 + 26(-2) - (-2)^2 &= 0 + 0 + C[2(-2) - 1] \\ 6 - 52 - 4 &= -5C \\ -50 &= -5C \\ 10 &= C \end{aligned}$$

To find B we first simplify equation (1).

$$\begin{aligned} &6 + 26x - x^2 \\ &= A(x^2 + 4x + 4) + B(2x^2 + 3x - 2) + C(2x - 1) \\ &= Ax^2 + 4Ax + 4A + 2Bx^2 + 3Bx - 2B + 2Cx - C \\ &= (A + 2B)x^2 + (4A + 3B + 2C)x + (4A - 2B - C) \end{aligned}$$

Equate the coefficients of x^2:

$-1 = A + 2B$

Substituting 3 for A, we obtain $B = -2$.

The decomposition is as follows:

$$\frac{3}{2x - 1} - \frac{2}{x + 2} + \frac{10}{(x + 2)^2}$$

16.

$$\begin{aligned} &\frac{5x^3 + 6x^2 + 5x}{(x^2 - 1)(x + 1)^3} \\ &= \frac{5x^3 + 6x^2 + 5x}{(x - 1)(x + 1)^4} \\ &= \frac{A}{x-1} + \frac{B}{x+1} + \frac{C}{(x+1)^2} + \frac{D}{(x+1)^3} + \frac{E}{(x+1)^4} \\ &= [A(x+1)^4 + B(x-1)(x+1)^3 + C(x-1)(x+1)^2 + \\ &\qquad D(x-1)(x+1) + E(x-1)]/[(x - 1)(x + 1)^4] \end{aligned}$$

$5x^3 + 6x^2 + 5x =$

$A(x + 1)^4 + B(x - 1)(x + 1)^3 + C(x - 1)(x + 1)^2 + D(x - 1)(x + 1) + E(x - 1) \quad (1)$

Let $x = 1$: $5\cdot 1^3 + 6\cdot 1^2 + 5\cdot 1 = A(1 + 1)^4$

$$\begin{aligned} 16 &= 16A \\ A &= 1 \end{aligned}$$

Let $x = -1$:

$$5(-1)^3 + 6(-1)^2 + 5(-1) = E(-1-1)$$
$$-4 = -2E$$
$$2 = E$$

Simplify equation (1).

$$5x^3 + 6x^2 + 5x = (A+B)x^4 + (4A+2B+C)x^3 + (6A+C+D)x^2 + (4A-2B-C+E)x + (A-B-C-D-E)$$

$$0 = A + B$$
$$0 = 1 + B$$
$$-1 = B$$
$$5 = 4A + 2B + C$$
$$5 = 4\cdot 1 + 2(-1) + C$$
$$3 = C$$
$$0 = A - B - C - D - E$$
$$0 = 1 - (-1) - 3 - D - 2$$
$$D = -3$$

The decomposition is

$$\frac{1}{x-1} - \frac{1}{x+1} + \frac{3}{(x+1)^2} - \frac{3}{(x+1)^3} + \frac{2}{(x+1)^4}.$$

17. $\dfrac{6x^3 + 5x^2 + 6x - 2}{2x^2 + x - 1}$

Since the degree of the numerator is greater than the degree of the denominator, we divide.

$$\begin{array}{r} 3x + \ \ 1 \\ 2x^2 + x - 1 \overline{\big)\, 6x^3 + 5x^2 + 6x - 2} \\ \underline{6x^3 + 3x^2 - 3x} \\ 2x^2 + 9x - 2 \\ \underline{2x^2 + \ \ x - 1} \\ 8x - 1 \end{array}$$

The original expression is equivalent to

$$3x + 1 + \frac{8x-1}{2x^2 + x - 1}$$

We proceed to decompose the fraction.

$$\frac{8x-1}{2x^2+x-1} = \frac{8x-1}{(2x-1)(x+1)} \quad \text{Factoring the denominator}$$
$$= \frac{A}{2x-1} + \frac{B}{x+1}$$
$$= \frac{A(x+1) + B(2x-1)}{(2x-1)(x+1)} \quad \text{Adding}$$

Equate the numerators:

$$8x - 1 = A(x+1) + B(2x-1)$$

Let $x + 1 = 0$, or $x = -1$. Then we get

$$8(-1) - 1 = 0 + B[2(-1) - 1]$$
$$-8 - 1 = B(-2-1)$$
$$-9 = -3B$$
$$3 = B$$

Next let $2x - 1 = 0$, or $x = \dfrac{1}{2}$. We get

$$8\left(\frac{1}{2}\right) - 1 = A\left(\frac{1}{2} + 1\right) + 0$$
$$4 - 1 = A\left(\frac{3}{2}\right)$$
$$3 = \frac{3}{2}A$$
$$2 = A$$

The decomposition is

$$\frac{2}{2x-1} + \frac{3}{x+1}.$$

The original expression is equivalent to

$$3x + 1 + \frac{2}{2x-1} + \frac{3}{x+1}.$$

18. $\dfrac{2x^3 + 3x^2 - 11x - 10}{x^2 + 2x - 3} = 2x - 1 + \dfrac{-3x - 13}{x^2 + 2x - 3}$ Dividing

$$\frac{-3x-13}{x^2+2x-3} = \frac{A}{x+3} + \frac{B}{x-1}$$
$$= \frac{A(x-1) + B(x+3)}{(x+3)(x-1)}$$
$$-3x - 13 = A(x-1) + B(x+3)$$

Let $x = -3$: $-3(-3) - 13 = A(-3-1) + 0$

$$-4 = -4A$$
$$1 = A$$

Let $x = 1$: $-3\cdot 1 - 13 = 0 + B(1+3)$

$$-16 = 4B$$
$$-4 = B$$

The decomposition is $2x - 1 + \dfrac{1}{x+3} - \dfrac{4}{x-1}$.

19. $$\frac{2x^2 - 11x + 5}{(x-3)(x^2+2x-5)}$$
$$= \frac{A}{x-3} + \frac{Bx+C}{x^2+2x-5}$$
$$= \frac{A(x^2+2x-5) + (Bx+C)(x-3)}{(x-3)(x^2+2x-5)} \quad \text{Adding}$$

Equate the numerators:

$$2x^2 - 11x + 5 = A(x^2 + 2x - 5) + (Bx + C)(x - 3) \quad (1)$$

Let $x - 3 = 0$, or $x = 3$. Then, we get

$$2\cdot 3^2 - 11\cdot 3 + 5 = A(3^2 + 2\cdot 3 - 5) + 0$$
$$18 - 33 + 5 = A(9 + 6 - 5)$$
$$-10 = 10A$$
$$-1 = A$$

To find B and C, we first simplify equation (1).

$$2x^2 - 11x + 5 = Ax^2 + 2Ax - 5A + Bx^2 - 3Bx + Cx - 3C$$
$$= (A+B)x^2 + (2A - 3B + C)x + (-5A - 3C)$$

Equate the coefficients of x^2:

$2 = A + B$

Substituting -1 for A, we get $B = 3$.

Equate the constant terms:

$5 = -5A - 3C$

Substituting -1 for A, we get $C = 0$.

The decomposition is as follows:

$$-\frac{1}{x-3} + \frac{3x}{x^2+2x-5}$$

20. $$\frac{3x^2-3x-8}{(x-5)(x^2+x-4)}$$
$$= \frac{A}{x-5} + \frac{Bx+C}{x^2+x-4}$$
$$= \frac{A(x^2+x-4)+(Bx+C)(x-5)}{(x-5)(x^2+x-4)}$$

$$3x^2-3x-8 = A(x^2+x-4)+ (Bx+C)(x-5) \quad (1)$$

Let $x = 5$: $3\cdot 5^2 - 3\cdot 5 - 8 = A(5^2+5-4)+0$

$52 = 26A$

$2 = A$

Simplify equation (1).

$3x^2-3x-8$

$= Ax^2 + Ax - 4A + Bx^2 - 5Bx + Cx - 5C$

$= (A+B)x^2 + (A-5B+C)x + (-4A-5C)$

$3 = A + B$

$3 = 2 + B$

$1 = B$

$-8 = -4A - 5C$

$-8 = -4\cdot 2 - 5C$

$0 = C$

The decomposition is $\frac{2}{x-5} + \frac{x}{x^2+x-4}$.

21. $$\frac{-4x^2-2x+10}{(3x+5)(x+1)^2}$$

The decomposition looks like

$$\frac{A}{3x+5} + \frac{B}{x+1} + \frac{C}{(x+1)^2}.$$

Add and equate the numerators.

$-4x^2-2x+10$

$= A(x+1)^2 + B(3x+5)(x+1) + C(3x+5)$

$= A(x^2+2x+1) + B(3x^2+8x+5) + C(3x+5)$

or

$-4x^2-2x+10$

$= (A+3B)x^2 + (2A+8B+3C)x + (A+5B+5C)$

Then equate corresponding coefficients.

$-4 = A + 3B$ Coefficients of x^2-terms

$-2 = 2A + 8B + 3C$ Coefficients of x-terms

$10 = A + 5B + 5C$ Constant terms

We solve this system of three equations and find $A = 5$, $B = -3$, $C = 4$.

The decomposition is

$$\frac{5}{3x+5} - \frac{3}{x+1} + \frac{4}{(x+1)^2}.$$

22. $$\frac{26x^2-36x+22}{(x-4)(2x-1)^2} = \frac{A}{x-4} + \frac{B}{2x-1} + \frac{C}{(2x-1)^2}$$

Add and equate numerators.

$26x^2-36x+22$

$= A(2x-1)^2 + B(x-4)(2x-1) + C(x-4)$

$= A(4x^2-4x+1) + B(2x^2-9x+4) + C(x-4)$

or

$26x^2-36x+22 = (4A+2B)x^2 + (-4A-9B+C)x + (A+4B-4C)$

Solving the system of equations

$26 = 4A + 2B,$

$-36 = -4A - 9B + C,$

$22 = A + 4B - 4C$

we get $A = 6$, $B = 1$, and $C = -3$.

Then the decomposition is

$$\frac{6}{x-4} + \frac{1}{2x-1} - \frac{3}{(2x-1)^2}.$$

23. $$\frac{36x+1}{12x^2-7x-10} = \frac{36x+1}{(4x-5)(3x+2)}$$

The decomposition looks like

$$\frac{A}{4x-5} + \frac{B}{3x+2}.$$

Add and equate the numerators.

$36x+1 = A(3x+2) + B(4x-5)$

or $36x+1 = (3A+4B)x + (2A-5B)$

Then equate corresponding coefficients.

$36 = 3A + 4B$ Coefficients of x-terms

$1 = 2A - 5B$ Constant terms

We solve this system of equations and find $A = 8$ and $B = 3$.

The decomposition is

$$\frac{8}{4x-5} + \frac{3}{3x+2}.$$

24. $$\frac{-17x+61}{6x^2+39x-21} = \frac{A}{6x-3} + \frac{B}{x+7}$$

$-17x+61 = (A+6B)x + (7A-3B)$

$-17 = A + 6B,$

$61 = 7A - 3B$

Then $A = 7$ and $B = -4$.

The decomposition is

$$\frac{7}{6x-3} - \frac{4}{x+7}.$$

25. $\dfrac{-4x^2-9x+8}{(3x^2+1)(x-2)}$

The decomposition looks like

$\dfrac{Ax+B}{3x^2+1}+\dfrac{C}{x-2}$.

Add and equate the numerators.

$$\begin{aligned}&-4x^2-9x+8\\ =\ &(Ax+B)(x-2)+C(3x^2+1)\\ =\ &Ax^2-2Ax+Bx-2B+3Cx^2+C\end{aligned}$$

or

$$\begin{aligned}&-4x^2-9x+8\\ =\ &(A+3C)x^2+(-2A+B)x+(-2B+C)\end{aligned}$$

Then equate corresponding coefficients.

$-4=A+3C$ Coefficients of x^2-terms

$-9=-2A+B$ Coefficients of x-terms

$8=-2B+C$ Constant terms

We solve this system of equations and find

$A=2$, $B=-5$, $C=-2$.

The decomposition is

$\dfrac{2x-5}{3x^2+1}-\dfrac{2}{x-2}$.

26. $\dfrac{11x^2-39x+16}{(x^2+4)(x-8)}=\dfrac{Ax+B}{x^2+4}+\dfrac{C}{x-8}$

$11x^2-39x+16=(A+C)x^2+(-8A+B)x+(-8B+4C)$

$$\begin{aligned}11&=A+C,\\ -39&=-8A+B,\\ 16&=-8B+4C\end{aligned}$$

Then $A=5$, $B=1$, and $C=6$.

The decomposition is

$\dfrac{5x+1}{x^2+4}+\dfrac{6}{x-8}$.

27. See the procedure on page 416 of the text. One of the algebraic methods referred to in Step 5 involves substituting values for the variable that allow us to find the constants. The other method involves equating numerators, equating coefficients of the like terms, and then using a system of equations to find the constants. Answers will vary regarding the method preferred.

28. The denominator of the second fraction, x^2-5x+6, can be factored into linear factors with real coefficients: $(x-3)(x-2)$. Thus, the given expression is not a partial fraction decomposition.

29. The degree of the numerator is equal to the degree of the denominator, so the first step should be to divide the numerator by the denominator in order to express the fraction as a quotient + remainder/denominator.

30.
$$\begin{aligned}x^3-3x^2+x-3&=0\\ x^2(x-3)+(x-3)&=0\\ (x-3)(x^2+1)&=0\\ x-3=0\ \textit{or}\ x^2+1&=0\\ x=3\ \textit{or}\quad x^2&=-1\\ x=3\ \textit{or}\quad x&=\pm i\end{aligned}$$

The solutions are 3, i, and $-i$.

31. $f(x)=x^3+x^2-3x-2$

We use synthetic division to factor the polynomial. Using the possibilities found by the rational zeros theorem we find that $x+2$ is a factor:

$$\begin{array}{r|rrrr} -2 & 1 & 1 & -3 & -2\\ & & -2 & 2 & 2\\ \hline & 1 & -1 & -1 & 0\end{array}$$

We have $x^3+x^2-3x-2=(x+2)(x^2-x-1)$.

$$\begin{aligned}x^3+x^2-3x-2&=0\\ (x+2)(x^2-x-1)&=0\\ x+2=0\ \textit{or}\ x^2-x-1&=0\end{aligned}$$

The solution of the first equation is -2. We use the quadratic formula to solve the second equation.

$$\begin{aligned}x&=\frac{-b\pm\sqrt{b^2-4ac}}{2a}\\ &=\frac{-(-1)\pm\sqrt{(-1)^2-4\cdot1\cdot(-1)}}{2\cdot1}\\ &=\frac{1\pm\sqrt{5}}{2}\end{aligned}$$

The solutions are -2, $\dfrac{1+\sqrt{5}}{2}$ and $\dfrac{1-\sqrt{5}}{2}$.

32. $f(x)=x^4-x^3-5x^2-x-6$

$$\begin{array}{r|rrrrr} -2 & 1 & -1 & -5 & -1 & -6\\ & & -2 & 6 & -2 & 6\\ \hline & 1 & -3 & 1 & -3 & 0\end{array}$$

$$\begin{aligned}x^4-x^3-5x^2-x-6&=0\\ (x+2)(x^3-3x^2+x-3)&=0\\ (x+2)[x^2(x-3)+(x-3)]&=0\\ (x+2)(x-3)(x^2+1)&=0\end{aligned}$$

$x+2=0\ \textit{or}\ x-3=0\ \textit{or}\ x^2+1=0$

$x=-2\ \textit{or}\quad x=3\ \textit{or}\quad x=\pm i$

The solutions are -2, 3, i, and $-i$.

33. $f(x)=x^3+5x^2+5x-3$

$$\begin{array}{r|rrrr} -3 & 1 & 5 & 5 & -3\\ & & -3 & -6 & 3\\ \hline & 1 & 2 & -1 & 0\end{array}$$

$$\begin{aligned}x^3+5x^2+5x-3&=0\\ (x+3)(x^2+2x-1)&=0\\ x+3=0\ \textit{or}\ x^2+2x-1&=0\end{aligned}$$

The solution of the first equation is -3. We use the quadratic formula to solve the second equation.

$$x = \frac{-b \pm \sqrt{b^2 - 4ac}}{2a}$$
$$= \frac{-2 \pm \sqrt{2^2 - 4 \cdot 1 \cdot (-1)}}{2 \cdot 1} = \frac{-2 \pm \sqrt{8}}{2}$$
$$= \frac{-2 \pm 2\sqrt{2}}{2} = \frac{2(-1 \pm \sqrt{2})}{2}$$
$$= -1 \pm \sqrt{2}$$

The solutions are -3, $-1 + \sqrt{2}$, and $-1 - \sqrt{2}$.

34. $\dfrac{9x^3 - 24x^2 + 48x}{(x-2)^4(x+1)} = \dfrac{A}{x+1} + \dfrac{P(x)}{(x-2)^4}$

Add and equate numerators.

$$9x^3 - 24x^2 + 48x = A(x-2)^4 + P(x)(x+1)$$

Let $x = -1$:

$$9(-1)^3 - 24(-1)^2 + 48(-1) = A(-1-2)^4 + 0$$
$$-81 = 81A$$
$$-1 = A$$

Then

$$9x^3 - 24x^2 + 48x = -(x-2)^4 + P(x)(x+1)$$
$$9x^3 - 24x^2 + 48x = -x^4 + 8x^3 - 24x^2 + 32x - 16 + P(x)(x+1)$$
$$x^4 + x^3 + 16x + 16 = P(x)(x+1)$$
$$x^3 + 16 = P(x) \quad \text{Dividing by } x+1$$

Now decompose $\dfrac{x^3+16}{(x-2)^4}$.

$$\frac{x^3+16}{(x-2)^4} = \frac{B}{x-2} + \frac{C}{(x-2)^2} + \frac{D}{(x-2)^3} + \frac{E}{(x-2)^4}$$

Add and equate numerators.

$$x^3 + 16 = B(x-2)^3 + C(x+2)^2 + D(x-2) + E \quad (1)$$

Let $x = 2$: $2^3 + 16 = 0 + 0 + 0 + E$

$$24 = E$$

Simplify equation (1).

$$x^3 + 16 = Bx^3 + (-6B+C)x^2 + (12B-4C+D)x + (-8B + 4C - 2D + E)$$

$1 = B$

$0 = -6B + C$

$0 = -6 \cdot 1 + C$

$6 = C$

$0 = 12B - 4C + D$

$0 = 12 \cdot 1 - 4 \cdot 6 + D$

$12 = D$

The decomposition is

$$-\frac{1}{x+1} + \frac{1}{x-2} + \frac{6}{(x-2)^2} + \frac{12}{(x-2)^3} + \frac{24}{(x-2)^4}.$$

35. $\dfrac{x}{x^4 - a^4}$

$$= \frac{x}{(x^2+a^2)(x+a)(x-a)} \quad \text{Factoring the denominator}$$
$$= \frac{Ax+B}{x^2+a^2} + \frac{C}{x+a} + \frac{D}{x-a}$$
$$= [(Ax+B)(x+a)(x-a) + C(x^2+a^2)(x-a) + D(x^2+a^2)(x+a)]/[(x^2+a^2)(x+a)(x-a)]$$

Equate the numerators:

$$x = (Ax+B)(x+a)(x-a) + C(x^2+a^2)(x-a) + D(x^2+a^2)(x+a)$$

Let $x - a = 0$, or $x = a$. Then, we get

$$a = 0 + 0 + D(a^2+a^2)(a+a)$$
$$a = D(2a^2)(2a)$$
$$a = 4a^3D$$
$$\frac{1}{4a^2} = D$$

Let $x + a = 0$, or $x = -a$. We get

$$-a = 0 + C[(-a)^2 + a^2](-a-a) + 0$$
$$-a = C(2a^2)(-2a)$$
$$-a = -4a^3C$$
$$\frac{1}{4a^2} = C$$

Equate the coefficients of x^3:

$0 = A + C + D$

Substituting $\dfrac{1}{4a^2}$ for C and for D, we get

$A = -\dfrac{1}{2a^2}$.

Equate the constant terms:

$0 = -Ba^2 - Ca^3 + Da^3$

Substitute $\dfrac{1}{4a^2}$ for C and for D. Then solve for B.

$$0 = -Ba^2 - \frac{1}{4a^2} \cdot a^3 + \frac{1}{4a^2} \cdot a^3$$
$$0 = -Ba^2$$
$$0 = B$$

The decomposition is as follows:

$$-\frac{\frac{1}{2a^2}x}{x^2+a^2} + \frac{\frac{1}{4a^2}}{x+a} + \frac{\frac{1}{4a^2}}{x-a}$$

36. $\dfrac{1}{e^{-x} + 3 + 2e^x} = \dfrac{e^x}{1 + 3e^x + 2e^{2x}}$

Multiplying by e^x/e^x

Let $y = e^x$, decompose $\dfrac{y}{2y^2 + 3y + 1}$, and then substitute e^x for y. The result is $\dfrac{1}{e^x+1} - \dfrac{1}{2e^x+1}$.

37. $\dfrac{1 + \ln x^2}{(\ln x + 2)(\ln x - 3)^2} = \dfrac{1 + 2\ln x}{(\ln x + 2)(\ln x - 3)^2}$

Let $u = \ln x$. Then we have:

$$\frac{1+2u}{(u+2)(u-3)^2}$$

$$= \frac{A}{u+2} + \frac{B}{u-3} + \frac{C}{(u-3)^2}$$

$$= \frac{A(u3)^2 + B(u+2)(u-3) + C(u+2)}{(u+2)(u-3)^2}$$

Equate the numerators:

$1 + 2u = A(u-3)^2 + B(u+2)(u-3) + C(u+2)$

Let $u - 3 = 0$, or $u = 3$.

$$1 + 2 \cdot 3 = 0 + 0 + C(5)$$
$$7 = 5C$$
$$\frac{7}{5} = C$$

Let $u + 2 = 0$, or $u = -2$.

$$1 + 2(-2) = A(-2-3)^2 + 0 + 0$$
$$-3 = 25A$$
$$-\frac{3}{25} = A$$

To find B, we equate the coefficients of u^2:

$0 = A + B$

Substituting $-\frac{3}{25}$ for A and solving for B, we get $B = \frac{3}{25}$.

The decomposition of $\frac{1+2u}{(u+2)(u-3)^2}$ is as follows:

$$-\frac{3}{25(u+2)} + \frac{3}{25(u-3)} + \frac{7}{5(u-3)^2}$$

Substituting $\ln x$ for u we get

$$-\frac{3}{25(\ln x + 2)} + \frac{3}{25(\ln x - 3)} + \frac{7}{5(\ln x - 3)^2}.$$

Chapter 6

Conic Sections

Exercise Set 6.1

1. Graph (f) is the graph of $x^2 = 8y$.

2. Graph (c) is the graph of $y^2 = -10x$.

3. Graph (b) is the graph of $(y-2)^2 = -3(x+4)$.

4. Graph (e) is the graph of $(x+1)^2 = 5(y-2)$.

5. Graph (d) is the graph of $13x^2 - 8y - 9 = 0$.

6. Graph (a) is the graph of $41x + 6y^2 = 12$.

7. $x^2 = 20y$

   $x^2 = 4 \cdot 5 \cdot y$ Writing $x^2 = 4py$

   Vertex: $(0,0)$

   Focus: $(0,5)$ $[(0,p)]$

   Directrix: $y = -5$ $(y = -p)$

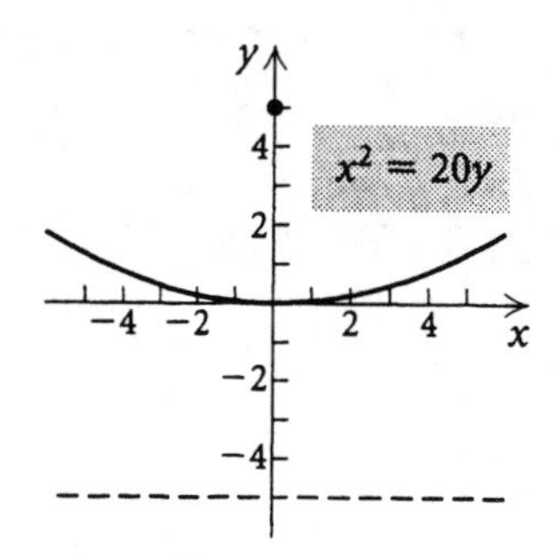

8. $x^2 = 16y$

   $x^2 = 4 \cdot 4 \cdot y$

   $V: (0,0),\ F: (0,4),\ D: y = -4$

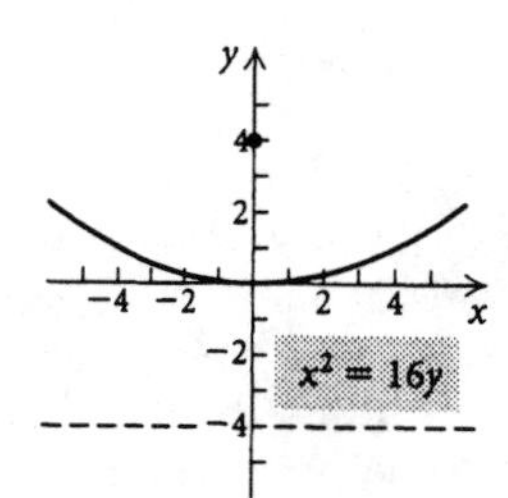

9. $y^2 = -6x$

   $y^2 = 4\left(-\dfrac{3}{2}\right)x$ Writing $y^2 = 4px$

   Vertex: $(0,0)$

   Focus: $\left(-\dfrac{3}{2}, 0\right)$ $[(p,0)]$

   Directrix: $x = -\left(-\dfrac{3}{2}\right) = \dfrac{3}{2}$ $(x = -p)$

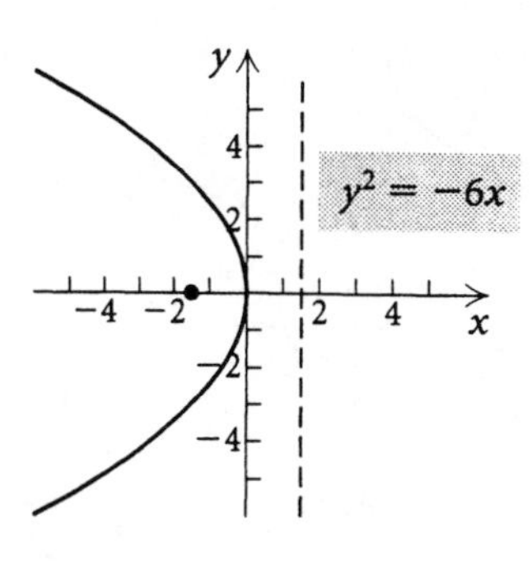

10. $y^2 = -2x$

    $y^2 = 4\left(-\dfrac{1}{2}\right)x$

    $V: (0,0), F: \left(-\dfrac{1}{2}, 0\right), D: x = \dfrac{1}{2}$

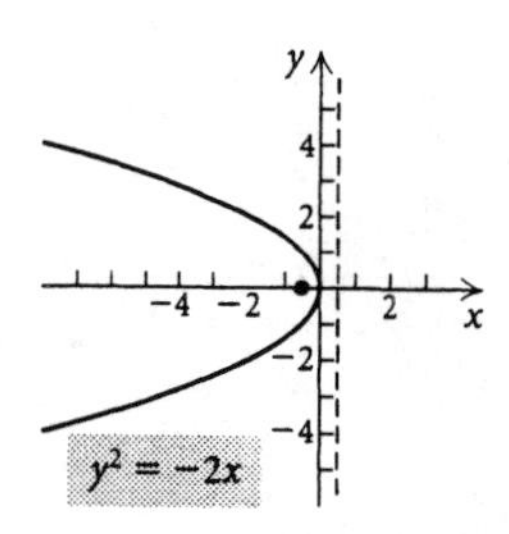

11. $x^2 - 4y = 0$

    $x^2 = 4y$

    $x^2 = 4 \cdot 1 \cdot y$ Writing $x^2 = 4py$

    Vertex: $(0,0)$

    Focus: $(0,1)$ $[(0,p)]$

    Directrix: $y = -1$ $(y = -p)$

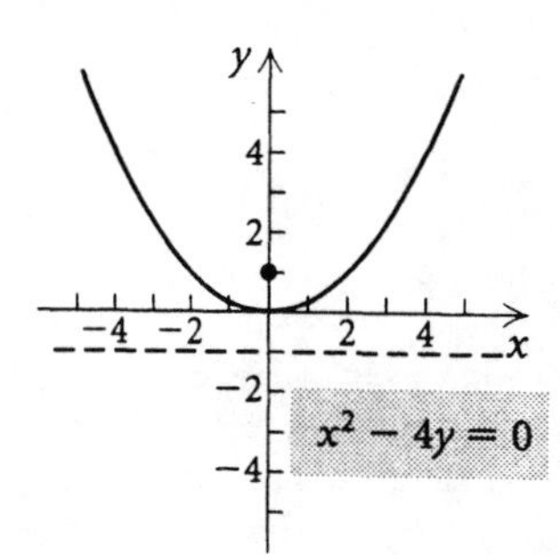

12. $y^2 + 4x = 0$

    $y^2 = -4x$

    $y^2 = 4(-1)x$

    $V: (0,0), F: (-1,0), D: x = 1$

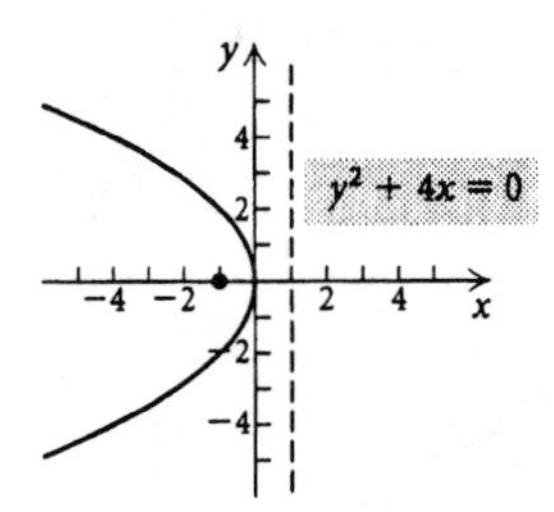

13. $x = 2y^2$

$y^2 = \frac{1}{2}x$

$y^2 = 4 \cdot \frac{1}{8} \cdot x$ Writing $y^2 = 4px$

Vertex: $(0, 0)$

Focus: $\left(\frac{1}{8}, 0\right)$

Directrix: $x = -\frac{1}{8}$

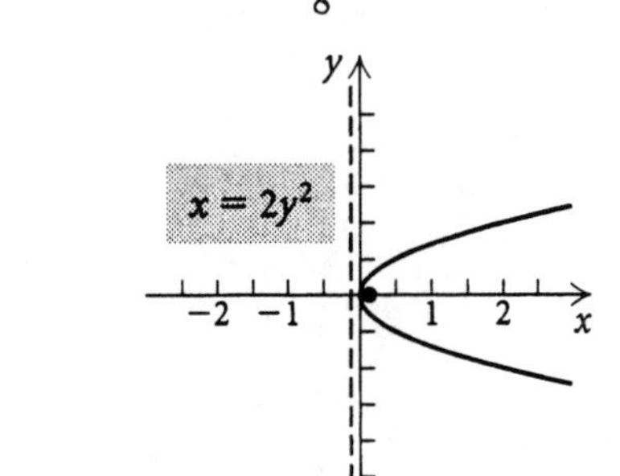

14. $y = \frac{1}{2}x^2$

$V: (0, 0)$, $F: \left(0, \frac{1}{2}\right)$, $D: y = -\frac{1}{2}$

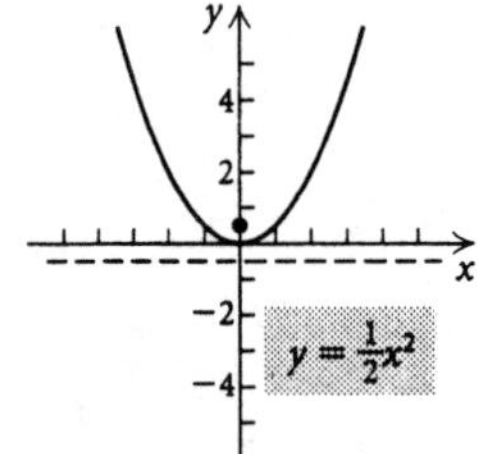

15. Since the directrix, $x = -4$, is a vertical line, the equation is of the form $(y - k)^2 = 4p(x - h)$. The focus, $(4, 0)$, is on the x-axis so the line of symmetry is the x-axis and $p = 4$. The vertex, (h, k), is the point on the x-axis midway between the directrix and the focus. Thus, it is $(0, 0)$. We have

$$(y - k)^2 = 4p(x - h)$$
$$(y - 0)^2 = 4 \cdot 4(x - 0) \quad \text{Substituting}$$
$$y^2 = 16x.$$

16. $(x - h)^2 = 4p(y - k)$

$$(x - 0)^2 = 4 \cdot \frac{1}{4}(y - 0)$$
$$x^2 = y$$

17. Since the directrix, $y = \pi$, is a horizontal line, the equation is of the form $(x - h)^2 = 4p(y - k)$. The focus, $(0, -\pi)$, is on the y-axis so the line of symmetry is the y-axis and $p = -\pi$. The vertex (h, k) is the point on the y-axis midway between the directrix and the focus. Thus, it is $(0, 0)$. We have

$$(x - h)^2 = 4p(y - k)$$
$$(x - 0)^2 = 4(-\pi)(y - 0) \quad \text{Substituting}$$
$$x^2 = -4\pi y$$

18. $(y - k)^2 = 4p(x - h)$

$$(y - 0)^2 = 4(-\sqrt{2})(x - 0)$$
$$y^2 = -4\sqrt{2}x$$

19. Since the directrix, $x = -4$, is a vertical line, the equation is of the form $(y - k)^2 = 4p(x - h)$. The focus, $(3, 2)$, is on the horizontal line $y = 2$, so the line of symmetry is $y = 2$. The vertex is the point on the line $y = 2$ that is midway between the directrix and the focus. That is, it is the midpoint of the segment from $(-4, 2)$ to $(3, 2)$: $\left(\frac{-4 + 3}{2}, \frac{2 + 2}{2}\right)$, or $\left(-\frac{1}{2}, 2\right)$. Then $h = -\frac{1}{2}$ and the directrix is $x = h - p$, so we have

$$x = h - p$$
$$-4 = -\frac{1}{2} - p$$
$$-\frac{7}{2} = -p$$
$$\frac{7}{2} = p.$$

Now we find the equation of the parabola.

$$(y - k)^2 = 4p(x - h)$$
$$(y - 2)^2 = 4\left(\frac{7}{2}\right)\left[x - \left(-\frac{1}{2}\right)\right]$$
$$(y - 2)^2 = 14\left(x + \frac{1}{2}\right)$$

20. Since the directrix, $y = -3$, is a horizontal line, the equation is of the form $(x - h)^2 = 4p(y - k)$. The focus $(-2, 3)$, is on the vertical line $x = -2$, so the line of symmetry is $x = -2$. The vertex is the point on the line $x = -2$ that is midway between the directrix and the focus. That is, it is the midpoint of the segment from $(-2, 3)$ to $(-2, -3)$: $\left(\frac{-2 - 2}{3}, \frac{3 - 3}{2}\right)$, or $(-2, 0)$. Then $k = 0$ and the directrix is $y = k - p$, so we have

$$y = k - p$$
$$-3 = 0 - p$$
$$3 = p$$

Now we find the equation of the parabola.

$$(x - h)^2 = 4p(y - k)$$
$$[x - (-2)]^2 = 4 \cdot 3(y - 0)$$
$$(x + 2)^2 = 12y$$

21. $(x+2)^2 = -6(y-1)$

$[x-(-2)]^2 = 4\left(-\frac{3}{2}\right)(y-1)$ $\quad [(x-h)^2 = 4p(y-k)]$

Vertex: $(-2, 1)$ $\quad [(h,k)]$

Focus: $\left(-2, 1+\left(-\frac{3}{2}\right)\right)$, or $\left(-2, -\frac{1}{2}\right)$

$[(h, k+p)]$

Directrix: $y = 1 - \left(-\frac{3}{2}\right) = \frac{5}{2}$ $\quad (y = k-p)$

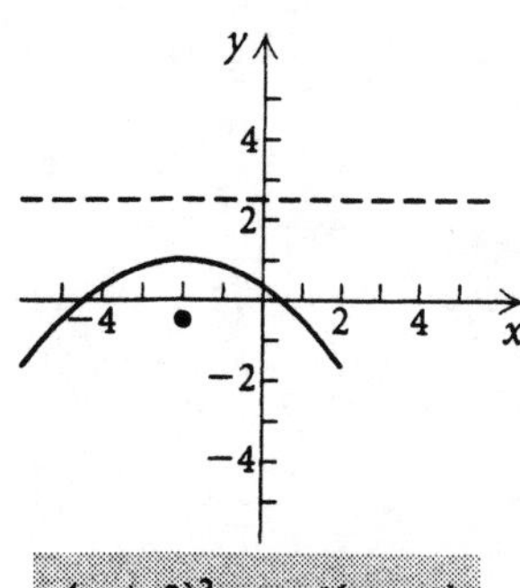

22. $(y-3)^2 = -20(x+2)$

$(y-3)^2 = 4(-5)[x-(-2)]$

V: $(-2, 3)$

F: $(-2-5, 3)$, or $(-7, 3)$

D: $x = -2-(-5) = 3$

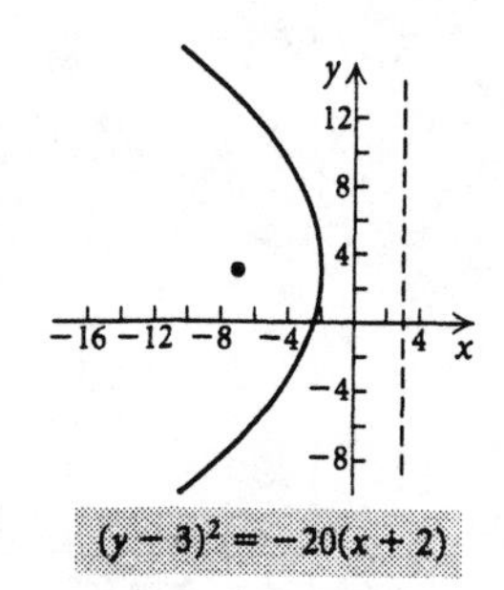

23. $x^2 + 2x + 2y + 7 = 0$

$x^2 + 2x = -2y - 7$

$(x^2 + 2x + 1) = -2y - 7 + 1 = -2y - 6$

$(x+1)^2 = -2(y+3)$

$[x-(-1)]^2 = 4\left(-\frac{1}{2}\right)[y-(-3)]$

$[(x-h)^2 = 4p(y-k)]$

Vertex: $(-1, -3)$ $\quad [(h,k)]$

Focus: $\left(-1, -3+\left(-\frac{1}{2}\right)\right)$, or $\left(-1, -\frac{7}{2}\right)$

$[(h, k+p)]$

Directrix: $y = -3 - \left(-\frac{1}{2}\right) = -\frac{5}{2}$ $\quad (y = k-p)$

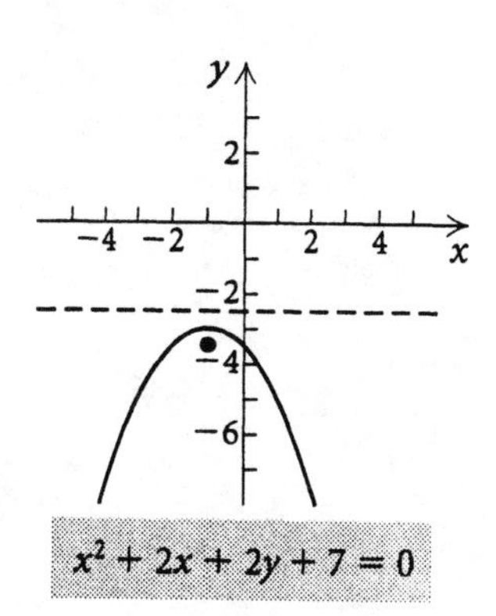

24. $y^2 + 6y - x + 16 = 0$

$y^2 + 6y + 9 = x - 16 + 9$

$(y+3)^2 = x - 7$

$[y-(-3)]^2 = 4\left(\frac{1}{4}\right)(x-7)$

V: $(7, -3)$

F: $\left(7+\frac{1}{4}, -3\right)$, or $\left(\frac{29}{4}, -3\right)$

D: $x = 7 - \frac{1}{4} = \frac{27}{4}$

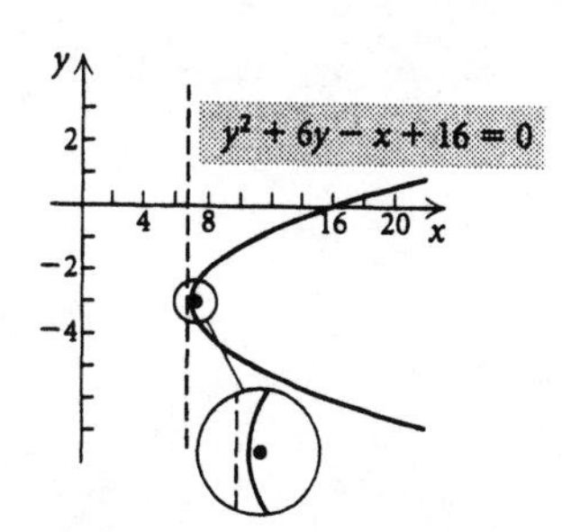

25. $x^2 - y - 2 = 0$

$x^2 = y + 2$

$(x-0)^2 = 4 \cdot \frac{1}{4} \cdot [y-(-2)]$

$[(x-h)^2 = 4p(y-k)]$

Vertex: $(0, -2)$ $\quad [(h,k)]$

Focus: $\left(0, -2+\frac{1}{4}\right)$, or $\left(0, -\frac{7}{4}\right)$ $\quad [(h, k+p)]$

Directrix: $y = -2 - \frac{1}{4} = -\frac{9}{4}$ $\quad (y = k-p)$

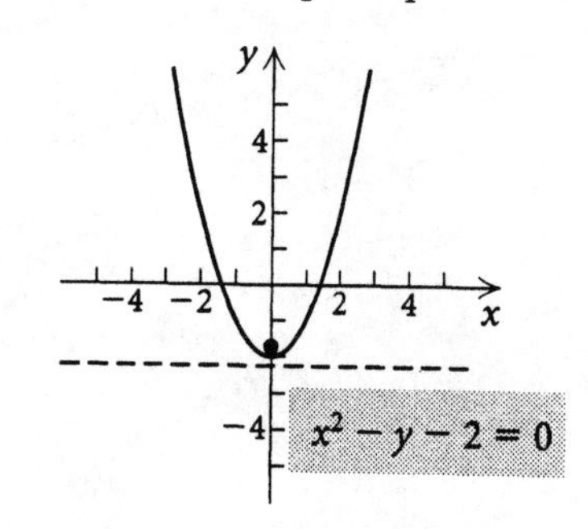

26. $x^2 - 4x - 2y = 0$

$$x^2 - 4x + 4 = 2y + 4$$
$$(x-2)^2 = 2(y+2)$$
$$(x-2)^2 = 4\left(\frac{1}{2}\right)[y-(-2)]$$

$V:\ (2,-2)$

$F:\ \left(2, -2+\frac{1}{2}\right)$, or $\left(2, -\frac{3}{2}\right)$

$D:\ y = -2 - \frac{1}{2} = -\frac{5}{2}$

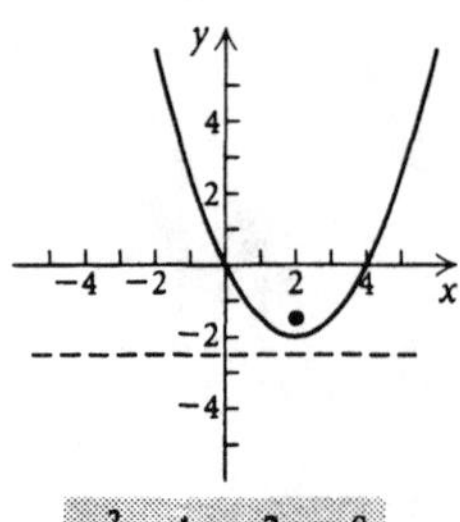

27.

$$y = x^2 + 4x + 3$$
$$y - 3 = x^2 + 4x$$
$$y - 3 + 4 = x^2 + 4x + 4$$
$$y + 1 = (x+2)^2$$
$$4 \cdot \frac{1}{4} \cdot [y - (-1)] = [x - (-2)]^2$$

$[(x-h)^2 = 4p(y-k)]$

Vertex: $(-2,-1)$ $\quad [(h,k)]$

Focus: $\left(-2, -1+\frac{1}{4}\right)$, or $\left(-2, -\frac{3}{4}\right)$ $\quad [(h, k+p)]$

Directrix: $y = -1 - \frac{1}{4} = -\frac{5}{4}$ $\quad (y = k - p)$

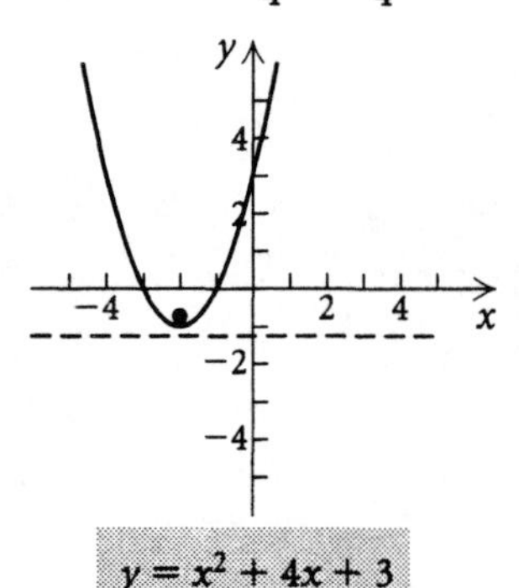

28.

$$y = x^2 + 6x + 10$$
$$y - 10 + 9 = x^2 + 6x + 9$$
$$y - 1 = (x+3)^2$$
$$4\left(\frac{1}{4}\right)(y-1) = [x-(-3)]^2$$

$V:\ (-3,1)$

$F:\ \left(-3, 1+\frac{1}{4}\right)$, or $\left(-3, \frac{5}{4}\right)$

$D:\ y = 1 - \frac{1}{4} = \frac{3}{4}$

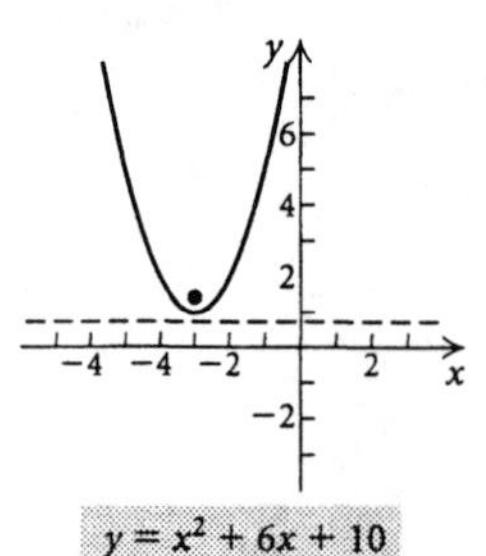

29. $y^2 - y - x + 6 = 0$

$$y^2 - y = x - 6$$
$$y^2 - y + \frac{1}{4} = x - 6 + \frac{1}{4}$$
$$\left(y - \frac{1}{2}\right)^2 = x - \frac{23}{4}$$
$$\left(y - \frac{1}{2}\right)^2 = 4 \cdot \frac{1}{4}\left(x - \frac{23}{4}\right)$$

$[(y-k)^2 = 4p(x-h)]$

Vertex: $\left(\frac{23}{4}, \frac{1}{2}\right)$ $\quad [(h,k)]$

Focus: $\left(\frac{23}{4} + \frac{1}{4}, \frac{1}{2}\right)$, or $\left(6, \frac{1}{2}\right)$ $\quad [(h+p, k)]$

Directrix: $x = \frac{23}{4} - \frac{1}{4} = \frac{22}{4}$ or $\frac{11}{2}$ $\quad (x = h - p)$

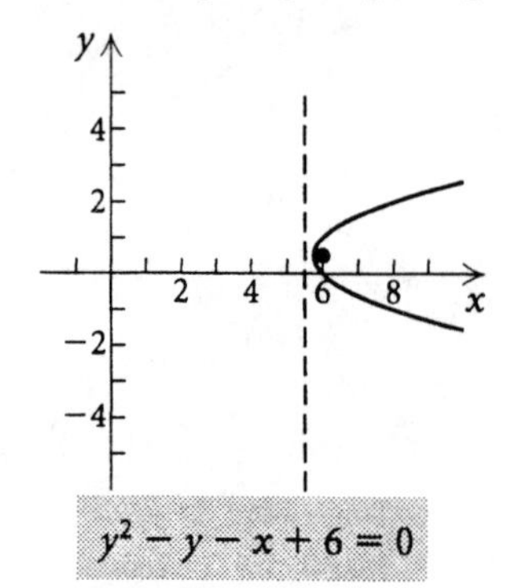

30. $y^2 + y - x - 4 = 0$

$$y^2 + y = x + 4$$
$$y^2 + y + \frac{1}{4} = x + 4 + \frac{1}{4}$$
$$\left(y + \frac{1}{2}\right)^2 = x + \frac{17}{4}$$
$$\left[y - \left(-\frac{1}{2}\right)\right]^2 = 4 \cdot \frac{1}{4}\left[x - \left(-\frac{17}{4}\right)\right]$$

$V: \left(-\frac{17}{4}, -\frac{1}{2}\right)$

$F: \left(-\frac{17}{4}+\frac{1}{4}, -\frac{1}{2}\right)$, or $\left(-4, -\frac{1}{2}\right)$

$D: x = -\frac{17}{4} - \frac{1}{4} = -\frac{9}{2}$

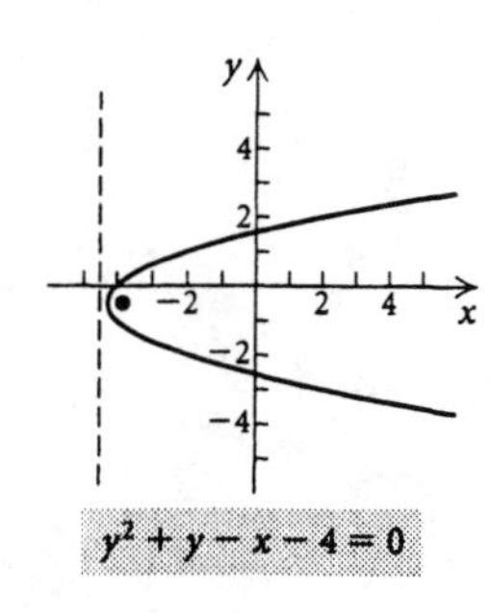

31. a) The vertex is $(0,0)$. The focus is $(4,0)$, so $p = 4$. The parabola has a horizontal axis of symmetry so the equation is of the form $y^2 = 4px$. We have

$$y^2 = 4px$$
$$y^2 = 4 \cdot 4 \cdot x$$
$$y^2 = 16x$$

b) We make a drawing.

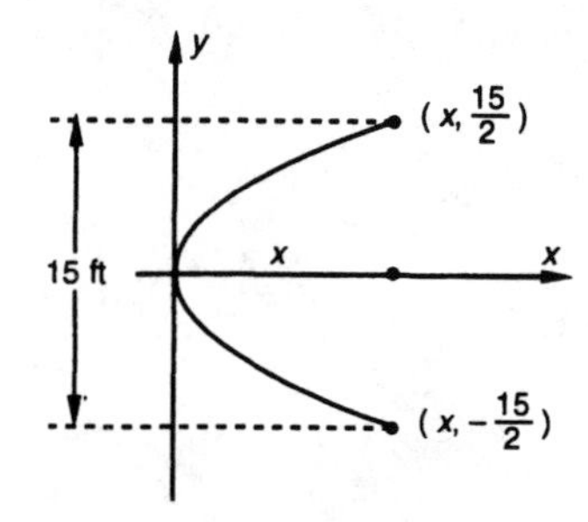

The depth of the satellite dish at the vertex is x where $\left(x, \frac{15}{2}\right)$ is a point on the parabola.

$$y^2 = 16x$$
$$\left(\frac{15}{2}\right)^2 = 16x \quad \text{Substituting } \frac{15}{2} \text{ for } y$$
$$\frac{225}{4} = 16x$$
$$\frac{225}{64} = x, \text{ or}$$
$$3\frac{33}{64} = x$$

The depth of the satellite dish at the vertex is $3\frac{33}{64}$ ft.

32. a) The parabola is of the form $y^2 = 4px$. A point on the parabola is $\left(1, \frac{6}{2}\right)$, or $(1,3)$.

$$y^2 = 4px$$
$$3^2 = 4 \cdot p \cdot 1$$
$$\frac{9}{4} = p$$

Then the equation of the parabola is $y^2 = 4 \cdot \frac{9}{4}x$, or $y^2 = 9x$.

b) The focus is at $(p,0)$, or $\left(\frac{9}{4}, 0\right)$, so the bulb should be placed $\frac{9}{4}$ in., or $2\frac{1}{4}$ in., from the vertex.

33. We position a coordinate system with the origin at the vertex and the x-axis on the parabola's axis of symmetry.

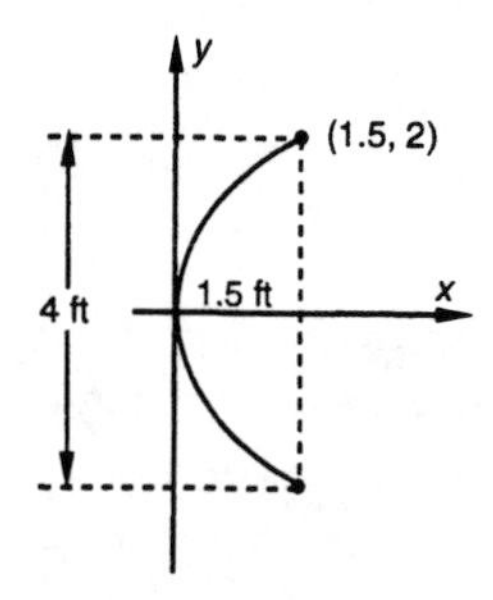

The parabola is of the form $y^2 = 4px$ and a point on the parabola is $\left(1.5, \frac{4}{2}\right)$, or $(1.5, 2)$.

$$y^2 = 4px$$
$$2^2 = 4 \cdot p \cdot (1.5) \quad \text{Substituting}$$
$$4 = 6p$$
$$\frac{4}{6} = p, \text{ or}$$
$$\frac{2}{3} = p$$

Since the focus is at $(p,0)$, or $\left(\frac{2}{3}, 0\right)$, the focus is $\frac{2}{3}$ ft, or 8 in., from the vertex.

34.

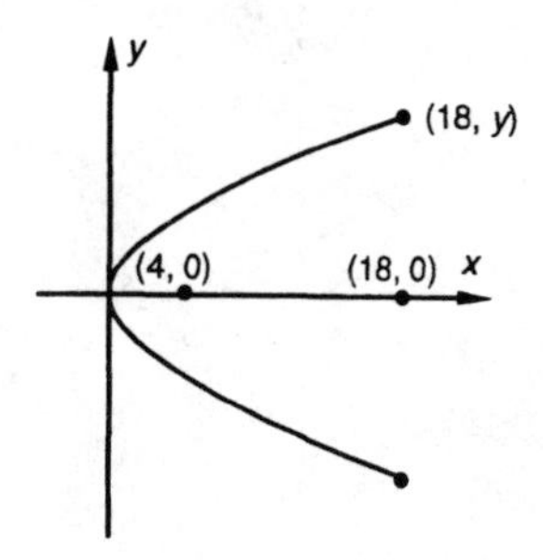

$$y^2 = 4px$$
$$y^2 = 4 \cdot 4 \cdot 18$$
$$y^2 = 288$$
$$y = 12\sqrt{2}$$

The width at the opening is $2 \cdot 12\sqrt{2}$, or $24\sqrt{2}$ in. $\approx$ 34 in.

35. No; parabolas with a horizontal axis of symmetry fail the vertical-line test.

36. See page 429 of the text.

37. $x^2+10x = x^2+10x+25 \quad (\frac{1}{2}\cdot 10 = 5 \text{ and } 5^2 = 25)$
$= (x+5)^2$

38. $y^2 - 9y = y^2 - 9y + \frac{81}{4} = \left(y - \frac{9}{2}\right)^2$

39. $(x-1)^2 + (y+2)^2 = 9$
$(x-1)^2 + [y-(-2)]^2 = 3^2$
Center: $(1,-2)$
Radius: 3

40. $[x-(-3)]^2 + (y-5)^2 = 6^2$
Center: $(-3,5)$
Radius: 6

41. A parabola with a vertical axis of symmetry has an equation of the type $(x-h)^2 = 4p(y-k)$.

Solve for p substituting $(-1,2)$ for (h,k) and $(-3,1)$ for (x,y).

$$[-3-(-1)]^2 = 4p(1-2)$$
$$4 = -4p$$
$$-1 = p$$

The equation of the parabola is
$[x-(-1)]^2 = 4(-1)(y-2)$, or
$(x+1)^2 = -4(y-2)$.

42. A parabola with a horizontal axis of symmetry has an equation of the type $(y-k)^2 = 4p(x-h)$.

Find p by substituting $(-2,1)$ for (h,k) and $(-3,5)$ for (x,y).

$$(5-1)^2 = 4p[-3-(-2)]$$
$$16 = 4p(-1)$$
$$16 = -4p$$
$$-4 = p$$

The equation of the parabola is
$(y-1)^2 = 4(-4)[x-(-2)]$, or
$(y-1)^2 = -16(x+2)$.

43. Vertex: $(0.867, 0.348)$
Focus: $(0.867, -0.191)$
Directrix: $y = 0.887$

44. Vertex: $(7.126, 1.180)$
Focus: $(7.045, 1.180)$
Directrix: $x = 7.207$

45. Position a coordinate system as shown below with the y-axis on the parabola's axis of symmetry.

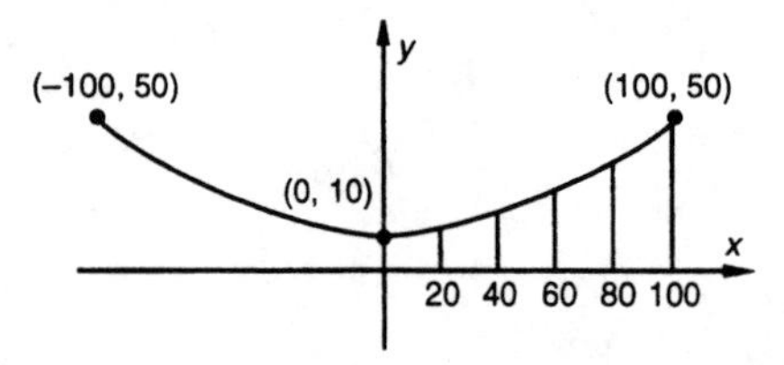

The equation of the parabola is of the form $(x-h)^2 = 4p(y-k)$. Substitute 100 for x, 50 for y, 0 for h, and 10 for k and solve for p.

$$(x-h)^2 = 4p(y-k)$$
$$(100-0)^2 = 4p(50-10)$$
$$10,000 = 160p$$
$$\frac{250}{4} = p$$

Then the equation is
$$x^2 = 4\left(\frac{250}{4}\right)(y-10), \text{ or}$$
$$x^2 = 250(y-10).$$

To find the lengths of the vertical cables, find y when $x = 0$, 20, 40, 60, 80, and 100.

When $x = 0$: $0^2 = 250(y-10)$
$0 = y - 10$
$10 = y$

When $x = 20$: $20^2 = 250(y-10)$
$400 = 250(y-10)$
$1.6 = y-10$
$11.6 = y$

When $x = 40$: $40^2 = 250(y-10)$
$1600 = 250(y-10)$
$6.4 = y-10$
$16.4 = y$

When $x = 60$: $60^2 = 250(y-10)$
$3600 = 250(y-10)$
$14.4 = y-10$
$24.4 = y$

When $x = 80$: $80^2 = 250(y-10)$
$6400 = 250(y-10)$
$25.6 = y-10$
$35.6 = y$

When $x = 100$, we know from the given information that $y = 50$.

The lengths of the vertical cables are 10 ft, 11.6 ft, 16.4 ft, 24.4 ft, 35.6 ft, and 50 ft.

Exercise Set 6.2

1. Graph (b) is the graph of $x^2 + y^2 = 5$.

2. Graph (f) is the graph of $y^2 = 20 - x^2$.

3. Graph (d) is the graph of $x^2 + y^2 - 6x + 2y = 6$.

4. Graph (c) is the graph of $x^2 + y^2 + 10x - 12y = 3$.

5. Graph (a) is the graph of $x^2 + y^2 - 5x + 3y = 0$.

6. Graph (e) is the graph of $x^2 + 4x - 2 = 6y - y^2 - 6$.

7. Complete the square twice.

$$x^2 + y^2 - 14x + 4y = 11$$
$$x^2 - 14x + y^2 + 4y = 11$$
$$x^2 - 14x + 49 + y^2 + 4y + 4 = 11 + 49 + 4$$
$$(x-7)^2 + (y+2)^2 = 64$$
$$(x-7)^2 + [y-(-2)]^2 = 8^2$$

Center: $(7, -2)$

Radius: 8

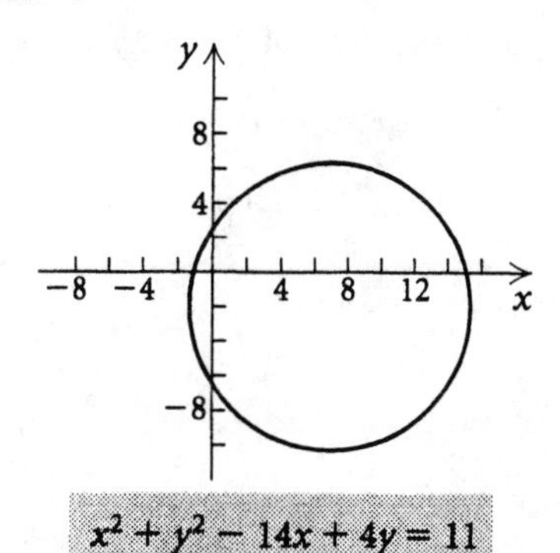

$x^2 + y^2 - 14x + 4y = 11$

8.

$$x^2 + y^2 + 2x - 6y = -6$$
$$x^2 + 2x + 1 + y^2 - 6y + 9 = -6 + 1 + 9$$
$$(x+1)^2 + (y-3)^2 = 4$$
$$[x-(-1)]^2 + (y-3)^2 = 2^2$$

Center: $(-1, 3)$

Radius: 2

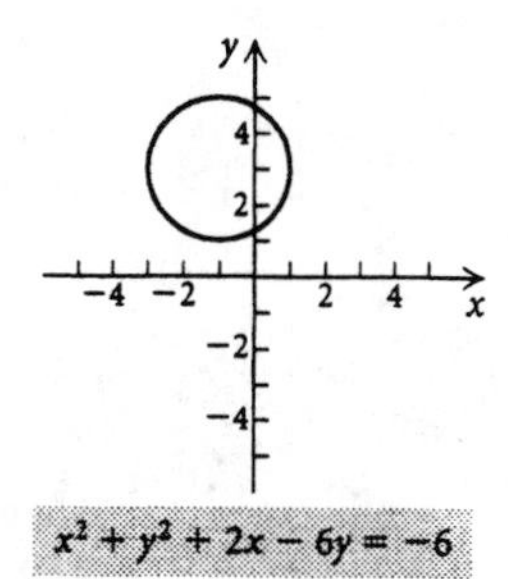

$x^2 + y^2 + 2x - 6y = -6$

9. Complete the square twice.

$$x^2 + y^2 + 4x - 6y - 12 = 0$$
$$x^2 + 4x + y^2 - 6y = 12$$
$$x^2 + 4x + 4 + y^2 - 6y + 9 = 12 + 4 + 9$$
$$(x+2)^2 + (y-3)^2 = 25$$
$$[x-(-2)]^2 + (y-3)^2 = 5^2$$

Center: $(-2, 3)$

Radius: 5

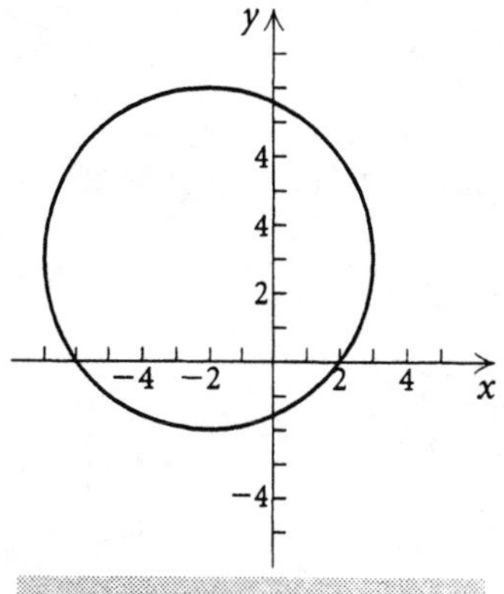

$x^2 + y^2 + 4x - 6y - 12 = 0$

10.

$$x^2 + y^2 - 8x - 2y - 19 = 0$$
$$x^2 - 8x + y^2 - 2y = 19$$
$$x^2 - 8x + 16 + y^2 - 2y + 1 = 19 + 16 + 1$$
$$(x-4)^2 + (y-1)^2 = 36$$
$$(x-4)^2 + (y-1)^2 = 6^2$$

Center: $(4, 1)$

Radius: 6

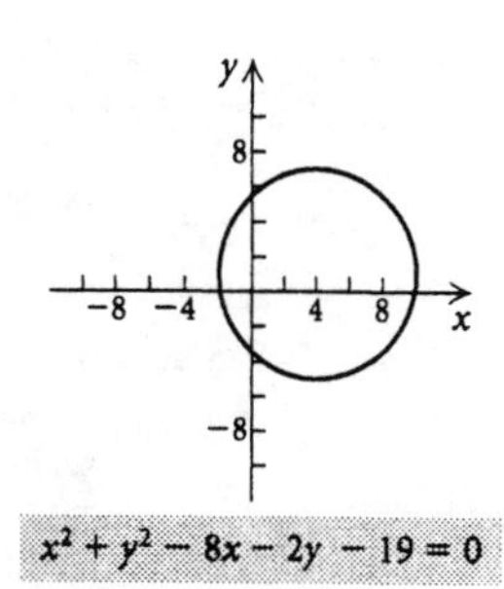

$x^2 + y^2 - 8x - 2y - 19 = 0$

11. Complete the square twice.

$$x^2 + y^2 + 6x - 10y = 0$$
$$x^2 + 6x + y^2 - 10y = 0$$
$$x^2 + 6x + 9 + y^2 - 10y + 25 = 0 + 9 + 25$$
$$(x+3)^2 + (y-5)^2 = 34$$
$$[x-(-3)]^2 + (y-5)^2 = (\sqrt{34})^2$$

Center: $(-3, 5)$

Radius: $\sqrt{34}$

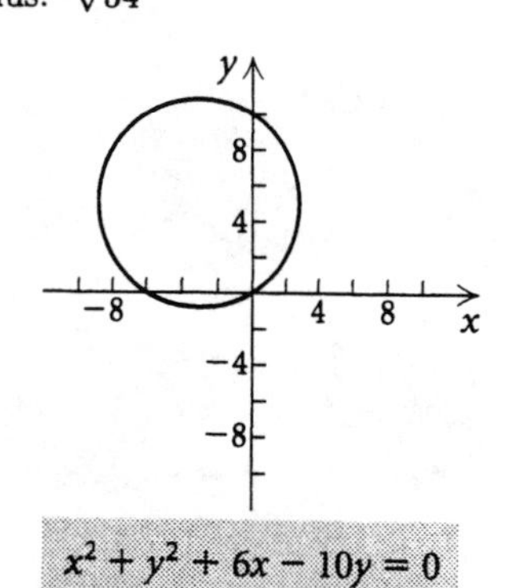

$x^2 + y^2 + 6x - 10y = 0$

12.
$$x^2 + y^2 - 7x - 2y = 0$$
$$x^2 - 7x + \frac{49}{4} + y^2 - 2y + 1 = \frac{49}{4} + 1$$
$$\left(x - \frac{7}{2}\right)^2 + (y-1)^2 = \frac{53}{4}$$
$$\left(x - \frac{7}{2}\right)^2 + (y-1)^2 = \left(\frac{\sqrt{53}}{2}\right)^2$$

Center: $\left(\frac{7}{2}, 1\right)$

Radius: $\frac{\sqrt{53}}{2}$

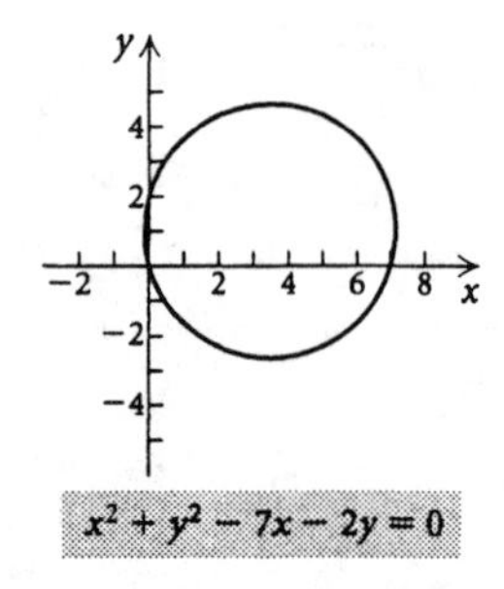

13. Complete the square twice.
$$x^2 + y^2 - 9x = 7 - 4y$$
$$x^2 - 9x + y^2 + 4y = 7$$
$$x^2 - 9x + \frac{81}{4} + y^2 + 4y + 4 = 7 + \frac{81}{4} + 4$$
$$\left(x - \frac{9}{2}\right)^2 + (y+2)^2 = \frac{125}{4}$$
$$\left(x - \frac{9}{2}\right)^2 + [y - (-2)]^2 = \left(\frac{5\sqrt{5}}{2}\right)^2$$

Center: $\left(\frac{9}{2}, -2\right)$

Radius: $\frac{5\sqrt{5}}{2}$

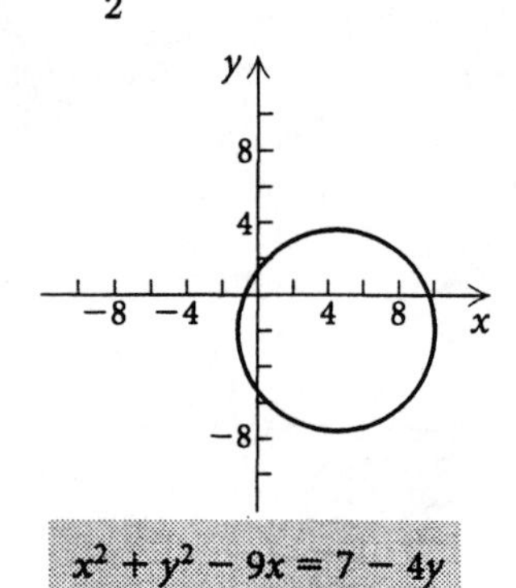

14.
$$y^2 - 6y - 1 = 8x - x^2 + 3$$
$$x^2 - 8x + y^2 - 6y = 4$$
$$x^2 - 8x + 16 + y^2 - 6y + 9 = 4 + 16 + 9$$
$$(x-4)^2 + (y-3)^2 = 29$$
$$(x-4)^2 + (y-3)^2 = (\sqrt{29})^2$$

Center: $(4, 3)$

Radius: $\sqrt{29}$

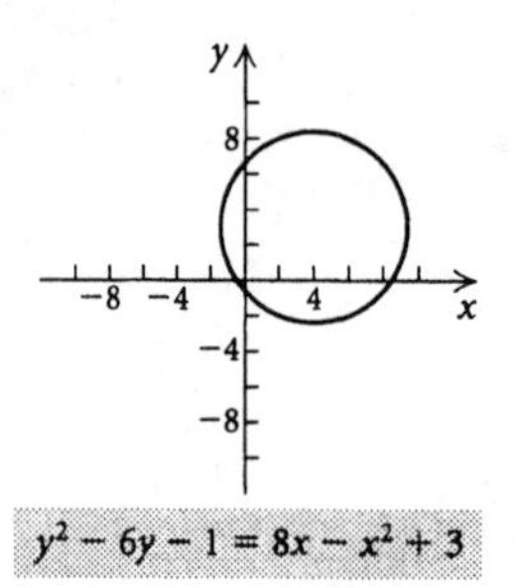

15. Graph (c) is the graph of $16x^2 + 4y^2 = 64$.

16. Graph (a) is the graph of $4x^2 + 5y^2 = 20$.

17. Graph (d) is the graph of $x^2 + 9y^2 - 6x + 90y = -225$.

18. Graph (b) is the graph of $9x^2 + 4y^2 + 18x - 16y = 11$.

19. $\frac{x^2}{4} + \frac{y^2}{1} = 1$

$\frac{x^2}{2^2} + \frac{y^2}{1^2} = 1$ Standard form

$a = 2$, $b = 1$

The major axis is horizontal, so the vertices are $(-2, 0)$ and $(2, 0)$. Since we know that $c^2 = a^2 - b^2$, we have $c^2 = 4 - 1 = 3$, so $c = \sqrt{3}$ and the foci are $(-\sqrt{3}, 0)$ and $(\sqrt{3}, 0)$.

To graph the ellipse, plot the vertices. Note also that since $b = 1$, the y-intercepts are $(0, -1)$ and $(0, 1)$. Plot these points as well and connect the four plotted points with a smooth curve.

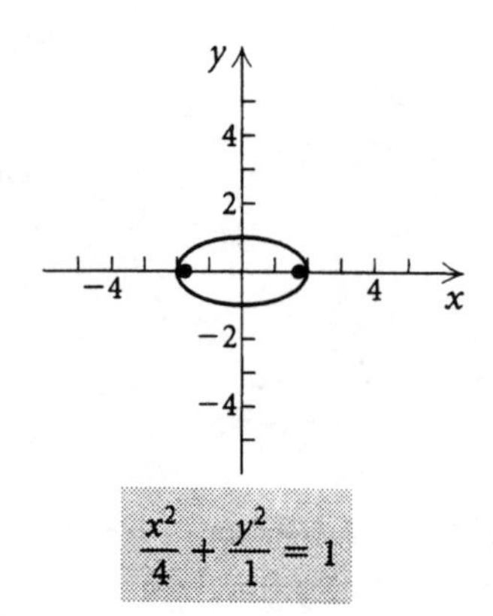

20. $\frac{x^2}{25} + \frac{y^2}{36} = 1$, or $\frac{x^2}{5^2} + \frac{y^2}{6^2} = 1$

$a = 6$, $b = 5$

The major axis is vertical, so the vertices are $(0, -6)$ and $(0, 6)$. Since $c^2 = a^2 - b^2$, we have $c^2 = 36 - 25 = 11$, so $c = \sqrt{11}$ and the foci are $(0, -\sqrt{11})$ and $(0, \sqrt{11})$.

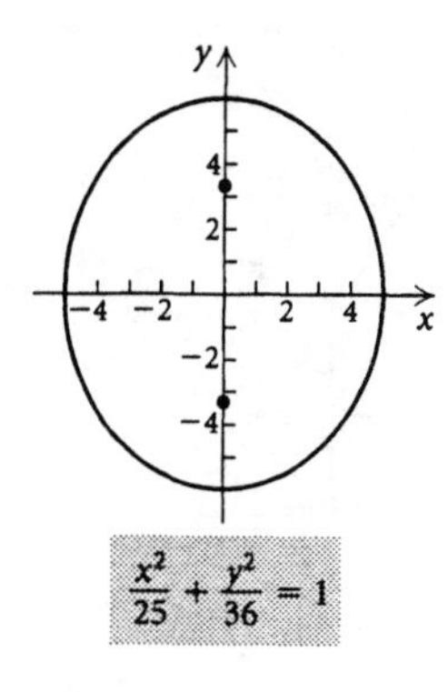

21. $16x^2 + 9y^2 = 144$

$\frac{x^2}{9} + \frac{y^2}{16} = 1$ Dividing by 144

$\frac{x^2}{3^2} + \frac{y^2}{4^2} = 1$ Standard form

$a = 4$, $b = 3$

The major axis is vertical, so the vertices are $(0, -4)$ and $(0, 4)$. Since $c^2 = a^2 - b^2$, we have $c^2 = 16 - 9 = 7$, so $c = \sqrt{7}$ and the foci are $(0, -\sqrt{7})$ and $(0, \sqrt{7})$.

To graph the ellipse, plot the vertices. Note also that since $b = 3$, the x-intercepts are $(-3, 0)$ and $(3, 0)$. Plot these points as well and connect the four plotted points with a smooth curve.

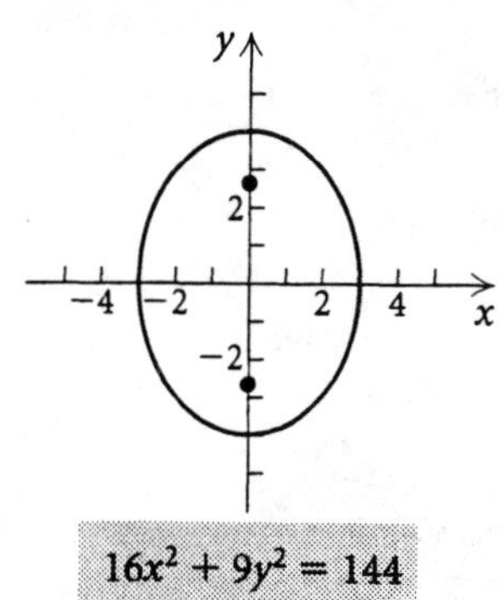

22. $9x^2 + 4y^2 = 36$

$\frac{x^2}{4} + \frac{y^2}{9} = 1$

$\frac{x^2}{2^2} + \frac{y^2}{3^2} = 1$

$a = 3$, $b = 2$

The major axis is vertical, so the vertices are $(0, -3)$ and $(0, 3)$. Since $c^2 = a^2 - b^2$, we have $c^2 = 9 - 4 = 5$, so $c = \sqrt{5}$ and the foci are $(0, -\sqrt{5})$ and $(0, \sqrt{5})$.

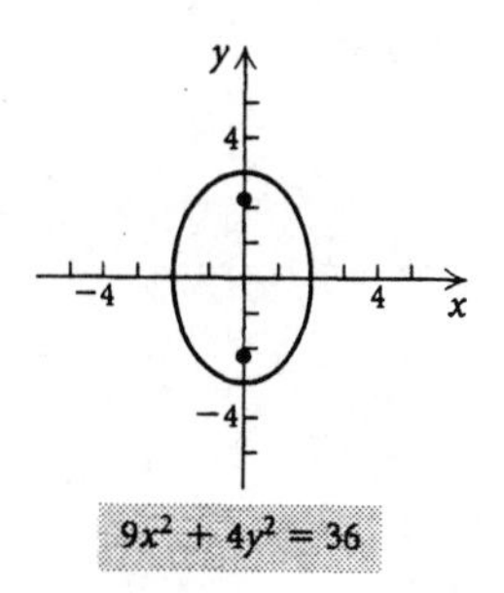

23. $2x^2 + 3y^2 = 6$

$\frac{x^2}{3} + \frac{y^2}{2} = 1$

$\frac{x^2}{(\sqrt{3})^2} + \frac{y^2}{(\sqrt{2})^2} = 1$

$a = \sqrt{3}$, $b = \sqrt{2}$

The major axis is horizontal, so the vertices are $(-\sqrt{3}, 0)$ and $(\sqrt{3}, 0)$. Since $c^2 = a^2 - b^2$, we have $c^2 = 3 - 2 = 1$, so $c = 1$ and the foci are $(-1, 0)$ and $(1, 0)$.

To graph the ellipse, plot the vertices. Note also that since $b = \sqrt{2}$, the y-intercepts are $(0, -\sqrt{2})$ and $(0, \sqrt{2})$. Plot these points as well and connect the four plotted points with a smooth curve.

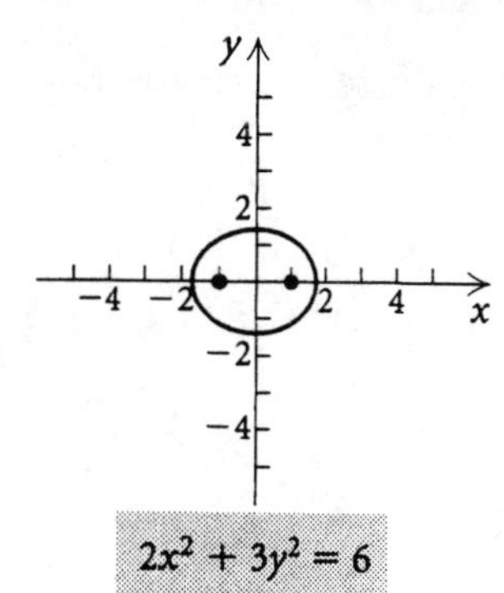

24. $5x^2 + 7y^2 = 35$

$\frac{x^2}{7} + \frac{y^2}{5} = 1$

$\frac{x^2}{(\sqrt{7})^2} + \frac{y^2}{(\sqrt{5})^2} = 1$

$a = \sqrt{7}$, $b = \sqrt{5}$

The major axis is horizontal, so the vertices are $(-\sqrt{7}, 0)$ and $(\sqrt{7}, 0)$. Since $c^2 = a^2 - b^2$, we have $c^2 = 7 - 5 = 2$, so $c = \sqrt{2}$ and the foci are $(-\sqrt{2}, 0)$ and $(\sqrt{2}, 0)$.

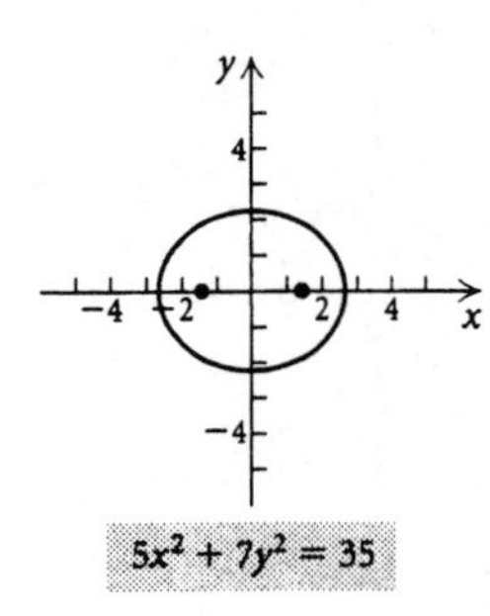

25.
$$4x^2 + 9y^2 = 1$$
$$\frac{x^2}{\frac{1}{4}} + \frac{y^2}{\frac{1}{9}} = 1$$
$$\frac{x^2}{\left(\frac{1}{2}\right)^2} + \frac{y^2}{\left(\frac{1}{3}\right)^2} = 1$$
$$a = \frac{1}{2},\ b = \frac{1}{3}$$

The major axis is horizontal, so the vertices are $\left(-\frac{1}{2}, 0\right)$ and $\left(\frac{1}{2}, 0\right)$. Since $c^2 = a^2 - b^2$, we have $c^2 = \frac{1}{4} - \frac{1}{9} = \frac{5}{36}$, so $c = \frac{\sqrt{5}}{6}$ and the foci are $\left(-\frac{\sqrt{5}}{6}, 0\right)$ and $\left(\frac{\sqrt{5}}{6}, 0\right)$.

To graph the ellipse, plot the vertices. Note also that since $b = \frac{1}{3}$, the y-intercepts are $\left(0, -\frac{1}{3}\right)$ and $\left(0, \frac{1}{3}\right)$. Plot these points as well and connect the four plotted points with a smooth curve.

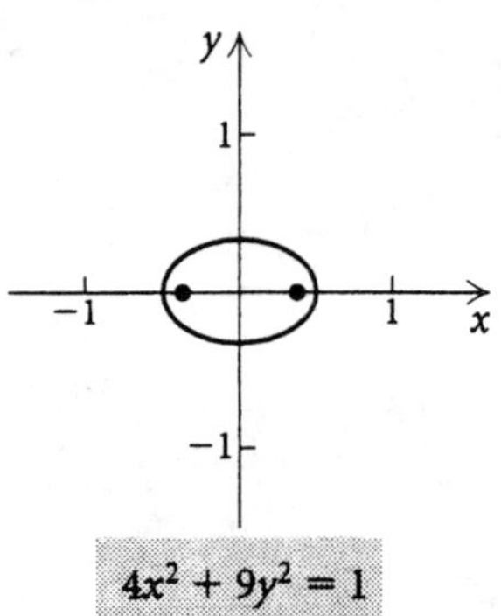

26.
$$25x^2 + 16y^2 = 1$$
$$\frac{x^2}{\frac{1}{25}} + \frac{y^2}{\frac{1}{16}} = 1$$
$$\frac{x^2}{\left(\frac{1}{5}\right)^2} + \frac{y^2}{\left(\frac{1}{4}\right)^2} = 1$$
$$a = \frac{1}{4},\ b = \frac{1}{5}$$

The major axis is vertical, so the vertices are $\left(0, -\frac{1}{4}\right)$ and $\left(0, \frac{1}{4}\right)$. Since $c^2 = a^2 - b^2$, we have $c^2 = \frac{1}{16} - \frac{1}{25} = \frac{9}{400}$, so $c = \frac{3}{20}$ and the foci are $\left(0, -\frac{3}{20}\right)$ and $\left(0, \frac{3}{20}\right)$.

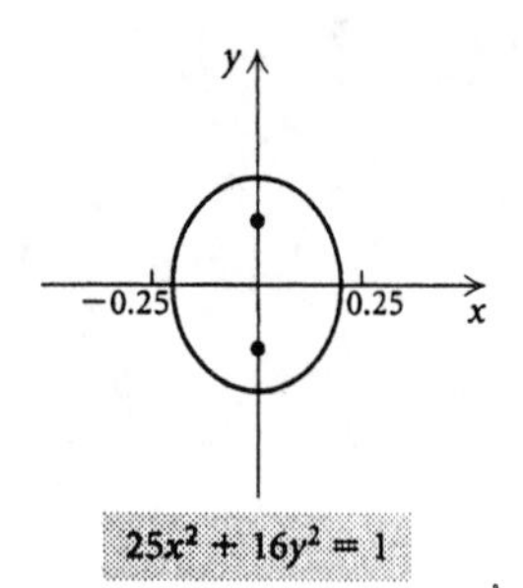

27. The vertices are on the x-axis, so the major axis is horizontal. We have $a = 7$ and $c = 3$, so we can find b^2:

$$c^2 = a^2 - b^2$$
$$3^2 = 7^2 - b^2$$
$$b^2 = 49 - 9 = 40$$

Write the equation:

$$\frac{x^2}{a^2} + \frac{y^2}{b^2} = 1$$
$$\frac{x^2}{49} + \frac{y^2}{40} = 1$$

28. The major axis is vertical; $a = 6$ and $c = 4$.

$$c^2 = a^2 - b^2$$
$$16 = 36 - b^2$$
$$b^2 = 20$$

The equation is $\frac{x^2}{20} + \frac{y^2}{36} = 1$.

29. The vertices, $(0, -8)$ and $(0, 8)$, are on the y-axis, so the major axis is vertical and $a = 8$. Since the vertices are equidistant from the origin, the center of the ellipse is at the origin. The length of the minor axis is 10, so $b = 10/2$, or 5.

Write the equation:

$$\frac{x^2}{b^2} + \frac{y^2}{a^2} = 1$$
$$\frac{x^2}{5^2} + \frac{y^2}{8^2} = 1$$
$$\frac{x^2}{25} + \frac{y^2}{64} = 1$$

30. The vertices, $(-5, 0)$ and $(5, 0)$ are on the x-axis, so the major axis is horizontal and $a = 5$. Since the vertices are equidistant from the origin, the center of the ellipse is at the origin. The length of the minor axis is 6, so $b = 6/2$, or 3. The equation is $\frac{x^2}{25} + \frac{y^2}{9} = 1$.

31. The foci, $(-2, 0)$ and $(2, 0)$ are on the x-axis, so the major axis is horizontal and $c = 2$. Since the foci are equidistant from the origin, the center of the ellipse is at the origin. The length of the major axis is 6, so $a = 6/2$, or 3. Now we find b^2:

$$c^2 = a^2 - b^2$$
$$2^2 = 3^2 - b^2$$
$$4 = 9 - b^2$$
$$b^2 = 5$$

Write the equation:

$$\frac{x^2}{a^2} + \frac{y^2}{b^2} = 1$$
$$\frac{x^2}{9} + \frac{y^2}{5} = 1$$

32. The foci, $(0, -3)$ and $(0, 3)$, are on the y-axis, so the major axis is vertical. The foci are equidistant from the origin, so the center of the ellipse is at the origin. The length of the major axis is 10, so $a = 10/2$, or 5. Find b^2:

$$c^2 = a^2 - b^2$$
$$9 = 25 - b^2$$
$$b^2 = 16$$

The equation is $\frac{x^2}{16} + \frac{y^2}{25} = 1$.

33. $\frac{(x-1)^2}{9} + \frac{(y-2)^2}{4} = 1$

$\frac{(x-1)^2}{3^2} + \frac{(y-2)^2}{2^2} = 1$ Standard form

The center is $(1, 2)$. Note that $a = 3$ and $b = 2$. The major axis is horizontal so the vertices are 3 units left and right of the center:

$(1-3, 2)$ and $(1+3, 2)$, or $(-2, 2)$ and $(4, 2)$.

We know that $c^2 = a^2 - b^2$, so $c^2 = 9 - 4 = 5$ and $c = \sqrt{5}$. Then the foci are $\sqrt{5}$ units left and right of the center:

$$(1 - \sqrt{5}, 2) \text{ and } (1 + \sqrt{5}, 2).$$

To graph the ellipse, plot the vertices. Since $b = 2$, two other points on the graph are 2 units below and above the center:

$(1, 2-2)$ and $(1, 2+2)$ or $(1, 0)$ and $(1, 4)$

Plot these points also and connect the four plotted points with a smooth curve.

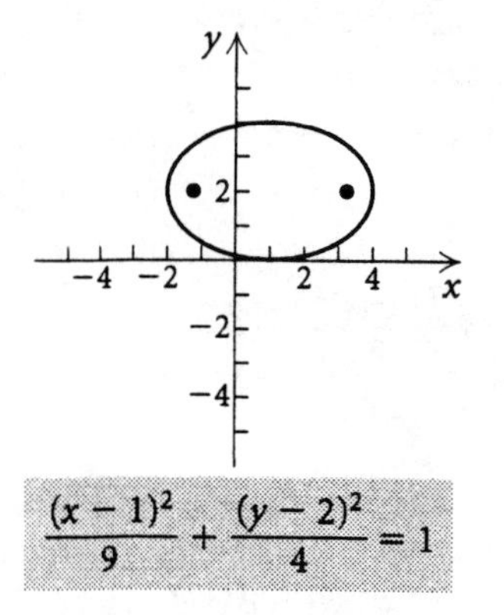

34. $\frac{(x-1)^2}{1} + \frac{(y-2)^2}{4} = 1$

$\frac{(x-1)^2}{1^2} + \frac{(y-2)^2}{2^2} = 1$ Standard form

The center is $(1, 2)$. Note that $a = 2$ and $b = 1$. The major axis is vertical so the vertices are 2 units below and above the center:

$(1, 2-2)$ and $(1, 2+2)$, or $(1, 0)$ and $(1, 4)$.

We know that $c^2 = a^2 - b^2$, so $c^2 = 4 - 1 = 3$ and $c = \sqrt{3}$. Then the foci are $\sqrt{3}$ units below and above the center:

$$(1, 2 - \sqrt{3}) \text{ and } (1, 2 + \sqrt{3}).$$

Since $b = 1$, two points on the graph other than the vertices are $(1-1, 2)$ and $(1+1, 2)$ or $(0, 2)$ and $(2, 2)$.

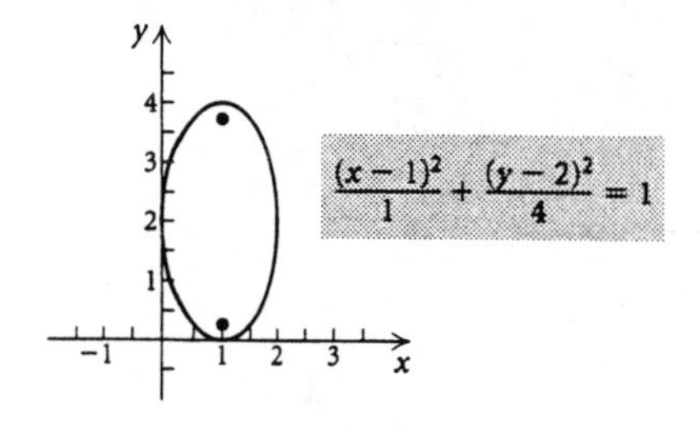

35. $\frac{(x+3)^2}{25} + \frac{(y-5)^2}{36} = 1$

$\frac{[x-(-3)]^2}{5^2} + \frac{(y-5)^2}{6^2} = 1$ Standard form

The center is $(-3, 5)$. Note that $a = 6$ and $b = 5$. The major axis is vertical so the vertices are 6 units below and above the center:

$(-3, 5-6)$ and $(-3, 5+6)$, or $(-3, -1)$ and $(-3, 11)$.

We know that $c^2 = a^2 - b^2$, so $c^2 = 36 - 25 = 11$ and $c = \sqrt{11}$. Then the foci are $\sqrt{11}$ units below and above the vertex:

$$(-3, 5 - \sqrt{11}) \text{ and } (-3, 5 + \sqrt{11}).$$

To graph the ellipse, plot the vertices. Since $b = 5$, two other points on the graph are 5 units left and right of the center:

$(-3-5, 5)$ and $(-3+5, 5)$, or $(-8, 5)$ and $(2, 5)$

Plot these points also and connect the four plotted points with a smooth curve.

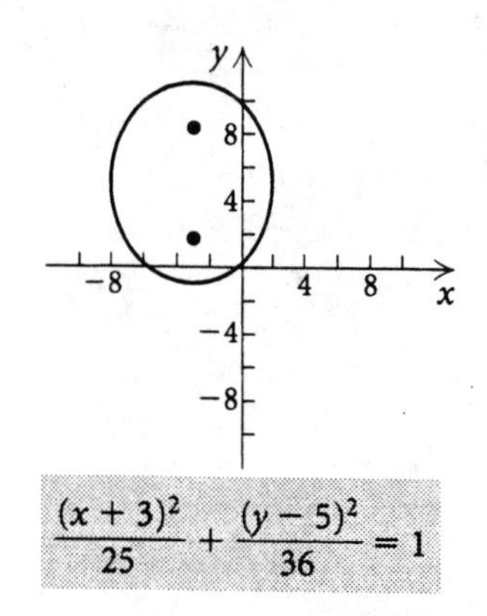

36. $\dfrac{(x-2)^2}{16}+\dfrac{(y+3)^2}{25}=1$

$\dfrac{(x-2)^2}{4^2}+\dfrac{[y-(-3)]^2}{5^2}=1$ Standard form

The center is $(2,-3)$. Note that $a=5$ and $b=4$. The major axis is vertical so the vertices are 5 units below and above the center:

$(2,-3-5)$ and $(2,-3+5)$, or $(2,-8)$ and $(2,2)$.

We know that $c^2=a^2-b^2$, so $c^2=25-16=9$ and $c=3$. Then the foci are 3 units below and above the center:

$(2,-3-3)$ and $(2,-3+3)$, or $(2,-6)$ and $(2,0)$.

Since $b=4$, two points on the graph other than the vertices are $(2-4,-3)$ and $(2+4,-3)$, or $(-2,-3)$ and $(6,-3)$.

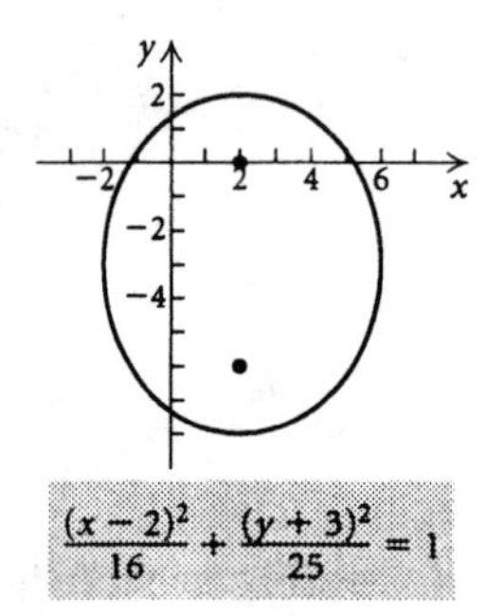

37. $3(x+2)^2+4(y-1)^2=192$

$\dfrac{(x+2)^2}{64}+\dfrac{(y-1)^2}{48}=1$ Dividing by 192

$\dfrac{[x-(-2)]^2}{8^2}+\dfrac{(y-1)^2}{(\sqrt{48})^2}=1$ Standard form

The center is $(-2,1)$. Note that $a=8$ and $b=\sqrt{48}$, or $4\sqrt{3}$. The major axis is horizontal so the vertices are 8 units left and right of the center:

$(-2-8,1)$ and $(-2+8,1)$, or $(-10,1)$ and $(6,1)$.

We know that $c^2=a^2-b^2$, so $c^2=64-48=16$ and $c=4$. Then the foci are 4 units left and right of the center:

$(-2-4,1)$ and $(-2+4,1)$ or $(-6,1)$ and $(2,1)$.

To graph the ellipse, plot the vertices. Since $b=4\sqrt{3}\approx 6.928$, two other points on the graph are about 6.928 units below and above the center:

$(-2,1-6.928)$ and $(-2,1+6.928)$, or

$(-2,-5.928)$ and $(-2,7.928)$.

Plot these points also and connect the four plotted points with a smooth curve.

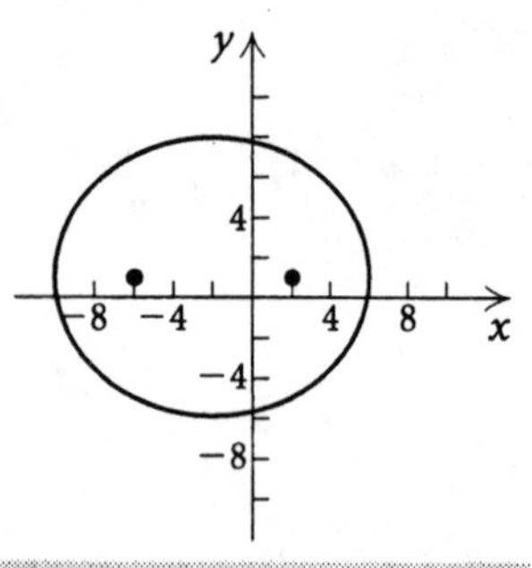

38. $4(x-5)^2+3(y-4)^2=48$

$\dfrac{(x-5)^2}{12}+\dfrac{(y-4)^2}{16}=1$

$\dfrac{(x-5)^2}{(\sqrt{12})^2}+\dfrac{(y-4)^2}{4^2}=1$ Standard form

The center is $(5,4)$. Note that $a=16$ and $b=\sqrt{12}$, or $2\sqrt{3}$. The major axis is vertical so the vertices are 4 units below and above the center:

$(5,4-4)$ and $(5,4+4)$, or $(5,0)$ and $(5,8)$.

We know that $c^2=a^2-b^2$, so $c^2=16-12=4$ and $c=2$. Then the foci are 2 units below and above the center:

$(5,4-2)$ and $(5,4+2)$, or $(5,2)$ and $(5,6)$.

Since $b=2\sqrt{3}\approx 3.464$, two points on the graph other than the vertices are about $(5-3.464,4)$ and $(5+3.464,4)$, or $(1.536,4)$ and $(8.464,4)$.

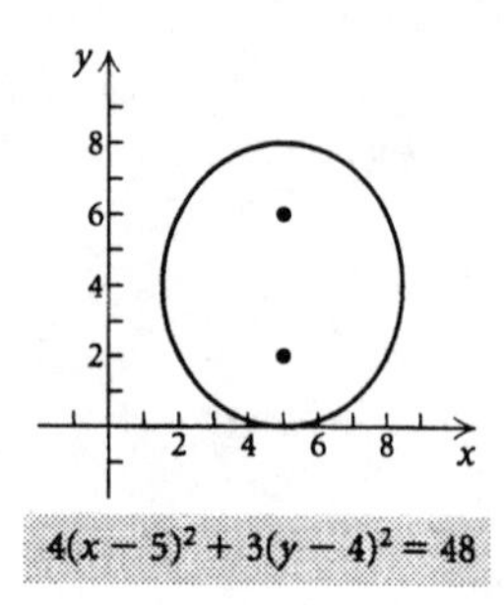

39. Begin by completing the square twice.

$$4x^2+9y^2-16x+18y-11=0$$
$$4x^2-16x+9y^2+18y=11$$
$$4(x^2-4x)+9(y^2+2y)=11$$
$$4(x^2-4x+4)+9(y^2+2y+1)=11+4\cdot 4+9\cdot 1$$
$$4(x-2)^2+9(y+1)^2=36$$
$$\frac{(x-2)^2}{9}+\frac{(y+1)^2}{4}=1$$
$$\frac{(x-2)^2}{3^2}+\frac{[y-(-1)]^2}{2^2}=1$$

The center is $(2,-1)$. Note that $a=3$ and $b=2$. The major axis is horizontal so the vertices are 3 units left and right of the center:

$(2-3, -1)$ and $(2+3, -1)$, or $(-1, -1)$ and $(5, -1)$.

We know that $c^2 = a^2 - b^2$, so $c^2 = 9 - 4 = 5$ and $c = \sqrt{5}$. Then the foci are $\sqrt{5}$ units left and right of the center:

$$(2-\sqrt{5}, -1) \text{ and } (2+\sqrt{5}, -1).$$

To graph the ellipse, plot the vertices. Since $b = 2$, two other points on the graph are 2 units below and above the center:

$(2, -1-2)$ and $(2, -1+2)$, or $(2, -3)$ and $(2, 1)$.

Plot these points also and connect the four plotted points with a smooth curve.

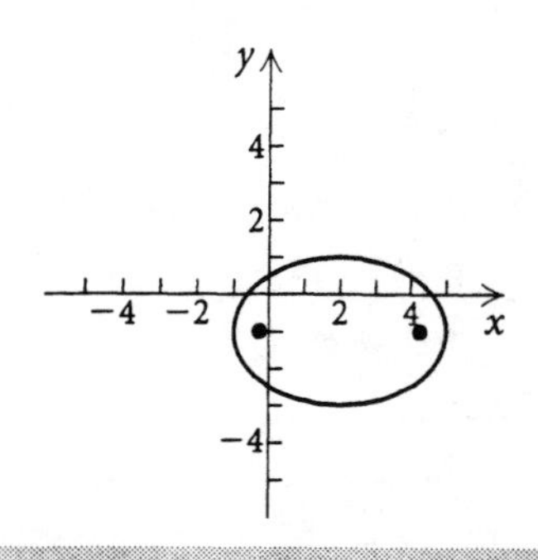

$4x^2 + 9y^2 - 16x + 18y - 11 = 0$

40. Begin by completing the square twice.

$$x^2 + 2y^2 - 10x + 8y + 29 = 0$$
$$x^2 - 10x + 2(y^2 + 4y) = -29$$
$$x^2 - 10x + 25 + 2(y^2 + 4y + 4) = -29 + 25 + 2 \cdot 4$$
$$(x-5)^2 + 2(y+2)^2 = 4$$
$$\frac{(x-5)^2}{4} + \frac{(y+2)^2}{2} = 1$$
$$\frac{(x-5)^2}{2^2} + \frac{[y-(-2)]^2}{(\sqrt{2})^2} = 1$$

The center is $(5, -2)$. Note that $a = 2$ and $b = \sqrt{2}$. The major axis is horizontal so the vertices are 2 units left and right of the center:

$(5-2, -2)$ and $(5+2, -2)$, or $(3, -2)$ and $(7, -2)$.

We know that $c^2 = a^2 - b^2$, so $c^2 = 4 - 2 = 2$ and $c = \sqrt{2}$. Then the foci are $\sqrt{2}$ units left and right of the center:

$$(5-\sqrt{2}, -2) \text{ and } (5+\sqrt{2}, -2).$$

Since $b = \sqrt{2} \approx 1.414$, two points on the graph other than the vertices are about

$(5, -2-1.414)$ and $(5, -2+1.414)$, or

$(5, -3.414)$ and $(5, -0.586)$.

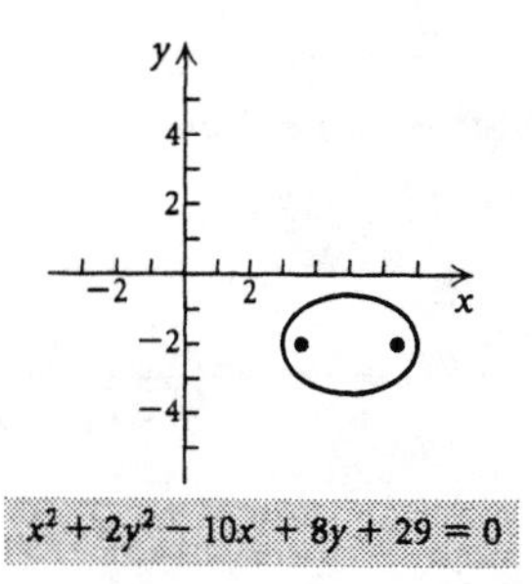

$x^2 + 2y^2 - 10x + 8y + 29 = 0$

41. Begin by completing the square twice.

$$4x^2 + y^2 - 8x - 2y + 1 = 0$$
$$4x^2 - 8x + y^2 - 2y = -1$$
$$4(x^2 - 2x) + y^2 - 2y = -1$$
$$4(x^2 - 2x + 1) + y^2 - 2y + 1 = -1 + 4 \cdot 1 + 1$$
$$4(x-1)^2 + (y-1)^2 = 4$$
$$\frac{(x-1)^2}{1} + \frac{(y-1)^2}{4} = 1$$
$$\frac{(x-1)^2}{1^2} + \frac{(y-1)^2}{2^2} = 1$$

The center is $(1, 1)$. Note that $a = 2$ and $b = 1$. The major axis is vertical so the vertices are 2 units below and above the center:

$(1, 1-2)$ and $(1, 1+2)$, or $(1, -1)$ and $(1, 3)$.

We know that $c^2 = a^2 - b^2$, so $c^2 = 4 - 1 = 3$ and $c = \sqrt{3}$. Then the foci are $\sqrt{3}$ units below and above the center:

$$(1, 1-\sqrt{3}) \text{ and } (1, 1+\sqrt{3}).$$

To graph the ellipse, plot the vertices. Since $b = 1$, two other points on the graph are 1 unit left and right of the center:

$(1-1, 1)$ and $(1+1, 1)$ or $(0, 1)$ and $(2, 1)$.

Plot these points also and connect the four plotted points with a smooth curve.

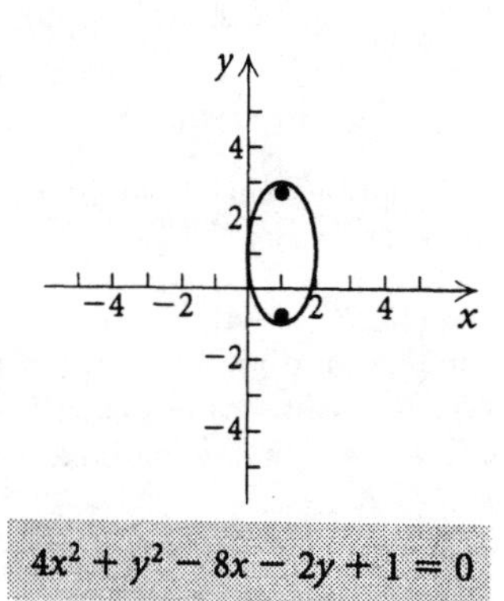

$4x^2 + y^2 - 8x - 2y + 1 = 0$

42. Begin by completing the square twice.

$$9x^2 + 4y^2 + 54x - 8y + 49 = 0$$
$$9(x^2 + 6x) + 4(y^2 - 2y) = -49$$
$$9(x^2 + 6x + 9) + 4(y^2 - 2y + 1) = -49+9\cdot 9+4\cdot 1$$
$$9(x + 3)^2 + 4(y - 1)^2 = 36$$
$$\frac{(x - 3)^2}{4} + \frac{(y - 1)^2}{9} = 1$$
$$\frac{[x - (-3)]^2}{2^2} + \frac{(y - 1)^2}{3^2} = 1$$

The center is $(-3, 1)$. Note that $a = 3$ and $b = 2$. The major axis is vertical so the vertices are 3 units below and above the center:

$(-3, 1 - 3)$ and $(-3, 1 + 3)$, or $(-3, -2)$ and $(-3, 4)$.

We know that $c^2 = a^2 - b^2$, so $c^2 = 9 - 4 = 5$ and $c = \sqrt{5}$. Then the foci are $\sqrt{5}$ units below and above the center:

$(-3, 1 - \sqrt{5})$ and $(-3, 1 + \sqrt{5})$.

Since $b = 2$, two points on the graph other than the vertices are

$(-3 - 2, 1)$ and $(-3 + 2, 1)$, or $(-5, 1)$ and $(-1, 1)$.

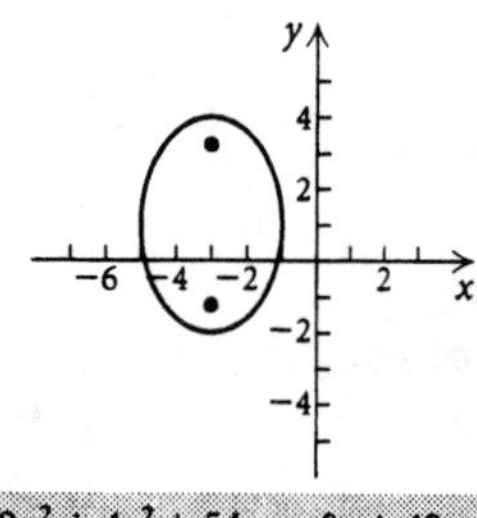

43. The ellipse in Example 4 is flatter than the one in Example 2, so the ellipse in Example 2 has the smaller eccentricity.

We compute the the eccentricities: In Example 2, $c = 3$ and $a = 5$, so $e = c/a = 3/5 = 0.6$. In Example 4, $c = 2\sqrt{3}$ and $a = 4$, so $e = c/a = 2\sqrt{3}/4 \approx 0.866$. These computations confirm that the ellipse in Example 2 has the smaller eccentricity.

44. Ellipse (b) is flatter than ellipse (a), so ellipse (a) has the smaller eccentricity.

45. Since the vertices, $(0, -4)$ and $(0, 4)$ are on the y-axis and are equidistant from the origin, we know that the major axis of the ellipse is vertical, its center is at the origin, and $a = 4$. Use the information that $e = 1/4$ to find c:

$$e = \frac{c}{a}$$
$$\frac{1}{4} = \frac{c}{4} \quad \text{Substituting}$$
$$c = 1$$

Now $c^2 = a^2 - b^2$, so we can find b^2:

$$1^2 = 4^2 - b^2$$
$$1 = 16 - b^2$$
$$b^2 = 15$$

Write the equation of the ellipse:

$$\frac{x^2}{b^2} + \frac{y^2}{a^2} = 1$$
$$\frac{x^2}{15} + \frac{y^2}{16} = 1$$

46. Since the vertices, $(-3, 0)$ and $(3, 0)$, are on the x-axis and are equidistant from the origin, we know that the major axis of the ellipse is horizontal, its center is at the origin, and $a = 3$.

Find c:

$$\frac{c}{a} = \frac{7}{10}$$
$$\frac{c}{3} = \frac{7}{10}$$
$$c = \frac{21}{10}$$

Now find b^2:

$$c^2 = a^2 - b^2$$
$$\left(\frac{21}{10}\right)^2 = 3^2 - b^2$$
$$b^2 = \frac{459}{100}$$

The equation of the ellipse is $\dfrac{x^2}{9} + \dfrac{y^2}{459/100} = 1$.

47. From the figure in the text we see that the center of the ellipse is $(0, 0)$, the major axis is horizontal, the vertices are $(-50, 0)$ and $(50, 0)$, and one y-intercept is $(0, 12)$. Then $a = 50$ and $b = 12$. The equation is

$$\frac{x^2}{a^2} + \frac{y^2}{b^2} = 1$$
$$\frac{x^2}{50^2} + \frac{y}{12^2} = 1$$
$$\frac{x^2}{2500} + \frac{y^2}{144} = 1.$$

48. Find the equation of the ellipse with center $(0, 0)$, $a = 1048/2 = 524$, $b = 898/2 = 449$, and a horizontal major axis:

$$\frac{x^2}{524^2} + \frac{y^2}{449^2} = 1$$
$$\frac{x^2}{274,576} + \frac{y^2}{201,601} = 1$$

49. Position a coordinate system as shown below where 1 unit $= 10^7$ mi.

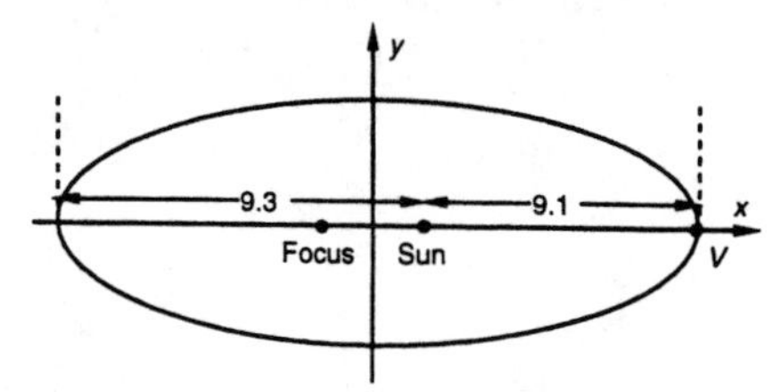

The length of the major axis is $9.3 + 9.1$, or 18.4. Then the distance from the center of the ellipse (the origin) to V is 18.4/2, or 9.2. Since the distance from the sun to V is 9.1, the distance from the sun to the center is $9.2 - 9.1$, or 0.1. Then the distance from the sun to the other focus is twice this distance:

$$2(0.1 \times 10^7 \text{ mi}) = 0.2 \times 10^7 \text{ mi}$$
$$= 2 \times 10^6 \text{ mi}$$

50. Position a coordinate system as shown.

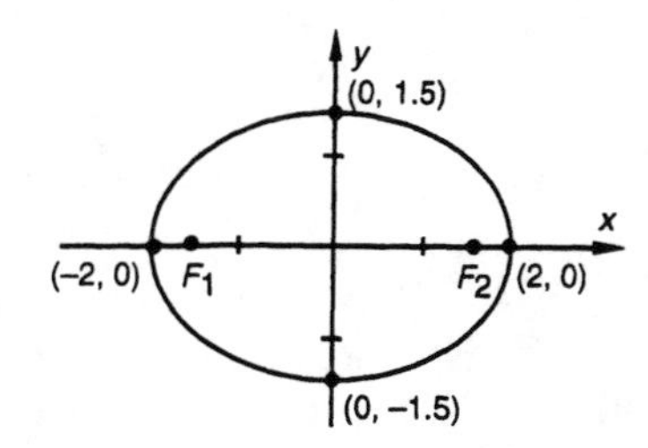

a) We have an ellipse with $a = 2$ and $b = 1.5$. The foci are at $(\pm c, 0)$ where $c^2 = a^2 - b^2$, so $c^2 = 4 - 2.25 = 1.75$ and $c = \sqrt{1.75}$. Then the string should be attached $2 - \sqrt{1.75}$ ft, or about 0.7 ft from the ends of the board.

b) The string should be the length of the major axis, 4 ft.

51. Circles and ellipses are not functions.

52. Answers will vary. Students who have graphers that require an equation to be expressed in "$y =$" form will probably prefer to graph the ellipse by hand.

53. a) The graph of $f(x) = 2x - 3$ is shown below.

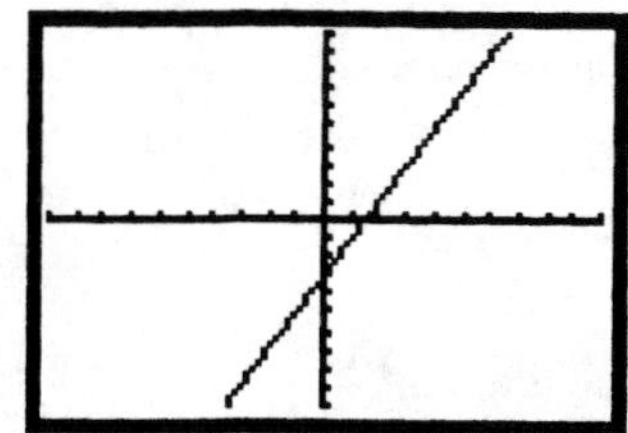

Since there is no horizontal line that crosses the graph more than once, the function is one-to-one.

b) Replace $f(x)$ with y: $y = 2x - 3$

Interchange x and y: $x = 2y - 3$

Solve for y: $x + 3 = 2y$

$$\frac{x+3}{2} = y$$

Replace y with $f^{-1}(x)$: $f^{-1}(x) = \dfrac{x+3}{2}$

54. a) The graph of $f(x) = x^3 + 2$ is shown below. It passes the horizontal line test, so it is one-to-one.

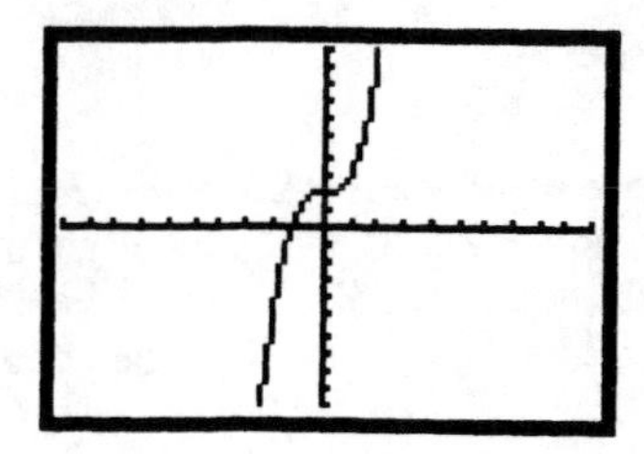

b) Replace $f(x)$ with y: $y = x^3 + 2$

Interchange x and y: $x = y^3 + 2$

Solve for y: $x - 2 = y^3$

$$\sqrt[3]{x-2} = y$$

Replace y with $f^{-1}(x)$: $f^{-1}(x) = \sqrt[3]{x-2}$

55. a) The graph of $f(x) = \dfrac{5}{x-1}$ is shown below.

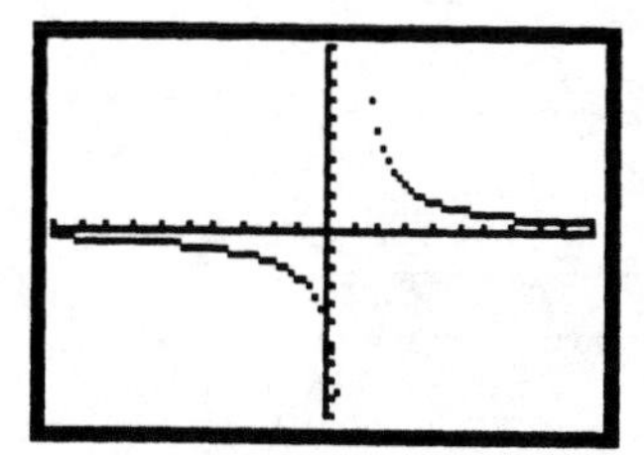

Since there is no horizontal line that crosses the graph more than once, the function is one-to-one.

b) Replace $f(x)$ with y: $y = \dfrac{5}{x-1}$

Interchange x and y: $x = \dfrac{5}{y-1}$

Solve for y: $x(y-1) = 5$

$$y - 1 = \frac{5}{x}$$
$$y = \frac{5}{x} + 1$$

Replace y with $f^{-1}(x)$: $f^{-1} = \dfrac{5}{x} + 1$, or $\dfrac{5+x}{x}$

56. a) The graph of $f(x) = \sqrt{x+4}$ is shown below. It passes the horizontal line test, so it is one-to-one.

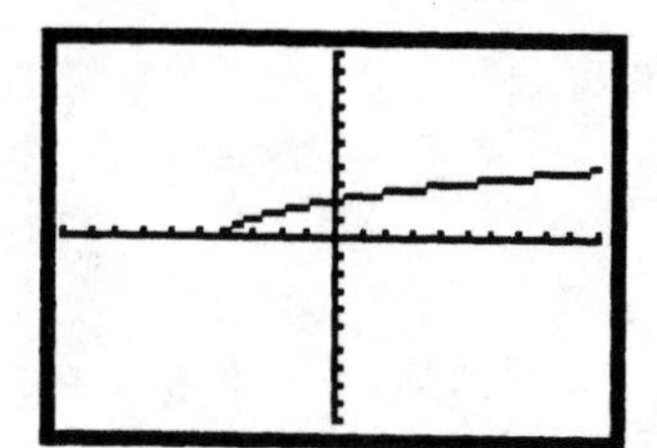

b) Replace $f(x)$ with y: $y = \sqrt{x+4}$

Interchange x and y: $x = \sqrt{y+4}$

Solve for y: $x^2 = y + 4$

$$x^2 - 4 = y$$

Replace y with $f^{-1}(x)$: $f^{-1}(x) = x^2 - 4$, $x \geq 0$

57. The center of the ellipse is the midpoint of the segment connecting the vertices:

$$\left(\frac{3+3}{2}, \frac{-4+6}{2}\right), \text{ or } (3, 1).$$

Now a is the distance from the origin to a vertex. We use the vertex $(3, 6)$.

$$a = \sqrt{(3-3)^2 + (6-1)^2} = 5$$

Also b is one-half the length of the minor axis.

$$b = \frac{\sqrt{(5-1)^2+(1-1)^2}}{2} = \frac{4}{2} = 2$$

The vertices lie on the vertical line $x = 3$, so the major axis is vertical. We write the equation of the ellipse.

$$\frac{(x-h)^2}{b^2} + \frac{(y-k)^2}{a^2} = 1$$

$$\frac{(x-3)^2}{4} + \frac{(y-1)^2}{25} = 1$$

58. Center: $\left(\frac{-1-1}{2}, \frac{-1+5}{2}\right)$, or $(-1, 2)$

$$a = \sqrt{[1-(-1)]^2+(-1-2)^2} = 3$$

$$b = \frac{\sqrt{(-3-1)^2+(2-2)^2}}{2} = \frac{4}{2} = 2$$

The vertices are on the line $x = -1$, so the major axis is vertical. The equation is

$$\frac{(x+1)^2}{4} + \frac{(y-2)^2}{9} = 1.$$

59. The center is the midpoint of the segment connecting the vertices:

$\left(\frac{-3+3}{0}, \frac{0+0}{0}\right)$, or $(0, 0)$.

Then $a = 3$ and since the vertices are on the x-axis, the major axis is horizontal. The equation is of the form $\frac{x^2}{a^2} + \frac{y^2}{b^2} = 1$.

Substitute 3 for a, 2 for x, and $\frac{22}{3}$ for y and solve for b^2.

$$\frac{4}{9} + \frac{\frac{484}{9}}{b^2} = 1$$

$$\frac{4}{9} + \frac{484}{9b^2} = 1$$

$$4b^2 + 484 = 9b^2$$

$$484 = 5b^2$$

$$\frac{484}{5} = b^2$$

Then the equation is $\frac{x^2}{9} + \frac{y^2}{484/5} = 1$.

60. $a = 4/2 = 2$; $b = 1/2$

The equation is $\frac{(x+2)^2}{1/4} + \frac{(y-3)^2}{4} = 1$.

61. Center: $(2.003, -1.005)$

Vertices: $(-1.017, -1.005)$, $(5.023, -1.005)$

62. Center: $(-3.004, 1.002)$

Vertices: $(-3.004, -1.970)$, $(-3.004, 3.974)$

63. Position a coordinate system as shown.

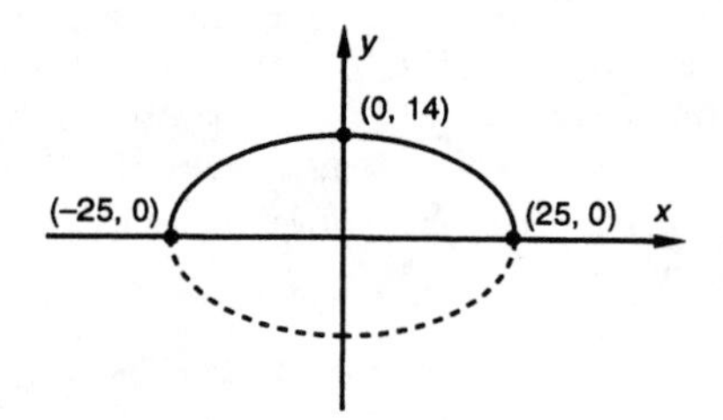

The equation of the ellipse is

$$\frac{x^2}{25^2} + \frac{y^2}{14^2} = 1$$

$$\frac{x^2}{625} + \frac{y^2}{196} = 1.$$

A point 6 ft from the riverbank corresponds to $(25-6, 0)$, or $(19, 0)$ or to $(-25+6, 0)$, or $(-19, 0)$. Substitute either 19 or -19 for x and solve for y, the clearance.

$$\frac{19^2}{625} + \frac{y^2}{196} = 1$$

$$\frac{y^2}{196} = 1 - \frac{361}{625}$$

$$y^2 = 196\left(1 - \frac{361}{625}\right)$$

$$y \approx 9.1$$

The clearance 6 ft from the riverbank is about 9.1 ft.

Exercise Set 6.3

1. Graph (b) is the graph of $\frac{x^2}{25} - \frac{y^2}{9} = 1$.

2. Graph (e) is the graph of $\frac{y^2}{4} - \frac{x^2}{36} = 1$.

3. Graph (c) is the graph of $\frac{(y-1)^2}{16} - \frac{(x+3)^2}{1} = 1$.

4. Graph (f) is the graph of $\frac{(x+4)^2}{100} - \frac{(y-2)^2}{81} = 1$.

5. Graph (a) is the graph of $25x^2 - 16y^2 = 400$.

6. Graph (d) is the graph of $y^2 - x^2 = 9$.

7. The vertices are equidistant from the origin and are on the y-axis, so the center is at the origin and the transverse axis is vertical. Since $c^2 = a^2 + b^2$, we have $5^2 = 3^2 + b^2$ so $b^2 = 16$.

The equation is of the form $\frac{y^2}{a^2} - \frac{x^2}{b^2} = 1$, so we have $\frac{y^2}{9} - \frac{x^2}{16} = 1$.

8. The vertices are equidistant from the origin and are on the x-axis, so the center is at the origin and the transverse axis is horizontal. Since $c^2 = a^2 + b^2$, we have $2^2 = 1^2 + b^2$ so $b^2 = 3$. The equation is $\frac{x^2}{1} - \frac{y^2}{3} = 1$.

9. The asymptotes pass through the origin, so the center is the origin. The given vertex is on the x-axis, so the transverse axis is horizontal. Since $\frac{b}{a}x = \frac{3}{2}x$ and $a = 2$, we have $b = 3$. The equation is of the form $\frac{x^2}{a^2} - \frac{y^2}{b^2} = 1$, so we have $\frac{x^2}{2^2} - \frac{y^2}{3^2} = 1$, or $\frac{x^2}{4} - \frac{y^2}{9} = 1$.

10. The asymptotes pass through the origin, so the center is the origin. The given vertex is on the y-axis, so the transverse axis is vertical. We use the equation of an asymptote to find b.

$$\frac{a}{b}x = \frac{5}{4}x$$

$$\frac{3}{b}x = \frac{5}{4}x \quad \text{Substituting 3 for } a$$

$$\frac{12}{5} = b$$

The equation is $\frac{y^2}{9} - \frac{x^2}{144/25} = 1$.

11. $\frac{x^2}{4} - \frac{y^2}{4} = 1$

$\frac{x^2}{2^2} - \frac{y^2}{2^2} = 1$ Standard form

The center is $(0,0)$; $a = 2$ and $b = 2$. The transverse axis is horizontal so the vertices are $(-2,0)$ and $(2,0)$. Since $c^2 = a^2+b^2$, we have $c^2 = 4+4 = 8$ and $c = \sqrt{8}$, or $2\sqrt{2}$. Then the foci are $(-2\sqrt{2},0)$ and $(2\sqrt{2},0)$.

Find the asymptotes:

$$y = \frac{b}{a}x \text{ and } y = -\frac{b}{a}x$$

$$y = \frac{2}{2}x \text{ and } y = -\frac{2}{2}x$$

$$y = x \quad \text{and} \quad y = -x$$

To draw the graph sketch the asymptotes, plot the vertices, and draw the branches of the hyperbola outward from the vertices toward the asymptotes.

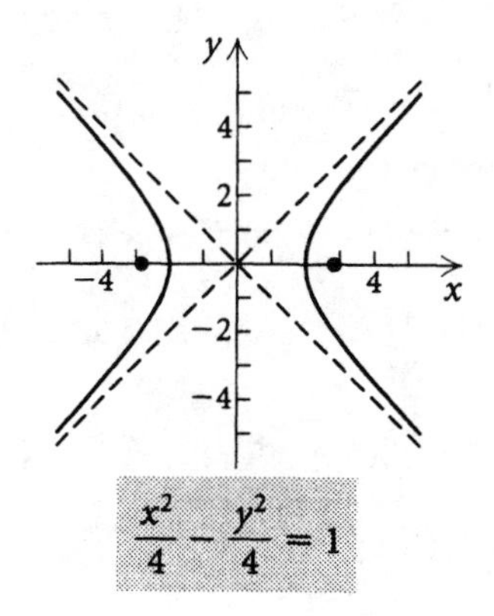

12. $\frac{x^2}{1} - \frac{y^2}{9} = 1$

$\frac{x^2}{1^2} - \frac{y^2}{3^2} = 1$ Standard form

The center is $(0,0)$; $a = 1$ and $b = 3$. The transverse axis is horizontal so the vertices are $(-1,0)$ and $(1,0)$. Since $c^2 = a^2+b^2$, we have $c^2 = 1+9 = 10$ and $c = \sqrt{10}$. Then the foci are $(-\sqrt{10},0)$ and $(\sqrt{10},0)$.

Find the asymptotes:

$$y = \frac{b}{a}x \text{ and } y = -\frac{b}{a}x$$

$$y = \frac{3}{1}x \text{ and } y = -\frac{3}{1}x$$

$$y = 3x \quad \text{and} \quad y = -3x$$

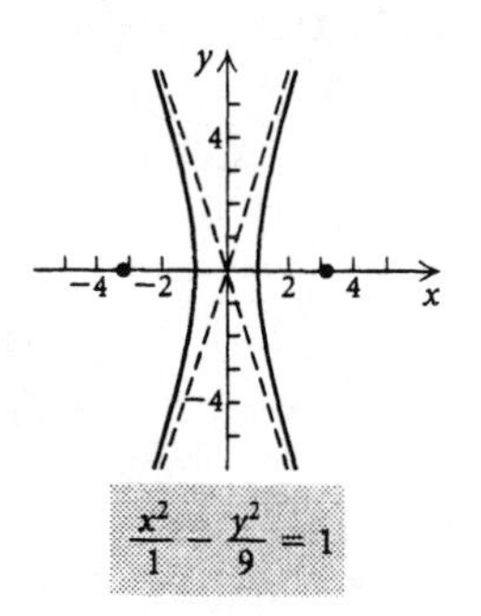

13. $\frac{(x-2)^2}{9} - \frac{(y+5)^2}{1} = 1$

$\frac{(x-2)^2}{3^2} - \frac{[y-(-5)]^2}{1^2} = 1$ Standard form

The center is $(2,-5)$; $a = 3$ and $b = 1$. The transverse axis is horizontal, so the vertices are 3 units left and right of the center:

$(2-3,-5)$ and $(2+3,-5)$, or $(-1,-5)$ and $(5,-5)$.

Since $c^2 = a^2 + b^2$, we have $c^2 = 9 + 1 = 10$ and $c = \sqrt{10}$. Then the foci are $\sqrt{10}$ units left and right of the center:

$$(2-\sqrt{10},-5) \text{ and } (2+\sqrt{10},-5).$$

Find the asymptotes:

$$y - k = \frac{b}{a}(x-h) \text{ and } \quad y - k = -\frac{b}{a}(x-h)$$

$$y - (-5) = \frac{1}{3}(x-2) \text{ and } y - (-5) = -\frac{1}{3}(x-2)$$

$$y + 5 = \frac{1}{3}(x-2) \text{ and } \quad y + 5 = -\frac{1}{3}(x-2), \text{ or}$$

$$y = \frac{1}{3}x - \frac{17}{3} \text{ and } \quad y = -\frac{1}{3}x - \frac{13}{3}$$

Sketch the asymptotes, plot the vertices, and draw the graph.

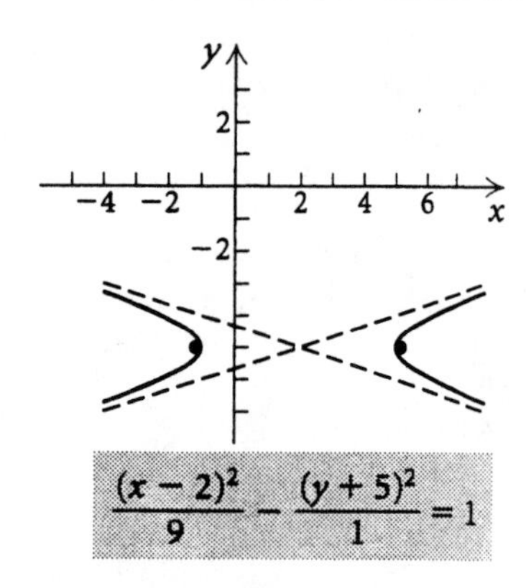

14. $$\frac{(x-5)^2}{16} - \frac{(y+2)^2}{9} = 1$$

$$\frac{(x-5)^2}{4^2} - \frac{[y-(-2)]^2}{3^2} = 1 \quad \text{Standard form}$$

The center is $(5, -2)$; $a = 4$ and $b = 3$. The transverse axis is horizontal, so the vertices are 4 units left and right of the center:

$(5-4, -2)$ and $(5+4, -2)$, or $(1, -2)$ and $(9, -2)$.

Since $c^2 = a^2 + b^2$, we have $c^2 = 16 + 9 = 25$ and $c = 5$. Then the foci are 5 units left and right of the center:

$(5-5, -2)$ and $(5+5, -2)$, or $(0, -2)$ and $(10, -2)$.

Find the asymptotes:

$$y - k = \frac{b}{a}(x-h) \text{ and } y - k = -\frac{b}{a}(x-h)$$

$$y + 2 = \frac{3}{4}(x-5) \text{ and } y + 2 = -\frac{3}{4}(x-5), \text{ or}$$

$$y = \frac{3}{4}x - \frac{23}{4} \quad \text{and} \quad y = -\frac{3}{4}x + \frac{7}{4}$$

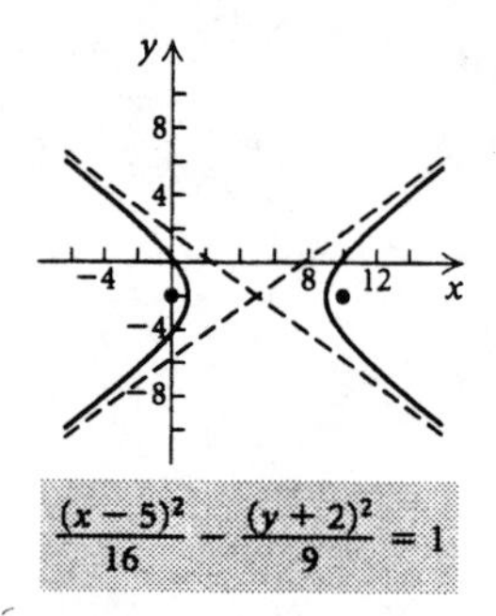

15. $$\frac{(y+3)^2}{4} - \frac{(x+1)^2}{16} = 1$$

$$\frac{[y-(-3)]^2}{2^2} - \frac{[x-(-1)]^2}{4^2} = 1 \quad \text{Standard form}$$

The center is $(-1, -3)$; $a = 2$ and $b = 4$. The transverse axis is vertical, so the vertices are 2 units below and above the center:

$(-1, -3-2)$ and $(1, -3+2)$, or $(-1, -5)$ and $(-1, -1)$.

Since $c^2 = a^2 + b^2$, we have $c^2 = 4 + 16 = 20$ and $c = \sqrt{20}$, or $2\sqrt{5}$. Then the foci are $2\sqrt{5}$ units below and above of the center:

$$(-1, -3 - 2\sqrt{5}) \text{ and } (-1, -3 + 2\sqrt{5}).$$

Find the asymptotes:

$$y - k = \frac{a}{b}(x-h) \quad \text{and} \quad y - k = -\frac{a}{b}(x-h)$$

$$y-(-3) = \frac{2}{4}(x-(-1)) \text{ and } y-(-3) = -\frac{2}{4}(x-(-1))$$

$$y+3 = \frac{1}{2}(x+1) \quad \text{and} \quad y+3 = -\frac{1}{2}(x+1), \text{ or}$$

$$y = \frac{1}{2}x - \frac{5}{2} \quad \text{and} \quad y = -\frac{1}{2}x - \frac{7}{2}$$

Sketch the asymptotes, plot the vertices, and draw the graph.

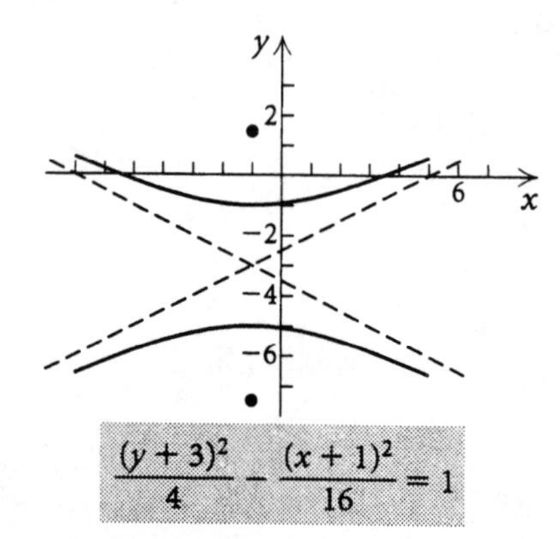

16. $$\frac{(y+4)^2}{25} - \frac{(x+2)^2}{16} = 1$$

$$\frac{[y-(-4)]^2}{5^2} - \frac{[x-(-2)]^2}{4^2} = 1 \quad \text{Standard form}$$

The center is $(-2, -4)$; $a = 5$ and $b = 4$. The transverse axis is vertical, so the vertices are 5 units below and above the center:

$(-2, -4-5)$ and $(-2, -4+5)$, or $(-2, -9)$ and $(-2, 1)$.

Since $c^2 = a^2 + b^2$, we have $c^2 = 25 + 16 = 41$ and $c = \sqrt{41}$. Then the foci are $\sqrt{41}$ units below and above the center:

$$(-2, -4 - \sqrt{41}) \text{ and } (-2, -4 + \sqrt{41}).$$

Find the asymptotes:

$$y - k = \frac{a}{b}(x-h) \text{ and } y - k = -\frac{a}{b}(x-h)$$

$$y + 4 = \frac{5}{4}(x+2) \text{ and } y + 4 = -\frac{5}{4}(x+2), \text{ or}$$

$$y = \frac{5}{4}x - \frac{3}{2} \quad \text{and} \quad y = -\frac{5}{4}x - \frac{13}{2}$$

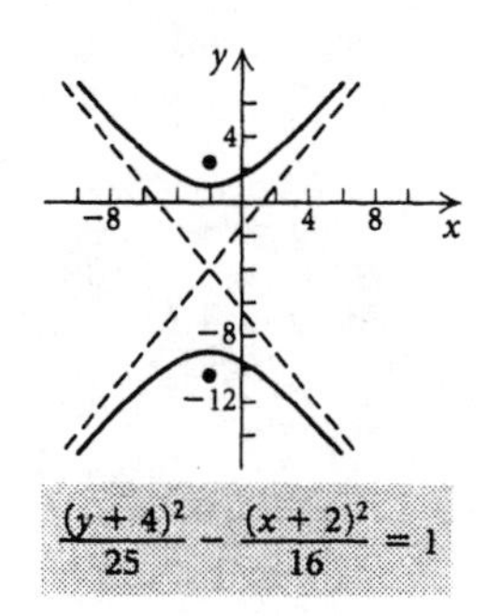

17. $x^2 - 4y^2 = 4$

$$\frac{x^2}{4} - \frac{y^2}{1} = 1$$

$$\frac{x^2}{2^2} - \frac{y^2}{1^2} = 1 \quad \text{Standard form}$$

The center is $(0,0)$; $a = 2$ and $b = 1$. The transverse axis is horizontal, so the vertices are $(-2,0)$ and $(2,0)$. Since $c^2 = a^2 + b^2$, we have $c^2 = 4+1 = 5$ and $c = \sqrt{5}$. Then the foci are $(-\sqrt{5},0)$ and $(\sqrt{5},0)$.

Find the asymptotes:

$$y = \frac{b}{a}x \text{ and } y = -\frac{b}{a}x$$

$$y = \frac{1}{2}x \text{ and } y = -\frac{1}{2}x$$

Sketch the asymptotes, plot the vertices, and draw the graph.

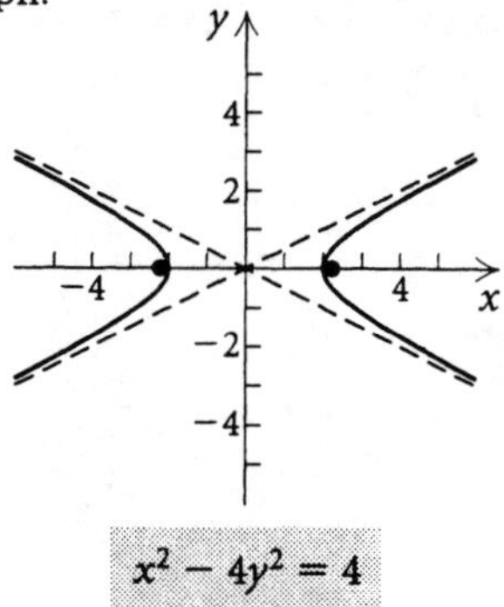

18. $4x^2 - y^2 = 16$

$$\frac{x^2}{4} - \frac{y^2}{16} = 1$$

$$\frac{x^2}{2^2} - \frac{y^2}{4^2} = 1 \quad \text{Standard form}$$

The center is $(0,0)$; $a = 2$ and $b = 4$. The transverse axis is horizontal, so the vertices are $(-2,0)$ and $(2,0)$. Since $c^2 = a^2 + b^2$, we have $c^2 = 4 + 16 = 20$ and $c = \sqrt{20}$, or $2\sqrt{5}$. Then the foci are $(-2\sqrt{5},0)$ and $(2\sqrt{5},0)$.

Find the asymptotes:

$$y = \frac{b}{a}x \text{ and } y = -\frac{b}{a}x$$

$$y = \frac{4}{2}x \text{ and } y = -\frac{4}{2}x$$

$$y = 2x \text{ and } y = -2x$$

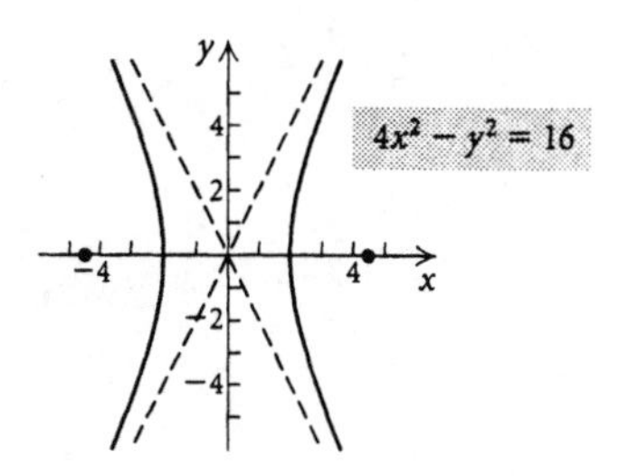

19. $9y^2 - x^2 = 81$

$$\frac{y^2}{9} - \frac{x^2}{81} = 1$$

$$\frac{y^2}{3^2} - \frac{x^2}{9^2} = 1 \quad \text{Standard form}$$

The center is $(0,0)$; $a = 3$ and $b = 9$. The transverse axis is vertical, so the vertices are $(0,-3)$ and $(0,3)$. Since $c^2 = a^2 + b^2$, we have $c^2 = 9 + 81 = 90$ and $c = \sqrt{90}$, or $3\sqrt{10}$. Then the foci are $(0,-3\sqrt{10})$ and $(0,3\sqrt{10})$.

Find the asymptotes:

$$y = \frac{a}{b}x \text{ and } y = -\frac{a}{b}x$$

$$y = \frac{3}{9}x \text{ and } y = -\frac{3}{9}x$$

$$y = \frac{1}{3}x \text{ and } y = -\frac{1}{3}x$$

Sketch the asymptotes, plot the vertices, and draw the graph.

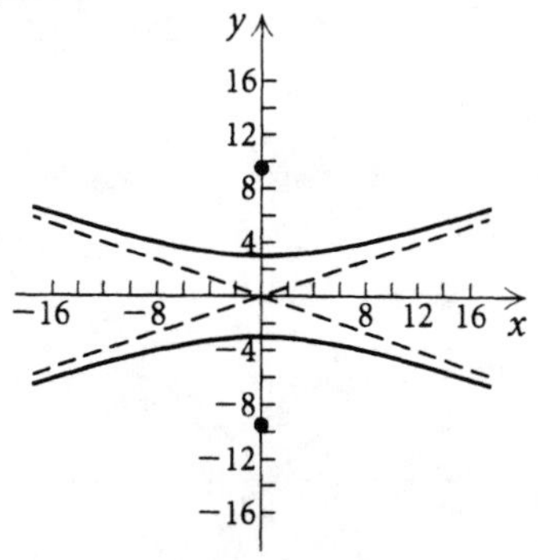

20. $y^2 - 4x^2 = 4$

$$\frac{y^2}{4} - \frac{x^2}{1} = 1$$

$$\frac{y^2}{2^2} - \frac{x^2}{1^2} = 1 \quad \text{Standard form}$$

The center is $(0,0)$; $a = 2$ and $b = 1$. The transverse axis is vertical, so the vertices are $(0,-2)$ and $(0,2)$. Since $c^2 = a^2 + b^2$, we have $c^2 = 4 + 1 = 5$ and $c = \sqrt{5}$. Then the foci are $(0,-\sqrt{5})$ and $(0,\sqrt{5})$.

Find the asymptotes:

$$y = \frac{a}{b}x \quad \text{and} \quad y = -\frac{a}{b}x$$
$$y = \frac{2}{1}x \quad \text{and} \quad y = -\frac{2}{1}x$$
$$y = 2x \quad \text{and} \quad y = -2x$$

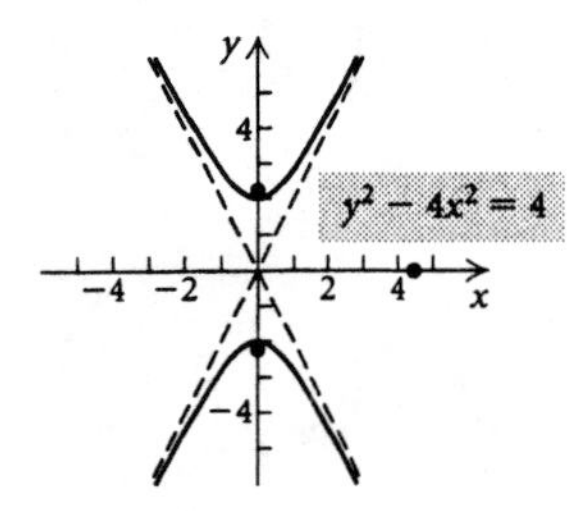

21.
$$x^2 - y^2 = 2$$
$$\frac{x^2}{2} - \frac{y^2}{2} = 1$$
$$\frac{x^2}{(\sqrt{2})^2} - \frac{y^2}{(\sqrt{2})^2} = 1 \quad \text{Standard form}$$

The center is $(0, 0)$; $a = \sqrt{2}$ and $b = \sqrt{2}$. The transverse axis is horizontal, so the vertices are $(-\sqrt{2}, 0)$ and $(\sqrt{2}, 0)$. Since $c^2 = a^2 + b^2$, we have $c^2 = 2+2 = 4$ and $c = 2$. Then the foci are $(-2, 0)$ and $(2, 0)$.

Find the asymptotes:

$$y = \frac{b}{a}x \quad \text{and} \quad y = -\frac{b}{a}x$$
$$y = \frac{\sqrt{2}}{\sqrt{2}}x \quad \text{and} \quad y = -\frac{\sqrt{2}}{\sqrt{2}}x$$
$$y = x \quad \text{and} \quad y = -x$$

Sketch the asymptotes, plot the vertices, and draw the graph.

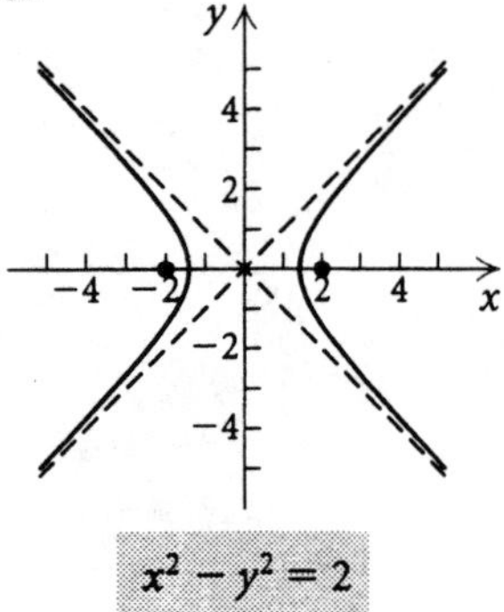

22.
$$x^2 - y^2 = 3$$
$$\frac{x^2}{3} - \frac{y^2}{3} = 1$$
$$\frac{x^2}{(\sqrt{3})^2} - \frac{y^2}{(\sqrt{3})^2} = 1 \quad \text{Standard form}$$

The center is $(0, 0)$; $a = \sqrt{3}$ and $b = \sqrt{3}$. The transverse axis is horizontal, so the vertices are $(-\sqrt{3}, 0)$ and $(\sqrt{3}, 0)$. Since $c^2 = a^2 + b^2$, we have $c^2 = 3+3 = 6$ so $c = \sqrt{6}$. Then the foci are $(-\sqrt{6}, 0)$ and $(\sqrt{6}, 0)$.

Find the asymptotes:

$$y = \frac{b}{a}x \quad \text{and} \quad y = -\frac{b}{a}x$$
$$y = \frac{\sqrt{3}}{\sqrt{3}}x \quad \text{and} \quad y = -\frac{\sqrt{3}}{\sqrt{3}}x$$
$$y = x \quad \text{and} \quad y = -x$$

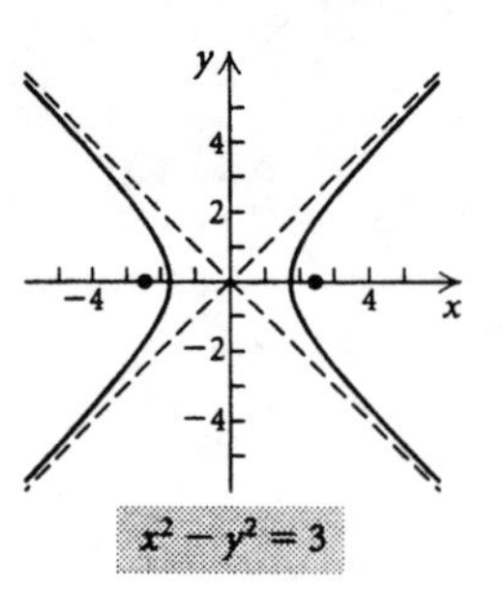

23.
$$y^2 - x^2 = \frac{1}{4}$$
$$\frac{y^2}{1/4} - \frac{x^2}{1/4} = 1$$
$$\frac{y^2}{(1/2)^2} - \frac{x^2}{(1/2)^2} = 1 \quad \text{Standard form}$$

The center is $(0, 0)$; $a = \frac{1}{2}$ and $b = \frac{1}{2}$. The transverse axis is vertical, so the vertices are $\left(0, -\frac{1}{2}\right)$ and $\left(0, \frac{1}{2}\right)$. Since $c^2 = a^2 + b^2$, we have $c^2 = \frac{1}{4} + \frac{1}{4} = \frac{1}{2}$ and $c = \sqrt{\frac{1}{2}}$, or $\frac{\sqrt{2}}{2}$. Then the foci are $\left(0, -\frac{\sqrt{2}}{2}\right)$ and $\left(0, \frac{\sqrt{2}}{2}\right)$.

Find the asymptotes:

$$y = \frac{a}{b}x \quad \text{and} \quad y = -\frac{a}{b}x$$
$$y = \frac{1/2}{1/2}x \quad \text{and} \quad y = -\frac{1/2}{1/2}x$$
$$y = x \quad \text{and} \quad y = -x$$

Sketch the asymptotes, plot the vertices, and draw the graph.

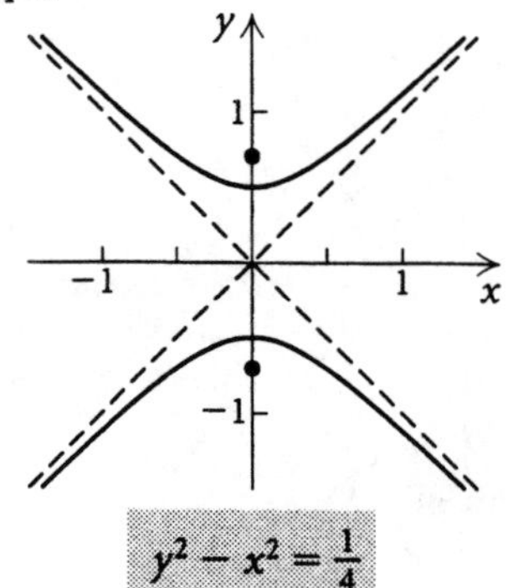

24.
$$y^2 - x^2 = \frac{1}{9}$$
$$\frac{y^2}{1/9} - \frac{x^2}{1/9} = 1$$
$$\frac{y^2}{(1/3)^2} - \frac{x^2}{(1/3)^2} = 1 \quad \text{Standard form}$$

The center is $(0,0)$; $a = \frac{1}{3}$ and $b = \frac{1}{3}$. The transverse axis is vertical, so the vertices are $\left(0, -\frac{1}{3}\right)$ and $\left(0, \frac{1}{3}\right)$. Since $c^2 = a^2 + b^2$, we have $c^2 = \frac{1}{9} + \frac{1}{9} = \frac{2}{9}$ and $c = \frac{\sqrt{2}}{3}$. Then the foci are $\left(0, -\frac{\sqrt{2}}{3}\right)$ and $\left(0, \frac{\sqrt{2}}{3}\right)$.

Find the asymptotes:

$$y = \frac{a}{b}x \quad \text{and} \quad y = -\frac{a}{b}x$$
$$y = \frac{1/3}{1/3}x \quad \text{and} \quad y = -\frac{1/3}{1/3}x$$
$$y = x \quad \text{and} \quad y = -x$$

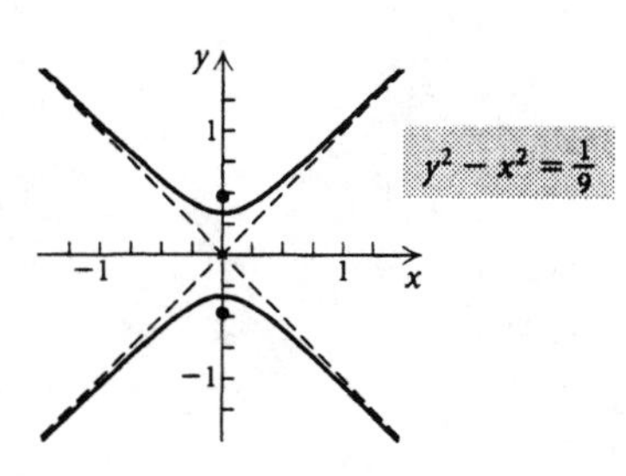

25. Begin by completing the square twice.
$$x^2 - y^2 - 2x - 4y - 4 = 0$$
$$(x^2 - 2x) - (y^2 + 4y) = 4$$
$$(x^2 - 2x + 1) - (y^2 + 4y + 4) = 4 + 1 - 1 \cdot 4$$
$$(x-1)^2 - (y+2)^2 = 1$$
$$\frac{(x-1)^2}{1^2} - \frac{[y-(-2)]^2}{1^2} = 1 \quad \text{Standard form}$$

The center is $(1, -2)$; $a = 1$ and $b = 1$. The transverse axis is horizontal, so the vertices are 1 unit left and right of the center:

$(1-1, -2)$ and $(1+1, -2)$ or $(0, -2)$ and $(2, -2)$

Since $c^2 = a^2 + b^2$, we have $c^2 = 1 + 1 = 2$ and $c = \sqrt{2}$. Then the foci are $\sqrt{2}$ units left and right of the center:

$(1-\sqrt{2}, -2)$ and $(1+\sqrt{2}, -2)$.

Find the asymptotes:

$$y - k = \frac{b}{a}(x-h) \quad \text{and} \quad y - k = -\frac{b}{a}(x-h)$$
$$y - (-2) = \frac{1}{1}(x-1) \quad \text{and} \quad y - (-2) = -\frac{1}{1}(x-1)$$
$$y + 2 = x - 1 \quad \text{and} \quad y + 2 = -(x-1), \text{ or}$$
$$y = x - 3 \quad \text{and} \quad y = -x - 1$$

Sketch the asymptotes, plot the vertices, and draw the graph.

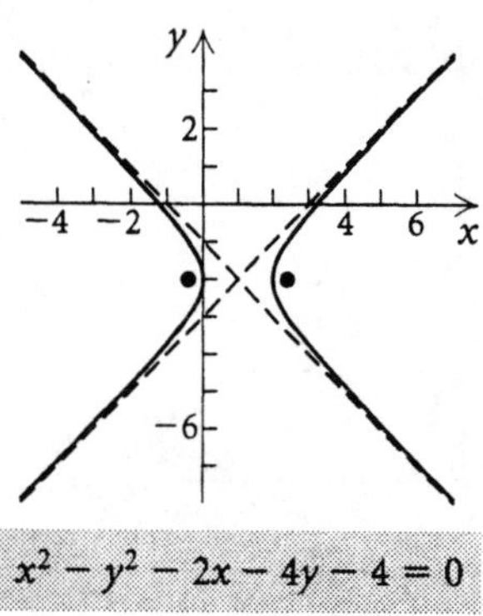

26. Begin by completing the square twice.
$$4x^2 - y^2 + 8x - 4y - 4 = 0$$
$$4(x^2 + 2x) - (y^2 + 4y) = 4$$
$$4(x^2 + 2x + 1) - (y^2 + 4y + 4) = 4 + 4 \cdot 1 - 1 \cdot 4$$
$$4(x+1)^2 - (y+2)^2 = 4$$
$$\frac{(x+1)^2}{1} - \frac{(y+2)^2}{4} = 1$$
$$\frac{[x-(-1)]^2}{1^2} - \frac{[y-(-2)]^2}{2^2} = 1 \quad \text{Standard form}$$

The center is $(-1, -2)$; $a = 1$ and $b = 2$. The transverse axis is horizontal, so the vertices are 1 unit left and right of the center:

$(-1-1, -2)$ and $(-1+1, -2)$ or $(-2, -2)$ and $(0, -2)$

Since $c^2 = a^2 + b^2$, we have $c^2 = 1 + 4 = 5$ and $c = \sqrt{5}$. Then the foci are $\sqrt{5}$ units left and right of the center:

$(-1-\sqrt{5}, -2)$ and $(1+\sqrt{5}, -2)$.

Find the asymptotes:

$$y - k = \frac{b}{a}(x-h) \quad \text{and} \quad y - k = -\frac{b}{a}(x-h)$$
$$y + 2 = \frac{2}{1}(x+1) \quad \text{and} \quad y + 2 = -\frac{2}{1}(x+1)$$
$$y + 2 = 2(x+1) \quad \text{and} \quad y + 2 = -2(x+1), \text{ or}$$
$$y = 2x \quad \text{and} \quad y = -2x - 4$$

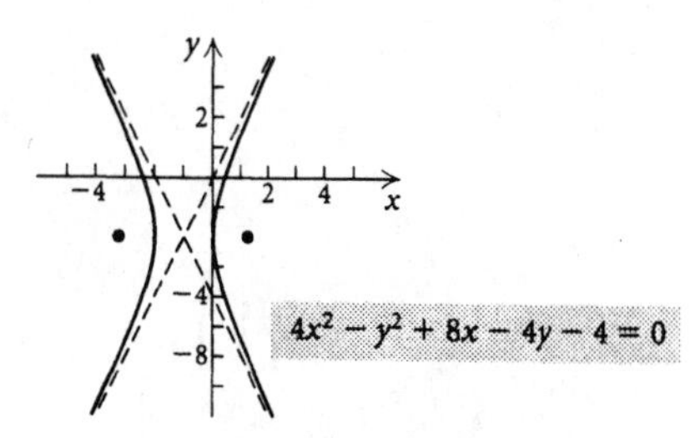

27. Begin by completing the square twice.

$$36x^2 - y^2 - 24x + 6y - 41 = 0$$
$$(36x^2 - 24x) - (y^2 - 6y) = 41$$
$$36\left(x^2 - \frac{2}{3}x\right) - (y^2 - 6y) = 41$$
$$36\left(x^2 - \frac{2}{3}x + \frac{1}{9}\right) - (y^2 - 6y + 9) = 41 + 36\cdot\frac{1}{9} - 1\cdot 9$$
$$36\left(x - \frac{1}{3}\right)^2 - (y-3)^2 = 36$$
$$\frac{\left(x - \frac{1}{3}\right)^2}{1} - \frac{(y-3)^2}{36} = 1$$
$$\frac{\left(x - \frac{1}{3}\right)^2}{1^2} - \frac{(y-3)^2}{6^2} = 1 \quad \text{Standard form}$$

The center is $\left(\frac{1}{3}, 3\right)$; $a = 1$ and $b = 6$. The transverse axis is horizontal, so the vertices are 1 unit left and right of the center:

$\left(\frac{1}{3} - 1, 3\right)$ and $\left(\frac{1}{3} + 1, 3\right)$ or $\left(-\frac{2}{3}, 3\right)$ and $\left(\frac{4}{3}, 3\right)$.

Since $c^2 = a^2 + b^2$, we have $c^2 = 1 + 36 = 37$ and $c = \sqrt{37}$. Then the foci are $\sqrt{37}$ units left and right of the center:

$$\left(\frac{1}{3} - \sqrt{37}, 3\right) \text{ and } \left(\frac{1}{3} + \sqrt{37}, 3\right).$$

Find the asymptotes:

$$y - k = \frac{b}{a}(x-h) \quad \text{and} \quad y - k = -\frac{b}{a}(x-h)$$
$$y - 3 = \frac{6}{1}\left(x - \frac{1}{3}\right) \quad \text{and} \quad y - 3 = -\frac{6}{1}\left(x - \frac{1}{3}\right)$$
$$y - 3 = 6\left(x - \frac{1}{3}\right) \quad \text{and} \quad y - 3 = -6\left(x - \frac{1}{3}\right), \text{ or}$$
$$y = 6x + 1 \quad \text{and} \quad y = -6x + 5$$

Sketch the asymptotes, plot the vertices, and draw the graph.

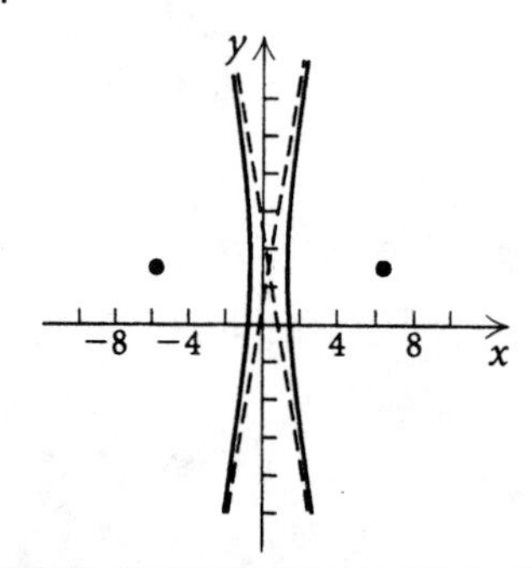

28. Begin by completing the square twice.

$$9x^2 - 4y^2 + 54x + 8y + 41 = 0$$
$$9(x^2 + 6x) - 4(y^2 - 2y) = -41$$
$$9(x^2 + 6x + 9) - 4(y^2 - 2y + 1) = -41 + 9\cdot 9 - 4\cdot 1$$
$$9(x+3)^2 - 4(y-1)^2 = 36$$
$$\frac{(x+3)^2}{4} - \frac{(y-1)^2}{9} = 1$$
$$\frac{[x-(-3)]^2}{2^2} - \frac{(y-1)^2}{3^2} = 1 \quad \text{Standard form}$$

The center is $(-3, 1)$; $a = 2$ and $b = 3$. The transverse axis is horizontal, so the vertices are 2 units left and right of the center:

$(-3-2, 1)$ and $(-3+2, 1)$, or $(-5, 1)$ and $(-1, 1)$.

Since $c^2 = a^2 + b^2$, we have $c^2 = 4 + 9 = 13$ and $c = \sqrt{13}$. Then the foci are $\sqrt{13}$ units left and right of the center:

$$(-3 - \sqrt{13}, 1) \text{ and } (-3 + \sqrt{13}, 1).$$

Find the asymptotes:

$$y - k = \frac{b}{a}(x-h) \text{ and } y - k = -\frac{b}{a}(x-h)$$
$$y - 1 = \frac{3}{2}(x+3) \text{ and } y - 1 = -\frac{3}{2}(x+3), \text{ or}$$
$$y = \frac{3}{2}x + \frac{11}{2} \quad \text{and} \quad y = -\frac{3}{2}x - \frac{7}{2}$$

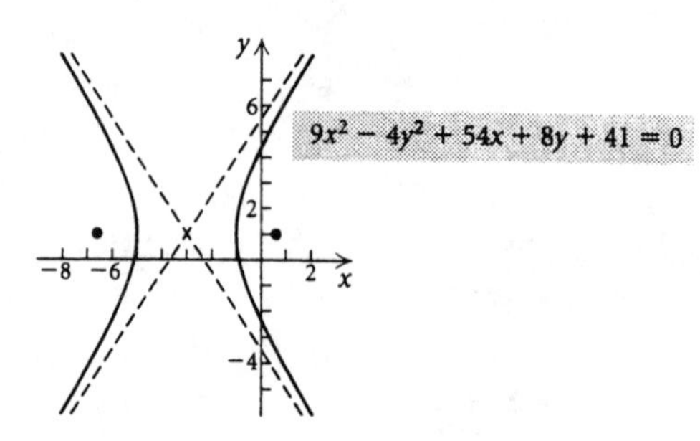

29. Begin by completing the square twice.

$$9y^2 - 4x^2 - 18y + 24x - 63 = 0$$
$$9(y^2 - 2y) - 4(x^2 - 6x) = 63$$
$$9(y^2 - 2y + 1) - 4(x^2 - 6x + 9) = 63 + 9\cdot 1 - 4\cdot 9$$
$$9(y-1)^2 - 4(x-3)^2 = 36$$
$$\frac{(y-1)^2}{4} - \frac{(x-3)^2}{9} = 1$$
$$\frac{(y-1)^2}{2^2} - \frac{(x-3)^2}{3^2} = 1 \quad \text{Standard form}$$

The center is $(3, 1)$; $a = 2$ and $b = 3$. The transverse axis is vertical, so the vertices are 2 units below and above the center:

$(3, 1-2)$ and $(3, 1+2)$, or $(3, -1)$ and $(3, 3)$.

Since $c^2 = a^2 + b^2$, we have $c^2 = 4 + 9 = 13$ and $c = \sqrt{13}$. Then the foci are $\sqrt{13}$ units below and above the center:

$$(3, 1 - \sqrt{13}) \text{ and } (3, 1 + \sqrt{13}).$$

Find the asymptotes:

$$y-k=\frac{a}{b}(x-h) \text{ and } y-k=-\frac{a}{b}(x-h)$$

$$y-1=\frac{2}{3}(x-3) \text{ and } y-1=-\frac{2}{3}(x-3), \text{ or}$$

$$y=\frac{2}{3}x-1 \quad \text{and} \quad y=-\frac{2}{3}x+3$$

Sketch the asymptotes, plot the vertices, and draw the graph.

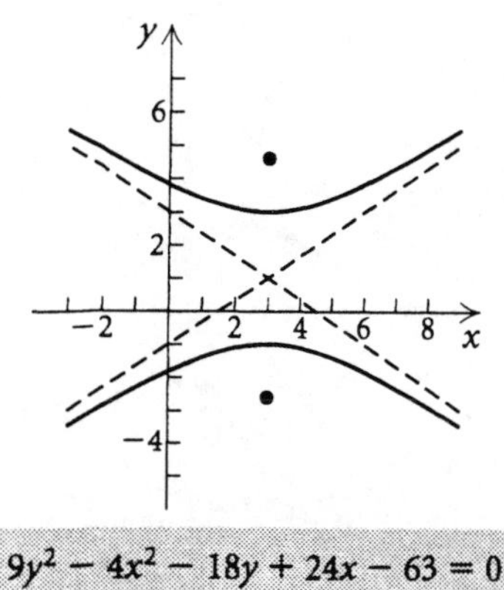

$9y^2 - 4x^2 - 18y + 24x - 63 = 0$

30. Begin by completing the square twice.

$$x^2-25y^2+6x-50y=41$$

$$x^2+6x+9-25(y^2+2y+1)=41+9-25\cdot 1$$

$$(x+3)^2-25(y+1)^2=25$$

$$\frac{(x+3)^2}{25}-\frac{(y+1)^2}{1}=1$$

$$\frac{[x-(-3)]^2}{5^2}-\frac{[y-(-1)]^2}{1^2}=1$$

The center is $(-3,-1)$; $a=5$ and $b=1$. The transverse axis is horizontal, so the vertices are 5 units left and right of the center:

$(-3-5,-1)$ and $(-3+5,-1)$, or $(-8,-1)$ and $(2,-1)$.

Since $c^2=a^2+b^2$, we have $c^2=25+1=26$ and $c=\sqrt{26}$. Then the foci are $\sqrt{26}$ units left and right of the center:

$$(-3-\sqrt{26},-1) \text{ and } (-3+\sqrt{26},-1).$$

Find the asymptotes:

$$y-k=\frac{b}{a}(x-h) \text{ and } y-k=-\frac{b}{a}(x-h)$$

$$y+1=\frac{1}{5}(x+3) \text{ and } y+1=-\frac{1}{5}(x+3), \text{ or}$$

$$y=\frac{1}{5}x-\frac{2}{5} \quad \text{and} \quad y=-\frac{1}{5}x-\frac{8}{5}$$

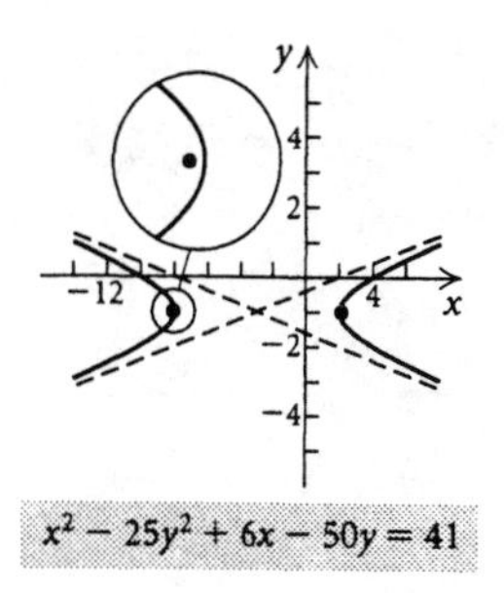

$x^2 - 25y^2 + 6x - 50y = 41$

31. Begin by completing the square twice.

$$x^2-y^2-2x-4y=4$$

$$(x^2-2x+1)-(y^2+4y+4)=4+1-4$$

$$(x-1)^2-(y+2)^2=1$$

$$\frac{(x-1)^2}{1^2}-\frac{[y-(-2)]^2}{1^2}=1 \quad \text{Standard form}$$

The center is $(1,-2)$; $a=1$ and $b=1$. The transverse axis is horizontal, so the vertices are 1 unit left and right of the center:

$(1-1,-2)$ and $(1+1,-2)$, or $(0,-2)$ and $(2,-2)$.

Since $c^2=a^2+b^2$, we have $c^2=1+1=2$ and $c=\sqrt{2}$. Then the foci are $\sqrt{2}$ units left and right of the center:

$$(1-\sqrt{2},-2) \text{ and } (1+\sqrt{2},-2).$$

Find the asymptotes:

$$y-k=\frac{b}{a}(x-h) \text{ and } y-k=-\frac{b}{a}(x-h)$$

$$y-(-2)=\frac{1}{1}(x-1) \text{ and } y-(-2)=-\frac{1}{1}(x-1)$$

$$y+2=x-1 \quad \text{and} \quad y+2=-(x-1), \text{ or}$$

$$y=x-3 \quad \text{and} \quad y=-x-1$$

Sketch the asymptotes, plot the vertices, and draw the graph.

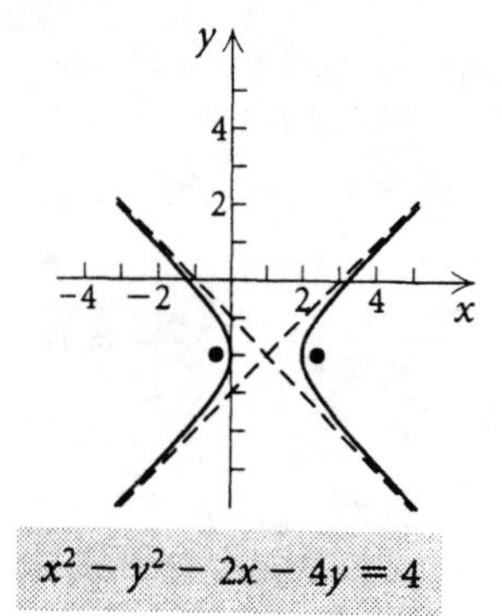

$x^2 - y^2 - 2x - 4y = 4$

32. Begin by completing the square twice.

$$9y^2 - 4x^2 - 54y - 8x + 41 = 0$$
$$9(y^2 - 6y + 9) - 4(x^2 + 2x + 1) = -41 + 9 \cdot 9 - 4 \cdot 1$$
$$9(y-3)^2 - 4(x+1)^2 = 36$$
$$\frac{(y-3)^2}{4} - \frac{(x+1)^2}{9} = 1$$
$$\frac{(y-3)^2}{2^2} - \frac{[x-(-1)]^2}{3^2} = 1 \quad \text{Standard form}$$

The center is $(-1, 3)$; $a = 2$ and $b = 3$. The transverse axis is vertical, so the vertices are 2 units below and above the center:

$(-1, 3-2)$ and $(-1, 3+2)$, or $(-1, 1)$ and $(-1, 5)$.

Since $c^2 = a^2 + b^2$, we have $c^2 = 4 + 9$ and $c = \sqrt{13}$. Then the foci are $\sqrt{13}$ units below and above the center:

$(-1, 3-\sqrt{13})$ and $(-1, 3+\sqrt{13})$.

Find the asymptotes:

$$y - k = \frac{a}{b}(x-h) \text{ and } y - k = -\frac{a}{b}(x-h)$$
$$y - 3 = \frac{2}{3}(x+1) \text{ and } y - 3 = -\frac{2}{3}(x+1), \text{ or}$$
$$y = \frac{2}{3}x + \frac{11}{3} \text{ and } y = -\frac{2}{3}x + \frac{7}{3}$$

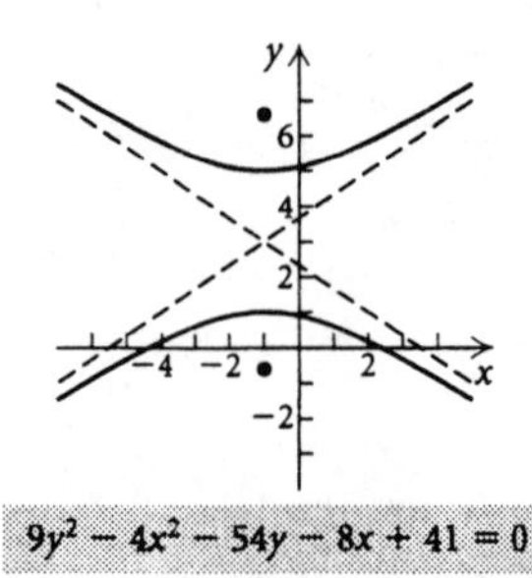

$9y^2 - 4x^2 - 54y - 8x + 41 = 0$

33. Begin by completing the square twice.

$$y^2 - x^2 - 6x - 8y - 29 = 0$$
$$(y^2 - 8y + 16) - (x^2 + 6x + 9) = 29 + 16 - 9$$
$$(y-4)^2 - (x+3)^2 = 36$$
$$\frac{(y-4)^2}{36} - \frac{(x+3)^2}{36} = 1$$
$$\frac{(y-4)^2}{6^2} - \frac{[x-(-3)]^2}{6^2} = 1 \quad \text{Standard form}$$

The center is $(-3, 4)$; $a = 6$ and $b = 6$. The transverse axis is vertical, so the vertices are 6 units below and above the center:

$(-3, 4-6)$ and $(-3, 4+6)$, or $(-3, -2)$ and $(-3, 10)$.

Since $c^2 = a^2 + b^2$, we have $c^2 = 36 + 36 = 72$ and $c = \sqrt{72}$, or $6\sqrt{2}$. Then the foci are $6\sqrt{2}$ units below and above the center:

$(-3, 4-6\sqrt{2})$ and $(-3, 4+6\sqrt{2})$.

Find the asymptotes:

$$y - k = \frac{a}{b}(x-h) \quad \text{and} \quad y - k = -\frac{a}{b}(x-h)$$
$$y - 4 = \frac{6}{6}(x-(-3)) \quad \text{and} \quad y - 4 = -\frac{6}{6}(x-(-3))$$
$$y - 4 = x + 3 \quad \text{and} \quad y - 4 = -(x+3), \text{ or}$$
$$y = x + 7 \quad \text{and} \quad y = -x + 1$$

Sketch the asymptotes, plot the vertices, and draw the graph.

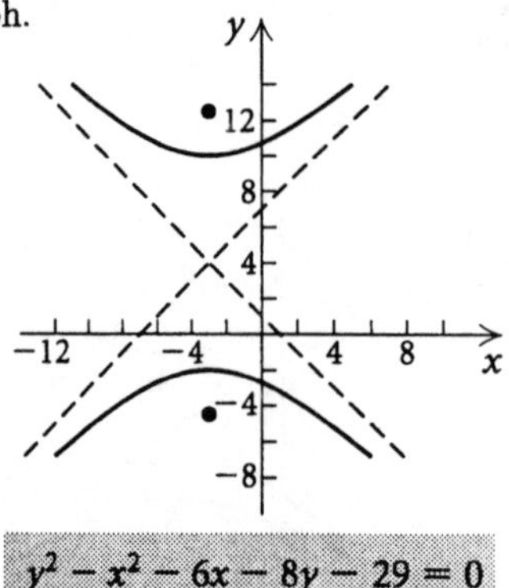

$y^2 - x^2 - 6x - 8y - 29 = 0$

34. Begin by completing the square twice.

$$x^2 - y^2 = 8x - 2y - 13$$
$$x^2 - 8x - y^2 + 2y = 13$$
$$x^2 - 8x + 16 - (y^2 - 2y + 1) = -13 + 16 - 1$$
$$(x-4)^2 - (y-1)^2 = 2$$
$$\frac{(x-4)^2}{(\sqrt{2})^2} - \frac{(y-1)^2}{(\sqrt{2})^2} = 1 \quad \text{Standard form}$$

The center is $(4, 1)$; $a = \sqrt{2}$ and $b = \sqrt{2}$. The transverse axis is horizontal, so the vertices are $\sqrt{2}$ units left and right of the center:

$(4-\sqrt{2}, 1)$ and $(4+\sqrt{2}, 1)$.

Since $c^2 = a^2 + b^2$, we have $c^2 = 2 + 2 = 4$ and $c = 2$. Then the foci are 2 units left and right of the center:

$(4-2, 1)$ and $(4+2, 1)$, or $(2, 1)$ and $(6, 1)$.

Find the asymptotes:

$$y - k = \frac{b}{a}(x-h) \quad \text{and} \quad y - k = -\frac{b}{a}(x-h)$$
$$y - 1 = \frac{\sqrt{2}}{\sqrt{2}}(x-4) \text{ and } y - 1 = -\frac{\sqrt{2}}{\sqrt{2}}(x-4)$$
$$y - 1 = x - 4 \quad \text{and} \quad y - 1 = -(x-4), \text{ or}$$
$$y = x - 3 \quad \text{and} \quad y = -x + 5$$

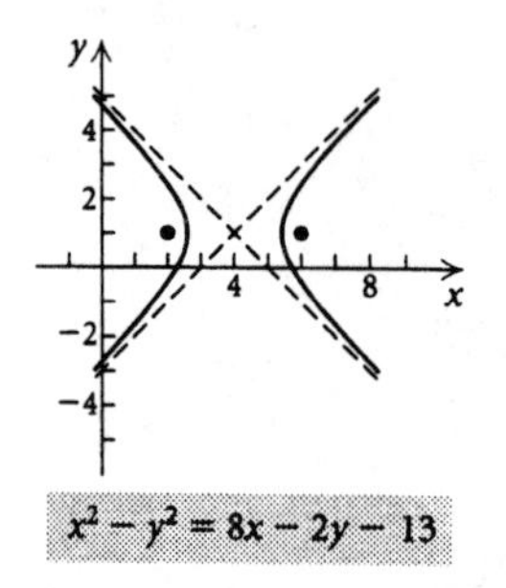

$x^2 - y^2 = 8x - 2y - 13$

35. The hyperbola in Example 3 is wider than the one in Example 2, so the hyperbola in Example 3 has the larger eccentricity.

Compute the eccentricities: In Example 2, $c = 5$ and $a = 4$, so $e = 5/4$, or 1.25. In Example 3, $c = \sqrt{5}$ and $a = 1$, so $e = \sqrt{5}/1 \approx 2.24$. These computations confirm that the hyperbola in Example 3 has the larger eccentricity.

36. Hyperbola (*b*) is wider so it has the larger eccentricity.

37. The center is the midpoint of the segment connecting the vertices:

$$\left(\frac{3-3}{2}, \frac{7+7}{2}\right), \text{ or } (0,7).$$

The vertices are on the horizontal line $y = 7$, so the transverse axis is horizontal. Since the vertices are 3 units left and right of the center, $a = 3$.

Find c:

$$e = \frac{c}{a} = \frac{5}{3}$$

$$\frac{c}{3} = \frac{5}{3} \quad \text{Substituting 3 for } a$$

$$c = 5$$

Now find b^2:

$$c^2 = a^2 + b^2$$
$$5^2 = 3^2 + b^2$$
$$16 = b^2$$

Write the equation:

$$\frac{(x-h)^2}{a^2} - \frac{(y-k)^2}{b^2} = 1$$

$$\frac{x^2}{9} - \frac{(y-7)^2}{16} = 1$$

38. The center is the midpoint of the segment connecting the vertices:

$$\left(\frac{-1-1}{2}, \frac{3+7}{2}\right), \text{ or } (-1,5).$$

The vertices are on the vertical line $x = -1$, so the transverse axis is vertical. Since the vertices are 2 units below and above the center, $a = 2$.

Find c:

$$e = \frac{c}{a} = 4$$

$$\frac{c}{2} = 4$$

$$c = 8$$

Now find b^2:

$$c^2 = a^2 + b^2$$
$$64 = 4 + b^2$$
$$60 = b^2$$

The equation is $\dfrac{(y-5)^2}{4} - \dfrac{(x+1)^2}{60} = 1$.

39.

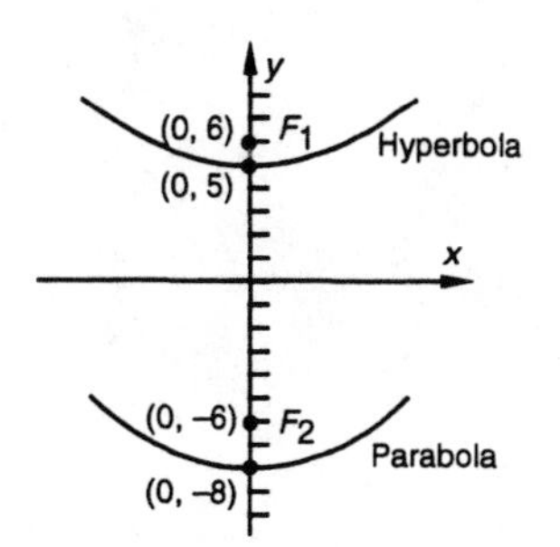

One focus is 6 units above the center of the hyperbola, so $c = 6$. One vertex is 5 units above the center, so $a = 5$. Find b^2:

$$c^2 = a^2 + b^2$$
$$6^2 = 5^2 + b^2$$
$$11 = b^2$$

Write the equation:

$$\frac{y^2}{a^2} - \frac{x^2}{b^2} = 1$$

$$\frac{y^2}{25} - \frac{x^2}{11} = 1$$

40.

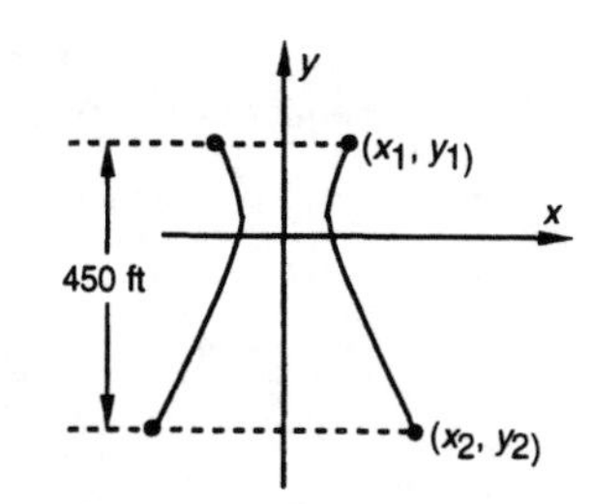

$y_1 = \frac{1}{2}y_2$, so $y_2 + \frac{1}{2}y_2 = 450$. Then $y_2 = 300$ and $y_1 = \frac{1}{2} \cdot 300 = 150$.

Find x_1:

$$\frac{x_1^2}{90^2} - \frac{150^2}{130^2} = 1$$

$$x_1^2 = 90^2\left(1 + \frac{150^2}{130^2}\right)$$

$$x_1 \approx 137.4$$

Then the diameter of the top of the tower is $2x_1 \approx 2(137.4) \approx 275$ ft.

Find x_2:

$$\frac{x_2^2}{90^2} - \frac{300^2}{130^2} = 1$$

$$x_2^2 = 90^2\left(1 + \frac{300^2}{130^2}\right)$$

$$x_2 \approx 226.4$$

Then the diameter of the bottom of the tower is $2x_2 \approx 2(226.4) \approx 453$ ft.

41. See the figure on page 428 of the text.

42. Answers will vary. Students who have graphers that require equations to be entered in "$y =$ " form will probably prefer to graph the hyperbola by hand.

43. $\begin{aligned} x + y &= 5, \quad (1) \\ x - y &= 7 \quad (2) \\ \hline 2x \quad &= 12 \quad \text{Adding} \\ x &= 6 \end{aligned}$

Back-substitute in either equation (1) or (2) and solve for y. We use equation (1).

$$6 + y = 5$$
$$y = -1$$

The solution is $(6, -1)$.

44. $\begin{aligned} 3x - 2y &= 5, \quad (1) \\ 5x + 2y &= 3 \quad (2) \\ \hline 8x \quad &= 8 \quad \text{Adding} \\ x &= 1 \end{aligned}$

Back-substitute and solve for y.

$$5 \cdot 1 + 2y = 3 \quad \text{Using equation (2)}$$
$$2y = -2$$
$$y = -1$$

The solution is $(1, -1)$.

45. $2x - 3y = 7, \quad (1)$

$3x + 5y = 1 \quad (2)$

Multiply equation (1) by 5 and equation (2) by 3 and add to eliminate y.

$$\begin{aligned} 10x - 15y &= 35 \\ 9x + 15y &= 3 \\ \hline 19x \quad &= 38 \\ x &= 2 \end{aligned}$$

Back-substitute and solve for y.

$$3 \cdot 2 + 5y = 1 \quad \text{Using equation (2)}$$
$$5y = -5$$
$$y = -1$$

The solution is $(2, -1)$.

46. $3x + 2y = -1 \quad (1)$

$2x + 3y = 6 \quad (2)$

Multiply equation (1) by 3 and equation (2) by -2 and add.

$$\begin{aligned} 9x + 6y &= -3 \\ -4x - 6y &= -12 \\ \hline 5x \quad &= -15 \\ x &= -3 \end{aligned}$$

Back-substitute and solve for y.

$$3(-3) + 2y = -1 \quad \text{Using equation (1)}$$
$$2y = 8$$
$$y = 4$$

The solution is $(-3, 4)$.

47. The center is the midpoint of the segment connecting $(3, -8)$ and $(3, -2)$:

$$\left(\frac{3+3}{2}, \frac{-8-2}{2}\right), \text{ or } (3, -5).$$

The vertices are on the vertical line $x = 3$ and are 3 units above and below the center so the transverse axis is vertical and $a = 3$. Use the equation of an asymptote to find b:

$$y - k = \frac{a}{b}(x - h)$$
$$y + 5 = \frac{3}{b}(x - 3)$$
$$y = \frac{3}{b}x - \frac{9}{b} - 5$$

This equation corresponds to the asymptote $y = 3x - 14$, so $\frac{3}{b} = 3$ and $b = 1$.

Write the equation of the hyperbola:

$$\frac{(y-k)^2}{a^2} - \frac{(x-h)^2}{b^2} = 1$$
$$\frac{(y+5)^2}{9} - \frac{(x-3)^2}{1} = 1$$

48. The center is the midpoint of the segment connecting the vertices:

$$\left(\frac{-9-5}{2}, \frac{4+4}{2}\right), \text{ or } (-7, 4).$$

The vertices are on the horizontal line $y = 4$ and are 2 units left and right of the center, so the transverse axis is horizontal and $a = 2$. Use the equation of an asymptote to find b:

$$y - k = \frac{b}{a}(x - h)$$
$$y - 4 = \frac{b}{2}(x + 7)$$
$$y = \frac{b}{2}x + \frac{7}{2}b + 4$$

This equation corresponds to the asymptote $y = 3x + 25$, so $\frac{b}{2} = 3$ and $b = 6$.

Write the equation of the hyperbola:

$$\frac{(x+7)^2}{4} - \frac{(y-4)^2}{36} = 1$$

49. Center: $(-1.460, -0.957)$

Vertices: $(-2.360, -0.957)$, $(-0.560, -0.957)$

Asymptotes: $y = -1.429x - 3.043$, $y = 1.429x + 1.129$

50. Center: $(1.023, -2.044)$

Vertices: $(2.07, -2.044)$, $(-0.024, -2.044)$

Asymptotes: $y = x - 3.067$, $y = -x - 1.021$

51. S and T are the foci of the hyperbola, so $c = 300/2 = 150$.

200 microseconds $\cdot \dfrac{0.186 \text{ mi}}{1 \text{ microsecond}} = 37.2$ mi, the difference of the ships' distances from the foci. That is, $2a = 37.2$, so $a = 18.6$.

Find b^2:

$$c^2 = a^2 + b^2$$
$$150^2 = 18.6^2 + b^2$$
$$22{,}154.04 = b^2$$

Then the equation of the hyperbola is

$$\frac{x^2}{18.6^2} - \frac{y^2}{22,154.04} = 1, \text{ or } \frac{x^2}{345.96} - \frac{y^2}{22,154.04} = 1.$$

Exercise Set 6.4

1. The correct graph is (e).

2. The correct graph is (a).

3. The correct graph is (c).

4. The correct graph is (f).

5. The correct graph is (b).

6. The correct graph is (d).

7. $x^2 + y^2 = 25,$ (1)

$y - x = 1$ (2)

First solve equation (2) for y.

$y = x + 1$ (3)

Then substitute $x+1$ for y in equation (1) and solve for x.

$$x^2 + y^2 = 25$$
$$x^2 + (x+1)^2 = 25$$
$$x^2 + x^2 + 2x + 1 = 25$$
$$2x^2 + 2x - 24 = 0$$
$$x^2 + x - 12 = 0 \quad \text{Multiplying by } \frac{1}{2}$$
$$(x+4)(x-3) = 0 \quad \text{Factoring}$$
$$x + 4 = 0 \quad \text{or} \quad x - 3 = 0 \quad \text{Principle of zero products}$$
$$x = -4 \quad \text{or} \quad x = 3$$

Now substitute these numbers into equation (3) and solve for y.

$y = -4 + 1 = -3$

$y = 3 + 1 = 4$

The pairs $(-4, -3)$ and $(3, 4)$ check, so they are the solutions.

8. $x^2 + y^2 = 100,$

$y - x = 2$

$y = x + 2$

$$x^2 + (x+2)^2 = 100$$
$$x^2 + x^2 + 4x + 4 = 100$$
$$2x^2 + 4x - 96 = 0$$
$$x^2 + 2x - 48 = 0$$
$$(x+8)(x-6) = 0$$

$x = -8$ or $x = 6$

$y = -8 + 2 = -6$

$y = 6 + 2 = 8$

The pairs $(-8, -6)$ and $(6, 8)$ check.

9. $4x^2 + 9y^2 = 36,$ (1)

$3y + 2x = 6$ (2)

First solve equation (2) for y.

$3y = -2x + 6$

$y = -\frac{2}{3}x + 2$ (3)

Then substitute $-\frac{2}{3}x + 2$ for y in equation (1) and solve for x.

$$4x^2 + 9y^2 = 36$$
$$4x^2 + 9\left(-\frac{2}{3}x + 2\right)^2 = 36$$
$$4x^2 + 9\left(\frac{4}{9}x^2 - \frac{8}{3}x + 4\right) = 36$$
$$4x^2 + 4x^2 - 24x + 36 = 36$$
$$8x^2 - 24x = 0$$
$$x^2 - 3x = 0$$
$$x(x-3) = 0$$

$x = 0$ or $x = 3$

Now substitute these numbers in equation (3) and solve for y.

$y = -\frac{2}{3} \cdot 0 + 2 = 2$

$y = -\frac{2}{3} \cdot 3 + 2 = 0$

The pairs $(0, 2)$ and $(3, 0)$ check, so they are the solutions.

10. $9x^2 + 4y^2 = 36,$

$3x + 2y = 6$

$$x = 2 - \frac{2}{3}y$$
$$9\left(2 - \frac{2}{3}y\right)^2 + 4y^2 = 36$$
$$9\left(4 - \frac{8}{3}y + \frac{4}{9}y^2\right) + 4y^2 = 36$$
$$36 - 24y + 4y^2 + 4y^2 = 36$$
$$8y^2 - 24y = 0$$
$$y^2 - 3y = 0$$
$$y(y-3) = 0$$

$y = 0$ or $y = 3$

$x = 2 - \frac{2}{3}(0) = 2$

$x = 2 - \frac{2}{3}(3) = 0$

The pairs $(2, 0)$ and $(0, 3)$ check.

11. $x^2 + y^2 = 25,$ (1)

$y^2 = x + 5$ (2)

We substitute $x + 5$ for y^2 in equation (1) and solve for x.

$$x^2 + y^2 = 25$$
$$x^2 + (x+5) = 25$$
$$x^2 + x - 20 = 0$$
$$(x+5)(x-4) = 0$$
$$x+5=0 \quad \text{or} \quad x-4=0$$
$$x=-5 \quad \text{or} \quad x=4$$

We substitute these numbers for x in either equation (1) or equation (2) and solve for y. Here we use equation (2).

$y^2 = -5+5 = 0$ and $y = 0$.

$y^2 = 4+5 = 9$ and $y = \pm 3$.

The pairs $(-5,0)$, $(4,3)$ and $(4,-3)$ check. They are the solutions.

12. $y = x^2$,
$x = y^2$

$$x = (x^2)^2$$
$$x = x^4$$
$$0 = x^4 - x$$
$$0 = x(x^3-1)$$
$$0 = x(x-1)(x^2+x+1)$$
$$x=0 \quad \text{or} \quad x=1 \quad \text{or} \quad x = \frac{-1 \pm \sqrt{1^2 - 4\cdot 1\cdot 1}}{2}$$
$$x=0 \quad \text{or} \quad x=1 \quad \text{or} \quad x = -\frac{1}{2} \pm \frac{\sqrt{3}}{2}i$$
$$y = 0^2 = 0$$
$$y = 1^2 = 1$$
$$y = \left(-\frac{1}{2} + \frac{\sqrt{3}}{2}i\right)^2 = -\frac{1}{2} - \frac{\sqrt{3}}{2}i$$
$$y = \left(-\frac{1}{2} - \frac{\sqrt{3}}{2}i\right)^2 = -\frac{1}{2} + \frac{\sqrt{3}}{2}i$$

The pairs $(0,0)$, $(1,1)$, $\left(-\frac{1}{2} + \frac{\sqrt{3}}{2}i, -\frac{1}{2} - \frac{\sqrt{3}}{2}i\right)$, and $\left(-\frac{1}{2} - \frac{\sqrt{3}}{2}i, -\frac{1}{2} + \frac{\sqrt{3}}{2}i\right)$ check.

13. $x^2 + y^2 = 9$, (1)
$x^2 - y^2 = 9$ (2)

Here we use the elimination method.

$$\begin{aligned} x^2 + y^2 &= 9 \quad (1)\\ x^2 - y^2 &= 9 \quad (2)\\ \hline 2x^2 &= 18 \quad \text{Adding}\\ x^2 &= 9\\ x &= \pm 3 \end{aligned}$$

If $x = 3$, $x^2 = 9$, and if $x = -3$, $x^2 = 9$, so substituting 3 or -3 in equation (1) gives us

$$x^2 + y^2 = 9$$
$$9 + y^2 = 9$$
$$y^2 = 0$$
$$y = 0.$$

The pairs $(3,0)$ and $(-3,0)$ check. They are the solutions.

14. $y^2 - 4x^2 = 4$ (1)
$4x^2 + y^2 = 4$ (2)

$$\begin{aligned} -4x^2 + y^2 &= 4 \quad (1)\\ 4x^2 + y^2 &= 4 \quad (2)\\ \hline 2y^2 &= 8 \quad \text{Adding}\\ y^2 &= 4\\ y &= \pm 2 \end{aligned}$$

Substitute for y in equation (2).

$$4x^2 + 4 = 4$$
$$4x^2 = 0$$
$$x = 0$$

The pairs $(0,2)$ and $(0,-2)$ check.

15. $y^2 - x^2 = 9$ (1)
$2x - 3 = y$ (2)

Substitute $2x-3$ for y in equation (1) and solve for x.

$$y^2 - x^2 = 9$$
$$(2x-3)^2 - x^2 = 9$$
$$4x^2 - 12x + 9 - x^2 = 9$$
$$3x^2 - 12x = 0$$
$$x^2 - 4x = 0$$
$$x(x-4) = 0$$

$x = 0$ or $x = 4$

Now substitute these numbers into equation (2) and solve for y.

If $x = 0$, $y = 2\cdot 0 - 3 = -3$.

If $x = 4$, $y = 2\cdot 4 - 3 = 5$.

The pairs $(0,-3)$ and $(4,5)$ check. They are the solutions.

16. $x + y = -6$,
$xy = -7$

$$y = -x - 6$$
$$x(-x-6) = -7$$
$$-x^2 - 6x = -7$$
$$0 = x^2 + 6x - 7$$
$$0 = (x+7)(x-1)$$

$x = -7$ or $x = 1$

$y = -(-7) - 6 = 1$

$y = -1 - 6 = -7$

The pairs $(-7,1)$ and $(1,-7)$ check.

17. $y^2 = x + 3$, (1)
$2y = x + 4$ (2)

First solve equation (2) for x.

$$2y - 4 = x \qquad (3)$$

Then substitute $2y-4$ for x in equation (1) and solve for y.

$$y^2 = x + 3$$
$$y^2 = (2y - 4) + 3$$
$$y^2 = 2y - 1$$
$$y^2 - 2y + 1 = 0$$
$$(y-1)(y-1) = 0$$
$$y - 1 = 0 \quad \text{or} \quad y - 1 = 0$$
$$y = 1 \quad \text{or} \quad y = 1$$

Now substitute 1 for y in equation (3) and solve for x.

$$2 \cdot 1 - 4 = x$$
$$-2 = x$$

The pair $(-2, 1)$ checks. It is the solution.

18. $y = x^2$,

$3x = y + 2$

$$y = 3x - 2$$

$$3x - 2 = x^2$$
$$0 = x^2 - 3x + 2$$
$$0 = (x-2)(x-1)$$
$$x = 2 \text{ or } x = 1$$

$$y = 3 \cdot 2 - 2 = 4$$
$$y = 3 \cdot 1 - 2 = 1$$

The pairs $(2, 4)$ and $(1, 1)$ check.

19. $x^2 + y^2 = 25, \quad (1)$

$xy = 12 \qquad (2)$

First we solve equation (2) for y.

$$xy = 12$$
$$y = \frac{12}{x}$$

Then we substitute $\frac{12}{x}$ for y in equation (1) and solve for x.

$$x^2 + y^2 = 25$$
$$x^2 + \left(\frac{12}{x}\right)^2 = 25$$
$$x^2 + \frac{144}{x^2} = 25$$
$$x^4 + 144 = 25x^2 \quad \text{Multiplying by } x^2$$
$$x^4 - 25x^2 + 144 = 0$$
$$u^2 - 25u + 144 = 0 \quad \text{Letting } u = x^2$$
$$(u-9)(u-16) = 0$$
$$u = 9 \quad \text{or} \quad u = 16$$

We now substitute x^2 for u and solve for x.

$$x^2 = 9 \quad \text{or} \quad x^2 = 16$$
$$x = \pm 3 \quad \text{or} \quad x = \pm 4$$

Since $y = 12/x$, if $x = 3$, $y = 4$; if $x = -3$, $y = -4$; if $x = 4$, $y = 3$; and if $x = -4$, $y = -3$. The pairs $(3, 4)$, $(-3, -4)$, $(4, 3)$, and $(-4, -3)$ check. They are the solutions.

20. $x^2 - y^2 = 16, \quad (1)$

$\underline{x + y^2 = 4} \quad (2)$

$x^2 + x = 20 \quad$ Adding

$$x^2 + x - 20 = 0$$
$$(x+5)(x-4) = 0$$
$$x = -5 \quad \text{or} \quad x = 4$$
$$y^2 = 4 - x \quad \text{Solving equation (2) for } y^2$$
$$y^2 = 4 - (-5) = 9 \text{ and } y = \pm 3$$
$$y^2 = 4 - 4 = 0 \text{ and } y = 0$$

The pairs $(-5, 3)$, $(-5, -3)$, and $(4, 0)$ check.

21. $x^2 + y^2 = 4, \quad (1)$

$16x^2 + 9y^2 = 144 \quad (2)$

$-9x^2 - 9y^2 = -36 \quad$ Multiplying (1) by -9

$\underline{16x^2 + 9y^2 = 144}$

$7x^2 = 108 \quad$ Adding

$$x^2 = \frac{108}{7}$$
$$x = \pm\sqrt{\frac{108}{7}} = \pm 6\sqrt{\frac{3}{7}}$$
$$x = \pm\frac{6\sqrt{21}}{7} \quad \text{Rationalizing the denominator}$$

Substituting $\frac{6\sqrt{21}}{7}$ or $-\frac{6\sqrt{21}}{7}$ for x in equation (1) gives us

$$\frac{36 \cdot 21}{49} + y^2 = 4$$
$$y^2 = 4 - \frac{108}{7}$$
$$y^2 = -\frac{80}{7}$$
$$y = \pm\sqrt{-\frac{80}{7}} = \pm 4i\sqrt{\frac{5}{7}}$$
$$y = \pm\frac{4i\sqrt{35}}{7}. \quad \text{Rationalizing the denominator}$$

The pairs $\left(\frac{6\sqrt{21}}{7}, \frac{4i\sqrt{35}}{7}\right)$, $\left(\frac{6\sqrt{21}}{7}, -\frac{4i\sqrt{35}}{7}\right)$, $\left(-\frac{6\sqrt{21}}{7}, \frac{4i\sqrt{35}}{7}\right)$, and $\left(-\frac{6\sqrt{21}}{7}, -\frac{4i\sqrt{35}}{7}\right)$ check. They are the solutions.

22. $x^2 + y^2 = 25, \quad (1)$

$25x^2 + 16y^2 = 400 \quad (2)$

$-16x^2 - 16y^2 = -400 \quad$ Multiplying (1) by -16

$\underline{25x^2 + 16y^2 = 400}$

$9x^2 = 0 \quad$ Adding

$x = 0$

$0^2 + y^2 = 25$ Substituting in (1)

$y = \pm 5$

The pairs $(0, 5)$ and $(0, -5)$ check.

23. $x^2 + 4y^2 = 25$, (1)

$x + 2y = 7$ (2)

First solve equation (2) for x.

$x = -2y + 7$ (3)

Then substitute $-2y + 7$ for x in equation (1) and solve for y.

$$x^2 + 4y^2 = 25$$
$$(-2y + 7)^2 + 4y^2 = 25$$
$$4y^2 - 28y + 49 + 4y^2 = 25$$
$$8y^2 - 28y + 24 = 0$$
$$2y^2 - 7y + 6 = 0$$
$$(2y - 3)(y - 2) = 0$$

$y = \frac{3}{2}$ or $y = 2$

Now substitute these numbers in equation (3) and solve for x.

$x = -2 \cdot \frac{3}{2} + 7 = 4$

$x = -2 \cdot 2 + 7 = 3$

The pairs $\left(4, \frac{3}{2}\right)$ and $(3, 2)$ check, so they are the solutions.

24. $y^2 - x^2 = 16$,

$2x - y = 1$

$y = 2x - 1$

$$(2x - 1)^2 - x^2 = 16$$
$$4x^2 - 4x + 1 - x^2 = 16$$
$$3x^2 - 4x - 15 = 0$$
$$(3x + 5)(x - 3) = 0$$

$x = -\frac{5}{3}$ or $x = 3$

$y = 2\left(-\frac{5}{3}\right) - 1 = -\frac{13}{3}$

$y = 2(3) - 1 = 5$

The pairs $\left(-\frac{5}{3}, -\frac{13}{3}\right)$ and $(3, 5)$ check.

25. $x^2 - xy + 3y^2 = 27$, (1)

$x - y = 2$ (2)

First solve equation (2) for y.

$x - 2 = y$ (3)

Then substitute $x - 2$ for y in equation (1) and solve for x.

$$x^2 - xy + 3y^2 = 27$$
$$x^2 - x(x - 2) + 3(x - 2)^2 = 27$$
$$x^2 - x^2 + 2x + 3x^2 - 12x + 12 = 27$$
$$3x^2 - 10x - 15 = 0$$

$$x = \frac{-(-10) \pm \sqrt{(-10)^2 - 4(3)(-15)}}{2 \cdot 3}$$

$$x = \frac{10 \pm \sqrt{100 + 180}}{6} = \frac{10 \pm \sqrt{280}}{6}$$

$$x = \frac{10 \pm 2\sqrt{70}}{6} = \frac{5 \pm \sqrt{70}}{3}$$

Now substitute these numbers in equation (3) and solve for y.

$$y = \frac{5 + \sqrt{70}}{3} - 2 = \frac{-1 + \sqrt{70}}{3}$$

$$y = \frac{5 - \sqrt{70}}{3} - 2 = \frac{-1 - \sqrt{70}}{3}$$

The pairs $\left(\frac{5 + \sqrt{70}}{3}, \frac{-1 + \sqrt{70}}{3}\right)$ and $\left(\frac{5 - \sqrt{70}}{3}, \frac{-1 - \sqrt{70}}{3}\right)$ check, so they are the solutions.

26. $2y^2 + xy + x^2 = 7$,

$x - 2y = 5$

$x = 2y + 5$

$$2y^2 + (2y + 5)y + (2y + 5)^2 = 7$$
$$2y^2 + 2y^2 + 5y + 4y^2 + 20y + 25 = 7$$
$$8y^2 + 25y + 18 = 0$$
$$(8y + 9)(y + 2) = 0$$

$y = -\frac{9}{8}$ or $y = -2$

$x = 2\left(-\frac{9}{8}\right) + 5 = \frac{11}{4}$

$x = 2(-2) + 5 = 1$

The pairs $\left(\frac{11}{4}, -\frac{9}{8}\right)$ and $(1, -2)$ check.

27. $x^2 + y^2 = 16$, $\quad$ $x^2 + y^2 = 16$, (1)

or

$y^2 - 2x^2 = 10$ $\quad$ $-2x^2 + y^2 = 10$ (2)

Here we use the elimination method.

$$\begin{aligned} 2x^2 + 2y^2 &= 32 \quad \text{Multiplying (1) by 2} \\ -2x^2 + y^2 &= 10 \\ \hline 3y^2 &= 42 \quad \text{Adding} \\ y^2 &= 14 \\ y &= \pm\sqrt{14} \end{aligned}$$

Substituting $\sqrt{14}$ or $-\sqrt{14}$ for y in equation (1) gives us

$$x^2 + 14 = 16$$
$$x^2 = 2$$
$$x = \pm\sqrt{2}$$

The pairs $(-\sqrt{2}, -\sqrt{14})$, $(-\sqrt{2}, \sqrt{14})$, $(\sqrt{2}, -\sqrt{14})$, and $(\sqrt{2}, \sqrt{14})$ check. They are the solutions.

28. $x^2 + y^2 = 14$, (1)
$x^2 - y^2 = 4$ (2)

$$2x^2 = 18 \quad \text{Adding}$$
$$x^2 = 9$$
$$x = \pm 3$$

$9 + y^2 = 14$ Substituting in equation (1)

$$y^2 = 5$$
$$y = \pm\sqrt{5}$$

The pairs $(-3, -\sqrt{5})$, $(-3, \sqrt{5})$, $(3, -\sqrt{5})$, and $(3, \sqrt{5})$ check.

29. $x^2 + y^2 = 5$, (1)
$xy = 2$ (2)

First we solve equation (2) for y.

$$xy = 2$$
$$y = \frac{2}{x}$$

Then we substitute $\frac{2}{x}$ for y in equation (1) and solve for x.

$$x^2 + y^2 = 5$$
$$x^2 + \left(\frac{2}{x}\right)^2 = 5$$
$$x^2 + \frac{4}{x^2} = 5$$
$$x^4 + 4 = 5x^2 \quad \text{Multiplying by } x^2$$
$$x^4 - 5x^2 + 4 = 0$$
$$u^2 - 5u + 4 = 0 \quad \text{Letting } u = x^2$$
$$(u-4)(u-1) = 0$$
$$u = 4 \text{ or } u = 1$$

We now substitute x^2 for u and solve for x.

$$x^2 = 4 \quad \text{or} \quad x^2 = 1$$
$$x = \pm 2 \qquad x = \pm 1$$

Since $y = 2/x$, if $x = 2$, $y = 1$; if $x = -2$, $y = -1$; if $x = 1$, $y = 2$; and if $x = -1$, $y = -2$. The pairs $(2, 1)$, $(-2, -1)$, $(1, 2)$, and $(-1, -2)$ check. They are the solutions.

30. $x^2 + y^2 = 20$,
$xy = 8$

$$y = \frac{8}{x}$$
$$x^2 + \left(\frac{8}{x}\right)^2 = 20$$
$$x^2 + \frac{64}{x^2} = 20$$
$$x^4 + 64 = 20x^2$$
$$x^4 - 20x^2 + 64 = 0$$
$$u^2 - 20u + 64 = 0 \qquad \text{Letting } u = x^2$$
$$(u-16)(u-4) = 0$$
$$u = 16 \quad \text{or} \quad u = 4$$
$$x^2 = 16 \quad \text{or} \quad x^2 = 4$$
$$x = \pm 4 \quad \text{or} \quad x = \pm 2$$

$y = 8/x$, so if $x = 4$, $y = 2$; if $x = -4$, $y = -2$; if $x = 2$, $y = 4$; if $x = -2$, $y = -4$. The pairs $(4, 2)$, $(-4, -2)$, $(2, 4)$, and $(-2, -4)$ check.

31. $3x + y = 7$ (1)
$4x^2 + 5y = 56$ (2)

First solve equation (1) for y.

$$3x + y = 7$$
$$y = 7 - 3x \quad (3)$$

Next substitute $7 - 3x$ for y in equation (2) and solve for x.

$$4x^2 + 5y = 56$$
$$4x^2 + 5(7 - 3x) = 56$$
$$4x^2 + 35 - 15x = 56$$
$$4x^2 - 15x - 21 = 0$$

Using the quadratic formula, we find that

$$x = \frac{15 - \sqrt{561}}{8} \text{ or } x = \frac{15 + \sqrt{561}}{8}.$$

Now substitute these numbers into equation (3) and solve for y.

If $x = \frac{15 - \sqrt{561}}{8}$, $y = 7 - 3\left(\frac{15 - \sqrt{561}}{8}\right)$, or $\frac{11 + 3\sqrt{561}}{8}$.

If $x = \frac{15 + \sqrt{561}}{8}$, $y = 7 - 3\left(\frac{15 + \sqrt{561}}{8}\right)$, or $\frac{11 - 3\sqrt{561}}{8}$.

The pairs $\left(\frac{15 - \sqrt{561}}{8}, \frac{11 + 3\sqrt{561}}{8}\right)$ and $\left(\frac{15 + \sqrt{561}}{8}, \frac{11 - 3\sqrt{561}}{8}\right)$ check and are the solutions.

32. $2y^2 + xy = 5$,
$4y + x = 7$

$$x = -4y + 7$$
$$2y^2 + (-4y + 7)y = 5$$
$$2y^2 - 4y^2 + 7y = 5$$
$$0 = 2y^2 - 7y + 5$$
$$0 = (2y - 5)(y - 1)$$
$$y = \frac{5}{2} \text{ or } y = 1$$
$$x = -4\left(\frac{5}{2}\right) + 7 = -3$$
$$x = -4(1) + 7 = 3$$

The pairs $\left(-3, \frac{5}{2}\right)$ and $(3, 1)$ check.

33. $a+b=7,\quad (1)$

$ab=4\qquad (2)$

First solve equation (1) for a.

$a=-b+7\quad (3)$

Then substitute $-b+7$ for a in equation (2) and solve for b.

$$(-b+7)b=4$$
$$-b^2+7b=4$$
$$0=b^2-7b+4$$
$$b=\frac{-(-7)\pm\sqrt{(-7)^2-4\cdot 1\cdot 4}}{2\cdot 1}$$
$$b=\frac{7\pm\sqrt{33}}{2}$$

Now substitute these numbers in equation (3) and solve for a.

$$a=-\left(\frac{7+\sqrt{33}}{2}\right)+7=\frac{7-\sqrt{33}}{2}$$
$$a=-\left(\frac{7-\sqrt{33}}{2}\right)+7=\frac{7+\sqrt{33}}{2}$$

The pairs $\left(\frac{7-\sqrt{33}}{2},\frac{7+\sqrt{33}}{2}\right)$ and $\left(\frac{7+\sqrt{33}}{2},\frac{7-\sqrt{33}}{2}\right)$ check, so they are the solutions.

34. $p+q=-4,$

$pq=-5$

$$p=-q-4$$
$$(-q-4)q=-5$$
$$-q^2-4q=-5$$
$$0=q^2+4q-5$$
$$0=(q+5)(q-1)$$
$$q=-5 \text{ or } q=1$$
$$p=-(-5)-4=1$$
$$p=-1-4=-5$$

The pairs $(1,-5)$ and $(-5,1)$ check.

35. $x^2+y^2=13,\qquad (1)$

$xy=6\qquad (2)$

First we solve Equation (2) for y.

$$xy=6$$
$$y=\frac{6}{x}$$

Then we substitute $\frac{6}{x}$ for y in equation (1) and solve for x.

$$x^2+y^2=13$$
$$x^2+\left(\frac{6}{x}\right)^2=13$$
$$x^2+\frac{36}{x^2}=13$$
$$x^4+36=13x^2\qquad \text{Multiplying by } x^2$$
$$x^4-13x^2+36=0$$
$$u^2-13u+36=0\qquad \text{Letting } u=x^2$$
$$(u-9)(u-4)=0$$
$$u=9 \quad\text{or}\quad u=4$$

We now substitute x^2 for u and solve for x.

$$x^2=9 \quad\text{or}\quad x^2=4$$
$$x=\pm 3 \quad\text{or}\quad x=\pm 2$$

Since $y=6/x$, if $x=3$, $y=2$; if $x=-3$, $y=-2$; if $x=2$, $y=3$; and if $x=-2$, $y=-3$. The pairs $(3,2)$, $(-3,-2)$, $(2,3)$, and $(-2,-3)$ check. They are the solutions.

36. $x^2+4y^2=20,$

$xy=4$

$$y=\frac{4}{x}$$
$$x^2+4\left(\frac{4}{x}\right)^2=20$$
$$x^2+\frac{64}{x^2}=20$$
$$x^4+64=20x^2$$
$$x^4-20x^2+64=0$$
$$u^2-20u+64=0\qquad \text{Letting } u=x^2$$
$$(u-16)(u-4)=0$$
$$u=16 \quad\text{or}\quad u=4$$
$$x^2=16 \quad\text{or}\quad x^2=4$$
$$x=\pm 4 \quad\text{or}\quad x=\pm 2$$

$y=4/x$, so if $x=4$, $y=1$; if $x=-4$, $y=-1$; if $x=2$, $y=2$; and if $x=-2$, $y=-2$. The pairs $(4,1)$, $(-4,-1)$, $(2,2)$, and $(-2,-2)$ check.

37. $x^2+y^2+6y+5=0\qquad (1)$

$x^2+y^2-2x-8=0\qquad (2)$

Using the elimination method, multiply equation (2) by -1 and add the result to equation (1).

$$\begin{array}{rl} x^2+y^2+6y+5=0 & (1)\\ -x^2-y^2+2x+8=0 & (2)\\ \hline 2x+6y+13=0 & (3)\end{array}$$

Solve equation (3) for x.

$$2x+6y+13=0$$
$$2x=-6y-13$$
$$x=\frac{-6y-13}{2}$$

Substitute $\frac{-6y-13}{2}$ for x in equation (1) and solve for y.

$$x^2 + y^2 + 6y + 5 = 0$$
$$\left(\frac{-6y-13}{2}\right)^2 + y^2 + 6y + 5 = 0$$
$$\frac{36y^2 + 156y + 169}{4} + y^2 + 6y + 5 = 0$$
$$36y^2 + 156y + 169 + 4y^2 + 24y + 20 = 0$$
$$40y^2 + 180y + 189 = 0$$

Using the quadratic formula, we find that $y = \frac{-45 \pm 3\sqrt{15}}{20}$. Substitute $\frac{-45 \pm 3\sqrt{15}}{20}$ for y in $x = \frac{-6y-13}{2}$ and solve for x.

If $y = \frac{-45+3\sqrt{15}}{20}$, then

$$x = \frac{-6\left(\frac{-45+3\sqrt{15}}{20}\right) - 13}{2} = \frac{5-9\sqrt{15}}{20}.$$

If $y = \frac{-45-3\sqrt{15}}{20}$, then

$$x = \frac{-6\left(\frac{-45-3\sqrt{15}}{20}\right) - 13}{2} = \frac{5+9\sqrt{15}}{20}.$$

The pairs $\left(\frac{5+9\sqrt{15}}{20}, \frac{-45-3\sqrt{15}}{20}\right)$ and $\left(\frac{5-9\sqrt{15}}{20}, \frac{-45+3\sqrt{15}}{20}\right)$ check and are the solutions.

38. $2xy + 3y^2 = 7, \quad (1)$
$3xy - 2y^2 = 4 \quad (2)$

$$\begin{aligned} 6xy + 9y^2 &= 21 && \text{Multiplying (1) by 3} \\ -6xy + 4y^2 &= -8 && \text{Multiplying (2) by } -2 \\ \hline 13y^2 &= 13 \\ y^2 &= 1 \\ y &= \pm 1 \end{aligned}$$

Substitute for y in equation (1) and solve for x.

When $y = 1$: $2 \cdot x \cdot 1 + 3 \cdot 1^2 = 7$
$2x = 4$
$x = 2$

When $y = -1$: $2 \cdot x \cdot (-1) + 3(-1)^2 = 7$
$-2x = 4$
$x = -2$

The pairs $(2, 1)$ and $(-2, -1)$ check.

39. $2a + b = 1, \quad (1)$
$b = 4 - a^2 \quad (2)$

Equation (2) is already solved for b. Substitute $4-a^2$ for b in equation (1) and solve for a.

$2a + 4 - a^2 = 1$
$0 = a^2 - 2a - 3$
$0 = (a-3)(a+1)$
$a = 3$ or $a = -1$

Substitute these numbers in equation (2) and solve for b.

$b = 4 - 3^2 = -5$
$b = 4 - (-1)^2 = 3$

The pairs $(3, -5)$ and $(-1, 3)$ check. They are the solutions.

40. $4x^2 + 9y^2 = 36,$
$x + 3y = 3$

$x = -3y + 3$

$$4(-3y+3)^2 + 9y^2 = 36$$
$$4(9y^2 - 18y + 9) + 9y^2 = 36$$
$$36y^2 - 72y + 36 + 9y^2 = 36$$
$$45y^2 - 72y = 0$$
$$5y^2 - 8y = 0$$
$$y(5y-8) = 0$$

$y = 0$ or $y = \frac{8}{5}$
$x = -3 \cdot 0 + 3 = 3$
$x = -3\left(\frac{8}{5}\right) + 3 = -\frac{9}{5}$

The pairs $(3, 0)$ and $\left(-\frac{9}{5}, \frac{8}{5}\right)$ check.

41. $a^2 + b^2 = 89, \quad (1)$
$a - b = 3 \quad (2)$

First solve equation (2) for a.

$a = b + 3 \quad (3)$

Then substitute $b+3$ for a in equation (1) and solve for b.

$$(b+3)^2 + b^2 = 89$$
$$b^2 + 6b + 9 + b^2 = 89$$
$$2b^2 + 6b - 80 = 0$$
$$b^2 + 3b - 40 = 0$$
$$(b+8)(b-5) = 0$$

$b = -8$ or $b = 5$

Substitute these numbers in equation (3) and solve for a.

$a = -8 + 3 = -5$
$a = 5 + 3 = 8$

The pairs $(-5, -8)$ and $(8, 5)$ check. They are the solutions.

42. $xy = 4,$
$x + y = 5$

$x = -y + 5$
$(-y+5)y = 4$
$-y^2 + 5y = 4$
$0 = y^2 - 5y + 4$
$0 = (y-4)(y-1)$

$y = 4$ or $y = 1$

$x = -4 + 5 = 1$

$x = -1 + 5 = 4$

The pairs $(1, 4)$ and $(4, 1)$ check.

43. $xy - y^2 = 2,\quad (1)$

$2xy - 3y^2 = 0\quad (2)$

$$\begin{aligned} -2xy + 2y^2 &= -4 \quad \text{Multiplying (1) by } -2 \\ 2xy - 3y^2 &= 0 \\ \hline -y^2 &= -4 \quad \text{Adding} \\ y^2 &= 4 \\ y &= \pm 2 \end{aligned}$$

We substitute for y in equation (1) and solve for x.

When $y = 2$: $x \cdot 2 - 2^2 = 2$

$$\begin{aligned} 2x - 4 &= 2 \\ 2x &= 6 \\ x &= 3 \end{aligned}$$

When $y = -2$: $x(-2) - (-2)^2 = 2$

$$\begin{aligned} -2x - 4 &= 2 \\ -2x &= 6 \\ x &= -3 \end{aligned}$$

The pairs $(3, 2)$ and $(-3, -2)$ check. They are the solutions.

44. $4a^2 - 25b^2 = 0,\quad (1)$

$2a^2 - 10b^2 = 3b + 4\quad (2)$

$$\begin{aligned} 4a^2 - 25b^2 &= 0 \\ -4a^2 + 20b^2 &= -6b - 8 \quad \text{Multiplying (2) by } -2 \\ \hline -5b^2 &= -6b - 8 \\ 0 &= 5b^2 - 6b - 8 \\ 0 &= (5b + 4)(b - 2) \end{aligned}$$

$b = -\frac{4}{5}$ or $b = 2$

Substitute for b in equation (1) and solve for a.

When $b = -\frac{4}{5}$: $4a^2 - 25\left(-\frac{4}{5}\right)^2 = 0$

$$\begin{aligned} 4a^2 &= 16 \\ a^2 &= 4 \\ a &= \pm 2 \end{aligned}$$

When $b = 2$: $4a^2 - 25(2)^2 = 0$

$$\begin{aligned} 4a^2 &= 100 \\ a^2 &= 25 \\ a &= \pm 5 \end{aligned}$$

The pairs $\left(2, -\frac{4}{5}\right)$, $\left(-2, -\frac{4}{5}\right)$, $(5, 2)$ and $(-5, 2)$ check.

45. $m^2 - 3mn + n^2 + 1 = 0,\quad (1)$

$3m^2 - mn + 3n^2 = 13\quad (2)$

$m^2 - 3mn + n^2 = -1\quad (3)$ Rewriting (1)

$3m^2 - mn + 3n^2 = 13\quad (2)$

$$\begin{aligned} -3m^2 + 9mn - 3n^2 &= 3 \quad \text{Multiplying (3) by } -3 \\ 3m^2 - mn + 3n^2 &= 13 \\ \hline 8mn &= 16 \\ mn &= 2 \\ n &= \frac{2}{m} \quad (4) \end{aligned}$$

Substitute $\frac{2}{m}$ for n in equation (1) and solve for m.

$$\begin{aligned} m^2 - 3m\left(\frac{2}{m}\right) + \left(\frac{2}{m}\right)^2 + 1 &= 0 \\ m^2 - 6 + \frac{4}{m^2} + 1 &= 0 \\ m^2 - 5 + \frac{4}{m^2} &= 0 \\ m^4 - 5m^2 + 4 &= 0 \quad \text{Multiplying by } m^2 \end{aligned}$$

Substitute u for m^2.

$$\begin{aligned} u^2 - 5u + 4 &= 0 \\ (u - 4)(u - 1) &= 0 \end{aligned}$$

$u = 4$ *or* $u = 1$

$m^2 = 4$ *or* $m^2 = 1$

$m = \pm 2$ *or* $m = \pm 1$

Substitute for m in equation (4) and solve for n.

When $m = 2$, $n = \frac{2}{2} = 1$.

When $m = -2$, $n = \frac{2}{-2} = -1$.

When $m = 1$, $n = \frac{2}{1} = 2$.

When $m = -1$, $n = \frac{2}{-1} = -2$.

The pairs $(2, 1)$, $(-2, -1)$, $(1, 2)$, and $(-1, -2)$ check. They are the solutions.

46. $ab - b^2 = -4,\quad (1)$

$ab - 2b^2 = -6\quad (2)$

$$\begin{aligned} ab - b^2 &= -4 \\ -ab + 2b^2 &= 6 \quad \text{Multiplying (2) by } -1 \\ \hline b^2 &= 2 \\ b &= \pm\sqrt{2} \end{aligned}$$

Substitute for b in equation (1) and solve for a.

When $b = \sqrt{2}$: $a(\sqrt{2}) - (\sqrt{2})^2 = -4$

$$\begin{aligned} a\sqrt{2} &= -2 \\ a &= -\frac{2}{\sqrt{2}} = -\sqrt{2} \end{aligned}$$

When $b = -\sqrt{2}$: $a(-\sqrt{2}) - (-\sqrt{2})^2 = -4$

$$\begin{aligned} -a\sqrt{2} &= -2 \\ a &= \frac{-2}{-\sqrt{2}} = \sqrt{2} \end{aligned}$$

The pairs $(-\sqrt{2}, \sqrt{2})$ and $(\sqrt{2}, -\sqrt{2})$ check.

47. $x^2 + y^2 = 5,\quad (1)$

$x - y = 8\quad (2)$

First solve equation (2) for x.

$x = y + 8\quad (3)$

Then substitute $y+8$ for x in equation (1) and solve for y.

$$(y+8)^2 + y^2 = 5$$
$$y^2 + 16y + 64 + y^2 = 5$$
$$2y^2 + 16y + 59 = 0$$
$$y = \frac{-16 \pm \sqrt{(16)^2 - 4(2)(59)}}{2 \cdot 2}$$
$$y = \frac{-16 \pm \sqrt{-216}}{4}$$
$$y = \frac{-16 \pm 6i\sqrt{6}}{4}$$
$$y = -4 \pm \frac{3}{2}i\sqrt{6}$$

Now substitute these numbers in equation (3) and solve for x.

$$x = -4 + \frac{3}{2}i\sqrt{6} + 8 = 4 + \frac{3}{2}i\sqrt{6}$$
$$x = -4 - \frac{3}{2}i\sqrt{6} + 8 = 4 - \frac{3}{2}i\sqrt{6}$$

The pairs $\left(4 + \frac{3}{2}i\sqrt{6}, -4 + \frac{3}{2}i\sqrt{6}\right)$ and $\left(4 - \frac{3}{2}i\sqrt{6}, -4 - \frac{3}{2}i\sqrt{6}\right)$ check. They are the solutions.

48. $4x^2 + 9y^2 = 36,$

$y - x = 8$

$y = x + 8$

$$4x^2 + 9(x+8)^2 = 36$$
$$4x^2 + 9(x^2 + 16x + 64) = 36$$
$$4x^2 + 9x^2 + 144x + 576 = 36$$
$$13x^2 + 144x + 540 = 0$$
$$x = \frac{-144 \pm \sqrt{(144)^2 - 4(13)(540)}}{2 \cdot 13}$$
$$x = \frac{-72 \pm 6i\sqrt{51}}{13} = -\frac{72}{13} \pm \frac{6}{13}i\sqrt{51}$$
$$y = -\frac{72}{13} + \frac{6}{13}i\sqrt{51} + 8 = \frac{32}{13} + \frac{6}{13}i\sqrt{51}$$
$$y = -\frac{72}{13} - \frac{6}{13}i\sqrt{51} + 8 = \frac{32}{13} - \frac{6}{13}i\sqrt{51}$$

The pairs $\left(-\frac{72}{13} + \frac{6}{13}i\sqrt{51}, \frac{32}{13} + \frac{6}{13}i\sqrt{51}\right)$ and $\left(-\frac{72}{13} - \frac{6}{13}i\sqrt{51}, \frac{32}{13} - \frac{6}{13}i\sqrt{51}\right)$ check.

49. $a^2 + b^2 = 14,\quad (1)$

$ab = 3\sqrt{5}\quad (2)$

Solve equation (2) for b.

$$b = \frac{3\sqrt{5}}{a}$$

Substitute $\frac{3\sqrt{5}}{a}$ for b in equation (1) and solve for a.

$$a^2 + \left(\frac{3\sqrt{5}}{a}\right)^2 = 14$$
$$a^2 + \frac{45}{a^2} = 14$$
$$a^4 + 45 = 14a^2$$
$$a^4 - 14a^2 + 45 = 0$$
$$u^2 - 14u + 45 = 0 \quad \text{Letting } u = a^2$$
$$(u-9)(u-5) = 0$$
$$u = 9 \quad or \quad u = 5$$
$$a^2 = 9 \quad or \quad a^2 = 5$$
$$a = \pm 3 \quad or \quad a = \pm\sqrt{5}$$

Since $b = 3\sqrt{5}/a$, if $a = 3$, $b = \sqrt{5}$; if $a = -3$, $b = -\sqrt{5}$; if $a = \sqrt{5}$, $b = 3$; and if $a = -\sqrt{5}$, $b = -3$. The pairs $(3, \sqrt{5})$, $(-3, -\sqrt{5})$, $(\sqrt{5}, 3)$, $(-\sqrt{5}, -3)$ check. They are the solutions.

50. $x^2 + xy = 5,\quad (1)$

$2x^2 + xy = 2\quad (2)$

$$\begin{aligned} -x^2 - xy &= -5 \quad \text{Multiplying (1) by } -1 \\ 2x^2 + xy &= 2 \\ \hline x^2 &= -3 \\ x &= \pm i\sqrt{3} \end{aligned}$$

Substitute for x in equation (1) and solve for y.

When $x = i\sqrt{3}$: $(i\sqrt{3})^2 + (i\sqrt{3})y = 5$

$$i\sqrt{3}y = 8$$
$$y = \frac{8}{i\sqrt{3}}$$
$$y = -\frac{8i\sqrt{3}}{3}$$

When $x = -i\sqrt{3}$: $(-i\sqrt{3})^2 + (-i\sqrt{3})y = 5$

$$-i\sqrt{3}y = 8$$
$$y = -\frac{8}{i\sqrt{3}}$$
$$y = -\frac{8i\sqrt{3}}{3}$$

The pairs $\left(i\sqrt{3}, -\frac{8i\sqrt{3}}{3}\right)$ and $\left(-i\sqrt{3}, \frac{8i\sqrt{3}}{3}\right)$.

51. $x^2 + y^2 = 25,\quad (1)$

$9x^2 + 4y^2 = 36\quad (2)$

$-4x^2 - 4y^2 = -100$ Multiplying (1) by -4

$$\begin{aligned} 9x^2 + 4y^2 &= 36 \\ \hline 5x^2 &= -64 \\ x^2 &= -\frac{64}{5} \\ x &= \pm\sqrt{\frac{-64}{5}} = \pm\frac{8i}{\sqrt{5}} \\ x &= \pm\frac{8i\sqrt{5}}{5} \quad \text{Rationalizing the denominator} \end{aligned}$$

Substituting $\frac{8i\sqrt{5}}{5}$ or $-\frac{8i\sqrt{5}}{5}$ for x in equation (1) and solving for y gives us

$$\begin{aligned} -\frac{64}{5} + y^2 &= 25 \\ y^2 &= \frac{189}{5} \\ y &= \pm\sqrt{\frac{189}{5}} = \pm 3\sqrt{\frac{21}{5}} \\ y &= \pm\frac{3\sqrt{105}}{5}. \quad \text{Rationalizing the denominator} \end{aligned}$$

The pairs $\left(\frac{8i\sqrt{5}}{5}, \frac{3\sqrt{105}}{5}\right)$, $\left(-\frac{8i\sqrt{5}}{5}, \frac{3\sqrt{105}}{5}\right)$, $\left(\frac{8i\sqrt{5}}{5}, -\frac{3\sqrt{105}}{5}\right)$, and $\left(-\frac{8i\sqrt{5}}{5}, -\frac{3\sqrt{105}}{5}\right)$ check.

They are the solutions.

52. $x^2 + y^2 = 1,$ (1)

$9x^2 - 16y^2 = 144$ (2)

$16x^2 + 16y^2 = 16$ Multiplying (1) by 16

$$\begin{aligned} 9x^2 - 16y^2 &= 144 \\ \hline 25x^2 &= 160 \\ x^2 &= \frac{160}{25} \\ x &= \pm\frac{4\sqrt{10}}{5} \end{aligned}$$

Substituting for x in equation (1) and solving for y gives us

$$\begin{aligned} \frac{160}{25} + y^2 &= 1 \\ y^2 &= -\frac{135}{25} \\ y &= \pm\sqrt{-\frac{135}{25}} \\ y &= \pm\frac{3i\sqrt{15}}{5}. \end{aligned}$$

The pairs $\left(\frac{4\sqrt{10}}{5}, \frac{3i\sqrt{15}}{5}\right)$, $\left(\frac{4\sqrt{10}}{5}, -\frac{3i\sqrt{15}}{5}\right)$, $\left(-\frac{4\sqrt{10}}{5}, \frac{3i\sqrt{15}}{5}\right)$, and $\left(-\frac{4\sqrt{10}}{5}, -\frac{3i\sqrt{15}}{5}\right)$ check.

53. $5y^2 - x^2 = 1,$ (1)

$xy = 2$ (2)

Solve equation (2) for x.

$$x = \frac{2}{y}$$

Substitute $\frac{2}{y}$ for x in equation (1) and solve for y.

$$\begin{aligned} 5y^2 - \left(\frac{2}{y}\right)^2 &= 1 \\ 5y^2 - \frac{4}{y^2} &= 1 \\ 5y^4 - 4 &= y^2 \\ 5y^4 - y^2 - 4 &= 0 \\ 5u^2 - u - 4 &= 0 \quad \text{Letting } u = y^2 \\ (5u + 4)(u - 1) &= 0 \end{aligned}$$

$$\begin{aligned} 5u + 4 = 0 \quad &\textit{or} \quad u - 1 = 0 \\ u = -\frac{4}{5} \quad &\textit{or} \quad u = 1 \\ y^2 = -\frac{4}{5} \quad &\textit{or} \quad y^2 = 1 \\ y = \pm\frac{2i}{\sqrt{5}} \quad &\textit{or} \quad y = \pm 1 \\ y = \pm\frac{2i\sqrt{5}}{5} \quad &\textit{or} \quad y = \pm 1 \end{aligned}$$

Since $x = 2/y$, if $y = \frac{2i\sqrt{5}}{5}$, $x = \frac{2}{\frac{2i\sqrt{5}}{5}} = \frac{5}{i\sqrt{5}} = \frac{5}{i\sqrt{5}} \cdot \frac{-i\sqrt{5}}{-i\sqrt{5}} = -i\sqrt{5}$; if $y = -\frac{2i\sqrt{5}}{5}$, $x = \frac{2}{-\frac{2i\sqrt{5}}{5}} = i\sqrt{5}$;

if $y = 1$, $x = 2/1 = 2$; if $y = -1$, $x = 2/-1 = -2$.

The pairs $\left(-i\sqrt{5}, \frac{2i\sqrt{5}}{5}\right)$, $\left(i\sqrt{5}, -\frac{2i\sqrt{5}}{5}\right)$, $(2, 1)$ and $(-2, -1)$ check. They are the solutions.

54. $x^2 - 7y^2 = 6,$

$xy = 1$

$$y = \frac{1}{x}$$

$$\begin{aligned} x^2 - 7\left(\frac{1}{x}\right)^2 &= 6 \\ x^2 - \frac{7}{x^2} &= 6 \\ x^4 - 7 &= 6x^2 \\ x^4 - 6x^2 - 7 &= 0 \\ u^2 - 6u - 7 &= 0 \quad \text{Letting } u = x^2 \\ (u - 7)(u + 1) &= 0 \end{aligned}$$

$u = 7 \quad or \quad u = -1$

$x^2 = 7 \quad or \quad x^2 = -1$

$x = \pm\sqrt{7} \quad or \quad x = \pm i$

Since $y = 1/x$, if $x = \sqrt{7}$, $y = 1/\sqrt{7} = \sqrt{7}/7$; if $x = -\sqrt{7}$, $y = 1/(-\sqrt{7}) = -\sqrt{7}/7$; if $x = i$, $y = 1/i = -i$, if $x = -i$, $y = 1/(-i) = i$.

The pairs $\left(\sqrt{7}, \frac{\sqrt{7}}{7}\right)$, $\left(-\sqrt{7}, -\frac{\sqrt{7}}{7}\right)$, $(i, -i)$, and $(-i, i)$ check.

55. ***Familiarize.*** We first make a drawing. We let l and w represent the length and width, respectively.

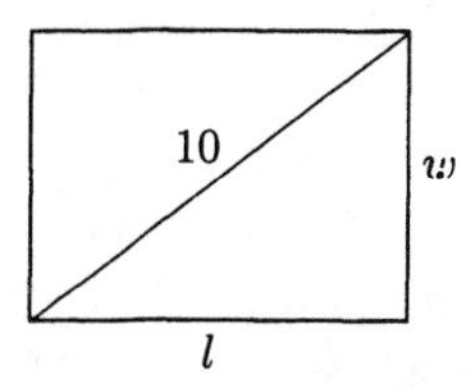

Translate. The perimeter is 28 cm.

$2l + 2w = 28$, or $l + w = 14$

Using the Pythagorean theorem we have another equation.

$l^2 + w^2 = 10^2$, or $l^2 + w^2 = 100$

Carry out. We solve the system:

$l + w = 14, \quad (1)$

$l^2 + w^2 = 100 \quad (2)$

First solve equation (1) for w.

$w = 14 - l \quad (3)$

Then substitute $14-l$ for w in equation (2) and solve for l.

$$l^2 + w^2 = 100$$
$$l^2 + (14 - l)^2 = 100$$
$$l^2 + 196 - 28l + l^2 = 100$$
$$2l^2 - 28l + 96 = 0$$
$$l^2 - 14l + 48 = 0$$
$$(l - 8)(l - 6) = 0$$

$l = 8$ or $l = 6$

If $l = 8$, then $w = 14 - 8$, or 6. If $l = 6$, then $w = 14 - 6$, or 8. Since the length is usually considered to be longer than the width, we have the solution $l = 8$ and $w = 6$, or $(8, 6)$.

Check. If $l = 8$ and $w = 6$, then the perimeter is $2 \cdot 8 + 2 \cdot 6$, or 28. The length of a diagonal is $\sqrt{8^2 + 6^2}$, or $\sqrt{100}$, or 10. The numbers check.

State. The length is 8 cm, and the width is 6 cm.

56. Let l and w represent the length and width, respectively. Solve the system:

$2l + 2w = 6,$

$l^2 + w^2 = (\sqrt{5})^2$, or

$l + w = 3$

$l^2 + w^2 = 5$

The solutions are $(1, 2)$ and $(2, 1)$. Choosing the larger number as the length, we have the solution. The length is 2 m, and the width is 1 m.

57. ***Familiarize.*** We first make a drawing. Let l = the length and w = the width of the rectangle.

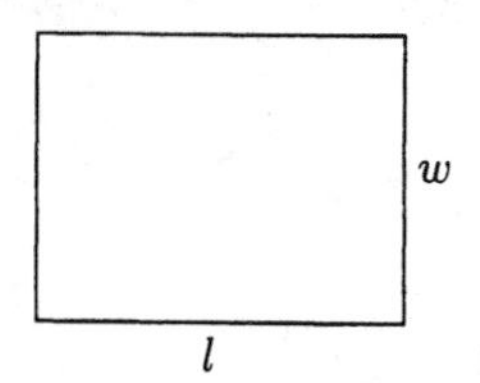

Translate.

Area: $lw = 20$

Perimeter: $2l + 2w = 18$, or $l + w = 9$

Carry out. We solve the system:

Solve the second equation for l: $l = 9 - w$

Substitute $9 - w$ for l in the first equation and solve for w.

$$(9 - w)w = 20$$
$$9w - w^2 = 20$$
$$0 = w^2 - 9w + 20$$
$$0 = (w - 5)(w - 4)$$

$w = 5$ or $w = 4$

If $w = 5$, then $l = 9 - w$, or 4. If $w = 4$, then $l = 9 - 4$, or 5. Since length is usually considered to be longer than width, we have the solution $l = 5$ and $w = 4$, or $(5, 4)$.

Check. If $l = 5$ and $w = 4$, the area is $5 \cdot 4$, or 20. The perimeter is $2 \cdot 5 + 2 \cdot 4$, or 18. The numbers check.

State. The length is 5 in. and the width is 4 in.

58. Let l and w represent the length and width, respectively. We solve the system:

$lw = 2,$

$2l + 2w = 6$

The solutions are $(1, 2)$ and $(2, 1)$. We choose the larger number to be the length, so the length is 2 yd and the width is 1 yd.

59. ***Familiarize.*** We first make a drawing. Let l = the length and w = the width.

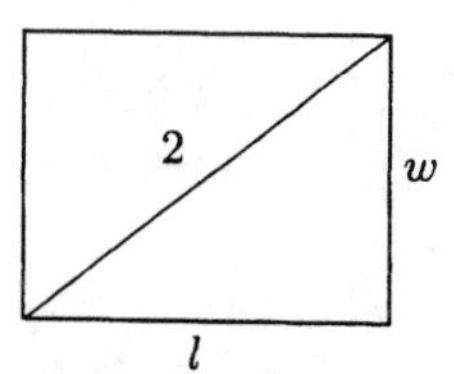

Translate.

Area: $lw = \sqrt{3} \quad (1)$

From the Pythagorean theorem: $l^2 + w^2 = 2^2 \quad (2)$

Carry out. We solve the system of equations.

We first solve equation (1) for w.

$$lw = \sqrt{3}$$
$$w = \frac{\sqrt{3}}{l}$$

Then we substitute $\frac{\sqrt{3}}{l}$ for w in equation 2 and solve for l.

$$l^2 + \left(\frac{\sqrt{3}}{l}\right)^2 = 4$$
$$l^2 + \frac{3}{l^2} = 4$$
$$l^4 + 3 = 4l^2$$
$$l^4 - 4l^2 + 3 = 0$$
$$u^2 - 4u + 3 = 0 \quad \text{Letting } u = l^2$$
$$(u-3)(u-1) = 0$$
$$u = 3 \quad \text{or} \quad u = 1$$

We now substitute l^2 for u and solve for l.

$$l^2 = 3 \quad \text{or} \quad l^2 = 1$$
$$l = \pm\sqrt{3} \quad \text{or} \quad l = \pm 1$$

Measurements cannot be negative, so we only need to consider $l = \sqrt{3}$ and $l = 1$. Since $w = \sqrt{3}/l$, if $l = \sqrt{3}$, $w = 1$ and if $l = 1$, $w = \sqrt{3}$. Length is usually considered to be longer than width, so we have the solution $l = \sqrt{3}$ and $w = 1$, or $(\sqrt{3}, 1)$.

Check. If $l = \sqrt{3}$ and $w = 1$, the area is $\sqrt{3}\cdot 1 = \sqrt{3}$. Also $(\sqrt{3})^2 + 1^2 = 3 + 1 = 4 = 2^2$. The numbers check.

State. The length is $\sqrt{3}$ m, and the width is 1 m.

60. Let l = the length and w = the width. Solve the system

$$lw = \sqrt{2},$$
$$l^2 + w^2 = (\sqrt{3})^2.$$

The solutions are $(\sqrt{2}, 1)$, $(-\sqrt{2}, -1)$, $(1, \sqrt{2})$, and $(-1, -\sqrt{2})$. Only the pairs $(\sqrt{2}, 1)$ and $(1, \sqrt{2})$ have meaning in this problem. Since length is usually considered to be longer than width, the length is $\sqrt{2}$ m, and the width is 1 m.

61. ***Familiarize***. We make a drawing of the dog run. Let l = the length and w = the width.

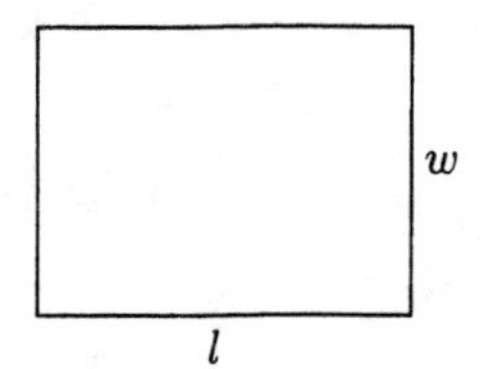

Since it takes 210 yd of fencing to enclose the run, we know that the perimeter is 210 yd.

Translate.

Perimeter: $2l + 2w = 210$, or $l + w = 105$

Area: $lw = 2250$

Carry out. We solve the system:

Solve the first equation for l: $l = 105 - w$

Substitute $105 - w$ for l in the second equation and solve for w.

$$(105 - w)w = 2250$$
$$105w - w^2 = 2250$$
$$0 = w^2 - 105w + 2250$$
$$0 = (w - 30)(w - 75)$$
$$w = 30 \quad \text{or} \quad w = 75$$

If $w = 30$, then $l = 105 - 30$, or 75. If $w = 75$, then $l = 105 - 75$, or 30. Since length is usually considered to be longer than width, we have the solution $l = 75$ and $w = 30$, or $(75, 30)$.

Check. If $l = 75$ and $w = 30$, the perimeter is $2\cdot 75 + 2\cdot 30$, or 210. The area is 75(30), or 2250. The numbers check.

State. The length is 75 yd and the width is 30 yd.

62. Let l and w represent the length and width, respectively. Solve the system:

$$\sqrt{l^2 + w^2} = l + 1,$$
$$\sqrt{l^2 + w^2} = 2w + 3$$

The solutions are $(12, 5)$ and $(0, -1)$. Only $(12, 5)$ has meaning in this problem. It checks. The length is 12 ft and the width is 5 ft.

63. ***Familiarize***. We let x = the length of a side of one test plot and y = the length of a side of the other plot. Make a drawing.

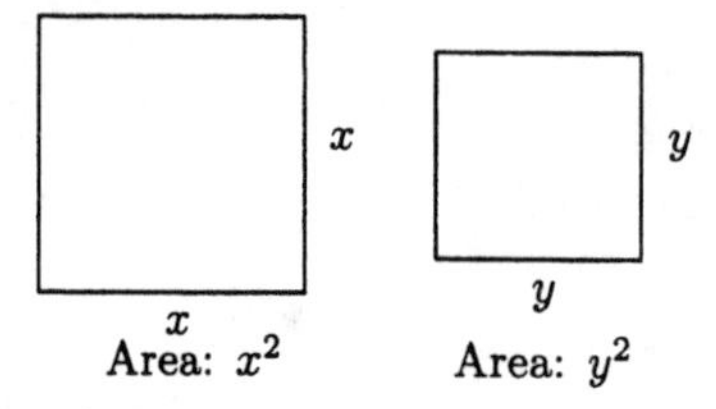

Translate.

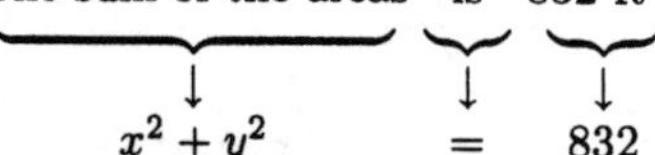

The difference of the areas is 320 ft².

$$x^2 - y^2 \quad = \quad 320$$

Carry out. We solve the system of equations.

$$x^2 + y^2 = 832$$
$$x^2 - y^2 = 320$$
$$2x^2 = 1152 \quad \text{Adding}$$
$$x^2 = 576$$
$$x = \pm 24$$

Since measurements cannot be negative, we consider only $x = 24$. Substitute 24 for x in the first equation and solve for y.

$$24^2 + y^2 = 832$$
$$576 + y^2 = 832$$
$$y^2 = 256$$
$$y = \pm 16$$

Again, we consider only the positive value, 16. The possible solution is $(24, 16)$.

Check. The areas of the test plots are 24^2, or 576, and 16^2, or 256. The sum of the areas is $576 + 256$, or 832. The difference of the areas is $576 - 256$, or 320. The values check.

State. The lengths of the test plots are 24 ft and 16 ft.

64. Let $p =$ the principal and $r =$ the interest rate. Solve the system:

$$pr = 7.5,$$
$$(p + 25)(r - 0.01) = 7.5$$

The solutions are $(125, 0.06)$ and $(-1.50, -0.05)$. Only $(125, 0.06)$ has meaning in this problem. The principal was \$125 and the interest rate was 0.06, or 6%.

65. a) $I(x) = 21.15508287x + 198.9151934$

b) $E(x) = 197.3380928(1.053908223)^x$

c) Find the first coordinate of the point of intersection of the graphs of $y_1 = I(x)$ and $y_2 = E(x)$. It is approximately 25, so income will equal expenditures about 25 years after 1985.

66. a) $M(x) = 65.50776523(0.9888648008)^x$

b) $F(x) = 0.6080952381x + 35.05833333$

c) Find the first coordinate of the point of intersection of the graphs of $y_1 = M(x)$ and $y_2 = F(x)$. It is approximately 24, so equal percents of men and women earned degrees about 24 years after 1960.

67. We can only find the real-number solutions graphically.

68. Answers will vary.

69.
$$2^{3x} = 64$$
$$2^{3x} = 2^6$$
$$3x = 6$$
$$x = 2$$

The solution is 2.

70.
$$5^x = 27$$
$$\ln 5^x = \ln 27$$
$$x \ln 5 = \ln 27$$
$$x = \frac{\ln 27}{\ln 5}$$
$$x \approx 2.048$$

71.
$$\log_3 x = 4$$
$$x = 3^4$$
$$x = 81$$

The solution is 81.

72.
$$\log(x - 3) + \log x = 1$$
$$\log(x - 3)(x) = 1$$
$$x^2 - 3x = 10$$
$$x^2 - 3x - 10 = 0$$
$$(x - 5)(x + 2) = 0$$
$$x = 5 \quad or \quad x = -2$$

Only 5 checks.

73. $(x - h)^2 + (y - k)^2 = r^2$

If $(2, 4)$ is a point on the circle, then

$(2 - h)^2 + (4 - k)^2 = r^2$.

If $(3, 3)$ is a point on the circle, then

$(3 - h)^2 + (3 - k)^2 = r^2$.

Thus

$$(2 - h)^2 + (4 - k)^2 = (3 - h)^2 + (3 - k)^2$$
$$4 - 4h + h^2 + 16 - 8k + k^2 =$$
$$9 - 6h + h^2 + 9 - 6k + k^2$$
$$-4h - 8k + 20 = -6h - 6k + 18$$
$$2h - 2k = -2$$
$$h - k = -1$$

If the center (h, k) is on the line $3x - y = 3$, then $3h - k = 3$.

Solving the system

$$h - k = -1,$$
$$3h - k = 3$$

we find that $(h, k) = (2, 3)$.

Find r^2, substituting $(2, 3)$ for (h, k) and $(2, 4)$ for (x, y). We could also use $(3, 3)$ for (x, y).

$$(x - h)^2 + (y - k)^2 = r^2$$
$$(2 - 2)^2 + (4 - 3)^2 = r^2$$
$$0 + 1 = r^2$$
$$1 = r^2$$

The equation of the circle is 1 is $(x-2)^2+(y-3)^2 = 1$.

74. Let (h, k) represent the point on the line $5x+8y = -2$ which is the center of a circle that passes through the points $(-2, 3)$ and $(-4, 1)$. The distance between (h, k) and $(-2, 3)$ is the same as the distance between (h, k) and $(-4, 1)$. This gives us one equation:

$$\sqrt{[h-(-2)]^2+(k-3)^2} = \sqrt{[h-(-4)]^2+(k-1)^2}$$
$$(h + 2)^2 + (k - 3)^2 = (h + 4)^2 + (k - 1)^2$$
$$h^2+4h+4+k^2-6k+9 = h^2+8h+16+k^2-2k+1$$
$$4h - 6k + 13 = 8h - 2k + 17$$
$$-4h - 4k = 4$$
$$h + k = -1$$

We get a second equation by substituting (h, k) in $5x + 8y = -2$.

$$5h + 8k = -2$$

We now solve the following system:

$$h + k = -1,$$
$$5h + 8k = -2$$

The solution, which is the center of the circle, is $(-2, 1)$.

Next we find the length of the radius. We can find the distance between either $(-2, 3)$ or $(-4, 1)$ and the center $(-2, 1)$. We use $(-2, 3)$.

$$r = \sqrt{[-2-(-2)]^2 + (1-3)^2}$$
$$r = \sqrt{0^2 + (-2)^2}$$
$$r = \sqrt{4} = 2$$

We can write the equation of the circle with center $(-2, 1)$ and radius 2.

$$(x-h)^2 + (y-k)^2 = r^2$$
$$[x-(-2)]^2 + (y-1)^2 = 2^2$$
$$(x+2)^2 + (y-1)^2 = 4$$

75. The equation of the ellipse is of the form $\frac{x^2}{a^2} + \frac{y^2}{b^2} = 1$. Substitute $\left(1, \frac{\sqrt{3}}{2}\right)$ and $\left(\sqrt{3}, \frac{1}{2}\right)$ for (x, y) to get two equations.

$$\frac{1^2}{a^2} + \frac{\left(\frac{\sqrt{3}}{2}\right)^2}{b^2} = 1, \text{ or } \frac{1}{a^2} + \frac{3}{4b^2} = 1$$

$$\frac{(\sqrt{3})^2}{a^2} + \frac{\left(\frac{1}{2}\right)^2}{b^2} = 1, \text{ or } \frac{3}{a^2} + \frac{1}{4b^2} = 1$$

Substitute u for $\frac{1}{a^2}$ and v for $\frac{1}{b^2}$.

$$u + \frac{3}{4}v = 1, \qquad 4u + 3v = 4,$$
$$\text{or}$$
$$3u + \frac{1}{4}v = 1 \qquad 12u + v = 4$$

Solving for u and v, we get $u = \frac{1}{4}$, $v = 1$. Then $u = \frac{1}{a^2} = \frac{1}{4}$, so $a^2 = 4$; $v = \frac{1}{b^2} = 1$, so $b^2 = 1$.

Then the equation of the ellipse is

$$\frac{x^2}{4} + \frac{y^2}{1} = 1, \text{ or } \frac{x^2}{4} + y^2 = 1.$$

76. $\frac{x^2}{a^2} - \frac{y^2}{b^2} = 1$

Substitute each ordered pair for (x, y).

$$\frac{(-3)^2}{a^2} - \frac{\left(-\frac{3\sqrt{5}}{2}\right)^2}{b^2} = 1,$$

$$\frac{(-3)^2}{a^2} - \frac{\left(\frac{3\sqrt{5}}{b^2}\right)^2}{b^2} = 1,$$

$$\frac{\left(-\frac{3}{2}\right)^2}{a^2} - \frac{0^2}{b^2} = 1$$

$$\frac{9}{a^2} - \frac{45}{4b^2} = 1, \qquad (1)$$

$$\frac{9}{a^2} - \frac{45}{4b^2} = 1, \qquad (2)$$

$$\frac{9}{4a^2} = 1 \qquad (3)$$

Note that equation (1) and equation (2) are identical. Multiply both sides of equation (3) by 4:

$$\frac{9}{a^2} = 4$$

Substitute 4 for $\frac{9}{a^2}$ in equation (1) and solve for b^2.

$$4 - \frac{45}{4b^2} = 1$$
$$16b^2 - 45 = 4b^2$$
$$12b^2 = 45$$
$$b^2 = \frac{45}{12}, \text{ or } \frac{15}{4}$$

Solve equation (3) for a^2.

$$\frac{9}{4a^2} = 1$$
$$\frac{9}{4} = a^2$$

The equation of the hyperbola is $\frac{x^2}{9/4} - \frac{y^2}{15/4} = 1$.

77. $(x-h)^2 + (y-k)^2 = r^2$ Standard form

Substitute $(4, 6)$, $(-6, 2)$, and $(1, -3)$ for (x, y).

$$(4-h)^2 + (6-k)^2 = r^2 \qquad (1)$$
$$(-6-h)^2 + (2-k)^2 = r^2 \qquad (2)$$
$$(1-h)^2 + (-3-k)^2 = r^2 \qquad (3)$$

Thus

$(4-h)^2 + (6-k)^2 = (-6-h)^2 + (2-k)^2$, or

$5h + 2k = 3$

and

$(4-h)^2 + (6-k)^2 = (1-h)^2 + (-3-k)^2$, or

$h + 3k = 7$.

We solve the system

$$5h + 2k = 3,$$
$$h + 3k = 7.$$

Solving we get $h = -\frac{5}{13}$ and $k = \frac{32}{13}$. Substituting these values in equation (1), (2), or (3), we find that $r^2 = \frac{5365}{169}$.

The equation of the circle is
$\left(x+\frac{5}{13}\right)^2+\left(y-\frac{32}{13}\right)^2=\frac{5365}{169}$.

78. Using $(x-h)^2+(y-k)^2=r^2$ and the given points, we have

$$(2-h)^2+(3-k)^2=r^2 \quad (1)$$
$$(4-h)^2+(5-k)^2=r^2 \quad (2)$$
$$(0-h)^2+(-3-k)^2=r^2 \quad (3)$$

Then equation (1) − equation (2) gives $h+k=7$ and equation (2) − equation (3) gives $h+2k=4$. We solve this system:

$$h+k=7,$$
$$h+2k=4.$$

Then $h=10$, $k=-3$, $r=10$ and the equation of the circle is $(x-10)^2+(y+3)^2=10^2$.

79. See the answer section in the text.

80. Let x and y represent the numbers. Solve:

$$xy=2,$$
$$\frac{1}{x}+\frac{1}{y}=\frac{33}{8}.$$

The solutions are $\left(\frac{1}{4},8\right)$ and $\left(8,\frac{1}{4}\right)$. In either case the numbers are $\frac{1}{4}$ and 8.

81. ***Familiarize***. Let x and y represent the numbers.

Translate.

The square of a certain number exceeds twice the square of another number by $\frac{1}{8}$.

$$x^2=2y^2+\frac{1}{8}$$

The sum of the squares is $\frac{5}{16}$.

$$x^2+y^2=\frac{5}{16}$$

Carry out. We solve the system.

$$x^2-2y^2=\frac{1}{8}, \quad (1)$$
$$x^2+y^2=\frac{5}{16} \quad (2)$$

$$x^2-2y^2=\frac{1}{8},$$
$$2x^2+2y^2=\frac{5}{8} \quad \text{Multiplying (2) by 2}$$
$$3x^2=\frac{6}{8}$$
$$x^2=\frac{1}{4}$$
$$x=\pm\frac{1}{2}$$

Substitute $\pm\frac{1}{2}$ for x in (2) and solve for y.

$$\left(\pm\frac{1}{2}\right)^2+y^2=\frac{5}{16}$$
$$\frac{1}{4}+y^2=\frac{5}{16}$$
$$y^2=\frac{1}{16}$$
$$y=\pm\frac{1}{4}$$

We get $\left(\frac{1}{2},\frac{1}{4}\right)$, $\left(-\frac{1}{2},\frac{1}{4}\right)$, $\left(\frac{1}{2},-\frac{1}{4}\right)$ and $\left(-\frac{1}{2},-\frac{1}{4}\right)$.

Check. It is true that $\left(\pm\frac{1}{2}\right)^2$ exceeds twice $\left(\pm\frac{1}{4}\right)^2$ by $\frac{1}{8}$: $\frac{1}{4}=2\left(\frac{1}{16}\right)+\frac{1}{8}$

Also $\left(\pm\frac{1}{2}\right)^2+\left(\pm\frac{1}{4}\right)^2=\frac{5}{16}$. The pairs check.

State. The numbers are $\frac{1}{2}$ and $\frac{1}{4}$ or $-\frac{1}{2}$ and $\frac{1}{4}$ or $\frac{1}{2}$ and $-\frac{1}{4}$ or $-\frac{1}{2}$ and $-\frac{1}{4}$.

82. Make a drawing.

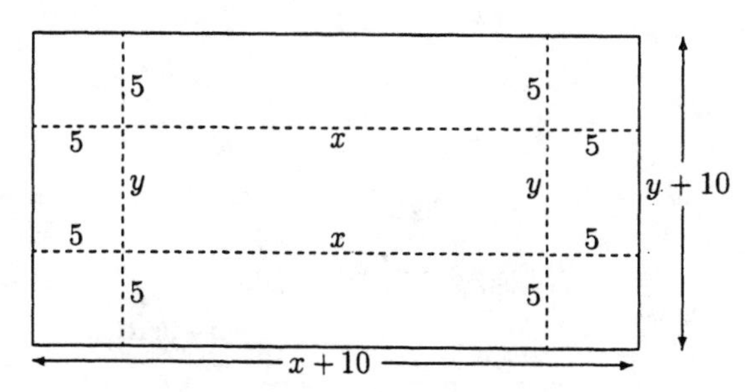

We let x and y represent the length and width of the base of the box, respectively. Then the dimensions of the metal sheet are $x+10$ and $y+10$.

Solve the system

$$(x+10)(y+10)=340,$$
$$x\cdot y\cdot 5=350.$$

The solutions are $(10,7)$ and $(7,10)$. The dimensions of the box are 10 in. by 7 in. by 5 in.

83. See the answer section in the text.

84. $x^2-y^2=a^2-b^2, \quad (1)$

$x-y=a-b \quad (2)$

Solve equation (2) for x.

$x=y+a-b \quad (3)$

Substitute for x in equation (1) and solve for y.

$$(y+a-b)^2-y^2=a^2-b^2$$
$$y^2+2ay-2by+a^2-2ab+b^2-y^2=a^2-b^2$$
$$2ay-2by=2ab-2b^2$$
$$2y(a-b)=2b(a-b)$$
$$y=b$$

Substitute for y in equation (3) and solve for x.

$x = b + a - b = a$

The pair (a, b) checks.

85. $x^3 + y^3 = 72$, (1)

$x + y = 6$ (2)

Solve equation (2) for y: $y = 6 - x$

Substitute for y in equation (1) and solve for x.

$$x^3 + (6 - x)^3 = 72$$
$$x^3 + 216 - 108x + 18x^2 - x^3 = 72$$
$$18x^2 - 108x + 144 = 0$$
$$x^2 - 6x + 8 = 0$$

Multiplying by $\frac{1}{18}$

$$(x - 4)(x - 2) = 0$$

$x = 4$ or $x = 2$

If $x = 4$, then $y = 6 - 4 = 2$.

If $x = 2$, then $y = 6 - 2 = 4$.

The pairs $(4, 2)$ and $(2, 4)$ check.

86. $a + b = \frac{5}{6}$, (1)

$\frac{a}{b} + \frac{b}{a} = \frac{13}{6}$ (2)

$b = \frac{5}{6} - a = \frac{5 - 6a}{6}$ Solving equation (1) for b

$$\frac{a}{\frac{5-6a}{6}} + \frac{\frac{5-6a}{6}}{a} = \frac{13}{6}$$ Substituting for b in equation (2)

$$\frac{6a}{5 - 6a} + \frac{5 - 6a}{6a} = \frac{13}{6}$$
$$36a^2 + 25 - 60a + 36a^2 = 65a - 78a^2$$
$$150a^2 - 125a + 25 = 0$$
$$6a^2 - 5a + 1 = 0$$
$$(3a - 1)(2a - 1) = 0$$

$a = \frac{1}{3}$ or $a = \frac{1}{2}$

Substitute for a and solve for b.

When $a = \frac{1}{3}$, $b = \frac{5 - 6\left(\frac{1}{3}\right)}{6} = \frac{1}{2}$.

When $a = \frac{1}{2}$, $b = \frac{5 - 6\left(\frac{1}{2}\right)}{6} = \frac{1}{3}$.

The pairs $\left(\frac{1}{3}, \frac{1}{2}\right)$ and $\left(\frac{1}{2}, \frac{1}{3}\right)$ check. They are the solutions.

87. $p^2 + q^2 = 13$, (1)

$\frac{1}{pq} = -\frac{1}{6}$ (2)

Solve equation (2) for p.

$$\frac{1}{q} = -\frac{p}{6}$$
$$-\frac{6}{q} = p$$

Substitute $-6/q$ for p in equation (1) and solve for q.

$$\left(-\frac{6}{q}\right)^2 + q^2 = 13$$
$$\frac{36}{q^2} + q^2 = 13$$
$$36 + q^4 = 13q^2$$
$$q^4 - 13q^2 + 36 = 0$$
$$u^2 - 13u + 36 = 0 \quad \text{Letting } u = q^2$$
$$(u - 9)(u - 4) = 0$$
$$u = 9 \text{ or } u = 4$$
$$x^2 = 9 \text{ or } x^2 = 4$$
$$x = \pm 3 \text{ or } x = \pm 2$$

Since $p = -6/q$, if $q = 3$, $p = -2$; if $q = -3$, $p = 2$; if $q = 2$, $p = -3$; and if $q = -2$, $p = 3$. The pairs $(-2, 3)$, $(2, -3)$, $(-3, 2)$, and $(3, -2)$ check. They are the solutions.

88. $x^2 + y^2 = 4$, (1)

$(x - 1)^2 + y^2 = 4$ (2)

Solve equation (1) for y^2.

$y^2 = 4 - x^2$ (3)

Substitute for $4 - x^2$ for y^2 in equation (2) and solve for x.

$$(x - 1)^2 + (4 - x^2) = 4$$
$$x^2 - 2x + 1 + 4 - x^2 = 4$$
$$-2x = -1$$
$$x = \frac{1}{2}$$

Substitute $\frac{1}{2}$ for x in equation (3) and solve for y.

$$y^2 = 4 - \left(\frac{1}{2}\right)^2$$
$$y^2 = 4 - \frac{1}{4}$$
$$y^2 = \frac{15}{4}$$
$$y = \pm\frac{\sqrt{15}}{2}$$

The pairs $\left(\frac{1}{2}, \frac{\sqrt{15}}{2}\right)$ and $\left(\frac{1}{2}, -\frac{\sqrt{15}}{2}\right)$ check. They are the solutions.

89. $5^{x+y} = 100$,

$3^{2x-y} = 1000$

$(x + y) \log 5 = 2$, Taking logarithms and

$(2x - y) \log 3 = 3$ simplifying

$x\log 5 + y\log 5 = 2,\quad (1)$

$2x\log 3 - y\log 3 = 3\quad (2)$

Multiply equation (1) by $\log 3$ and equation (2) by $\log 5$ and add.

$$\begin{array}{l} x\log 3\cdot\log 5 + y\log 3\cdot\log 5 = 2\log 3 \\ \underline{2x\log 3\cdot\log 5 - y\log 3\cdot\log 5 = 3\log 5} \\ 3x\log 3\cdot\log 5 = 2\log 3 + 3\log 5 \end{array}$$

$$x = \frac{2\log 3 + 3\log 5}{3\log 3\cdot\log 5}$$

Substitute in (1) to find y.

$$\frac{2\log 3 + 3\log 5}{3\log 3\cdot\log 5}\cdot\log 5 + y\log 5 = 2$$

$$y\log 5 = 2 - \frac{2\log 3 + 3\log 5}{3\log 3}$$

$$y\log 5 = \frac{6\log 3 - 2\log 3 - 3\log 5}{3\log 3}$$

$$y\log 5 = \frac{4\log 3 - 3\log 5}{3\log 3}$$

$$y = \frac{4\log 3 - 3\log 5}{3\log 3\cdot\log 5}$$

The pair $\left(\frac{2\log 3 + 3\log 5}{3\log 3\cdot\log 5}, \frac{4\log 3 - 3\log 5}{3\log 3\cdot\log 5}\right)$ checks. It is the solution.

90. $e^x - e^{x+y} = 0,\quad (1)$

$e^y - e^{x-y} = 0\quad (2)$

Factor (1): $e^x(1 - e^y) = 0$

$e^x = 0$ or $1 - e^y = 0$

No solution $\quad y = 0$

Substitute in (2).

$$e^0 - e^{x-0} = 0$$

$$1 - e^x = 0$$

$$x = 0$$

The solution is $(0, 0)$.

91. Find the points of intersection of $y_1 = \ln x + 2$ and $y_2 = x^2$. They are $(1.564, 2.448)$ and $(0.138, 0.019)$.

92. Graph $y_1 = \ln(x+4)$, $y_2 = \sqrt{6 - x^2}$, and $y_3 = -\sqrt{6 - x^2}$ and find the points of intersection. They are $(1.720, 1.744)$ and $(-2.405, 0.467)$.

93. Find the point of intersection of $y_1 = e^x - 1$ and $y_2 = -3x + 4$. It is $(0.871, 1.388)$.

94. Find the point of intersection of $y_1 = e^{-x} + 1$ and $y_2 = 2x + 5$. It is $(-0.841, 3.318)$.

95. Find the points of intersection of $y_1 = e^x$ and $y_2 = x + 2$. They are $(1.146, 3.146)$ and $(-1.841, 0.159)$.

96. Find the points of intersection of $y_1 = e^{-x}$ and $y_2 = 3 - x$. They are $(2.948, 0.052)$ and $(-1.505, 4.505)$.

97. Graph $y_1 = \sqrt{19,380,510.36 - x^2}$, $y_2 = -\sqrt{19,380,510.36 - x^2}$, and $y_3 = 27,941.25x/6.125$ and find the points of intersection. They are $(0.965, 4402.33)$ and $(-0.965, -4402.33)$.

98. Find the points of intersection of $y = (1660 - 2x)/2$ (or $y = 830 - x$) and $y = 35,325/x$. They are $(785, 45)$ and $(45, 785)$.

99. Graph $y_1 = \sqrt{\frac{14.5x^2 - 64.5}{13.5}}$, $y_2 = -\sqrt{\frac{14.5x^2 - 64.5}{13.5}}$, and $y_3 = (5.5x - 12.3)/6.3$ and find the points of intersection. They are $(2.112, -0.109)$ and $(-13.041, -13.337)$.

100. Find the points of intersection of $y = -15.6/(13.5x)$ and $y_2 = (5.6x - 42.3)/6.7$. They are $(7.366, -0.157)$ and $(0.188, -6.157)$.

101. Graph $y_1 = \sqrt{\frac{56,548 - 0.319x^2}{2688.7}}$, $y_2 = -\sqrt{\frac{56,548 - 0.319x^2}{2688.7}}$, $y_3 = \sqrt{\frac{0.306x^2 - 43,452}{2688.7}}$, and $y_4 = -\sqrt{\frac{0.306x^2 - 43,452}{2688.7}}$ and find the points of intersection. They are $(400, 1.431)$, $(-400, 1.431)$, $(400, -1.431)$, and $(-400, -1.431)$.

102. Graph $y_1 = \sqrt{\frac{6408 - 18.465x^2}{788.723}}$, $y_2 = \sqrt{\frac{6408 - 18.465x^2}{788.723}}$, $y_3 = \sqrt{\frac{106.535x^2 - 2692}{788.723}}$, and $y_4 = \sqrt{\frac{106.535x^2 - 2692}{788.723}}$ and find the points of intersection. They are $(8.532, 2.534)$, $(8.532, -2.534)$, $(-8.532, 2.534)$, and $(-8.532, -2.534)$.

Chapter 7

Sequences, Series, and Combinatorics

Exercise Set 7.1

1. $a_n = 4n - 1$
$a_1 = 4 \cdot 1 - 1 = 3,$
$a_2 = 4 \cdot 2 - 1 = 7,$
$a_3 = 4 \cdot 3 - 1 = 11,$
$a_4 = 4 \cdot 4 - 1 = 15;$
$a_{10} = 4 \cdot 10 - 1 = 39;$
$a_{15} = 4 \cdot 15 - 1 = 59$

2. $a_1 = (1-1)(1-2)(1-3) = 0,$
$a_2 = (2-1)(2-2)(2-3) = 0,$
$a_3 = (3-1)(3-2)(3-3) = 0,$
$a_4 = (4-1)(4-2)(4-3) = 3 \cdot 2 \cdot 1 = 6;$
$a_{10} = (10-1)(10-2)(10-3) = 9 \cdot 8 \cdot 7 = 504;$
$a_{15} = (15-1)(15-2)(15-3) = 14 \cdot 13 \cdot 12 = 2184$

3. $a_n = \dfrac{n}{n-1},\ n \geq 2$
The first 4 terms are a_2, a_3, a_4, and a_5:
$a_2 = \dfrac{2}{2-1} = 2,$
$a_3 = \dfrac{3}{3-1} = \dfrac{3}{2},$
$a_4 = \dfrac{4}{4-1} = \dfrac{4}{3},$
$a_5 = \dfrac{5}{5-1} = \dfrac{5}{4};$
$a_{10} = \dfrac{10}{10-1} = \dfrac{10}{9};$
$a_{15} = \dfrac{15}{15-1} = \dfrac{15}{14}$

4. $a_3 = 3^2 - 1 = 8,$
$a_4 = 4^2 - 1 = 15,$
$a_5 = 5^2 - 1 = 24,$
$a_6 = 6^2 - 1 = 35;$
$a_{10} = 10^2 - 1 = 99;$
$a_{15} = 15^2 - 1 = 224$

5. $a_n = \dfrac{n^2-1}{n^2+1},$
$a_1 = \dfrac{1^2-1}{1^2+1} = 0,$
$a_2 = \dfrac{2^2-1}{2^2+1} = \dfrac{3}{5},$
$a_3 = \dfrac{3^2-1}{3^2+1} = \dfrac{8}{10} = \dfrac{4}{5},$
$a_4 = \dfrac{4^2-1}{4^2+1} = \dfrac{15}{17};$
$a_{10} = \dfrac{10^2-1}{10^2+1} = \dfrac{99}{101};$
$a_{15} = \dfrac{15^2-1}{15^2+1} = \dfrac{224}{226} = \dfrac{112}{113}$

6. $a_1 = \left(-\dfrac{1}{2}\right)^{1-1} = 1,$
$a_2 = \left(-\dfrac{1}{2}\right)^{2-1} = -\dfrac{1}{2},$
$a_3 = \left(-\dfrac{1}{2}\right)^{3-1} = \dfrac{1}{4},$
$a_4 = \left(-\dfrac{1}{2}\right)^{4-1} = -\dfrac{1}{8},$
$a_{10} = \left(-\dfrac{1}{2}\right)^{10-1} = -\dfrac{1}{512};$
$a_{15} = \left(-\dfrac{1}{2}\right)^{15-1} = \dfrac{1}{16,384}$

7. $a_n = (-1)^n n^2$
$a_1 = (-1)^1 1^2 = -1,$
$a_2 = (-1)^2 2^2 = 4,$
$a_3 = (-1)^3 3^2 = -9,$
$a_4 = (-1)^4 4^2 = 16;$
$a_{10} = (-1)^{10} 10^2 = 100;$
$a_{15} = (-1)^{15} 15^2 = -225$

8. $a_1 = (-1)^{1-1}(3 \cdot 1 - 5) = -2,$
$a_2 = (-1)^{2-1}(3 \cdot 2 - 5) = -1,$
$a_3 = (-1)^{3-1}(3 \cdot 3 - 5) = 4,$
$a_4 = (-1)^{4-1}(3 \cdot 4 - 5) = -7;$
$a_{10} = (-1)^{10-1}(3 \cdot 10 - 5) = -25,$
$a_{15} = (-1)^{15-1}(3 \cdot 15 - 5) = 40$

9. $a_n = 5 + \frac{(-2)^{n+1}}{2^n}$

$a_1 = 5 + \frac{(-2)^{1+1}}{2^1} = 5 + \frac{4}{2} = 7,$

$a_2 = 5 + \frac{(-2)^{2+1}}{2^2} = 5 + \frac{-8}{4} = 3,$

$a_3 = 5 + \frac{(-2)^{3+1}}{2^3} = 5 + \frac{16}{8} = 7,$

$a_4 = 5 + \frac{(-2)^{4+1}}{2^4} = 5 + \frac{-32}{16} = 3;$

$a_{10} = 5 + \frac{(-2)^{10+1}}{2^{10}} = 5 + \frac{-1 \cdot 2^{11}}{2^{10}} = 3;$

$a_{15} = 5 + \frac{(-2)^{15+1}}{2^{15}} = 5 + \frac{2^{16}}{2^{15}} = 7$

10. $a_1 = \frac{2 \cdot 1 - 1}{1^2 + 2 \cdot 1} = \frac{1}{3},$

$a_2 = \frac{2 \cdot 2 - 1}{2^2 + 2 \cdot 2} = \frac{3}{8},$

$a_3 = \frac{2 \cdot 3 - 1}{3^2 + 2 \cdot 3} = \frac{5}{15} = \frac{1}{3},$

$a_4 = \frac{2 \cdot 4 - 1}{4^2 + 2 \cdot 4} = \frac{7}{24};$

$a_{10} = \frac{2 \cdot 10 - 1}{10^2 + 2 \cdot 10} = \frac{19}{120};$

$a_{15} = \frac{2 \cdot 15 - 1}{15^2 + 2 \cdot 15} = \frac{29}{255}$

11. $a_n = 5n - 6$

$a_8 = 5 \cdot 8 - 6 = 40 - 6 = 34$

12. $a_7 = (3 \cdot 7 - 4)(2 \cdot 7 + 5) = 17 \cdot 19 = 323$

13. $a_n = (2n+3)^2$

$a_6 = (2 \cdot 6 + 3)^2 = 225$

14. $a_{12} = (-1)^{12-1}[4.6(12) - 18.3] = -36.9$

15. $a_n = 5n^2(4n - 100)$

$a_{11} = 5(11)^2(4 \cdot 11 - 100) = 5(121)(-56) =$
$-33,880$

16. $a_{80} = \left(1 + \frac{1}{80}\right)^2 = \left(\frac{81}{80}\right)^2 = \frac{6561}{6400}$

17. $a_n = \ln e^n$

$a_{67} = \ln e^{67} = 67$

18. $a_{100} = 2 - \frac{1000}{100} = 2 - 10 = -8$

19. 2, 4, 6, 8, 10,. . .

These are the even integers, so the general term might be $2n$.

20. 3^n

21. $-2, 6, -18, 54, \ldots$

We can see a pattern if we write the sequence as

$-1 \cdot 2 \cdot 1, 1 \cdot 2 \cdot 3, -1 \cdot 2 \cdot 9, 1 \cdot 2 \cdot 27, \ldots$

The general term might be $(-1)^n 2(3)^{n-1}$.

22. $5n - 7$

23. $\frac{2}{3}, \frac{3}{4}, \frac{4}{5}, \frac{5}{6}, \frac{6}{7}, \ldots$

These are fractions in which the denominator is 1 greater than the numerator. Also, each numerator is 1 greater than the preceding numerator. The general term might be $\frac{n+1}{n+2}$.

24. $\sqrt{2n}$

25. $1 \cdot 2, 2 \cdot 3, 3 \cdot 4, 4 \cdot 5, \ldots$

These are the products of pairs of consecutive natural numbers. The general term might be $n(n+1)$.

26. $-1 - 3(n-1)$, or $-3n + 2$, or $-(3n - 2)$

27. $0, \log 10, \log 100, \log 1000, \ldots$

We can see a pattern if we write the sequence as

$\log 1, \log 10, \log 100, \log 1000, \ldots$

The general term might be $\log 10^{n-1}$. This is equivalent to $n - 1$.

28. $\ln e^{n+1}$, or $n + 1$

29. 1, 2, 3, 4, 5, 6, 7, . . .

$S_3 = 1 + 2 + 3 = 6$

$S_7 = 1 + 2 + 3 + 4 + 5 + 6 + 7 = 28$

30. $S_2 = 1 - 3 = -2$

$S_8 = 1 - 3 + 5 - 7 + 9 = 5$

31. 2, 4, 6, 8, . . .

$S_4 = 2 + 4 + 6 + 8 = 20$

$S_5 = 2 + 4 + 6 + 8 + 10 = 30$

32. $S_1 = 1$

$S_5 = 1 + \frac{1}{4} + \frac{1}{9} + \frac{1}{16} + \frac{1}{25} = \frac{5269}{3600}$

33.
$$\sum_{k=1}^{5} \frac{1}{2k} = \frac{1}{2 \cdot 1} + \frac{1}{2 \cdot 2} + \frac{1}{2 \cdot 3} + \frac{1}{2 \cdot 4} + \frac{1}{2 \cdot 5}$$
$$= \frac{1}{2} + \frac{1}{4} + \frac{1}{6} + \frac{1}{8} + \frac{1}{10}$$
$$= \frac{60}{120} + \frac{30}{120} + \frac{20}{120} + \frac{15}{120} + \frac{12}{120}$$
$$= \frac{137}{120}$$

34. $\frac{1}{3} + \frac{1}{5} + \frac{1}{7} + \frac{1}{9} + \frac{1}{11} + \frac{1}{13} = \frac{43,024}{45,045}$

35. $\sum_{i=0}^{6} 2^i = 2^0+2^1+2^2+2^3+2^4+2^5+2^6$
$= 1+2+4+8+16+32+64$
$= 127$

36. $\sqrt{7}+\sqrt{9}+\sqrt{11}+\sqrt{13} \approx 12.5679$

37. $\sum_{k=7}^{10} \ln k = \ln 7 + \ln 8 + \ln 9 + \ln 10 =$
$\ln(7\cdot 8\cdot 9\cdot 10) = \ln 5040 \approx 8.5252$

38. $\pi + 2\pi + 3\pi + 4\pi = 10\pi \approx 31.4159$

39. $\sum_{k=1}^{8} \frac{k}{k+1} = \frac{1}{1+1}+\frac{2}{2+1}+\frac{3}{3+1}+\frac{4}{4+1}+$
$\frac{5}{5+1}+\frac{6}{6+1}+\frac{7}{7+1}+\frac{8}{8+1}$
$= \frac{1}{2}+\frac{2}{3}+\frac{3}{4}+\frac{4}{5}+\frac{5}{6}+\frac{6}{7}+\frac{7}{8}+\frac{8}{9}$
$= \frac{15,551}{2520}$

40. $\frac{1-1}{1+3}+\frac{2-1}{2+3}+\frac{3-1}{3+3}+\frac{4-1}{4+3}+\frac{5-1}{5+3}$
$= 0+\frac{1}{5}+\frac{2}{6}+\frac{3}{7}+\frac{4}{8}$
$= 0+\frac{1}{5}+\frac{1}{3}+\frac{3}{7}+\frac{1}{2}$
$= \frac{307}{210}$

41. $\sum_{i=1}^{5} (-1)^i$
$= (-1)^1+(-1)^2+(-1)^3+(-1)^4+(-1)^5$
$= -1+1-1+1-1$
$= -1$

42. $-1+1-1+1-1+1=0$

43. $\sum_{k=1}^{8} (-1)^{k+1}3k$
$= (-1)^2 3\cdot 1 + (-1)^3 3\cdot 2 + (-1)^4 3\cdot 3+$
$(-1)^5 3\cdot 4 + (-1)^6 3\cdot 5 + (-1)^7 3\cdot 6+$
$(-1)^8 3\cdot 7 + (-1)^9 3\cdot 8$
$= 3-6+9-12+15-18+21-24$
$= -12$

44. $4-4^2+4^3-4^4+4^5-4^6+4^7-4^8 = -52,428$

45. $\sum_{k=0}^{6} \frac{2}{k^2+1} = \frac{2}{0^2+1}+\frac{2}{1^2+1}+\frac{2}{2^2+1}+\frac{2}{3^2+1}+$
$\frac{2}{4^2+1}+\frac{2}{5^2+1}+\frac{2}{6^2+1}$
$= 2+1+\frac{2}{5}+\frac{2}{10}+\frac{2}{17}+\frac{2}{26}+\frac{2}{37}$
$= 2+1+\frac{2}{5}+\frac{1}{5}+\frac{2}{17}+\frac{1}{13}+\frac{2}{37}$
$= \frac{157,351}{40,885}$

46. $1\cdot2+2\cdot3+3\cdot4+4\cdot5+5\cdot6+6\cdot7+7\cdot8+8\cdot9+9\cdot10+10\cdot11 = 440$

47. $\sum_{k=0}^{5} (k^2-2k+3)$
$= (0^2-2\cdot0+3)+(1^2-2\cdot1+3)+$
$(2^2-2\cdot2+3)+(3^2-2\cdot3+3)+$
$(4^2-2\cdot4+3)+(5^2-2\cdot5+3)$
$= 3+2+3+6+11+18$
$= 43$

48. $\frac{1}{1\cdot2}+\frac{1}{2\cdot3}+\frac{1}{3\cdot4}+\frac{1}{4\cdot5}+\frac{1}{5\cdot6}+$
$\frac{1}{6\cdot7}+\frac{1}{7\cdot8}+\frac{1}{8\cdot9}+\frac{1}{9\cdot10}+\frac{1}{10\cdot11}$
$= \frac{1}{2}+\frac{1}{6}+\frac{1}{12}+\frac{1}{20}+\frac{1}{30}+\frac{1}{42}+\frac{1}{56}+$
$\frac{1}{72}+\frac{1}{90}+\frac{1}{110}$
$= \frac{10}{11}$

49. $\sum_{i=0}^{10} \frac{2i}{2^i+1}$
$= \frac{2^0}{2^0+1}+\frac{2^1}{2^1+1}+\frac{2^2}{2^2+1}+\frac{2^3}{2^3+1}+\frac{2^4}{2^4+1}+$
$\frac{2^5}{2^5+1}+\frac{2^6}{2^6+1}+\frac{2^7}{2^7+1}+\frac{2^8}{2^8+1}+\frac{2^9}{2^9+1}+$
$\frac{2^{10}}{2^{10}+1}$
$= \frac{1}{2}+\frac{2}{3}+\frac{4}{5}+\frac{8}{9}+\frac{16}{17}+\frac{32}{33}+\frac{64}{65}+\frac{128}{129}+$
$\frac{256}{257}+\frac{512}{513}+\frac{1024}{1025}$
≈ 9.736

50. $(-2)^0+(-2)^2+(-2)^4+(-2)^6 = 1+4+16+64 = 85$

51. $5+10+15+20+25+\ldots$

This is a sum of multiples of 5, and it is an infinite series. Sigma notation is

$$\sum_{k=1}^{\infty} 5k.$$

52. $\sum_{k=1}^{\infty} 7k$

53. $2-4+8-16+32-64$

This is a sum of powers of 2 with alternating signs. Sigma notation is

$$\sum_{k=1}^{6} (-1)^{k+1}2k, \text{ or } \sum_{k=1}^{6} (-1)^{k-1}2k$$

54. $\sum_{k=1}^{5} 3k$

55. $-\frac{1}{2}+\frac{2}{3}-\frac{3}{4}+\frac{4}{5}-\frac{5}{6}+\frac{6}{7}$

This is a sum of fractions in which the denominator is one greater than the numerator. Also, each numerator is 1 greater than the preceding numerator and the signs alternate. Sigma notation is

$$\sum_{k=1}^{6} (-1)^k \frac{k}{k+1}.$$

56. $\sum_{k=1}^{5} \frac{1}{k^2}$

57. $4-9+16-25+\ldots+(-1)^n n^2$

This is a sum of terms of the form $(-1)^k k^2$, beginning with $k=2$ and continuing through $k=n$. Sigma notation is

$$\sum_{k=2}^{n} (-1)^k k^2.$$

58. $\sum_{k=3}^{n} (-1)^{k+1} k^2$, or $\sum_{k=3}^{n} (-1)^{k-1} k^2$

59. $\frac{1}{1\cdot 2}+\frac{1}{2\cdot 3}+\frac{1}{3\cdot 4}+\frac{1}{4\cdot 5}+\ldots$

This is a sum of fractions in which the numerator is 1 and the denominator is a product of two consecutive integers. The larger integer in each product is the smaller integer in the succeeding product. It is an infinite series. Sigma notation is

$$\sum_{k=1}^{\infty} \frac{1}{k(k+1)}.$$

60. $\sum_{k=1}^{\infty} \frac{1}{k(k+1)^2}$

61. $a_1 = 4, \qquad a_{k+1} = 1 + \frac{1}{a_k}$

$a_2 = 1 + \frac{1}{4} = 1\frac{1}{4}$, or $\frac{5}{4}$

$a_3 = 1 + \frac{1}{\frac{5}{4}} = 1 + \frac{4}{5} = 1\frac{4}{5}$, or $\frac{9}{5}$

$a_4 = 1 + \frac{1}{\frac{9}{5}} = 1 + \frac{5}{9} = 1\frac{5}{9}$, or $\frac{14}{9}$

62. $a_1 = 256,\ a_2 = \sqrt{256} = 16,\ a_3 = \sqrt{16} = 4,$
$a_4 = \sqrt{4} = 2$

63. $a_1 = 6561, \qquad a_{k+1} = (-1)^k \sqrt{a_k}$

$a_2 = (-1)^1\sqrt{6561} = -81$

$a_3 = (-1)^2\sqrt{-81} = 9i$

$a_4 = (-1)^3\sqrt{9i} = -3\sqrt{i}$

64. $a_1 = e^Q,\ a_2 = \ln e^Q = Q,\ a_3 = \ln Q,\ a_4 = \ln(\ln Q)$

65. $a_1 = 2, \qquad a_{k+1} = a_k + a_{k-1}$

$a_2 = 3$

$a_3 = 3 + 2 = 5$

$a_4 = 5 + 3 = 8$

66. $a_1 = -10,\ a_2 = 8,\ a_3 = 8 - (-10) = 18,$
$a_4 = 18 - 8 = 10$

67. See the answer section in the text.

68.

n	Un
1	.41421
2	.31784
3	.26795
4	.23607
5	.21342
6	.19626
7	.18268
8	.17157
9	.16228
10	.15435

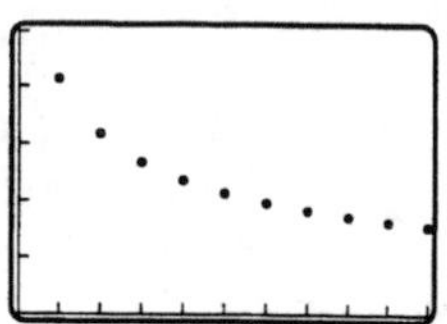

69. See the answer section in the text.

70.

n	Un
1	2
2	1.5
3	1.4167
4	1.4142
5	1.4142
6	1.4142
7	1.4142
8	1.4142
9	1.4142
10	1.4142

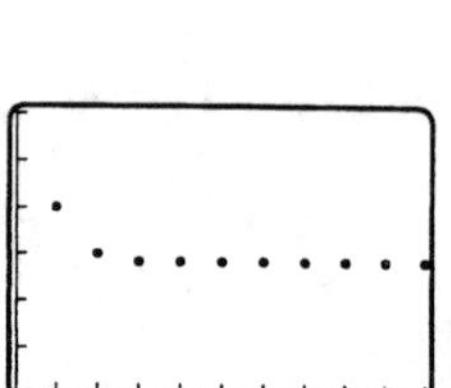

71. a) $a_1 = \$1000(1.062)^1 = \$1062$

$a_2 = \$1000(1.062)^2 \approx \$1127.84$

$a_3 = \$1000(1.062)^3 \approx \$1197.77$

$a_4 = \$1000(1.062)^4 \approx \$1272.03$

$a_5 = \$1000(1.062)^5 \approx \$1350.90$

$a_6 = \$1000(1.062)^6 \approx \$1434.65$

$a_7 = \$1000(1.062)^7 \approx \$1523.60$

$a_8 = \$1000(1.062)^8 \approx \$1618.07$

$a_9 = \$1000(1.062)^9 \approx \$1718.39$

$a_{10} = \$1000(1.062)^{10} \approx \$1824.93$

b) $a_{20} = \$1000(1.062)^{20} \approx \$3330.35$

72. Find each term by multiplying the preceding term by 0.75:

$5200, $3900, $2925, $2193.75, $1645.31,

$1233.98, $925.49, $694.12, $520.59, $390.44

73. Find each term by multiplying the preceding term by 2. Find 17 terms, beginning with $a_1 = 1$, since there are 16 fifteen minute periods in 4 hr.

1, 2, 4, 8, 16, 32, 64, 128, 256, 512, 1024,

2048, 4096, 8192, 16,384, 32,768, 65,536

74. Find each term by adding $0.40 to the preceding term:

$6.30, $6.70, $7.10, $7.50, $7.90, $8.30,

$8.70, $9.10, $9.50, $9.90

75. a)

n	Un
15	27.25
16	26.108
17	25.002
18	23.932
19	22.898
20	21.9
21	20.938
22	20.012
23	19.122
24	18.268

b)

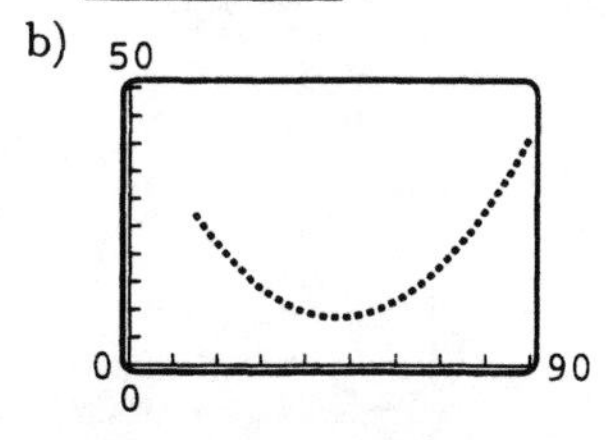

c) From the table in part (a), we know that $a_{16} = 26.108 \approx 26$ and $a_{23} = 19.122 \approx 19$. Using a table set in Ask mode we find that $a_{50} = 8.7 \approx 9$, $a_{75} = 22.45 \approx 22$, and $a_{85} = 34.25 \approx 34$.

76. a) Using the quadratic regression operation on a grapher we get

$a_n = -17.748n^2 + 1784.945n - 23,768.848$

b) $a \approx \$246.78$

$a_{20} \approx \$4830.85$

$a_{40} \approx \$19,232.15$

$a_{60} \approx \$19,435.05$

$a_{75} \approx \$10,269.53$

(Answers may vary slightly due to rounding differences.)

77. $a_1 = 1$ (Given)

$a_2 = 1$ (Given)

$a_3 = a_2 + a_1 = 1 + 1 = 2$

$a_4 = a_3 + a_2 = 2 + 1 = 3$

$a_5 = a_4 + a_3 = 3 + 2 = 5$

$a_6 = a_5 + a_4 = 5 + 3 = 8$

$a_7 = a_6 + a_5 = 8 + 5 = 13$

78. a) $a_1 = 41$, $a_2 = 43$, $a_3 = 47$, $a_4 = 53$; the terms fit the pattern $41, 41+2, 41+6, 41+12, \cdots$, or $41+1\cdot 0, 41+2\cdot 1, 41+3\cdot 2, 41+4\cdot 3, \cdots$, or $41 + n(n-1)$.

b) Yes; $n^2 - n + 41 = n(n-1) + 41 = 41 + n(n-1)$.

79. For a Fibonacci sequence, as $n \to \infty$, $a_{n+1}/a_n \to \tau \approx 1.618033989$ where τ is a ratio called the golden section. It is the ratio that results when a line is divided so that the ratio of the larger part to the smaller part is equal to the ratio of the whole to the larger part.

80. $3x - 2y = 3$, (1)

$2x + 3y = -11$ (2)

Multiply equation (1) by 3 and equation (2) by 2 and add.

$$\begin{aligned} 9x - 6y &= 9 \\ 4x + 6y &= -22 \\ \hline 13x \quad &= -13 \\ x &= -1 \end{aligned}$$

Back-substitute to find y.

$2(-1) + 3y = -11$ Using equation (2)

$-2 + 3y = -11$

$3y = -9$

$y = -3$

The solution is $(-1, -3)$.

81. ***Familiarize***. Let $x =$ the amount spent on original research and development and $y =$ the amount spent on purchased research and development.

Translate. A total of $1.19 billion was spent, so we have one equation:

$x + y = 1.19$

We also know that the amount spent on original research and development is $0.837 billion more than the amount spent on purchased research and development. This gives us another equation:

$x = y + 0.837$

We have a system of equations:

$x + y = 1.19$,

$x = y + 0.837$

Carry out. We use the substitution method.

$$(y+0.837)+y=1.19$$
$$2y+0.837=1.19$$
$$2y=0.352$$
$$y=0.1765$$

Back-substitute to find x.

$$x=0.1765+0.837=1.0135$$

Check. The total spent is \$1.0135 billion + \$0.837 billion, or \$1.19 billion. Also, \$1.0135 billion is \$0.837 billion more than \$0.1765 billion. The answer checks.

State. In 1998 \$1.0135 billion was spent on original research and development and \$0.1765 was spent on purchased research and development.

82.
$$x^2+y^2-6x+4y=3$$
$$x^2-6x+9+y^2+4y+4=3+9+4$$
$$(x-3)^2+(y+2)^2=16$$

Center: $(3,-2)$; radius: 4

83. We complete the square twice.

$$x^2+y^2+5x-8y=2$$
$$x^2+5x+y^2-8y=2$$
$$x^2+5x+\frac{25}{4}+y^2-8y+16=2+\frac{25}{4}+16$$
$$\left(x+\frac{5}{2}\right)^2+(y-4)^2=\frac{97}{4}$$
$$\left[x-\left(-\frac{5}{2}\right)\right]^2+(y-4)^2=\left(\frac{\sqrt{97}}{2}\right)^2$$

The center is $\left(-\frac{5}{2},4\right)$ and the radius is $\frac{\sqrt{97}}{2}$.

84. $a_n=\frac{1}{2^n}\log\ 1000^n$

$a_1=\frac{1}{2^1}\log\ 1000^1=\frac{1}{2}\log\ 10^3=\frac{1}{2}\cdot 3=\frac{3}{2}$

$a_2=\frac{1}{2^2}\log\ 1000^2=\frac{1}{4}\log\ (10^3)^2=\frac{1}{4}\log\ 10^6=\frac{1}{4}\cdot 6=\frac{3}{2}$

$a_3=\frac{1}{2^3}\log\ 1000^3=\frac{1}{8}\log\ (10^3)^3=\frac{1}{8}\log\ 10^9=\frac{1}{8}\cdot 9=\frac{9}{8}$

$a_4=\frac{1}{2^4}\log\ 1000^4=\frac{1}{16}\log\ (10^3)^4=\frac{1}{16}\log\ 10^{12}=\frac{1}{16}\cdot 12=\frac{3}{4}$

$a_5=\frac{1}{2^5}\log\ 1000^5=\frac{1}{32}\log\ (10^3)^5=\frac{1}{32}\log\ 10^{15}=\frac{1}{32}\cdot 15=\frac{15}{32}$

$S_5=\frac{3}{2}+\frac{3}{2}+\frac{9}{8}+\frac{3}{4}+\frac{15}{32}=\frac{171}{32}$

85. $a_n=i^n$

$a_1=i$

$a_2=i^2=-1$

$a_3=i^3=-i$

$a_4=i^4=1$

$a_5=i^5=i^4\cdot i=i$

$S_5=i-1-i+1+i=i$

86. $a_n=\ln(1\cdot 2\cdot 3\cdots n)$

$a_1=\ln\ 1=0$

$a_2=\ln(1\cdot 2)=\ln\ 2$

$a_3=\ln(1\cdot 2\cdot 3)=\ln\ 6$

$a_4=\ln(1\cdot 2\cdot 3\cdot 4)=\ln\ 24$

$a_5=\ln(1\cdot 2\cdot 3\cdot 4\cdot 5)=\ln\ 120$

$S_5=0+\ln\ 2+\ln\ 6+\ln\ 24+\ln\ 120$
$=\ln(2\cdot 6\cdot 24\cdot 120)=\ln\ 34,560\approx 10.450$

87. $S_n=\ln 1+\ln+2\ln 3+\cdots+\ln n$
$=\ln(1\cdot 2\cdot 3\cdots n)$

88. $S_n=\left(1-\frac{1}{2}\right)+\left(\frac{1}{2}-\frac{1}{3}\right)+\left(\frac{1}{3}-\frac{1}{4}\right)+\cdots+\left(\frac{1}{n-1}-\frac{1}{n}\right)+\left(\frac{1}{n}-\frac{1}{n+1}\right)$
$=1-\frac{1}{n+1}=\frac{n}{n+1}$

Exercise Set 7.2

1. 3, 8, 13, 18, . . .

$a_1=3$

$d=5\quad (8-3=5,\ 13-8=5,\ 18-13=5)$

2. $a_1=\$1.08$, $d=\$0.08$, $(\$1.16-\$1.08=\$0.08)$

3. 9, 5, 1, −3, . . .

$a_1=9$

$d=-4\quad (5-9=-4,\ 1-5=-4,\ -3-1=-4)$

4. $a_1=-8$, $d=3$ $(-5-(-8)=3)$

5. $\frac{3}{2},\frac{9}{4},3,\frac{15}{4},\ldots$

$a_1=\frac{3}{2}$

$d=\frac{3}{4}\quad \left(\frac{9}{4}-\frac{3}{2}=\frac{3}{4}, 3-\frac{9}{4}=\frac{3}{4}\right)$

6. $a_1=\frac{3}{5}$, $d=-\frac{1}{2}$ $\left(\frac{1}{10}-\frac{3}{5}=-\frac{1}{2}\right)$

7. $a_1=\$316$

$d=-\$3$ $(\$313-\$316=-\$3,$
$\$310-\$313=-\$3,\ \$307-\$310=-\$3)$

8. $a_1 = 0.07,\ d = 0.05,$ and $n = 11$

$a_{11} = 0.07 + (11-1)(0.05) = 0.07 + 0.5 = 0.57$

9. 2, 6, 10, . . .

$a_1 = 2,\ d = 4,$ and $n = 12$

$a_n = a_1 + (n-1)d$

$a_{12} = 2 + (12-1)4 = 2 + 11\cdot 4 = 2 + 44 = 46$

10. $a_1 = 7,\ d = -3,$ and $n = 17$

$a_{17} = 7 + (17-1)(-3) = 7 + 16(-3) = 7 - 48 = -41$

11. $3, \frac{7}{3}, \frac{5}{3}, \ldots$

$a_1 = 3,\ d = -\frac{2}{3},$ and $n = 14$

$a_n = a_1 + (n-1)d$

$a_{14} = 3 + (14-1)\left(-\frac{2}{3}\right) = 3 - \frac{26}{3} = -\frac{17}{3}$

12. $a_1 = \$1200,\ d = \$964.32 - \$1200 = -\$235.68,$ and $n = 13$

$a_{13} = \$1200 + (13-1)(-\$235.68) = \$1200 + 12(-\$235.68) = \$1200 - \$2828.16 = -\$1628.16$

13. \$2345.78, \$2967.54, \$3589.30, . . .

$a_1 = \$2345.78,\ d = \$621.76,$ and $n = 10$

$a_n = a_1 + (n-1)d$

$a_{10} = \$2345.78 + (10-1)(\$621.76) = \$7941.62$

14.
$$106 = 2 + (n-1)(4)$$
$$106 = 2 + 4n - 4$$
$$108 = 4n$$
$$27 = n$$

The 27th term is 106.

15. $a_1 = 0.07,\ d = 0.05$

$a_n = a_1 + (n-1)d$

Let $a_n = 1.67$, and solve for n.

$$1.67 = 0.07 + (n-1)(0.05)$$
$$1.67 = 0.07 + 0.05n - 0.05$$
$$1.65 = 0.05n$$
$$33 = n$$

The 33rd term is 1.67.

16.
$$-296 = 7 + (n-1)(-3)$$
$$-296 = 7 - 3n + 3$$
$$-306 = -3n$$
$$102 = n$$

The 102nd term is -296.

17. $a_1 = 3,\ d = -\frac{2}{3}$

$a_n = a_1 + (n-1)d$

Let $a_n = -27$, and solve for n.

$$-27 = 3 + (n-1)\left(-\frac{2}{3}\right)$$
$$-81 = 9 + (n-1)(-2)$$
$$-81 = 9 - 2n + 2$$
$$-92 = -2n$$
$$46 = n$$

The 46th term is -27.

18. $a_{20} = 14 + (20-1)(-3) = 14 + 19(-3) = -43$

19. $a_n = a_1 + (n-1)d$

$33 = a_1 + (8-1)4$ Substituting 33 for a_8, 8 for n, and 4 for d

$33 = a_1 + 28$

$5 = a_1$

(Note that this procedure is equivalent to subtracting d from a_8 seven times to get a_1: $33 - 7(4) = 33 - 28 = 5$)

20.
$$26 = 8 + (11-1)d$$
$$26 = 8 + 10d$$
$$18 = 10d$$
$$1.8 = d$$

21.
$$a_n = a_1 + (n-1)d$$
$$-507 = 25 + (n-1)(-14)$$
$$-507 = 25 - 14n + 14$$
$$-546 = -14n$$
$$39 = n$$

22. We know that $a_{17} = -40$ and $a_{28} = -73$. We would have to add d eleven times to get from a_{17} to a_{28}. That is,

$$-40 + 11d = -73$$
$$11d = -33$$
$$d = -3.$$

Since $a_{17} = -40$, we subtract d sixteen times to get to a_1.

$$a_1 = -40 - 16(-3) = -40 + 48 = 8$$

We write the first five terms of the sequence:

8, 5, 2, -1, -4

23.
$$\frac{25}{3} + 15d = \frac{95}{6}$$
$$15d = \frac{45}{6}$$
$$d = \frac{1}{2}$$

$$a_1 = \frac{25}{3} - 16\left(\frac{1}{2}\right) = \frac{25}{3} - 8 = \frac{1}{3}$$

The first five terms of the sequence are $\frac{1}{3}, \frac{5}{6}, \frac{4}{3}, \frac{11}{6}, \frac{7}{3}$.

24. $a_{14} = 11 + (14-1)(-4) = 11 + 13(-4) = -41$

$$S_{14} = \frac{14}{2}[11 + (-41)] = 7(-30) = -210$$

25. $5 + 8 + 11 + 14 + \ldots$

Note that $a_1 = 5$, $d = 3$, and $n = 20$. First we find a_{20}:

$$\begin{aligned} a_n &= a_1 + (n-1)d \\ a_{20} &= 5 + (20-1)3 \\ &= 5 + 19 \cdot 3 = 62 \end{aligned}$$

Then

$$\begin{aligned} S_n &= \frac{n}{2}(a_1 + a_n) \\ S_{20} &= \frac{20}{2}(5 + 62) \\ &= 10(67) = 670. \end{aligned}$$

26. $1 + 2 + 3 + \ldots + 299 + 300.$

$$S_{300} = \frac{300}{2}(1 + 300) = 150(301) = 45,150$$

27. The sum is $2 + 4 + 6 + \ldots + 798 + 800$. This is the sum of the arithmetic sequence for which $a_1 = 2$, $a_n = 800$, and $n = 400$.

$$\begin{aligned} S_n &= \frac{n}{2}(a_1 + a_n) \\ S_{400} &= \frac{400}{2}(2 + 800) = 200(802) = 160,400 \end{aligned}$$

28. $1 + 3 + 5 + \ldots + 197 + 199.$

$$S_{100} = \frac{100}{2}(1 + 199) = 50(200) = 10,000$$

29. The sum is $7 + 14 + 21 + \ldots + 91 + 98$. This is the sum of the arithmetic sequence for which $a_1 = 7$, $a_n = 98$, and $n = 14$.

$$\begin{aligned} S_n &= \frac{n}{2}(a_1 + a_n) \\ S_{14} &= \frac{14}{2}(7 + 98) = 7(105) = 735 \end{aligned}$$

30. $16 + 20 + 24 + \ldots + 516 + 520$

$$S_{127} = \frac{127}{2}(16 + 520) = 34,036$$

31. First we find a_{20}:

$$\begin{aligned} a_n &= a_1 + (n-1)d \\ a_{20} &= 2 + (20-1)5 \\ &= 2 + 19 \cdot 5 = 97 \end{aligned}$$

Then

$$\begin{aligned} S_n &= \frac{n}{2}(a_1 + a_n) \\ S_{20} &= \frac{20}{2}(2 + 97) \\ &= 10(99) = 990. \end{aligned}$$

32. $a_{32} = 7 + (32-1)(-3) = 7 + (31)(-3) = -86$

$$S_{32} = \frac{32}{2}[7 + (-86)] = 16(-79) = -1264$$

33. $\sum_{k=1}^{40} (2k+3)$

Write a few terms of the sum:

$$5 + 7 + 9 + \ldots + 83$$

This is a series coming from an arithmetic sequence with $a_1 = 5$, $n = 40$, and $a_{40} = 83$. Then

$$\begin{aligned} S_n &= \frac{n}{2}(a_1 + a_n) \\ S_{40} &= \frac{40}{2}(5 + 83) \\ &= 20(88) = 1760 \end{aligned}$$

34. $\sum_{k=5}^{20} 8k$

$$40 + 48 + 56 + 64 + \ldots + 160$$

This is equivalent to a series coming from an arithmetic sequence with $a_1 = 40$, $n = 16$, and $n_{16} = 160$.

$$S_{16} = \frac{16}{2}(40 + 160) = 1600$$

35. $\sum_{k=0}^{19} \frac{k-3}{4}$

Write a few terms of the sum:

$$-\frac{3}{4} - \frac{1}{2} - \frac{1}{4} + 0 + \frac{1}{4} + \ldots + 4$$

Since k goes from 0 through 19, there are 20 terms. Thus, this is equivalent to a series coming from an arithmetic sequence with $a_1 = -\frac{3}{4}$, $n = 20$, and $a_{20} = 4$. Then

$$\begin{aligned} S_n &= \frac{n}{2}(a_1 + a_n) \\ S_{20} &= \frac{20}{2}\left(-\frac{3}{4} + 4\right) \\ &= 10 \cdot \frac{13}{4} = \frac{65}{2}. \end{aligned}$$

36. $\sum_{k=2}^{50} (2000 - 3k)$

$$1994 + 1991 + 1988 + \ldots + 1850$$

This is equivalent to a series coming from an arithmetic sequence with $a_1 = 1994$, $n = 49$, and $n_{49} = 1850$.

$$S_{49} = \frac{49}{2}(1994 + 1850) = 94,178$$

37. $\sum_{k=12}^{57} \frac{7-4k}{13}$

Write a few terms of the sum:

$$-\frac{41}{13} - \frac{45}{13} - \frac{49}{13} - \ldots - \frac{221}{13}$$

Since k goes from 12 through 57, there are 46 terms. Thus, this is equivalent to a series coming from an arithmetic sequence with $a_1 = -\frac{41}{13}$, $n = 46$, and $a_{46} = -\frac{221}{13}$. Then

$$S_n = \frac{n}{2}(a_1 + a_n)$$

$$S_{46} = \frac{46}{2}\left(-\frac{41}{13} - \frac{221}{13}\right)$$

$$= 23\left(-\frac{262}{13}\right) = -\frac{6026}{13}.$$

38. First find $\sum_{k=101}^{200} (1.14k - 2.8)$.

$$112.34 + 113.48 + \ldots + 225.2$$

This is equivalent to a series coming from an arithmetic sequence with $a_1 = 112.34$, $n = 100$, and $a_{100} = 225.2$.

$$S_{100} = \frac{100}{2}(112.34 + 225.2) = 16,877$$

Next find $\sum_{k=1}^{5} \left(\frac{k+4}{10}\right)$.

$$S_5 = \frac{5}{2}\left(\frac{1}{2} + \frac{9}{10}\right) = 3.5$$

Then $16,877 - 3.5 = 16,873.5$.

39. ***Familiarize***. We go from 50 poles in a row, down to six poles in the top row, so there must be 45 rows. We want the sum $50 + 49 + 48 + \ldots + 6$. Thus we want the sum of an arithmetic sequence. We will use the formula $S_n = \frac{n}{2}(a_1 + a_n)$.

Translate. We want to find the sum of the first 45 terms of an arithmetic sequence with $a_1 = 50$ and $a_{45} = 6$.

Carry out. Substituting into the formula, we have

$$S_{45} = \frac{45}{2}(50 + 6)$$

$$= \frac{45}{2} \cdot 56 = 1260$$

Check. We can do the calculation again, or we can do the entire addition:

$50 + 49 + 48 + \ldots + 6$.

State. There will be 1260 poles in the pile.

40. $a_{25} = 5000 + (25 - 1)(1125) = 5000 + 24 \cdot 1125 = 32,000$

$S_{25} = \frac{25}{2}(5000 + 32,000) = \frac{25}{2}(37,000) = \$462,500$

41. We first find how many plants will be in the last row.

Familiarize. The sequence is 35, 31, 27, ... It is an arithmetic sequence with $a_1 = 35$ and $d = -4$. Since each row must contain a positive number of plants, we must determine how many times we can add -4 to 35 and still have a positive result.

Translate. We find the largest integer x for which $35 + x(-4) > 0$. Then we evaluate the expression $35 - 4x$ for that value of x.

Carry out. We solve the inequality.

$$35 - 4x > 0$$

$$35 > 4x$$

$$\frac{35}{4} > x$$

$$8\frac{3}{4} > x$$

The integer we are looking for is 8. Thus $35 - 4x = 35 - 4(8) = 3$.

Check. If we add -4 to 35 eight times we get 3, a positive number, but if we add -4 to 35 more than eight times we get a negative number.

State. There will be 3 plants in the last row.

Next we find how many plants there are altogether.

Familiarize. We want to find the sum $35 + 31 + 27 + \ldots + 3$. We know $a_1 = 35$ $a_n = 3$, and, since we add -4 to 35 eight times, $n = 9$. (There are 8 terms after a_1, for a total of 9 terms.) We will use the formula $S_n = \frac{n}{2}(a_1 + a_n)$.

Translate. We want to find the sum of the first 9 terms of an arithmetic sequence in which $a_1 = 35$ and $a_9 = 3$.

Carry out. Substituting into the formula, we have

$$S_9 = \frac{9}{2}(35 + 3)$$

$$= \frac{9}{2} \cdot 38 = 171$$

Check. We can check the calculations by doing them again. We could also do the entire addition:

$35 + 31 + 27 + \ldots + 3$.

State. There are 171 plants altogether.

42. $a_{25} = 14 + (25 - 1)(2) = 14 + 24 \cdot 2 = 62$

$S_{25} = \frac{25}{2}(14 + 62) = \frac{25}{2}(76) = 950$

43. ***Familiarize***. We have a sequence 10, 20, 30, . . . It is an arithmetic sequence with $a_1 = 10$, $d = 10$, and $n = 31$.

Translate. We want to find $S_n = \frac{n}{2}(a_1 + a_n)$ where $a_n = a_1 + (n - 1)d$, $a_1 = 10$, $d = 10$, and $n = 31$.

Carry out. First we find a_{31}.

$a_{31} = 10 + (31 - 1)10 = 10 + 30 \cdot 10 = 310$

Then $S_{31} = \frac{31}{2}(10 + 310) = \frac{31}{2} \cdot 320 = 4960$.

Check. We can do the calculation again, or we can do the entire addition:

$10 + 20 + 30 + \ldots + 310$.

State. A total of 4960¢, or \$49.60 is saved.

44. Yes; $d = 48 - 16 = 80 - 48 = 112 - 80 = 144 - 112 = 32$.

$a_{10} = 16 + (10 - 1)32 = 304$

$S_{10} = \frac{10}{2}(16 + 304) = 1600$ ft

45. ***Familiarize***. We have arithmetic sequence with $a_1 = 28$, $d = 4$, and $n = 20$.

Translate. We want to find $S_n = \frac{n}{2}(a_1 + a_n)$ where $a_n = a_1 + (n - 1)d$, $a_1 = 28$, $d = 4$, and $n = 20$.

Carry out. First we find a_{20}.

$a_{20} = 28 + (20 - 1)4 = 104$

Then $S_{20} = \frac{20}{2}(28 + 104) = 10 \cdot 132 = 1320$.

Check. We can do the calculations again, or we can do the entire addition:

$28 + 32 + 36 + \ldots 104.$

State. There are 1320 seats in the first balcony.

46. Yes; $d = 0.6080 - 0.5908 = 0.6252 - 0.6080 = \ldots = 0.7112 - 0.6940 = 0.0172$

47. Yes; $d = 6 - 3 = 9 - 6 = 3n - 3(n - 1) = 3$.

48. The first formula can be derived from the second by substituting $a_1 + (n-1)d$ for a_n. When the first and last terms of the sum are known, the second formula is the better one to use. If the last term is not known, the first formula allows us to compute the sum in one step without first finding a_n.

49.

$$\begin{aligned}&1 + 2 + 3 + \ldots + 100\\ &= (1 + 100) + (2 + 99) + (3 + 98) + \ldots + (50 + 51)\\ &= \underbrace{101 + 101 + 101 + \ldots + 101}_{\text{50 addends of 101}}\\ &= 50 \cdot 101\\ &= 5050\end{aligned}$$

A formula for the first n natural numbers is $\frac{n}{2}(1 + n)$.

50. $7x - 2y = 4$, (1)

$x + 3y = 17$ (2)

Multiply equation (1) by 3 and equation (2) by 2 and add.

$$\begin{aligned}21x - 6y &= 12\\ 2x + 6y &= 34\\ \hline 23x \quad &= 46\\ x &= 2\end{aligned}$$

Back-substitute to find y.

$2 + 3y = 17$ Using equation (2)

$3y = 15$

$y = 5$

The solution is $(2, 5)$.

51.

$$\begin{aligned}2x + y + 3z &= 12\\ x - 3y - 2z &= -1\\ 5x + 2y - 4z &= -4\end{aligned}$$

We will use Gauss-Jordan elimination with matrices. First we write the augmented matrix.

$$\left[\begin{array}{ccc|c} 2 & 1 & 3 & 12\\ 1 & -3 & -2 & -1\\ 5 & 2 & -4 & -4\end{array}\right]$$

Next we interchange the first two rows.

$$\left[\begin{array}{ccc|c} 1 & -3 & -2 & -1\\ 2 & 1 & 3 & 12\\ 5 & 2 & -4 & -4\end{array}\right]$$

Now multiply the first row by -2 and add it to the second row. Also multiply the first row by -5 and add it to the third row.

$$\left[\begin{array}{ccc|c} 1 & -3 & -2 & -1\\ 0 & 7 & 7 & 14\\ 0 & 17 & 6 & 1\end{array}\right]$$

Multiply the second row by $\frac{1}{7}$.

$$\left[\begin{array}{ccc|c} 1 & -3 & -2 & -1\\ 0 & 1 & 1 & 2\\ 0 & 17 & 6 & 1\end{array}\right]$$

Multiply the second row by 3 and add it to the first row. Also multiply the second row by -17 and add it to the third row.

$$\left[\begin{array}{ccc|c} 1 & 0 & 1 & 5\\ 0 & 1 & 1 & 2\\ 0 & 0 & -11 & -33\end{array}\right]$$

Multiply the third row by $-\frac{1}{11}$.

$$\left[\begin{array}{ccc|c} 1 & 0 & 1 & 5\\ 0 & 1 & 1 & 2\\ 0 & 0 & 1 & 3\end{array}\right]$$

Multiply the third row by -1 and add it to the first row and also to the second row.

$$\left[\begin{array}{ccc|c} 1 & 0 & 0 & 2\\ 0 & 1 & 0 & -1\\ 0 & 0 & 1 & 3\end{array}\right]$$

Now we can read the solution from the matrix. It is $(2, -1, 3)$.

52. $9x^2 + 16y^2 = 144$

$\frac{x^2}{16} + \frac{y^2}{9} = 1$

Vertices: $(-4, 0)$, $(4, 0)$

$c^2 = a^2 - b^2 = 16 - 9 = 7$

$c = \sqrt{7}$

Foci: $(-\sqrt{7}, 0)$, $(\sqrt{7}, 0)$

53. The vertices are on the y-axis, so the transverse axis is vertical and $a = 5$. The length of the minor axis is 4, so $b = 4/2 = 2$. The equation is

$$\frac{x^2}{4} + \frac{y^2}{25} = 1.$$

54. Let x = the first number in the sequence and let d = the common difference. Then the three numbers in the sequence are x, $x + d$, and $x + 2d$. Solve:

$$x + x + 2d = 10,$$
$$x(x + d) = 15.$$

We get $x = 3$ and $d = 2$ so the numbers are 3, $3 + 2$, and $3 + 2 \cdot 2$, or 3, 5, and 7.

55. $S_n = \frac{n}{2}(1 + 2n - 1) = n^2$

56.

$$a_1 = \$8760$$
$$a_2 = \$8760 + (-\$798.23) = \$7961.77$$
$$a_3 = \$8760 + 2(-\$798.23) = \$7163.54$$
$$a_4 = \$8760 + 3(-\$798.23) = \$6365.31$$
$$a_5 = \$8760 + 4(-\$798.23) = \$5567.08$$
$$a_6 = \$8760 + 5(-\$798.23) = \$4768.85$$
$$a_7 = \$8760 + 6(-\$798.23) = \$3970.62$$
$$a_8 = \$8760 + 7(-\$798.23) = \$3172.39$$
$$a_9 = \$8760 + 8(-\$798.23) = \$2374.16$$
$$a_{10} = \$8760 + 9(-\$798.23) = \$1575.93$$
$$S_{10} = \frac{10}{2}(\$8760 + \$1575.93) = \$51,679.65$$

57. Let d = the common difference. Then $a_4 = a_2 + 2d$, or

$$10p + q = 40 - 3q + 2d$$
$$10p + 4q - 40 = 2d$$
$$5p + 2q - 20 = d.$$

Also, $a_1 = a_2 - d$, so we have

$$a_1 = 40 - 3q - (5p + 2q - 20)$$
$$= 40 - 3q - 5p - 2q + 20$$
$$= 60 - 5p - 5q.$$

58. $P(x) = x^4 + 4x^3 - 84x^2 - 176x + 640$ has at most 4 zeros because $P(x)$ is of degree 4. By the rational roots theorem, the possible zeros are

$\pm1, \pm2, \pm4, \pm5, \pm8, \pm10, \pm16, \pm20, \pm32, \pm40, \pm64, \pm80, \pm128, \pm160, \pm320, \pm640$

Using the graph of the function and synthetic division, we find that two zeros are -4 and 2.

Also by synthetic division we determine that ±1 and -2 are not zeros. Therefore we determine that $d = 6$ in the arithmetic sequence.

Possible arithmetic sequences:

a) $-4,\ 2,\ 8,\ 14$

b) $-4,\ 2,\ 8$

c) $-16,\ -10,\ -4,\ 2$

The solution cannot be (a) because 14 is not a possible zero. Checking -16 by synthetic division we find that -16 is not a zero. Thus (*b*) is the only possible arithmetic sequence which contains all four zeros. Synthetic division confirms that -10 and 8 are also zeros. The zeros are -10, -4, 2, and 8.

59. 4, m_1, m_2, m_3, 1

We look for m_1, m_2, and m_3 such that 4, m_1, m_2, m_3, 13 is an arithmetic sequence. In this case, $a_1 = 4$, $n = 5$, and $a_5 = 12$. First we find d:

$$a_n = a_1 + (n - 1)d$$
$$12 = 4 + (5 - 1)d$$
$$12 = 4 + 4d$$
$$8 = 4d$$
$$2 = d$$

Then we have

$$m_1 = a_1 + d = 4 + 2 = 6$$
$$m_2 = m_1 + d = 6 + 2 = 8$$
$$m_3 = m_2 + d = 8 + 2 = 10$$

60. 4, m_1, m_2, m_3, m_4, 13

$a_1 = 4$, $n = 6$, $a_6 = 13$

$$13 = 4 + (6 - 1)d$$
$$9 = 5d$$
$$1\frac{4}{5} = d$$

$$m_1 = 4 + 1\frac{4}{5} = 5\frac{4}{5}$$
$$m_2 = 5\frac{4}{5} + 1\frac{4}{5} = 6\frac{8}{5} = 7\frac{3}{5}$$
$$m_3 = 7\frac{3}{5} + 1\frac{4}{5} = 8\frac{7}{5} = 9\frac{2}{5}$$
$$m_4 = 9\frac{2}{5} + 1\frac{4}{5} = 10\frac{6}{5} = 11\frac{1}{5}$$

61. -3, m_1, m_2, m_3, 5

We look for m_1, m_2, and m_3 such that -3, m_1, m_2, m_3, 5 is an arithmetic sequence. In this case, $a_1 = -3$, $n = 5$, and $a_5 = 5$. First we find d:

$$a_n = a_1 + (n - 1)d$$
$$5 = -3 + (5 - 1)d$$
$$8 = 4d$$
$$2 = d$$

Then we have

$$m_1 = a_1 + d = -3 + 2 = -1,$$
$$m_2 = m_1 + d = -1 + 2 = 1,$$
$$m_3 = m_2 + d = 1 + 2 = 3.$$

62. 27, m_1, m_2, . . . ,m_9, m_{10}, 300

$$300 = 27 + (12 - 1)d$$
$$273 = 11d$$
$$\frac{273}{11} = d, \text{ or}$$
$$24\frac{9}{11} = d$$
$$m_1 = 27 + 24\frac{9}{11} = 51\frac{9}{11}$$
$$m_2 = 51\frac{9}{11} + 24\frac{9}{11} = 76\frac{7}{11}$$
$$m_3 = 76\frac{7}{11} + 24\frac{9}{11} = 101\frac{5}{11},$$
$$m_4 = 101\frac{5}{11} + 24\frac{9}{11} = 126\frac{3}{11},$$
$$m_5 = 126\frac{3}{11} + 24\frac{9}{11} = 151\frac{1}{11},$$
$$m_6 = 151\frac{1}{11} + 24\frac{9}{11} = 175\frac{10}{11},$$
$$m_7 = 175\frac{10}{11} + 24\frac{9}{11} = 200\frac{8}{11},$$
$$m_8 = 200\frac{8}{11} + 24\frac{9}{11} = 225\frac{6}{11},$$
$$m_9 = 225\frac{6}{11} + 24\frac{9}{11} = 250\frac{4}{11},$$
$$m_{10} = 250\frac{4}{11} + 24\frac{9}{11} = 275\frac{2}{11}$$

63. 1, $1+d$, $1+2d$, . . . , 50 has n terms and $S_n = 459$.

Find n:

$$459 = \frac{n}{2}(1+50)$$
$$18 = n$$

Find d:

$$50 = 1 + (18-1)d$$
$$\frac{49}{17} = d$$

The sequence has a total of 18 terms, so we insert 16 arithmetic means between 1 and 50 with $d = \frac{49}{17}$.

64. a) $a_t = \$5200 - t\left(\frac{\$5200 - \$1100}{8}\right)$

$a_t = \$5200 - \$512.50t$

b) $a_0 = \$5200 - \$512.50(0) = \$5200$

$a_1 = \$5200 - \$512.50(1) = \$4687.50$

$a_2 = \$5200 - \$512.50(2) = \$4175$

$a_3 = \$5200 - \$512.50(3) = \$3662.50$

$a_4 = \$5200 - \$512.50(4) = \$3150$

$a_7 = \$5200 - \$512.50(7) = \$1612.50$

$a_8 = \$5200 - \$512.50(8) = \$1100$

65.
$$m = p + d$$
$$m = q - d$$
$$2m = p + q \quad \text{Adding}$$
$$m = \frac{p+q}{2}$$

Exercise Set 7.3

1. 2, 4, 8, 16, . . .

$\frac{4}{2} = 2,\ \frac{8}{4} = 2,\ \frac{16}{8} = 2$

$r = 2$

2. $r = -\frac{6}{18} = -\frac{1}{3}$

3. 1, −1, 1, −1, . . .

$\frac{-1}{1} = -1,\ \frac{1}{-1} = -1,\ \frac{-1}{1} = -1$

$r = -1$

4. $r = \frac{-0.8}{8} = 0.1$

5. $\frac{2}{3}, -\frac{4}{3}, \frac{8}{3}, -\frac{16}{3}, \ldots$

$\frac{-\frac{4}{3}}{\frac{2}{3}} = -2,\ \frac{\frac{8}{3}}{-\frac{4}{3}} = -2,\ \frac{-\frac{16}{3}}{\frac{8}{3}} = -2$

$r = -2$

6. $r = \frac{15}{75} = \frac{1}{5}$

7. $\frac{0.6275}{6.275} = 0.1,\ \frac{0.06275}{0.6275} = 0.1$

$r = 0.1$

8. $r = \frac{\frac{1}{x^2}}{\frac{1}{x}} = \frac{1}{x}$

9. $\frac{\frac{5a}{2}}{5} = \frac{a}{2},\ \frac{\frac{5a^2}{4}}{\frac{5a}{2}} = \frac{a}{2},\ \frac{\frac{5a^3}{8}}{\frac{5a}{4}} = \frac{a}{2}$

$r = \frac{a}{2}$

10. $r = \frac{\$858}{\$780} = 1.1$

11. 2, 4, 8, 16, . . .

$a_1 = 2$, $n = 7$, and $r = \frac{4}{2}$, or 2.

We use the formula $a_n = a_1r^{n-1}$.

$$a_7 = 2(2)^{7-1} = 2 \cdot 2^6 = 2 \cdot 64 = 128$$

12. $a_9 = 2(-5)^{9-1} = 781,250$

13. $2, 2\sqrt{3}, 6, \ldots$

$a_1 = 2,\ n = 9,\ \text{and } r = \dfrac{2\sqrt{3}}{2},\ \text{or } \sqrt{3}$

$a_n = a_1 r^{n-1}$

$a_9 = 2(\sqrt{3})^{9-1} = 2(\sqrt{3})^8 = 2 \cdot 81 = 162$

14. $a_{57} = 1(-1)^{57-1} = 1$

15. $\dfrac{7}{625}, -\dfrac{7}{25}, \ldots$

$a_1 = \dfrac{7}{625},\ n = 23,\ \text{and } r = \dfrac{-\frac{7}{25}}{\frac{7}{625}} = -25.$

$a_n = a_1 r^{n-1}$

$a_{23} = \dfrac{7}{625}(-25)^{23-1} = \dfrac{7}{625}(-25)^{22}$

$= \dfrac{7}{25^2} \cdot 25^2 \cdot 25^{20} = 7(25)^{20},\ \text{or } 7(5)^{40}$

16. $a_5 = \$1000(1.06)^{5-1} \approx \$1262.48$

17. $1, 3, 9, \ldots$

$a_1 = 1 \text{ and } r = \dfrac{3}{1},\ \text{or } 3$

$a_n = a_1 r^{n-1}$

$a_n = 1(3)^{n-1} = 3^{n-1}$

18. $a_n = 25\left(\dfrac{1}{5}\right)^{n-1} = \dfrac{5^2}{5^{n-1}} = 5^{3-n}$

19. $1, -1, 1, -1, \ldots$

$a_1 = 1 \text{ and } r = \dfrac{-1}{1} = -1$

$a_n = a_1 r^{n-1}$

$a_n = 1(-1)^{n-1} = (-1)^{n-1}$

20. $a_n = (-2)^n$

21. $\dfrac{1}{x}, \dfrac{1}{x^2}, \dfrac{1}{x^2}, \ldots$

$a_1 = \dfrac{1}{x} \text{ and } r = \dfrac{\frac{1}{x^2}}{\frac{1}{x}} = \dfrac{1}{x}$

$a_n = a_1 r^{n-1}$

$a_n = \dfrac{1}{x}\left(\dfrac{1}{x}\right)^{n-1} = \dfrac{1}{x} \cdot \dfrac{1}{x^{n-1}} = \dfrac{1}{x^{1+n-1}} = \dfrac{1}{x^n}$

22. $a_n = 5\left(\dfrac{a}{2}\right)^{n-1}$

23. $6 + 12 + 24 + \ldots$

$a_1 = 6,\ n = 7,\ \text{and } r = \dfrac{12}{6},\ \text{or } 2$

$S_n = \dfrac{a_1(1 - r^n)}{1 - r}$

$S_7 = \dfrac{6(1 - 2^7)}{1 - 2} = \dfrac{6(1 - 128)}{-1} = \dfrac{6(-127)}{-1} = 762$

24. $S_{10} = \dfrac{16\left[1 - \left(-\frac{1}{2}\right)^{10}\right]}{1 - \left(-\frac{1}{2}\right)} = \dfrac{16\left(1 - \frac{1}{1024}\right)}{\frac{3}{2}} =$

$\dfrac{16\left(\frac{1023}{1024}\right)}{\frac{3}{2}} = \dfrac{341}{32},\ \text{or } 10\dfrac{21}{32}$

25. $\dfrac{1}{18} - \dfrac{1}{6} + \dfrac{1}{2} - \ldots$

$a_1 = \dfrac{1}{18},\ n = 9,\ \text{and } r = \dfrac{-\frac{1}{6}}{\frac{1}{18}} = -\dfrac{1}{6} \cdot \dfrac{18}{1} = -3$

$S_n = \dfrac{a_1(1 - r^n)}{1 - r}$

$S_9 = \dfrac{\frac{1}{18}\left[1 - (-3)^9\right]}{1 - (-3)} = \dfrac{\frac{1}{18}(1 + 19{,}683)}{4}$

$\dfrac{\frac{1}{18}(19{,}684)}{4} = \dfrac{1}{18}(19{,}684)\left(\dfrac{1}{4}\right) = \dfrac{4921}{18}$

26. $a_n = a_1 r^{n-1}$

$-\dfrac{1}{32} = (-8) \cdot \left(-\dfrac{1}{2}\right)^{n-1}$

$n = 9$

$S_9 = \dfrac{-8\left[\left(-\frac{1}{2}\right)^9 - 1\right]}{-\frac{1}{2} - 1} = -\dfrac{171}{32}$

27. $4 + 2 + 1 + \ldots$

$|r| = \left|\dfrac{2}{4}\right| = \left|\dfrac{1}{2}\right| = \dfrac{1}{2}$, and since $|r| < 1$, the series does have a sum.

$S_\infty = \dfrac{a_1}{1 - r} = \dfrac{4}{1 - \frac{1}{2}} = \dfrac{4}{\frac{1}{2}} = 4 \cdot \dfrac{2}{1} = 8$

28. $|r| = \left|\dfrac{3}{7}\right| = \dfrac{3}{7} < 1$, so the series has a sum.

$S_\infty = \dfrac{7}{1 - \frac{3}{7}} = \dfrac{7}{\frac{4}{7}} = \dfrac{49}{4}$

29. $25 + 20 + 16 + \ldots$

$|r| = \left|\dfrac{20}{25}\right| = \left|\dfrac{4}{5}\right| = \dfrac{4}{5}$, and since $|r| < 1$, the series does have a sum.

$S_\infty = \dfrac{a_1}{1 - r} = \dfrac{25}{1 - \frac{4}{5}} = \dfrac{25}{\frac{1}{5}} = 25 \cdot \dfrac{5}{1} = 125$

30. $|r| = \left|\dfrac{-10}{100}\right| = \dfrac{1}{10} < 1$, so the series has a sum.

$S_\infty = \dfrac{100}{1 - \left(-\frac{1}{10}\right)} = \dfrac{100}{\frac{11}{10}} = \dfrac{1000}{11}$

31. $8 + 40 + 200 + \dots$

$|r| = \left|\frac{40}{8}\right| = |5| = 5$, and since $|r| > 1$ the series does not have a sum.

32. $|r| = \left|\frac{3}{-6}\right| = \left|-\frac{1}{2}\right| = \frac{1}{2} < 1$, so the series has a sum.

$$S_\infty = \frac{-6}{1-\left(-\frac{1}{2}\right)} = \frac{-6}{\frac{3}{2}} = -6 \cdot \frac{2}{3} = -4$$

33. $0.6 + 0.06 + 0.006 + \dots$

$|r| = \left|\frac{0.06}{0.6}\right| = |0.1| = 0.1$, and since $|r| < 1$, the series does have a sum.

$$S_\infty = \frac{a_1}{1-r} = \frac{0.6}{1-0.1} = \frac{0.6}{0.9} = \frac{6}{9} = \frac{2}{3}$$

34. $\sum_{k=0}^{10} 3^k$

$a_1 = 1,\ |r| = 3,\ n = 11$

$$S_{11} = \frac{1(1-3^{11})}{1-3} = 88,573$$

35. $\sum_{k=1}^{11} 15\left(\frac{2}{3}\right)^k$

$a_1 = 15 \cdot \frac{2}{3}$ or 10; $|r| = \left|\frac{2}{3}\right| = \frac{2}{3},\ n = 11$

$$S_{11} = \frac{10\left[1-\left(\frac{2}{3}\right)^{11}\right]}{1-\frac{2}{3}} = \frac{10\left[1-\frac{2048}{177,147}\right]}{\frac{1}{3}}$$

$$= 10 \cdot \frac{175,099}{177,147} \cdot 3$$

$$= \frac{1,750,990}{59,049}, \text{ or } 29\frac{38,569}{59,049}$$

36. $\sum_{k=0}^{50} 200(1.08)^k$

$a_1 = 200,\ |r| = 1.08,\ n = 51$

$$S_{51} = \frac{200[1-(1.08)^{51}]}{1-1.08} \approx 124,134.354$$

37. $\sum_{k=1}^{\infty} \left(\frac{1}{2}\right)^{k-1}$

$a_1 = 1,\ |r| = \left|\frac{1}{2}\right| = \frac{1}{2}$

$$S_\infty = \frac{a_1}{1-r} = \frac{1}{1-\frac{1}{2}} = \frac{1}{\frac{1}{2}} = 2$$

38. Since $|r| = |2| > 1$, the sum does not exist.

39. $\sum_{k=1}^{\infty} 12.5^k$

Since $|r| = 12.5 > 1$, the sum does not exist.

40. Since $|r| = 1.0625 > 1$, the sum does not exist.

41. $\sum_{k=1}^{\infty} \$500(1.11)^{-k}$

$a_1 = \$500(1.11)^{-1}$, or $\frac{\$500}{1.11}$; $|r| = |1.11^{-1}| = \frac{1}{1.11}$

$$S_\infty = \frac{a_1}{1-r} = \frac{\frac{\$500}{1.11}}{1-\frac{1}{1.11}} = \frac{\frac{\$500}{1.11}}{\frac{0.11}{1.11}} \approx \$4545.\overline{45}$$

42. $\sum_{k=1}^{\infty} \$1000(1.06)^{-k}$

$a_1 = \frac{\$1000}{1.06},\ |r| = \frac{1}{1.06}$

$$S_\infty = \frac{\frac{\$1000}{1.06}}{1-\frac{1}{1.06}} = \frac{\frac{\$1000}{1.06}}{\frac{0.06}{1.06}} \approx \$16,666.\overline{66}$$

43. $\sum_{k=1}^{\infty} 16(0.1)^{k-1}$

$a_1 = 16,\ |r| = |0.1| = 0.1$

$$S_\infty = \frac{a_1}{1-r} = \frac{16}{1-0.1} = \frac{16}{0.9} = \frac{160}{9}$$

44. $\sum_{k=1}^{\infty} \frac{8}{3}\left(\frac{1}{2}\right)^{k-1}$

$a_1 = \frac{8}{3},\ |r| = \frac{1}{2}$

$$S_\infty = \frac{\frac{8}{3}}{1-\frac{1}{2}} = \frac{\frac{8}{3}}{\frac{1}{2}} = \frac{16}{3}$$

45. $0.131313\ldots = 0.13 + 0.0013 + 0.000013 + \dots$

This is an infinite geometric series with $a_1 = 0.13$.

$|r| = \left|\frac{0.0013}{0.13}\right| = |0.01| = 0.01 < 1$, so the series has a limit.

$$S_\infty = \frac{a_1}{1-r} = \frac{0.13}{1-0.01} = \frac{0.13}{0.99} = \frac{13}{99}$$

46. $0.2222 = 0.2 + 0.02 + 0.002 + 0.0002 + \dots$

$|r| = \left|\frac{0.02}{0.2}\right| = |0.1| = 0.1$

$$S_\infty = \frac{0.2}{1-0.1} = \frac{0.2}{0.9} = \frac{2}{9}$$

47. We will find fractional notation for $0.999\overline{9}$ and then add 8.

$0.999\overline{9} = 0.9 + 0.09 + 0.009 + 0.0009 + \dots$

This is an infinite geometric series with $a_1 = 0.9$.

$|r| = \left|\frac{0.09}{0.9}\right| = |0.1| = 0.1 < 1$, so the series has a limit.

$$S_\infty = \frac{a_1}{1-r} = \frac{0.9}{1-0.1} = \frac{0.9}{0.9} = 1$$

Then $8.999\overline{9} = 8 + 1 = 9$.

48. $0.1\overline{6} = 0.16 + 0.0016 + 0.000016 + \ldots$

$$|r| = \left|\frac{0.0016}{0.16}\right| = |0.01| = 0.01$$

$$S_\infty = \frac{0.16}{1-0.01} = \frac{0.16}{0.99} = \frac{16}{99}$$

Then $6.1\overline{6} = 6 + \frac{16}{99} = \frac{610}{99}$.

49. $3.4125\overline{125} = 3.4 + 0.0125\overline{125}$

We will find fractional notation for $0.0125\overline{125}$ and then add

3.4, or $\frac{34}{10}$, or $\frac{17}{5}$.

$0.0125\overline{125} = 0.0125 + 0.0000125 + \ldots$

This is an infinite geometric series with $a_1 = 0.0125$.

$|r| = \left|\frac{0.0000125}{0.0125}\right| = |0.001| = 0.001 < 1$, so the series has a limit.

$$S_\infty = \frac{a_1}{1-r} = \frac{0.0125}{1-0.001} = \frac{0.0125}{0.999} = \frac{125}{9990}$$

Then $\frac{17}{5} + \frac{125}{9990} = \frac{33,966}{9990} + \frac{125}{9990} = \frac{34,091}{9990}$

50. $12.7809\overline{809} = 12.7 + 0.0809\overline{809}$

$0.0809\overline{809} = 0.0809 + 0.0000809 + \ldots$

$$|r| = \left|\frac{0.0000809}{0.0809}\right| = |0.001| = 0.001$$

$$S_\infty = \frac{0.0809}{1-0.001} = \frac{0.0809}{0.999} = \frac{809}{9990}$$

Then $12.7 + \frac{809}{9990} = \frac{127}{10} + \frac{809}{9990} = \frac{127,682}{9990} = \frac{63,841}{4995}$

51. a) ***Familiarize***. The rebound distances form a geometric sequence:

$$\frac{1}{4} \times 16,\ \left(\frac{1}{4}\right)^2 \times 16,\ \left(\frac{1}{4}\right)^3 \times 16, \ldots,$$

or $4,\quad \frac{1}{4} \times 4,\quad \left(\frac{1}{4}\right)^2 \times 4, \ldots$

The height of the 6th rebound is the 6th term of the sequence.

Translate. We will use the formula $a_n = a_1 r^{n-1}$, with $a_1 = 4$, $r = \frac{1}{4}$, and $n = 6$:

$$a_6 = 4\left(\frac{1}{4}\right)^{6-1}$$

Carry out. We calculate to obtain $a_6 = \frac{1}{256}$.

Check. We can do the calculation again.

State. It rebounds $\frac{1}{256}$ ft the 6th time.

b) $S_\infty = \frac{a}{1-r} = \frac{4}{1-\frac{1}{4}} = \frac{4}{\frac{3}{4}} = \frac{16}{3}$ ft, or $5\frac{1}{3}$ ft

52. $a_1 = \$0.01$, $r = 2$, $n = 28$

$$S_{28} = \frac{\$0.01(1-2^{28})}{1-2} \approx \$2,684,355$$

53. a) ***Familiarize***. The rebound distances form a geometric sequence:

$$0.6 \times 200,\ (0.6)^2 \times 200,\ (0.6)^3 \times 200, \ldots,$$

or $120,\ 0.6 \times 120,\ (0.6)^2 \times 120, \ldots$

The total rebound distance after 9 rebounds is the sum of the first 9 terms of this sequence.

Translate. We will use the formula $S_n = \frac{a_1(1-r^n)}{1-r}$ with $a_1 = 120$, $r = 0.6$, and $n = 9$.

Carry out.

$$S_9 = \frac{120[1-(0.6)^9]}{1-0.6} \approx 297$$

Check. We repeat the calculation.

State. The bungee jumper has traveled about 297 ft upward after 9 rebounds.

b) $S_\infty = \frac{a_1}{1-r} = \frac{120}{1-0.6} = 300$ ft

54. a) $a_1 = 100,000$, $r = 1.03$. The population in 15 years will be the 16th term of the sequence $100,000,\ (1.03)100,000,\ (1.03)^2 100,000, \ldots$

$$a_{16} = 100,000(1.03)^{16-1} \approx 155,797$$

b) Solve: $200,000 = 100,000(1.03)^{n-1}$

$n \approx 24$ yr

55. ***Familiarize***. The amount of the annuity is the geometric series

$\$1000 + \$1000(1.062) + \$1000(1.062)^2 + \ldots + \$1000(1.062)^{17}$, where $a_1 = \$1000$, $r = 1.062$, and $n = 18$.

Translate. Using the formula

$$S_n = \frac{a_1(1-r^n)}{1-r}$$

we have

$$S_{18} = \frac{\$1000[1-(1.062)^{18}]}{1-1.062}$$

Carry out. We carry out the computation and get $S_{18} \approx \$31,497.57$.

Check. Repeat the calculations.

State. The amount of the annuity is \$31,497.57.

56. a) We have a geometric sequence with $a_1 = P$ and $r = 1 + i$. Then

$$S_N = V = \frac{P[1-(1+i)^N]}{1-(1+i)}$$

$$= \frac{P[1-(1+i)^N]}{-i}$$

$$= \frac{P[(1+i)^N - 1]}{i}$$

b) We have a geometric sequence with $a_1 = P$ and $r = 1 + \frac{i}{n}$.

The number of terms is nN. Then

$$S_{nN} = V = \frac{P\left[1 - \left(1 + \frac{i}{n}\right)^{nN}\right]}{1 - \left(1 + \frac{i}{n}\right)}$$

$$= \frac{P\left[1 - \left(1 + \frac{i}{n}\right)^{nN}\right]}{-\frac{i}{n}}$$

$$= \frac{P\left[\left(1 + \frac{i}{n}\right)^{nN} - 1\right]}{\frac{i}{n}}$$

57. ***Familiarize***. The amounts owed at the beginning of successive years form a geometric sequence:

$\$120,000,\ (1.12)\$120,000,\ (1.12)^2\$120,000,\ \ldots$

The amount to be repaid at the end of 13 years is the amount owed at the beginning of the 14th year.

Translate. Use the formula $a_n = a_1 r^{n-1}$ with $a_1 = 120,000$, $r = 1.12$, and $n = 14$:

$$a_{14} = 120,000(1.12)^{14-1}$$

Carry out. We perform the calculation, obtaining $a_{14} \approx \$523,619.17$.

Check. Repeat the calculation.

State. At the end of 13 years, \$523,619.17 will be repaid.

58. We have a sequence $0.01,\ 2(0.01),\ 2^2(0.01),\ 2^3(0.01),\ \ldots$. The thickness after 20 folds is given by the 21st term of the sequence.

$$a_{21} = 0.01(2)^{21-1} = 10,485.76 \text{ in.}$$

59. ***Familiarize***. The total effect on the economy is the sum of an infinite geometric series

$\$13,000,000,000 + \$13,000,000,000(0.85) + \$13,000,000,000(0.85)^2 + \ldots$

with $a_1 = \$13,000,000,000$ and $r = 0.85$.

Translate. Using the formula

$$S_\infty = \frac{a_1}{1 - r}$$

we have

$$S_\infty = \frac{\$13,000,000,000}{1 - 0.85}.$$

Carry out. Perform the calculation:

$S_\infty \approx \$86,666,666,667$.

Check. Repeat the calculation.

State. The total effect on the economy is \$86,666,666,667.

60. $S_\infty = \dfrac{5,000,000(0.3)}{1 - 0.7} \approx 2,142,857$

$\dfrac{2,142,857}{5,000,000} \approx 0.429$, so this is about 42.9% of the population.

61. Answers may vary. One possibility is given. Casey invests \$900 at 8% interest, compounded annually. How much will be in the account at the end of 40 years?

62. $S_1 = 2,\ S_2 = 2.5,\ S_3 = 2.\overline{6},\ S_4 = 2.708\overline{3},$

$S_5 = 2.71\overline{6},\ S_6 = 2.7180\overline{5}$

$2.718 < S_\infty < 2.719$ since the terms from a_7 on will cause changes in S_n in the fourth decimal place and beyond.

63. $f(x) = x^2,\ g(x) = 4x + 5$

$(f \circ g)(x) = f(g(x)) = f(4x + 5) = (4x + 5)^2 = 16x^2 + 40x + 25$

$(g \circ f)(x) = g(f(x)) = g(x^2) = 4x^2 + 5$

64. $f(x) = x - 1,\ g(x) = x^2 + x + 3$

$(f \circ g)(x) = f(g(x)) = f(x^2 + x + 3) = x^2 + x + 3 - 1 = x^2 + x + 2$

$(g \circ f)(x) = g(f(x)) = g(x - 1) = (x-1)^2 + (x-1) + 3 = x^2 - 2x + 1 + x - 1 + 3 = x^2 - x + 3$

65.

$$5^x = 35$$
$$\ln 5^x = \ln 35$$
$$x \ln 5 = \ln 35$$
$$x = \frac{\ln 35}{\ln 5}$$
$$x \approx 2.209$$

66. $\log_2 x = -4$

$$x = 2^{-4} = \frac{1}{2^4}$$
$$x = \frac{1}{16}$$

67. See the answer section in the text.

68. The sequence is not geometric; $a_4/a_3 \neq a_3/a_2$.

69. a) If the sequence is arithmetic, then $a_2 - a_1 = a_3 - a_2$.

$$x + 7 - (x + 3) = 4x - 2 - (x + 7)$$
$$x = \frac{13}{3}$$

The three given terms are $\frac{13}{3} + 3 = \frac{22}{3}$, $\frac{13}{3} + 7 = \frac{34}{3}$, and $4 \cdot \frac{13}{3} - 2 = \frac{46}{3}$.

Then $d = \frac{12}{3}$, or 4, so the fourth term is $\frac{46}{3} + \frac{12}{3} = \frac{58}{3}$.

b) If the sequence is geometric, then $a_2/a_1 = a_3/a_2$.

$$\frac{x+7}{x+3} = \frac{4x-2}{x+7}$$

$$x = -\frac{11}{3} \text{ or } x = 5$$

For $x = -\frac{11}{3}$: The three given terms are

$-\frac{11}{3} + 3 = -\frac{2}{3}$, $-\frac{11}{3} + 7 = \frac{10}{3}$, and

$4\left(-\frac{11}{3}\right) - 2 = -\frac{50}{3}$.

Then $r = -5$, so the fourth term is

$-\frac{50}{3}(-5) = \frac{250}{3}$.

For $x = 5$: The three given terms are $5 + 3 = 8$, $5 + 7 = 12$, and $4 \cdot 5 - 2 = 18$. Then $r = \frac{3}{2}$, so the fourth term is $18 \cdot \frac{3}{2} = 27$.

70. $S_n = \frac{1(1-x^n)}{1-x} = \frac{1-x^n}{1-x}$

71. $x^2 - x^3 + x^4 - x^5 + \ldots$

This is a geometric series with $a_1 = x^2$ and $r = -x$.

$$S_n = \frac{a_1(1-r^n)}{1-r} = \frac{x^2(1-(-x)^n)}{1-(-x)} = \frac{x^2(1-(-x)^n)}{1+x}$$

72. $\frac{a_n+1}{a_n} = r$, so $\frac{(a_n+1)^2}{(a_n)^2} = r^2$; Thus $a_1^2, a_2^2, a_3^2, \ldots$ is a geometric sequence with the common ratio r^2.

73. See the answer section in the text.

74. Let the arithmetic sequence have the common difference $d = a_{n+1} - a_n$. Then for the sequence $5^{a_1}, 5^{a_2}, 5^{a_3}, \ldots,$

we have $\frac{5^{a_{n+1}}}{5^{a_n}} = 5^{a_{n+1}-a_n} = 5^d$. Thus, we have a geometric sequence with the common ratio 5^d.

75. ***Familiarize***. The length of a side of the first square is 16 cm. The length of a side of the next square is the length of the hypotenuse of a right triangle with legs 8 cm and 8 cm, or $8\sqrt{2}$ cm. The length of a side of the next square is the length of the hypotenuse of a right triangle with legs $4\sqrt{2}$ cm and $4\sqrt{2}$ cm, or 8 cm. The areas of the squares form a sequence:

$(16)^2$, $(8\sqrt{2})^2$, $(8)^2, \ldots,$ or

$256, \quad 128, \quad 64, \ldots.$

This is a geometric sequence with $a_1 = 256$ and $r = \frac{1}{2}$.

Translate. We find the sum of the infinite geometric series $256 + 128 + 64 + \ldots$.

$$S_\infty = \frac{a_1}{1-r}$$

$$S_\infty = \frac{256}{1-\frac{1}{2}}$$

Carry out. We calculate to obtain $S_\infty = 512$.

Check. We can do the calculation again.

State. The sum of the areas is 512 cm^2.

Exercise Set 7.4

1. $n^2 < n^3$

$1^2 < 1^3$, $2^2 < 2^3$, $3^2 < 3^3$, $4^2 < 4^3$, $5^2 < 5^3$

The first statement is false, and the others are true.

2. $1^2 - 1 + 41$ is prime, $2^2 - 2 + 41$ is prime, $3^3 - 3 + 41$ is prime, $4^2 - 4 + 41$ is prime, $5^2 - 5 + 41$ is prime. Each of these statements is true.

The statement is false for $n = 41$; $41^2 - 41 + 41$ is not prime.

3. A polygon of n sides has $\frac{n(n-3)}{2}$ diagonals.

A polygon of 3 sides has $\frac{3(3-3)}{2}$ diagonals.

A polygon of 4 sides has $\frac{4(4-3)}{2}$ diagonals.

A polygon of 5 sides has $\frac{5(5-3)}{2}$ diagonals.

A polygon of 6 sides has $\frac{6(6-3)}{2}$ diagonals.

A polygon of 7 sides has $\frac{7(7-3)}{2}$ diagonals.

Each of these statements is true.

4. The sum of the angles of a polygon of 3 sides is $(3-2) \cdot 180°$.

The sum of the angles of a polygon of 4 sides is $(4-2) \cdot 180°$.

The sum of the angles of a polygon of 5 sides is $(5-2) \cdot 180°$.

The sum of the angles of a polygon of 6 sides is $(6-2) \cdot 180°$.

The sum of the angles of a polygon of 7 sides is $(7-2) \cdot 180°$.

Each of these statements is true.

5. See the answer section in the text.

6. $S_n: \quad 4 + 8 + 12 + \ldots + 4n = 2n(n+1)$

$S_1: \quad 4 = 2 \cdot 1 \cdot (1+1)$

$S_k: \quad 4 + 8 + 12 + \ldots + 4k = 2k(k+1)$

$S_{k+1}: \quad 4 + 8 + 12 + \ldots + 4k + 4(k+1) = 2(k+1)(k+2)$

1) *Basis step:* Since $2 \cdot 1 \cdot (1+1) = 2 \cdot 2 = 4$, S_1 is true.

2) *Induction step:* Let k be any natural number. Assume S_k. Deduce S_{k+1}. Starting with the left side of S_{k+1}, we have

$$\underbrace{4+8+12+\ldots+4k}\ +4(k+1)$$
$$= \quad 2k(k+1) \qquad +4(k+1) \qquad \text{By } S_k$$
$$= (k+1)(2k+4)$$
$$= 2(k+1)(k+2)$$

7. See the answer section in the text.

8. $S_n: \ 3+6+9+\ldots+3n = \dfrac{3n(n+1)}{2}$

$S_1: \ 3 = \dfrac{3\cdot 1(1+1)}{2}$

$S_k: \ 3+6+9+\ldots+3k = \dfrac{3k(k+1)}{2}$

$S_{k+1}: \ 3+6+9+\ldots+3k+3(k+1) = \dfrac{3(k+1)(k+2)}{2}$

1) *Basis step:* Since $\dfrac{3\cdot 1(1+1)}{2} = \dfrac{3\cdot 2}{2} = 3$, S_1 is true.

2) *Induction step:* Let k be any natural number. Assume S_k. Deduce S_{k+1}. Starting with the left side of S_{k+1}, we have

$$\underbrace{3+6+9+\ldots+3k}\ +3(k+1)$$
$$= \frac{3k(k+1)}{2} \qquad +3(k+1) \qquad \text{By } S_k$$
$$= \frac{3k(k+1)+6(k+1)}{2}$$
$$= \frac{(k+1)(3k+6)}{2}$$
$$= \frac{3(k+1)(k+2)}{2}$$

9. See the answer section in the text.

10. $S_n: \ 2 \le 2^n$

$S_1: \ 2 \le 2^1$

$S_k: \ 2 \le 2^k$

$S_{k+1}: \ 2 \le 2^{k+1}$

1) *Basis step:* Since $2 \le 2$, S_1 is true.

2) *Induction step:* Let k be any natural number. Assume S_k. Deduce S_{k+1}.

$$2 \le 2^k \qquad S_k$$
$$2\cdot 2 \le 2^k\cdot 2 \qquad \text{Multiplying by 2}$$
$$2 < 2\cdot 2 \le 2^{k+1} \qquad (2 < 2\cdot 2)$$
$$2 \le 2^{k+1}$$

11. See the answer section in the text.

12. $S_n: \ 3^n < 3^{n+1}$

$S_1: \ 3^1 < 3^{1+1}$

$S_k: \ 3^k < 3^{k+1}$

$S_{k+1}: \ 3^{k+1} < 3^{k+2}$

1) *Basis step:* Since $3^1 < 3^{1+1}$, or $3 < 9$, S_1 is true.

2) *Induction step:* Let k be any natural number. Assume S_k. Deduce S_{k+1}.

$$3^k < 3^{k+1} \qquad S_k$$
$$3^k\cdot 3 < 3^{k+1}\cdot 3 \qquad \text{Multiplying by 3}$$
$$3^{k+1} < 3^{k+2}$$

13. See the answer section in the text.

14. $S_n: \ \dfrac{1}{1\cdot 2}+\dfrac{1}{2\cdot 3}+\ldots+\dfrac{1}{n(n+1)} = \dfrac{n}{n+1}$

$S_1: \ \dfrac{1}{1\cdot 2} = \dfrac{1}{1+1}$

$S_k: \ \dfrac{1}{1\cdot 2}+\dfrac{1}{2\cdot 3}+\ldots+\dfrac{1}{k(k+1)} = \dfrac{k}{k+1}$

$S_{k+1}: \ \dfrac{1}{1\cdot 2}+\dfrac{1}{2\cdot 3}+\ldots+\dfrac{1}{k(k+1)}+\dfrac{1}{(k+1)(k+2)} = \dfrac{k+1}{k+2}$

1) *Basis step:* Since $\dfrac{1}{1+1} = \dfrac{1}{2} = \dfrac{1}{1\cdot 2}$, S_1 is true.

2) *Induction step:* Let k be any natural number. Assume S_k. Deduce S_{k+1}.

$$\frac{1}{1\cdot 2}+\frac{1}{2\cdot 3}+\ldots+\frac{1}{k(k+1)} = \frac{k}{k+1} \qquad (S_k)$$
$$\frac{1}{1\cdot 2}+\frac{1}{2\cdot 3}+\ldots+\frac{1}{k(k+1)}+\frac{1}{(k+1)(k+2)} = \frac{k}{k+1}+\frac{1}{(k+1)(k+2)}$$
$$\left(\text{Adding } \frac{1}{(k+1)(k+2)} \text{ on both sides}\right)$$
$$= \frac{k(k+2)+1}{(k+1)(k+2)}$$
$$= \frac{k^2+2k+1}{(k+1)(k+2)}$$
$$= \frac{(k+1))(k+1)}{(k+1)(k+2)}$$
$$= \frac{k+1}{k+2}$$

15. See the answer section in the text.

16. $S_n: \ x \le x^n$

$S_1: \ x \le x$

$S_k: \ x \le x^k$

$S_{k+1}: \ x \le x^{k+1}$

1) *Basis step:* Since $x = x$, S_1 is true.

2) *Induction step:* Let k be any natural number. Assume S_k. Deduce S_{k+1}.

$$x \le x^k \qquad S_k$$
$$x\cdot x \le x^k\cdot x \qquad \text{Multiplying by } x,\ x > 1$$
$$x \le x\cdot x \le x^k\cdot x$$
$$x \le x^{k+1}$$

17. See the answer section in the text.

18. $S_n:\ 1^2+2^2+3^2+\ldots+n^2=\dfrac{n(n+1)(2n+1)}{6}$

$S_1:\ 1^2=\dfrac{1(1+1)(2\cdot 1+1)}{6}$

$S_k:\ 1^2+2^2+3^2+\ldots+k^2=\dfrac{k(k+1)(2k+1)}{6}$

$S_{k+1}:\ 1^2+2^2+\ldots+k^2+(k+1)^2=\dfrac{(k+1)(k+1+1)(2(k+1)+1)}{6}$

1) *Basis step:* $1^2=\dfrac{1(1+1)(2\cdot 1+1)}{6}$ is true.

2) *Induction step:* Let k be any natural number. Assume S_k. Deduce S_{k+1}.

$$1^2+2^2+\ldots+k^2=\frac{k(k+1)(2k+1)}{6}$$

$$\begin{aligned}1^2+2^2+\ldots+k^2+(k+1)^2 &= \frac{k(k+1)(2k+1)}{6}+(k+1)^2\\ &= \frac{k(k+1)(2k+1)+6(k+1)^2}{6}\\ &= \frac{(k+1)(2k^2+7k+6)}{6}\\ &= \frac{(k+1)(k+2)(2k+3)}{6}\\ &= \frac{(k+1)(k+1+1)(2(k+1)+1)}{6}\end{aligned}$$

19. See the answer section in the text.

20. $S_n:\ 1^4+2^4+3^4+\ldots+n^4=\dfrac{n(n+1)(2n+1)(3n^2+3n-1)}{30}$

$S_1:\ 1^4=\dfrac{1(1+1)(2\cdot 1+1)(3\cdot 1^2+3\cdot 1-1)}{30}$

$S_k:\ 1^4+2^4+3^4+\ldots+k^4=\dfrac{k(k+1)(2k+1)(3k^2+3k-1)}{30}$

$S_{k+1}:\ 1^4+2^4+\ldots+k^4+(k+1)^4=\dfrac{(k+1)(k+1+1)(2(k+1)+1)(3(k+1)^2-3(k+1)-1)}{30}$

1) *Basis step:*

$1^4=\dfrac{1(1+1)(2\cdot 1+1)(3\cdot 1^2+3\cdot 1-1)}{30}$ is true.

2) *Induction step:*

$$1^4+2^4+\ldots+k^4=\frac{k(k+1)(2k+1)(3k^2+3k-1)}{30}$$

$$\begin{aligned}1^4+2^4+\ldots+k^4+(k+1)^4 &= \frac{k(k+1)(2k+1)(3k^2+3k-1)}{30}+(k+1)^4\\ &= \frac{k(k+1)(2k+1)(3k^2+3k-1)+30(k+1)^4}{30}\\ &= \frac{(k+1)(6k^4+39k^3+91k^2+89k+30)}{30}\\ &= \frac{(k+1)(k+2)(2k+3)(3k^2+9k+5)}{30}\\ &= [(k+1)(k+1+1)(2(k+1)+1)(3(k+1)^2+3(k+1)-1)]/30\end{aligned}$$

21. See the answer section in the text.

22. $S_n:\ 2+5+8+\ldots+3n-1=\dfrac{n(3n+1)}{2}$

$S_1:\ 2=\dfrac{1(3\cdot 1+1)}{2}$

$S_k:\ 2+5+8+\ldots+3k-1=\dfrac{k(3k+1)}{2}$

$S_{k+1}:\ 2+5+\ldots+(3k-1)+(3(k+1)-1)=\dfrac{(k+1)(3(k+1)+1)}{2}$

1) *Basis step:* $2=\dfrac{1(3\cdot 1+1)}{2}$ is true.

2) *Induction step:* Let k be any natural number. Assume S_k. Deduce S_{k+1}.

$$2+5+\ldots+3k-1=\frac{k(3k+1)}{2}$$

$$\begin{aligned}2+5+\ldots+(3k-1)+(3(k+1)-1) &= \frac{k(3k+1)}{2}+(3(k+1)-1)\\ &= \frac{3k^2+k+6k+6-2}{2}\\ &= \frac{3k^2+7k+4}{2}\\ &= \frac{(k+1)(3k+4)}{2}\\ &= \frac{(k+1)(3(k+1)+1)}{2}\end{aligned}$$

23. See the answer section in the text.

24. S_n $\left(1+\frac{1}{1}\right)\left(1+\frac{1}{2}\right)\left(1+\frac{1}{3}\right)\cdots\left(1+\frac{1}{n}\right)=$ $n+1$

$S_1:\ 1+\frac{1}{1}=1+1$

S_k $\left(1+\frac{1}{1}\right)\cdots\left(1+\frac{1}{k}\right)=k+1$

S_{k+1} $\left(1+\frac{1}{1}\right)\cdots\left(1+\frac{1}{k}\right)\left(1+\frac{1}{k+1}\right)=(k+1)+1$

1) *Basis step:* $\left(1+\frac{1}{1}\right)=1+1$ is true.

2) *Induction step:* Let k be any natural number. Assume S_k. Deduce S_{k+1}.

$$\left(1+\frac{1}{1}\right)\cdots\left(1+\frac{1}{k}\right)=k+1$$

$$\left(1+\frac{1}{1}\right)\cdots\left(1+\frac{1}{k}\right)\left(1+\frac{1}{k+1}\right)=(k+1)\left(1+\frac{1}{k+1}\right)$$

Multiplying by $\left(1+\frac{1}{k+1}\right)$

$$=(k+1)\left(\frac{k+1+1}{k+1}\right)$$

$$=(k+1)+1$$

25. See the answer section in the text.

26. We can prove an infinite sequence of statements S_n by showing that a basis statement S_1 is true and then that for all natural numbers k, if S_k is true, then S_{k+1} is true.

27. Two possibilities are $n<n^2$ and $n^2\le 2^n$. The basis step is false for the first and the induction step fails for the second.

28. $2x-3y=1$, (1)

$3x-4y=3$ (2)

Multiply equation (1) by 4 and multiply equation (2) by -3 and add.

$$\begin{aligned} 8x-12y&=4\\ -9x+12y&=-9\\ \hline -x\qquad&=-5\\ x&=5\end{aligned}$$

Back-substitute to find y. We use equation (1).

$$\begin{aligned}2\cdot5-3y&=1\\10-3y&=1\\-3y&=-9\\y&=3\end{aligned}$$

The solution is $(5,3)$.

29. $x+y+z=3$, (1)

$2x-3y-2z=5$, (2)

$3x+2y+2z=8$ (3)

We will use Gaussian elimination. First multiply equation (1) by -2 and add it to equation (2). Also multiply equation (1) by -3 and add it to equation (3).

$$\begin{aligned}x+y+z&=3\\-5y-4z&=-1\\-y-z&=-1\end{aligned}$$

Now multiply the last equation above by 5 to make the y-coefficient a multiple of the y-coefficient in the equation above it.

$$\begin{aligned}x+y+z&=3 \quad (1)\\-5y-4z&=-1 \quad (4)\\-5y-5z&=-5 \quad (5)\end{aligned}$$

Multiply equation (4) by -1 and add it to equation (3).

$$\begin{aligned}x+y+z&=3 \quad (1)\\-5y-4z&=-1 \quad (4)\\-z&=-4 \quad (6)\end{aligned}$$

Now solve equation (6) for z.

$$\begin{aligned}-z&=-4\\z&=4\end{aligned}$$

Back-substitute 4 for z in equation (4) and solve for y.

$$\begin{aligned}-5y-4\cdot4&=-1\\-5y-16&=-1\\-5y&=15\\y&=-3\end{aligned}$$

Finally, back-substitute -3 for y and 4 for z in equation (1) and solve for x.

$$\begin{aligned}x-3+4&=3\\x+1&=3\\x&=2\end{aligned}$$

The solution is $(2,-3,4)$.

30. Let h = the number of hardback books sold and p = the number of paperback books sold.

Solve: $h+p=80$,

$24.95h+9.95p=1546$

$h=50$, $p=30$

31. ***Familiarize***. Let x, y, and z represent the amounts invested at 6%, 8%, and 10%, respectively.

Translate. We know that simple interest for one year was \$376. This gives us one equation:

$$0.06x+0.08y+0.1z=376$$

The amount invested at 8% is twice the amount invested at 6%:

$$y=2x,\ \text{or}\ -2x+y=0$$

There is \$400 more invested at 10% than at 8%:

$$z=y+400,\ \text{or}\ -y+z=400$$

We have a system of equations:

$$\begin{aligned} 0.06x + 0.08y + 0.1z &= 376, \\ -2x + \quad y \quad &= 0 \\ - \quad y + \quad z &= 400 \end{aligned}$$

Carry out. Solving the system of equations, we get (800, 1600, 2000).

Check. Simple interest for one year would be 0.06($800)+0.08($1600)+0.1($2000), or $48+$128+$200, or $376. The amount invested at 8%, $1600, is twice $800, the amount invested at 6%. The amount invested at 10%, $2000, is $400 more than $1600, the amount invested at 8%. The answer checks.

State. Martin invested $800 at 6%, $1600 at 8%, and $2000 at 10%.

32. $S_n: \; a_1 + a_1r + a_1r^2 + ... + a_1r^{n-1} = \dfrac{a_1 - a_1r^n}{1-r}$

$S_1: \; a_1 = \dfrac{a_1 - a_1r}{1-r}$

$S_k: \; a_1 + a_1r + a_2r^2 + ... + a_1r^{k-1} = \dfrac{a_1 - a_1r^k}{1-r}$

$S_{k+1}: \; a_1 + a_1r + ... + a_1r^{k-1} + a_1r^{(k+1))-1} = \dfrac{a_1 - a_1r^{k+1}}{1-r}$

1) *Basis step:* $a_1 = \dfrac{a_1(1-r)}{1-r} = \dfrac{a_1 - a_1r}{1-r}$ is true.

2) *Induction step:* Let n be any natural number. Assume S_k. Deduce S_{k+1}.

$$a_1 + a_1r + ... + a_1r^{k-1} = \frac{a_1 - a_1r^k}{1-r}$$

$$a_1 + a +_1 r + ... + a_1r^{k-1} + a_1r^k = \frac{a_1 - a_1r^k}{1-r} + a_1r^k \quad \text{Adding } a_1r^k$$

$$= \frac{a_1 - a_1r^k + a_1r^k - a_1r^{k+1}}{1-r}$$

$$= \frac{a_1 - a_1r^{k+1}}{1-r}$$

33. See the answer section in the text.

34. $S_n: \; 2n + 1 < 3^n$

$S_2: \; 2 \cdot 2 + 1 < 3^2$

$S_k: \; 2k + 1 < 3^k$

$S_{k+1}: \; 2(k+1) + 1 = 3^{k+1}$

1) *Basis step:* $2 \cdot 2 + 1 < 3^2$ is true.

2) *Induction step:* Let k be any natural number greater than or equal to 2. Assume S_k. Deduce S_{k+1}.

$$2k + 1 < 3^k$$

$$3(2k+1) < 3 \cdot 3^k \quad \text{Multiplying by 3}$$

$$6k + 3 < 3^{k+1}$$

$$2k + 3 < 6k + 3 < 3^{k+1} \quad (2k < 6k)$$

$$2(k+1) + 1 < 3^{k+1}$$

35. See the answer section in the text.

36. $S_1: \; \overline{z^1} = \overline{z}^1$ If $z = a + bi$, $\overline{z} = a - bi$.

$S_k: \; \overline{z^k} = \overline{z}^k$

$\overline{z^k} \cdot \overline{z} = \overline{z}^k \cdot \overline{z}$ Multiplying both sides of S_k by $\overline{z}$

$\overline{z^k} \cdot \overline{z} = \overline{z}^{k+1}$

$\overline{z^{k+1}} = \overline{z}^{k+1}$

37. See the answer section in the text.

38. $S_2: \; \overline{z_1z_2} = \overline{z_1} \cdot \overline{z_2}$

$S_k: \; \overline{z_1z_2 \cdots z_k} = \overline{z}_1\overline{z}_2 \cdots \overline{z}_k$

Starting with the left side of S_{k+1}, we have

$\overline{z_1z_2 \cdots z_kz_{k+1}} = \overline{z_1z_2 \cdots z_k} \cdot \overline{z_{k+1}}$ By S_2

$= \overline{z}_1\overline{z}_2 \cdots \overline{z}_k \cdot \overline{z}_{k+1}$ By S_k

39. See the answer section in the text.

40. S_1 : 2 is a factor of $1^2 + 1$.

S_k : 2 is a factor of $k^2 + k$.

$$\begin{aligned} (k+1)^2 + (k+1) &= k^2 + 2k + 1 + k + 1 \\ &= k^2 + k + 2(k+1) \end{aligned}$$

By S_k, 2 is a factor of k^2+k; hence 2 is a factor of the right-hand side, so 2 is a factor of $(k+1)^2 + (k+1)$.

41. See the answer section in the text.

42. a) The least number of moves for

1 disk(s) is $1 = 2^1 - 1$,

2 disk(s) is $3 = 2^2 - 1$,

3 disk(s) is $7 = 2^3 - 1$,

4 disk(s) is $15 = 2^4 - 1$; etc.

b) Let P_n be the least number of moves for n disks. We conjecture and must show:

$S_n: \; P_n = 2^n - 1$.

1) *Basis step:* S_1 is true by substitution.

2) *Induction step:* Assume S_k for k disks: $P_k = 2^k - 1$. Show: $P_{k+1} = 2^{k+1} - 1$. Now suppose there are $k+1$ disks on one peg. Move k of them to another peg in $2^k - 1$ moves (by S_k) and move the remaining disk to the free peg (1 move). Then move the k disks onto it in (another) $2^k - 1$ moves. Thus the total moves P_{k+1} is $2(2^k - 1) + 1 = 2^{k+1} - 1$: $P_{k+1} = 2^{k+1} - 1$.

Exercise Set 7.5

1. ${}_6P_6 = 6! = 6 \cdot 5 \cdot 4 \cdot 3 \cdot 2 \cdot 1 = 720$

2. ${}_4P_3 = 4 \cdot 3 \cdot 2 = 24$, or

$${}_4P_3 = \frac{4!}{(4-3)!} = \frac{4!}{1!} = \frac{4 \cdot 3 \cdot 2 \cdot 1}{1} = 24$$

3. Using formula (1), we have
$_{10}P_7 = 10 \cdot 9 \cdot 8 \cdot 7 \cdot 6 \cdot 5 \cdot 4 = 604,800.$
Using formula (2), we have
$$_{10}P_7 = \frac{10!}{(10-7)!} = \frac{10!}{3!} = \frac{10 \cdot 9 \cdot 8 \cdot 7 \cdot 6 \cdot 5 \cdot 4 \cdot 3!}{3!} = 604,800.$$

4. $_{10}P_3 = 10 \cdot 9 \cdot 8 = 720$, or
$$_{10}P_3 = \frac{10!}{(10-3)!} = \frac{10!}{7!} = \frac{10 \cdot 9 \cdot 8 \cdot 7!}{7!} = 720$$

5. $5! = 5 \cdot 4 \cdot 3 \cdot 2 \cdot 1 = 120$

6. $7! = 7 \cdot 6 \cdot 5 \cdot 4 \cdot 3 \cdot 2 \cdot 1 = 5040$

7. 0! is defined to be 1.

8. $1! = 1$

9. $\frac{9!}{5!} = \frac{9 \cdot 8 \cdot 7 \cdot 6 \cdot 5!}{5!} = 9 \cdot 8 \cdot 7 \cdot 6 = 3024$

10. $\frac{9!}{4!} = \frac{9 \cdot 8 \cdot 7 \cdot 6 \cdot 5 \cdot 4!}{4!} = 9 \cdot 8 \cdot 7 \cdot 6 \cdot 5 = 15,120$

11. $(8-3)! = 5! = 5 \cdot 4 \cdot 3 \cdot 2 \cdot 1 = 120$

12. $(8-5)! = 3! = 3 \cdot 2 \cdot 1 = 6$

13. $\frac{10!}{7!3!} = \frac{10 \cdot 9 \cdot 8 \cdot 7!}{7!3 \cdot 2 \cdot 1} = \frac{10 \cdot 3 \cdot 3 \cdot 4 \cdot 2}{3 \cdot 2 \cdot 1} = 10 \cdot 3 \cdot 4 = 120$

14. $\frac{7!}{(7-2)!} = \frac{7!}{5!} = \frac{7 \cdot 6 \cdot 5!}{5!} = 42$

15. Using formula (2), we have
$$_8P_0 = \frac{8!}{(8-0)!} = \frac{8!}{8!} = 1.$$

16. $_{13}P_1 = 13$ (Using formula (1))

17. Using a grapher, we find $_{52}P_4 = 6,497,400$

18. $_{52}P_5 = 311,875,200$

19. Using formula (1), we have $_nP_3 = n(n-1)(n-2)$.
Using formula (2), we have
$$_nP_3 = \frac{n!}{(n-3)!} = \frac{n(n-1)(n-2)(n-3)!}{(n-3)!} = n(n-1)(n-2).$$

20. $_nP_2 = n(n-1)$

21. Using formula (1), we have $_nP_1 = n$.
Using formula (2), we have
$$_nP_1 = \frac{n!}{(n-1)!} = \frac{n(n-1)!}{(n-1)!} = n.$$

22. $_nP_0 = \frac{n!}{(n-0)!} = \frac{n!}{n!} = 1$

23. $_6P_6 = 6! = 720$

24. $_4P_4 = 4! = 24$

25. $_9P_9 = 9! = 362,880$

26. $_8P_8 = 8! = 40,320$

27. $_9P_4 = 9 \cdot 8 \cdot 7 \cdot 6 = 3024$

28. $_8P_5 = 8 \cdot 7 \cdot 6 \cdot 5 \cdot 4 = 6720$

29. Without repetition: $_5P_5 = 5! = 120$
With repetition: $5^5 = 3125$

30. $_7P_7 = 7! = 5040$

31. BUSINESS: 1 B, 1 U, 3 S's, 1 I, 1 N, 1 E, a total of 8.
$$= \frac{8!}{1! \cdot 1! \cdot 3! \cdot 1! \cdot 1! \cdot 1!}$$
$$= \frac{8!}{3!} = \frac{8 \cdot 7 \cdot 6 \cdot 5 \cdot 4 \cdot 3!}{3!} = 8 \cdot 7 \cdot 6 \cdot 5 \cdot 4 = 6720$$
BIOLOGY: 1 B, 1 I, 2 0's, 1 L, 1 G, 1 Y, a total of 7.
$$= \frac{7!}{1! \cdot 1! \cdot 2! \cdot 1! \cdot 1! \cdot 1!}$$
$$= \frac{7!}{2!} = \frac{7 \cdot 6 \cdot 5 \cdot 4 \cdot 3 \cdot 2!}{2!} = 7 \cdot 6 \cdot 5 \cdot 4 \cdot 3 = 2520$$
MATHEMATICS: 2 M's, 2 A's, 2 T's, 1 H, 1 E, 1 I, 1 C, 1 S, a total of 11.
$$= \frac{11!}{2! \cdot 2! \cdot 2! \cdot 1! \cdot 1! \cdot 1! \cdot 1! \cdot 1!}$$
$$= \frac{11!}{2! \cdot 2! \cdot 2!} = \frac{11 \cdot 10 \cdot 9 \cdot 8 \cdot 7 \cdot 6 \cdot 5 \cdot 4 \cdot 3 \cdot 2!}{2! \cdot 2! \cdot 2!}$$
$$= \frac{11 \cdot 10 \cdot 9 \cdot 8 \cdot 7 \cdot 6 \cdot 5 \cdot 4 \cdot 3}{2 \cdot 1 \cdot 2 \cdot 1}$$
$$= 4,989,600$$

32. $\frac{24!}{3!5!9!4!3!} = 16,491,024,950,400$

33. The first number can be any of the eight digits other than 0 and 1. The remaining 6 numbers can each be any of the ten digits 0 through 9. We have
$$8 \cdot 10^6 = 8,000,000$$
Accordingly, there can be 8,000,000 telephone numbers within a given area code before the area needs to be split with a new area code.

34. $_5P_5 \cdot_4 P_4 = 5!4! = 2880$

35. $a^2b^3c^4 = a \cdot a \cdot b \cdot b \cdot b \cdot c \cdot c \cdot c \cdot c$
There are 2 a's, 3 b's, and 4 c's, for a total of 9. We have
$$\frac{9!}{2! \cdot 3! \cdot 4!}$$
$$= \frac{9 \cdot 8 \cdot 7 \cdot 6 \cdot 5 \cdot 4!}{2 \cdot 1 \cdot 3 \cdot 2 \cdot 1 \cdot 4!} = \frac{9 \cdot 8 \cdot 7 \cdot 6 \cdot 5}{2 \cdot 3 \cdot 2} = 1260.$$

36. a) ${}_4P_4 = 4! = 24$

b) There are 4 choices for the first coin and 2 possibilities (head or tail) for each choice. This results in a total of 8 choices for the first selection.

Likewise there are 6 choices for the second selection, 4 for the third, and 2 for the fourth. Then the number of ways in which the coins can be lined up is $8 \cdot 6 \cdot 4 \cdot 2$, or 384.

37. a) ${}_6P_5 = 6 \cdot 5 \cdot 4 \cdot 3 \cdot 2 = 720$

b) $6^5 = 7776$

c) The first letter can only be D. The other four letters are chosen from A, B, C, E, F without repetition. We have

$$1 \cdot {}_5P_4 = 1 \cdot 5 \cdot 4 \cdot 3 \cdot 2 = 120.$$

d) The first letter can only be D. The second letter can only be E. The other three letters are chosen from A, B, C, F without repetition. We have

$$1 \cdot 1 \cdot {}_4P_3 = 1 \cdot 1 \cdot 4 \cdot 3 \cdot 2 = 24.$$

38. There are 80 choices for the number of the county, 26 choices for the letter of the alphabet, and 9999 choices for the number that follows the letter. By the fundamental counting principle we know there are $80 \cdot 26 \cdot 9999$, or 20,797,920 possible license plates.

39. a) Since repetition is allowed, each of the 5 digits can be chosen in 10 ways. The number of zip-codes possible is $10 \cdot 10 \cdot 10 \cdot 10 \cdot 10$, or 100,000.

b) Since there are 100,000 possible zip-codes, there could be 100,000 post offices.

40. $10^9 = 1,000,000,000$

41. a) Since repetition is allowed, each digit can be chosen in 10 ways. There can be $10 \cdot 10 \cdot 10 \cdot 10 \cdot 10 \cdot 10 \cdot 10 \cdot 10 \cdot 10$, or 1,000,000,000 social security numbers.

b) Since more than 275 million social security numbers are possible, each person can have a social security number.

42. Put the following in the form of a paragraph.

First find the number of seconds in a year (365 days):

$$365 \text{ days} \cdot \frac{24 \text{ hr}}{1 \text{ day}} \cdot \frac{60 \text{ min}}{1 \text{ hr}} \cdot \frac{60 \text{ sec}}{1 \text{ min}} =$$

31,536,000 sec.

The number of arrangements possible is 15!.

The time is $\dfrac{15!}{31,536,000} \approx 41,466$ yr.

43. For each circular arrangement of the numbers on a clock face there are 12 distinguishable ordered arrangements on a line. The number of arrangements of 12 objects on a line is ${}_{12}P_{12}$, or 12!. Thus, the number of circular permutations is $\dfrac{{}_{12}P_{12}}{12} = \dfrac{12!}{12} = 11! = 39,916,800$.

In general, for each circular arrangement of n objects, there are n distinguishable ordered arrangements on a line. The total number of arrangements of n objects on a line is ${}_nP_n$, or $n!$. Thus, the number of circular permutations is $\dfrac{n!}{n} = \dfrac{n(n-1)!}{n} = (n-1)!$.

44. $4x - 9 = 0$

$$4x = 9$$

$$x = \frac{9}{4}, \text{ or } 2.25$$

We could also graph $y = 4x - 9$ and use the Zero feature. The solution is $\dfrac{9}{4}$, or 2.25.

45.

$$x^2 + x - 6 = 0$$
$$(x+3)(x-2) = 0$$
$$x + 3 = 0 \quad \textit{or} \quad x - 2 = 0$$
$$x = -3 \quad \textit{or} \quad x = 2$$

We could also graph $y = x^2 + x - 6$ and use the Zero feature twice. The solutions are -3 and 2.

46. $2x^2 - 3x - 1 = 0$

$$x = \frac{-(-3) \pm \sqrt{(-3)^2 - 4 \cdot 2 \cdot (-1)}}{2 \cdot 2}$$
$$= \frac{3 \pm \sqrt{17}}{4}$$

47. $f(x) = x^3 - 4x^2 - 7x + 10$

We use synthetic division to find one factor of the polynomial. We try $x - 1$.

$$\begin{array}{r|rrrr} 1 & 1 & -4 & -7 & 10 \\ & & 1 & -3 & -10 \\ \hline & 1 & -3 & -10 & 0 \end{array}$$

$$x^3 - 4x^2 - 7x + 10 = 0$$
$$(x-1)(x^2 - 3x - 10) = 0$$
$$(x-1)(x-5)(x+2) = 0$$
$$x - 1 = 0 \quad \textit{or} \quad x - 5 = 0 \quad \textit{or} \quad x + 2 = 0$$
$$x = 1 \quad \textit{or} \quad x = 5 \quad \textit{or} \quad x = -2$$

We could also graph $y = x^3 - 4x^2 - 7x + 10$ and use the Zero feature three times. The solutions are -2, 1, and 5.

48.

$${}_nP_5 = 7 \cdot {}_nP_4$$
$$\frac{n!}{(n-5)!} = 7 \cdot \frac{n!}{(n-4)!}$$
$$\frac{n!}{7(n-5)!} = \frac{n!}{(n-4)!}$$
$$7(n-5)! = (n-4)!$$

The denominators must be the same.

$$7(n-5)! = (n-4)(n-5)!$$
$$7 = n - 4$$
$$11 = n$$

49. $_nP_4 = 8 \cdot {}_{n-1}P_3$

$\frac{n!}{(n-4)!} = 8 \cdot \frac{(n-1)!}{(n-1-3)!}$

$\frac{n!}{(n-4)!} = 8 \cdot \frac{(n-1)!}{(n-4)!}$

$n! = 8 \cdot (n-1)!$ Multiplying by $(n-4)!$

$n(n-1)! = 8 \cdot (n-1)!$

$n = 8$ Dividing by $(n-1)!$

50. $_nP_5 = 9 \cdot {}_{n-1}P_4$

$\frac{n!}{(n-5)!} = 9 \cdot \frac{(n-1)!}{(n-1-4)!}$

$\frac{n!}{(n-5)!} = 9 \cdot \frac{(n-1)!}{(n-5)!}$

$n! = 9(n-1)!$

$n(n-1)! = 9(n-1)!$

$n = 9$

51. $_nP_4 = 8 \cdot {}_nP_3$

$\frac{n!}{(n-4)!} = 8 \cdot \frac{n!}{(n-3)!}$

$(n-3)! = 8(n-4)!$ Multiplying by $\frac{(n-4)!(n-3)!}{n!}$

$(n-3)(n-4)! = 8(n-4)!$

$n-3 = 8$ Dividing by $(n-4)!$

$n = 11$

52. $n! = n(n-1)(n-2)(n-3)(n-4)\cdots 1 = n(n-1)(n-2)[(n-3)(n-4)\cdots 1] = n(n-1)(n-2)(n-3)!$

53. There is one losing team per game. In order to leave one tournament winner there must be $n-1$ losers produced in $n-1$ games.

54. 2 losses for each of $(n-1)$ losing teams means $2n-2$ losses. The tournament winner will have lost at most 1 game; thus at most there are $(2n-2)+1$ or $(2n-1)$ losses requiring $2n-1$ games.

Exercise Set 7.6

1. $_{13}C_2 = \frac{13!}{2!(13-2)!}$

$= \frac{13!}{2!11!} = \frac{13 \cdot 12 \cdot 11!}{2 \cdot 1 \cdot 11!}$

$= \frac{13 \cdot 12}{2 \cdot 1} = \frac{13 \cdot 6 \cdot 2}{2 \cdot 1}$

$= 78$

2. $_9C_6 = \frac{9!}{6!(9-6)!}$

$= \frac{9!}{6!3!} = \frac{9 \cdot 8 \cdot 7 \cdot 6!}{6! \cdot 3 \cdot 2 \cdot 1}$

$= 84$

3. $\binom{13}{11} = \frac{13!}{11!(13-11)!}$

$= \frac{13!}{11!2!}$

$= 78$ (See Exercise 1.)

4. $\binom{9}{3} = \frac{9!}{3!(9-3)!}$

$= \frac{9!}{3!6!}$

$= 84$ (See Exercise 2.)

5. $\binom{7}{1} = \frac{7!}{1!(7-1)!}$

$= \frac{7!}{1!6!} = \frac{7 \cdot 6!}{1 \cdot 6!}$

$= 7$

6. $\binom{8}{8} = \frac{8!}{8!(8-8)!}$

$= \frac{8!}{8!0!} = \frac{8!}{8! \cdot 1}$

$= 1$

7. $\frac{_5P_3}{3!} = \frac{5 \cdot 4 \cdot 3}{3!}$

$= \frac{5 \cdot 4 \cdot 3}{3 \cdot 2 \cdot 1} = \frac{5 \cdot 2 \cdot 2 \cdot 3}{3 \cdot 2 \cdot 1}$

$= 5 \cdot 2 = 10$

8. $\frac{_{10}P_5}{5!} = \frac{10 \cdot 9 \cdot 8 \cdot 7 \cdot 6}{5 \cdot 4 \cdot 3 \cdot 2 \cdot 1} = 252$

9. $\binom{6}{0} = \frac{6!}{0!(6-0)!}$

$= \frac{6!}{0!6!} = \frac{6!}{6! \cdot 1}$

$= 1$

10. $\binom{6}{1} = \frac{6}{1} = 6$

11. $\binom{6}{2} = \frac{6 \cdot 5}{2 \cdot 1} = 15$

12. $\binom{6}{3} = \frac{6!}{3!(6-3)!}$

$= \frac{6!}{3!3!} = \frac{6 \cdot 5 \cdot 4 \cdot 3!}{3! \cdot 3 \cdot 2 \cdot 1}$

$= 20$

13. $\binom{n}{r} = \binom{n}{n-r}$, so

$$\binom{7}{0}+\binom{7}{1}+\binom{7}{2}+\binom{7}{3}+\binom{7}{4}+\binom{7}{5}+\binom{7}{6}+\binom{7}{7}$$
$$= 2\left[\binom{7}{0}+\binom{7}{1}+\binom{7}{2}+\binom{7}{3}\right]$$
$$= 2\left[\frac{7!}{7!0!}+\frac{7!}{6!1!}+\frac{7!}{5!2!}+\frac{7!}{4!3!}\right]$$
$$= 2(1+7+21+35) = 2\cdot 64 = 128$$

14. $$\binom{6}{0}+\binom{6}{1}+\binom{6}{2}+\binom{6}{3}+\binom{6}{4}+\binom{6}{5}+\binom{6}{6}$$
$$= 2\left[\binom{6}{0}+\binom{6}{1}+\binom{6}{2}\right]+\binom{6}{3}$$
$$= 2(1+6+15)+20 = 64$$

15. Use a grapher.

$_{52}C_4 = 270,725$

16. $_{52}C_5 = 2,598,960$

17. Use a grapher.

$\binom{27}{11} = {_{27}C_{11}} = 13,037,895$

18. $\binom{37}{8} = 38,608,020$

19. $\binom{n}{1} = \frac{n!}{1!(n-1)!} = \frac{n(n-1)!}{1!(n-1)!} = n$

20. $\binom{n}{3} = \frac{n!}{3!(n-3)!} =$

$\frac{n(n-1)(n-2)(n-3)!}{3\cdot 2\cdot 1\cdot (n-3)!} = \frac{n(n-1)(n-2)}{6}$

21. $\binom{m}{m} = \frac{m!}{m!(m-m)!} = \frac{m!}{m!0!} = 1$

22. $\binom{t}{4} = \frac{t!}{4!(t-4)!} = \frac{t(t-1)(t-2)(t-3)(t-4)!}{4\cdot 3\cdot 2\cdot 1\cdot (t-4)!} =$

$\frac{t(t-1)(t-2)(t-3)}{12}$

23. $$_{23}C_4 = \frac{23!}{4!(23-4)!}$$
$$= \frac{23!}{4!19!} = \frac{23\cdot 22\cdot 21\cdot 20\cdot 19!}{4\cdot 3\cdot 2\cdot 1\cdot 19!}$$
$$= \frac{23\cdot 22\cdot 21\cdot 20}{4\cdot 3\cdot 2\cdot 1} = \frac{23\cdot 2\cdot 11\cdot 3\cdot 7\cdot 4\cdot 5}{4\cdot 3\cdot 2\cdot 1}$$
$$= 8855$$

24. Playing all other teams once: $_9C_2 = 36$

Playing all other teams twice: $2\cdot {_9C_2} = 72$

25. $$_{13}C_{10} = \frac{13!}{10!(13-10)!}$$
$$= \frac{13!}{10!3!} = \frac{13\cdot 12\cdot 11\cdot 10!}{10!\cdot 3\cdot 2\cdot 1}$$
$$= \frac{13\cdot 12\cdot 11}{3\cdot 2\cdot 1} = \frac{13\cdot 3\cdot 2\cdot 2\cdot 11}{3\cdot 2\cdot 1}$$
$$= 286$$

26. $_{10}C_7 \cdot {_5C_3} = \binom{10}{7}\cdot\binom{5}{3}$ Using the fundamental counting principle

$$= \frac{10!}{7!(10-7)!}\cdot\frac{5!}{3!(5-3)!}$$
$$= \frac{10\cdot 9\cdot 8\cdot 7!}{7!\cdot 3!}\cdot\frac{5\cdot 4\cdot 3!}{3!\cdot 2!}$$
$$= \frac{10\cdot 9\cdot 8}{3\cdot 2\cdot 1}\cdot\frac{5\cdot 4}{2\cdot 1} = 120\cdot 10 = 1200$$

27. Since two points determine a line and no three of these 8 points are colinear, we need to find the number of combinations of 8 points taken 2 at a time, $_8C_2$.

$$_8C_2 = \binom{8}{2} = \frac{8!}{2!(8-2)!}$$
$$= \frac{8\cdot 7\cdot 6!}{2\cdot 1\cdot 6!} = \frac{4\cdot 2\cdot 7}{2\cdot 1}$$
$$= 28$$

Thus 28 lines are determined.

Since three noncolinear points determine a triangle, we need to find the number of combinations of 8 points taken 3 at a time, $_8C_3$.

$$_8C_3 = \binom{8}{3} = \frac{8!}{3!(8-3)!}$$
$$= \frac{8\cdot 7\cdot 6\cdot 5!}{3\cdot 2\cdot 1\cdot 5!} = \frac{8\cdot 7\cdot 3\cdot 2}{3\cdot 2\cdot 1}$$
$$= 56$$

Thus 56 triangles are determined.

28. Using the fundamental counting principle, we have $_{58}C_6 \cdot {_{42}C_4}$.

29. $_{52}C_5 = 2,598,960$

30. $_{52}C_{13} = 635,013,559,600$

31. a) $_{31}P_2 = 930$

b) $31\cdot 31 = 961$

c) $_{31}C_2 = 465$

32. Order is considered in a combination lock.

33. Choosing k objects from a set of n objects is equivalent to not choosing the other $n-k$ objects.

34. $$3x - 7 = 5x + 10$$
$$-17 = 2x$$
$$-\frac{17}{2} = x, \text{ or}$$
$$-8.5 = x$$

We could also graph $y_1 = 3x-7$ and $y_2 = 5x+10$ and use the Intersect feature to find the first coordinate of the point of intersection of the graphs. The solution is $-\frac{17}{2}$, or -8.5.

35.
$$\begin{aligned} 2x^2 - x &= 3 \\ 2x^2 - x - 3 &= 0 \\ (2x-3)(x+1) &= 0 \\ 2x - 3 = 0 \quad &or \quad x+1 = 0 \\ 2x = 3 \quad &or \quad x = -1 \\ x = \frac{3}{2} \quad &or \quad x = -1 \end{aligned}$$

We could also graph $y_1 = 2x^2 - x$ and $y_2 = 3$ and use the Intersect feature twice to find the first coordinates of the points of intersection of the graphs. The solutions are $\frac{3}{2}$ and -1, or 1.5 and -1.

36. $x^2 + 5x + 1 = 0$
$$\begin{aligned} x &= \frac{-5 \pm \sqrt{5^2 - 4 \cdot 1 \cdot 1}}{2 \cdot 1} \\ &= \frac{-5 \pm \sqrt{21}}{2} \end{aligned}$$

37.
$$\begin{aligned} x^3 + 3x^2 - 10x &= 24 \\ x^3 + 3x^2 - 10x - 24 &= 0 \end{aligned}$$

We use synthetic division to find one factor of the polynomial on the left side of the equation. We try $x - 3$.

3	1	3	−10	−24
		3	18	24
	1	6	8	0

Now we have:
$$\begin{aligned} (x-3)(x^2 + 6x + 8) &= 0 \\ (x-3)(x+2)(x+4) &= 0 \end{aligned}$$
$$x - 3 = 0 \quad or \quad x + 2 = 0 \quad or \quad x + 4 = 0$$
$$x = 3 \quad or \quad x = -2 \quad or \quad x = -4$$

We could also graph $y_1 = x^3 + 3x^2 - 10x$ and $y_2 = 24$ and use the Intersect feature three times to find the first coordinates of the points of intersection of the graphs. The solutions are -4, -2, and 3.

38. ${}_4C_3 \cdot {}_4C_2 = 24$

39. There are 13 diamonds, and we choose 5. We have ${}_{13}C_5 = 1287$.

40. ${}_nC_4$

41. Playing once: ${}_nC_2$

Playing twice: $2 \cdot {}_nC_2$

42.
$$\begin{aligned} \binom{n+1}{3} &= 2 \cdot \binom{n}{2} \\ \frac{(n+1)!}{(n+1-3)!3!} &= 2 \cdot \frac{n!}{(n-2)!2!} \\ \frac{(n+1)!}{(n-2)!3!} &= 2 \cdot \frac{n!}{(n-2)!2!} \\ \frac{(n+1)(n)(n-1)(n-2)!}{(n-2)!3 \cdot 2 \cdot 1} &= 2 \cdot \frac{n(n-1)(n-2)!}{(n-2)! \cdot 2 \cdot 1} \\ \frac{(n+1)(n)(n-1)}{6} &= n(n-1) \\ \frac{n^3 - n}{6} &= n^2 - n \\ n^3 - n &= 6n^2 - 6n \\ n^3 - 6n^2 + 5n &= 0 \\ n(n^2 - 6n + 5) &= 0 \\ n(n-5)(n-1) &= 0 \end{aligned}$$
$$n = 0 \text{ or } n = 5 \text{ or } n = 1$$

Only 5 checks. The solution is 5.

43.
$$\begin{aligned} \binom{n}{n-2} &= 6 \\ \frac{n!}{(n-(n-2))!(n-2)!} &= 6 \\ \frac{n!}{2!(n-2)!} &= 6 \\ \frac{n(n-1)(n-2)!}{2 \cdot 1 \cdot (n-2)!} &= 6 \\ \frac{n(n-1)}{2} &= 6 \\ n(n-1) &= 12 \\ n^2 - n &= 12 \\ n^2 - n - 12 &= 0 \\ (n-4)(n+3) &= 0 \end{aligned}$$
$$n = 4 \text{ or } n = -3$$

Only 4 checks. The solution is 4.

44.
$$\begin{aligned} \binom{n}{3} &= 2 \cdot \binom{n-1}{2} \\ \frac{n!}{(n-3)!3!} &= 2 \cdot \frac{(n-1)!}{(n-1-2)!2!} \\ \frac{n!}{(n-3)!3!} &= 2 \cdot \frac{(n-1)!}{(n-3)!2!} \\ \frac{n!}{3!} &= 2 \cdot \frac{(n-1)!}{2!} \\ n! &= 3!(n-1)! \\ \frac{n(n-1)!}{(n-1)!} &= 6 \\ n &= 6 \end{aligned}$$

This number checks. The solution is 6.

45.
$$\binom{n+2}{4} = 6 \cdot \binom{n}{2}$$
$$\frac{(n+2)!}{(n+2-4)!4!} = 6 \cdot \frac{n!}{(n-2)!2!}$$
$$\frac{(n+2)!}{(n-2)!4!} = 6 \cdot \frac{n!}{(n-2)!2!}$$
$$\frac{(n+2)!}{4!} = 6 \cdot \frac{n!}{2!} \quad \text{Multiplying by } (n-2)!$$
$$4! \cdot \frac{(n+2)!}{4!} = 4! \cdot 6 \cdot \frac{n!}{2!}$$
$$(n+2)! = 72 \cdot n!$$
$$(n+2)(n+1)n! = 72 \cdot n!$$
$$(n+2)(n+1) = 72 \quad \text{Dividing by } n!$$
$$n^2 + 3n + 2 = 72$$
$$n^2 + 3n - 70 = 0$$
$$(n+10)(n-7) = 0$$
$$n = -10 \text{ or } n = 7$$

Only 7 checks. The solution is 7.

46. Line segments: $_nC_2 = \dfrac{n!}{2!(n-2)!} = \dfrac{n(n-1)(n-2)!}{2 \cdot 1 \cdot (n-2)!} = \dfrac{n(n-1)}{2}$

Diagonals: The n line segments that form the sides of the n-agon are not diagonals. Thus, the number of diagonals is $_nC_2 - n = \dfrac{n(n-1)}{2} - n = \dfrac{n^2-n-2n}{2} = \dfrac{n^2-3n}{2} = \dfrac{n(n-3)}{2}$, $n \geq 4$.

Let D_n be the number of diagonals on an n-agon. Prove the result above for diagonals using mathematical induction.

$$S_n: \ D_n = \frac{n(n-3)}{2}, \text{ for } n = 4, 5, 6, ...$$
$$S_4: \ D_4 = \frac{4 \cdot 1}{2}$$
$$S_k: \ D_k = \frac{k(k-3)}{2}$$
$$S_{k+1}: \ D_{k+1} = \frac{(k+1)(k-2)}{2}$$

1) *Basis step:* S_4 is true (a quadrilateral has 2 diagonals).

2) *Induction step:* Assume S_k. Observe that when an additional vertex V_{k+1} is added to the k-gon, we gain k segments, 2 of which are sides of the $(k+1)$-gon], and a former side $\overline{V_1V_k}$ becomes a diagonal. Thus the additional number of diagonals is $k-2+1$, or $k-1$. Then the new total of diagonals is $D_k + (k-1)$, or

$$D_{k+1} = D_k + (k-1)$$
$$= \frac{k(k-3)}{2} + (k-1) \quad (\text{by } S_k)$$
$$= \frac{(k+1)(k-2)}{2}$$

47. See the answer section in the text.

Exercise Set 7.7

1. Expand: $(x+5)^4$.

We have $a = x$, $b = 5$, and $n = 4$.

Pascal's triangle method: Use the fifth row of Pascal's triangle.

1 4 6 4 1

$$(x+5)^4$$
$$= 1 \cdot x^4 + 4 \cdot x^3 \cdot 5 + 6 \cdot x^2 \cdot 5^2 + 4 \cdot x \cdot 5^3 + 1 \cdot 5^4$$
$$= x^4 + 20x^3 + 150x^2 + 500x + 625$$

Factorial notation method:

$$(x+5)^4$$
$$= \binom{4}{0}x^4 + \binom{4}{1}x^3 \cdot 5 + \binom{4}{2}x^2 \cdot 5^2 + \binom{4}{3}x \cdot 5^3 + \binom{4}{4}5^4$$
$$= \frac{4!}{0!4!}x^4 + \frac{4!}{1!3!}x^3 \cdot 5 + \frac{4!}{2!2!}x^2 \cdot 5^2 + \frac{4!}{3!1!}x \cdot 5^3 + \frac{4!}{4!0!}5^4$$
$$= x^4 + 20x^3 + 150x^2 + 500x + 625$$

2. Expand: $(x-1)^4$.

Pascal's triangle method: Use the 5th row of Pascal's triangle.

1 4 6 4 1

$$(x-1)^4$$
$$= 1 \cdot x^4 + 4 \cdot x^3(-1) + 6x^2(-1)^2 + 4x(-1)^3 + 1 \cdot (-1)^4$$
$$= x^4 - 4x^3 + 6x^2 - 4x + 1$$

Factorial notation method:

$$(x-1)^4$$
$$= \binom{4}{0}x^4 + \binom{4}{1}x^3(-1) + \binom{4}{2}x^2(-1)^2 + \binom{4}{3}x(-1)^3 + \binom{4}{4}(-1)^4$$
$$= x^4 - 4x^3 + 6x^2 - 4x + 1$$

3. Expand: $(x-3)^5$.

We have $a = x$, $b = -3$, and $n = 5$.

Pascal's triangle method: Use the sixth row of Pascal's triangle.

1 5 10 10 5 1

$$(x-3)^5$$
$$= 1 \cdot x^5 + 5x^4(-3) + 10x^3(-3)^2 + 10x^2(-3)^3 + 5x(-3)^4 + 1 \cdot (-3)^5$$
$$= x^5 - 15x^4 + 90x^3 - 270x^2 + 405x - 243$$

Factorial notation method:

$$(x-3)^5$$
$$= \binom{5}{0}x^5 + \binom{5}{1}x^4(-3) + \binom{5}{2}x^3(-3)^2 + \binom{5}{3}x^2(-3)^3 + \binom{5}{4}x(-3)^4 + \binom{5}{5}(-3)^5$$
$$= \frac{5!}{0!5!}x^5 + \frac{5!}{1!4!}x^4(-3) + \frac{5!}{2!3!}x^3(9) + \frac{5!}{3!2!}x^2(-27) + \frac{5!}{4!1!}x(81) + \frac{5!}{5!0!}(-243)$$
$$= x^5 - 15x^4 + 90x^3 - 270x^2 + 405x - 243$$

4. Expand: $(x+2)^9$.

Pascal's triangle method: Use the 10th row of Pascal's triangle.

1 9 36 84 126 126 84 36 9 1

$$(x+2)^9$$
$$= 1\cdot x^9 + 9x^8\cdot 2 + 36x^7\cdot 2^2 + 84x^6\cdot 2^3 + 126x^5\cdot 2^4 + 126x^4\cdot 2^5 + 84x^3\cdot 2^6 + 36x^2\cdot 2^7 + 9x\cdot 2^8 + 1\cdot 2^9$$
$$= x^9 + 18x^8 + 144x^7 + 672x^6 + 2016x^5 + 4032x^4 + 5376x^3 + 4608x^2 + 2304x + 512$$

Factorial notation method:

$$(x+2)^9$$
$$= \binom{9}{0}x^9 + \binom{9}{1}x^8\cdot 2 + \binom{9}{2}x^7\cdot 2^2 + \binom{9}{3}x^6\cdot 2^3 + \binom{9}{4}x^5\cdot 2^4 + \binom{9}{5}x^4\cdot 2^5 + \binom{9}{6}x^3\cdot 2^6 + \binom{9}{7}x^2\cdot 2^7 + \binom{9}{8}x\cdot 2^8 + \binom{9}{9}2^9$$
$$= x^9 + 18x^8 + 144x^7 + 672x^6 + 2016x^5 + 4032x^4 + 5376x^3 + 4608x^2 + 2304x + 512$$

5. Expand: $(x-y)^5$.

We have $a = x$, $b = -y$, and $n = 5$.

Pascal's triangle method: We use the sixth row of Pascal's triangle.

1 5 10 10 5 1

$$(x-y)^5$$
$$= 1\cdot x^5 + 5x^4(-y) + 10x^3(-y)^2 + 10x^2(-y)^3 + 5x(-y)^4 + 1\cdot(-y)^5$$
$$= x^5 - 5x^4y + 10x^3y^2 - 10x^2y^3 + 5xy^4 - y^5$$

Factorial notation method:

$$(x-y)^5$$
$$= \binom{5}{0}x^5 + \binom{5}{1}x^4(-y) + \binom{5}{2}x^3(-y)^2 + \binom{5}{3}x^2(-y)^3 + \binom{5}{4}x(-y)^4 + \binom{5}{5}(-y)^5$$
$$= \frac{5!}{0!5!}x^5 + \frac{5!}{1!4!}x^4(-y) + \frac{5!}{2!3!}x^3(y^2) + \frac{5!}{3!2!}x^2(-y^3) + \frac{5!}{4!1!}x(y^4) + \frac{5!}{5!0!}(-y^5)$$
$$= x^5 - 5x^4y + 10x^3y^2 - 10x^2y^3 + 5xy^4 - y^5$$

6. Expand: $(x+y)^8$.

Pascal's triangle method: Use the ninth row of Pascal's triangle.

1 8 28 56 70 56 28 8 1

$$(x+y)^8$$
$$= x^8 + 8x^7y + 28x^6y^2 + 56x^5y^3 + 70x^4y^4 + 56x^3y^5 + 28x^2y^6 + 8xy^7 + y^8$$

Factorial notation method:

$$(x+y)^8$$
$$= \binom{8}{0}x^8 + \binom{8}{1}x^7y + \binom{8}{2}x^6y^2 + \binom{8}{3}x^5y^3 + \binom{8}{4}x^4y^4 + \binom{8}{5}x^3y^5 + \binom{8}{6}x^2y^6 + \binom{8}{7}xy^7 + \binom{8}{8}y^8$$
$$= x^8 + 8x^7y + 28x^6y^2 + 56x^5y^3 + 70x^4y^4 + 56x^3y^5 + 28x^2y^6 + 8xy^7 + y^8$$

7. Expand: $(5x+4y)^6$.

We have $a = 5x$, $b = 4y$, and $n = 6$.

Pascal's triangle method: Use the seventh row of Pascal's triangle.

1 6 15 20 15 6 1

$$(5x+4y)^6$$
$$= 1\cdot(5x)^6 + 6\cdot(5x)^5(4y) + 15(5x)^4(4y)^2 + 20(5x)^3(4y)^3 + 15(5x)^2(4y)^4 + 6(5x)(4y)^5 + 1\cdot(4y)^6$$
$$= 15,625x^6 + 75,000x^5y + 150,000x^4y^2 + 160,000x^3y^3 + 96,000x^2y^4 + 30,720xy^5 + 4096y^6$$

Factorial notation method:

$$(5x+4y)^6$$
$$= \binom{6}{0}(5x)^6 + \binom{6}{1}(5x)^5(4y) + \binom{6}{2}(5x)^4(4y)^2 + \binom{6}{3}(5x)^3(4y)^3 + \binom{6}{4}(5x)^2(4y)^4 + \binom{6}{5}(5x)(4y)^5 + \binom{6}{6}(4y)^6$$

$= \frac{6!}{0!6!}(15,625x^6) + \frac{6!}{1!5!}(3125x^5)(4y) +$
$\frac{6!}{2!4!}(625x^4)(16y^2) + \frac{6!}{3!3!}(125x^3)(64y^3) +$
$\frac{6!}{4!2!}(25x^2)(256y^4) + \frac{6!}{5!1!}(5x)(1024y^5) +$
$\frac{6!}{6!0!}(4096y^6)$
$= 15,625x^6 + 75,000x^5y + 150,000x^4y^2 +$
$160,000x^3y^3 + 96,000x^2y^4 + 30,720xy^5 +$
$4096y^6$

8. Expand: $(2x - 3y)^5$.

Pascal's triangle method: Use the sixth row of Pascal's triangle.

1 5 10 10 5 1

$(2x - 3y)^5$
$= 1 \cdot (2x)^5 + 5(2x)^4(-3y) + 10(2x)^3(-3y)^2 +$
$10(2x)^2(-3y)^3 + 5(2x)(-3y)^4 + 1 \cdot (-3y)^5$
$= 32x^5 - 240x^4y + 720x^3y^2 - 1080x^2y^3 +$
$810xy^4 - 243y^5$

Factorial notation method:

$(2x - 3y)^5$
$= \binom{5}{0}(2x)^5 + \binom{5}{1}(2x)^4(-3y) +$
$\binom{5}{2}(2x)^3(-3y)^2 + \binom{5}{3}(2x)^2(-3y)^3 +$
$\binom{5}{4}(2x)(-3y)^4 + \binom{5}{5}(-3y)^5$
$= 32x^5 - 240x^4y + 720x^3y^2 - 1080x^2y^3 +$
$810xy^4 - 243y^5$

9. Expand: $\left(2t + \frac{1}{t}\right)^7$.

We have $a = 2t$, $b = \frac{1}{t}$, and $n = 7$.

Pascal's triangle method: Use the eighth row of Pascal's triangle.

1 7 21 35 35 21 7 1

$\left(2t + \frac{1}{t}\right)^7$
$= 1 \cdot (2t)^7 + 7(2t)^6\left(\frac{1}{t}\right) + 21(2t)^5\left(\frac{1}{t}\right)^2 +$
$35(2t)^4\left(\frac{1}{t}\right)^3 + 35(2t)^3\left(\frac{1}{t}\right)^4 + 21(2t)^2\left(\frac{1}{t}\right)^5 +$
$7(2t)\left(\frac{1}{t}\right)^6 + 1 \cdot \left(\frac{1}{t}\right)^7$
$= 128t^7 + 7 \cdot 64t^6 \cdot \frac{1}{t} + 21 \cdot 32t^5 \cdot \frac{1}{t^2} +$
$35 \cdot 16t^4 \cdot \frac{1}{t^3} + 35 \cdot 8t^3 \cdot \frac{1}{t^4} + 21 \cdot 4t^2 \cdot \frac{1}{t^5} +$
$7 \cdot 2t \cdot \frac{1}{t^6} + \frac{1}{t^7}$
$= 128t^7 + 448t^5 + 672t^3 + 560t + 280t^{-1} +$
$84t^{-3} + 14t^{-5} + t^{-7}$

Factorial notation method:

$\left(2t + \frac{1}{t}\right)^7$
$= \binom{7}{0}(2t)^7 + \binom{7}{1}(2t)^6\left(\frac{1}{t}\right) +$
$\binom{7}{2}(2t)^5\left(\frac{1}{t}\right)^2 + \binom{7}{3}(2t)^4\left(\frac{1}{t}\right)^3 +$
$\binom{7}{4}(2t)^3\left(\frac{1}{t}\right)^4 + \binom{7}{5}(2t)^2\left(\frac{1}{t}\right)^5 +$
$\binom{7}{6}(2t)\left(\frac{1}{t}\right)^6 + \binom{7}{7}\left(\frac{1}{t}\right)^7$
$= \frac{7!}{0!7!}(128t^7) + \frac{7!}{1!6!}(64t^6)\left(\frac{1}{t}\right) + \frac{7!}{2!5!}(32t^5)\left(\frac{1}{t^2}\right) +$
$\frac{7!}{3!4!}(16t^4)\left(\frac{1}{t^3}\right) + \frac{7!}{4!3!}(8t^3)\left(\frac{1}{t^4}\right) +$
$\frac{7!}{5!2!}(4t^2)\left(\frac{1}{t^5}\right) + \frac{7!}{6!1!}(2t)\left(\frac{1}{t^6}\right) + \frac{7!}{7!0!}\left(\frac{1}{t^7}\right)$
$= 128t^7 + 448t^5 + 672t^3 + 560t + 280t^{-1} +$
$84t^{-3} + 14t^{-5} + t^{-7}$

10. Expand: $\left(3y - \frac{1}{y}\right)^4$.

Pascal's triangle method: Use the fifth row of Pascal's triangle.

1 4 6 4 1

$\left(3y - \frac{1}{y}\right)^4$
$= 1 \cdot (3y)^4 + 4(3y)^3\left(-\frac{1}{y}\right) + 6(3y)^2\left(-\frac{1}{y}\right)^2 +$
$4(3y)\left(-\frac{1}{y}\right)^3 + 1 \cdot \left(-\frac{1}{y}\right)^4$
$= 81y^4 - 108y^2 + 54 - 12y^{-2} + y^{-4}$

Factorial notation method:

$\left(3y - \frac{1}{y}\right)^4$
$= \binom{4}{0}(3y)^4 + \binom{4}{1}(3y)^3\left(-\frac{1}{y}\right) +$
$\binom{4}{2}(3y)^2\left(-\frac{1}{y}\right)^2 + \binom{4}{3}(3y)\left(-\frac{1}{y}\right)^3 +$
$\binom{4}{4}\left(-\frac{1}{y}\right)^4$
$= 81y^4 - 108y^2 + 54 - 12y^{-2} + y^{-4}$

11. Expand: $(x^2-1)^5$.

We have $a=x^2$, $b=-1$, and $n=5$.

Pascal's triangle method: Use the sixth row of Pascal's triangle.

1 5 10 10 5 1

$$\begin{aligned}&(x^2-1)^5\\&=1\cdot(x^2)^5+5(x^2)^4(-1)+10(x^2)^3(-1)^2+\\&\quad 10(x^2)^2(-1)^3+5(x^2)(-1)^4+1\cdot(-1)^5\\&=x^{10}-5x^8+10x^6-10x^4+5x^2-1\end{aligned}$$

Factorial notation method:

$$\begin{aligned}&(x^2-1)^5\\&=\binom{5}{0}(x^2)^5+\binom{5}{1}(x^2)^4(-1)+\\&\quad\binom{5}{2}(x^2)^3(-1)^2+\binom{5}{3}(x^2)^2(-1)^3+\\&\quad\binom{5}{4}(x^2)(-1)^4+\binom{5}{5}(-1)^5\\&=\frac{5!}{0!5!}(x^{10})+\frac{5!}{1!4!}(x^8)(-1)+\frac{5!}{2!3!}(x^6)(1)+\\&\quad\frac{5!}{3!2!}(x^4)(-1)+\frac{5!}{4!1!}(x^2)(1)+\frac{5!}{5!0!}(-1)\\&=x^{10}-5x^8+10x^6-10x^4+5x^2-1\end{aligned}$$

12. Expand: $(1+2q^3)^8$.

Pascal's triangle method: Use the ninth row of Pascal's triangle.

1 8 28 56 70 56 28 8 1

$$\begin{aligned}&(1+2q^3)^8\\&=1\cdot1^8+8\cdot1^7(2q^3)+28\cdot1^6(2q^3)^2+56\cdot1^5(2q^3)^3+\\&\quad70\cdot1^4(2q^3)^4+56\cdot1^3(2q^3)^5+28\cdot1^2(2q^3)^6+\\&\quad8\cdot1(2q^3)^7+1\cdot(2q^3)^8\\&=1+16q^3+112q^6+448q^9+1120q^{12}+1792q^{15}+\\&\quad1792q^{18}+1024q^{21}+256q^{24}\end{aligned}$$

Factorial notation method:

$$\begin{aligned}&(1+2q^3)^8\\&=\binom{8}{0}(1)^8+\binom{8}{1}(1)^7(2q^3)+\binom{8}{2}(1)^6(2q^3)^2+\\&\quad\binom{8}{3}(1)^5(2q^3)^3+\binom{8}{4}(1)^4(2q^3)^4+\\&\quad\binom{8}{5}(1)^3(2q^3)^5+\binom{8}{6}(1)^2(2q^3)^6+\\&\quad\binom{8}{7}(1)(2q^3)^7+\binom{8}{8}(2q^3)^8\\&=1+16q^3+112q^6+448q^9+1120q^{12}+1792q^{15}+\\&\quad1792q^{18}+1024q^{21}+256q^{24}\end{aligned}$$

13. Expand: $(\sqrt{5}+t)^6$.

We have $a=\sqrt{5}$, $b=t$, and $n=6$.

Pascal's triangle method: We use the seventh row of Pascal's triangle:

1 6 15 20 15 6 1

$$\begin{aligned}(\sqrt{5}+t)^6&=1\cdot(\sqrt{5})^6+6(\sqrt{5})^5(t)+\\&\quad15(\sqrt{5})^4(t^2)+20(\sqrt{5})^3(t^3)+\\&\quad15(\sqrt{5})^2(t^4)+6\sqrt{5}t^5+1\cdot t^6\\&=125+150\sqrt{5}\,t+375t^2+100\sqrt{5}\,t^3+\\&\quad75t^4+6\sqrt{5}\,t^5+t^6\end{aligned}$$

Factorial notation method:

$$\begin{aligned}(\sqrt{5}+t)^6&=\binom{6}{0}(\sqrt{5})^6+\binom{6}{1}(\sqrt{5})^5(t)+\\&\quad\binom{6}{2}(\sqrt{5})^4(t^2)+\binom{6}{3}(\sqrt{5})^3(t^3)+\\&\quad\binom{6}{4}(\sqrt{5})^2(t^4)+\binom{6}{5}(\sqrt{5})(t^5)+\\&\quad\binom{6}{6}(t^6)\\&=\frac{6!}{0!6!}(125)+\frac{6!}{1!5!}(25\sqrt{5})t+\frac{6!}{2!4!}(25)(t^2)+\\&\quad\frac{6!}{3!3!}(5\sqrt{5})(t^3)+\frac{6!}{4!2!}(5)(t^4)+\\&\quad\frac{6!}{5!1!}(\sqrt{5})(t^5)+\frac{6!}{6!0!}(t^6)\\&=125+150\sqrt{5}\,t+375t^2+100\sqrt{5}\,t^3+\\&\quad75t^4+6\sqrt{5}\,t^5+t^6\end{aligned}$$

14. Expand: $(x-\sqrt{2})^6$.

Pascal's triangle method: Use the seventh row of Pascal's triangle.

1 6 15 20 15 6 1

$$\begin{aligned}&(x-\sqrt{2})^6\\&=1\cdot x^6+6x^5(-\sqrt{2})+15x^4(-\sqrt{2})^2+20x^3(-\sqrt{2})^3+\\&\quad15x^2(-\sqrt{2})^4+6x(-\sqrt{2})^5+1\cdot(-\sqrt{2})^6\\&=x^6-6\sqrt{2}x^5+30x^4-40\sqrt{2}x^3+60x^2-24\sqrt{2}x+8\end{aligned}$$

Factorial notation method:

$$\begin{aligned}&(x-\sqrt{2})^6\\&=\binom{6}{0}x^6+\binom{6}{1}(x^5)(-\sqrt{2})+\\&\quad\binom{6}{2}(x^4)(-\sqrt{2})^2+\binom{6}{3}(x^3)(-\sqrt{2})^3+\\&\quad\binom{6}{4}(x^2)(-\sqrt{2})^4+\binom{6}{5}(x)(-\sqrt{2})^5+\\&\quad\binom{6}{6}(-\sqrt{2})^6\\&=x^6-6\sqrt{2}x^5+30x^4-40\sqrt{2}x^3+60x^2-24\sqrt{2}x+8\end{aligned}$$

15. Expand: $\left(a-\frac{2}{a}\right)^9$.

We have $a=a$, $b=-\frac{2}{a}$, and $n=9$.

Pascal's triangle method: Use the tenth row of Pascal's triangle.

1 9 36 84 126 126 84 36 9 1

$$\left(a-\frac{2}{a}\right)^9 = 1\cdot a^9 + 9a^8\left(-\frac{2}{a}\right) + 36a^7\left(-\frac{2}{a}\right)^2 +$$
$$84a^6\left(-\frac{2}{a}\right)^3 + 126a^5\left(-\frac{2}{a}\right)^4 +$$
$$126a^4\left(-\frac{2}{a}\right)^5 + 84a^3\left(-\frac{2}{a}\right)^6 +$$
$$36a^2\left(-\frac{2}{a}\right)^7 + 9a\left(-\frac{2}{a}\right)^8 + 1\cdot\left(-\frac{2}{a}\right)^9$$
$$= a^9 - 18a^7 + 144a^5 - 672a^3 + 2016a -$$
$$4032a^{-1} + 5376a^{-3} - 4608a^{-5} +$$
$$2304a^{-7} - 512a^{-9}$$

Factorial notation method:

$$\left(a-\frac{2}{a}\right)^9$$
$$= \binom{9}{0}a^9 + \binom{9}{1}a^8\left(-\frac{2}{a}\right) + \binom{9}{2}a^7\left(-\frac{2}{a}\right)^2 +$$
$$\binom{9}{3}a^6\left(-\frac{2}{a}\right)^3 + \binom{9}{4}a^5\left(-\frac{2}{a}\right)^4 +$$
$$\binom{9}{5}a^4\left(-\frac{2}{a}\right)^5 + \binom{9}{6}a^3\left(-\frac{2}{a}\right)^6 +$$
$$\binom{9}{7}a^2\left(-\frac{2}{a}\right)^7 + \binom{9}{8}a\left(-\frac{2}{a}\right)^8 +$$
$$\binom{9}{9}\left(-\frac{2}{a}\right)^9$$
$$= \frac{9!}{9!0!}a^9 + \frac{9!}{8!1!}a^8\left(-\frac{2}{a}\right) + \frac{9!}{7!2!}a^7\left(\frac{4}{a^2}\right) +$$
$$\frac{9!}{6!3!}a^6\left(-\frac{8}{a^3}\right) + \frac{9!}{5!4!}a^5\left(\frac{16}{a^4}\right) +$$
$$\frac{9!}{4!5!}a^4\left(-\frac{32}{a^5}\right) + \frac{9!}{3!6!}a^3\left(\frac{64}{a^6}\right) +$$
$$\frac{9!}{2!7!}a^2\left(-\frac{128}{a^7}\right) + \frac{9!}{1!8!}a\left(\frac{256}{a^8}\right) +$$
$$\frac{9!}{0!9!}\left(-\frac{512}{a^9}\right)$$
$$= a^9 - 9(2a^7) + 36(4a^5) - 84(8a^3) + 126(16a) -$$
$$126(32a^{-1}) + 84(64a^{-3}) - 36(128a^{-5}) +$$
$$9(256a^{-7}) - 512a^{-9}$$
$$= a^9 - 18a^7 + 144a^5 - 672a^3 + 2016a - 4032a^{-1} +$$
$$5376a^{-3} - 4608a^{-5} + 2304a^{-7} - 512a^{-9}$$

16. Expand: $(1+3)^n$

Use the factorial notation method.

$$(1+3)^n$$
$$= \binom{n}{0}(1)^n + \binom{n}{1}(1)^{n-1}3 + \binom{n}{2}(1)^{n-2}3^2 +$$
$$\binom{n}{3}(1)^{n-3}3^3 + \cdots + \binom{n}{n-2}(1)^2 3^{n-2} +$$
$$\binom{n}{n-1}(1)3^{n-1} + \binom{n}{n}3^n$$
$$= 1 + 3n + \binom{n}{2}3^2 + \binom{n}{3}3^3 + \cdots +$$
$$\binom{n}{n-2}3^{n-2} + 3^{n-1}n + 3^n$$

17. $(\sqrt{2}+1)^6 - (\sqrt{2}-1)^6$

First, expand $(\sqrt{2}+1)^6$.

$$(\sqrt{2}+1)^6 = \binom{6}{0}(\sqrt{2})^6 + \binom{6}{1}(\sqrt{2})^5(1) +$$
$$\binom{6}{2}(\sqrt{2})^4(1)^2 + \binom{6}{3}(\sqrt{2})^3(1)^3 +$$
$$\binom{6}{4}(\sqrt{2})^2(1)^4 + \binom{6}{5}(\sqrt{2})(1)^5 +$$
$$\binom{6}{6}(1)^6$$
$$= \frac{6!}{6!0!}\cdot 8 + \frac{6!}{5!1!}\cdot 4\sqrt{2} + \frac{6!}{4!2!}\cdot 4 +$$
$$\frac{6!}{3!3!}\cdot 2\sqrt{2} + \frac{6!}{2!4!}\cdot 2 + \frac{6!}{1!5!}\cdot\sqrt{2} + \frac{6!}{0!6!}$$
$$= 8 + 24\sqrt{2} + 60 + 40\sqrt{2} + 30 + 6\sqrt{2} + 1$$
$$= 99 + 70\sqrt{2}$$

Next, expand $(\sqrt{2}-1)^6$.

$$(\sqrt{2}-1)^6$$
$$= \binom{6}{0}(\sqrt{2})^6 + \binom{6}{1}(\sqrt{2})^5(-1) +$$
$$\binom{6}{2}(\sqrt{2})^4(-1)^2 + \binom{6}{3}(\sqrt{2})^3(-1)^3 +$$
$$\binom{6}{4}(\sqrt{2})^2(-1)^4 + \binom{6}{5}(\sqrt{2})(-1)^5 +$$
$$\binom{6}{6}(-1)^6$$
$$= \frac{6!}{6!0!}\cdot 8 - \frac{6!}{5!1!}\cdot 4\sqrt{2} + \frac{6!}{4!2!}\cdot 4 - \frac{6!}{3!3!}\cdot 2\sqrt{2} +$$
$$\frac{6!}{2!4!}\cdot 2 - \frac{6!}{1!5!}\cdot\sqrt{2} + \frac{6!}{0!6!}$$
$$= 8 - 24\sqrt{2} + 60 - 40\sqrt{2} + 30 - 6\sqrt{2} + 1$$
$$= 99 - 70\sqrt{2}$$
$$(\sqrt{2}+1)^6 - (\sqrt{2}-1)^6$$
$$= (99 + 70\sqrt{2}) - (99 - 70\sqrt{2})$$
$$= 99 + 70\sqrt{2} - 99 + 70\sqrt{2}$$
$$= 140\sqrt{2}$$

18. $(1-\sqrt{2})^4 = 1\cdot 1^4 + 4\cdot 1^3(-\sqrt{2}) + 6\cdot 1^2(-\sqrt{2})^2 + 4\cdot 1\cdot(-\sqrt{2})^3 + 1\cdot(-\sqrt{2})^4$

$$= 1 - 4\sqrt{2} + 12 - 8\sqrt{2} + 4$$
$$= 17 - 12\sqrt{2}$$

$(1+\sqrt{2})^4 = 1 + 4\sqrt{2} + 12 + 8\sqrt{2} + 4$ Using the result above

$$= 17 + 12\sqrt{2}$$

$$(1-\sqrt{2})^4 + (1+\sqrt{2})^4 = 17 - 12\sqrt{2} + 17 + 12\sqrt{2} = 34$$

19. Expand: $(x^{-2}+x^2)^4$.

We have $a = x^{-2}$, $b = x^2$, and $n = 4$.

Pascal's triangle method: Use the fifth row of Pascal's triangle.

1 4 6 4 1.

$$(x^{-2}+x^2)^4$$
$$= 1\cdot(x^{-2})^4 + 4(x^{-2})^3(x^2) + 6(x^{-2})^2(x^2)^2 + 4(x^{-2})(x^2)^3 + 1\cdot(x^2)^4$$
$$= x^{-8} + 4x^{-4} + 6 + 4x^4 + x^8$$

Factorial notation method:

$$(x^{-2}+x^2)^4$$
$$= \binom{4}{0}(x^{-2})^4 + \binom{4}{1}(x^{-2})^3(x^2) + \binom{4}{2}(x^{-2})^2(x^2)^2 + \binom{4}{3}(x^{-2})(x^2)^3 + \binom{4}{4}(x^2)^4$$
$$= \frac{4!}{4!0!}(x^{-8}) + \frac{4!}{3!1!}(x^{-6})(x^2) + \frac{4!}{2!2!}(x^{-4})(x^4) + \frac{4!}{1!3!}(x^{-2})(x^6) + \frac{4!}{0!4!}(x^8)$$
$$= x^{-8} + 4x^{-4} + 6 + 4x^4 + x^8$$

20. Expand: $\left(\frac{1}{\sqrt{x}} - \sqrt{x}\right)^6$.

Pascal's triangle method: We use the seventh row of Pascal's triangle:

1 6 15 20 15 6 1

$$\left(\frac{1}{\sqrt{x}} - \sqrt{x}\right)^6$$
$$= 1\cdot\left(\frac{1}{\sqrt{x}}\right)^6 + 6\left(\frac{1}{\sqrt{x}}\right)^5(-\sqrt{x}) + 15\left(\frac{1}{\sqrt{x}}\right)^4(-\sqrt{x})^2 + 20\left(\frac{1}{\sqrt{x}}\right)^3(-\sqrt{x})^3 + 15\left(\frac{1}{\sqrt{x}}\right)^2(-\sqrt{x})^4 + 6\left(\frac{1}{\sqrt{x}}\right)(-\sqrt{x})^5 + 1\cdot(\sqrt{x})^6$$
$$= x^{-3} - 6x^{-2} + 15x^{-1} - 20 + 15x - 6x^2 + x^3$$

Factorial notation method:

$$\left(\frac{1}{\sqrt{x}} - \sqrt{x}\right)^6$$
$$= \binom{6}{0}\left(\frac{1}{\sqrt{x}}\right)^6 + \binom{6}{1}\left(\frac{1}{\sqrt{x}}\right)^5(-\sqrt{x}) + \binom{6}{2}\left(\frac{1}{\sqrt{x}}\right)^4(-\sqrt{x})^2 + \binom{6}{3}\left(\frac{1}{\sqrt{x}}\right)^3(-\sqrt{x})^3 + \binom{6}{4}\left(\frac{1}{\sqrt{x}}\right)^2(-\sqrt{x})^4 + \binom{6}{5}\left(\frac{1}{\sqrt{x}}\right)(-\sqrt{x})^5 + \binom{6}{6}(-\sqrt{x})^6$$
$$= x^{-3} - 6x^{-2} + 15x^{-1} - 20 + 15x - 6x^2 + x^3$$

21. Find the 3rd term of $(a+b)^7$.

First, we note that $3 = 2+1$, $a = a$, $b = b$, and $n = 7$. Then the 3rd term of the expansion of $(a+b)^7$ is

$$\binom{7}{2}a^{7-2}b^2, \text{ or } \frac{7!}{2!5!}a^5b^2, \text{ or } 21a^5b^2.$$

22. $\binom{8}{5}x^3y^5 = 56x^3y^5$

23. Find the 6th term of $(x-y)^{10}$.

First, we note that $6 = 5+1$, $a = x$, $b = -y$, and $n = 10$. Then the 6th term of the expansion of $(x-y)^{10}$ is

$$\binom{10}{5}x^5(-y)^5, \text{ or } -252x^5y^5.$$

24. $\binom{9}{4}p^5(-2q)^4 = 2016p^5q^4$

25. Find the 12th term of $(a-2)^{14}$.

First, we note that $12 = 11+1$, $a = a$, $b = -2$, and $n = 14$. Then the 12th term of the expansion of $(a-2)^{14}$ is

$$\binom{14}{11}a^{14-11}\cdot(-2)^{11} = \frac{14!}{3!11!}a^3(-2048)$$
$$= 364a^3(-2048)$$
$$= -745,472a^3$$

26. $\binom{12}{10}x^{12-10}(-3)^{10} = 3,897,234x^2$

27. Find the 5th term of $(2x^3 - \sqrt{y})^8$.

First, we note that $5 = 4+1$, $a = 2x^3$, $b = -\sqrt{y}$, and $n = 8$. Then the 5th term of the expansion of $(2x^3 - \sqrt{y})^8$ is

$$\binom{8}{4}(2x^3)^{8-4}(-\sqrt{y})^4$$
$$= \frac{8!}{4!4!}(2x^3)^4(-\sqrt{y})^4$$
$$= 70(16x^{12})(y^2)$$
$$= 1120x^{12}y^2$$

28. $\binom{7}{3}\left(\frac{1}{b^2}\right)^{7-3}\left(\frac{b}{3}\right)^3 = \frac{35}{27}b^{-5}$

29. The expansion of $(2u-3v^2)^{10}$ has 11 terms so the 6th term is the middle term. Note that $6 = 5+1$, $a = 2u$, $b = -3v^2$, and $n = 10$. Then the 6th term of the expansion of $(2u-3v^2)^{10}$ is

$$\binom{10}{5}(2u)^{10-5}(-3v^2)^5$$
$$= \frac{10!}{5!5!}(2u)^5(-3v^2)^5$$
$$= 252(32u^5)(-243v^{10})$$
$$= -1,959,552u^5v^{10}$$

30. 3rd term: $\binom{5}{2}(\sqrt{x})^{5-2}(\sqrt{3})^2 = 30x\sqrt{x}$

4th term: $\binom{5}{3}(\sqrt{x})^{5-3}(\sqrt{3})^3 = 30x\sqrt{3}$

31. The number of subsets is 2^7, or 128

32. 2^6, or 64

33. The number of subsets is 2^{24}, or 16,777,216.

34. 2^{26}, or 67,108,864

35. The term of highest degree of $(x^5+3)^4$ is the first term, or

$\binom{4}{0}(x^5)^{4-0}3^0 = \frac{4!}{4!0!}x^{20} = x^{20}.$

Therefore, the degree of $(x^5+3)^4$ is 20.

36. The term of highest degree of $(2-5x^3)^7$ is the last term, or

$\binom{7}{7}(-5x^3)^7 = -78,125x^{21}.$

Therefore, the degree of $(2-5x^3)^7$ is 21.

37. We use factorial notation. Note that

$a=3$, $b=i$, and $n=5$.

$$\begin{aligned}
&(3+i)^5\\
&= \binom{5}{0}(3^5)+\binom{5}{1}(3^4)(i)+\binom{5}{2}(3^3)(i^2)+\\
&\quad\binom{5}{3}(3^2)(i^3)+\binom{5}{4}(3)(i^4)+\binom{5}{5}(i^5)\\
&= \frac{5!}{0!5!}(243)+\frac{5!}{1!4!}(81)(i)+\frac{5!}{2!3!}(27)(-1)+\\
&\quad\frac{5!}{3!2!}(9)(-i)+\frac{5!}{4!1!}(3)(1)+\frac{5!}{5!0!}(i)\\
&= 243+405i-270-90i+15+i\\
&= -12+316i
\end{aligned}$$

38.

$$\begin{aligned}
&(1+i)^6\\
&= \binom{6}{0}1^6+\binom{6}{1}1^5\cdot i+\binom{6}{2}1^4\cdot i^2+\\
&\quad\binom{6}{3}1^3\cdot i^3+\binom{6}{4}1^2\cdot i^4+\binom{6}{5}1\cdot i^5+\\
&\quad\binom{6}{6}i^6\\
&= 1+6i-15-20i+15+6i-1\\
&= -8i
\end{aligned}$$

39. We use factorial notation. Note that

$a=\sqrt{2}$, $b=-i$, and $n=4$.

$$\begin{aligned}
(\sqrt{2}-i)^4 &= \binom{4}{0}(\sqrt{2})^4+\binom{4}{1}(\sqrt{2})^3(-i)+\\
&\quad\binom{4}{2}(\sqrt{2})^2(-i)^2+\binom{4}{3}(\sqrt{2})(-i)^3+\\
&\quad\binom{4}{4}(-i)^4\\
&= \frac{4!}{0!4!}(4)+\frac{4!}{1!3!}(2\sqrt{2})(-i)+\\
&\quad\frac{4!}{2!2!}(2)(-1)+\frac{4!}{3!1!}(\sqrt{2})(i)+\\
&\quad\frac{4!}{4!0!}(1)\\
&= 4-8\sqrt{2}i-12+4\sqrt{2}i+1\\
&= -7-4\sqrt{2}i
\end{aligned}$$

40.

$$\begin{aligned}
&\left(\frac{\sqrt{3}}{2}-\frac{1}{2}i\right)^{11}\\
&= \binom{11}{0}\left(\frac{\sqrt{3}}{2}\right)^{11}+\binom{11}{1}\left(\frac{\sqrt{3}}{2}\right)^{10}\left(-\frac{1}{2}i\right)+\\
&\quad\binom{11}{2}\left(\frac{\sqrt{3}}{2}\right)^{9}\left(-\frac{1}{2}i\right)^{2}+\\
&\quad\binom{11}{3}\left(\frac{\sqrt{3}}{2}\right)^{8}\left(-\frac{1}{2}i\right)^{3}+\\
&\quad\binom{11}{4}\left(\frac{\sqrt{3}}{2}\right)^{7}\left(-\frac{1}{2}i\right)^{4}+\\
&\quad\binom{11}{5}\left(\frac{\sqrt{3}}{2}\right)^{6}\left(-\frac{1}{2}i\right)^{5}+\\
&\quad\binom{11}{6}\left(\frac{\sqrt{3}}{2}\right)^{5}\left(-\frac{1}{2}i\right)^{6}+\\
&\quad\binom{11}{7}\left(\frac{\sqrt{3}}{2}\right)^{4}\left(-\frac{1}{2}i\right)^{7}+\\
&\quad\binom{11}{8}\left(\frac{\sqrt{3}}{2}\right)^{3}\left(-\frac{1}{2}i\right)^{8}+\\
&\quad\binom{11}{9}\left(\frac{\sqrt{3}}{2}\right)^{2}\left(-\frac{1}{2}i\right)^{9}+\\
&\quad\binom{11}{10}\left(\frac{\sqrt{3}}{2}\right)\left(-\frac{1}{2}i\right)^{10}+\binom{11}{11}\left(-\frac{1}{2}i\right)^{11}\\
&= \frac{243\sqrt{3}}{2048}-\frac{2673}{2048}i-\frac{4455\sqrt{3}}{2048}+\frac{13,365}{2048}i+\\
&\quad\frac{8910\sqrt{3}}{2048}-\frac{12,474}{2048}i-\frac{4158\sqrt{3}}{2048}+\frac{2970}{2048}i+\\
&\quad\frac{495\sqrt{3}}{2048}-\frac{165}{2048}i-\frac{11\sqrt{3}}{2048}+\frac{1}{2048}i\\
&= \frac{1024\sqrt{3}}{2048}+\frac{1024}{2048}i\\
&= \frac{\sqrt{3}}{2}+\frac{1}{2}i
\end{aligned}$$

41. $(a-b)^n = \binom{n}{0}a^n(-b)^0 + \binom{n}{1}a^{n-1}(-b)^1 +$
$\binom{n}{2}a^{n-2}(-b)^2 + \cdots +$
$\binom{n}{n-1}a^1(-b)^{n-1} + \binom{n}{n}a^0(-b)^n$
$= \binom{n}{0}(-1)^0a^nb^0 + \binom{n}{1}(-1)^1a^{n-1}b^1 +$
$\binom{n}{2}(-1)^2a^{n-2}b^2 + \cdots +$
$\binom{n}{n-1}(-1)^{n-1}a^1b^{n-1} +$
$\binom{n}{n}(-1)^na^0b^n$
$= \sum_{k=0}^{n}\binom{n}{k}(-1)^ka^{n-k}b^k$

42. $\frac{(x+h)^{13}-x^{13}}{h}$
$= (x^{13} + 13x^{12}h + 78x^{11}h^2 + 286x^{10}h^3 +$
$715x^9h^4 + 1287x^8h^5 + 1716x^7h^6 + 1716x^6h^7 +$
$1287x^5h^8 + 715x^4h^9 + 286x^3h^{10} + 78x^2h^{11} +$
$13xh^{12} + h^{13} - x^{13})/h$
$= 13x^{12} + 78x^{11}h + 286x^{10}h^2 + 715x^9h^3 +$
$1287x^8h^4 + 1716x^7h^5 + 1716x^6h^6 + 1287x^5h^7 +$
$715x^4h^8 + 286x^3h^9 + 78x^2h^{10} + 13xh^{11} + h^{12}$

43. $\frac{(x+h)^n - x^n}{h}$
$= \frac{\binom{n}{0}x^n + \binom{n}{1}x^{n-1}h + \cdots + \binom{n}{n}h^n - x^n}{h}$
$= \binom{n}{1}x^{n-1} + \binom{n}{2}x^{n-2}h + \cdots + \binom{n}{n}h^{n-1}$
$= \sum_{k=1}^{n}\binom{n}{k}x^{n-k}h^{k-1}$

44. In expanding $(a+b)^n$, it would probably be better to use Pascal's triangle when n is relatively small. When n is large, and many rows of Pascal's triangle must be computed to get to the $(n+1)$st row, it would probably be better to use factorial notation. In addition, factorial notation allows us to write a particular term of the expansion more efficiently than Pascal's triangle.

45. The array of numbers that is known as Pascal's triangle appeared as early as 1303 in a work of the Chinese algebraist Chu Shï-kié. Because Pascal developed many of the triangle's properties and then found many applications of these properties, this array became known as Pascal's triangle.

46. $(f+g)(x) = f(x) + g(x) = (x^2+1) + (2x-3) =$
$x^2 + 2x - 2$

47. $(fg)(x) = f(x)g(x) = (x^2+1)(2x-3) =$
$2x^3 - 3x^2 + 2x - 3$

48. $(f \circ g)(x) = f(g(x)) = f(2x-3) = (2x-3)^2 + 1 =$
$4x^2 - 12x + 9 + 1 = 4x^2 - 12x + 10$

49. $(g \circ f)(x) = g(f(x)) = g(x^2+1) = 2(x^2+1) - 3 =$
$2x^2 + 2 - 3 = 2x^2 - 1$

50. $\sum_{k=0}^{8}\binom{8}{k}x^{8-k}3^k = 0$

The left side of the equation is sigma notation for $(x+3)^8$, so we have:

$(x+3)^8 = 0$
$x + 3 = 0$ Taking the 8th root on both sides
$x = -3$

51. $\sum_{k=0}^{4}\binom{4}{k}5^{4-k}x^k = 64$

The left side of the equation is sigma notation for $(5+x)^4$, so we have:

$(5+x)^4 = 64$
$5 + x = \pm 2\sqrt{2}$ Taking the 4th root on both sides
$x = -5 \pm 2\sqrt{2}$

The real solutions are $-5 \pm 2\sqrt{2}$.

If we also observe that $(2\sqrt{2}i)^4 = 64$, we also find the imaginary solutions $-5 \pm 2\sqrt{2}i$.

52. $\sum_{k=0}^{5}\binom{5}{k}(-1)^kx^{5-k}3^k = \sum_{k=0}^{5}\binom{5}{k}x^{5-k}(-3)^k$, so the left side of the equation is sigma notation for $(x-3)^5$. We have:

$(x-3)^5 = 32$
$x - 3 = 2$ Taking the 5th root on both sides
$x = 5$

53. $\sum_{k=0}^{4}\binom{4}{k}(-1)^kx^{4-k}6^k = \sum_{k=0}^{4}\binom{4}{k}x^{4-k}(-6)^k$, so the left side of the equation is sigma notation for $(x-6)^4$. We have:

$(x-6)^4 = 81$
$x - 6 = \pm 3$ Taking the 4th root on both sides
$x - 6 = 3$ or $x - 6 = -3$
$x = 9$ or $x = 3$

The solutions are 9 and 3.

If we also observe that $(3i)^4 = 81$, we also find the imaginary solutions $6 \pm 3i$.

54. The $(k+1)$st term of $\left(\frac{3x^2}{2}-\frac{1}{3x}\right)^{12}$ is

$\binom{12}{k}\left(\frac{3x^2}{2}\right)^{12-k}\left(-\frac{1}{3x}\right)^k$. In the term which does not contain x, the exponent of x in the numerator is equal to the exponent of x in the denominator.

$$2(12-k)=k$$
$$24-2k=k$$
$$24=3k$$
$$8=k$$

Find the $(8+1)$st, or 9th term:

$$\binom{12}{8}\left(\frac{3x^2}{2}\right)^4\left(-\frac{1}{3x}\right)^8=\frac{12!}{4!8!}\left(\frac{3^4x^8}{2^4}\right)\left(\frac{1}{3^8x^8}\right)=\frac{55}{144}$$

55. The expansion of $(x^2-6y^{3/2})^6$ has 7 terms, so the 4th term is the middle term.

$$\binom{6}{3}(x^2)^3(-6y^{3/2})^3=\frac{6!}{3!3!}(x^6)(-216y^{9/2})=$$
$$-4320x^6y^{9/2}$$

56. $$\frac{\binom{5}{3}(p^2)^2\left(-\frac{1}{2}p\sqrt[3]{q}\right)^3}{\binom{5}{2}(p^2)^3\left(-\frac{1}{2}p\sqrt[3]{q}\right)^2}=\frac{-\frac{1}{8}p^7q}{\frac{1}{4}p^8\sqrt[3]{q^2}}=-\frac{\sqrt[3]{q}}{2p}$$

57. The $(k+1)$st term of $\left(\sqrt[3]{x}-\frac{1}{\sqrt{x}}\right)^7$ is

$\binom{7}{k}(\sqrt[3]{x})^{7-k}\left(-\frac{1}{\sqrt{x}}\right)^k$. The term containing $\frac{1}{x^{1/6}}$ is the term in which the sum of the exponents is $-1/6$. That is,

$$\left(\frac{1}{3}\right)(7-k)+\left(-\frac{1}{2}\right)(k)=-\frac{1}{6}$$
$$\frac{7}{3}-\frac{k}{3}-\frac{k}{2}=-\frac{1}{6}$$
$$-\frac{5k}{6}=-\frac{15}{6}$$
$$k=3$$

Find the $(3+1)$st, or 4th term.

$$\binom{7}{3}(\sqrt[3]{x})^4\left(-\frac{1}{\sqrt{x}}\right)^3=\frac{7!}{4!3!}(x^{4/3})(-x^{-3/2})=$$
$$-35x^{-1/6},\text{ or }-\frac{35}{x^{1/6}}.$$

58. The total number of subsets of a set of 7 bills is 2^7. This includes the empty set. Thus, 2^7-1, or 127, different sums of money can be formed.

59. ${}_{100}C_0+{}_{100}C_1+\cdots+{}_{100}C_{100}$ is the total number of subsets of a set with 100 members, or 2^{100}.

60. ${}_nC_0+{}_nC_1+\ldots+{}_nC_n$ is the total number of subsets of a set with n members, or 2^n.

61. $$\sum_{k=0}^{23}\binom{23}{k}(\log_a x)^{23-k}(\log_a t)^k=$$
$$(\log_a x+\log_a t)^{23}=[\log_a(xt)]^{23}$$

62. $$\sum_{k=0}^{15}\binom{15}{k}i^{30-2k}.$$
$$=i^{30}+15i^{28}+105i^{26}+455i^{24}+1365i^{22}+3003i^{20}+5005i^{18}+6435i^{16}+6435i^{14}+5005i^{12}+3003i^{10}+1365i^8+455i^6+105i^4+15i^2+1$$
$$=-1+15-105+455-1365+3003-5005+6435-6435+5005-3003+1365-455+105-15+1$$
$$=0$$

63. See the answer section in the text.

Exercise Set 7.8

1. a) We use Principle P.

For 1: $P=\frac{18}{100}$, or 0.18

For 2: $P=\frac{24}{100}$, or 0.24

For 3: $P=\frac{23}{100}$, or 0.23

For 4: $P=\frac{23}{100}$, or 0.23

For 5: $P=\frac{12}{100}$, or 0.12

b) Opinions may vary, but it seems that people tend not to select the first or last numbers.

2. The total number of gumdrops is $7+8+9+4+5+6$, or 39.

Lemon: $\frac{8}{39}$

Lime: $\frac{5}{39}$

Orange: $\frac{9}{39}=\frac{3}{13}$

Grape: $\frac{6}{39}=\frac{2}{13}$

Strawberry: $\frac{7}{39}$

Licorice: $\frac{0}{39}=0$

3. The company can expect 78% of the 15,000 pieces of advertising to be opened and read. We have:

$78\%(15,000)=0.78(15,000)=11,700.$

4. a) B: $136/9136\approx 1.5\%$

C: $273/9136\approx 3.0\%$

D: $286/9136\approx 3.1\%$

E: $1229/9136\approx 13.5\%$

F: $173/9136\approx 1.9\%$

G: $190/9136\approx 2.1\%$

H: $399/9136 \approx 4.4\%$

I: $539/9136 \approx 5.9\%$

J: $21/9136 \approx 0.2\%$

K: $57/9136 \approx 0.6\%$

L: $417/9136 \approx 4.6\%$

M: $231/9136 \approx 2.5\%$

N: $597/9136 \approx 6.5\%$

O: $705/9136 \approx 7.7\%$

P: $238/9136 \approx 2.6\%$

Q: $4/9136 \approx 0.04\%$

R: $609/9136 \approx 6.7\%$

S: $745/9136 \approx 8.2\%$

T: $789/9136 \approx 8.6\%$

U: $240/9136 \approx 2.6\%$

V: $113/9136 \approx 1.2\%$

W: $127/9136 \approx 1.4\%$

X: $20/9136 \approx 0.2\%$

Y: $124/9136 \approx 1.4\%$

b) $\frac{853 + 1229 + 539 + 705 + 240}{9136} = \frac{3566}{9136} \approx 39\%$

c) In part (b) we found that there are 3566 vowels. Thus, there are $9136 - 3566$, or 5570 consonants.

$$P = \frac{5570}{9136} \approx 61\%$$

(We could also find this by subtracting the probability of a vowel occurring from 100%: $100\% - 39\% = 61\%$)

5. a) The consonants with the 5 greatest numbers of occurrences are the 5 consonants with the greatest probability of occurring. They are T, S, R, N, and L.

b) E is the vowel with the greatest number of occurrences, so E is the vowel with the greatest probability of occurring.

c) Yes

6. a) 52

b) $\frac{4}{52}$, or $\frac{1}{13}$

c) $\frac{13}{52}$, or $\frac{1}{4}$

d) $\frac{4}{52}$, or $\frac{1}{13}$

e) $\frac{26}{52}$, or $\frac{1}{2}$

f) $\frac{4+4}{52}$, or $\frac{2}{13}$

g) $\frac{2}{52}$, or $\frac{1}{26}$

7. a) Since there are 14 equally likely ways of selecting a marble from a bag containing 4 red marbles and 10 green marbles, we have, by Principle P,

$P(\text{selecting a red marble}) = \frac{4}{14} = \frac{2}{7}$.

b) Since there are 14 equally likely ways of selecting a marble from a bag containing 4 red marbles and 10 green marbles, we have, by Principle P,

$P(\text{selecting a green marble}) = \frac{10}{14} = \frac{5}{7}$.

c) Since there are 14 equally likely ways of selecting a marble from a bag containing 4 red marbles and 10 green marbles, we have, by Principle P,

$P(\text{selecting a purple marble}) = \frac{0}{14} = 0$.

d) Since there are 14 equally likely ways of selecting a marble from a bag containing 4 red marbles and 10 green marbles, we have, by Principle P,

$P(\text{selecting a red or a green marble}) =$
$\frac{4+10}{14} = 1$.

8. $\frac{{}_{10}C_2 \cdot {}_{10}C_2}{{}_{20}C_4} = \frac{45 \cdot 45}{4845} = \frac{135}{323}$

9. The total number of coins is $7 + 5 + 10$, or 22 and the total number of coins to be drawn is $4 + 3 + 1$, or 8. The number of ways of selecting 8 coins from a group of 22 is ${}_{22}C_8$. Four dimes can be selected in ${}_7C_4$ ways, 3 nickels in ${}_5C_3$ ways, and 1 quarter in ${}_{10}C_1$ ways.

$P(\text{selecting 4 dimes, 3 nickels, and 1 quarter}) =$
$\frac{{}_7C_4 \cdot {}_5C_3 \cdot {}_{10}C_1}{{}_{22}C_8}$, or $\frac{350}{31,977}$

10. a) ${}_{49}C_6 = 13,983,816$

b) About 0.000007%

11. The number of ways of selecting 5 cards from a deck of 52 cards is ${}_{52}C_5$. Three sevens can be selected in ${}_4C_3$ ways and 2 kings in ${}_4C_2$ ways.

$P(\text{drawing 3 sevens and 2 kings}) = \frac{{}_4C_3 \cdot {}_4C_2}{{}_{52}C_5}$, or $\frac{1}{108,290}$.

12. Since there are only 4 aces in the deck, $P(5 \text{ aces}) = 0$.

13. The number of ways of selecting 5 cards from a deck of 52 cards is ${}_{52}C_5$. Since 13 of the cards are spades, then 5 spades can be drawn in ${}_{13}C_5$ ways

$P(\text{drawing 5 spades}) = \frac{{}_{13}C_5}{{}_{52}C_5} = \frac{1287}{2,598,960} =$
$\frac{33}{66,640}$

14. $\frac{{}_4C_4 \cdot {}_4C_1}{{}_{52}C_5} = \frac{1}{649,740}$

15. a), b) Answers will vary.

16. a) HHH, HHT, HTH, HTT, THH, THT, TTH, TTT

b) Three of the 8 outcomes have exactly one head. Thus, $P(\text{exactly one head}) = \frac{3}{8}$.

c) Seven of the 8 outcomes have exactly 0, 1, or 2 heads. Thus, $P(\text{at most two heads}) = \frac{7}{8}$.

d) Seven of the 8 outcomes have 1, 2, or 3 heads. Thus, $P(\text{at least one head}) = \frac{7}{8}$.

e) Three of the 8 outcomes have exactly two tails. Thus, $P(\text{exactly two tails}) = \frac{3}{8}$.

17. The roulette wheel contains 38 equally likely slots. Eighteen of the 38 slots are colored black. Thus, by Principle P,

$P(\text{the ball falls in a black slot}) = \frac{18}{38} = \frac{9}{19}$.

18. $\frac{18}{38}$, or $\frac{9}{19}$

19. The roulette wheel contains 38 equally likely slots. Thirty-six of the slots are colored red or black. Then, by Principle P,

$P(\text{the ball falls in a red or a black slot}) = \frac{36}{38} = \frac{18}{19}$.

20. $\frac{1}{38}$

21. The roulette wheel contains 38 equally likely slots. Only 1 slot is numbered 0. Then, by Principle P,

$P(\text{the ball falls in the 0 slot}) = \frac{1}{38}$.

22. $\frac{2}{38} = \frac{1}{19}$

23. The roulette wheel contains 38 equally likely slots. Eighteen of the slots are odd-numbered. Then, by Principle P,

$P(\text{the ball falls in a an odd-numbered slot}) =$

$\frac{18}{38} = \frac{9}{19}$.

24. $\frac{1}{38}$

25. The dartboard can be thought of as having 18 areas of equal size. Of these, 6 are red, 4 are green, 3 are blue, and 5 are yellow.

$P(\text{red}) = \frac{6}{18} = \frac{1}{3}$

$P(\text{green}) = \frac{4}{18} = \frac{2}{9}$

$P(\text{blue}) = \frac{3}{18} = \frac{1}{6}$

$P(\text{yellow}) = \frac{5}{18}$

26. From the choices of the 26 letters of the alphabet or a space, the probability of selecting each of the 50 letters or spaces is $\frac{1}{27}$. Then the probability that the given passage could have been written by a monkey is $\left(\frac{1}{27}\right)^{50} \approx 2.7 \times 10^{-72}$. Thus, it is extremely unlikely that the given passage, much less an entire novel, could be written by a monkey.

27. Answers may vary.

28.

$$3x^2 - 4x = 3$$
$$3x^2 - 4x - 3 = 0$$
$$x = \frac{-(-4) \pm \sqrt{(-4)^2 - 4\cdot 3\cdot(-3)}}{2\cdot 3}$$
$$= \frac{4 \pm \sqrt{52}}{6} = \frac{4 \pm 2\sqrt{13}}{6}$$
$$= \frac{2 \pm \sqrt{13}}{3}$$

29. $2x^3 + 5x^2 - 4x - 3 = 0$

We use synthetic division to find a factor of the polynomial on the left side of the equation. We try $x - 1$.

$$\begin{array}{r|rrrr} 1 & 2 & 5 & -4 & -3 \\ & & 2 & 7 & 3 \\ \hline & 2 & 7 & 3 & 0 \end{array}$$

Then we have:

$$(x-1)(2x^2 + 7x + 3) = 0$$
$$(x-1)(2x+1)(x+3) = 0$$

$$x - 1 = 0 \quad or \quad 2x + 1 = 0 \quad or \quad x + 3 = 0$$
$$x = 1 \quad or \quad 2x = -1 \quad or \quad x = -3$$
$$x = 1 \quad or \quad x = -\frac{1}{2} \quad or \quad x = -3$$

We could also graph $y = 2x^3 + 5x^2 - 4x - 3$ and use the Zero feature three times. The solutions are -3, $-\frac{1}{2}$, and 1, or -3, -0.5, and 1.

30.

$$x - y = 1, \quad (1)$$
$$2x - 3y = 6 \quad (2)$$

Multiply equation (1) by -3 and add.

$$\begin{array}{rcl} -3x + 3y &=& -3 \\ 2x - 3y &=& 6 \\ \hline -x &=& 3 \\ x &=& -3 \end{array}$$

Back-substitute to find y. We use equation (1).

$$-3 - y = 1$$
$$-y = 4$$
$$y = -4$$

The solution is $(-3, -4)$.

31.

$$2x + y - 3z = 5,$$
$$3x + 3y - 5z = 4,$$
$$x - 2y + 2z = 11$$

We use Gauss-Jordan elimination with matrices. First we write the augmented matrix.

$$\left[\begin{array}{rrr|r} 2 & 1 & -3 & 5 \\ 3 & 3 & -5 & 4 \\ 1 & -2 & 2 & 11 \end{array}\right]$$

Now interchange the first and third rows.

$$\left[\begin{array}{ccc|c} 1 & -2 & 2 & 11 \\ 3 & 3 & -5 & 4 \\ 2 & 1 & -3 & 5 \end{array}\right]$$

Multiply the first row by -3 and add it to the second row. Also multiply the first row by -2 and add it to the third row.

$$\left[\begin{array}{ccc|c} 1 & -2 & 2 & 11 \\ 0 & 9 & -11 & -29 \\ 0 & 5 & -7 & -17 \end{array}\right]$$

Multiply the third row by -2 and add it to the second row.

$$\left[\begin{array}{ccc|c} 1 & -2 & 2 & 11 \\ 0 & -1 & 3 & 5 \\ 0 & 5 & -7 & -17 \end{array}\right]$$

Now multiply the second row by -1.

$$\left[\begin{array}{ccc|c} 1 & -2 & 2 & 11 \\ 0 & 1 & -3 & -5 \\ 0 & 5 & -7 & -17 \end{array}\right]$$

Multiply the second row by 2 and add it to the first row. Also multiply the second row by -5 and add it to the third row.

$$\left[\begin{array}{ccc|c} 1 & 0 & -4 & 1 \\ 0 & 1 & -3 & -5 \\ 0 & 0 & 8 & 8 \end{array}\right]$$

Multiply the third row by $\frac{1}{8}$.

$$\left[\begin{array}{ccc|c} 1 & 0 & -4 & 1 \\ 0 & 1 & -3 & -5 \\ 0 & 0 & 1 & 1 \end{array}\right]$$

Multiply the third row by 4 and add it to the first row. Also multiply the third row by 3 and add it to the second row.

$$\left[\begin{array}{ccc|c} 1 & 0 & 0 & 5 \\ 0 & 1 & 0 & -2 \\ 0 & 0 & 1 & 1 \end{array}\right]$$

Now we can read the solution from the matrix. We have $x = 5$, $y = -2$, $z = 1$. The solution is $(5, -2, 1)$.

32. a) 4

b) $\dfrac{4}{_{52}C_5} = \dfrac{4}{2,598,960} \approx 1.54 \times 10^{-6}$

33. Consider a suit

A K Q J 10 9 8 7 6 5 4 3 2

A straight flush can be any of the following combinations in the same suit.

K Q J 10 9
Q J 10 9 8
J 10 9 8 7
10 9 8 7 6
9 8 7 6 5
8 7 6 5 4
7 6 5 4 3
6 5 4 3 2
5 4 3 2 A

Remember a straight flush does not include A K Q J 10 which is a royal flush.

a) Since there are 9 straight flushes per suit, there are 36 straight flushes in all 4 suits.

b) Since 2,598,960, or $_{52}C_5$, poker hands can be dealt from a standard 52-card deck and 36 of those hands are straight flushes, the probability of getting a straight flush is $\dfrac{36}{2,598,960}$, or about 1.39×10^{-5}.

34. a) There are 13 ways to choose a denomination. Then there are 48 ways to choose one of the 48 cards remaining after 4 cards of the same denomination are chosen. Thus there are $13 \cdot 48$, or 624, four of a kind hands.

b) $\dfrac{624}{_{52}C_5} = \dfrac{624}{2,598,960} \approx 2.4 \times 10^{-4}$

35. a) There are 13 ways to select a denomination. Then from that denomination there are $_4C_3$ ways to pick 3 of the 4 cards in that denomination. Now there are 12 ways to select any one of the remaining 12 denominations and $_4C_2$ ways to pick 2 cards from the 4 cards in that denomination. Thus the number of full houses is $(13 \cdot {_4C_3}) \cdot (12 \cdot {_4C_2})$ or 3744.

b) $\dfrac{3744}{_{52}C_5} = \dfrac{3744}{2,598,960} \approx 0.00144$

36. a) There are 13 ways to select a denomination and then $\binom{4}{3}$ ways to choose 3 of the 4 cards in that denomination. Now there are $\binom{48}{2}$ ways to choose 2 cards from the 12 remaining denominations ($4 \cdot 12$, or 48 cards). But these combinations include the 3744 hands in a full house like Q-Q-Q-4-4 (Exercise 31), so these must be subtracted. Thus the number of three of a kind hands is $13 \cdot \binom{4}{3} \cdot \binom{48}{2} - 3744$, or 54,912.

b) $\dfrac{54,912}{_{52}C_5} = \dfrac{54,912}{2,598,960} \approx 0.0211$

37. a) There are 4 ways to select a suit and then $\binom{13}{5}$ ways to choose 5 cards from that suit. But these combinations include hands with all the cards in sequence, so we subtract the 4 royal flushes (Exercise 28) and the 36 straight flushes (Exercise 29). Thus there are $4 \cdot \binom{13}{5} - 4 - 36$, or 5108 flushes.

b) $\dfrac{5108}{_{52}C_5} = \dfrac{5108}{2,598,960} \approx 0.00197$

38. a) There are $\binom{13}{2}$ ways to select 2 denominations from the 13 denominations. Then in each denomination there are $\binom{4}{2}$ ways to choose 2 of the 4 cards. Finally there are $\binom{44}{1}$ ways to choose the fifth card from the 11 remaining denominations ($4 \cdot 11$, or 44 cards). Thus the number of two pairs hands is

$\binom{13}{2} \cdot \binom{4}{2} \cdot \binom{4}{2} \cdot \binom{44}{1}$, or 123,552.

b) $\dfrac{123,552}{_{52}C_5} = \dfrac{123,552}{2,598,960} \approx 0.0475$

39. a) There are 10 sets of 5 consecutive cards:

A K Q J 10
K Q J 10 9
.
.
.
5 4 3 2 A

In each of these 10 sets there are 4 ways to choose (from 4 suits) each of the 5 cards. These combinations include the 4 royal flushes and the 36 straight flushes, both of which consist of 5 cards of the *same suit* in sequence.

Thus there are $10 \cdot 4 \cdot 4 \cdot 4 \cdot 4 \cdot 4 - 4 - 36$, or 10,200 straights.

b) $\dfrac{10,200}{_{52}C_5} = \dfrac{10,200}{2,598,960} \approx 0.00392$

Sample Chapter Tests

The sample Chapter Tests and the sample Final Examination on the following pages are available to students in the Student's Solutions Manuals to accompany *College Algebra: Graphs & Models*, Second Edition, by Bittinger, Beecher, Ellenbogen, and Penna.

Chapter G Test

1. Make a hand-drawn graph of the set of points. Label each point.

   $\{(-3,0),(1,4),(-2,-3),(0,2),(3,-2),(-1,5)\}$

2. Determine whether each ordered pair is a solution of the equation $2x - 3y = 7$. Answer yes or no.

   a) $(4, 5)$ b) $(2, -1)$

3. Make a hand-drawn graph of each equation.

   a) $y = 2x - 3$ b) $y = x^2 + 1$.

Match the equations with graphs (a) - (d).

4. $y = |x - 3|$

5. $y = \sqrt{x + 2}$

6. $y = 5 - 3x$

7. $y = x^2 - 4$

a)

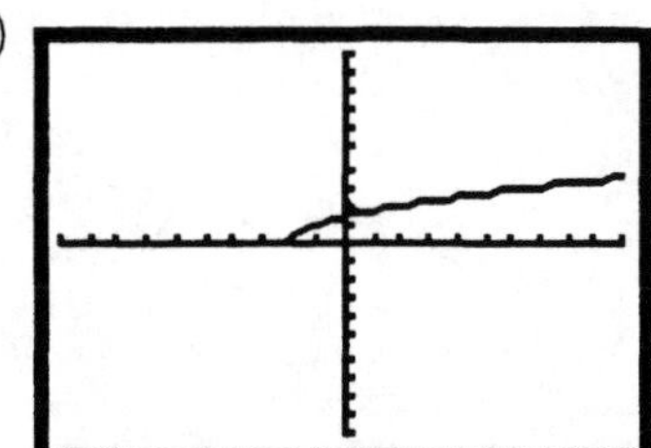

b)

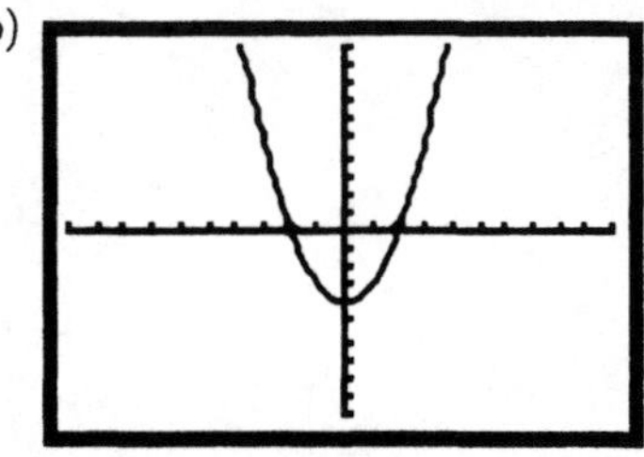

c)

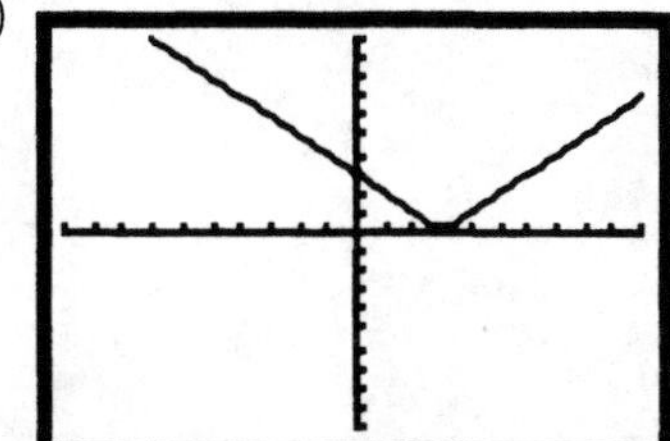

d)

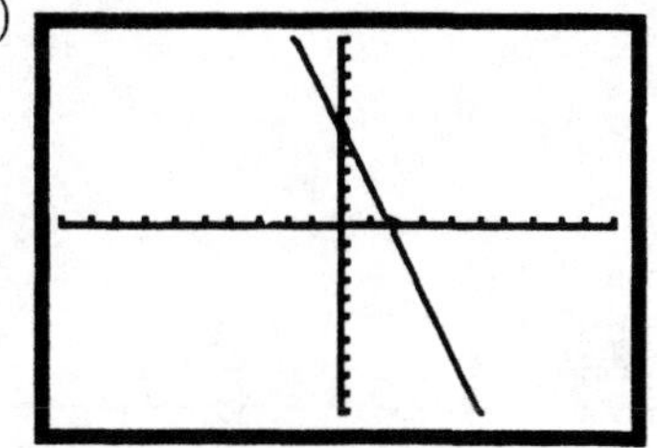

8. Use a grapher to complete the table for $y_1 = \sqrt{9 - x^2}$ and $y_2 = 3 - x$.

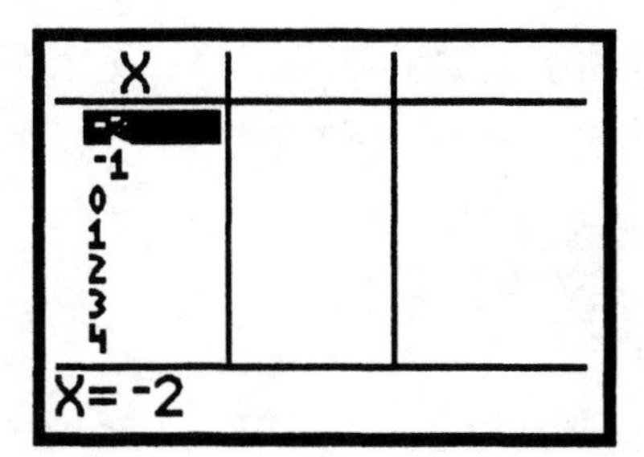

9. Graph the equation $y = x^3 - 2x + 2$ using the window $[-4, 4, -8, 8]$, Xscl = 1, Yscl = 1.

10. Graph the equation $y = x^2 - 10$ in the standard $[-10, 10, -10, 10]$ window and in the window $[-5, 5, -5, 5]$. Determine which window better shows the shape of the graph and where it crosses the x- and y-axes.

11. a) Graph the equations $y_1 = 2x^3 - 3x^2 + 3$ and $y_2 = 7 - x^2$ in the same viewing window using the dimensions $[-10, 10, -10, 10]$, Xscl = 1, Yscl = 1.

    b) Find the coordinates of the point of intersection of the graphs in part (a). Round each coordinate to two decimal places.

12. Solve $3.4x^3 - 1.9x = -4$. Round your answer to two decimal places.

Chapter R Test

1. Consider the numbers -8, $\frac{11}{3}$, $\sqrt{15}$, 0, -5.49, 36, $\sqrt[3]{7}$, $10\frac{1}{6}$.
   a) Which are integers?
   b) Which are rational numbers?

2. Simplify: $|-1.2xy|$.

3. Find the distance between -9 and 5 on a number line.

4. Write interval notation for $\{x \mid -3 < x \leq 6\}$.

5. Compute: $32 \div 2^3 - 12 \div 4 \cdot 3$.

6. Compute and write scientific notation for the answer: $\dfrac{2.7 \times 10^4}{3.6 \times 10^{-3}}$.

Simplify.

7. $(-3a^5b^{-4})(5a^{-1}b^3)$

8. $(3x^4 - 2x^2 + 6x) - (5x^3 - 3x^2 + x)$

9. $(x+3)(2x-5)$

10. $(2y-1)^2$

11. $3\sqrt{75} + 2\sqrt{27}$

12. $\dfrac{\frac{x}{y} - \frac{y}{x}}{x+y}$

Factor.

13. $2n^2 + 5n - 12$

14. $8x^2 - 18$

15. $m^3 - 8$

16. Use a grapher to graph the equation(s) you would use to check the factorization $x^2 + x - 6 = (x+3)(x-2)$. Show your graph in a window and write the equation(s) you graphed above the window.

17. Multiply and simplify: $\dfrac{x^2+x-6}{x^2+8x+15} \cdot \dfrac{x^2-25}{x^2-4x+4}$.

18. Subtract and simplify: $\dfrac{x}{x^2-1} - \dfrac{3}{x^2+4x-5}$.

19. Convert to radical notation: $t^{5/7}$.

20. Rationalize the denominator: $\dfrac{5}{7-\sqrt{3}}$.

Solve.

21. $6x + 7 = 1$

22. $(2x-1)(x+5) = 0$

23. $x^2 - 2x - 3 = 0$

24. $6x^2 - 36 = 0$

25. Solve $V = \frac{2}{3}\pi r^2 h$ for h.

26. How long is a guy wire that reaches from the top of a 12-ft pole to a point on the ground 5 feet from the pole?

SYNTHESIS

27. Solve: $(x+1)^3 = x^3 + 7$

Chapter 1 Test

1. a) Determine whether the relation

$$\{(-4,7),(3,0),(1,5),(0,7)\}$$

is a function. Answer yes or no.

b) Find the domain of the relation.

c) Find the range of the relation.

2. Given that $f(x) = 2x^2 - x + 5$, find

a) $f(-1)$; b) $f(a+2)$.

3. a) Use a grapher to graph $f(x) = |x-2| + 3$.

b) Visually estimate the domain of $f(x)$.

c) Visually estimate the range of $f(x)$.

4. Determine whether each graph is that of a function. Answer yes or no.

a)

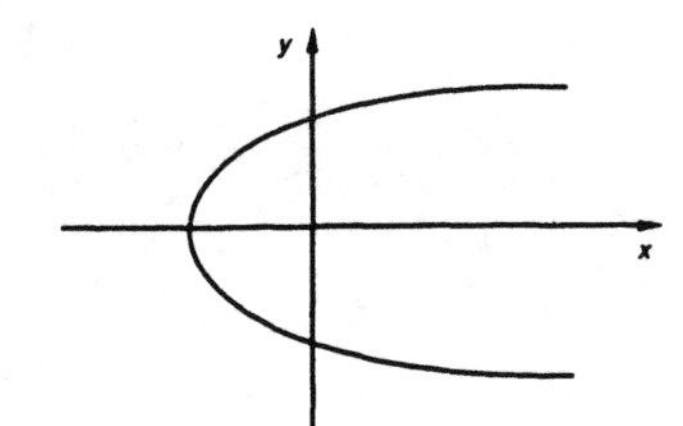

b)

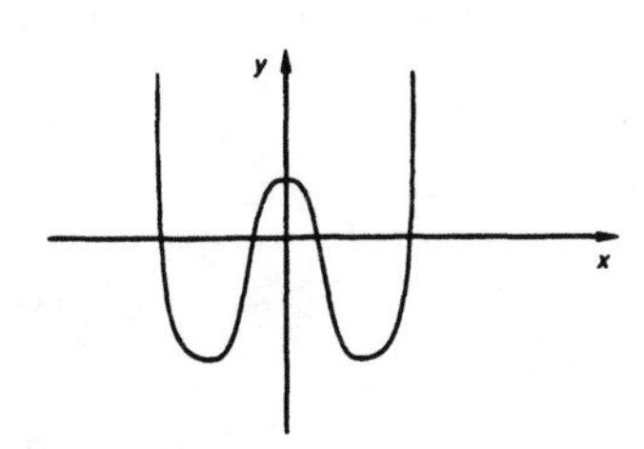

5. For the function $f(x) = x^3 - 4x^2 - 3x + 5$, use a grapher to find

a) the intervals on which the function is increasing or decreasing;

b) any relative maxima or minima.

6. Make a hand-drawn graph of $f(x)$.

$$f(x) = \begin{cases} x^2, & \text{for } x < -1, \\ |x|, & \text{for } -1 \le x \le 1, \\ \sqrt{x-1}, & \text{for } x > 1, \end{cases}$$

7. Find the slope and the y-intercept of the graph of $-3x + 2y = 5$.

8. Write an equation for the line that passes through $(-5,4)$ and $(3,-2)$.

9. Find an equation of the line containing the point $(-1,3)$ and parallel to the line $x + 2y = -6$.

10. The table below shows the per capita consumption of tea in the United States in several years.

Year, x	Per Capita Consumption of Tea, in gallons
0. '93	8.4
1. '94	8.2
2. '95	8.0
3. '96	8.0

a) Use a grapher to fit a regression line $y = ax + b$ to the data, where x is the number of years after 1993.

b) What is the correlation coefficient for the regression line?

c) Use the regression line in part (a) to predict the per capita consumption of tea in 2005.

11. Find the distance between (5,8) and $(-1,5)$.

12. Find the midpoint of the segment with endpoints $(-2,6)$ and $(-4,3)$.

13. Find an equation of the circle with center $(-1,2)$ and radius $\sqrt{5}$.

14. a) Use a grapher to graph $y = x^4 - 2x^2$ in the window $[-5,5,-5,5]$.

b) Determine whether the graph in part (a) is symmetric with respect to the x-axis, the y-axis, and/or the origin.

15. Test algebraically whether the function $f(x) = \dfrac{2x}{x^2+1}$ is even, odd, or neither even nor odd. Show your work.

16. Write an equation for a function that has the shape of $y = x^2$, but shifted right 2 units and down 1 unit.

17. Write an equation for a function that has the shape of $y = x^2$ but shifted left 2 units and down 3 units.

18. The graph of a function $y = f(x)$ is shown below. No formula for f is given. Make a hand-drawn graph of $y = -\frac{1}{2}f(x)$.

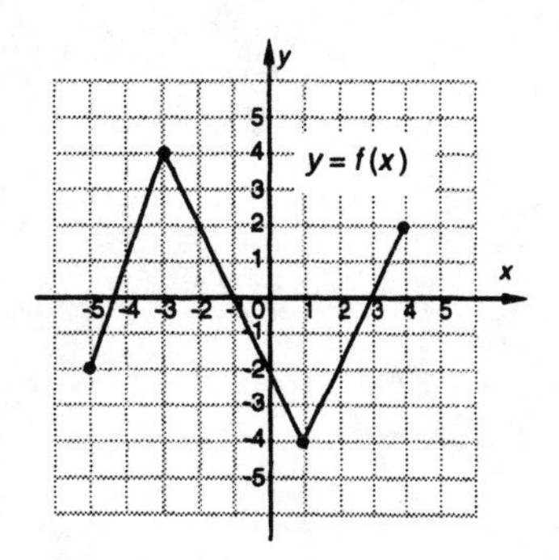

19. For $f(x) = x^2$ and $g(x) = \sqrt{x-3}$, find

a) the domain of f;

b) the domain of g;

c) $(f - g)(x)$;

d) $(fg)(x)$;

e) the domain of $(f/g)(x)$.

20. Find an equation of variation in which y varies inversely as x and $y = 5$ when $x = 6$.

21. The stopping distance d of a car after the brakes have been applied varies directly as the square of the speed r. If a car traveling 60 mph can stop in 200 ft, how long will it take a car traveling 30 mph to stop?

SYNTHESIS

22. If $(-3, 1)$ is a point on the graph of $y = f(x)$, what point do you know is on the graph of $y = f(3x)$?

Chapter 2 Test

Find the zero(s) of each function.

1. $f(x) = 3x + 9$

2. $f(x) = 4x^2 - 11x - 3$

3. $f(x) = 2x^2 - x - 7$

4. Solve $x^2 + 4x = 1$ by completing the square. Find exact solutions. Show your work.

Solve. Find exact solutions.

5. $2t^2 - 3t + 4 = 0$

6. $x + 5\sqrt{x} - 36 = 0$

7. $\dfrac{3}{3x+4} + \dfrac{2}{x-1} = 2$

8. $\sqrt{x+4} - 2 = 1$

9. $|4y - 3| = 5$

10. $-7 < 2x + 3 < 9$

11. $|x + 5| > 2$

12. Solve for n: $R = \sqrt{3np}$.

Express in terms of i.

13. $\sqrt{-43}$

14. $-\sqrt{-25}$

Simplify.

15. $(3 + 4i)(2 - i)$

16. i^{33}

17. For the graph of the function $f(x) = -x^2 + 2x + 8$,

a) find the vertex;

b) find the line of symmetry;

c) state whether there is a maximum or minimum value and find that value;

d) find the range.

18. *Maximum Area* A homeowner wants to fence a rectangular play yard using 60 ft of fencing. The side of the house will be used as one side of the rectangle. Find the dimensions for which the area is a maximum.

19. Use the regression feature on a grapher to fit a quadratic function to the data in the table.

x	y
−3	300
−1	167
1	98
3	136
5	310

SYNTHESIS

20. Find a such that $f(x) = ax^2 - 4x + 3$ has a maximum value of 12.

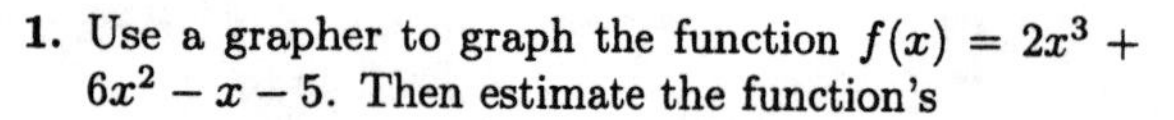

Chapter 3 Test

1. Use a grapher to graph the function $f(x) = 2x^3 + 6x^2 - x - 5$. Then estimate the function's
   a) real zeros;
   b) relative maxima;
   c) relative minima;
   d) domain;
   e) range.

2. *Interest compounded annually* When P dollars is invested at interest rate i, compounded annually, for t years, the investment grows to A dollars, where
   $$A = P(1+i)^t.$$
   Find the interest rate i if \$1500 grows to \$1858.24 in 4 years.

3. a) Use the regression feature on a grapher to fit a cubic function $y = ax^3 + bx^2 + cx + d$ to the data in the table.

x	y
−5	−200
−3	−66
0	−4
2	10
6	175

   b) Use your answer to part (a) to find the function value for $x = 9$.

4. Use long division to find the quotient and remainder. Show your work.
   $(x^4 + 3x^3 + 2x - 5) \div (x^2 - 1)$

5. Use synthetic division to find the quotient and remainder. Show your work.
   $(3x^3 - 12x + 7) \div (x - 5)$

6. Use synthetic division to determine whether -2 is a zero of $f(x) = x^3 + 4x^2 + x - 6$. Answer yes or no. Show your work.

7. Use synthetic division to find $P(-3)$ for $P(x) = 2x^3 - 6x^2 + x - 4$. Show your work.

8. Suppose that a polynomial function of degree 5 with rational coefficients has 1, $\sqrt{3}$, and $2 - i$ as zeros. Find the other zeros.

9. For the polynomial function $P(x) = x^4 + 2x^3 - 4x^2 - 5x + 6$,
   a) use a grapher to graph $P(x)$;
   b) solve $P(x) = 0$;
   c) express $P(x)$ as a product of linear factors.

10. Make a hand-drawn graph of $f(x) = \dfrac{2}{(x-3)^2}$. Label all the asymptotes.

11. Find a rational function that has vertical asymptotes $x = -1$ and $x = 2$ and x-intercept $(-4, 0)$.

Solve.

12. $2x^2 > 5x + 3$

13. $\dfrac{x+1}{x-4} \leq 3$

14. The function $S(t) = -16t^2 + 64t + 192$ gives the height S, in feet, of a model rocket launched from a hill that is 192 ft high with a velocity of 64 ft/sec.
    a) Determine how long it will take the rocket to reach the ground.
    b) Find the interval on which the height of the rocket is greater than 240 ft.

SYNTHESIS

15. Find the domain of $f(x) = \sqrt{x^2 + x - 12}$.

Chapter 4 Test

1. For $f(x) = x - 5$ and $g(x) = x^2 + 1$, find $(f \circ g)(x)$ and $(g \circ f)(x)$.

2. Find the inverse of the relation

$$\{(-2, 5),\ (4, 3),\ (0, -1),\ (-6, -3)\}.$$

3. Determine whether the function shown below is one-to-one. Answer yes or no.

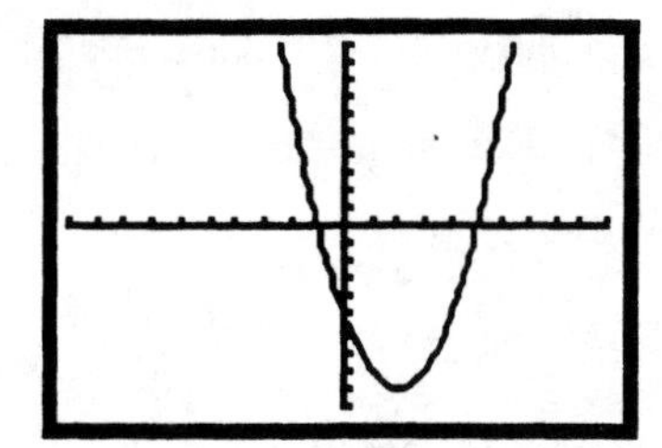

4. Find a formula for the inverse of the function $f(x) = x^3 + 1$.

Graph each of the following functions using a grapher.

5. $f(x) = e^x - 3$

6. $f(x) = \ln(x + 2)$

7. Convert to an exponential equation: $\ln x = 4$.

8. Convert to a logarithmic equation: $3^x = 5.4$.

Solve.

9. $\log_{25} 5 = x$

10. $\log_3 x + \log_3(x + 8) = 2$

11. $3^{4-x} = 27^x$

12. $e^x = 65$

13. Express in terms of sums and differences of logarithms: $\ln \sqrt[5]{x^2 y}$.

14. Given that $\log_a 2 = 0.328$ and $\log_a 8 = 0.984$, find $\log_a 4$.

15. Simplify: $\ln e^{-4t}$.

16. *Growth rate* The population of a country doubled in 45 yr. What was the exponential growth rate?

17. *Compound interest* Suppose \$1000 is invested at interest rate k, compounded continuously, and grows to \$1144.54 in 3 yr.

a) Find the interest rate.

b) Find the exponential growth function.

c) Find the balance after 8 yr.

d) Find the doubling time.

18. The following table contains data regarding the sales of a small business.

Year, x	Sales, y (in millions of dollars)
0. 1995	2.8
1. 1996	4.4
2. 1997	6.5
3. 1998	10.1
4. 1999	15.4
5. 2000	23.0

a) Create a scatterplot of the data.

b) Use regression to fit an exponential function $y = ab^x$ or $y = ae^{kx}$ to the data.

c) Use the function to predict the sales in 2003.

d) After how long will the sales be \$50 million?

SYNTHESIS

19. Solve: $4^{\sqrt[3]{x}} = 8$.

Chapter 5 Test

1. Graph the system of equations in the window $[-10, 10, -10, 10]$.

$$x - y = 5,$$
$$2x + 3y = 5.$$

Solve. Use any method.

2. $3x + 2y = 1,$
$2x - y = -11$

3. $2x - 3y = 8,$
$5x - 2y = 9$

4. $4x + 2y + z = 4,$
$3x - y + 5z = 4$
$5x + 3y - 3z = -2$

5. Classify the system of equations in Exercise 2 as consistent or inconsistent

6. Classify the system of equations in Exercise 3 as dependent or independent.

7. *Ticket Sales* One evening 750 tickets were sold for Shortridge Community College's spring musical. Tickets cost $3 for students and $5 for non-students. Total receipts were $3066. How many of each type of ticket were sold?

For Exercises 8 - 13, let

$$\boldsymbol{A} = \begin{bmatrix} 1 & -1 & 3 \\ -2 & 5 & 2 \end{bmatrix}, \boldsymbol{B} = \begin{bmatrix} -5 & 1 \\ -2 & 4 \end{bmatrix}, \text{ and } \boldsymbol{C} = \begin{bmatrix} 3 & -4 \\ -1 & 0 \end{bmatrix}.$$

Find each of the following, if possible.

8. $\boldsymbol{B} + \boldsymbol{C}$

9. $\boldsymbol{A} - \boldsymbol{C}$

10. $\boldsymbol{CB}$

11. $\boldsymbol{AB}$

12. $2\boldsymbol{A}$

13. $\boldsymbol{C}^{-1}$

14. *Food Service Management* The table below shows the cost per serving, in cents, for items on three lunch menus served at a senior-citizens' center.

Menu	Main Dish	Side Dish	Dessert
1	49	10	13
2	43	12	11
3	51	8	12

On a particular day 26 Menu 1 meals, 18 Menu 2 meals, and 23 Menu 3 meals are served.

a) Write the information in the table as a 3×3 matrix $\boldsymbol{M}$.

b) Write a row matrix $\boldsymbol{N}$ that represents the number of each menu served.

c) Find the product $\boldsymbol{NM}$.

d) State what the entries of $\boldsymbol{NM}$ represent.

15. Write a matrix equation equivalent to the system of equations

$$3x - 4y + 2z = -8,$$
$$2x + 3y + z = 7,$$
$$x - 5y - 3z = 3.$$

16. Make a hand-drawn graph of $3x + 4y \leq -12$.

17. Find the maximum and minimum values of $Q = 2x + 3y$ subject to

$$x + y \leq 6,$$
$$2x - 3y \geq -3,$$
$$x \geq 1,$$
$$y \geq 0.$$

18. *Maximizing Profit* Casey's Cakes prepares pound cakes and carrot cakes. In a given week, at most 100 cakes can be prepared of which 25 pound cakes and 15 carrot cakes are required by regular customers. The profit from each pound cake is $3 and the profit from each carrot cake is $4. How many of each type of cake should be prepared in order to maximize the profit?

19. Decompose into partial fractions: $\dfrac{3x - 11}{x^2 + 2x - 3}$.

SYNTHESIS

20. Three solutions of the equation $Ax - By = Cz - 8$ are $(2, -2, 2)$, $(-3, -1, 1)$, and $(4, 2, 9)$. Find A, B, and C.

Chapter 6 Test

Match each of the following with its graph.

1. $4x^2 - y^2 = 4$

2. $x^2 - 2x - 3y = 5$

3. $x^2 + 4x + y^2 - 2y - 4 = 0$

4. $9x^2 + 4y^2 = 36$

a)

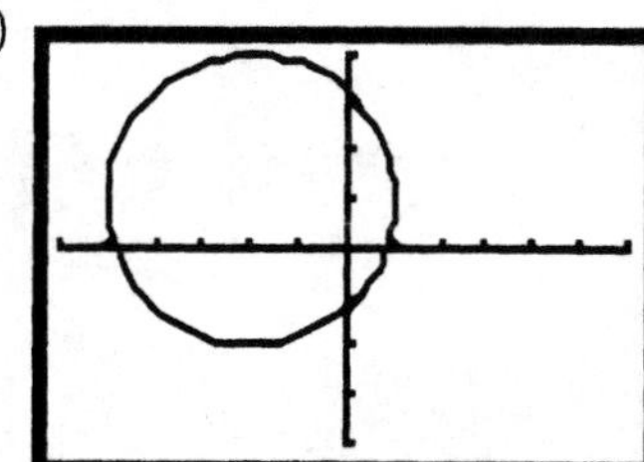

b)

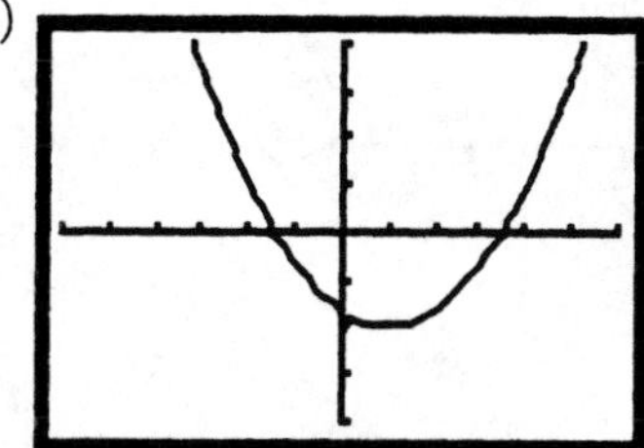

c)

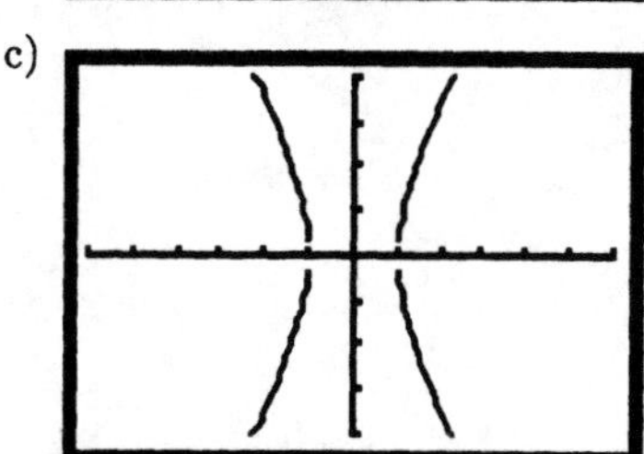

d)

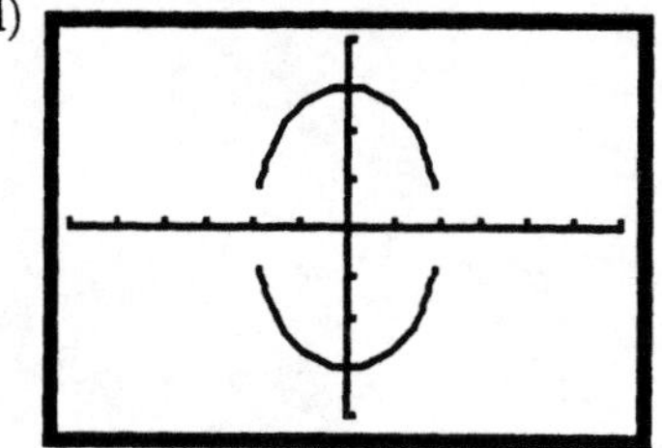

5. Find an equation of the parabola with focus (0,2) and directrix $y = -2$.

6. Find the vertex of the parabola given by $y^2 - 8y + 2x - 4 = 0$.

7. Find the center and the radius of the circle given by $x^2 + y^2 + 2x - 6y - 15 = 0$.

8. Find an equation of the ellipse having vertices $(0, -5)$ and (0,5) with minor axis of length 4.

9. Find the asymptotes of the hyperbola given by $2y^2 - x^2 = 18$.

10. *Satellite Dish* A satellite dish has a parabolic cross-section that is 18 in. wide at the opening and 6 in. deep at the vertex. How far from the vertex is the focus?

Solve.

11. $2x^2 - 3y^2 = -10,$
$x^2 + 2y^2 = 9$

12. $x^2 + y^2 = 13,$
$x + y = 1$

13. $x + y = 5,$
$xy = 6$

14. *Landscaping* Leisurescape is planting a rectangular flower garden with a perimeter of 18 ft and a diagonal of $\sqrt{41}$ ft. Find the dimensions of the garden.

SYNTHESIS

15. Find an equation of a circle that passes through the points (1,1) and $(5, -3)$ and whose center is on the line $x - y = 4$.

Chapter 7 Test

1. For the sequence whose nth term is $a_n = (-1)^n(2n+1)$, find a_{21}.

2. Find and evaluate: $\sum_{k=1}^{4}(k^2+1)$.

3. Use a grapher to graph the first 10 terms of the sequence with general term $a_n = \frac{n+1}{n+2}$.

4. Find the 15th term of the arithmetic sequence $2, 5, 8, \ldots$.

5. The 1st term of an arithmetic sequence is 8 and the 21st term is 108. Find the 7th term.

6. Find the sum of the first 20 terms of the series $17 + 13 + 9 + \ldots$.

7. For a geometric sequence, $r = 0.2$ and $S_4 = 1248$. Find a_1.

8. Find the sum, if it exists: $18 + 6 + 2 \cdots$.

9. Find fractional notation for $0.\overline{56}$.

10. *Amount of an Annuity* To create a college fund, a parent makes a sequence of 18 yearly deposits of $2500 each in a savings account on which interest is compounded annually at 5.6%. Find the amount of the annuity.

11. Use mathematical induction to prove that, for every natural number n, $2 + 5 + 8 + \cdots + (3n-1) = \frac{n(3n+1)}{2}$.

Evaluate.

12. ${}_{15}P_6$

13. ${}_{21}C_{10}$

14. $\binom{n}{4}$

15. How many 4-digit numbers can be formed using the digits 1, 3, 5, 6, 7, and 9 without repetition?

16. *Test Options* On a test with 20 questions, a student must answer 8 of the first 12 questions and 4 of the last 8. In how many ways can this be done?

17. Expand $(x+1)^5$.

18. Determine the number of subsets of a set containing 9 members.

19. *Marbles* Suppose we select, without looking, one marble from a bag containing 6 red marbles and 8 blue marbles. What is the probability of selecting a blue marble?

SYNTHESIS

20. Solve for n: ${}_nP_7 = 9 \cdot {}_nP_6$.

Final Examination

1. Make a hand-drawn graph of the equation $y = 4 - 2x$.

2. Find the coordinates of the point of intersection of $y_1 = x^3 - 5x + 7$ and $y_2 = -2x + 1$. (Round to three decimal places.)

3. Compute: $100 - 80 \div 2^2 \cdot 5 + 3$.

4. Find an equation of variation in which y varies directly as x and $y = 30$ when $x = 45$.

5. Solve. Write the answer in interval notation.

   $-16 \leq 3x + 2 < 5$

6. Given that $f(x) = 5x^3 + 4x^2 - 6x - 8$, find $f(-1)$.

7. Use a grapher to find the zeros of the function $f(x) = -x^3 + 2x^2 + 5x - 4$. (Round to three decimal places.)

8. Write an equation for the line that passes through $(1, -4)$ and $(3, -6)$.

9. The table below shows the per capita consumption of flour and cereal products in the United States in recent years.

Year, x	Per Capita Consumption of Flour and Cereal Products, in pounds
0. '93	191.0
1. '94	194.1
2. '95	192.5
3. '96	198.5

   a) Use a grapher to fit a regression line to the data.

   b) What is the correlation coefficient for the regression line?

10. a) Use a grapher to graph $y = 3x^3 - 2x$ in the window $[-5, 5, -5, 5]$.

    b) Determine whether the graph in part (a) is symmetric with respect to the x-axis, the y-axis, and/or the origin.

11. Solve $4x^2 - 3x + 2 = 0$.

12. For the graph of the function $f(x) = -x^2 + 3x + 5$,

    a) find the vertex;

    b) find the line of symmetry;

    c) state whether there is a maximum or minimum value and find that value;

    d) find the range.

13. For the polynomial function $P(x) = x^4 + 5x^3 + 3x^2 - 5x - 4$, solve $P(x) = 0$.

14. Solve $\dfrac{x-2}{x+3} < 2$.

15. Use the regression feature on a grapher to fit a quadratic function to the data in the table. (Round a, b, and c to three decimal places.)

x	y
0	252
1	216
3	198
7	243
9	278

16. Find a formula for the inverse of the function $f(x) = 3x - 5$.

17. Use a grapher to graph $f(x) = e^{x-1}$ in the window $[-10, 10, -10, 10]$.

18. Solve $16^x = 2^{3x-1}$.

19. Express in terms of sums and differences of logarithms of x, y, and z: $\log \dfrac{x^3 y}{z^2}$.

20. Suppose \$ 3000 is invested at interest rate k, compounded continuously, and grows to \$3635 in 4 years.

    a) Find the exponential growth function.

    b) Find the doubling time.

Solve. Use any method.

21. $$\begin{aligned} 2x + 3y &= 1, \\ 3x - 2y &= 21 \end{aligned}$$

22. $$\begin{aligned} 2x - 3y + z &= 11, \\ 3x + 2y + 2z &= -1, \\ 4x - 5y - 3z &= -1 \end{aligned}$$

23. For $\mathbf{A} = \begin{bmatrix} 2 & -5 & 4 \\ -3 & 1 & -1 \end{bmatrix}$ and $\mathbf{B} = \begin{bmatrix} 6 & -4 \\ -1 & 2 \\ -3 & 5 \end{bmatrix}$, find $\mathbf{AB}$, if possible.

24. Make a hand-drawn graph of $3x - 2y > 6$.

25. *Maximizing Profit* Henry's Bakery produces blueberry muffins and bran muffins. Each day at most 60 dozen muffins can be produced of which 12 dozen blueberry muffins and 6 dozen bran muffins are required by regular customers. The profit from 1 dozen blueberry muffins is \$3.50 and the profit from 1 dozen bran muffins is \$2.50. How many dozen of each type of muffin should be prepared in order to maximize the profit?

Match each of the following with its graph.

26. $2x^2 + 3x - 4y = 1$

27. $2y^2 - 3x^2 = 6$

28. $4x^2 + y^2 = 16$

29. $x^2 - 2x + y^2 - 4y - 4 = 0$

a)

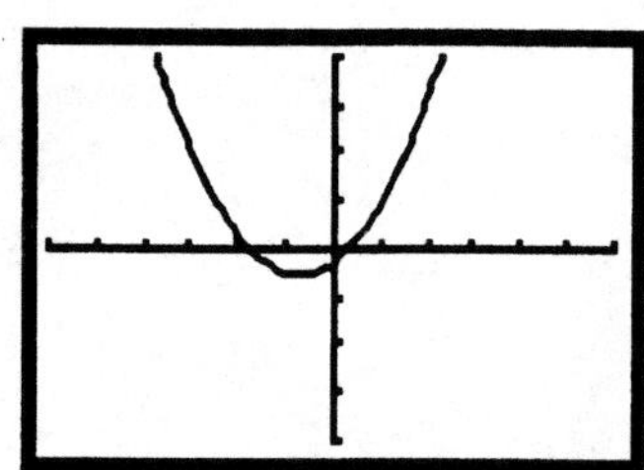

b)

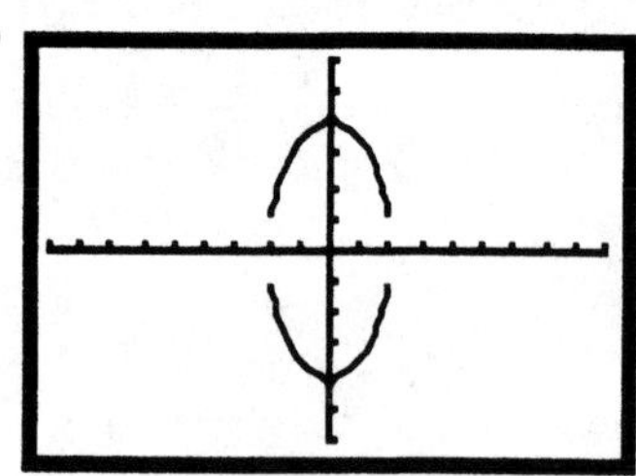

c)

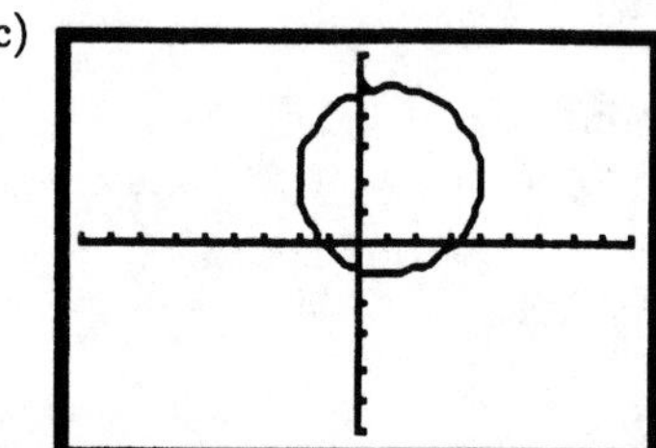

d)

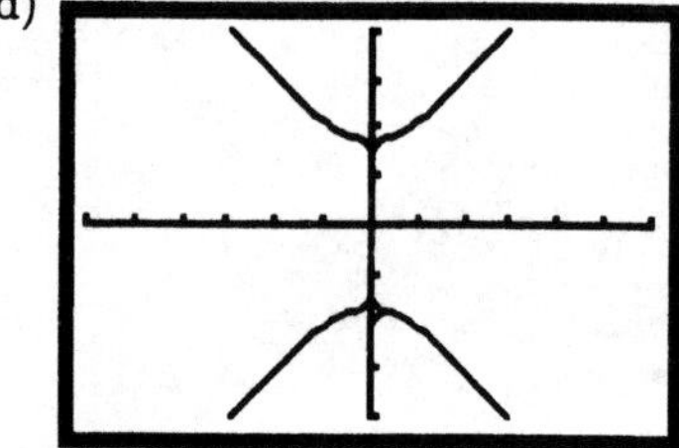

30. Solve: $x^2 - y^2 = 24,$,

$x + y = 2$

31. Find the 18th term of the arithmetic sequence $-7, -4, -1, \ldots$.

32. Find the sum, if it exists: $48 + 12 + 3 + \ldots$.

33. Use mathematical induction to prove that, for every natural number n, $2 + 7 + 12 + \cdots + (5n - 3) = \dfrac{n(5n-1)}{2}$.

34. In how many ways can 5 different books be arranged on a shelf?

35. From a bag containing 3 red marbles, 8 blue marbles, and 5 green marbles, 4 marbles are drawn all at once. What is the probability of getting 2 blue marbles and 2 green marbles?

SYNTHESIS

36. If $(5, -2)$ is a point on the graph of $y = f(x)$, what point do you know is on the graph of $y = f(x + 3)$?

37. Find the domain of $f(x) = \log_3(2x + 1)$.

CHAPTER TEST ANSWERS

Chapter G

1.

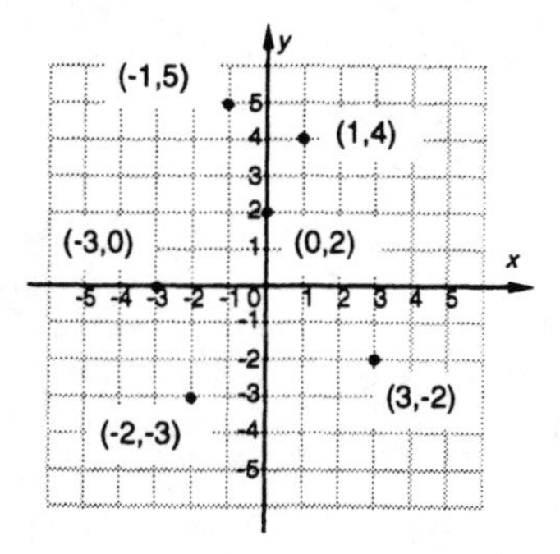

2. a) No; b) yes

3. a)

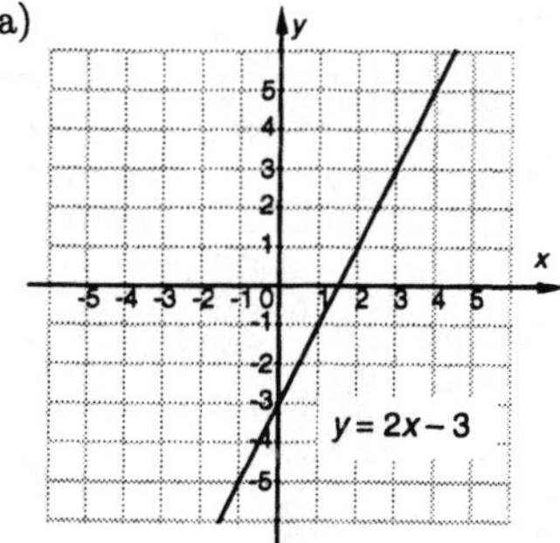

b)

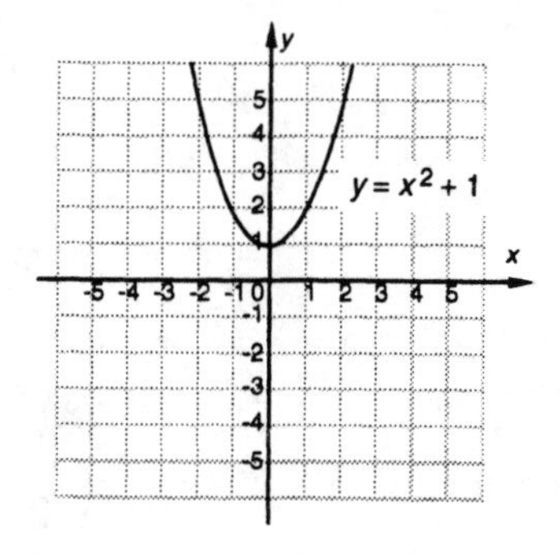

4. c **5.** a **6.** d **7.** b

8.

X	Y1	Y2
[illegible]	2.2361	5
-1	2.8284	4
0	3	3
1	2.8284	2
2	2.2361	1
3	0	0
4	ERROR	-1

X=-2

9.

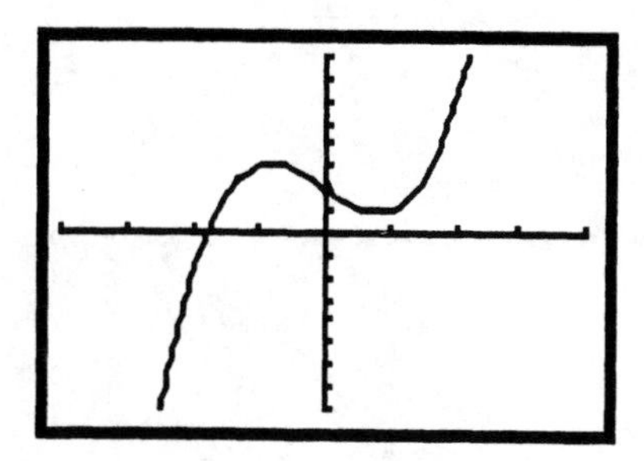

10. Standard window

11. a)

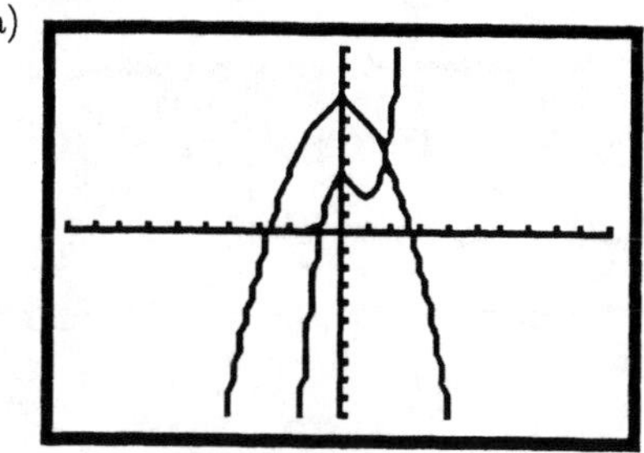

b) (1.70,4.12) **12.** −1.23

Chapter R

1. a) $-8, 0, 36$; b) $-8, \frac{11}{3}, 0, -5.49, 36, 10\frac{1}{6}$ **2.** $1.2|xy|$, or $1.2|x| \cdot |y|$ **3.** 14 **4.** $(-3, 6]$ **5.** -5 **6.** 7.5×10^6

7. $-15a^4b^{-1}$, or $-\frac{15a^4}{b}$ **8.** $3x^4 - 5x^3 + x^2 + 5x$

9. $2x^2 + x - 15$ **10.** $4y^2 - 4y + 1$ **11.** $21\sqrt{3}$

12. $\frac{x-y}{xy}$ **13.** $(2n-3)(n+4)$ **14.** $2(2x+3)(2x-3)$

15. $(m-2)(m^2+2m+4)$

16. $y_1 = x^2 + x - 6, y_2 = (x+3)(x-2)$

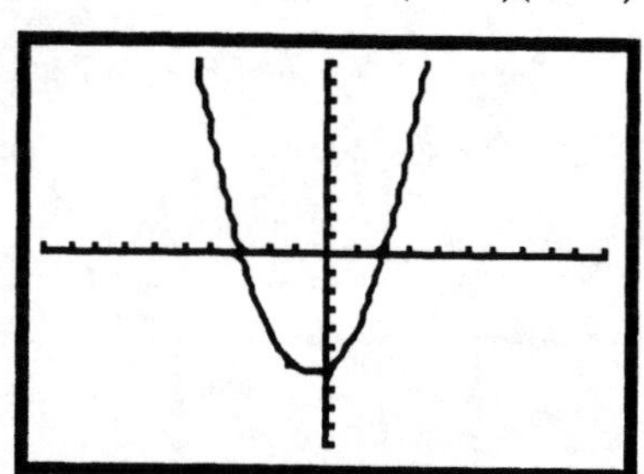

17. $\frac{x-5}{x-2}$ **18.** $\frac{x+3}{(x+1)(x+5)}$ **19.** $\sqrt[7]{t^5}$

20. $\frac{35+5\sqrt{3}}{46}$ **21.** -1 **22.** $-5, \frac{1}{2}$ **23.** $-1, 3$

24. $\pm\sqrt{6}$ **25.** $h = \frac{3V}{2\pi r^2}$ **26.** 13 ft **27.** -2,1

Chapter 1

1. a) Yes; b) $\{-4,3,1,0\}$; c) $\{7,0,5\}$ **2.** a) 8; b) $2a^2+7a+11$

3. a)

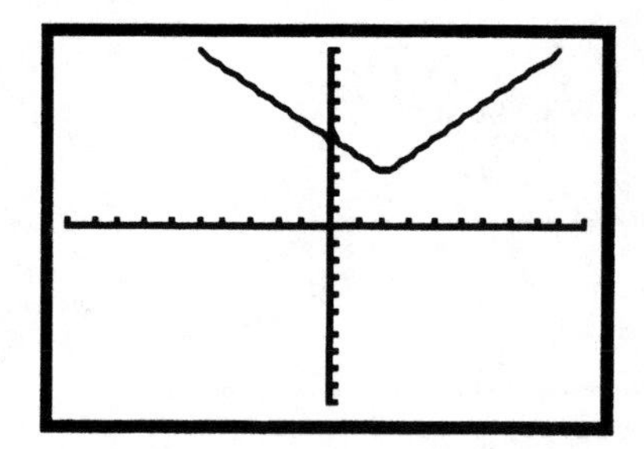

b) $(-\infty,\infty)$; c) $[3,\infty)$ **4.** a) No; b) yes

5. a) Increasing: $(-\infty,-0.33),(3,\infty)$, decreasing: $(-0.33,3)$; b) Maximum: 5.52 at $x=-0.33$, minimum: -13 at $x=3$

6.

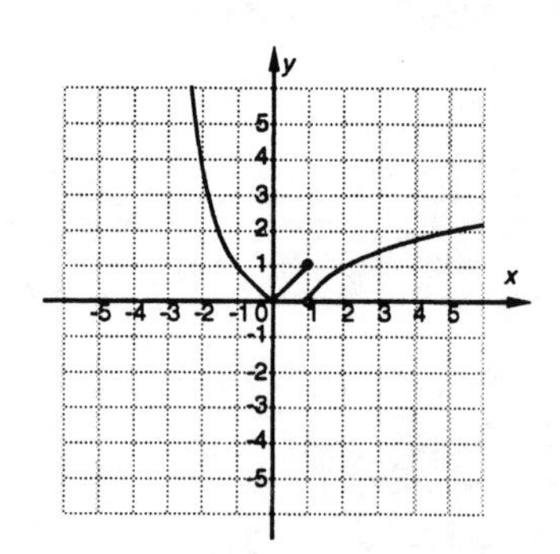

7. Slope: $\frac{3}{2}$; y-intercept: $\left(0,\frac{5}{2}\right)$

8. $y-4=-\frac{3}{4}(x-(-5))$, or $y-(-2)=-\frac{3}{4}(x-3)$, or $y=-\frac{3}{4}x+\frac{1}{4}$ **9.** $y-3=-\frac{1}{2}(x+1)$, or $y=-\frac{1}{2}x+\frac{5}{2}$

10. a) $y=-0.14x+8.36$; b) $r\approx-0.9439$; c) 6.7 gal

11. $\sqrt{45}\approx6.708$ **12.** $\left(-3,\frac{9}{2}\right)$

13. $(x+1)^2+(y-2)^2=5$

14. a)

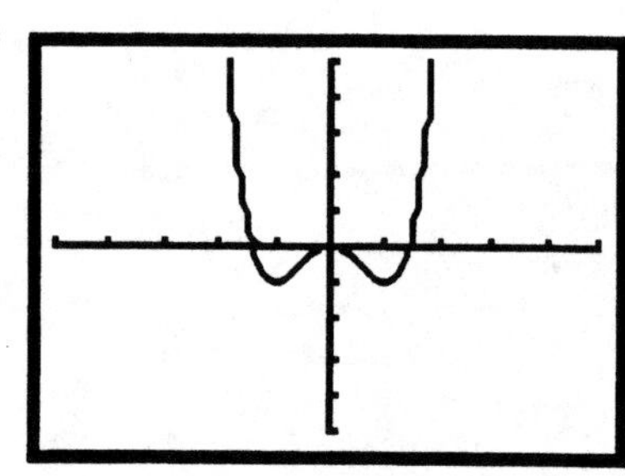

b) x-axis: no, y-axis: yes, origin: no **15.** Odd

16. $f(x)=(x-2)^2-1$ **17.** $f(x)=(x+2)^2-3$

18.

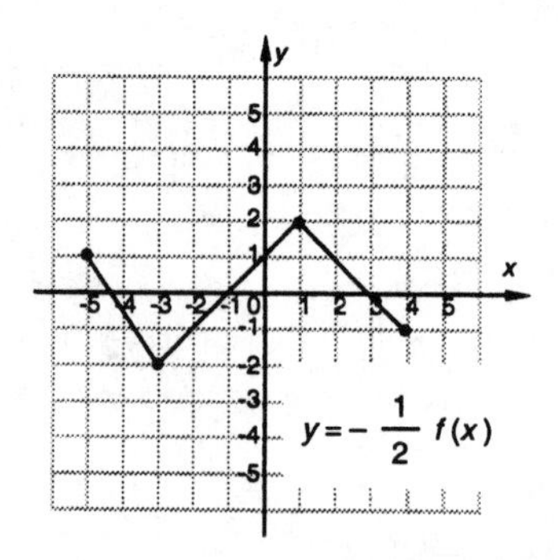

19. a) $(-\infty,\infty)$; b) $[3,\infty)$; c) $x^2-\sqrt{x-3}$; d) $x^2\sqrt{x-3}$; e) $(3,\infty)$ **20.** $y=\frac{30}{x}$ **21.** 50 ft

22. $(-1,1)$

Chapter 2

1. -3 **2.** $-\frac{1}{4}$ and 3, or -0.25 and 3 **3.** $\frac{1\pm\sqrt{57}}{4}$, or approximately -1.637 and 2.137 **4.** $-2\pm\sqrt{5}$

5. $\frac{3}{4}\pm\frac{\sqrt{23}}{4}i$ **6.** 16 **7.** $-1,\frac{13}{6}$ **8.** 5 **9.** $-\frac{1}{2},2$

10. $(-5,3)$ **11.** $(-\infty,-7)\cup(-3,\infty)$ **12.** $n=\frac{R^2}{3p}$

13. $\sqrt{43}i$ **14.** $-5i$ **15.** $10+5i$ **16.** i **17.** a) (1,9); b) $x=1$; c) maximum: 9; d) $(-\infty,9]$ **18.** 15 ft by 30 ft **19.** $y=12.875x^2-26.3x+112.625$ **20.** $\frac{1}{6}$

Chapter 3

1. a) $-2.87,-1,0.87$; b) 5.04 at $x=-2.08$; c) -5.04 at $x=0.08$; d) $(-\infty,\infty)$; e) $(-\infty,\infty)$ **2.** 5.5%

3. a) $y=0.9360221882x^3-1.786434004x^2+6.861222601x-4.039496246$; b) approximately 595

4. $x^2+3x+1+\frac{5x-4}{x^2-1}$ **5.** $3x^2+15x+63$, R 322

6. Yes **7.** -115 **8.** $-\sqrt{3},2+i$

9. a)

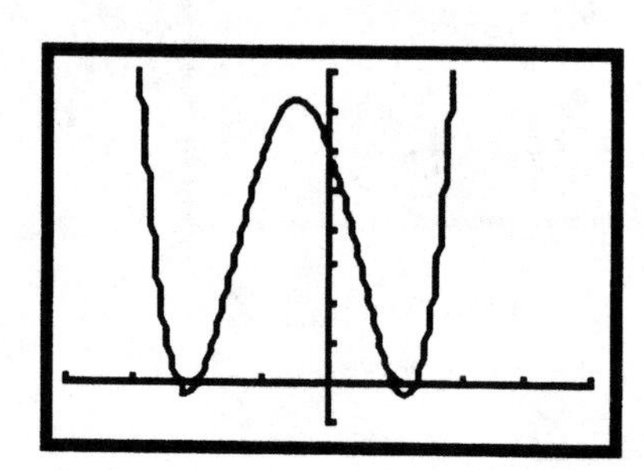

b) $1, -2, \frac{-1 \pm \sqrt{13}}{2}$;

c) $P(x) = (x-1)(x+2)\left(x - \frac{-1+\sqrt{13}}{2}\right)\left(x - \frac{-1-\sqrt{13}}{2}\right)$

10.

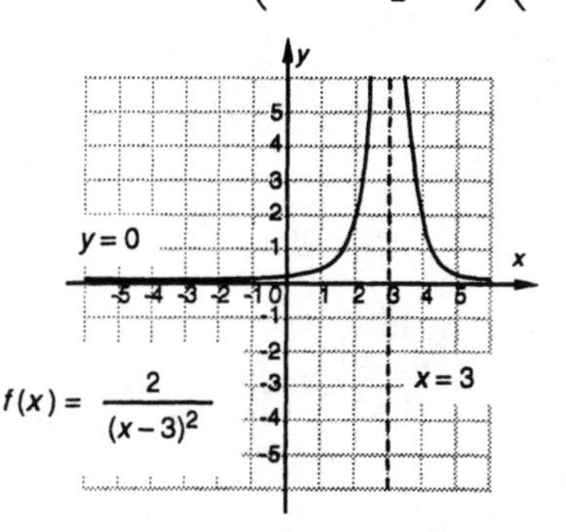

11. Answers may vary; $f(x) = \frac{x+4}{x^2 - x - 2}$

12. $\left(-\infty, -\frac{1}{2}\right) \cup (3, \infty)$ **13.** $(-\infty, 4) \cup \left[\frac{13}{2}, \infty\right)$

14. a) 6 sec; b) (1,3) **15.** $(-\infty, -4] \cup [3, \infty)$

Chapter 4

1. $(f \circ g)(x) = x^2 - 4$; $(g \circ f)(x) = x^2 - 10x + 26$

2. $\{(5, -2), (3, 4), (-1, 0), (-3, -6)\}$ **3.** No

4. $f^{-1}(x) = \sqrt[3]{x - 1}$

5.

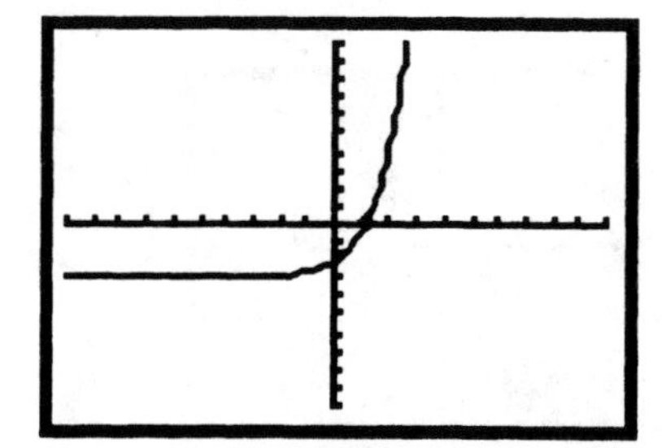

6.

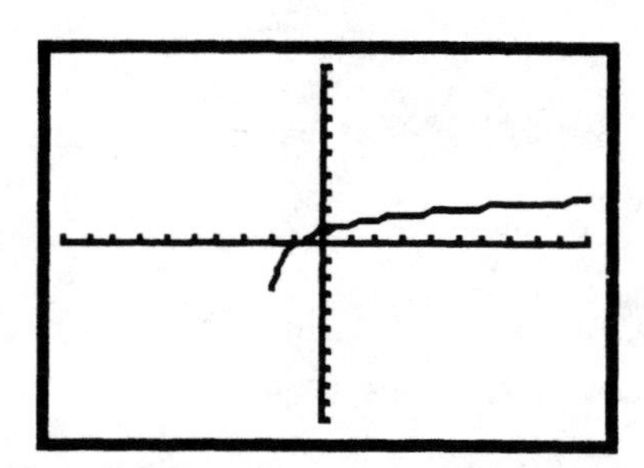

7. $x = e^4$ **8.** $x = \log_3 5.4$ **9.** $\frac{1}{2}$ **10.** 1 **11.** 1

12. 4.174 **13.** $\frac{2}{5} \ln x + \frac{1}{5} \ln y$ **14.** 0.656 **15.** $-4t$

16. 0.0154 **17.** a) 4.5%; b) $P(t) = 1000e^{0.045t}$; c) \$1433.33; d) 15.4 yr

18.

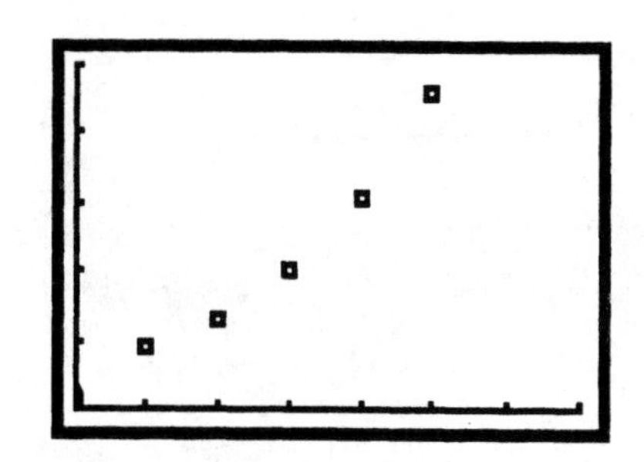

b) $y = 2.83547343(1.523196852)^x$, or $y = 2.83547343e^{0.4208113184x}$; c) \$82.2 million; d) about 6.8 yr **19.** $\frac{27}{8}$

Chapter 5

1.

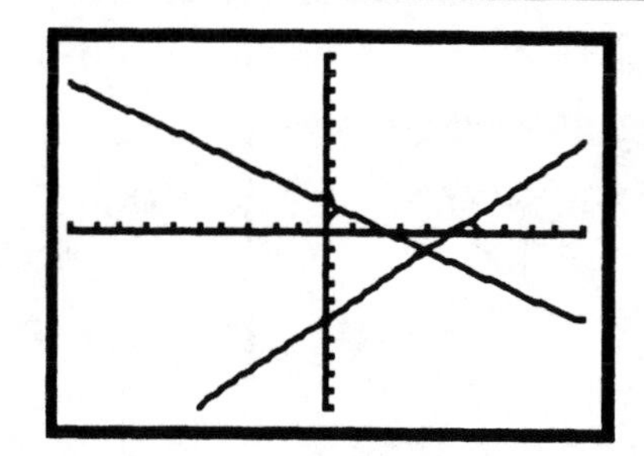

2. $(-3, 5)$ **3.** $(1, -2)$ **4.** $(-1, 3, 2)$ **5.** Consistent

6. Independent **7.** Student: 342, non-student: 408

8. $\begin{bmatrix} -2 & -3 \\ -3 & 4 \end{bmatrix}$ **9.** Not defined **10.** $\begin{bmatrix} -7 & -13 \\ 5 & -1 \end{bmatrix}$

11. Not defined **12.** $\begin{bmatrix} 2 & -2 & 6 \\ -4 & 10 & 4 \end{bmatrix}$

13. $\begin{bmatrix} 0 & -1 \\ -0.25 & -0.75 \end{bmatrix}$ **14.** a) $\begin{bmatrix} 49 & 10 & 13 \\ 43 & 12 & 11 \\ 51 & 8 & 12 \end{bmatrix}$;

b) $[26 \quad 18 \quad 23]$; c) $[3221 \quad 660 \quad 812]$; d) the total cost, in cents, for each type of menu item served on the given day **15.** $\begin{bmatrix} 3 & -4 & 2 \\ 2 & 3 & 1 \\ 1 & -5 & -3 \end{bmatrix} \begin{bmatrix} x \\ y \\ z \end{bmatrix} = \begin{bmatrix} -8 \\ 7 \\ 3 \end{bmatrix}$

16.

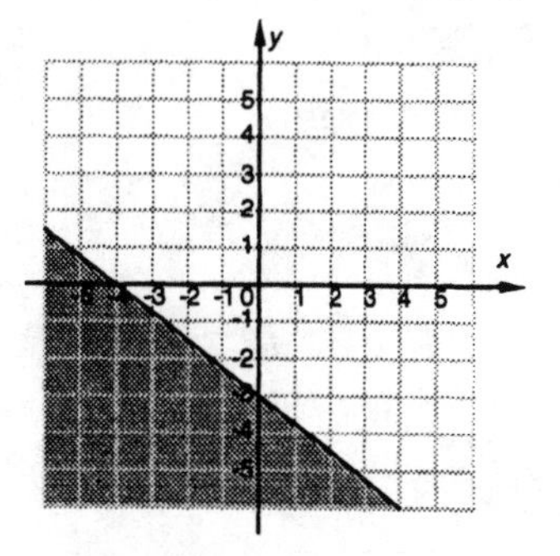

17. Maximum: 15 at (3,3), minimum: 2 at (1,0)

18. 25 pound cakes, 75 carrot cakes

19. $-\frac{2}{x-1}+\frac{5}{x+3}$ **20.** $A=1, B=-3, C=2$

Chapter 6

1. c **2.** b **3.** a **4.** d **5.** $x^2=8y$ **6.** (10,4)

7. Center: $(-1,3)$, radius: 5 **8.** $\frac{y^2}{25}+\frac{x^2}{4}=1$

9. $y=\frac{\sqrt{2}}{2}x,\ y=-\frac{\sqrt{2}}{2}x$ **10.** $\frac{27}{8}$ in.

11. $(1,2),\ (1,-2),\ (-1,2),\ (-1,-2)$

12. $(3,-2),\ (-2,3)$ **13.** (2,3), (3,2) **14.** 5 ft by 4 ft

15. $(x-1)^2+(y+3)^2=16$

Chapter 7

1. -43 **2.** $2+5+10+17=34$

3.

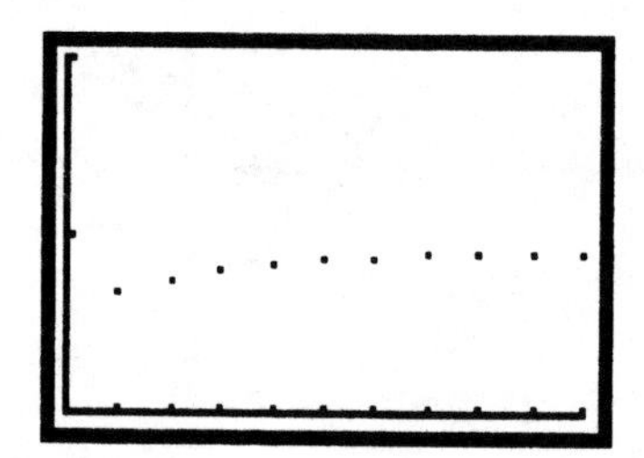

4. 44 **5.** 38 **6.** -420 **7.** 156,000 **8.** 27 **9.** $\frac{56}{99}$

10. \$74,399.77

11. $S_1: 2=\frac{1(3\cdot 1+1)}{2}$

$S_k: 2+5+8+\cdots+(3k-1)=\frac{k(3k+1)}{2}$

$S_{k+1}: 2+5+8+\cdots+(3k-1)+[3(k+1)-1]=\frac{(k+1)[3(k+1)+1]}{2}$

Basis step: $\frac{1(3\cdot 1+1)}{2}=\frac{1\cdot 4}{2}=2$, so S_1 is true.

Induction step:

$$2+5+8+\cdots+(3k-1)+[3(k+1)-1]$$
$$=\frac{k(3k+1)}{2}+[3k+3-1]$$
$$=\frac{3k^2}{2}+\frac{k}{2}+3k+2$$
$$=\frac{3k^2}{2}+\frac{7k}{2}+2$$
$$=\frac{3k^2+7k+4}{2}$$
$$=\frac{(k+1)(3k+4)}{2}$$
$$=\frac{(k+1)[3(k+1)+1]}{2}$$

12. 3,603,600 **13.** 352,716

14. $\frac{n(n-1)(n-2)(n-3)}{24}$ **15.** 360 **16.** 34,650

17. $x^5+5x^4+10x^3+10x^2+5x+1$ **18.** $2^9=512$

19. $\frac{4}{7}$ **20.** 15

Final Examination

1.

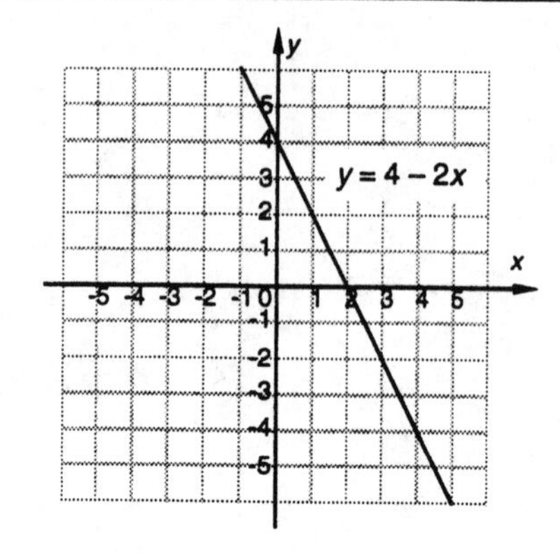

2. $(-2.355, 5.711)$ **3.** 3 **4.** $y=\frac{2}{3}x$ **5.** $[-6,1)$ **6.** -3

7. $-1.856,\ 0.678,\ 3.177$ **8.** $y-(-4)=-1(x-1)$, or $y-(-6)=-1(x-3)$, or $y=-x-3$

9. a) $y=2.09x+190.89$; b) $r\approx 0.8326$

10. a)

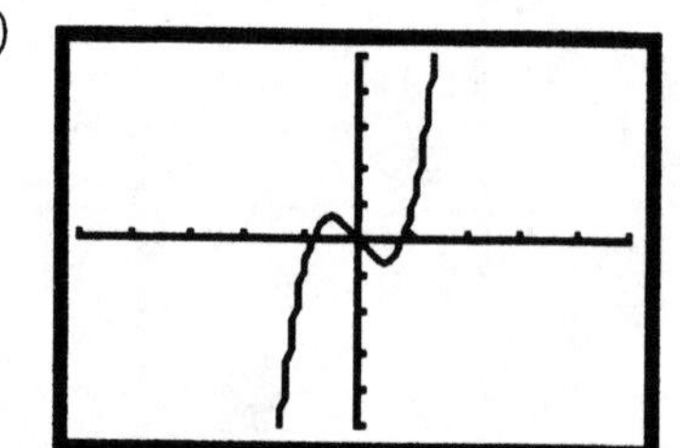

b) x-axis: no; y-axis: no; origin: yes **11.** $\frac{3}{8}\pm\frac{\sqrt{23}}{8}i$

12. a) $\left(\frac{3}{2},\frac{29}{4}\right)$; b) $x=\frac{3}{2}$; c) maximum: $\frac{29}{4}$ at $x=\frac{3}{2}$; d) $\left(-\infty,\frac{29}{4}\right]$ **13.** $-4,-1,1$

14. $(-\infty,-8)\cup(-3,\infty)$

15. $y=3.038x^2-22.924x+244.036$

16. $f^{-1}(x)=\frac{x+5}{3}$

17.

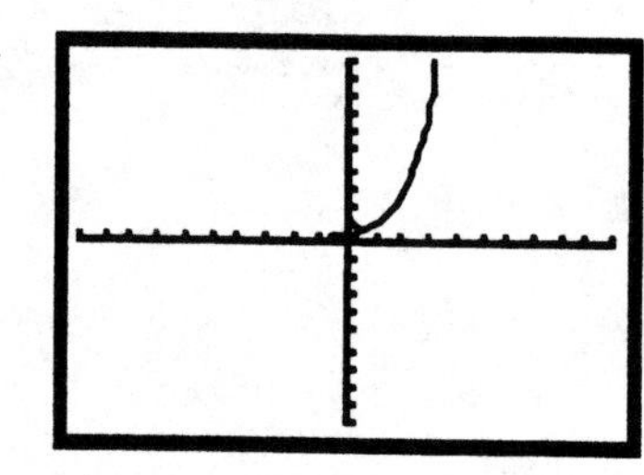

18. $-1s$ **19.** $3\log x + \log y - 2\log z$

20. a) $P(t) = 3000e^{0.048t}$; b) about 14.4 yr

21. $(5, -3)$ **22.** $(-1, -3, 4)$ **23.** $\begin{bmatrix} 5 & 2 \\ -16 & 9 \end{bmatrix}$

24.

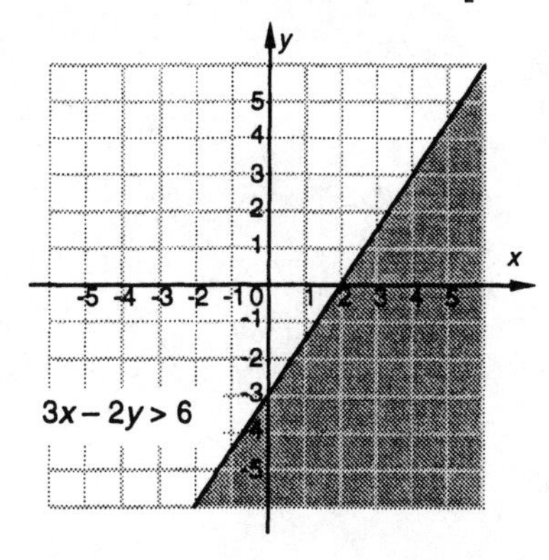

25. 54 dozen blueberry, 6 dozen bran **26.** a **27.** d

28. b **29.** c **30.** $(7, -5)$ **31.** 44 **32.** 64

33. $S_1 : 2 = \dfrac{1(5 \cdot 1 - 1)}{2}$

$S_k : 2 + 7 + 12 + \cdots + (5k - 3) = \dfrac{k(5k-1)}{2}$

$S_{k+1} : 2 + 7 + 12 + \cdots + (5k - 3) + [5(k+1) - 3] = \dfrac{(k+1)[5(k+1)-1]}{2}$

Basis step: $\dfrac{1(5 \cdot 1 - 1)}{2} = \dfrac{1 \cdot 4}{2} = 2$, so S_1 is true.

Induction step:

$$2 + 7 + 12 + \cdots + (5k - 3) + [5(k+1) - 3]$$

$$= \frac{k(5k-1)}{2} + [5k + 5 - 3]$$

$$= \frac{5k^2}{2} - \frac{k}{2} + 5k + 2$$

$$= \frac{5k^2 + 9k + 4}{2}$$

$$= \frac{(k+1)(5k+4)}{2}$$

$$= \frac{(k+1)[5(k+1)-1]}{2}$$

34. 120 **35.** $\dfrac{2}{13}$ **36.** $(2, -2)$ **37.** $\left(-\dfrac{1}{2}, \infty\right)$

Appendix Answers

Appendix A

1. 3 or 1; 0 **2.** 3 or 1; 0 **3.** 0; 3 or 1 **4.** 0; 1 **5.** 2 or 0; 2 or 0 **6.** 2 or 0; 4, 2, or 0 **7.** 1; 1 **8.** 2 or 0; 0 **9.** 1; 0 **10.** 1; 1 **11.** 2 or 0; 2 or 0 **12.** 3 or 1; 1 **13.** 3 or 1; 1 **14.** 1; 1 **15.** 1; 1 **16.** 2 or 0; 2 or 0 **17.** 2 or 0; 2 or 0 **18.** 0; 0 **19.** 0; 0 **20.** 3 or 1; 2 or 0 **21.** 2 or 0; 1 **22.** 1; 1

Appendix B

1. -11 **2.** $2\sqrt{5}+12$ **3.** x^3-4x **4.** $3y^2+2y$ **5.** -109 **6.** -9 **7.** $-x^4+x^2-5x$ **8.** $-2x^3$ **9.** The answer checks. **10.** The answer checks. **11.** The answer checks. **12.** The answer checks.

13. $M_{11}=6,\ M_{32}=-9,\ M_{22}=-29$

14. $M_{13}=4,\ M_{31}=12,\ M_{23}=18$

15. $A_{11}=6,\ A_{32}=9,\ A_{22}=-29$

16. $A_{13}=4,\ A_{31}=12,\ A_{23}=-18$ **17.** -10

18. -10 **19.** -10 **20.** -10

21. $M_{41}=-14,\ M_{33}=20$ **22.** $M_{12}=32,\ M_{44}=7$

23. $A_{24}=15,\ A_{43}=30$ **24.** $A_{22}=-10,\ A_{34}=1$

25. 110 **26.** 110 **27.** 110 **28.** -195

29. $\left(-\frac{25}{2},-\frac{11}{2}\right)$ **30.** $\left(\frac{9}{19},\frac{51}{38}\right)$ **31.** (3,1)

32. $\left(-\frac{10}{41},-\frac{13}{41}\right)$ **33.** $\left(\frac{1}{2},-\frac{1}{3}\right)$ **34.** $\left(-\frac{3}{2},\frac{2}{3}\right)$

35. (1,1) **36.** $(5,-4)$ **37.** $\left(\frac{3}{2},\frac{13}{14},\frac{33}{14}\right)$

38. $\left(-1,-\frac{6}{7},\frac{11}{7}\right)$ **39.** $(3,-2,1)$ **40.** $(-1,4,-2)$

41. $(1,3,-2)$ **42.** $(4,-3,2)$ **43.** $\left(\frac{1}{2},\frac{2}{3},-\frac{5}{6}\right)$

44. $\left(-\frac{31}{16},\frac{25}{16},-\frac{29}{8}\right)$ **45.** Discussion and Writing

46. Discussion and Writing **47.** ± 2 **48.** $3,-2$

49. $(-\infty,-\sqrt{3}]\cup[\sqrt{3},\infty)$ **50.** $(-\sqrt{10},\sqrt{10})$ **51.** -34

52. 3 **53.** 4 **54.** 0 **55.** Answers may vary. $\begin{vmatrix} L & -W \\ 2 & 2 \end{vmatrix}$ **56.** Answers may vary. $\begin{vmatrix} \pi & -\pi \\ h & r \end{vmatrix}$

57. Answers may vary. $\begin{vmatrix} a & b \\ -b & a \end{vmatrix}$ **58.** Answers may vary. $\begin{vmatrix} \frac{h}{2} & -\frac{h}{2} \\ b & a \end{vmatrix}$ **59.** Answers may vary. $\begin{vmatrix} 2\pi r & 2\pi r \\ -h & r \end{vmatrix}$

60. Answers may vary. $\begin{vmatrix} xy & Q \\ Q & xy \end{vmatrix}$

Appendix C

1. $y=12x-7,\ -\frac{1}{2}\le x\le 3$ **2.** $y=5-x,\ -2\le x\le 3$

3. $y=\sqrt[3]{x}-4,\ -1\le x\le 1000$

4. $y=2x^2+3,\ 0\le x\le 2\sqrt{2}$ **5.** $x=y^4,\ 0\le x\le 16$

6. $y=\sqrt[3]{x-1},\ -26\le x\le 28$ **7.** $y=\frac{1}{x},\ 1\le x\le 5$

8. $y=x-2,\ -127\le x\le 129$

9. $y=\frac{1}{4}(x+1)^2,\ -7\le x\le 5$

10. $y=3x,\ -\frac{5}{3}\le x\le\frac{5}{3}$ **11.** $y=\frac{1}{x},\ x>0$

12. $y=e^x, -\infty<x<\infty$ **13.** Answers may vary. $x=t,\ y=4t-3;\ x=\frac{t}{4}+3,\ y=t+9$ **14.** Answers may vary. $x=t,\ y=t^2-1;\ x=t-2,\ y=t^2-4t+3$

15. Answers may vary. $x=t,\ y=(t-2)^2-6t;$ $x=t+2,\ y=t^2-6t-12$ **16.** Answers may vary. $x=t,\ y=t^3+3;\ x=\sqrt[3]{t},\ y=t+3$ **17.** Answers may vary. $x=\frac{1}{4}t,\ y=-t-4$ **18.** Answers may vary. $x=t,\ y=2\pm\sqrt{25-(t+1)^2}$